DENDRITIC CELLS IN FUNDAMENTAL AND CLINICAL IMMUNOLOGY

ADVANCES IN EXPERIMENTAL MEDICINE AND BIOLOGY

Recent Volumes in this Series

Volume 321
PANCREATIC ISLET CELL REGENERATION AND GROWTH
Edited by Aaron I. Vinik

Volume 322
EXERCISE, CALORIES, FAT, AND CANCER
Edited by Maryce M. Jacobs

Volume 323
MECHANISMS OF LYMPHOCYTE ACTIVATION AND IMMUNE REGULATION IV: Cellular Communications
Edited by Sudhir Gupta and Thomas A. Waldmann

Volume 324
PROSTATE CANCER AND BONE METASTASIS
Edited by James P. Karr and Hidetoshi Yamanaka

Volume 325
RECOVERY FROM BRAIN DAMAGE: Reflections and Directions
Edited by F. D. Rose and D. A. Johnson

Volume 326
THE USE OF RESEALED ERYTHROCYTES AS CARRIERS AND BIOREACTORS
Edited by Mauro Magnani and John R. DeLoach

Volume 327
GENETICALLY ENGINEERED VACCINES
Edited by Joseph E. Ciardi, Jerry R. McGhee, and Jerry M. Keith

Volume 328
ENZYMOLOGY AND MOLECULAR BIOLOGY OF CARBONYL METABOLISM 4
Edited by Henry Weiner, David W. Crabb, and T. Geoffrey Flynn

Volume 329
DENDRITIC CELLS IN FUNDAMENTAL AND CLINICAL IMMUNOLOGY
Edited by Eduard W. A. Kamperdijk, Paul Nieuwenhuis, and Elisabeth C. M. Hoefsmit

---

A Continuation Order Plan is available for this series. A continuation order will bring delivery of each new volume immediately upon publication. Volumes are billed only upon actual shipment. For further information please contact the publisher.

DENDRITIC CELLS IN FUNDAMENTAL AND CLINICAL IMMUNOLOGY

Edited by

Eduard W. A. Kamperdijk
Vrije Universiteit
Amsterdam, The Netherlands

Paul Nieuwenhuis
University of Groningen
Groningen, The Netherlands

and

Elisabeth C. M. Hoefsmit
Vrije Universiteit
Amsterdam, The Netherlands

PLENUM PRESS • NEW YORK AND LONDON

Library of Congress Cataloging in Publication Data

Dendritic cells in fundamental and clinical immunology / edited by Eduard W. A. Kamperdijk, Paul Nieuwenhuis, and Elisabeth C. M. Hoefsmit.
p. cm.—(Advances in experimental medicine and biology; v. 329)
"Proceedings of the Second International Symposium on Dendritic Cells in Fundamental and Clinical Immunology, held June 21–25, 1992, in Amsterdam, the Netherlands"—T.p. verso.
Includes bibliographical references and index.
ISBN 0-306-44407-0
1. Dendritic cells—Congresses. I. Kamperdijk, Eduard W. A. II. Nieuwenhuis, Paul. III. Hoefsmit, Elisabeth, C. M. IV. International Symposium on Dendritic Cells in Fundamental and Clinical Immunology (2nd: 1992: Amsterdam, Netherlands) V. Series.
[DNLM: 1. Dendritic Cells—immunology—congresses. QW 568 D391 1992]
QR185.8.D45D46 1993
591.2'9—dc20
DNLM/DLC 92-48429
for Library of Congress CIP

Proceedings of the Second International Symposium on Dendritic Cells in Fundamental and Clinical Immunology, held June 21–25, 1992, in Amsterdam, The Netherlands

ISBN 0-306-44407-0

A Division of Plenum Publishing Corporation
233 Spring Street, New York, N.Y. 10013

Printed in the United States of America

PREFACE

These Proceedings contain the contributions of the participants of the Second International Symposium on Dendritic Cells that was held from the 1st to 25th of June 1992 in Amsterdam, the Netherlands.

The First International Symposium on Dendritic Cells was organized as a Satellite symposium at the occasion of the 30th anniversary of the Japanese Reticuloendothelial Society by Dr. Y. Imai in Yamagata (Japan), in 1990. It was entitled "Dendritic Cells in Lymphoid Tissues," and focused primarily on the Interdigitating Cells (IDC), Epidermal Langerhans cells (LC) and Follicular Dendritic Cells (FDC), from the point of view of human pathology. However, the concept of Dendritic Cell System, comprising the bone marrow derived IDC and LC but not the FDC, was based on animal experiments and mainly on in vitro experiments on isolated cells. In a report from the Reticuloendothelial Society Committee on Nomenclature in 1982, Tew, Thorbecke and Steinman had already characterized these different types of DC, but the gap between in vivo and in vitro function remained.

In Amsterdam, the Symposium focused on the Role of Dendritic Cells in Fundamental and Clinical Immunology. First, recent developments in molecular biology of antigen presentation and cell biological aspects of signal transduction were discussed, in relation to the potential of DC to stimulate lymphocytes and to trigger their in vitro differentiation. Next, the immunological role of DC in vivo, as studied in animal experiments, was expertly dealt with, highlighting origin and kinetics of DC as well as their potential, upon in vitro exposure to antigen, to induce vigorous immune responses in vivo. Finally, the role of DC in Clinical Immunology was illustrated in presentations dealing with isolation and functional characterization of DC and other accessory cells from normal and pathologic human tissues. DC may play a pivotal role in infection diseases such as AIDS, chronic inflammatory reactions, allergy and autoimmune diseases, transplantation reactions and tumor rejection. During a session on DC nomenclature further consensus on the distinctive roles of FDC and DC was reached. In contrast to the FDC, the origin of which still remains a mystery, DC clearly are bone marrow derived and most likely should be considered as belonging to the monocyte-macrophage system. On the other hand, the gap between the functional characterization of DC, studied ex vivo as isolated cells and their morphological and phenotypical identification as epidermal LC and IDC in situ, has yet to be closed.

The meeting was attended by some 220 participants and 15 invited speakers. The format of the respective sessions was designed so as to facilitate personal communication to advantage. The core of each session consisted of ample time for poster viewing which was preceded and followed by short oral presentations illustrating major issues of the poster data, a format which was highly appreciated by all participants.

Finally we would like to express our gratitude to the British Society for Immunology, the Dutch Society for Immunology, the European Federation of Immunological Societies, Koninklijke Nederlandse Academie voor Wetenschappen, Nederlands Astma Fonds, and Nier Stichting Nederland. Without the generous financial support of a number of companies listed beneath, it would have been impossible to organize this symposium.

Preparations for a follow-up of this successful meeting, which is to take place sometime in 1994, are already under way.

For the organizing committee,

Prof.dr. E.C.M. Hoefsmit
Chairman

ACKNOWLEDGEMENTS

B.V. Clean Air Techniek Woerden
Coulter Electronics Nederland
Diagnostic Products Corporation Nederland BV
Eurocetus
European Immunodermatology Society
Gambro B.V.
Greiner Nederland
Harlan C.P.B.
Instruchemie B.V.
Miltenyi Biotec GmbH
Novo Nordisk Farma B.V.
Nycomed B.V.
Pharmacia Diagnostics BV
Sanbio B.V.
The Wellcome Trust

CONTENTS

General Introduction 1
Dendritic cells: Antigen presentation, accessory function and clinical relevance
Ralph M. Steinman, Margit Witmer-Pack, and Kayo Inaba

CELLULAR AND MOLECULAR BIOLOGY OF ANTIGEN PRESENTATION

Endocytic activity of dendritic cells is similar to other antigen presenting cells . . 11
Timothy P. Levine and Benjamin M. Chain

Immunocytochemical characterization of dendritic cells 17
Joanne M.S. Arkema, Inge L. Schadee-Eestermans, Donna M. Broekhuis-Fluitsma and Elisabeth C.M. Hoefsmit

Requirements of exogenous protein antigens for presentation to $CD4^+$ T lymphocytes by MHC Class II-positive APC 23
Gernot Gradehandt, Johannes Hampl, Nadja Kleber, Christina Lobron, Silke Milbradt, Burkhard Schmidt, and Erwin Rüde

Modulation of MHC Class II determinants on rat Langerhans cells during short term culture 29
Ursula Neiß, Detlef Becker, Jürgen Knop, and Konrad Reske

A dendritic cell specific determinant present in endosomes is involved in the presentation of protein antigens 35
Toshiyuki Maruyama, Elisabeth C.M. Hoefsmit and Georg Kraal

Dendritic cells are potent antigen-presenting cells for microbial superantigen . . . 41
Sharifah Iqball, Reshma Bhatt, Penelope Bedford, Peter Borriello and Stella Knight

Divergent T-cell cytokine profiles induced by dendritic cells from different tissues 47
Michael P. Everson, William J. Koopman and Kenneth W. Beagley

Adhesion molecules: Co-stimulators and Co-mitogens in dendritic cell - T cell interaction 53
Philip D. King, Mohammad A.A. Ibrahim and David R. Katz

Dendritic cell dependent expression of IgA by clones in T/B microcultures 59
Carol E. Schrader and John J. Cebra

Adhesion molecules in tonsil DC-T cell interactions 65
Derek N.J. Hart and Timothy C.R. Prickett

Dendritic cells have reduced cell surface membrane glycoproteins including CD43 determinants 71
William Egner, Barry D. Hock, and Derek N.J. Hart

The effect of human dendritic cells on the lectin-induced responsiveness of $CD4^+$T cells to IL-2 and IL-4 75
Sergiusz Markowicz and Anita Mehta

Analysis of cytokine and cytokine receptor production by human dendritic cells .. 81
Victoria L. Calder, Timothy C.R. Prickett, Judith L. McKenzie and Derek N.J. Hart

Costimulating factors and signals relevant for antigen presenting cell function ... 87
Marc K. Jenkins, Dimuthu R. DeSilva, Julia G. Johnson, and Steven D. Norton

ANTIGEN PRESENTING CELLS IN T CELL REGULATION AND REPERTOIRE SELECTION

Distinct T cell stimulation mechanism and phenotype of human blood dendritic cells 93
Hui Xu, Ulrike Friedrichs, Robert K.H. Gieseler, Jörg Ruppert, Göran Ocklind, and J.Hinrich Peters

The role of dendritic cells in the regulation of T cell cytokine synthesis 99
Joanna Ellis, Mohammad A. A. Ibrahim, Benjamin M. Chain, and David R. Katz

The role of dendritic cells as co-stimulators in tolerance induction 105
Michal Dabrowski, Mohammad A.A. Ibrahim, Benjamin M. Chain and David R. Katz

The influence of dendritic cells on T-cell cytokine production 111
Erna van Wilsem, John Brevé, Ingrid van Hoogstraten, Huub Savelkoul and Georg Kraal

Phenotypical and functional characterization of dendritic cells in the human peritoneal cavity 117
Michiel G.H. Betjes, Cees W. Tuk, and Robert H.J. Beelen

Dendritic cells isolated from rat and human non-lymphoid tissue are very potent accessory cells 123
Robert H.J. Beelen, Ellen van Vugt, Joke J.E. Steenbergen, Michiel G.H. Betjes, Carin E.G. Havenith and Eduard W.A. Kamperdijk

Antigen presenting capacity of peritoneal macrophages and dendritic cells 129
Ellen van Vugt, Marina A.M. Verdaasdonk, Eduard W.A. Kamperdijk, and Robert H.J. Beelen

T-cell repertoire development in MHC class II deficient humans 135
Marja van Eggermond, Marijke Lambert, Françoise Mascart, Etienne Dupont and Peter van den Elsen

Rat thymic dendritic cells 141
Ed. W.A. Kamperdijk, Joanne M.S. Arkema, Marina A.M. Verdaasdonk, Robert H.J. Beelen and Ellen van Vugt

Rat thymic dendriticCells: flow cytometry analysis 147
Maria Purificacion Bañuls, Alberto Alvarez, Isabel Ferrero, Agustin Zapata and Carlos Ardavin

Ultrastructure of interdigitating cells in the rat thymus during Cyclosporin A treatment 153
Eric J. de Waal, Louk H.P.M. Rademakers, Henk-Jan Schuurman and Henk van Loveren

T cell tolerance and antigen presenting cell function in the thymus 159
David J. Izon, John D. Nieland, Lori A. Jones and Ada M. Kruisbeek

DENDRITIC CELLS: IN VITRO PHENOTYPE AND FUNCTION

Tolerizing mice to human leukocytes: A step toward the production of monoclonal antibodies specific for human dendritic cells 165
Una O'Doherty, William J. Swiggard, Kayo Inaba, Yasunori Yamaguchi, Iris Kopeloff, Nina Bhardwaj, and Ralph M. Steinman

Morphological and functional differences between HLA-Dr$^+$ peripheral blood dendritic cells and HLA Dr$^+$IFN-alpha producing cells 173
Stuart E. Starr, Santu Bandyopodhyay, Vedapuri Shanmugam, Nassef Hassan, Steven Douglas, Stephanie J. Jackson, Giorgio Trinchieri and Jihed Chehimi

Three monoclonal antibodies to antigen presenting cells in the rat with differential influence on cellular interactions 179
Jan Damoiseaux, Ed Döpp, Marina Verdaasdonk, Ed Kamperdijk and Christine Dijkstra

The MHC expression of dendritic cells from mouse spleen isolated by centrifugal elutriation is upregulated during short term culture 185
Miriam A. Ossevoort, René E.M. Toes, Marloes L.H. De Bruijn, Cornelis J.M. Melief, Carl G. Figdor and W. Martin Kast

Il-6 and its high affinity receptor during differentiation of monocytes into Langerhans cells 191
Gertrud Rossi, Markus Schmitt, Martin Owsianowski, Bernhard Thiele, and Harald Gollnick

Phagocytosis of antigens by Langerhans cells 199
Caetano Reis e Sousa and Jonathan M. Austyn

Dissection of human Langerhans cell allostimulatory function. Modulation by interferon-γ 205
J. Péguet-Navarro, C. Dalbiez-Gauthier, C. Dezutter-Dambuyant and D. Schmitt

Human *in vitro* T cell sensitization using hapten-modified epidermal Langerhans cells 209
Corinne Moulon, Josette Péguet-Navarro, Pascal Courtellemont, Gerard Redziniak and Daniel Schmitt

Monocyte-derived Langerhans cells from different species - morphological and functional characterization 213
Falko Steinbach and Bernhard Thiele

A serial section study of mice Langerhans cell granules after DNFB painting . . 219
Akira Senoo, Nobuo Imazeki, Yoshifusa Matsuura and Yusuke Fuse

Skin dendritic cell-lymphocyte interactions in autologous system 225
Hanna Galkowska and Waldemar L. Olszewski

Induction of the low affinity receptor for IgE (FcεRII/CD23) on human blood dendritic cells by interleukin-4 231
Susanne Krauss, Elfriede Mayer, Gerti Rank and E.Peter Rieber

Fc_{ϵ}RI mediates IGE-binding to human epidermal Langerhans cells 237
Oliver Kilgus, Binghe Wang, Armin Rieger, Birgit Osterhoff, Kenichi Ochiai, Dieter Maurer, Dagmar Födinger, Jean-Pierre Kinet and Georg Stingl

Murine epidermal Langerhans cells as a model to study tissue dendritic cells . . 243
Gerold Schuler, Franz Koch, Christine Heufler, Eckhardt Kämpgen, Gerda Topar, and Nikolaus Romani

DENDRITIC CELLS IN VIVO: PHENOTYPE AND FUNCTION

Differentiation of dendritic cells in cultures of rat and mouse bone marrow cells 251
William E. Bowers, Shengqiang Yu, and Syeedur Khandkar

TNF and GM-CSF dependent growth of an early progenitor of dendritic Langerhans cells in human bone marrow 257
Cecil D.L. Reid, Arthur Stackpoole and Jaak Tikerpae

Human bone marrow contains potent stimulatory cells for the allogeneic MLR with the phenotype of dendritic cells 263
William Egner, Judith L. McKenzie, Stewart M. Smith, Michael E.J. Beard, and Derek N.J. Hart

Recombinant GM-CSF induces in vitro differentiation of dendritic cells from mouse bone marrow 269
Christoph Scheicher, Maria Mehlig, Reinhard Zecher, Friedrich Seiler, Petra Hintz-Obertreis and Konrad Reske

Signals required for differentiating dendritic cells from human monocytes in vitro 275
J. Hinrich Peters, Hui Xu, Jörg Ruppert, Dorothea Ostermeier, Detlef Friedrichs, Robert K.H. Gieseler

Down-regulation and release of CD14 on human monocytes by IL-4 depends on the presence of serum or GM-CSF 281
Jörg Ruppert, Christiane Schütt, Dorothea Ostermeier and J.Hinrich Peters

Serum-free differentiation of rat and human dendritic cells, accompanied by acquisition of the nuclear lamins A/C as differentiation markers 287
Robert K.H. Gieseler, Hui Xu, Rolf Schlemminger and J.Hinrich Peters

Immunophenotypic and ultrastructural differentiation and maturation of nonlymphoid dendritic cells in osteopetrotic (op) mice with the total absence of macrophage colony stimulating factor activity 293
Kiyoshi Takahashi, Makoto Naito, Yuki Morioka and Leonard D. Shultz

Loading of dendritic cells with antigen in vitro or in vivo by immunotargeting can replace the need for adjuvant 299
T. Sornasse, V. Flamand, G. De Becker, K. Thielemans, J. Urbain, O. Leo and M. Moser

Migration of alveolar macrophages from alveolar space to paracortical T cell area of the draining lymph node 305
Theo Thepen, Eric Claassen, Karin Hoeben, John Brevé and Georg Kraal

Comparison of Langerhans cells and interdigitating reticulum cells 311
Mikihiro Shamoto, Satoru Hosokawa, Masanori Shinzato and Chiyuki Kaneko

Migration of dendritic cells during contact sensitization 315
S. Hill, S. Griffiths, I. Kimber and S.C. Knight

Peritoneal cell labelling: A study on the migration of macrophages and dendritic cells towards the gut 321
Marsetyawan Soesatyo, Theo Thepen, Muhammad Ghufron, Jeike Biewenga, and Taede Sminia

Dendritic Cells "in vivo": Migration and antigen handling 327
G. Gordon MacPherson and Liming Liu

FOLLICULAR DENDRITIC CELLS: PHYSIOLOGY AND PATHOLOGY

Follicular dendritic cells: Isolation procedures, short and long term cultures . . . 333
Ernst Heinen, Rikyia Tsunoda, Christian Marcoty, Nadine Antoine, Alain Bosseloir, Nadine Cormann and Léon Simar

Heterogeneity and cellular origin of follicular dendritic cells 339
Yutaka Imai, Kunihiko Maeda, Mitsunori Yamakawa, Yasuaki Karube, Mikio Matsuda, Michio Dobashi, Hironobu Sato and Kazuo Terashima

Differential uptake and trapping of TI-2 antigens: An unexpected role for follicular dendritic cells in the induction of TI-2 immune responses 345
A.J.M. Van den Eertwegh, M.M. Schellekens, W.J.A. Boersma and E. Claassen

Ultrastructural heterogeneity of follicular dendritic cells in the human tonsil ... 353
Louk H.P.M. Rademakers and Henk-Jan Schuurman

Ultrastructural analysis of human lymph node follicles after HIV-1 infection ... 359
Louk H.P.M. Rademakers, Henk-Jan Schuurman and Jack F. de Frankrijker

Emperipolesis of lymphoid cells by human follicular dendritic cells in vitro ... 365
Rikiya Tsunoda, Masayuki Nakayama, Ernst Heinen, Katsuya Miyake, Kazunori Suzuki, Hirooki Okamura, Naonori Sugai and Mizu Kojima

The localization of lymphokines in murine germinal centers 371
T. Kasajima, A. Andoh, Y. Takeo and T. Nishikawa

Two different mechanisms of immune-complex trapping in the mouse spleen during immune responses 377
Kikuyoshi Yoshida, Timo K. van den Berg and Christine D. Dijkstra

Cellular requirements for functional reconstitution of follicular dendritic cells in SCID mice 383
Zoher F. Kapasi, Gregory F. Burton, Leonard D. Shultz, John G. Tew and Andras K.Szakal

Interaction through the LFA-1/ICAM-1 pathway prevents programmed cell death of germinal center B cells 387
Gerrit Koopman, Robert M.J. Keehnen and Steven T. Pals

Membrane expression of Fcϵ RII/CD23 and release of soluble CD23 by follicular dendritic cells 393
E.Peter Rieber, Gerti Rank, Ingrid Köhler and Susanne Krauss

Follicular dendritic cells in malignant lymphomas - Distribution, phenotypes & ultrastructures 399
Kunihiko Maeda, Mikio Matsuda, Masaru Narabayashi, Ryuichi Nagashia, Noriyuki Degawa and Yutaka Imai

Lymphoid follicles in Cynomolgus monkeys after infection with simian immunodeficiency virus 405
Piet Joling, Peter Biberfeld, Henk K. Parmentier, Dick F. van Wichen, Timo Meerloo, Louk H.P.M. Rademakers, Jürg Tschopp, Jaap Goudsmit, and Henk-Jan Schuurman

Destruction of follicular dendritic cells in murine acquired immunodeficiency syndrome (MAIDS) 411
Akihiro Masuda, Gregory F. Burton, Bruse A. Fuchs, Andras K. Szakal and John G. Tew

Changes in follicular dendritic cell and CD8+ cell function in macaque lymph nodes following infection with SIV_{251} 417
Yvonne J. Rosenberg, Marie H. Kosco, Mark G. Lewis, Enrique C. Leon, Jack J. Greenhouse, Kyle E. Bieg, Gerald A. Eddy, and Philip M. Zack

Rapid and selective isolation of follicular dendritic cells by low speed centrifugations on discontinuous BSA gradients 425
Christian Marcoty, Ernst Heinen, Nadine Antoine, Rikiya Tsunoda and L.J. Simar

Follicular dendritic cells do not produce TNF-α nor its receptor 431
Isabelle Mancini, Alain Bosseloir, Elisabeth Hooghe-Peters, Ernst Heinen, and Léon Simar

Splenic lesions in hypogammaglobulinaemia 437
J. Weston, B.M. Balfour, W. Tsohas, N. English, J. Farrant and A.D.B. Webster

Ontogenic study on the bronchus associated lymphoid tissue (BALT) in the rat, with special reference to dendritic cells 443
Yoshifusa Matsuura, Nobuo Imazeki, Akira Senoo, and Yusuke Fuse

DRC1 expression on lymphoid normal and pathological cells 449
Nadine Antoine, Ernst Heinen, Christian Marcoty, Alain Bosseloir and Léon Simar

Binding of HIV-1 to human follicular dendritic cells 455
Piet Joling, Leendert J. Bakker, Dick F. van Wichen, Loek de Graaf, Timo Meerloo, Maarten R. Visser, Jos A.G. van Strijp, Jaap Goudsmit, Jan Verhoef, and Henk-Jan Schuurman

Follicular dendritic cells in germinal center reactions 461
John G. Tew, Gregory F. Burton, Leo I. Kupp and Andras Szakal

PANEL DISCUSSION ON DENDRITIC CELL NOMENCLATURE

Follicular dendritic cells and dendritic cell nomenclature 467
John G. Tew

Langerhans cells as outposts of the dendritic cell system 469
A.H. Warfel, G.J. Thorbecke and D.V. Belsito

Heterogeneity of dendritic cells and nomenclature 481
Elisabeth C.M. Hoefsmit, Joanne M.S. Arkema, Michiel G.H. Betjes, Carin E.G. Havenith, Ellen van Vugt, Robert H.J. Beelen and Eduard W.A. Kamperdijk.

Report of the panel discussion 487
Elisabeth C.M. Hoefsmit

DENDRITIC CELLS IN CLINICAL IMMUNOLOGY

Dendritic cells in transplantation 489
Jonathan M. Austyn

Down-regulation of MHC-expression on dendritic cells in rat kidney grafts by PUVA pretreatment 495
B. v. Gaudecker, R. Petersen, M. Epstein, J. Kaden, and H. Oesterwitz

Cytokine mediators of non-lymphoid dendritic cell migration 501
Justin A. Roake, Abdul S. Rao, Christian P. Larsen, Deborah F. Hankins, Peter J. Morris, and Jonathan M. Austyn

Isolation of dendritic leukocytes from non-lymphoid organs 507
Abdul S. Rao, Justin A. Roake, Christian P. Larsen, Deborah F. Hankins, Peter J.Morris and Jonathan M. Austyn

RTIB/D$^+$ non-lymphoid DC in early GVHD and Hg-induced autoimmunity of rat salivary and lacrimal glands 513
Ake Larsson, Kouji Fujuwara and Michael Peszkowski

In-vitro infection of peripheral blood dendritic cells with human immunodeficiency virus-1 causes impairment of accessory functions . . . 521
Jihed Chehimi, Kesh Prakash, Vedapuri Shanmugam, Stephanie J. Jackson, Santu Bandyopadhyay, and Stuart E. Starr

Simian immunodeficiency virus (SIV) induced alterations of thymus IDCs 527
V. Krenn, J. Müller, W. Mosgöller, S. Czub, C. Schindler, C. Stahl-Hennig, C. Coulibaly, G. Hunsmann, and H.K. Müller-Hermelink

Murine leukaemia virus infections as models for retroviral disease in humans . . 533
Mary S. Roberts, Jennifer J. Harvey, Steven E. Macatonia, and Stella C. Knight

Langerhans cells and interdigitating cells in HIV infection 539
Andrea M.R. von Stemm, Julia Ramsauer, Klara Tenner-Racz, Heidemarie F. Schmidt, Irma Gigli, and Paul Racz

Dendritic cells in HIV-1 and HTLV-1 infection 545
Stella C. Knight, Steven E. Macatonia, Kennedy Cruickshank, Peter Rudge, and Steven Patterson

Dendritic cells in allergic and chronic inflammatory responses 551
Leonard W. Poulter and George Janossy

Pulmonary dendritic cell populations 557
Patrick G. Holt

Blood dendritic cells are highly adherent to untreated and cytokine-treated cultured endothelium 563
K.Alun Brown, Frances LeRoy, Penelope A. Bedford, Stella C. Knight and Dudley C. Dumonde

Antigen specific T cell priming *in vivo* by intratracheal injection of antigen presenting cells 571
Carin E.G. Havenith, Annette J. Breedijk, Wim Calame, Robert H.J. Beelen and Elisabeth C.M. Hoefsmit

Histology and immunophenotype of dendritic cells in the human lung 577
Jan Maarten W. van Haarst, Harm J. de Wit, Hemmo A. Drexhage and Henk C. Hoogsteden

Acquisition of Chlamydial antigen by dendritic cells and monocytes 581
Andrew J. Stagg, Arthur Stackpoole, William J. Elsley and Stella C. Knight

Experimental cutaneous leishmaniasis: Langerhans cells internalize Leishmania major and induce an antigen-specific T-cell response 587
Heidrun Moll

Endocytosis of potential contact sensitizers by human dendritic cells 593
Marie Larsson and Urban Forsum

Dendritic cells and "dendritic" macrophages in the uveal tract 599
John V. Forrester, Paul G. McMenamin, Janet Liversidge and Lynne Lumsden

Macrophages and dendritic cells in rat colon in experimental inflammatory bowel disease 605
Emmelien P. van Rees, Marsetyawan Soesatyo, Marja van der Ende, and Taede Sminia

Vaccination with tumor antigen-pulsed dendritic cells induces in vivo resistance to a B cell lymphoma 611
V. Flamand, T. Sornasse, K. Thielemans, C. Demanet, O. Leo, J. Urbain and M. Moser

Studies on Langerhans cells in the tracheal squamous metaplasia of vitamin A deficient rats 617
Satoru Hosokawa, Masanori Shinzato, Chiyuki Kaneko and Mikihiro Shamoto

Depletion of Langerhans cells following carcinogen treatment is partly due to antigenicity 623
Gregory M. Woods, Imogen H. Liew and H. Konrad Muller

A proportion of patients with premature ovarian failure show lowered percentages of blood monocyte derived dendritic cells capable of forming clusters with lymphocytes 629
Annemieke Hoek, Yvonne van Kasteren, Meeny de Haan-Meulman, Joop Schoemaker, and Hemmo A. Drexhage

Thyroid hormones and their iodinated breakdown products enhance the capability of monocytes to mature into veiled cells. Blocking effects of α-GM-CSF 633
P. Mooij, M. de Haan-Meulman, H.J. de Wit and H.A. Drexhage

Relationship between dendritic cells and folliculo-stellate cells in the pituitary: Immuno-histochemical comparison between mouse, rat and human pituitaries 637
W. Allaerts, P.H.M. Jeucken, F.T. Bosman and H.A. Drexhage

Dendritic cells in tumor growth and endocrine diseases 643
H.A. Drexhage, P. Mooy, A. Jansen, J. Kerrebijn, W. Allaerts and M.P.R. Tas

Index 651

DENDRITIC CELLS: ANTIGEN PRESENTATION, ACCESSORY FUNCTION AND CLINICAL RELEVANCE

Ralph M. Steinman[1], Margit Witmer-Pack[1], and Kayo Inaba[2]

[1] Laboratory of Cellular Physiology and Immunology, The Rockefeller University, New York, NY 10021, USA
[2] Department of Zoology, Kyoto University, Kyoto 693, Japan

INTRODUCTION

Laboratories around the world, many represented here, have now elucidated the broad outlines of the dendritic cell system. Dendritic cells occupy three compartments: a) nonlymphoid tissues such as the Langerhans cells of the skin and the interstitial dendritic cells of lung and heart, b) the circulation especially the veiled cells of the afferent lymph, and c) lymphoid organs primarily the interdigitating cells of the T cell areas and thymic medulla. These pools of dendritic cells are interconnected, since nonlymphoid cells can pick up antigens and migrate into the blood or lymph and to lymphoid organs.

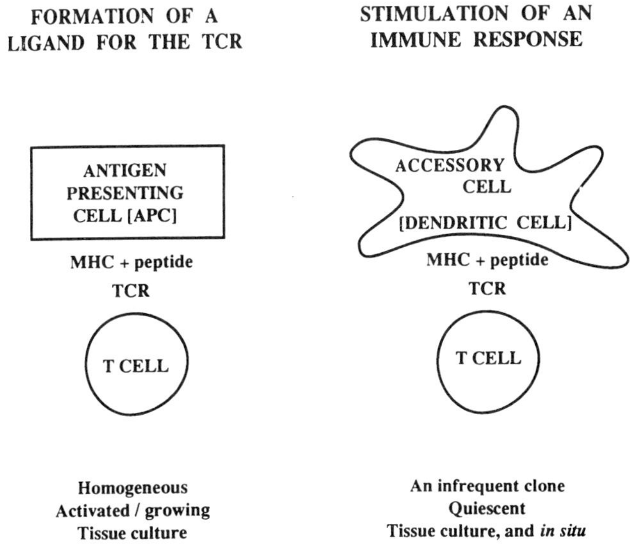

Fig 1. Typical assays for antigen presenting and accessory cells

Dendritic Cells in Fundamental and Clinical Immunology,
Edited by Kamperdijk *et al.*, Plenum Press, New York, 1993

The function of dendritic cells in the immune system is to act as highly specialized *accessory cells* for inducing immunity and most likely tolerance [Fig 1]. Dendritic cells are not simply *antigen presenting cells* [APCs]. All cells that express MHC products can use these products for presentation i.e., to bind and present peptide fragments of an antigen to MHC-restricted receptors on T cells. However dendritic cells have many distinct "accessory" properties that relate to the acquisition and retention of antigens, the binding and stimulation of infrequent antigen-specific clones, and the generation of immunity within the confines of a whole animal. Accessory function for primary T cell populations, especially in situ, is best studied with dendritic cells, whereas in vitro, antigen presentation can be studied using cell lines as APCs and T lymphocytes [Fig 1].

METHODS AND MARKERS

One would know very little about the properties of dendritic cells if it were not feasible to isolate them for analysis, either in tissue culture or upon reinfusion into animals. However it is not easy to work with these APCs. Human blood, perhaps the most demanding tissue to study, illustrates how one goes about enriching this trace cell type. The principle is to remove other leukocytes, using physical methods and/or cell-specific markers. For blood, we successively remove granulocytes, T cells, monocytes, and B cells [Fig 2]. The dendritic cells are found in a low density fraction that represents but 1% of blood mononuclear cells [1]. A further step entailing cell sorting or immunoadsorption is required for high purity, particularly for the depletion of residual $CD4^+$ T cells in AIDS-related work. The large, low density, "null" fraction is almost entirely dendritic cells with an unusual shape and motility [next section], high levels of MHC products, and potent stimulating activity for T cell-dependent responses.

This negative selection approach [Fig 2] contrasts with methods in which APCs are positively selected with antibodies to MHC class II products. Expression of class II is not cell specific. There are enormous difference that become apparent by isolating different, class II positive APCs such as dendritic cells, B lymphocytes, or macrophages. It would be difficult to appreciate the presence of specialized accessories like dendritic cells, if one simply isolated cells on the basis of MHC class II expression.

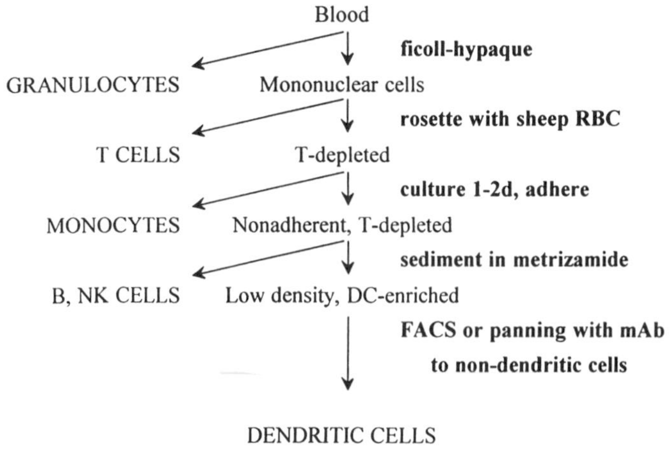

Figure 2. Enrichment of dendritic cells from human blood.

Dendritic cell-restricted reagents have been difficult to obtain. This also is the case with other myeloid cells such as granulocytes and monocytes. In contrast, B or T lymphocytes when injected into mice elicit many different lymphocyte-restricted monoclonals. There should be ways to overcome the difficulties in obtaining more dendritic cell-restricted reagents, one of these being the tolerization approaches outlined by O'Doherty et al in this meeting.

MORPHOLOGY AND MOTILITY

The morphologic features of dendritic cells are distinct in two respects. One is a peculiar endocytic apparatus. Dendritic cells lack abundant lysosomes but do possess a set of endocytic vacuoles that may be specialized for antigen presentation rather than antigen scavenging [see below]. The other feature is that dendritic cells are actively motile, constantly forming and retracting long cell processes, veils, or dendrites. This motility raises several questions [Table 1].

Table 1 Motility of dendritic cells

What stimulates the formation and retraction of large lamellipodia?
What is occurring at the level of the cytoskeleton?
What are the consequences? (capture of T cells, migration in situ)

MOVEMENT AND MIGRATION IN SITU

Beginning with the insights of Balfour, Knight, Drexhage, and MacPherson, dendritic cells were realized to be a constituent of the afferent lymph [2-4]. These "veiled cells" can carry antigens to the lymphoid tissues [5]. Dendritic cells pulsed with a soluble protein and placed into the foot pad, sensitize $CD4^+$ MHC-restricted, helper cells in draining lymph nodes [6]. Presumably migration to the T cell areas [below] is critical for the sensitization that is observed.

Several recent studies, especially those of J.Austyn et al, have established with some directness the prevalance of dendritic cell migration [7]. Two approaches have been taken. One is to look for the movement of dendritic cells out of transplants. Indeed, APCs leave organs like skin and heart in large numbers. This migration may consistently accompany the transplantation event. Recent findings by Starzl et al indicate that migration to the thymus is also possible [8]. The other approach is to isolate dendritic cells, label with a vital dye, inject back into a recipient, and then observe that the APCs home via the afferent lymph or blood to the T cell areas.

We have verified the second approach using the following technique. Dendritic cells are labeled with carboxyfluorescein succinimide ester [Molecular Probes] and injected into the foot pad or i.v. A day later, the draining lymph node or spleen is taken. Sections are stained with peroxidase-anti-fluorescein to localize the dendritic cells, and an alkaline phosphatase system to localize specific cell types. Below are spleen sections double labeled for the dendritic cells [arrows indicate the discrete, darker, injected dendritic cells] and with monoclonals to SER-4 [marginal zone metallophil or macrophage], F4/80 anti-macrophage, GK1.5 anti-CD4, and RA3-3.1 anti-B cell. The dendritic cells clearly home to the T cell areas of the white pulp. The mobilization and movement of dendritic cells could be a key event in initiating immunity. Several questions come to mind [Table 2].

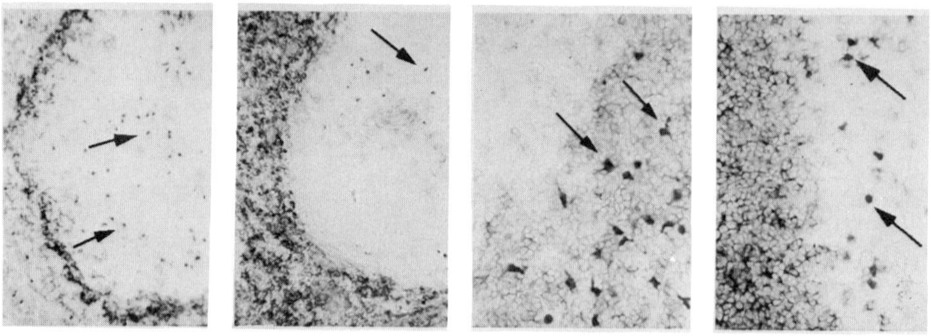

Fig.3 Homing of dendritic cells to splenic T areas [see text].

Table 2 Movement of dendritic cells in situ

What triggers movement of dendritic cells out of nonlymphoid tissues?
How do dendritic cells enter the afferent lymph? the blood?
What directs dendritic cells to the T-dependent areas?
What is the life span in the T-cell area?
Is movement essential for immunogenicity?

MATURATION OF DENDRITIC CELLS

White blood cells, following production in the bone marrow or thymus, undergo extensive further diffentiation before full immunologic function is expressed. For example, B and T lymphocytes undergo programs that lead to the production of antibodies, lymphokines and cytolysins. Monocytes, upon encountering cytokines and other stimuli, become activated to express many secretory and cidal activities. Dendritic cells also exist as less mature elements that later acquire the characteristic phenotype and function of stimulatory APCs, apparently as the end stage of their development.

The maturation process was encountered by Schuler and colleagues in cultures of skin dendritic cells [9] and has since been described in some detail [10-13]. Maturation likely occurs in dendritic cells from rat lung, mouse spleen, and human blood as well. In the case of skin, maturation entails many changes, some critical ones being listed in Table 3. Maturation nicely segregates the two main functions of accessory cells: presentation and T cell sensitization. The former is expressed primarily in the first day in culture, and the latter develops on the second and third days.

Table 3 Maturation of epidermal Langerhans cells

	0-12h	12-72h
Presentation of soluble proteins antigens	+++	+/-
Presentation of peptides	+++	+++
Presentation of allo MHC/mitogens	+	+++
Biosynthesis of MHC class I/II, invariant chain	+++	-
Acidic endocytic vacuoles	+++	+
Surface adhesion and costimulatory molecules	-	+++
Fc receptors	+	+/-

Witmer-Pack, Heufler and colleagues found that GM-CSF is a major mediator of dendritic cell maturation in culture [14,15]. Other cytokines like IL-1 can play a role, possibly by enhancing GM-CSF production and responsiveness as described by G.Schuler in this volume. The cytokines do not induce proliferation of epidermal dendritic cells, however. Several questions emerge from these findings [Table 4].

Table 4 Maturation of dendritic cells

Do tissues other than epidermis have immature dendritic cells?
Does maturation underlie the initiation of immunity e.g., contact allergy, transplantation
How are antigen processing and accessory functions carried out?
What are the regulatory roles of individual cytokines, e.g., IL-1, TNFα, GM-CSF?

MARROW ORIGIN AND RELATIONSHIP TO OTHER MYELOID CELLS

Dendritic cells in several tissues are known to arise from marrow progenitors. This holds for spleen [16], afferent lymph [4], epidermis [17], thymus [18], and the interstitial compartment of most organs [19]. Recently a distinct proliferating progenitor was identified in mouse blood by Inaba and Schuler [20]. They found that the Ia-negative, nonadherent fraction of mouse blood gives rise to distinctive aggregates upon culture in GM-CSF but not other CSF's. The aggregates have large numbers of proliferating cells and can be subcultured to produce additional aggregates. Around the periphery of the aggregate there appear typical dendritic cells with sheet-like processes and with markers such as abundant MHC class II, the NLDC-145 antigen, and the M342/2A1 intracellular antigens [below]. Single, nonproliferating dendritic cells are released, and these exhibit the typical functional properties of motility, movement to the T cell areas in situ, and strong T cell stimulating activity. The system has now been extended to mouse marrow which clearly can generate large numbers of dendritic cells in the presence of exogenous GM-CSF. It will be important to extend the findings to answer several questions [Table 5].

Table 5 Myeloid origin of dendritic cells

What organs harbour precursors to dendritic cells?
Can distinct CFU for dendritic cells be identified?
Where, in the myeloid pathway, do dendritic cells diverge?

It is likely that dendritic cells are a separate myeloid lineage given their distinct and stable features, a lack of interconversion with phagocytes and lymphocytes, the dependence on GM-CSF but not the more lineage restricted M- and G-CSF's, and the occurrence of dendritic cells in scid, nude, and op/op mice. The study of dendritic cell development in colony forming assays is now underway.

MECHANISM OF DENDRITIC CELL FUNCTION

Antigen presentation

One specialization of dendritic cells is the coordination of two activities, the capacity to acquire antigens and the synthesis of class II products and associated invariant chain [see Table 3][11,12]. There is considerable data that the invariant chain protects the peptide binding groove during biosynthesis. Upon reaching the endocytic system, the invariant chain is removed allowing access of the processed antigens to the MHC class II. The coordination of antigen uptake and class II biosynthesis may allow dendritic cells to charge the peptide binding grooves with large amounts of exogenously-derived antigen.

A second specialization is that the endocytic vacuolar apparatus of dendritic cells differs from other APCs. Although dendritic cells appear to be weakly endocytic, this really means weak *accumulation* of substrates for digestion in lysosomes. While dendritic cells are not scavenging cells, there must be sufficient internalization to generate MHC-peptide complexes, since processing and *prolonged retention of antigens* occurs. The distinctiveness of the dendritic cell vacuolar system has become apparent with new monoclonals, e.g., MIDC-8 and M342. These bind to antigens that are more abundant in dendritic cell vacuoles [21,22]. The M342 positive vacuoles are a kind of multivesicular body by EM. The M342 reagent that we have raised also react with the other specialized APC, the stimulated B cell, but not with mononuclear phagocytes in many tissues.

A current mystery is to understand how dendritic cells acquire particulate antigens. We favor the idea that during multiplication and maturation, small but sufficient levels of phagocytic activity occur. Another suggestion, but there is evidence to the contrary [23], is that phagocytes regurgitate peptide fragments for presentation by dendritic cells. Possibly, dendritic cells exhibit extracellular processing.

Accessory properties in T cell stimulation

Dendritic cells express very high levels of MHC class II. It has always been assumed that their potent activity as APCs simply reflects the presence of larger amounts of MHC-peptide complexes or signal one. In fact, recent data [[24] and Bhardwaj et al, in preparation] suggests that small amounts of ligand suffice. Therefore the large amounts of dendritic cell MHC products actually may serve two other functions: retention of a sufficient amount of "signal one" for long periods of time [11,23], and display of many different kinds of peptides especially in the case of transplants and infectious agents.

The regulation of class II MHC expression appears to be distinct in dendritic cells relative to other white cells. Whereas lymphokines upregulate biosynthesis of MHC class II in macrophages [25] and in B cells [26], this is not known to be the case in dendritic cells. For example, marked upregulation of MHC class I and II expression on dendritic cells occurs in culture in the apparent absence of T cells. This feature, T cell independent biosynthesis of MHC products, may enable dendritic cells to capture peptides during the afferent limb of the immune response.

Expressed on the dendritic cell surface are many of the known adhesion and costimulatory molecules such as ICAM-1, LFA-3, B7/BB1, the heat stable antigen. Young et al have shown that monoclonals directed to CD2, CD4, CD11a, and CD28 each blocks T cell responses to dendritic cells in the MLR [27]. Combinations of two or more antibodies ablate the response by 90-95% or more. Since each antibody has a blocking effect, it appears that each ligand-receptor interaction has a distinct role in T cell stimulation that is not readily bypassed by another ligand-receptor pair. However, all of the known adhesion and costimulatory systems are not cell specific. It remains possible that dendritic cells will express new and cell-restricted ligands for T cells.

Specializations of dendritic cells in situ

In spite of the above specializations for antigen presentation and costimulation, the distinctiveness of dendritic cells may relate primarily to all of the signals that control function in situ. For example, one needs to pinpoint how dendritic cells proliferate, mature, migrate to lymphoid organs, and home to the T areas. Such information may lead to new ways to manipulate immunogenicity and tolerance, i.e., to amplify or dampen the immune response at very early stages. If peripheral tolerance can occur when antigens are presented on nondendritic cells, then blocking dendritic cell function in situ may lead to tolerance and not just a lack of an immune response. Alternatively, inactivation of specific dendritic cell signals could provide a direct route to inducing peripheral tolerance.

MEDICINE

Since so many areas in clinical medicine involve T cells [Table 6], it is likely that new understanding can originate from additional study at the level of APCs. A recent example is AIDS, even though the precise contribution of dendritic cells to pathogenesis is controversial. The findings of Cameron et al on 25 patients at different stages of infection [asymptomatic, persistent generalized lymphadenopathy, AIDS] are that APC numbers and function are not significantly reduced relative to other non-T cells, and that the dendritic cells in blood are not heavily infected with HIV-1 [contrary data exist; reviewed in [28]]. Likewise there have been initial reports that epidermal Langerhans cells are altered and infected, yet recent studies are not confirmatory [29].

Table 6 Can unknowns in clinical immunology be studied at the level of APCs?

Rejection of transplants and in some situations, tumors
Resistance to microbial infections: killer cells, lymphokines, antibodies
Contact allergy and other delayed type hypersensitivies
Diseases attributed to autoimmunity, e.g.,
    Multiple sclerosis, juvenile diabetes, psoriasis, rheumatoid arthritis
AIDS

We have recently come upon evidence whereby dendritic cells transmit a vigorous cytopathic infection to $CD4^+$ T cells but without an infection of the APCs [30]. A longstanding enigma is to explain how ostensibly small amounts of virus and/or virus-infected cells lead to such profound $CD4^+$ depletion. Cameron and Freudenthal have found that HIV-1 simply takes advantage of the dendritic cell, nature's adjuvant for $CD4^+$ T cell immunity. If dendritic cells are exposed to HIV-1 and then present antigens to $CD4^+$ T cells, the antigen-responding T cells become infected and are killed.

The findings are as follows. Highly enriched, blood dendritic cells are not readily infected with HIV-1 [using primers for either gag or LTR sequences and PCR to look for reverse transcripts]. HIV-1 pulsed dendritic cells nonetheless transmit a vigorous cytopathic infection to those T cells that are responding to allogeneic MHC products [the mixed leukocyte reaction] or to superantigens. The dendritic cells are not even infected *during* the interaction with T cells that are producing large amounts of virus. Syncytium formation and death is prominent during immune responses to HIV-1 pulsed dendritic cells, but again these APCs do not participate in either manifestation of cytotoxicity.

In essence dendritic cells are a *catalyst for the cytopathic effects* of HIV-1. Dendritic cells allow the virus to gain access to a key site of replication i.e., T blasts that are juxtaposed to one another for several days. These findings may be the tip of the iceberg. The blood dendritic cells we have studied have very low levels of CD4. With higher levels of CD4, as occurs with epidermal Langerhans cells, transmission of a cytopathic infection may be even more vigorous.

DISCUSSION

If one considers the work on dendritic cells relative to other APCs, the key thrust is their many distinct "accessory cell functions" [Fig 1]. While much of the effort in fundamental immunology profitably uses cell lines as both APCs and as T lymphocytes, the accessory cell functions of dendritic cells provide a physiological dimension, allowing one to study the stimulation of resting lymphocytes that occur in very low frequencies and within whole animals. A macrophage or a B cell has yet to be used to elicit graft rejection [31], to induce an antibody response [32], or to sensitize an MHC-restricted helper T cell [6]. It will be important to use dendritic cells to decipher the molecular mechanisms underlying such events as antigen processing, APC-T cell binding, immunogenicity vs. tolerance, and the juxtaposition of APCs and T cells in situ. In each of these areas, dendritic cells are quantitatively and possibly qualitatively distinct from other APCs.

Obviously there is the difficulty in obtaining this trace cell type. This relates to their low frequency, the lack of many cell-restricted markers, and the multistep procedures required for their isolation. This difficulty is an experimental rather than a physiologic one since antigen specific T cells are far less numerous. Also a) one dendritic cell can activate 10-20 T cells in a day [24]; b) dendritic precursors can proliferate extensively [20], so that if this proliferation is coupled to the initiation of an immune response, many APCs can be provided; c) nonproliferating dendritic cells are poised as sentinels along body surfaces where they are specialized to capture antigens and to move to the T area to select out antigen-reactive T cell clones. Even the experimental obstacles of isolating dendritic cells may be solved with new methods to induce considerable growth of this lineage.

Clinically, dendritic cells hold clues to several areas [Table 6]. A few dendritic cells appear capable of inducing transplant rejection, and dendritic cells act as "nature's adjuvant" for a variety of other antigens. In the presence of low levels of HIV-1, these accessory cells catalyze the destruction of many T cells[30], thereby modeling the major immunologic deficit in patients. Allergy, vaccination, and autoimmunity are other areas where the study of dendritic cells should be fruitful. So dendritic cell experimentation has obstacles, but the impacts for fundamental and clinical immunology are considerable.

SUMMARY

Because of difficulties in isolation, it has taken some time to arrive at a reasonable outline of the dendritic cell system. With the international effort that is assembled here, the main features of this system are apparent. There are now several criteria that allow for dendritic cell identification, there is understanding of tissue distribution and the interconnections between different compartments, there is new data on the production and maturation components of this system, and there are many observations that help explain antigen presentation, T cell stimulatory function, and behaviour in situ.

The contributions of our Dutch hosts should be stressed. Many have energized the study of lymphoid, mononuclear phagocyte, and dendritic cell systems. The beginnings were made by Koenig, Langevoort, Thorbecke and van Furth, continued with Veldman, Nieuwenhuis, Hoefsmit, Drexhage, van Ewijk, Dijkstra, Kraal, Kamperdijk, and now there are many investigators in the biology of antigen presentation, one understands why it is appropriate to be in Holland. Holland even geographically is a "dendritic cell" [Fig 4].

Fig 4. Dendritic cells: cell biology and clinical immunology

REFERENCES

1. Freudenthal, P. S. & Steinman, R. M. *Proc.Natl.Acad.Sci.* 87, 7698-7702 (1990).
2. Drexhage, H. A. et al. *Cell.Tiss.Res.* **202**, 407-430 (1979).
3. Knight, S. C. et al. *Eur. J. Immunol.* **12**, 1057-1060 (1982).
4. Pugh, C. W., MacPherson, G. G. & Steer, H. W. *J.Exp.Med.* **157**, 1758-1779 (1983).
5. Bujdoso, R. et al *J. Exp. Med.* **170**, 1285-1302 (1989).
6. Inaba, K. et al. *J. Exp. Med.* **172**, 631-640 (1990).
7. Austyn, J. M. & Larsen, C. P. *Transpl.* **49**, 1-7 (1990).
8. Demetris, A. J., Murase, N. & Starzl, T. E. *Lancet* **In Press**, (1992).
9. Schuler, G. & Steinman, R. M. *J. Exp. Med.* **161**, 526-546 (1985).
10. Romani, N. et al. *J. Exp. Med.* **169**, 1169-1178 (1989).
11. Pure', E. et al. *J. Exp. Med.* **172**, 1459-1469 (1990).
12. Kampgen, E. et al. *Proc. Natl. Acad. Sci.* **88**, 3014-3018 (1991).
13. Romani, N. et al. *J. Invest. Dermatol.* **93**, 600-609 (1989).
14. Witmer-Pack, M. D. et al. *J. Exp. Med.* **166**, 1484-1498 (1987).
15. Heufler, C., Koch, F. & Schuler, G. *J. Exp. Med.* **167**, 700-705 (1987).
16. Steinman, R. M., Lustig, D. S. & Cohn, Z. A. *J. Exp. Med.* **139**, 1431-1445 (1974).
17. Katz, S. I., Tamaki, K. & Sachs, D. H. *Nature* **282**, 324-326 (1979).
18. Barclay, A. N. & Mayrhofer, G. *J. Exp. Med.* **153**, 1666-1671 (1981).
19. Hart, D. N. J. & Fabre, J. W. *J. Exp. Med.* **154**, 347-361 (1981).
20. Inaba, K. et al. *J. Exp. Med.* **175**, 1157-1167 (1992).
21. Breel, M., Mebius, R. E. & Kraal, G. *Eur. J. Immunol.* **17**, 1555-1559 (1987).
22. Agger, R. et al. *J. Leuk. Biol.* **52**, 34-42 (1992).
23. Crowley, M., Inaba, K. & Steinman, R. M. *J. Exp. Med.* **172**, 383-386 (1990).
24. Romani, N. et al. *J. Exp. Med.* **169**, 1153-1168 (1989).
25. Steinman, R. M. et al. *J. Exp. Med.* **152**, 1248-1261 (1980).
26. Roehm, N. W. et al *J. Exp. Med.* **160**, 679-694 (1984).
27. Young, J. W. et al. *J. Clin. Invest.* **90**, 229-237 (1992).
28. Cameron, P. U. et al. *Exp. Immunol.* **88**, 1-11 (1992).
29. Kalter, D. C. et al. *J. Immunol.* **146**, 3396-3404 (1991).
30. Cameron, P. U. et al. *Science In Press* (1992).
31. Lechler, R. I. & Batchelor, J. R. *J. Exp. Med.* **155**, 31-41 (1982).
32. Sornasse, T. et al. *J. Exp. Med.* **175**, 15-21 (1992).

ENDOCYTIC ACTIVITY OF DENDRITIC CELLS IS SIMILAR TO OTHER ANTIGEN PRESENTING CELLS

Timothy P. Levine and Benjamin M. Chain

Dept. of Biology, Medawar Building, University College London, Gower Street, London WC1E 6BT, UK

INTRODUCTION

Antigen processing occurs prior to presentation of an exogenous antigen in association with class II MHC, and requires endocytosis of antigen followed by partial digestion. The intra-cellular pathways taken by antigen and class II may vary between cells,[1] but most evidence suggests that antigen-derived peptides bind class II in organelles related to late endosomes.[2] Antigens enter the endocytic pathway either in the fluid phase or bound to internalized plasma membrane. Processing is enhanced by antigen receptors which are endocytosed from the plasma membrane, examples of which are found on all APC types except dendritic cells: surface immunoglobulin (sIg) on B cells,[3] opsonin receptors (FcR/C3R/C4R) on B cells and macrophages[4] and scavenger receptors on macrophages.[5] Therefore, endocytosis in dendritic cells occurs by the less efficient means of fluid-phase endocytosis and non-specific membrane absorption. These mechanisms appear to be sufficient to allow processing of at least some exogenous antigens by dendritic cells.[6,7]

Previous studies of dendritic cells have indicated poor uptake of endocytic markers compared to macrophages.[8,9] However, uptake was measured over long periods, and so may have represented accumulation in lysosomes. In order to study traffic through endosomes in dendritic cells, we have developed an assay of endocytosis which requires small numbers of non-adherent cells.[10] Here, we show that this assay can be applied to a marker which is taken up by a combination of fluid-phase endocytosis and non-specific membrane absorption. Endocytic traffic of this marker in dendritic cells is similar to B cells, with equivalent entry of marker into the late endosome during a one hour pulse.

METHODS

1-5*10^6 murine splenic dendritic cells, resting B lymphocytes, activated B cells, or peritoneal macrophages were pulsed with the endocytic marker rhodamine dextran (RD, 70 kDa, Sigma, final concentration 5-10mg/ml) for 0, 2, 6, 20 and 60 minutes, and the cells washed repeatedly by centrifugation in cold medium. For dendritic cell experiments, contaminating B cells were stained with anti-mouse sIg conjugated to

Dendritic Cells in Fundamental and Clinical Immunology,
Edited by Kamperdijk *et al.*, Plenum Press, New York, 1993

phycoerythrin-Texas Red. In all experiments some cells were labelled not with RD but with Di.O (1,1'-dihexadecyloxacarbocyanine perchlorate, Molecular Probes, Oregon, USA; final concentration 13μg/ml) and added to each group after the wash steps.

5000 events were acquired using a FACScan (Becton Dickinson, UK) with 3 fluorescence channels detecting RD, Di.O and phycoerythrin-Texas Red separately. Dead and clumped cells were excluded by a forward and side scatter gate. In dendritic cell preparations, B cells were excluded with a gate on phycoerythrin-Texas Red positive events.

The fluorescence due to RD was calculated from:
$[RD^+ - RD^-]_{2,6,20 \text{ or } 60} - [RD^+ - RD^-]_0$, where RD^+_t and RD^-_t were the linearized median fluorescences on the RD axis in flow cytometer units after pulse for t minutes of RD positive cells and RD negative (Di.O positive) cells respectively. The cells pulsed for 0 minutes together with the cells labelled with Di.O provided internal controls for variations in background fluorescence.[10]

Exocytosis was then commenced by re-incubation of cells at 37°C, the chase period, during which samples were regularly taken for flow cytometric analysis, as above.

RESULTS

Dendritic Cell Endocytosis/Exocytosis, Two Compartment Model

Murine splenic dendritic cells were pulsed with rhodamine dextran (RD) for 2, 6, 20 and 60 minutes, washed into cold marker-free medium, and analyzed by flow cytometry for RD content (figure 1). Contaminating B cells were excluded from the analysis by positive staining for sIg. Endocytosis was initially rapid, then slower.

Each group of dendritic cells was chased at 37°C to analyze exocytosis of RD during a chase period of upto 120 minutes (figure 1). Exocytosis of RD was initially rapid, then slower. The amount of RD exocytosed rapidly plateaued at 6-20 minutes pulse, while the amount of RD exocytosed slowly increased linearly throughout 60 minutes.

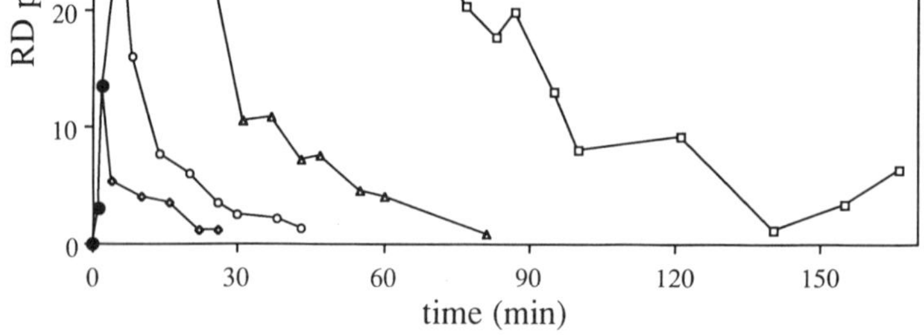

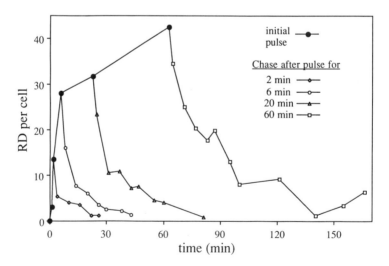

Figure 1. Endocytosis and exocytosis of RD by dendritic cells. Cells were pulsed with RD for upto 60 minutes and then chased for upto 120 minutes. Data are from a single representative of two experiments.

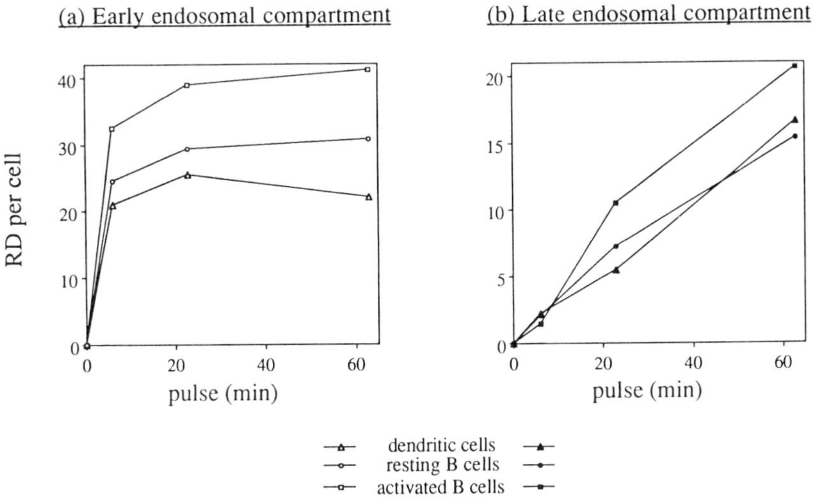

Figure 2. RD endocytosed by dendritic cells, resting B cells and activated B cells separated into early and late endosomal compartments. Data are the mean of two experiments.

The data were further analyzed using a two compartment model of the endocytic pathway,[10,11] which assumes (i) that there were two types of endosomes, with exponential kinetics of filling and emptying (early, rapid; late, slow), (ii) that entry into lysosomes was minimal within the short duration of the assay, and (iii) uptake of RD was proportional to its concentration in the medium. The amount of RD in early endosomes and late endosomes were estimated using this model (figure 2).

Comparison with B Cells

Data on endocytosis of RD followed by exocytosis were obtained for resting B cells and activated B cells (figure 2). After 60 minutes pulse the amount of RD in late endosomes was similar for all 3 APCs, whereas the amount of RD in early endosomes varied to a greater extent. From these data, the traffic of RD through late endosomes, which can be determined using the exponential model (see ref. 10), was found to be similar in dendritic cells and B cells.

After 60 minutes pulse, but not with shorter pulse times, dendritic cells had a broader RD profile than the other cell types (figure 3), indicating late endosomal heterogeneity in dendritic cells.

DISCUSSION

Uptake of the endocytic marker rhodamine dextran (RD) into early and late endosomes has been measured in various APC types, including dendritic cells, by a flow cytometric assay which excludes contaminating cells by positive immuno-staining, and which requires only small numbers of cells. All APC types showed biphasic kinetics of endocytosis and exocytosis, fitting a two compartment model with rapidly filling early endosomes and slowly filling late endosomes.

The importance of antigen traffic though APCs' late endosomes for antigen processing is founded on the proposition that the processing compartment is

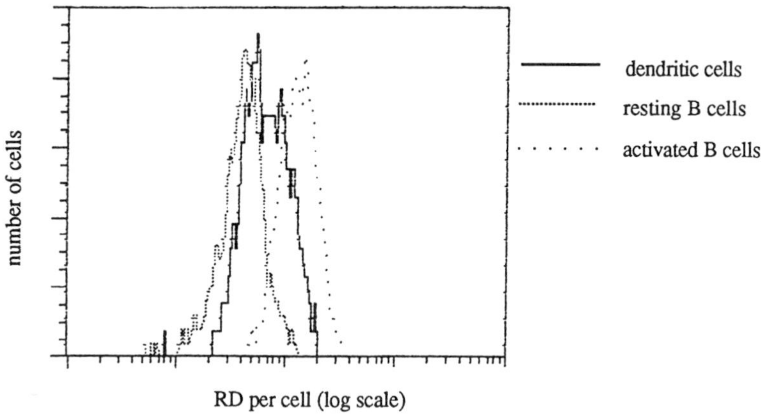

Figure 3. Profiles of RD endocytosed by dendritic cells, resting B cells and activated B cells during a 60 minute pulse.

related to late endosomes.[2] We have shown here that traffic of RD through dendritic cells' late endosomes is similar to B cells' late endosomes. We have previously shown (see ref. 10) that an endocytic marker which is taken up in the fluid-phase without appreciable membrane binding, lucifer yellow, traffics to a similar extent through dendritic cells' late endosomes as in other APCs, and that the late endosomes of dendritic cells are heterogeneous with respect to this endocytic marker. In addition, we have shown that RD is taken up by a mixture of fluid-phase endocytosis and, non-specific membrane adsorption.[10] The latter route of uptake is important in dendritic cells, as it is the only mechanism available to enhance fluid-phase entry,[3-5,12] which may be too inactive to be significant in antigen processing (C.A. Janeway, personal communication). Since RD has produced essentially the same findings as lucifer yellow, we therefore conclude that dendritic cells maintain a similar level of traffic through late

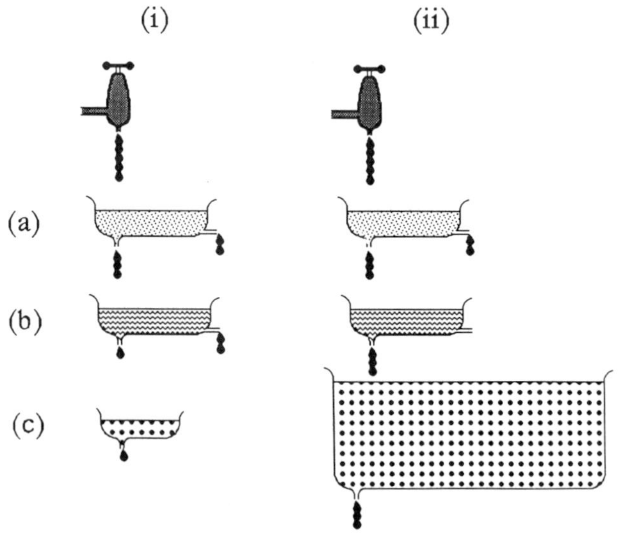

Figure 4. Endosomal traffic can be independent of lysosomal size. The three successive baths represent three compartments in series: (a) early and (b) late endosomes and (c) lysosomes. (i) Dendritic cells may direct a more late endosomal traffic for recycling compared to (ii) Macrophages which may direct more traffic into lysosomes.

endosomes as B cells, whether uptake is purely in the fluid-phase or via non-specific membrane adsorption.

Our findings contrast with previous findings of poor uptake of endocytic markers by dendritic cells.[8,9] These studies used pulses of antigen up to 18 hours long in dendritic cells and macrophages to demonstrate that dendritic cells do not accumulate marker in lysosomes to the same extent as macrophages. The larger number of lysosomes in macrophages may be exaggerated in these experiments by the use of an indigestible endocytic marker, which is retained indefinitely. However, the smaller size of the lysosomal compartment in dendritic cells does not necessarily reflect reduced endosomal activity. Traffic through late endosomes may be similar, while a much smaller proportion of traffic exiting dendritic cells' late endosomes is routed into lysosomes (figure 4). Preliminary studies of peritoneal macrophages with RD have showed that they take up more marker than other APCs (data not shown), but the differences were not as absolute as has been previously described.[7] The macrophages were not analyzed by the two compartment model, because of their adherent phenotype.

In summary, the present study has measured traffic through the early and late endosomes of dendritic cells, while previous studies have emphasized lysosomal size. We have found that endocytic traffic in dendritic cells is similar to that of B cells. As illustrated in figure 4, this does not preclude wide variation between APCs in the size and function of the lysosomal compartment.

Acknowledgement T. L. was supported by the Medical Research Council, U.K. (Training Fellowship).

References

1. F.M. Brodsky, and L.E. Guagliardi, Antigen processing and presentation: close encounters in the endocytic pathway, *Trends in Cell. Biol.* 2:109 (1992).
2. H.L. Pleogh, and J.J. Neefjes, Intracellular transport of MHC class II molecules, *Immunol. Today* 92:179 (1992).
3. A. Lanzavecchia, Receptor-mediated antigen uptake and its effect on antigen presentation to class II-restricted T lymphocytes, *Annu. Rev. Immunol.* 8:773 (1990).
4. J. Arvieux, H. Yssel, and M.G. Colomb, Antigen-bound C3b and C4b enhance antigen presenting cell function in activation of human T cell clones, *Immunology* 65:229 (1988).
5. M. Naito, T. Kodama, A. Matsumoto, T. Doi, and K. Takahashi, Tissue distribution, intracellular localization, and in vitro expression of bovine macrophage scavenger receptors, *Am. J. Pathol.* 139:1411 (1991).
6. R. Nonacs, C. Humborg, and R.M. Steinman, Stimulation of class I-restricted influenza-specific cytotoxic T cells by dendritic cells, *J. Cell. Biochem.* 16D:19 (abstract).
7. B.M. Chain, P.M. Kay, and M. Feldmann, The cellular pathway of antigen presentation: biochemical and functional analysis of antigen processing in dendritic cells and macrophages, *Immunology* 58:271 (1986).
8. M.L. Kapsenberg, M.C.B.M. Teunissen, F.E.M. Stiekema, and H.G. Keizer, Antigen-presenting cell function of dendritic cells and macrophages in proliferative T cell responses to soluble and particulate antigens, *Eur. J. Immunol.* 16:345 (1986).
9. K. Inaba, J.P. Metlay, M.T. Crowley, and R.M. Steinman, Dendritic cells pulsed with protein antigens in vitro can prime antigen-specific, MHC-restricted T cells in situ, *J. Exp. Med.* 172:631 (1990).
10. T.P. Levine, and B.M. Chain, Endocytosis by antigen presenting cells: dendritic cells are as endocytically active as other antigen presenting cells, *Proc. Natl. Acad. Sci. (USA)* in press.
11. J.A. Swanson, B.D. Yirinec, and S.C. Silverstein, Phorbol esters and horse radish peroxidase stimulate pinocytosis and redirect the flow of pinocytosed fluid in macrophages, *J. Cell Biol.* 100:851 (1985).
12. R.M. Steinman, The dendritic cell system and its role in immunogenicity, *Annu. Rev. Immunol.* 9:271 (1991).

IMMUNOCYTOCHEMICAL CHARACTERIZATION OF DENDRITIC CELLS

Joanne M.S. Arkema, Inge L. Schadee-Eestermans, Donna M. Broekhuis-Fluitsma and Elisabeth C.M. Hoefsmit

Dept. of Cell Biology, Div. E.M., Medical Faculty, Vrije Universiteit, NL-1081 BT Amsterdam, The Netherlands

INTRODUCTION

Dendritic cells (DC) are very efficient antigen presenting cells (APC) for various T-cell dependent immune responses (Inaba and Steinman, 1984; Macatonia et al., 1989; Metlay et al., 1989) in comparison with other types of APC (macrophages, B cells, B blasts or activated T cells).

The aim of our study was to investigate the MHC class II mediated antigen presentation immuno-cytochemically in DC. As the generation of MHC class II-peptide complexes is a crucial step in antigen presentation, we studied the intracellular localization of MHC class II molecules. Moreover, we studied the epidermal Langerhans cell (LC) as *in situ* representative of the DC (Hoefsmit et al., 1982).

We conclude that DC and LC contain a characteristic intracellular storage compartment for MHC class II molecules, which has connections to the lysosomal apparatus in the same region of the cell. We hypothesize that, depending on the functional phenotype of DC, the storage vesicles can deliver MHC class II molecules to different sites.

MATERIAL AND METHODS

Isolation of Dendritic Cells from Human Blood

DC were isolated from human peripheral blood using the procedure according to Knight et al. (1986). This procedure is based on adherence of monocytes on plastic petri dishes and gradient centrifugation over a 14.5% metrizamide gradient. Additionally, we performed panning on petri dishes coated with human IgG to deplete remaining monocytes. The remaining cells consist of 30-60% DC, as can be concluded from their strong HLA-DR-positivity and the presence of acid phosphatase in a spot near the nucleus (Kamperdijk et al., 1987).

Preparation of Dendritic Cells for Immuno-electron Microscopy

Samples of DC were fixed in a mixture of 2% paraformaldehyde, 0.01 M sodium periodate and 0.075 M lysine (PLP), in 0.0375 M sodium phosphate buffer. After peletting in 1% gelatin and cryo-protection in 2.3 M sucrose, ultrathin cryosections were made. On these cryosections double immunolabeling was performed using anti-HLA-DRα-β chain (a kind gift from Dr. H. Ploegh) and anti MHC class I heavy chain or anti-lamp-1 (a kind gift from Dr. M. Fukuda), recognizing a lysosomal membrane protein. A protein-A goldprobe of 10 or 15 nm was used to visualize the MHC class I, class II and lamp-1 molecules.

Preparation of skin biopsies for Immuno-electron Microscopy to study Langerhans Cells *in situ*

Small tissue blocks of skin biopsies were fixed in a mixture of 2% paraformaldehyde, 0.01 M sodium periodate and 0.075 M lysine (PLP), in 0.0375 M sodium phosphate buffer. Samples were cryoprotected, and substituted in methanol that contained 0.5% uranylacetate at -90° C for 30 hr, in the CS-Auto (Reichert). Than the temperature was raised to -45° C, for impregnation in HM_{20}, and polymerization. Labeling was performed using the anti-HLA-DRα- and -β chain polyclonal antibody, the anti-mannose 6-phosphate receptor (a kind gift from Dr. K. von Figura) (M6PR, a marker for prelysosomes) and goat-anti-rabbit and goat-anti-mouse goldprobe 10 nm respectively (Aurion, Wageningen).

RESULTS

Class II Molecules in Dendritic Cells

DC showed a cluster of vesicles in a juxtanuclear position which was MHC class II positive and class I and lamp-1 negative (Figure 1 and 2). However, only a few class II positive vesicles showed lysosomal markers such lamp-1 (Figure 2).

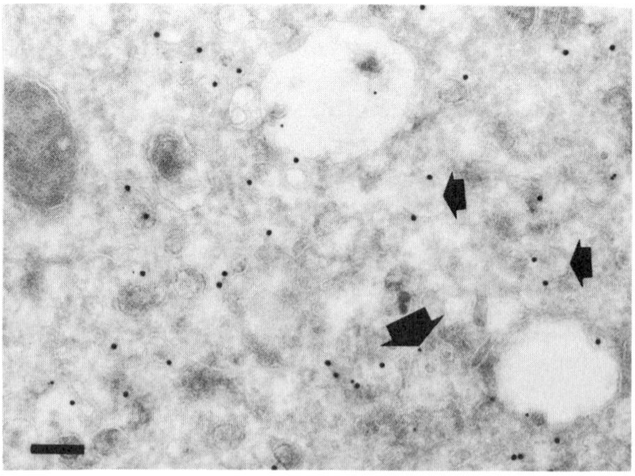

Figure 1. Double labeling for MHC class I (15 nm gold) and MHC class II (10 nm gold) on human blood derived dendritic cells. Many vesicles positive for MHC class II molecules can be found in the cell centre (small arrows), whereas only a few MHC class I positive vesicles can be found in this area (large arrows). Original magnification: X56420. Bar=0.2 μm.

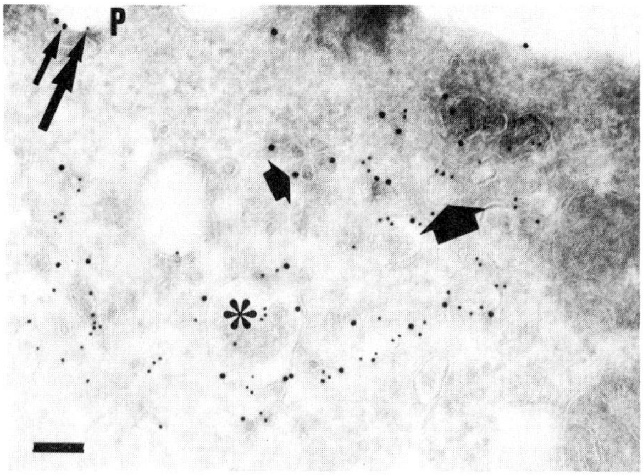

Figure 2. Double labeling for MHC class II molecules (15 nm gold) and the lysosomal integral membrane protein lamp-1 (10 nm gold) on human blood derived dendritic cells. Many MHC class II molecules (single arrow), but only a few lamp-1 molecules (double arrow) are present on the plasma membrane (P). In the cell centre vesicles only positive for MHC class II (small arrow), lamp-1 (large arrow), or both MHC class II and lamp-1 (asterix), can be observed. Original magnification: X56420. Bar=0.2 μm.

Class II molecules in Langerhans Cells

In Langerhans cells *in situ* MHC class II molecules can be found in peripherally located, multilammelar vesicles (Figure 3A) and in centrically located vesicles (Figure 3B). Peripherally located, multilamellar vesicles were also positive for the mannose 6-phosphate receptor (figure 4), strongly suggesting that the class II positive vesicles in peripheral position are prelysosomes. Birbeck granules, the characteristic LC granules, were neither positive for MHC class II molecules, nor for the M6PR.

DISCUSSION

In the rat system DC can be characterized by a cluster of acid phosphatase (APh) positive vesicles in the juxtanuclear area (Kamperdijk et al., 1987). More recently, the presence of a cluster of MHC class II positive vesicles in the same area was described in DC isolated from human peripheral blood (Arkema et al., 1990).

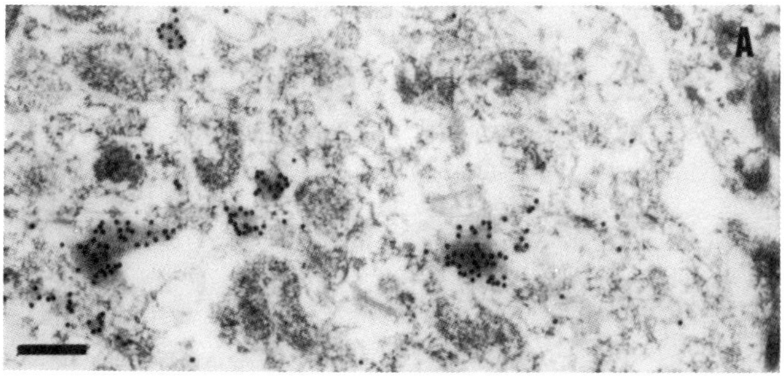

Figure 3A

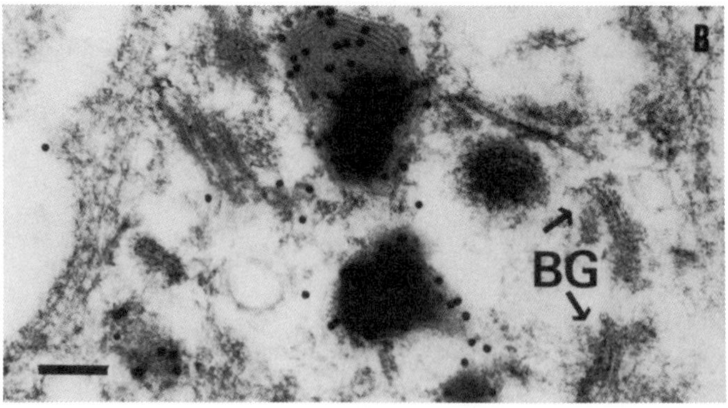

Figure 3B

Figure 3. Ultrathin Lowikryl HM_{20} section of a skin biopsy labelled for HLA-DRα and -β. Label (15 nm gold) is present in centrically located vesicles (A). Original magnification: X31500. Bar=0.3 μm. Label is also present in multilamellar vesicles (B). BG=Birbeck granulum. Original magnification X85000. Bar=0.2 μm.

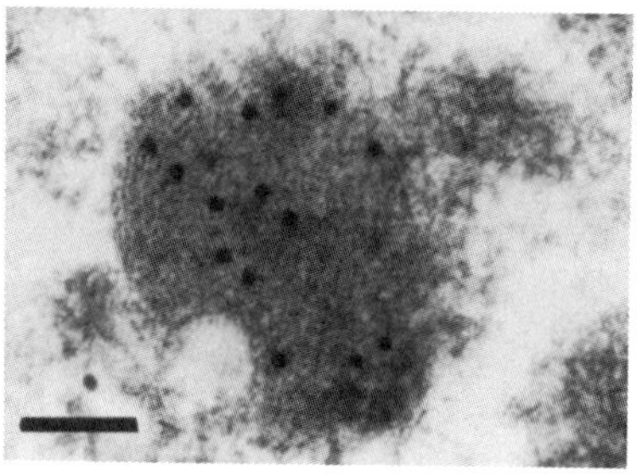

Figure 4. Ultrathin Lowikryl HM_{20} section of a skin biopsy labelled for the M6PR. Peripherally located multilamellar vesicles are haevily labelled. Original magnification X96000. Bar=0.1 μm.

After incubation of DC with BSA-gold, this endocytotic tracer could rarely be found in the MHC class II positive vesicles. Moreover, these vesicles disappeared after treatment with the protein synthesis inhibitor cycloheximide (Arkema et al., 1991). These findinges suggest that the MHC class II positive vesicles represent a storage compartment for MHC class II molecules.

We now conclude that DC and LC both contain a characteristic intracellular storage compartment for MHC class II molecules, which has connections to the lysosomal apparatus in the same region of the cell. We hypothesize that, depending on the functional phenotype of DC, the storage vesicles can deliver MHC class II molecules to different sites. A schematic model of the route of antigen and MHC class II molecules is given in figue 5.

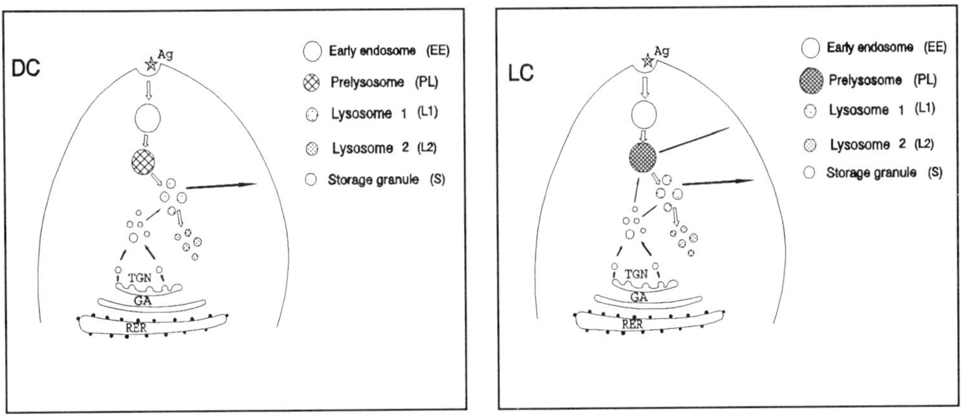

Figure 5. Schematic model of the endocytotic route of antigen (Ag) (white arrows) and MHC class II molecules (black arrows) in DC and LC.

REFERENCES

Arkema, J.M.S., Broekhuis-Fluitsma, D.M., de Laat, P.A.J.M., and Hoefsmit, E.C.M., 1990, Human peripheral blood dendritic cells concentrate in contrast to monocytes intracellular HLA class II molecules in a juxtanuclear position, *Immunobiology* 181:335.

Arkema, J.M.S., Schadee-Eestermans, I.L., Broekhuis-Fluitsma, D.M., and Hoefsmit, E.C.M., 1991, Localization of class II molecules in storage vesicles, endosomes and lysosomes in human dendritic cells, *Immunobiology* 183:396.

Hoefsmit, E.C.M, Duijvestijn, A.M., and Kamperdijk, E.W.A., 1982, Relation between Langerhans cells, veiled cells, and interdigitating cells, *Immunobiology* 161:255.

Inaba, K., and Steinman, R.M., 1984, Resting and sensitized T lymphocytes exhibit distinct stimulatory (antigen presenting cell) requirements for growth and lymphokine release, *J. Exp. Med.* 160:1717.

Kamperdijk, E.W.A., Kapsenberg, M.L., van den Berg, M., and Hoefsmit, E.C.M., 1985, Characterization of dendritic cells, isolated from normal and stimulated lymph nodes of the rat, *Cell Tissue Res.* 242:469.

Knight, S.C., Farrant, J., Bryant, A., Edwards, A.J., Burman, B., Lever, A., Clarke, J., and Webster, A.D.B., 1986, Non-adherent, low-density cells from human peripheral blood contain dendritic cells and monocytes, both with veiled morphology, *Immunology* 57:595.

Macatonia, S.E., Taylor, P.M., Knight, S.C., and Askonas, B.A., 1989, Primary stimulation by dendritic cells induces antiviral proliferative and cytotoxic T cell responses in vitro, *J. Exp. Med.* 169:1255.

Metlay, J.P., Pure, E., and Steinman, R.M., 1989, Distinct features of dendritic cells and anti-Ig activated B cells as stimulators of the primary mixed leucocyte reaction, *J. Exp. Med.* 169:239.

REQUIREMENTS OF EXOGENOUS PROTEIN ANTIGENS FOR PRESENTATION TO $CD4^+$ T LYMPHOCYTES BY MHC CLASS II-POSITIVE APC

Gernot Gradehandt, Johannes Hampl, Nadja Kleber, Christina Lobron, Silke Milbradt, Burkhard Schmidt, and Erwin Rüde

Institut für Immunologie, Johannes Gutenberg-Universität, W-6500 Mainz, Germany

INTRODUCTION

The antigen-specific activation of CD4-positive T helper cells depends on the recognition of a complex of MHC class II molecules and an antigen-derived peptide on the surface of antigen-presenting cells (APC). For most antigens generation of this MHC/peptide complex requires the uptake of the respective antigen by APC, followed by intracellular processing. The latter leads to suitable peptides of the antigen which are able to bind to MHC class II-molecules. Subsequently the resulting complexes are transported to the cell surface. Evidence supporting this concept came mainly from the finding that agents such as chloroquine[1], interfering with the function of endosomes and lysosomes, can block antigen presentation. The importance of proteolysis was demonstrated by the observation that certain preformed peptides do not require further processing and can also be presented by metabolically inactivated APC[2]. Nevertheless, the enzymes that are responsible for antigen processing are only known in few cases. Here we will describe the processing requirements for different protein antigens, all shown to need processing by APC prior to binding to MHC class II molecules: bovine insulin (BI), its S-sulfonated chymotryptic A-chain fragment (BI-A1-14), porcine insulin (PI), human insulin (HuI), ovalbumin (OVA), and conalbumin (CA).

PROCESSING STEPS INVOLVED IN THE PRESENTATION OF VARIOUS ANTIGENS

Processing by proteolytic cleavage

Evidence for the concept of proteolytic degradation of antigens to MHC-binding peptides by endo/lysosomal proteases came from the treatment of APC with protease inhibitors such as leupeptin[3] or pepstatin A[4] during antigen processing. We used these and other protease-inhibitors, among them Cbz-Phe-Ala-CHN_2, a selective inhibitor of cathepsin B and to a lesser extent of cathepsin H and L, to identify the proteases involved in the generation of antigenic peptides of CA, OVA, and BI. As shown in Fig. 1A, all these compounds failed to inhibit the presentation of BI or its A-chain fragment whose presentation, despite its small size, still depends on uptake and alteration by

Dendritic Cells in Fundamental and Clinical Immunology,
Edited by Kamperdijk *et al.*, Plenum Press, New York, 1993

APC[5]. This finding suggested that BI may be an antigen which needs processing by APC but not by proteolytic degradation[6]. In constrast, inhibition of cellular proteases enhanced the efficiency of insulin presentation (Fig. 1A), suggesting that intracellular proteolysis of insulin reduces the concentration of the antigenic peptides. Similar observations were reported from OVA by Vidard et al.[7]. In Fig. 1A it can also be seen that blocking of the endo/lysosomal thiol protease cathepsin B interfered with the processing of CA and OVA, while the presentation of an MHC-binding OVA-peptide remained unaffected and the presentation of insulin was enhanced. The latter argues against a general inhibitory effect of Cbz-Phe-Ala-CHN_2 on antigen presentation, for instance by interference with the degradation of the invariant chain (Ii) which may lead to alterations of the intracellular traffic of MHC class II-molecules[8].

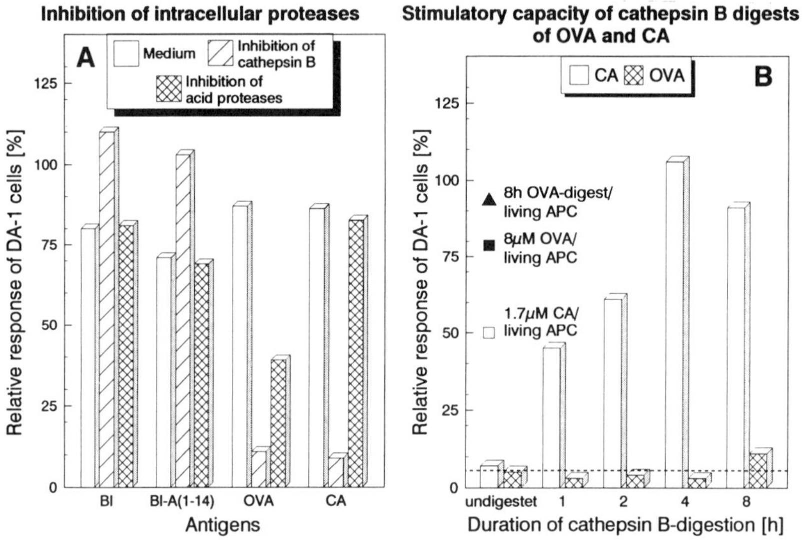

Figure 1: Panel A: B-hybridoma cells were pulsed with suboptimal concentrations of the antigens in presence or absence of 10 μg/ml pepstatin A or 8 μM Cbz-Phe-Ala-CHN_2. After fixation with 0.05 % glutaraldehyde (GA) they served as stimulator cells for specific T cell clones. Panel B: Fixed B cell-hybridomas were cocultivated with cathepsin B digests of CA and OVA and specific T cells. In both experiments supernatants of the cultures were harvested after 48 h and tested for their content of IL-3 and GM-CSF by using the DA-1 indicator cell line. Results are expressed as relative response of DA-1 cells compared to proliferation with saturating concentrations of IL-3.

For further characterization of the mode of action by which cathepsin B influences the presentation of OVA and CA, the two antigens were digested with the isolated enzyme. It was found that cathepsin B generates small (< 5000 kD) proteolytic CA-fragments[9] but in the case of OVA only 20-30 kD-fragments could be observed after extensive digestion (data not shown). In agreement with these results only the cathepsin B-derived

CA-digests were able to stimulate a CA-specific T cell clone together with fixed APC, while the OVA-digest was not stimulatory for OVA-specific T cells. Interestingly, these digests could be presented by living APC, indicating the involvement of other proteases which generate the antigenic OVA-peptide, for instance the acid proteases cathepsin D[4] or cathepsin E[10]. Possibly processing of OVA involves a combination of both types of proteases as indicated by the inhibitory effect of Cbz-Phe-Ala-CHN_2 (cathepsin B) as well as pepstatin A (cathepsin H,E) on OVA-presentation (Fig. 1A).

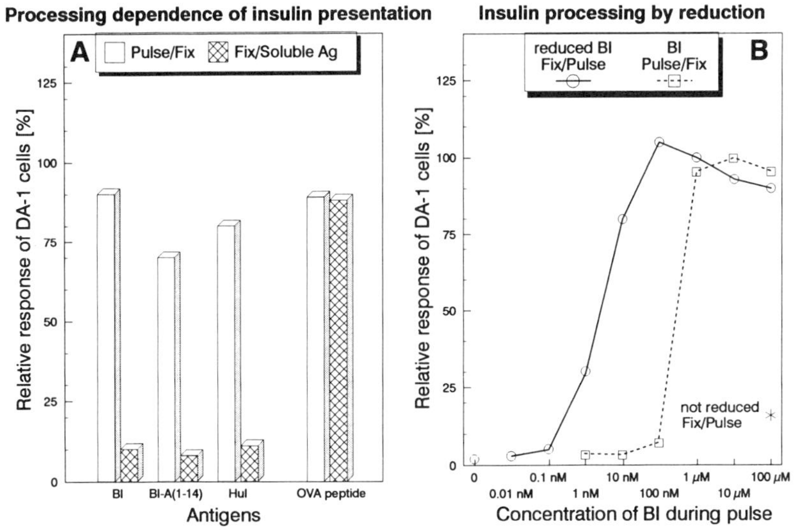

Figure 2: Panel A: LBK fibroblasts were pulsed with 5 μM of the antigens for 3 h, then fixed with GA (pulse/fix), and cocultured with specific T cells. Alternatively, the APC were prefixed with GA and the antigens were added in soluble form to the stimulation cultures (fix/soluble antigen). Panel B: BI was reduced with 10 mM glutathione. APC were pulsed with the reduced antigen under acidic conditions and were added to specific T cells. For comparison live APC were pulsed with BI, then fixed, and cocultured with T cells. The IL-3 content in 48 h supernatants was determined as described in Fig. 1.

Processing by reduction of disulfide bonds

Insulin is a model antigen to demonstrate the importance of disulfide reduction during intracellular processing. It consists of 51 amino acid residues arranged on two polypeptide chains (A and B) which are linked by two disulfide bonds (A7/B7 and A20/B19). The A chain contains an additional intrachain disulfide loop (A6/A11). Our insulin-specific T cells recognize an epitope within the chymotryptic A-chain peptide A1-14. This peptide could not be presented efficiently by fixed APC despite its relatively small size. Therefore A1-14 as well as native insulin need processing (Fig. 2)[11].

Upon reduction of insulin by glutathione this antigen could be presented by fixed APC in a highly efficient way (Fig. 3B, fix/pulse). In comparison, presentation by pulsing living APC with insulin followed by fixation wa about 100-fold less efficient (Fig. 3B, pulse/fix). Further characterization of this processing event using peptides with different substitutions of cysteine by serine residues showed that only CysA7 (which connects the A- and B-chains in insulin) is essential for activation by specific T cells [12,13]. This indicates that the reductive opening of disulfide bonds and generation of CysSH residues is a sufficient processing step for the presentation of insulin.

In summary, these data show that in addition to proteolytic fragmentation of proteins by thiol[9] and acidic proteases[4] the reduction of disulfide bonds play a key role in the processing of exogenous antigens.

Table 1. Processing requirements of the examined antigens

Antigen	Proteolysis	Reduction
BI, PI, HuI, BI-A(1-14)	no	yes
CA	cathepsin B	?
OVA	Cathepsin D[4] or E[10]	
	+ Cathepsin B (?)	yes[14]

REFERENCES

1 H.K. Ziegler and E.R. Unanue. Decrease in macrophage antigen catabolism caused by ammonia and chloroquine is associated with inhibition of antigen presentation to T cells, *Proc.Natl.Acad.Sci.U.S.A.* 79:175 (1982)

2 R. Shimonkevitz, J. Kappler, P. Marrack and H. Grey. Antigen recognition by H-2-restricted T cells. I. Cell-free antigen processing, *J.Exp.Med.* 158:303 (1983)

3 J. Puri and Y. Factorovich. Selective inhibition of antigen presentation to cloned T cells by protease inhibitors, *J.Immunol.* 141:3313 (1988)

4 S. Diment. Different roles for thiol and aspartyl proteases in antigen presentation of ovalbumin, *J.Immunol.* 145:417 (1990)

5 G. Gradehandt, J. Hampl, D. Plachov, K. Reske and E. Rude. Processing requirements for the recognition of insulin fragments by murine T cells, *Immunol.Rev.* 106:59 (1988)

6 G. Gradehandt, J. Hampl, S. Milbradt and E. Rude. Processing without proteolytic cleavage is required for recognition of insulin by T cells, *Eur.J.Immunol.* 20:2637 (1990)

7 L. Vidard, K.L. Rock and B. Benacerraf. The generation of immunogenic peptides can be selectively increased or decreased by proteolytic enzyme inhibitors, *J.Immunol.* 147:1786 (1991)

8 V.E. Reyes, S. Lu and R.E. Humphreys. Cathepsin B cleavage of Ii from class II MHC alpha- and beta- chains, *J.Immunol.* 146:3877 (1991)

9 G. Gradehandt and E. Ruede. The endo/lysosomal protease cathepsin B is able to process conalbumin fragments for presentation to T cells, *Immunology* 74:393 (1991)

10 K. Bennet, T. Levine, E.S. Ellis, R.J. Peanasky, I.M. Samloff, K. Kay and B.M. Chain. Antigen processing for presentation by class II major histocompatibility complex requires cleavage by cathepsin E, *Eur.J.Immunol.* 22:1519 (1992)

11 J. Hampl, G. Gradehandt, D. Plachov, H.G. Gattner, H. Kalbacher, W. Voelter, M. Meyer Delius and E. Rude. Presentation of insulin and insulin A chain peptides to mouse T cells: involvement of cysteine residues, *Mol.Immunol.* 28:479 (1991)

12 P.E. Jensen. Reduction of disulfide bonds during antigen processing: evidence from a thiol-dependent insulin determinant, *J.Exp.Med.* 174:1121 (1991)
13 J. Hampl, G. Gradehandt, H. Kalbacher and E. Rüde. In vitro processing of insulin for recognition by murine T cells results in the generation of A-chains with free CysSH, *J.Immunol.* 148:2664 (1992)
14 D.S. Collins, E.R. Unanue and C.V. Harding. Reduction of disulfide bonds within lysosomes is a key step in antigen processing, *J.Immunol.* 147:4054 (1991)

MODULATION OF MHC CLASS II DETERMINANTS ON RAT LANGERHANS CELLS DURING SHORT TERM CULTURE

Ursula Neiß,[1] Detlef Becker,[2] Jürgen Knop,[2] and Konrad Reske[1]

[1]Institut für Immunologie
[2]Hautklinik der Johannes Gutenberg Universität
D-6500 Mainz, FRG

INTRODUCTION

Epidermal Langerhans cells (LC) are regarded as the most peripheral outpost of the immune system. They play a pivotal role during the onset of an immune response in the skin. One of the principal functions of LC is reflected in their extraordinary potency to present antigen to high activation requiring naive T cells.

Prominent levels of cell surface expressed class II molecules are well documented for freshly isolated mouse and human LC[1,2]. As short term culture of LC results in further increase in expression of class II determinants, including several phenotypical and functional changes, it has been suggested that LC can adopt two functionally distinct states[3,4]. Freshly prepared LC can process and present protein antigens to previously primed T cells. Upon short term culture LC loose their processing capacity but instead gain costimulatory potential, that converts them into highly efficient accessory cells for activation of unprimed T cells. It has been suggested that fresh LC are mirrored by intraepidermal LC whereas cultured LC represent in vitro equivalents of LC, that have migrated from skin to the draining lymph nodes[5].

Work from our laboratory revealed the existance of mAb defined folding isomers of MHC class II molecules derived from various accessory cells of Lewis rats[6,7]. These and studies from other groups demonstrated that class II molecules undergo profound conformational changes along their biosynthetic route to the cell surface[8,9]. Consistent with these reports work by Murphy et al.[10] suggested that MHC class II molecules differ in conformation in a tissue-specific fashion.

Since the possibility existed that distinct conformational forms of class II restriction elements might be involved in cutaneous antigen presentation we investigated the constitutive synthesis and expression of class II in Lewis rat LC. Our further interest focused on invariant chain expression in these cells.

Dendritic Cells in Fundamental and Clinical Immunology,
Edited by Kamperdijk *et al.*, Plenum Press, New York, 1993

RESULTS AND DISCUSSION

Cell Surface Expression of MHC Class II Molecules and Invariant Chain on freshly prepared Rat LC

LC are known to be the principal cell population expressing MHC class II molecules constitutively within the epidermis[1,2]. Thus in order to identify LC in epidermal cell (EC) suspensions of Lewis rats a series of experiments was performed assessing the cell surface expression of Ia-molecules, by use of the two noncrossreactive mAb OX3 and OX6. FACS-analysis revealed distinct class II molecule expression on 3-5% of EC (Fig. 1A). Notably, as earlier shown for B cell class II[6] comparable numbers of OX3- and OX6-determinant expressing cells were detectable, suggesting that LC carry the two conformational forms defined by the mAb on their cell surface.

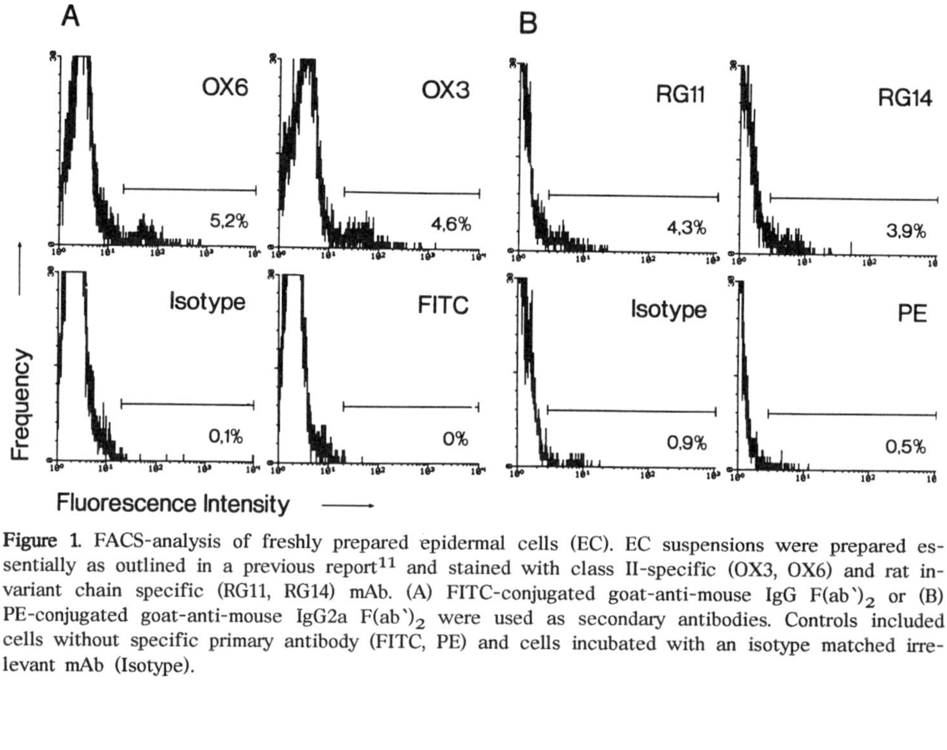

Figure 1. FACS-analysis of freshly prepared epidermal cells (EC). EC suspensions were prepared essentially as outlined in a previous report[11] and stained with class II-specific (OX3, OX6) and rat invariant chain specific (RG11, RG14) mAb. (A) FITC-conjugated goat-anti-mouse IgG F(ab')$_2$ or (B) PE-conjugated goat-anti-mouse IgG2a F(ab')$_2$ were used as secondary antibodies. Controls included cells without specific primary antibody (FITC, PE) and cells incubated with an isotype matched irrelevant mAb (Isotype).

Several mAb directed against the extracytoplasmic C-terminus of rat invariant chain were raised in our lab[12] and have been used to investigate γ chain expression on rat LC. mAb RG11 and RG14 that demonstrated cell surface display of the invariant chain on rat spleen cells also stained a small quantity of cells within the EC suspensions (Fig. 1B). In contrast mAb RG10 which was shown to recognize immature intracellular forms of the invariant chain preferentially failed to stain epidermal cells (Fig. 1B Isotype).

As class II- and γ chain-specific mAb identified roughly the same proportion of cells we conclude that MHC class II molecules and the invariant chain are coexpressed at the cell surface of freshly prepared rat LC.

Biochemical Properties of LC-derived Class II Molecules

Our further interest focused on the composition and chemical nature of class II molecules and the relationship between polymorphic and invariant chains.

To augment the fraction of LC, EC were subjected to a round of mAb panning employing the RT1.B-specific mAb OX6. Positively selected cells contained 18% LC. Biosynthetic labelling, immunoprecipitation analysis and 2D gel electrophoresis were performed with this cell fraction. Spleen cells were included as a source of constitutively Ia synthesizing B cells because 2D spot patterns obtained with the mAb used have been explored extensively in previous studies.

As shown in Fig. 2 A,B both mAb precipitate newly synthezied class II complexes in the LC fraction. Consistent with earlier observations with B cells[6] mAb OX3 recognizes a biosynthetically mature, terminally glycosylated α,β-heterodimer without invariant chains while the class II peptide chain complexes precipitated by mAb OX6 contain in addition the proteins of the invariant γ chain group.

The failure of mAb OX6 to precipitate in LC but not in spleen cells immature cytoplasmic precursor forms of the polymorphic chains might implicate differences in the turnover of class II molecules between these cells.

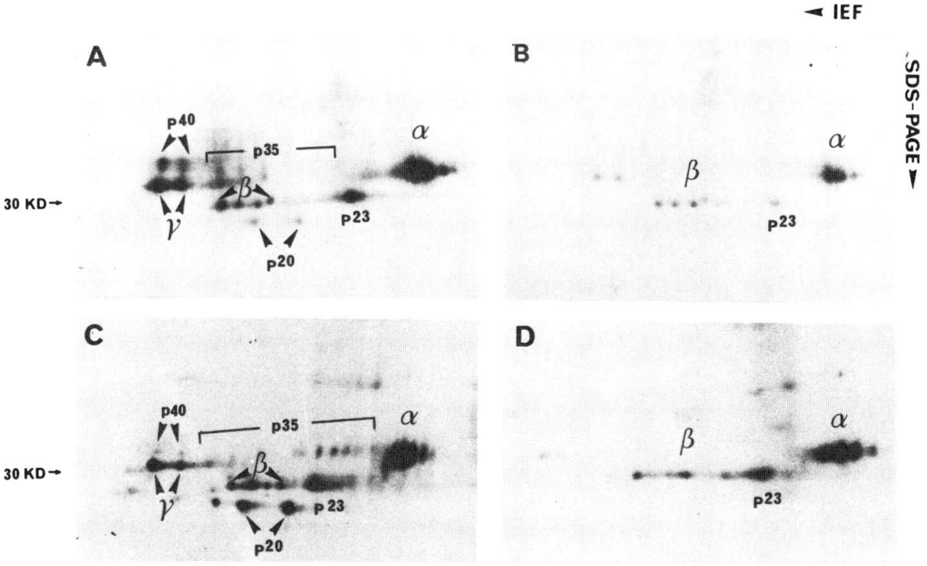

Figure 2. Comparative 2D-PAGE analysis of metabolically labelled RT1.B (I-A-equivalent) molecules from Lewis rat LC-enriched epidermal cell preparations. Cells were labelled with [^{35}S]-methionine for 4 h, solubilized with NP-40 and subjected to immunoprecipitation using mAb OX6 (A,C) and OX3 (B,D). Specific precipitates were resolved by twodimensional gel electrophoresis (1D: IEF; 2D: SDS-PAGE). A,B: Freshly prepared LC-enriched epidermal cell suspension (18% Ia$^+$ cells) obtained by panning. C,D: Lewis rat spleen cells used as positive control.

As described for mouse LC[11,13] rat LC synthesize high amounts of invariant chain form p40. However the excessively sialylated forms p35 and p45 of γ and p40, which have been discussed to be an outstanding feature of mouse LC, were absent in rat LC. This might be due to a high turnover rate of invariant chain products in general in rat LC, in particular of the highly glycosylated mature forms.

Alternatively poor sialylation of the invariant chain might be a prominent feature of rat LC. A low degree of sialylation of surface MHC class I and II molecules on dendritic cells has been discussed to contribute to the excellent immunostimulatory capacity of this type of accessory cells[14]. Although conspicuous differences between LC and control spleen cells in the glycosylation pattern of the polymorphic chains were not detectable, the deficient sialylation of the γ chain is remarkable. Since cell surface expression of the invariant polypeptide has been documented in this study (see Fig. 1B) it seems reasonable to consider the possibility that the low degree of γ-sialylation might influence APC function of rat LC.

Phenotypic Alterations of LC aquired during Short Term Culture

A number of studies indicate that in terms of antigen processing and presentation LC can occur in two fundamentally different states[3,4], reflected by phenotypical and functional changes observed between freshly isolated and short term cultured LC. In accordance with results published for mouse[3,11] and human LC[15] a significant upregulation of MHC class II determinants on short term cultured rat LC was revealed by FACS-analysis (data not shown). However invariant chain expression on these cells showed a quite different behaviour.

Whereas cell surface display of γ chain on freshly prepared human LC has been reported[16], a decrease of intracellular invariant chain for short term cultured mouse LC was detected by immunocytological staining of permeabilized cells[13,17]. When we performed double labelling experiments using class II- and γ chain-specific mAb we could clearly detect invariant chain molecules at the cell surface of 3 day cultured rat LC (Fig. 3). However in contrast to the elevated class II expression an increase in invariant chain expression upon short term culture could not be demonstrated.

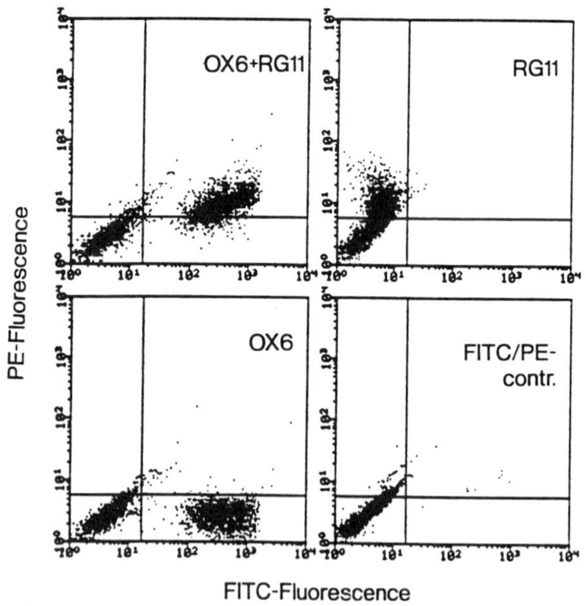

Figure 3. Double labelling analysis of 3 day cultured rat LC. Unseparated EC were cultured in DMEM, 10% FCS, glutamine (2 mM), penicillin (100 μg/ml), gentamycin (50 μg/ml). After 3 days of culture cells were harvested and dead cells were removed by incubation in 0,1% trypsin, 150 U/ml DNase for 30 min at 37°C. Double labelling was performed using OX6-FITC-conjugate to detect Ia^{+} cells. Invariant chain expression was identified by mAb RG11 and a subclass-specific PE-conjugated secondary antibody.

Considering the down-regulation of class II and invariant chain synthesis in short term cultured mouse LC[11,17] the invariant chain seems to be stably expressed at the cell surface of rat LC, at least for 3 days. These findings are distinct from a recent report describing a rather short half-life of invariant chain products in LC[13].

Severel studies emphazise a profound effect of the invariant chain on the intracellular routing and expression of MHC class II molecules. Furthermore proteolytic cleavage of invariant chain with subsequent dissociation from the α,β-dimer complex within an endosomal compartment is thought to enable peptide binding to the restriction element, thus implicating a role of γ in antigen-presentation (for review see [18]).

Here we demonstrate binding of γ chain-specific mAb to the surface of rat LC and show that binding persists on 3 day cultured LC. Wether this binding detects intact γ chain or γ chain fragments originated from its endosomal breakdown remains to be elucidated. In any case, the presence of γ chain determinants at the cell surface of class II expressing cells might have attractive functional implications.

ACKNOWLEDMENTS

This work was supported by BMFT 01KC8804 and BMFT 50QV8626.

REFERENCES

1. L. Klareskog, U. Malmnäs Tjernlund, U. Forsum, and P.A. Peterson, Epidermal Langerhans cells express Ia antigens, Nature 268:248 (1977)
2. K. Tamaki, G. Stingl, M. Gullino, D.H. Sachs, and S.I. Katz, Ia antigens in mouse skin are predominantly expressed on Langerhans cells, J. Immunol. 123:784 (1979)
3. G. Schuler, and R.M. Steinman, Murine epidermal Langerhans cells mature into potent immunostimulatory dendritic cells in vitro, J. Exp. Med. 161:526 (1985)
4. J.W. Streilein, and S.F. Grammer, In vitro evidence that Langerhans cells can adopt two functionally distinct forms capable of antigen presentation to T lymphocytes, J. Immunol. 143:3925 (1989)
5. N. Romani, S. Koide, M. Crowley, M. Witmer-Pack, A.M. Livingstone, C.G. Fathman, K. Inaba, and R.M. Steinman, Presentation of exogenous protein antigens by dendritic cells to T cell clones, J. Exp. Med. 169:1169 (1989)
6. K. Reske, and R. Weitzel, Immunologically discrete conformation isomers of I-A locus-equivalent class II molecules detected in Lewis rats, Eur. J. Immunol. 15:1229 (1985)
7. K. Reske, U. Möhle, D. Sun, and H. Wekerle, Synthesis and cell surface display of class II determinants by long-term propagated rat T line cells, Eur. J. Immunol. 17:909 (1987)
8. M. Peterson, and J. Miller, Invariant chain influences the immunological recognition of MHC class II molecules, Nature 345:172 (1990)
9. R.N. Germain, and L.R. Hendrix, MHC class II structure, occupancy and surface expression determined by post-endoplasmic reticulum antigen binding, Nature 353:134 (1991)
10. D.B. Murphy, D. Lo, S. Rath, R.L. Brinster, R.A. Flavell, A. Slanetz and C.A. Janeway, Jr., A novel MHC class II epitope expressed in thymic medulla but not cortex, Nature 338:765 (1989)
11. D. Becker, A.B. Reske-Kunz, J. Knop, and K. Reske, Biochemical properties of MHC class II molecules endogenously synthesized and expressed by mouse Langerhans cells, Eur. J. Immunol. 21:1213 (1991)
12. A. Fisch, and K. Reske, Cell surface display of rat invariant chain: Detection by monoclonal antibodies directed against a C-terminal γ chain segment, Eur. J. Immunol. 22:1413 (1992)
13. E. Kämpgen, N. Koch, F. Koch, P. Stöger, C. Heufler, G. Schuler, and N. Romani, Class II major histocompatibility complex molecules of murine dendritic cells: Synthesis, sialylation of invariant chain, and antigen processing capacity are down-regulated upon culture, Proc. Natl. Acad. Sci. USA 88:3014 (1991)
14. C.H.P. Boog, J.J. Neefjes, J. Boes, H.L. Ploegh, and C.J.M. Melief, Specific immune responses restored by alteration in carbohydrate chains of surface molecules on antigen-presenting cells, Eur. J. Immunol. 19:537 (1989)
15. N. Romani, A. Lenz, H. Glassel, H. Stössel, U. Stanzl, O. Majdic, P. Fritsch, and G. Schuler, Cultured human Langerhans cells resemble lymphoid dendritic cells in phenotype and function, J. Invest. Dermatol. 93:600 (1989)
16. L. Claesson-Welsh, A. Scheynius, U. Tjernlund, and P.A. Peterson, Cell surface expression of invariant γ-chain of class II histocompatibility antigens in human skin, J. Immunol. 136:484 (1985)

17. E. Pure´, K. Inaba, M.T. Crowley, L. Tardelli, M.D. Witmer-Pack, G. Ruberti, G. Fathman, and R.M. Steinman, Antigen processing by epidermal Langerhans cells correlates with the level of biosynthesis of major histocompatibility complex class II molecules and expression of invariant chain, J. Exp. Med. 172:1459 (1990)
18. J.J. Neefjes, and H.L. Ploegh, Intracellular transport of MHC class II molecules, Immunol. Today 13:179 (1992)

A DENDRITIC CELL SPECIFIC DETERMINANT PRESENT IN ENDOSOMES IS INVOLVED IN THE PRESENTATION OF PROTEIN ANTIGENS

Toshiyuki Maruyama, Elisabeth C.M. Hoefsmit and Georg Kraal

Department of Cell Biology
Faculty of Medicine, Vrije Universiteit
Van der Boechorststraat 7
1081 BT Amsterdam, The Netherlands

INTRODUCTION

Antigen presenting cells (APC) such as dendritic cells (DC), macrophages (M), and B lymphocytes present exogenous antigens on their cell surface to T cells, not as native proteins but as peptides, complexed with MHC-class II molecules (1). In contrast to macrophages where antigen processing occurs in endosomes (2). It is still unclear how and by which organelles the processing takes place in DC or B-cells (3). Using fluorescein conjugated protein antigen it has been shown that isolated splenic DC can take up some amount of protein antigen in intra-cytoplasmic organelles after over-night culturing. Furthermore, these DC were now able to prime T cells in situ (4). Therefore DC must have a small but relevant intra-cytoplasmic system for antigen processing.

A monoclonal antibody, MIDC-8, has been described which specifically recognizes a cytoplasmic determinant in cells of the DC lineage in mice such as interdigitating cells (IDC) in T cell dependent areas of secondary lymphoid organs, veiled cells, Langerhans cells and isolated DC from lymph node, spleen and thymus (5). By light microscopy the determinant recognized by the antibody can be seen as distinct spots near the nucleus. In this study we wished to determine the sub-cellular distribution of the MIDC-8 antigen, its relationship with endocytosed antigen, and the possible effects of the antibody on the antigen presenting function of DC.

THE LOCALIZATION AND FUNCTION OF THE MIDC-8 ANTIGEN

Using a pre-embedding labeling procedure for electron microscopy MIDC-8 was exclusively present in vesicles in lymph node IDC (Fig.1). The peroxidase reac-

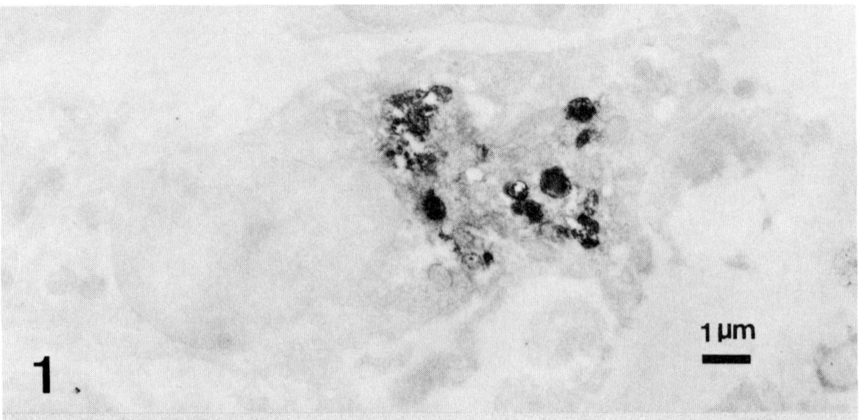

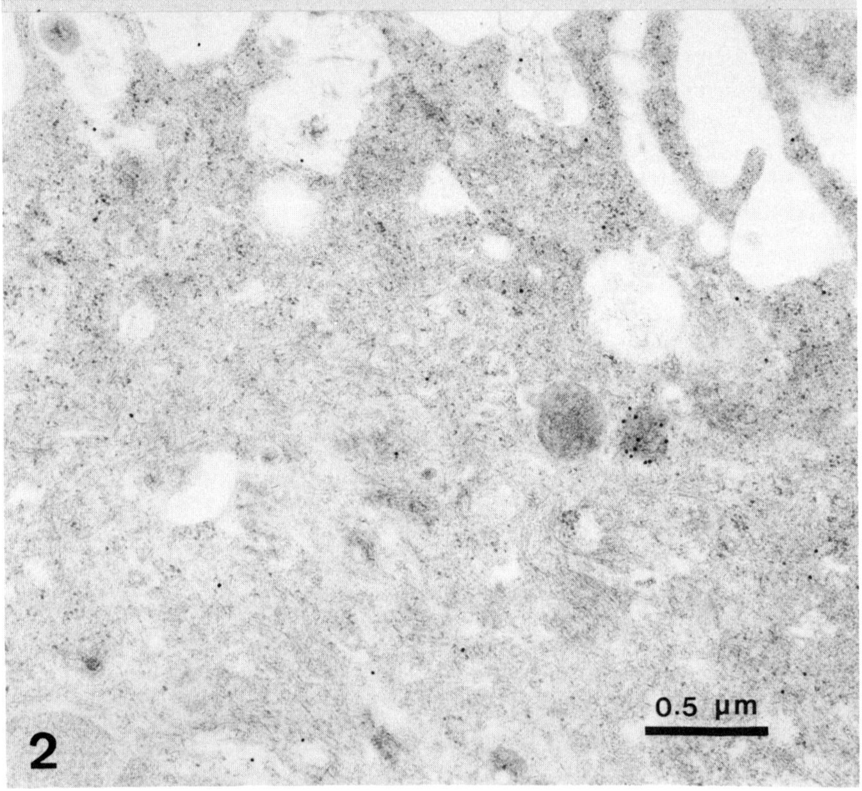

Figure 1. MIDC-8 in an interdigitating cell in situ.

A focal distribution of the reaction product is seen in endosomal organelles of a typical interdigitating cell after pre-embedding staining for immuno-electron microscopy. The cell is located in the paracortex, and contains an indented nucleus with unevenly distributed heterochromatin at the periphery of the nucleus. Small endosomes are found in the cytoplasm near the nucleus. The reaction product is seen inside and on the membrane of endosomal organelles (x 8300).

Figure 2. MIDC-8 in isolated dendritic cells. Post embedding staining

After post-embedding staining on ultrathin sections the MIDC-8 associated 10 nm gold particles can be observed in the endosomal vesicles of dendritic cells isolated from BALB/c lymph nodes (x 41 000).

tion product was distributed evenly throughout the positive vesicles. There was also some reaction product on the luminal side of the endosomal membrane.

When fluorescence double labeling was performed with the isolated dendritic cells (DC) we were able to confirm the uptake of TRITC-OVA by small vesicles as described by Inaba et al. (4). Furthermore, the majority of these vesicles were also positive for MIDC-8.

In addition to the pre-embedding labeling in situ, also post-embedding staining was performed on isolated DC using MIDC-8 in combination with gold particles (10nm). This confirmed that MIDC-8 staining was exclusively present in vesicles which could belong to the endocytotic or secretory route (Fig.2).

We then tested whether the structure recognized by the antibody could be involved in antigen processing, within the endosomal complex. Therefore, purified MIDC-8 antibody was added to isolated DC during over night pulsing with OVA, and these cells were thereafter added to OVA-primed T-cells.

This way it could be demonstrated that the MIDC-8 antibody interfered with antigen specific T-cell proliferation. This was most prominent when T-cell proliferation was performed in the continuous presence of MIDC-8 antibody (Table I). A similar effect was seen when in vitro OVA pulsed DC were injected into footpads of mice. When 5 days later T cells from draining lymph nodes were assayed for antigen-specific proliferation previous MIDC-8 incubation of the tranferred DC resulted in decrease of proliferation (Table II).

The molecular mass of the antigen recognized by MIDC-8 was determined by affinity chromography using a spleen DC lysate. The elution fluid obtained from the column showed a single band of 50 kD when it was run on SDS-PAGE under non-reducing conditions and of 66 kD under reducing conditions. This indicates that the MIDC-8 antibody recognizes a single chain protein, probably containing intra-chain sulphur bridges.

DISCUSSION

The above findings strongly suggest a role for the MIDC-8 molecule in antigen processing. So far we have only been able to identify the MIDC-8 antigen in cells of the DC lineage, which may be related to a concentration effect but is in line with recent studies in which it was emphasized that endosomes of cells of the DC lineage are more specialized in antigen processing , whereas in M antigen digestion and elimination is more important (6).

The molecular mass of the MIDC-8 antigen makes it unlikely that the antibody recognizes a DC specific epitope on invariant chains or MHC class II molecules. Since antigen processing is a multistep procedure involving proteolysis of antigen and binding to MHC class II antigens, but also proteolysis and degradation of invariant chain probably occuring in the endosomal or lysosomal compartment (7) it is unclear at what step the MIDC-8 antibody interferes. The mode of action of MIDC-8 resembles that of chloroquin, which lysosomotropic agent is thought to work via blockade of proteolytic enzyme activity. This is accomplished by neutralizing the pH of endosomes leading to inhibition of antigen processing and presentation (8).

In our experiments in which the antibody was washed away and in the in situ priming assay both MIDC-8 and chloroquin showed only partial inhibitory effects.

Table 1. Suppressive effect of MIDC-8 on antigen presentation in vitro

APC	Treatment		T cell proliferation	
Control DC	-		3.7	(0.6)
OVA pulsed DC	-		22.5	(2.2)
(100 μg /ml)	NLDC-145	0.1 mg/ml[a]	23.5	(1.0)
	Human IgG	0.1 mg/ml[a]	19.9	(4.6)
	Chloroquin	0.025 mM[a]	3.5	(0.1)**
	MIDC-8	0.1 mg/ml[a]	7.4	(0.8)**
	MIDC-8	0.02 mg/ml[a]	11.4	(0.3)**
	MIDC-8	0.01 mg/ml[a]	11.7	(0.8)**
	MIDC-8	0.005 mg/ml[a]	14.9	(0.8)**
	Cloroquin	0.025 mM[b]	16.0	(0.4)**
	MIDC-8	0.1 mg/ml[b]	15.5	(4.3)**
	MIDC-8	0.02 mg/ml[b]	18.6	(3.1)**
	MIDC-8	0.01 mg/ml[b]	16.4	(3.1)**
	MIDC-8	0.005 mg/ml[b]	17.9	(0.1)**

Purified OVA-primed T cells were prepared and 8×10^5 cells were added to 2×10^4 pulsed spleen dendritic cells. [^{3}H]TdR was added at 72-90 h to measure and the data are expressed as mean (SD) cpm x 10^{-3} from triplicate wells.

Dendritic cells were pulsed by culturing them for 18 h in 96 well flat bottom plates at 2×10^4 per well in the presence of 100 g/ml OVA. The plates were washed three times before further use.

a): Dendritic cells were pulsed with OVA in the presence of MIDC-8, NLDC-145 (9), human IgG or chloroquin (25 M) which was added to the cells 1 h previous to the OVA. The concentrations of antibodies used are indicated.

After washing the cells to remove free OVA the antibodies and chloroquin were added back to the wells, together with the primed T cells in a), but not in b). **: The statistical significance was computed using the student T test ($P < 0.01$).

Table 2. Suppressive effects of MIDC-8 on antigen presentation in vivo

Antigen presenting cell	challenge	cell proliferation	
Control DC	-	3.6	(0.9)
Control DC	TRITC-OVA	6.9	(1.7)
pulsed DC	-	9.4	(6.6)
pulsed DC	TRITC-OVA	74.3	(14.0)*
pulsed DC + MIDC-8	-	12.8	(8.2)
pulsed DC + MIDC-8	TRITC-OVA	57.5	(14.5)*

Spleen dendritic cells were incubated with MIDC-8 for 1 h and pulsed in vitro for 18 h with TRITC-OVA (0.1 mg/ml) in the continuous presence of MIDC-8. After washing, 3×10^5 cells were injected in the left front foot pad. After 5 days lymphocyte suspensions of the draining lymph nodes were cultured in 96 well flat bottom culture plates at 1×10^6 cells / well with or without TRITC-OVA (0.1 mg/ml). [^{3}H]TdR was added at 48-66 h to measure DNA synthesis. Data are mean cpm (SD) x 10^{-3} of the mean of three mice from triplicate wells in each animal. *: Statistical significance was determined using the student T test ($P<0.01$).

Since it is known that DC can retain antigen in an immunogenic form for several days in culture (4), it indicates that the effects of MIDC-8 and chloroquine are reversible and once the endosomal condition has been reestablished, normal antigen processing and presentation can occur.

The exact nature of the MIDC-8 antigen must come from further molecular analysis. In the meantime the data stress the uniqueness of dendritic cells in their antigen presenting capacity.

REFERENCES

1) T.J. Braciale and V.L. Braciale, Antigen presentation: Structural themes and functional variations, Immunol. Today 12:124 (1991).
2) E.R. Unanue, Antigen-presenting function of the macrophage, Annu. Rev. Immunol. 2:395 (1984).
3) P.D. King and R. Katz, Mechanisms of dendritic cell function, Immunol. Today 11: 206 (1990).
4) K. Inaba, P. Metlay, M.T. Crowley and R.M. Steinman, Dendritic cells pulsed with protein antigens in vitro can prime antigen-specific MHC-restricted T cell in situ. J. Exp. Med. 172: 631 (1990).
5) M. Breel, R.E. Mebius and G. Kraal, Dendritic cells of the mouse recognized by two monoclonal antibodies, Eur. J. Immunol. 17: 1555 (1987)
6) E. Puré, K. Inaba, M.T. Crowley, L. Tardelli, D.W. Witmer-Pack, G. Fathman and R.M. Steinman, antigen processing by epidermal Langerhans cells correlates with the level of biosynsthesis of major histocompatibility complex class II molecules and expression of invariant chain, J. Exp. Med. 172: 1459 (1990).
7) W.L. Elliott, C.J. Steille, L.J. Thomas and R.E. Humphreys, An hypothesis on the binding of an amphipathic, helical sequence in Ii to the desetope of class II antigens, J. Immunol. 138: 2949 (1987).
8) H.K. Zieglar and E.R. Unanue, Decrease in macrophage antigen catabolism caused by ammonia and chloroquin is associated with inhibition of antigen presentation to T cells, Proc. Natl. Acad. Sci. USA 79: 175 (1982).
9) G. Kraal, M. Breel, E.M. Janse and G. Bruin, Langerhans' cells, veiled cells, and interdigitating cells in the mouse recognized by a monoclonal antibody, J. Exp. Med. 163: 981 (1986).

DENDRITIC CELLS ARE POTENT ANTIGEN-PRESENTING CELLS FOR MICROBIAL SUPERANTIGEN

Sharifah Iqball[1], Reshma Bhatt[2], Penelope Bedford[1], Peter Borriello[2] and Stella Knight[1]

[1]Antigen Presentation Research Group
[2]Microbial Pathogenicity Research
Clinical Research Centre
Watford Road
Harrow Middlesex HA1 3UJ

SUMMARY

Dendritic cells (DC) were found to be more efficient than macrophages (MO) in activating T cell responses to Staphylococcal enterotoxin B (SEB) using the hanging drop techniques and DC as antigen presenting cells (APC). When superantigen was presented via DC, the activation of T cells was not dependent on antigen processing and MHC class II molecules IA and IE were involved.

INTRODUCTION

Microbial superantigens such as the Staphylococcal enterotoxins were first named for their association with the symptoms of food poisoning. The latter comprise of a group of structurally related but serologically distinct proteins produced by certain strains of Staphylococcus aureus. Staphylococcal enterotoxins are the most potent mitogens known and at extremely low concentrations[1] stimulate both human and murine lymphocytes. These superantigens require antigen presenting cells bearing major histocompatibility complex (MHC) class II molecules to stimulate T cells[2] and they appear to bind to sites on MHC class II molecules that are located outside the peptide binding groove[3]. The complex ligand so formed between the T cell and the MHC class II molecule has specificity for a particular part of T cell receptors vβ and by engaging vβ, can stimulate many T cells[4]. Superantigens can therefore be distinguished from each other by which vβ T cells they stimulate.

In this study, we have looked at the capacity of DC and MO to act as APC for the stimulation of T cells using SEB. DC were found to be more efficient than macrophages in initiating the T cell response.

MATERIALS AND METHODS

Mice

Female CBA mice (6-8 weeks old) were obtained from the specific pathogen-free unit at the Clinical Research Centre.

Antigen

SEB was kindly provided by R.D. Brehm (PHLS Centre for Applied Microbiology and Research, Salisbury)

Responder T Cells

Lymph nodes were taken from normal CBA mice. Single cell suspensions were prepared by pressing the tissue through a wire mesh and washing the cells in RPMI 1640 (Dutch modification; Flow Labs, Irvine, Ayrshire, U.K.) supplemented with 100 IU/ml penicillin, 100 μg/ml streptomycin, 10^{-5} M 2-mercaptoethanol and 10% heat-inactivated fetal calf serum (FCS). Enriched T cells (>80% pure) were obtained by passage of cell suspensions over nylon wool columns.

Antigen Presenting Cells

DC were isolated from non-adherent spleen cells after overnight culture in medium. The cell suspension at 5 x 10^6/ml was layered onto 2ml metrizamide (Nyegaard, Oslo, Norway; analytical grade, 14.5g added to 100ml of medium) and centrifuged for 10 min at 2000 rpm. These separated cells were >70% DC, assessed by morphology and the remaining cells were < 5% macrophages and small numbers of contaminating lymphocytes.

Peritoneal exudate cells (PEC) were taken from animals 3 days after intraperitoneal injection of 60μg Concanavalin A. These cells were used as a source of MO expressing class II MHC molecules. Morphologically, 60% of the PEC were MO.

DC or MO were either untreated or pulsed *in vitro* in medium with SEB (0.1μg/ml) for 2hrs and then washed twice in medium.

Antibodies to MHC class II molecules were prepared from tissue culture supernatant. H81 detects an Ia specificity on I-E^k and HB42 on I-A^k molecules. DC were incubated with these antibodies in a 1:1 dilution for 30 min after a 2hr pulse with medium or 0.1μg/ml SEB.

For experiments with Paraformaldehyde (PCHO), DC were either untreated or treated with 0.2% w/v PCHO for 30 min at room temperature subsequent to a 2hr incubation with SEB[5].

Lymphocyte Proliferation Assay

T cells were seeded at a density of 0.6-10 x 10^5 cells per well in a 60-well Terasaki plate in the presence of varying numbers of untreated or treated DC or MO. The cultures in 20μl volumes were inverted and placed over sterile saline in a 37^0 C, 5% CO_2 incubator[6]. After 3 days, cultures were pulsed with 1μl (^3H) thymidine (Amersham International, Amersham, Bucks, U.K.) at 2Ci/mmol (final concentration of 1μg of thymidine) for 2hrs and blotted onto filter discs. Filters were washed with saline and 5% trichloroacetic acid and dried with alcohol before being counted using a liquid scintillation counter. Significant differences between responses were assessed using analysis of variance[6].

RESULTS

Stimulation of T Cells by APC using SEB

We investigated the capacity of DC and M0 to act as APC in this system. The DC were reproducibly the more potent stimulators over the 1000 to 4000 APC cell range (Fig. 1). Responses with these cells peaked much earlier with a smaller number of T cells in comparison with responses with M0 which required a greater number of T cells to reach the same level of stimulation. The increased response of DC was found to be statistically significant ($P < 0.005$). DC were therefore more efficient than M0 in initiating the T cell response.

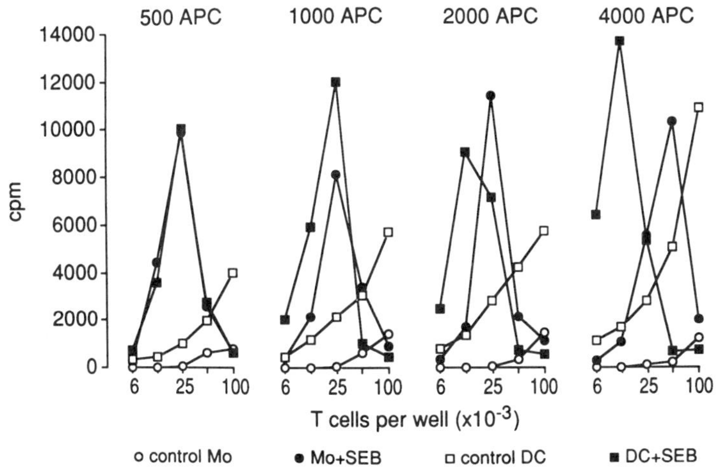

Figure 1. DC and M0 were pulsed with SEB (0.1μg/ml) for 2 hrs *in vitro* and cultured for 3 days with varying concentrations of T cells purified from lymph nodes. T cell proliferation was measured as (^3H) Tdr incorporation in counts per minute (average of triplicates).

Effect of anti-MHC Class II Antibodies on the Presentation of SEB pulsed DC to T Cells

Since presentation of superantigens by APC is dependent upon the presence of MHC class II molecules, we examined the involvement of these molecules in T cell proliferation induced by SEB. The stimulation of T cells by superantigens can be blocked by the addition of antibodies to the MHC class II molecules during culturing[7].

The anti-IA antibody was more inhibitory than those directed against IE (Fig. 2) in the 1000-4000 DC range, the difference in T cell proliferation being significant ($P < 0.025$). The combination of both antibodies further reduced the response to near control DC levels. Control DC when treated with the antibodies did not show any significant inhibition (data not shown).

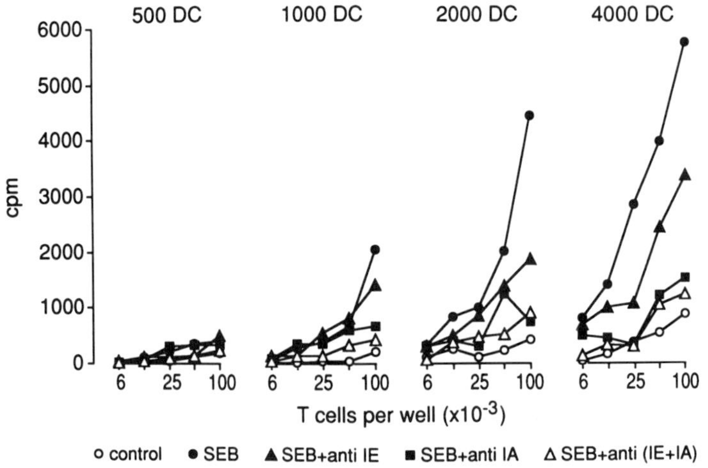

Figure 2. DC were pulsed with SEB and incubated with one volume of mAb H81 (anti IE^k), mAb HB42 (anti IA^k) or a combination of both for 30 min before being cultured for 3 days with T cells.

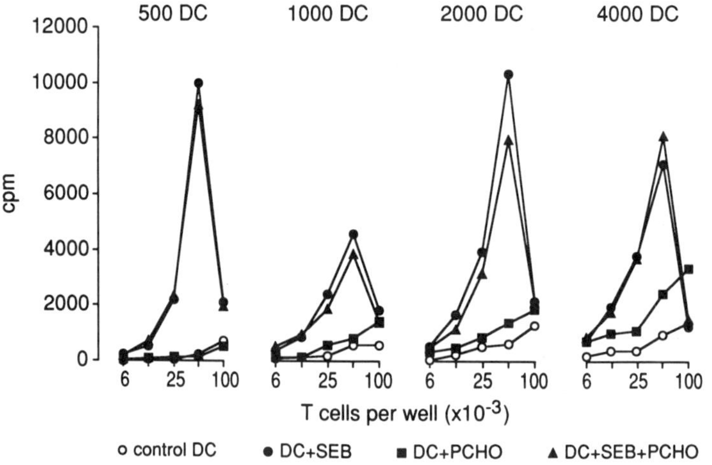

Figure 3. DC were treated with paraformaldehyde (0.2%w/v) for 30 min at room temperature subsequent to a 2hr incubation with 0.1µg/ml SEB. The DC were then cultured with T cells.

Effect of Paraformaldehyde on the Presenting Capacity of DC

Experiments with paraformaldehyde treated DC indicated that cellular processing of SEB was unnecessary. DC treated with PCHO were as effective in supporting T cell activation by SEB as were untreated DC (Fig. 3), the magnitude of T cell proliferation being similar in both cases.

DISCUSSION

Our data using the mouse system showed that DC are more efficient than MO in initiating the T cell response with SEB over a wide range of T cell concentrations. This confirms and extends observations by Bhardwaj *et al*[8] who have also shown that DC are potent APCs for several superantigens. DC were 10-50 fold more effective than monocytes or B cells isolated from human blood. The relative contribution of DC and other antigen presenting cells including MO to stimulation of T dependent immune responses has been controversial for many years. Evidence that DC may be the principal stimulators of primary responses has been obtained from functional studies of DC from mouse spleen and human blood[9-11] and similar findings have been made with rat DC[12]. Mouse DC were the most potent stimulators of immune responses on a per cell basis when compared to splenic MO even when Ia expression was induced to high levels on MO by culture with gamma interferon.

Fleischer *et al*[2] demonstrated that enterotoxin superantigens can arouse helper cells only if APC deliver the proteins to the T cells. Moreover, it is MHC class II molecules that do the presenting. The requirement for either Ia molecule for activation by SEB is most likely to be determined by the vβ expressed on responder T cells. It is possible that SEB bound to IA (or IE) molecules will activate different vβ TCR expressing T cells. Several studies[7,13,14] have shown that SEB required the IE molecule (equivalent to HLA-DR in humans) predominantly over the IA molecule in T cell lines. Our findings however suggest that SEB has a preference for the IA molecule rather than the IE. This may reflect the use of DC which express higher levels of all class II molecules constitutively. A mixture of anti-IA and anti-IE antibodies were more inhibitory than either one of the antibodies alone.

The internalization of superantigens by APC has not been well characterised. However, the conventional view of antigen presentation is that the antigen is internalized by the APC, processed and presented in a membrane bound form in association with MHC molecules to the specific T cell. Presentation of SEB to T cells is mediated by the direct binding of the toxin to MHC class II molecules present on the surface of the APC[14] without additional requirements from the APC such as antigen processing. The enhanced stimulation by DC does not appear to reflect any inhibition by processing of the molecules since PCHO treatment did not alter the level of stimulation significantly. Thus, for these superantigens, it is the intact toxin molecule and not processed fragments that is presented to T cells. This strange behaviour suggests that elucidation of the enterotoxin-binding sites on MHC molecules and T cells might reveal why more T cells react to superantigens than to conventional antigens.

REFERENCES

1. M.P. Langford, G.J. Stanton and H.M. Johnson. Biological effects of Staphylococcal enterotoxin A on human peripheral lymphocytes. *Infect. Immun.* 22:62(1978).
2. B. Fleischer and H. Schrezenmeier. T cell stimulation by Staphylococcal enterotoxins. Clonally variable response and requirement for major histocompatibility complex class II molecules on accessory or target cells. *J. exp. Med.* 167:1697(1988).
3. P. Dellabona, J. Pecoud, J. Kappler, P. Marrack, C. Benoist and D. Mathis. Superantigens interact with MHC class II molecules outside of the antigen groove. *Cell* 62:1115(1990).

4. Y. Choi, B. Kotzin, L. Herron, J. Callahan, P. Marrack and J. Kappler. Interaction of Staphylococcus aureus toxin 'superantigens' with human T cells. *Proc. Natl. Acad. Sci. USA.* 86:8941(1989).
5. P. K. Legaard, R.D. Legrand and M.L. Misfeldt. The superantigen Pseudomonas exotoxin A requires additional functions from accessory cells for T lymphocyte proliferation. *Cell. Immun.* 135:372(1991).
6. S.C. Knight. "Lymphocytes - A Practical Approach" G.G.B. Klaus, ed, IRL Press Ltd. Oxford (1987).
7. J. Yagi, J. Baron, S. Buxser and C.J. Janeway. Bacterial proteins that mediate the association of a defined subset of T cell receptor:CD4 complexes with class II MHC. *J. Immunol.* 144:892(1990).
8. N. Bhardwaj, S.M. Friedman, B.C. Cole and A.J. Nisanian. Dendritic cells are potent antigen-presenting cells for microbial superantigens. *J. exp. Med.* 175:267(1992).
9. R. Inaba and R.M. Steinman. Resting and sensitized T lymphocytes exhibit distinct stimulatory (antigen-presenting cell) requirements for growth and lymphokine release. *J. exp. Med.* 160:576(1984).
10. S.E. Macatonia, S. Patterson and S.C. Knight. Primary proliferative and cytotoxic T-cell responses to HIV induced *in vitro* by human dendritic cells. *Immunology* 74:399(1991).
11. S.E. Macatonia, P.M. Taylor, S.C. Knight and B.A. Askonas. Primary stimulation by dendritic cells induces antiviral proliferative and cytotoxic T cell responses *in vitro*. *J. Exp. Med.* 169:1255(1989).
12. W.E.F. Klinkert, J.H. Labadie and W.E. Bowers. Accessory and stimulating properties of dendritic cells and macrophages isolated from various rat tissues. *J. exp. Med.* 156:1(1982).
13. J. Yagi, S. Rath and C.A. Janeway Jr. Control of T cell responses to Staphylococcal enterotoxins by stimulator cell MHC class II polymorphism. *J. Immunol.* 147:1398(1991).
14. J.D. Fraser. High affinity binding of Staphylococcal enterotoxins A and B to HLA-DR. *Nature* 339:221(1989).

DIVERGENT T-CELL CYTOKINE PROFILES INDUCED BY DENDRITIC CELLS FROM DIFFERENT TISSUES

Michael P. Everson, William J. Koopman, and Kenneth W. Beagley

Department of Medicine
429 THT, UAB Station
The University of Alabama at Birmingham
Birmingham, AL 35294

INTRODUCTION

Dendritic cells (DC) were originally described by Steinman and Cohn in 1973.[1,2] Their methods of splenic DC isolation have involved the flotation of murine spleen (SP) cells on dense bovine serum albumin (BSA) gradients and differential glass or plastic adherence of SP cells. Nonadherent Peyer's patch (PP) DC were isolated in our laboratory by Spalding and coworkers in 1983.[3] These methods employed flotation of PP cells on dense BSA gradients, clustering of DC within the buoyant PP population with periodate-modified SP T cells, dissociation of clusters, and final BSA flotation of enriched PP DC.

The fact that DC from various tissues have been previously isolated under dissimilar conditions imposes certain limitations with regard to interpreting reported differences or similarities in the features and functions of these DC populations. Observed differences in SP and PP DC characteristics could be due to dissimilarities in isolation techniques, while similarities could be confounded by these different techniques. Therefore, ultimate elucidation of the role of DC in mucosal and systemic immune functions required development of a single isolation technique applicable to DC from both tissues. We combined novel techniques with previously published methods to develop a unified approach for the isolation of DC from SP and PP. This permitted a meaningful comparison of the function of DC from different tissues.

Previous data indicate that DC from different lymphoid tissues can possess similar functions; for example, SP DC and PP DC have been shown to possess similar stimulatory function in oxidative mitogenesis reactions using splenic T cells.[3] However, other data indicate that DC from different lymphoid tissues can possess divergent functions, i.e., SP DC induce preferential IgM production and PP DC induce preferential IgA production in B cells and pre-B cell lines.[4,5] We have previously shown that cytokines can influence Ig isotype production,[6] and, in particular, interleukin-6 (IL-6) can stimulate surface IgA-bearing B cells to preferentially secrete IgA.[7] Therefore, we sought to determine whether SP DC and PP DC, isolated under identical conditions, induced the production of similar levels of T cell-derived cytokines involved in regulation of Ig production. Specifically, DC-induced T cell cultures were tested for representative cytokines previously shown to be elaborated by T_H1 (gamma interferon [IFN-γ] and IL-2; termed T_H1 cytokines) and T_H2 (IL-4, IL-5, and IL-6; termed T_H2 cytokines) cell types.[8]

METHODS

Cells

Spleen (SP) and Peyer's patch (PP) cells were derived from 8-20 week old C3H/HeN mice. Thymus cells were derived from 6-9 week old C3H/HeN mice (Frederick Cancer Research Facility, Frederick, MD).

Cell Enrichment Procedures

We developed a procedure for the isolation of DC from SP and PP using identical conditions (to be published in detail elsewhere). Spleens and Peyer's patches were enzyme digested using Dispase® to yield single cell suspensions, cell suspensions were applied to bovine serum albumin (BSA) density gradients, and buoyant cells (spec. grav. $\leq$1.0735) were harvested from these gradients as previously described.[3,9] These buoyant cells were cocultured at a 1:5 to 1:20 ratio with gamma-irradiated (1500 rads), periodate-modified[3] dense thymocytes (obtained by mechanical dissociation through wire mesh) which had been harvested from the pelleted fraction of BSA density gradients. Following overnight (14-19 h) coculture, cell clusters containing thymocytes and DC were isolated using Percoll/fetal calf serum/phosphate-buffered saline density gradients as previously described.[10] Cell clusters were dissociated on ice with forceful pipetting in Ca^{2+}-/Mg^{2+}-free phosphate-buffered saline and buoyant cells were isolated using BSA density gradients. This procedure yielded approximately 0.3-0.6% of the original starting populations of SP and PP cells. These enriched populations were approximately 60-80% DC as determined by morphological characteristics. These methods provided a means for comparing the functional characteristics of DC derived from different lymphoid tissues.

Culture System

Preliminary experiments were performed using culture supernatants derived from bulk cultures of enzyme-dissociated, periodate-modified whole SP or PP cells cultured at 5×10^6 cells/mL in complete medium (RPMI 1640, 50 μg/ml gentamicin, 2 mM L-glutamine, 5% heat-inactivated fetal calf serum, 50 μM 2-ME, 100 U/mL penicillin, 100 μg/mL streptomycin, and 1 μg/ml indomethacin). Culture supernatants were harvested after 24 h of culture and filter-sterilized prior to cytokine analysis.

To more critically determine the role(s) for DC in stimulating production of T-cell cytokines, SP and PP DC were enriched by the unified procedure described above and mixed with periodate-modified SP or PP T cells. These T cells were enriched by passage over nylon-wool columns, followed by treatment with anti-Ia^k and complement as previously described.[9] Mixtures of DC and T cells (DC-T) were cocultured at a identical DC:T cell ratios (ratios ranged from 1:10 to 1:20) in complete medium in 24-well, flat-bottomed tissue culture wells at 5.0-5.25 $\times$ 10^6 cells/mL. For generation of conditioned medium, culture supernatants were harvested after 24 h of culture, filter-sterilized, and stored at 4°C until assayed for cytokine content.

Cytokine Determinations

Cytokine levels were determined using the cytokine-sensitive cell lines described below by comparison of supernatant dilutions with recombinant cytokine dose-response curves as previously described.[9] IL-2/-4, IFN-γ, and IL-6 levels were determined using CTLL-20,[9] WEHI-279[11] (provided by K. Bottomly, Yale University, New Haven, CT), and 7TD1[12] cells (provided by R. Nordan, National Institutes of Health, Bethesda, MD), respectively. IL-5 was determined by enzyme-linked immunosorbent assay (ELISA) as previously described[13] by using two anti-IL-5 monoclonal antibodies (TRFK4 and TRFK5; provided by R. Coffman, DNAX Research Institute, Palo Alto, CA).

RESULTS

Oxidative mitogenesis reactions using periodate as the mitogen have been demonstrated to elicit T cell activation by predominantly, and perhaps uniquely, stimulatory DC.[14] Therefore, periodate-modified cells and cell mixtures were used as a model for determining the potential of DC from SP and PP to induce cytokine production by T cells derived from these tissues.

Initial data were generated using bulk cultures of periodate-modified, unfractionated SP and PP cells. When 24-h culture supernatants were analyzed for IL-2 content, it was determined that SP cells produced approximately 2.5-fold more IL-2 (300 U/Ml) than PP cells (120 U/Ml) in cultures containing equal cell densities (data not shown). Data from similar oxidative mitogenesis proliferative cultures indicated that unfractionated SP cell cultures incorporated approximately 3- to 4-fold more tritiated thymidine than unfractionated PP cell cultures as determined by dose-response curve comparison (data not shown). The activity in the above supernatants was confirmed to be IL-2 by complete neutralization of supernatant-induced CTLL cell line proliferation by inclusion of S4B6 anti-IL-2 monoclonal antibody (provided by T. Mosmann, University of Alberta, Edmonton, Alberta, Canada) (data not shown). Similar results were obtained with an IL-4-non-responsive, IL-2-dependent HT-2AB cell line (derived from the parent HT-2 cell line[15] by H.A. Bouwer; Providence Medical Center, Portland, OR). To determine whether this contrast in cytokine levels was due to differences in the DC or T cell (or both) populations of these tissues, DC were enriched from SP and PP by a common technique and mixed with periodate-modified SP and PP T cells.

Twenty-four hour culture supernatants derived from homologous DC-T cell mixtures (i.e., SP DC + SP T cells or PP DC + PP T cells) and heterologous DC-T cell mixtures (i.e., SP DC + PP T cells or PP DC + SP T cells) were tested for IL-2, IFN-γ, IL-4, IL-5, and IL-6. The profile of IL-2 levels (not shown) produced by these different cell mixtures was similar to that of the other T_H1 cytokine tested, IFN-γ (Fig. 1). Homologous SP DC-T cell mixtures produced the greatest amount of IFN-γ while homologous PP DC-T cell mixtures produced the least IFN-γ, and these values were statistically different by Student's *t* test ($p < 0.005$). Results similar to these were obtained using buoyant SP or PP cells as stimulators, i.e., without further separation to enrich for DC populations. These marked differences in cytokine expression suggested that either the PP DC or PP T cell population was defective in induction or production of IFN-γ, respectively, or that the SP DC or SP T cell population was hyperactive in these functions. Therefore, the heterologous cell mixtures were analyzed to address these possibilities. Data derived from these control experiments indicate that the decreased levels of IFN-γ produced in the PP homologous cell mixture is not due to an inability of PP T cells to produce IFN-γ since SP DC were able to activate these cells to produce considerable quantities of IFN-γ. Using ELISPOT analysis to determine the frequency of cells producing IFN-γ as previously described,[16] the levels of IFN-γ protein (Fig. 1) were found to correlate with the number of cells from these different mixtures producing IFN-γ (data not shown).

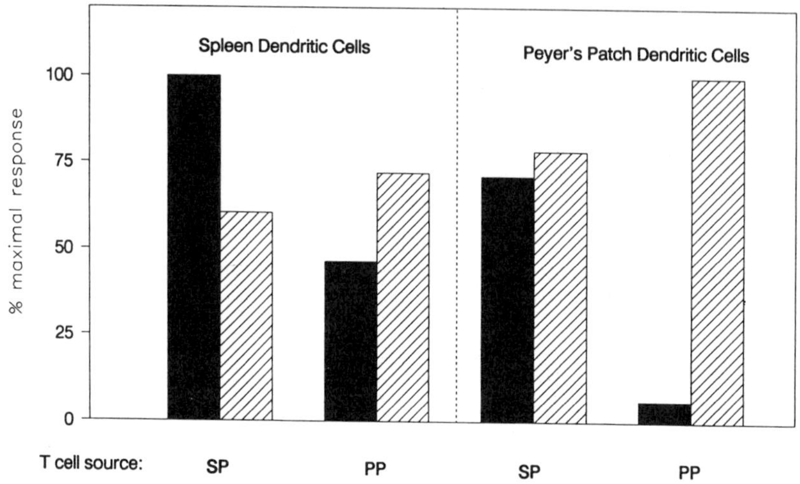

Figure 1. Divergent cytokines produced by DC and T cells from different tissues. DC from SP and PP were cultured for 24 h with periodate-modified SP or PP T cells, and culture supernatants were harvested and tested for cytokine levels. Results are depicted as % maximal response where maximal responses were 32-34 U/mL (SP DC + SP T cells) and 7-17 U/mL (PP DC + PP T cells)for IFN-γ and IL-6, respectively. Solid bars depict IFN-γ levels and hatched bars depict IL-6 levels for the means of 2 experiments. Standard deviations within an experimental group are routinely ≤15% and have been omitted here. Background levels for DC alone or T cells alone (<15% of stimulated response) have been subtracted from these data to yield DC-induced T cell cytokine levels. Indicator cells were WEHI-279 and 7TD1 for IFN-γ and IL-6, respectively.

Determination of T_H2 cytokine levels in cell mixture supernatants indicated that IL-4 (<10 U/mL) and IL-5 (<5 U/mL) concentrations were below detectable limits of the assays utilized (not shown). Nonetheless, the profile of the supernatant levels of IL-6, another T_H2 cytokine, contrasted with the T_H1 cytokine profiles (Fig. 1). Specifically, homologous PP DC-T cell mixtures produced the greatest quantity of IL-6 whereas homologous SP DC-T cell mixtures produced the least. Significant differences ($p < 0.005$ by Student's *t* test) existed between the IFN-γ (and IL-6) levels produced by the homologous DC-T cell mixtures from SP and PP. When DC or T cells were cultured alone, negligible levels (<15% of stimulated levels) of cytokines (IFN-γ and IL-6) were produced. Similar results were obtained when IL-2 was the cytokine of interest.[14] When detected, these background levels were subtracted from the reported levels to yield induced levels of the relevant T-cell cytokines.

DISCUSSION

Previous studies provide evidence for diversity of functional capabilities of antigen presenting (or accessory) cells. An example of macrophage diversity as a function of buoyant density is that density-defined alveolar macrophage subpopulations produce different quantities of tumor necrosis factor-α in pulmonary fibrosis.[17] Examples of DC diversity as a function of tissue origin are presented here and in the work of Spalding and colleagues.[4-5] For example, Spalding demonstrated that SP DC and PP DC induced different profiles of immunoglobulin regardless of the T cell and B cell origin. Experiments described above indicate that periodate-modified whole SP cells proliferate and produce IL-2 about threefold more readily than identically treated whole PP cells. Although provocative, these experiments did not use identically enriched DC. To further investigate this tissue-specific diversity, it was therefore necessary to develop a unified procedure for enriching DC from SP and PP.

Most SP DC isolation techniques use glass or plastic adherence as an enrichment step to remove lymphocytes and other nonadherent cells.[2,18] Although adherence is commonly used to isolate SP DC, its effects may not be fully appreciated. For example, adherence has been demonstrated to induce human monocytes to express IL-1β and TNF-α genes (personal communication, Ken Hardy, University of Alabama at Birmingham). Thus to obviate any bias possibly induced by adherence, we chose to isolate SP DC under nonadherent conditions. Since most PP DC are nonadherent,[3] this technique also would be applicable to PP DC isolation.

Spalding *et al.* used the capacity of periodate-modified SP T cells to cluster with DC in their procedure to enrich for PP DC.[3] Since our experimental design required the mixing of SP and PP DC with SP and PP T cells, we felt that the use of SP T cells in the DC isolation technique could possibly skew the data. Therefore, we decided to use a population of immature T cells that was derived from neither SP nor PP, i.e., thymocytes. Thymocytes have been demonstrated to be poorly responsive to DC.[19] Our data confirmed this finding (e.g., buoyant SP or PP cells induced periodate-modified, dense thymocytes to produce <0.1 U/mL IL-6). Nonetheless, once modified with sodium periodate, these rather inert thymocytes clustered well with DC from SP and PP. This step provided the necessary technique for enriching SP and PP DC in the absence of the previously used techniques of adherence and clustering with SP T cells.

Previous studies indicate that distinct T helper cell subpopulations may regulate humoral and cell-mediated immunity to foreign antigens. This regulation may be controlled in part by distinctly different patterns of cytokines produced by these different T_H populations, i.e., T_H1 cells produce IL-2 and IFN-γ, and T_H2 cells produce IL-4, IL-5, IL-6, and IL-10.[8] However, mechanisms underlying the induction and regulation of these T_H subsets *in vivo* are unclear. The possibility that different antigen presenting cells may induce divergent cytokine production profiles has been suggested by others but has not been demonstrated. The studies presented here suggest that distinct DC populations induce divergent T cell-derived cytokine production profiles.

Our data are consistent with the notion that DC are responsible for the cytokine profiles produced by T cells. These data also suggest that DC coordinate immune responses to at least some degree by controlling the cytokines produced in a given tissue or milieu. In this regard, further investigations are required to determine the molecular basis for the observed functional differences in DC derived from different tissues.

ACKNOWLEDGMENTS

The authors are greatly appreciative of the expert technical assistance provided by Mrs. Debbie McDuffie during the course of these studies.

REFERENCES

1. R.M. Steinman and Z.A. Cohn, Identification of a novel cell type in peripheral lymphoid organs of mice. I. Morphology, quantitation, tissue distribution, *J. Exp. Med.* 137:1142 (1973).
2. R.M. Steinman and M.C. Nussenzweig, Dendritic cells features and functions, *Immunol. Rev.* 53:127 (1980).
3. D.M. Spalding, W.J. Koopman, J.H. Eldridge, J.R. McGhee, and R.M. Steinman, Accessory cells in murine Peyer's patch. I. Identification and enrichment of a functional dendritic cell. *J. Exp. Med.* 157:1646 (1983).
4. D.M. Spalding, S.I. Williamson, W.J. Koopman, and J.R. McGhee, Preferential induction of polyclonal IgA secretion by murine Peyer's patch dendritic cell-T cell mixtures, *J. Exp. Med.* 160:941 (1984).
5. D.M. Spalding and J.A. Griffin, Different pathways of differentiation of pre-B cell lines are induced by dendritic cells and T cells from different lymphoid tissue, *Cell* 44:507 (1986).
6. K.W. Beagley, J.H. Eldridge, H. Kiyono, M.P. Everson, W.J.Koopman, T. Honjo, and J.R. McGhee, Recombinant murine IL-5 induces high rate IgA synthesis in cycling IgA-positive Peyer's patch B cells, *J. Immunol.* 141:2035 (1988).
7. K.W. Beagley, J.H. Eldridge, F. Lee, H. Kiyono, M.P. Everson, W.J. Koopman, T. Hirano, T. Kishimoto, and J.R. McGhee, Interleukins and IgA synthesis: human and murine interleukin 6 induce high rate IgA secretion in IgA-committed B cells, *J. Exp. Med.* 169:2133 (1989).
8. T.R. Mosmann, J.H Schumacher, N.F. Street, R. Budd, A. O'Garra, T.A.T. Fong, M.W.Bond, K.W.M. Moore, A. Sher, and D.F. Fiorentino, Diversity of cytokine synthesis and function of mouse $CD4^+$ T cells, *Immunol. Rev.* 123:209 (1991).
9. M.P. Everson, D.M. Spalding, and W.J. Koopman, Enhancement of IL-2-induced T cell proliferation by a novel factor(s) present in murine spleen dendritic cell-T cell culture supernatants, *J. Immunol.* 142:1183 (1989).
10. K. Inaba, A. Granelli-Piperno, and R.M. Steinman, Dendritic cells induce T lymphocytes to release B cell-stimulating factors by interleukin 2-dependent mechanism, *J. Exp. Med.* 158:2040 (1983).
11. D.S. Reynolds, W.H. Boom, and A.K. Abbas, Inhibition of B lymphocyte activation by interferon-γ, *J. Immunol.* 139:767 (1987).
12. J. Van Snick, S. Cayphas, A. Vink, C. Uyttenhove, P.G. Coulie, M.R. Rubira, and R. Simpson, Purification and NH_2-terminal amino acid sequence of a T-cell-derived lymphokine with growth factor activity for B-cell hybridomas, *Proc. Natl. Acad. Sci. USA* 83:9679 (1986).
13. J.H. Schumacher, A. O'Garra, B. Shrader, A. van Kimmenade, M.W. Bond, T.R. Mosmann, and R.L. Coffman, The characterization of four monoclonal antibodies specific for mouse IL-5 and development of mouse and human IL-5 enzyme-linked immunosorbent, *J.Immunol.* 141:1576 (1988).
14. J.M. Austyn, R.M. Steinman, D.E. Weinstein, A. Granelli-Piperno, and M.A. Pallidino, Dendritic cells initiate a two-stage mechanism of T lymphocyte proliferation, *J. Exp. Med.* 157:1101 (1983).
15. J. Watson, Continuous proliferation of murine antigen-specific helper T lymphocytes in culture, *J. Exp. Med.* 150:1510 (1979).
16. T. Taguchi, J.R. McGhee, R.L. Coffman, K.W. Beagley, J.H. Eldridge, K. Takatsu, and H. Kiyono, Analysis of Th1 and Th2 cells in murine gut-associated tissues. Frequencies of $CD4^+$ and $CD8^+$ T cells that secrete IFN-γ and IL-5, *J. Immunol.* 145:68 (1990).
17. M.P. Everson and D.B. Chandler, Changes in distribution, morphology, and tumor necrosis factor-α secretion of alveolar macrophage subpopulations during the development of bleomycin-induced pulmonary fibrosis, *Am. J. Pathol.* 140:503 (1992).
18. S.E. Macatonia, S.C. Knight, A. Edwards, P. Fryer, and S. Griffiths, Localization of antigen on lymph node dendritic cells after exposure to the contact sensitizer fluorescein isothiocyanate, *J. Exp. Med.* 166:1654 (1987).
19. M.C. Nussenzweig and R.M. Steinman, Contribution of dendritic cells to stimulation of the murine syngeneic mixed leukocyte reaction, *J. Exp. Med.* 151:1196 (1980).

ADHESION MOLECULES: CO-STIMULATORS AND CO-MITOGENS IN DENDRITIC CELL - T CELL INTERACTION

Philip D. King, Mohammad A. A. Ibrahim and David R. Katz

Depts of Histopathology, Immunology and Biology
University College and Middlesex School of Medicine
48, Riding House St., London W1P 7PN UK

INTRODUCTION

Dendritic cells (DC) are identified both by morphology and function (1-3). These cells are localised to T cell micro-environments and are the most potent known accessory inducer cells for T cell responses. Even when there is heterogeneity in in vitro comparative accessory cell function (i.e. when more than one cell type can present), then the DC almost invariably prove to be the most potent relative inducers (4,5). Thus the distinguishing molecular mechanisms involved in DC - T cell interaction that might account for this distinctive inducer capacity is obviously of considerable interest.

In previous studies we separated DC from solid human tissues (6) and used these cells to address questions about DC function that had only been examined previously using mouse DC. In these studies we demonstrated the role of the reciprocal receptor - ligand molecules CD11a/CD18 (LFA-1) and CD54 (ICAM1) in the initial bidirectional clustering interaction that occurs between DC and T cells. This role in DC - T cell clustering provided an explanation as to why antibodies against these CD molecules inhibited the induction of T cell proliferation in an oxidative mitogenesis assay. There were also other receptor-ligand pairs of molecules, such as CD58 (LFA-3) and CD2, which were implicated in the clustering stage, albeit in a unidirectional fashion. Furthermore, there were some antibodies, directed against molecules such as class II MHC and CD4 which inhibit proliferation, but do not affect clustering, suggesting that these molecules are involved in the signal transduction step rather than in the formation of the DC - T cell aggregates.

In subsequent studies we extended our analysis of the role of CD11a/CD18 - CD54 as accessory molecules in DC induced responses. We also used the same approach to examine the role of other cell surface molecules expressed at the DC - T cell interface, with particular emphasis on those that are expressed (or induced) on the T cell during the activation process. This approach has allowed us to begin to evaluate the possible role of these other molecules, and their ligands, as participants in the mechanism(s) of DC function.

MATERIALS AND METHODS

Isolation of human tonsillar cells

Tonsils digested with collagenase (Sigma) (90 min at 37^0C) were pushed through nylon meshes (Cadisch, London, GB) (125μ pore size) to separate lymphomedullary cells from surrounding stromal connective tissue. After 4 washes in RPMI 1640 medium (Gibco) cells were spun on a 7 step iso-osmolar percoll gradient (Pharmacia) at 600g for 15 min. Low density (LD) cells (30-50% percoll) were used to prepare DC; high density (HD) cells (50-70% percoll) to prepare T cells.

Dendritic cells

LD cells were depleted of macrophages (MO) and natural killer cells with 5 mM leucine methyl ester (LME) (Sigma) and incubated overnight in complete medium (CM) {RPMI 1640 containing 5% heat inactivated fetal calf serum (FCS) and antibiotics (Gibco), 10 mM HEPES buffer and 50μM 2-mercaptoethanol (Sigma)} on 140 mm^2 plastic petri dishes (Nunclon, Denmark) in a 5% $CO_2/37^0C$ humidified incubator, to remove adherent cells that had survived LME treatment.

Non-adherent cells were harvested and T cell depleted by sheep red blood cell (SRBC/E) rosetting (SRBC, 1 hr at 4^0C, followed by separation of E+ from E- non-T cells over ficoll - hypaque (Pharmacia) at 680g for 20 min.) The E- cells were B cell depleted in a second rosetting step using ox red blood cells (ORBC) pre - coupled to affinity purified rabbit anti-mouse immunoglobulin (RAM) (Dakopatts) After incubation with 3AC5 (a murine anti-CD45RA antibody) for 30 min at 4^0C, E- cells were washed x 2 and incubated with RAM - ORBC complexes for 5 min at 20^0C. Cells were pelleted, resuspended in their own supernatants and spun over ficoll-hypaque (680g / 20 min) to remove rosetted B cells.

The non-rosetted DC were washed twice before use in subsequent assays. These cells were >95% class II+, and do not express CD2, CD3, CD11b, CD14 and CD19 on their surface.

T cells

Responder T cells were prepared from HD fractions by SRBC rosetting as described above. Following removal of SRBC from rosettes by hypotonic shock, T cells were washed x 2 for use in the assay system. Previous studies had confirmed that these T cells are in G_o, and are >95% CD2+ CD3+.

Monoclonal antibodies

Reagents used and reported here were anti-CD5 (T169 and

Lo-Tact5); anti-CD11a (MHM24, MEM-25, MEM-30, 1524, BU17, 122-2A5, MEM-83, MEM-95, 25.3.1, 459, F110.22, GRS3, HI111, 2F12, CC51D7 and TMD3-1, all IgG1), anti-CD18 (MHM23, IgG1); anti-CD11b (Mo1, 44); anti-CD11c (KB23, KB43); anti-CD28 (KOLT2); anti-CD44 (GRHL1, IgG1; and F10-44-2, IgG2a) and anti-CD71 (BU54, BU55, JML-H9, MEM-75, 138-18, 120-2A3 and G11/71, IgG1).

Antibodies were dialysed before use {4-6 changes of phosphate buffered saline (PBS) over 3-4 days} and aggregates removed by high speed centrifugation (7000g / 30 min). All antibodies are of murine origin. Antibodies that were obtained from the Fourth Differentiation Workshop (Vienna, 1989) were used "blind" in these assays at 1:400 dilution. Concentrations of other antibodies were as indicated in the text.

Oxidative mitogenesis assays

Oxidative mitogenesis assays were performed in triplicate in 10% human AB serum (North London Blood Transfusion Centre) (i.e. CM containing heat inactivated human rather than FCS) T cells were modified with 2 mM sodium periodate (Sigma) for 30 min. at 4^0C, washed x 2 and 2 x 10^5 cells were incubated in a total volume of 0.2ml with 1 x 10^5 irradiated DC (3000 rads).

Monoclonal antibodies were included in DC plus T cell wells either at time zero or at different time points as indicated. After 48 hr cells were pulsed with 5-[^{125}I]-iodo-2-deoxyuridine (^{125}IdUrd) (Amersham International) at 0.75μCi/well and incubated for a further 16 hr. Cultures were terminated at 64 hr and incorporated radiolabel measured. Results are expressed either as mean cpm ^{125}IdUrd incorporation in triplicate assays, $\pm$ SD; as percentage change in proliferation in the presence of antibody compared to controls, calculated as [mean cpm (DC + T + antibody)] - [mean cpm (DC + T)] / [mean cpm (DC + T)] - [mean cpm T alone] x 100. The statistical significance of changes was calculated by comparing DC + T triplicates with DC + T + antibody triplicates in a students two sample t test.

RESULTS AND DISCUSSION

The role of a well - established anti-CD11a, MHM24, was explored in detail. For example, Figure 1a shows a representative dose - response curve, illustrating the inhibitory effect of the antibody on the DC induced oxidative mitogenesis assay. Control irrelevant isotype matched antibodies had no effect; and the MHM24 antibody itself had no effect when it was added after the first 24 hours of the assay (data not shown). Anti-CD18 had very a similar effect to anti-CD11a; and neither anti-CD11b nor anti-CD11c were inhibitory.

In subsequent studies a panel of IgG1 anti-CD11a antibodies were evaluated for their role in the response, and the results of a panel analysis are summarised in Figure 1b.

The use of a panel of CD11a antibodies thus confirms that there are indeed several isotype matched reagents which have a similar inhibitory effect to that seen with MHM24. However, a striking feature of the panel analysis was that there were also some antibodies that were not inhibitory. Rather, they induced a (statistically significant) increase in proliferation, despite the presence of an ongoing DC - induced response.

CD11a is regarded primarily as an adhesion molecule in cell-cell and cell-matrix interactions. Our earlier findings were consistent with this hypothesis, with antibody - mediated inhibition of the clustering phase required for T cell activation. The CD11a/CD18 molecule is also, however, an integrin molecule which has an intracellular signalling role in response to extracellular stimuli, and therefore the antibodies that increase proliferation probably recognise and trigger epitopes involved in this component of CD11a/CD18 function, rather than interfering with receptor-ligand interaction.

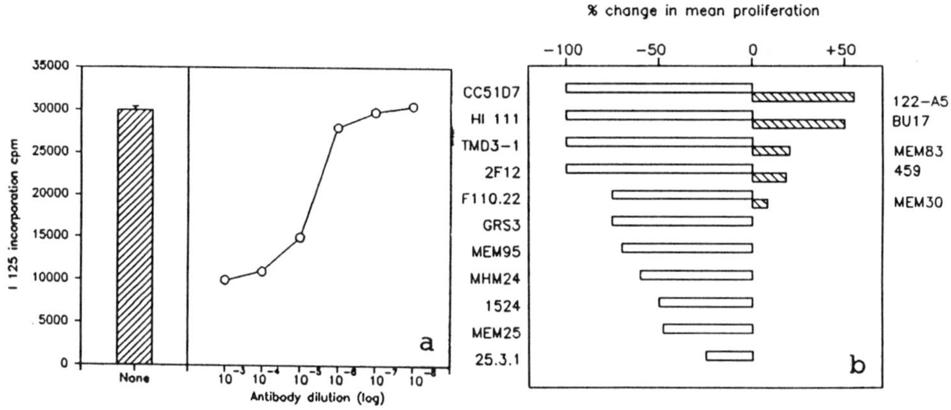

Figure 1. CD11a in DC induced T cell oxidative mitogenesis: (a) MHM24 and (b) the effects of different antibodies.

However, the other antibodies which had a significant enhancing effect on the DC - induced T cell oxidative mitogenesis reaction did not belong to known integrin groups. For example, Figure 2a shows that 6 of the 7 antibodies directed against the transferrin receptor (CD71) augment DC induced T cell proliferation.

Similarly the antibody against CD5 (2b), CD28 (2c) and CD44 (2d) increase the proliferative response.

Antibodies against the transferrin receptor have been reported previously as inhibitors of lymphocyte responses, interrupting a receptor-ligand interaction that is necessary for cell activation (7). This study suggests a possible alternative role for the anti-CD71 antibodies. Via a process of ligand mimicry, the antibody acts as an amplifying co-mitogenic agent for signal transduction in conjunction with DC.

It is possible that anti-CD28 may also act via a process of ligand mimicry. However, it has been suggested that CD28 -

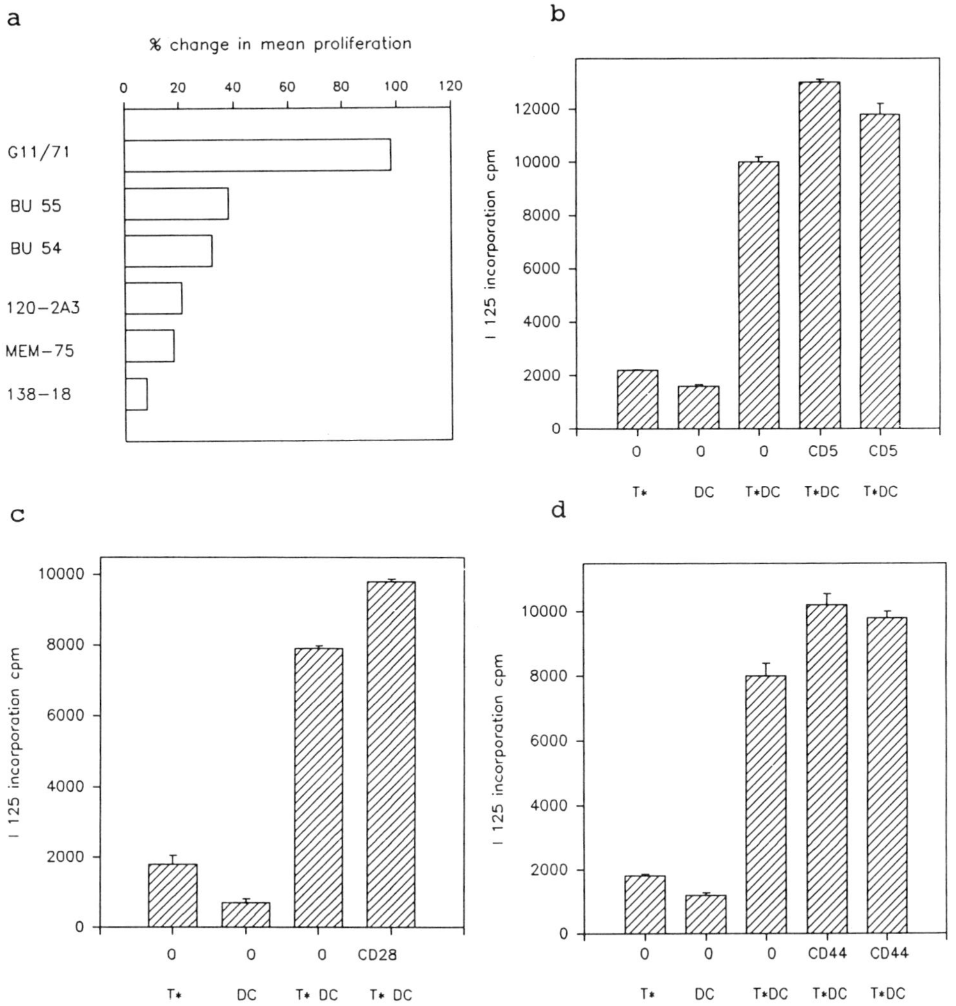

Figure 2. Effect of (a) CD71 (b) CD5 (c) CD28 and (d) CD44 antibodies on DC - induced oxidative mitogenesis.

B7, like CD11a/18 - CD54, represents a co-stimulatory complex that is required for antigen specific responses (8); and thus a similar pattern of either inhibition or amplification might be expected during a cell - cell interaction.

CD5 - anti-CD5 binding has been shown previously to lead to signal transduction, in synergy with other signals (9). The effects of the anti-CD5 resemble those seen with anti-CD71, and this is probably best regarded as mimicry acting as an amplifying co-mitogenic signal.

By morphology none of CD5, CD28 and CD71 antibodies show an effect on the DC - T cell clustering process (data not shown). However, both CD44 antibodies resulted in an increase in the size of clusters, analagous to our findings in the U937 - T cell assay, as previously reported (10). CD44 is known to be one of the heavily glycosylated surface molecules, which conveys a negative charge to the cell surface, and a possible explanation is thus that the abrogation of negative charge and increased clustering results in increased proliferation.

CONCLUSION

In summary, these studies suggest that:

1. There are two different roles for adhesion molecules during APC - T cell interaction:
    i. during clustering - as co-stimulators
    ii. in signal transduction - as co-mitogens
2. Other co-stimulatory molecules, such as CD28, may also have a dual role.
3. Some antibodies act via ligand mimicry, will synergise with DC in the induction of T cell proliferation, and are co-mitogenic.
4. Enhancement of clustering (e.g. by anti-CD44) may also represent a form of co-stimulation.

Acknowledgements

These studies were supported by grants from UK Medical Research Council, Arthritis and Rheumatism Council, Special Trustees of the Middlesex Hospital Medical School and Muirhead Trust.

REFERENCES

1. Austyn J.M. Immunology (1987) 62: 161
2. King P.D. and Katz D.R. Immunol To-day (1990) 11: 206
3. Steinman R.M. Ann Rev Immunol (1991) 9: 271
4. Sunshine G.H., Katz D.R. and Czitrom A.A. Eur J Immunol (1982) 12: 9
5. Katz D.R. et al Immunology (1986) 58: 167
6. King P.D. and Katz D.R. Eur J Immunol (1989) 19: 581
7. Trowbridge I.S. et al Methods in Enzymology (1987) 147: 265
8. Jenkins M.K. et al J Immunol (1991) 147: 2461
9. Ledbetter J.A. et al J Immunol (1985) 135: 2331
10. King P.D. et al Eur J Immunol (1991) 20: 363

DENDRITIC CELL DEPENDENT EXPRESSION OF IgA BY CLONES IN T/B MICROCULTURES

Carol E. Schrader and John J. Cebra

Department of Biology
University of Pennsylvania
Philadelphia, PA 19104-6018

INTRODUCTION

We sought to develop a T-dependent, clonal B cell microculture in order to assess changes in frequencies and Ig isotype potential of antigen (Ag)-specific B cells, associated with oral/gut mucosal exposure to Ag. Particularly, we were concerned with the development of IgA-memory B cells (1) as predictive of an in vivo, secondary mucosal IgA antibody (Ab) response (2) and with the role of germinal centers (GC) in Peyer's patches (PP) in the generation of these IgA-memory cells as well as IgA pre-plasmablasts. Our original T/B microculture was based on clonal culturing of B cells responsive to thymus-independent Ags, as practiced in the Nossal laboratory (3) using Ag-specific B cells enriched by panning on haptenated gelatin (4), except that we used cloned, Ag (conalbumin)-specific D10, T_H2 cells (5) and haptenated-Ag as stimuli (6). This system exhibited H-2 haplotype restriction and a requirement for linked recognition of hapten and carrier. An Ag-independent version of this T/B microculture utilized the alloreactivity of the D10 cells versus $I\text{-}A^b$ molecules and purified F_1, k × b, B cells (6). Both systems promoted a high frequency of productive B cell clones (1-50%) that exhibited intraclonal isotype switching. However, secreted IgA was rarely detected as a clonal product and B cells shown to include many IgA memory cells, such as PP B cells enriched on phosphocholine (PC)-gelatin and tested in splenic fragment cultures (7), failed to generate clones secreting only IgA in microculture. In an attempt to make T/B microcultures more supportive of IgA expression, we added either peritoneal macrophage, NIH/3T3, BALB/3T3, or dendritic cells (DC) as 'filler' or 'feeder' cells. Although all types of added cells increased the frequency of responding B cells, only DC prepared from either spleen by the Steinman procedure (8) or from PP by the Spaulding method (9) markedly potentiated IgA expression in both Ag-dependent and allo-stimulated clonal B cell microcultures (6). Addition of as few as 400 DC added to these T/B microcultures:

Dendritic Cells in Fundamental and Clinical Immunology,
Edited by Kamperdijk *et al.*, Plenum Press, New York, 1993

(a) support the expression of IgA among other isotypes by a high proportion of productive B cell clones (6);
(b) support the stimulation of IgA memory cells to generate clones exclusively secreting IgA (10);
(c) support the differentiation of pre-plasmablasts into secretory IgA plasma cells if their division is blocked by x-irradiation or with aphidicholin (10);
(d) support the expression of IgA among other isotypes by clones derived from IgD^+ B cells that undergo isotype switching in microculture (6);
(e) permit the reduction of input T_H cells (3×10^3 to 3×10^2) without diminishing appreciably the frequency of responding B cells (6);
(f) create more supportive cultures that permit clones to survive longer (12 d) and display increased output of Ig (6);
(g) does not require that the necessary T_H2 cells divide to support IgA expression by B cell clones (6), although the DC are far superior to splenic APC (100×) in stimulating Ag-dependent proliferation of D10 T_H2 cells. Thus, we sought to further analyze the characteristics of the T/B/DC microculture in an effort to understand the means by which DC effected IgA expression.

POSSIBLE DC EFFECTS AND INTERACTIONS IN CLONAL B CELL MICROCULTURES

DC Alone Cannot Effect IgA Expression in T-independent Clonal B Cell Microcultures

PC-specific B cells were enriched on PC-gelatin and were stimulated in clonal microculture with LPS, dextran sulfate (DXS) or both in the presence of DC. Although many highly productive clones secreting IgM and IgG isotypes developed, none secreted IgA. Further, addition of 10% conditioned medium (CM) from ConA-stimulated D10 T_H2 cells also failed to elicit IgA expression. Since the D10 cells themselves were capable of allostimulating a large proportion of clones from PC-specific B cells to express IgA in T/B/DC cultures in the absence of mitogens, we conclude that some sort of physical interactions among the cell types may be required for this effect.

DC Plus Ag Induces Rested D10 T_H2 Cells to Secrete IL-4 and IL-6 but Exogenous LK, D10 CM, or DC/D10 CM Cannot Replace DC for IgA Expression

Co-culturing DC with rested D10 cells results in an Ag dependent output of IL-4 measured using CTLL and CT.4S indicator cells, and IL-6, using B9 cells for bioassay. However, addition of exogenous IL-4, IL-5, IL-6, and combinations of these to T/B microcultures did not replace the effect of DC to promote IgA expression. Further, neither CM from ConA stimulated D10 cells nor 2-10% CM prepared from Ag-stimulated DC/D10 cultures effected IgA expression in T/B microculture. However, all exogenous lymphokines (LK) and CM described above increased the incidence of productive B cell clones -- as did the addition of various macrophage and fibroblast 'filler' cells. Thus the increase in cloning efficiency seems separable from the support of IgA expression. In the course of some of these assays we used cyclosporine A (CyA, 20 ng/ml) to eliminate the contributions of endogenous LK made by the D10 cells in T/B microcultures. Indeed, CyA prevented the development of productive B cell clones, but addition of exogenous LK in CM from ConA stimulated D10 cells or in CM from Ag/DC/D10 over-came its effect in T/B microculture, again without effecting IgA expression.

The Presence of DC in DC/T/B Microcultures Counteracts the Effect of CyA to Prevent Development of Ig-Producing B Cell Clones

Remarkably, we found that concentrations of CyA that sharply reduced or eliminated productive clones in T/B microcultures, had little effect on their incidence in DC/T/B

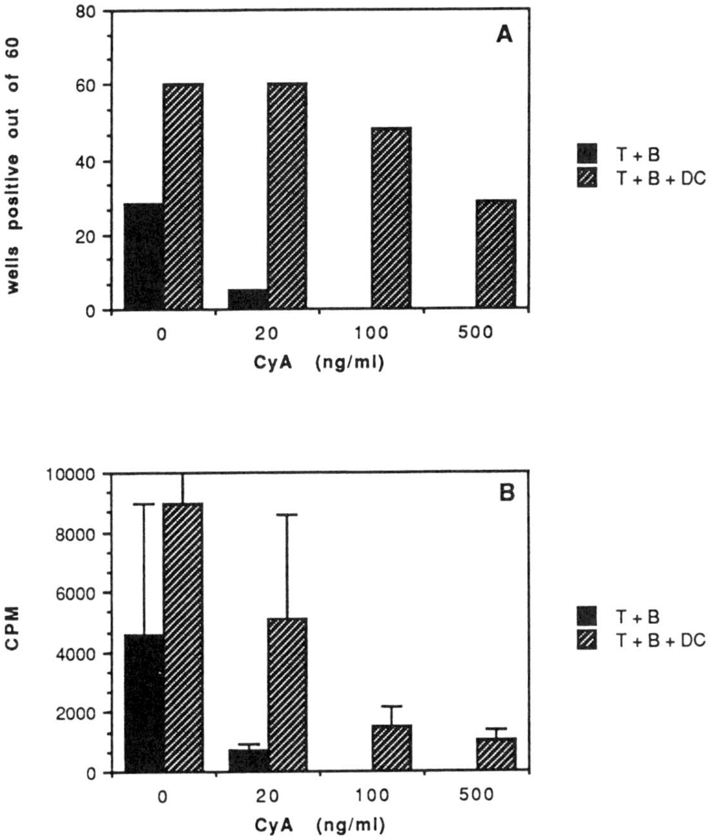

Figure 1. Increasing doses of cyclosporine A (CyA) were added at initiation of DC/T allogeneic microcultures. (A) # of wells scoring positive for total Ig. (B) relative amounts of Ig produced in these cultures, expressed as mean cpm of 60 wells.

microcultures (Fig. 1A). Higher concentrations of CyA did somewhat diminish the total output of Ig secreted by clones developing in the presence of DC (Fig. 1B), but not the proportion of these that expressed IgA. For instance, 20 ng/ml CyA had no effect on the frequency of responding B cells in DC/T/B microcultures and 87% versus 91% of productive clones expressed IgA in the presence or absence of CyA respectively.

TGFß Does Not Enhance IgA Expresssion in T/B or DC/T/B Microcultures

Transforming growth factor-ß (TGFß) has been proposed as an IgA 'switch factor' based on its effecting an increase in IgA expression in bulk B cell cultures stimulated by LPS (11). Table 1 shows that TGFß does not promote any IgA expression by clones grown in T/B microculture, with or without the CM from ConA stimulated D10 cells. Further, rather than enhancing the effect of DC to promote IgA expression, addition of TGFß both decreases the frequency of responding B cells in DC/T/B microculture and the incidence of IgA expression by the productive clones that do develop. Table 1 also shows that possible *in vivo* interactions with T cells prior to culture of B cells has no effect on their responsiveness to the presence of DC -- B cells from either euthymic or athymic mice generate productive clones displaying about the same incidence of IgA expression in the presence of DC.

Table 1. TGFß is not an IgA Switch Factor in Clonal Microcultures

			% producing	
	B cells	# clones	IgM	IgA
Medium	F_1	1	100	0
CM	F_1	2	100	0
DC	F_1	47	96	30
DC + TGFß	F_1	19	90	16
CM + TGFß	F_1	3	100	0
medium	nu/+	0	0	0
CM	nu/+	1	100	0
DC	nu/+	15	100	33
DC + TGFß	nu/+	8	100	0
CM + TGFß	nu/+	1	100	0
medium	nu/nu	2	100	0
CM	nu/nu	3	100	0
DC	nu/nu	44	98	43
DC + TGFß	nu/nu	11	100	9
CM + TGFß	nu/nu	3	100	0

Inoculum consisted of one B cell per well from strain of mice indicated: F_1 = k x b allotype. CM = 5% conditioned medium from D10 cultures. TGFß used at 5 ng/ml. Total Ig measured by reverse RIA; other isotypes not shown.

DISCUSSION

All of our observations are consistent with the requirement that T, B, and DC be physically co-cultured in order to enhance development of B cell clones that secrete IgA. Of course, we cannot rule out sequential interactions of DC with T cells, followed by T with B or of DC with B cells, followed by B with T. The simple propensity of DC to facilitate cell clustering and to nucleate cell aggregates can be observed in microculture and could account for a permissible decrease in T_H2 cell input. However, the remarkable ability of DC to overcome the effects of CyA to prevent the LK-dependent outgrowth of productive B cell clones suggests a more active role for these cells. CyA has generally been considered to effect a block in the signalling pathway leading to expres-sion of IL-2 and IL-2R that is ordinarily activated via the TCR (12). Recently, a CyA resistant signalling pathway leading to the expression of multiple LK by T cells has been described (13). The activation of this pathway is via CD28, constitutively expressed by most $CD4^+$ T cells, reacting with its ligand, B7, inducible on B cells (14). Of relevance to this present discussion was that finding that virtually all activated tonsillar $sIgG^+$ and $sIgA^+$ B cells (memory cells) expresssed B7 (14). It is tempting to surmise that DC may induce/provide B7 ligand for B/DC interaction with CD28 on T_H2 cells, leading to a CyA insensitive production and delivery of LK necessary for IgA expression by clonally-derived B cells that have undergone switching at the DNA-level to IgA.

Acknowledgements

The work reported here has been supported by a grant, AI-17997 from the National Institutes of Health. CES has been a Fellow of Training Grant CA-09140.

REFERENCES

1. D.A. Lebman, P.M. Griffin, and J.J. Cebra, Relationship between expression of IgA by Peyer's patch cells and functional IgA memory cells, *J. Exp. Med.* 166:1405 (1987).
2. J.A. Fuhrman and J.J. Cebra, Special features of the priming process for a secretory IgA response. B cell priming with Cholera toxin, *J. Exp. Med.* 153:534 (1981).
3. G.J.V. Nossal, B.L. Pike, and F.L. Battye, Sequential use of hapten-gelatin fractionation and fluorescence activated cell sorting in the enrichment of hapten-specific B lymphocytes, *Eur. J. Immunol.* 8:151 (1978).
4. P.A. Schweitzer and J.J. Cebra, Quality of antibodies secreted by clones in microcultures from B cells enriched on haptenated gelatin: isotypes and avidities, *Mol. Immunol.* 25:231 (1988).
5. J. Kaye, S. Porcelli, J. Tite, B. Jones, and C.A. Janeway, Jr., Both a monoclonal antibody and antisera specific for determinants unique to individual cloned helper T cell lines can substitute for antigen and antigen-presenting cells in the activation of T cells. *J. Exp. Med.* 158:836 (1983).

6. C.E. Schrader, A. George, R.L. Kerlin, and J.J. Cebra, Dendritic cells support production of IgA and other non-IgM isotypes in clonal microculture, *Int. Immunol.* 2:563 (1990).
7. D.B. Kotloff and J.J. Cebra, Effect of T_H-lines and clones on the growth and differentiation of B cells clones in microculture, *Mol. Immunol.* 25:147 (1988).
8. R.M. Steinman, G. Kaplan, M.D. Witmer, and Z.A. Cohn, Identification of a novel cell type in peripheral lymphoid organs of mice. V. Purification of spleen dendritic cells, new surface markers, and maintenance *in vitro. J. Exp. Med.* 149:1 (1979).
9. D.M. Spaulding, W.J. Koopman, J.H. Eldridge, J.R. McGhee, and R.M. Steinman, Accessory cells in murine Peyer's patch. I. Identification and enrichment of a functional dendritic cell. *J. Exp. Med.* 157:1646 (1983).
10. A. George and J.J. Cebra, Responses of single germinal-center B cells in T-cell-dependent microculture, *Proc. Natl. Acad. Sci. U.S.A.* 88:11 (1991).
11. R.L. Coffman, D.A. Lebman, and B. Shrader, Transforming growth factor-ß specifically enhances IgA production by lipopolysaccharide-stimulated murine B lymphocytes, *J. Exp. Med.* 170:1039 (1989).
12. J.F. Elliott, Y. Lin, S.B. Mizel, R.C. Bleackley, D.G. Harnish, and V. Paetkau, Induction of Interleukin 2 messenger RNA inhibited by cyclosporin A, *Science*, 226:1439 (1984).
13. C.B. Thompson, T. Lindsten, J.A. Ledbetter, S.L. Kunkel, H.A. Young, S.G. Emerson, J.M. Leiden, and C.H. June, CD28 activation pathway regulates the production of multiple T-cell-derived lymphokines/cytokines, *Proc. Natl. Acad. Sci. U.S.A.* 86:1333 (1989).
14. A. Vallé, J-P. Aubry, I. Durand,and J. Banchereau, IL-4 and IL-2 upregulate the expression of antigen B7, the B cell counterstructure to T cell CD28: an amplification mechanism for T-B interactions, *Int. Immunol.* 3:229 (1991).

ADHESION MOLECULES IN TONSIL DC-T CELL INTERACTIONS

Derek N.J. Hart and Timothy C.R. Prickett

Haematology/Immunology Research Group
Christchurch School of Medicine
Christchurch Hospital
Christchurch, New Zealand

INTRODUCTION

The specific activation of a T lymphocyte depends on the interaction of its T cell receptor with the antigenic peptide - MHC complex on the surface of an antigen presenting cells (APC). In order for full primary T cell activation to occur the APC must provide essential additional co-stimulatory activity.[1] The ability to provide this "second signal" is a distinguishing feature of the specialist APC known as dendritic cells (DC) but the exact nature of this cellular property remains uncertain.[2]

DC express a high level of MHC class II molecules, and their membrane processes may allow for increased area with the responding T lymphcoyte. At present it seems unlikely that DC produce activating cytokines such as IL-1,[3] nor have DC been shown to have unique costimulatory cell membrane molecules. DC form antigen independent clusters with T lymphocytes. These are stabilized in the presence of antigen to retain and activate the antigen specific T lymphocytes. These DC-T cell interactions were expected to involve several adhesion molecules known to be important in other cellular interactions.[4]

This report describes the expression of adhesion molecules on tonsil DC and their role in DC-stimulated allogeneic T lymphocyte responses.

MATERIALS AND METHODS

Tonsil DC preparation

Single cell suspensions were prepared from tonsils obtained at routine tonsillectomy (with Christchurch Hospital Ethical Commitee approval). The low density DC were enriched using a BSA density gradient as previously described.[5] The contaminating small B lymphocytes in these preparations do not appear to influence the allogeneic MLR.[5,6] Further purification was then undertaken using a final antibody mix of B1 (CD20), BA1 (CD24), OKT3 (CD3) and CMRF-21 (CD14) monoclonal antibodies, followed by goat anti-mouse Ig labelling for FACS IV separation (or incubation with coat anti-mouse Ig coated Dynabeads for magnetic separation). Tonsil DC were enriched in the mix-negative fraction.[6]

Dendritic Cells in Fundamental and Clinical Immunology,
Edited by Kamperdijk *et al.*, Plenum Press, New York, 1993

Mixed Leucocyte Reaction

Mixed leucocyte reactions (MLRs) were established using 10^5 peripheral blood T lymphocytes as responders, and 10^4 allogeneic tonsil DC. Test antibodies were present either in the MLR for the duration of incubation, or were preincubated with either responder or stimulatory fractions and washed off prior to initiating the MLR. MLR were harvested after five days following a four hour pulse with 0.5uCi ^{3}H thymidine. Clustering between DC and T lymphocytes was observed after 20 hours of MLR culture in flat- bottomed 96 well plates.

Monoclonal Antibodies to Adhesion Molecules

The monoclonal antibodies listed in Table 1 were kindly provided by colleagues, or made available via participation in the 3rd and 4th Leucocyte Differentiation Antigen Workshops.

Immunoperoxidase Staining

DC enriched preparations were stained using a sensitive immunoperoxidase technique.[6]

RESULTS

Expression of adhesion molecules on isolated tonsil DC

The results of the immunoperoxidase labelling are listed in Table 1. Tonsil DC stained with both anti-LFA alpha (CD11A) and anti-integrin beta$_2$ chain (LFA-1 beta, CD18) antibodies but no staining of the other integrin beta$_2$ associated molecules CD11b (CR3) or CD11c (p150) was detected. No other members of the integrin family were readily identifiable using the CD29 (anti-beta$_1$ chain or CD61 (anti-beta$_3$) chain) reagents, although very low density staining with the CD29 antibody QE-2ES was observed.

Adhesion molecules of the immunoglobulin gene superfamily were also present on DC. The anti-ICAM-1 (CD54), anti-ICAM-2 and the anti-LFA-3 (CD58) antibodies stained DC. Tonsil DC did not react with either the anti-LFA-2 (CD2) or CD56 reagents or the anti-VCAM-1 antibody. Weak CD31 staining was noted.

The LAM1.1 and LAM1.2 antibodies, recognising LEC-CAM-1 of the selectin family, did not stain DC. In contrast DC labelled intensely for CD44.

It was notable that the G19.1 epitope of CD43, a recently defined adhesion molecule, widely distributed on leucocytes, was not detected on DC.

Effect of the Antibodies to Adhesion Molecules on the MLR

The stimulation of T lymphocytes by tonsil DC was readily inhibited by the anti-class HLA-class II antibodies (positive control) but not by the mouse monoclonal antibodies or polyclonal mouse immunoglobulin (negative controls). Inhibition of the MLR was greatest using the anti-LFA-1 alpha and anti-LFA-1 beta reagents (see Table 2). No inhibition beyond the plateau occurred on adding these two reagents together. Less, but still significant inhibition (approximately 50%) was noted with the anti-LFA-3 reagents, similar to the inhibition observed with the CD2 antibody.

Interestingly, only relatively weak inhibition occurred with the anti-ICAM-1 antibodies, which had been selected for their ability to inhibit other ICAM-1 mediated adhesion interactions, even when added at high concentrations. A definite plateau of inhibition occurred with increasing concentrations of antibody. Furthermore, the antibody against the alternative LFA-1 ligand (ICAM-2), despite inhibiting other LFA-1 interactions, also failed to inhibit the DC stimulated MLR. The simultaneous addition of the anti-ICAM-1 and anti-ICAM-2 reagents failed to inhibit the MLR significantly beyond that occurring with the individual antibodies. Anti-ICAM-3 reagents have yet to be tested.[7]

Table 1. Reactivity of antibodies to known or potential adhesion molecules with isolated tonsil dendritic cells.

Cluster of Differentiation	Molecule	Antibody	Reactivity with tonsil DC
CD2	LFA-2	OKT11[a]	-
CD11a	LFA-1alpha	MHM24, CIMT	+
b	CR3	OKM1[a], 44	-
c	p150, 95	KB23, 3.9, Ki M1	-
CD18	LFA-1 beta (integrin B2)	MHM23, 60.3	+
CD29	VLA beta (integrin B1)	Goat Antisera	-
		QE-2E5	+/-
CD31	PECAM-1	SG134,TM3	+/-
CD36	GpIV	5F1	-
CD41	GpIIb/IIIa	ASH 1704	+/-
	GpIIb	HPL2	-
CD43	Leukosialin	G19-1	-
CD44	Hermes Antigen	GRHL1, F10-44-2,3E8	+++
CD46	MCP	122-2	+/-
CD54	ICAM-1	wCAM-1, 7F7,RRI/III	+
CD56	N-CAM	NKH-1[b]	-
CD58	LFA-3	BRIC5,TS2/9.1.4.3[a]	+
CD61	GpIIIa/VNR (integrin B3)	V1-P12	-
-	LEC-CAM-1	LAM-1.1,LAM-1.2	-
-	VCAM-1	VCAM-1	-

a Antibody from hybridomas obtained from the American Type Culture Collection.

b Purchased from Coulter Immunology, Hialeah, USA.

Contribution of Individual Adhesion Ligands to DC or T cell components of the MLR

The above experiments suggested that the LFA-1 - ICAM-1/ICAM-2/ICAM-3 interactions were important. Addition of LFA-1 and LFA-3 reagents simultaneously significantly increased the inhibition compared to either reagent alone but still permitted a significant MLR response (approximately 20% control).

Pre-incubation and washing the DC or T lymphocyte components prior to the MLR confirmed that DC LFA-3 interacted with T lymphocyte CD2. These experiments (see 6 for details) also indicated that the LFA-1 on T lymphocytes played the greater role, although the LFA-1 on DC clearly contributes significantly to DC-T lymphocyte interactions.

Adhesion Interactions in DC-T cell clusters

The anti-LFA alpha and LFA-1 beta reagents had the greatest effect on cluster stability (83 +/- 8 and 94 +/- 1% inhibition after pipetting) when compared to LFA-3 and CD2 reagents (12 +/- 8 and 24 +/- 11% inhibition). During these studies formation of clusters was actually increased by the CD44 reagents.

Table 2. Inhibition of the MLR by antibodies against known adhesion molecules.

Cluster of Differentiation	Molecular Specificity	Antibody	Mean inhibition (% +/- SE) EnrichedDC	/Purified DC
CD2	LFA-2	TS2/18.1.1	60 +/- 5	
CD11a	LFA-1alpha	MHM24	75 +/- 12	64 +/- 29
CD18	LFA-1beta	MHM23	74 +/- 8	59 +/- 28
		MHM24+23	79 +/- 7	
CD29	VLA beta	VLA beta Goat antisera	16 +/-51	11 +/- 1
CD31	PECAM-1	SG134	-2 +/- 6	
CD44	Hermes	GRHL1	-4 +/- 1	
	antigen	3E8	-4 +/- 6	
CD46	MCP	HuLym5	4 +/- 8	
CD54	ICAM-1	wCAM-1	24 +/- 4	25 +/- 26
	ICAM-1	7F7	14 +/- 12	
	ICAM-1	8F5	8 +/- 7	
CD58	LFA-3	G26	46 +/- 6	
	LFA-3	BRIC5	56 +/- 3	60 +/- 18
	ICAM-2	XBE-IC2/2	7.5 +/-1	

Results are expressed as the mean of three experiments each consisting of triplicate cultures involving 10^5 T cell responders and 10^4 mitomycin C treated DC enriched stimulators. The mean of two confirmatory experiments using 10^3 purified DC as stimulators is also tabulated.

DISCUSSION

Adhesion molecules on DC may have several functions. They may influence DC migration and distribution via interactions with endothelium, epithelial cells (eg Langerhans cells in the skin) and other connective tissue components. We have shown that adhesion molecules, notably LFA-1, are involved in the clustering interaction of tonsil DC with T lymphocytes as well as the primary activation of T lymphocytes. LFA-1 is present on DC and the documentation of both ICAM-1 and ICAM-2 on DC (again these are expressed on T lymphocytes) allows for these two ligand pairs to operate in reciprocal directions. Our functional data raises the probability that ICAM-3, the third ligand for LFA-1 recently defined,[7] may well be expressed on DC and be involved in DC mediated primary T cell activation. We have not as yet been able to demonstrate ICAM-1 or ICAM-2 on interstitial DC but expression of these molecules may be upregulated upon DC migration centrally and be relevant to endothelial adhesion and lymphatic drainage.

The presence of LFA-3 on DC clearly delivers a significant signal to T lymphocytes via CD2. As such the LFA-3 on DC may play a greater role than in other cellular APC systems.

These ligand pairs do not appear to account for all the adhesion molecule interactions between DC and T lymphocytes. As well as ICAM-3, it is possible that new members of the integrin family may be expressed selectively on DC[8] and contribute to the potent stimulatory properties of DC. The ability to induce adhesion molecules, mobilise them on the membrane, and possible changes in avidity of these molecules, dependent on cell activation will also merit consideration.[6]

Finally we would draw attention to the absence of the strongly charged leukosialin or CD43 molecule from DC. The absence of CD43 from tonsil DC (and blood DC) may enable intimate membrane association between DC and T lymphocyte by reducing electrostatic repulsion.

ACKNOWLEDGEMENTS

This work was supported by grants from the Canterbury Medical Research Foundation and the Health Research Council of New Zealand. D.N.J. Hart is also supported by the McClelland Trust.

REFERENCES

1. D.L. Mueller, M.K. Jenkins, and R.H. Schwartz, Clonal expansion versus functional clonal inactivation: a costimulatory signalling pathway determines the outcome of T cell antigen receptor occupancy, *Ann.Rev. Immunol.*.7:445(1989).
2. R.M. Steinman and M.C. Nussenweig, Dendritic cells: features and functions, *Immunol. Rev.*.53:127(198?).
3. J.L. McKenzie, T.C.R. Prickett and D.N.J. Hart, Human dendritic cells stimulate allogeneic T cells in the absence of interleukin 1, *Immunology*.67:290(1989).
4. D.N.J. Hart and J.L. McKenzie, Isolation and characterisation of human tonsil dendritic cells, *J.Exp.Med.* 168:157(1988).
5. T.A. Springer, Adhesion receptors of the immune system, *Nature*, 346:425(1990).
6. T.C.R. Prickett, J.L. McKenzie and D.N.J. Hart, Adhesion molecules on human tonsil dendritic cells, *Transplantation*, 53:4831(1992).
7. A.R. Fougerolles and T.A. Springer, Intercellular adhesion molecule 3, a third adhesion counter-receptor for lymphocyte function - associated molecule 1 on resting lymphocytes, *J.Exp.Med.* 175:185(1992).
8. Y.J. Rosenstein, J.K.Park, W.C.Hahn, F.S.Rosen, B.E.Bierer and S.J.Burakoff, CD43 - a molecule defective in Wiskott-Aldrich syndrome binds ICAM-1, *Nature*.354:233(1991).
9. M. Brenan and M. Puklavec, The MRC OX-62 antigen: a useful marker in the purification of rat veiled cells with the biochemical properties of an integrin, *J. Exp. Med.*175:1457(1992).

DENDRITIC CELLS HAVE REDUCED CELL SURFACE MEMBRANE GLYCOPROTEINS INCLUDING CD43 DETERMINANTS

William Egner, Barry D. Hock, and Derek N.J. Hart

Haematology/Immunology Research Group
Christchurch School of Medicine
Christchurch Hospital
Christchurch, New Zealand

INTRODUCTION

The molecular basis for the special ability of dendritic cells (DC) to stimulate primary T cell responses is at present unknown. Important accessory molecules such as the adhesion molecules LFA-3, LFA-1, ICAM-1 and ICAM-2 and probably BB1/B7 (the ligand for T cell CD28 and CTLA-4) are expressed on DC[1] and contribute to the antigen presenting function of DC. However no unique co-stimulatory cell membrane molecule or secreted cytokine has been identified on DC to date. Human blood, tonsil and bone marrow DC are potent stimulators of alloimmune T lymphocyte responses. Highly purified peripheral blood CD19 positive B cells and CD14 positive monocytes may also stimulate an alloresponse from peripheral blood T cells, including those carefully depleted of endogenous antigen presenting cells; albeit a less potent one.

The difference in stimulatory capacity between DC and other cell types may result from other membrane properties which favour close cellular association of the DC with T cells in cell clusters[2]. Studies have suggested that human DC may have a reduced level of cell surface molecule glycosylation compared with other cell types.

MATERIALS AND METHODS

Monocyte and Macrophage Preparation

Peripheral blood mononuclear cells (PBMC) from normal donors were prepared over Ficoll/hypaque (density 1.077), and used fresh or after Teflon bag culture. Monocytes were examined immediately after isolation, and after differentiation to macrophages by seven day Teflon bag culture (generously provided by Prof R Andreesen, Germany) in the presence of either 2.5% pooled human AB serum, recombinant M-CSF (Cetus Corp, USA) or GM-CSF (Boehringer Ingelheim, Germany) at 50 ug/ml.

Blood DCs

Peripheral blood DCs were enriched by culturing T cell depleted PBMC overnight, followed by separation on a bovine serum albumin gradient. Low density cells were then panned on human immunoglobulin coated plates, followed by panning with rabbit anti-human immunoglobulin to remove Fc-receptor positive cells and B lymphocytes. The remaining cells were incubated with a mixture of CD3 (OKT3), CD19 (FMC63), CD57

Dendritic Cells in Fundamental and Clinical Immunology,
Edited by Kamperdijk *et al.*, Plenum Press, New York, 1993

(HNK-1) and CD14 (CMRF-31) antibodies prior to negative selection using goat anti-mouse coated Dynabeads (Dynal, Norway).

Tonsil DCs

DC were prepared from tonsil mononuclear cell preparations as previously described[3].

Immunoperoxidase Staining

Monocyte, macrophage and DC preparations were labelled with control, CD11a, CD54, CD58 monoclonal antibodies, and the CD43 antibody, G19-1[4] using an indirect immunoperoxidase technique[3]. The intensity of staining was graded from weak (+/-) to strongly positive (+++).

Cell Membrane Glycoprotein Analysis

Cell surface sialoglycoproteins were selectively labelled with tritiated borohydride using the technique of Gahmberg and Andersson. Labelled cells were solubilised in 1% NP-40 and then examined using SDS-PAGE and fluorography as described previously[4].

RESULTS

Monocytes and macrophages stained strongly with the CD43 antibody (G19-1) and the expression of leukosialin was upregulated by Teflon bag culture as monocytes differentiated into macrophages. In contrast, DC isolated by conventional methods from blood and tonsil failed to stain for the leukosialin epitope recognised by the CD43 antibody (Table 1).

Table 1

CD43 and adhesion molecule expression on human dendritic cells, monocytes, and macrophages.

	DC	Mono	AB-MØ	M-MØ	GM-MØ
CD11a LFA-1α	++	+/-	++	++	++
CD43 leukosialin	-	+	++	++	+++
CD54 ICAM-1	+/-	+/-	+	+	+
CD58 LFA-3	++	+/-	+	+	+

Borohydride labelling of cell surface sialic acid residues in combination with fluorography was used to examine the cell surface glycoproteins of monocytes, macrophages and tonsil DC. Monocytes and macrophages labelled readily and the SDS-PAGE analysis showed a full size range of membrane glycoproteins. Lymphocytes have previously been shown to have a predominance of higher molecular weight bands[4] and these, plus increased labelling of lower molecular weight glycoproteins, were readily demonstrable on activated T and B lymphocytes. However attempts to borohydride label DC cell surface molecules failed to detect obviously sialated cell surface molecules.

DISCUSSION

The CD43 antigen or leukosialin is a cell membrane molecule which is extensively glycosylated and has a wide distribution on leukocyte subpopulations[5]. It is present on monocytes, T lymphocytes and a subpopulation of B lymphocytes and accounts for more than 50% of the negative charge on the cell surface of human peripheral blood T lymphocytes. The function of the molecule is not entirely clear although its absence in the Wiskott-Aldrich syndrome is associated with immunodeficiency[6]. Recently the molecule has been suggested to have an adhesion function and has been shown to bind to the ligand ICAM-1[7].The failure to stain DC with CD43 antibodies and our ability to readily identify cell surface sialated glycoproteins suggest that DC lack CD43. This should be confirmed by testing with antibodies against an alternative CD43 epitope and by testing for CD43 RNA expression. Two CD43 epitopes have been identified: the majority of antibodies, like G19-1, stain a neuraminidase sensitive epitope but a neuraminidase resistant epitope has also been described[4].

It has been suggested that DC MHC molecules are hyposialyated[8]. Our data indicates that this may be a more general phenomena affecting other DC cell surface molecules. If this is proven to be the case then the relative paucity of neuraminic acid residues on the DC membrane may in part explain the ability of DC to form the intimate cell membrane associations with T lymphocytes essential to initiate a primary T lymphocyte response.

Finally, the fact that monocytes express the CD43 antigenic epitopes, which show increased density of staining upon differentiation into macrophages, should be exploited in immunophenotypic analyses, which attempt to distinguish between DC and cells of the monocyte-macrophage lineage.

ACKNOWLEDGEMENT

This work was supported by grants from the Canterbury Medical Research Foundation and the Health Research Council of New Zealand. We thank colleagues for making monoclonal antibodies available, in particular Dr Ledbetter for providing G19-1.

REFERENCES

1. T.C.R. Prickett, J.L. McKenzie, and D.N.J. Hart, Adhesion molecules on human tonsil dendritic cells, *Transplantation.* 53:483(1992).
2. K. Inaba, M.D. Witmer, and R.M. Steinman, Clustering of dendritic cells, helper T lymphocytes and histocompatible B cells during primary antibody responses in vitro, *J. Exp. Med.* 260:858(1984).
3. D.N.J. Hart and J.L. McKenzie, Isolation and characterisation of human tonsil dendritic cells, *J. Exp. Med.* 168:157(1988).
4. B.D. Hock and D.N.J. Hart, Cellular protein profiles of the Hodgkin's disease cells lines L428, KM-H2 and HDLM-2: a comparative study, *Leuk Res.* 16:253(1992).
5. M. Stoll, R. Dalchau, and R.E. Schmidt, Cluster report: CD43, in *Leucocyte Typing IV. White Cell Differentiation Antigens*, W. Knapp,ed, Oxford University Press, Oxford (1989).
6. E. Remold-O'Donnell, C. Zimmerman, D. Kenney, and F.S. Rosen, Expression on blood cells of sialophorin, the surface glycoprotein that is defective in Wiskott-Aldrich syndrome, *Blood.* 70:104(1987).
7. Y.J. Rosenstein, J.K. Park, W.C. Hahn, F.S. Rosen, B.E. Bierer, and S.J. Burakoff, CD43- a molecule defective in Wiskott-Aldrich syndrome binds ICAM-1, *Nature.* 354:233(1991).
8. C.J.P. Boog, J.J. Neefjes, J. Boes, H.L. Ploegh and C.S.M. Melief, Specific immune responses restored by alteration in carbohydrate chains of surface molecules on antigen presenting cells, *Eur. J. Immunol.* 19:537(1989).

THE EFFECT OF HUMAN DENDRITIC CELLS ON THE LECTIN-INDUCED RESPONSIVENESS OF $CD4^+$T CELLS TO IL-2 AND IL-4

Sergiusz Markowicz[1] and Anita Mehta[2]

[1]Department of Immunology, The Maria Sklodowska-Curie Memorial Cancer Center and Institute of Oncology Wawelska 15, 02034 Warsaw, Poland

[2]Department of Pathology, Stanford University School of Medicine, Stanford, California 94305, USA

INTRODUCTION

Human peripheral blood dendritic cells (DC) are potent accessory cells (AC) for the lectin-induced proliferative response of T lymphocytes[1]. However, the contribution of DC to the responsiveness of lectin-induced T cells to IL-2 and IL-4 is not clearly understood. The AC-dependent induction of IL-2 responsiveness by mitogenic lectin has been reported[2]. It has been also shown that the proliferation of lectin-induced T cells is promoted by IL-2 or IL-4 in the absence of AC[3].

IL-2[4,5] and IL-4[6] at high concentrations induce proliferation of fresh human peripheral blood mononuclear cells in the absence of identifiable exogeneous mitogenic or antigenic stimuli. The proliferative response of resting $CD4^+$T cells to IL-2 or IL-4 in the absence exogeneous stimuli is DC-dependent and monocytes/macrophages do not support such a response (manuscript in preparation). It implies that DC can induce proliferation of autologous $CD4^+$T cells in the process which does not involve antigen presentation.

This study was aimed to determine whether DC regulate lectin-induced responsiveness of $CD4^+$T cells to IL-2 and IL-4. DC and monocytes were compared as accessory cells in the responses to IL-2 and IL-4. $CD4^+$T cells were stimulated with PHA: 1. in the presence of DC or monocytes, or 2. $CD4^+$T cells pulse-astivated with PHA in the absence of accessory cells were subsequently provided with DC or monocytes. The latter approach was undertaken to determine whether DC directly mediated the interaction of $CD4^+$T cells with PHA or provided $CD4^+$T cells with a distinct costimulatory signal.

The contribution of DC versus monocytes to the regulation of the responsiveness to IL-2 and IL-4 of naive and memory subsets of $CD4^+$T cells was examined with the use of suboptimal doses of PHA. Naive and memory subsets of $CD4^+$T cells were identified by isoforms of CD45 molecule[7].

It has been demonstrated[8] that the majority of human T cells proliferating in autologous mixed lymphocyte reaction (AMLR) were specific for xenogeneic antigens derived from sheep erythrocytes (SRBC) and xenogeneic sera. Therefore in this study $CD4^+$T cells, DC and monocytes were not exposed to xenogeneic sera and SRBC during the procedure of cell isolation or in the cultures.

MATERIALS AND METHODS

Human peripheral blood DC, monocytes and $CD4^+$T cells were obtained according to the previously described procedures[1,9,10] with modifications indicated below. Briefly,

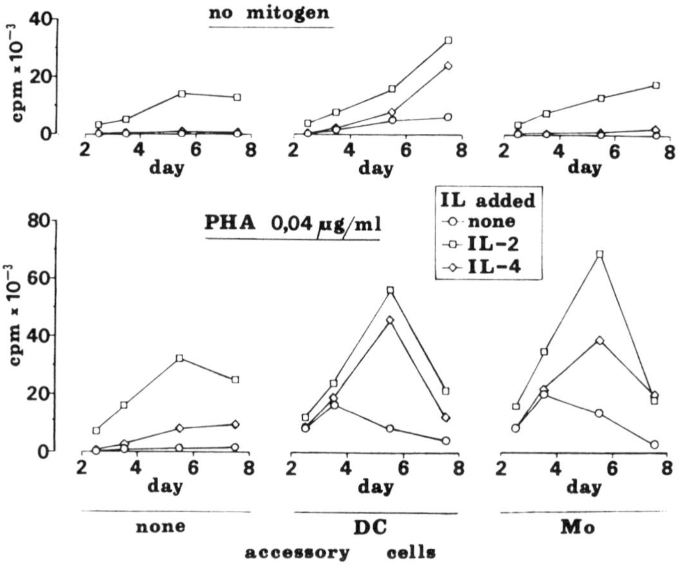

Figure 1. The effects of DC and monocytes (Mo) on the proliferative response of $CD4^+$T cells to exogeneous IL-2 and IL-4 in cultures unstimulated or stimulated with PHA at a submitogenic dose of 0,04 μg/ml. IL-2 was added at a final concentration of 12,5 U/ml, and IL-4 at a final concentration of 1000 U/ml. Results shown represent the mean counts per minute (cpm) for quadruplicate microcultures of 50 x 10^3 $CD4^+$T cells/well cultured alone or co-cultured with 10 x 10^3 of irradiated accessory cells (AC).

mononuclear cells were isolated by Ficoll-Uropoline gradient centrifugation. Low-density and high-density mononuclear cells were separated in a four-step discontinuous Percoll (Pharmacia LKB,Uppsala,Sweden) gradient.75%; 50,5%; 40%; and 30% dilutions of stock isoosmotic solution of Percoll (1,130 g/ml) in Dulbecco's calcium- and magnesium-free PBS containing 5% of heat-inactivated pooled human AB serum (DPBS/HS) were layered sequentially. Mononuclear cells suspended in DPBS/HS were overlayed onto the gradient and centrifuged at 1000 x g for 23 min. at 4°C. Low-density cells, mostly monocytes, were collected from the interface over 50,5% Percoll solution (density 1,065 g/ml), while high-density cells were collected from the interface between 75% and 50,5% Percoll solutions. High-density cells were suspended in RPMI-1640 medium supplemented with 2 mM L-glutamine, 50 μg/ml gentamycin , and 10% heat-inactivated human AB serum (hereafter designated complete medium [CM]) and cultured in Teflon vessels at 37°C in 5% CO_2 in air for 1 day. Thereafter DC and lymphocyte were separated in a two-step discontinuous Nycodenz gradient. Cells removed from Teflon vesels were suspended in DPBS/HS containing 5% of stock Na_2EDTA solution (2,7% w/v, pH 7,4), and thereafter 75% dilution of Nycodenz-monocytes (Nycomed,Oslo,Norway) in DPBS and undiluted Nycodenz were underlayered sequentially. Gradient was centrifuged at 400 x g for 13 min. at room temp. DC-enriched fraction was collected from the interface between Nycodenz layers. This fraction was directly used for cultures or further purified by paning procedure[11] simultaneously using LeuM3, Leu 12, Leu 11 and OKT3 moAbs. Lymphocyte fraction collected from the pellet was further purified by incubation on plastic and $CD4^+$T cells were isolated by a negative panning procedure using Leu2a moAb (anti-CD8), Leu-11 moAb and CA141 anti-HLA-DR moAb at the aim to remove $CD8^+$T cells, NK cells and remaining accessory cells. CD45RA $CD4^+$ naive and CD45RO $CD4^+$ memory T cell subsets were obtained by panning with UCHL-1 moAb (anti-CD45RO). By cytofluorographic analysis UCHL-1 negative naive cells stained brightly with Leu-18 moAb (anti-CD45RA), whereas dim staining was characteristic for UCHL-1 positive cells. Monocyte-enriched Percoll low-density fraction was further purified by overnight adherence to plastic yielding highly purified monocytes.

$CD4^+$T cells were cultured usually for 4 days in microwells containing 0,2 ml CM at 50 x 10^3 cells per well. 10 x 10^3, 7,5 x 10^3, or 5 x 10^3 of DC or monocytes were added per well. Accessory cells before the onset of the cultures were irradiated (30 Gy from ^{60}Co source). T cell proliferation was assessed by [^{3}H]-TdR incorporation. 0,5 μCi per well [^{3}H]-TdR (spec.activity 5,0 Ci/mmol) was added to the wells for the final 16 h of culture. PHA (HA 17, Wellcome,England) was used at a range from 0,01 to 1 μg/ml. Human rIL-2 (Genzyme, Boston, MA) had a sp.act. of approximately 2.5 x 10^6 BRMP units/mg, human rIL-4 (Genzyme) had a sp.act. of approximately 1 x 10^8 U/mg.

RESULTS

The proliferative response of human peripheral blood $CD4^+$T cells examined at a wide range of PHA doses was highly dependent on accessory cells. Both DC and monocytes were effective as accesory cells. The response increased when $CD4^+$T co-cultured with DC or monocytes were provided with exogeneous IL-2 or IL-4 (Fig.1). PHA-activated $CD4^+$T cells responded to IL-2 or IL-4 also in the absence of accesory cells even at low submitogenic doses of PHA. Synergistic effect of DC and IL-2 or IL-4 was observed at submitogenic concentrations of PHA. DC, in contrast to monocytes, augmented a weak proliferative response of $CD4^+$T cells induced by IL-2 in the absence of identifiable exogeneous stimuli. IL-4 did not induce proliferation of $CD4^+$T cells in the absence of identifiable exogeneous stimuli unless $CD4^+$T cells were cultured in the presence of DC. Monocytes failed to induce such a response. When the kinetics was concerned, the DC-dependent proliferative response of $CD4^+$T cells to IL-2 or IL-4 in the absence of identifiable foreign stimuli was retarded comparing to the proliferation induced in the presence of PHA.

The contact of $CD4^+T$ cells with DC was not prerequisite for their activation with PHA. However, $CD4^+T$ cells pulse-activated with submitogenic concentrations of PHA in the absence of DC required the interaction with DC in the subsequent culture to initiate proliferation unless provided with exogeneous interleukins (Fig.2). It implies that that the costimulatory effect of DC does not involve the presentation of PHA to $CD4^+T$ cells. DC augmented proliferative response of $CD4^+T$ cells to IL-2 and IL-4 when added subsequently to pulse-activation of $CD4^+T$ cells with PHA.

No clear-cut differences were found concerning the contribution of DC and monocytes to the responsiveness of $CD4^+T$ naive and memory subsets to IL-2 and IL-4. DC, in contrast to monocytes, were very potent as accessory cells for the proliferative response of CD45RA $CD4^+$ naive T cells pulse-activated with low submitogenic doses of PHA. However, differences between accessory cell activities of DC and monocytes were diminished when high doses of PHA were used (data not shown).

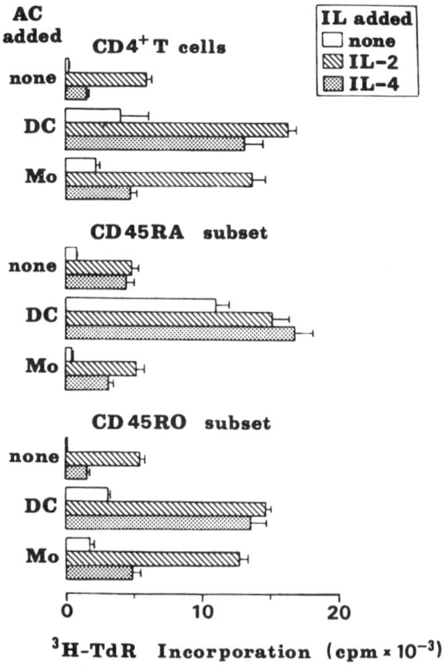

Figure 2. DC and monocytes (Mo) upregulate the responsiveness of $CD4^+T$ cells and their CD45RA naive and CD45RO memory subsets to exogeneous IL-2 and IL-4 when added subsequently to pulse-activation of $CD4^+T$ cells with PHA. $CD4^+T$ cells and their naive and memory subsets were incubated with PHA at a concentration of 0,2 μg/ml for 2 hours in the absence of accessory cells (AC), washed extensively and subsequently cultures in microwells were set up. 50×10^3 $CD4^+T$ cells were cultured alone or in the presence of $7{,}5 \times 10^3$ irradiated DC or monocytes. IL-2 was added to the cultures at a final concentration of 2 U/ml, and IL-4 at a final concentration of 500 U/ml. Results represent the mean counts per minute (cpm)±S.E. of five replicate cultures. The incorporation of [^{3}H]-TdR by $CD4^+T$ cells non-preincubated with PHA was < 400 cpm in cultures without exogeneous interleukins, and < 700 cpm or < 3000 cpm respectively in cultures supplemented with IL-4 or IL-2. This is a representative experiment from six performed.

DISCUSSION

PHA binds to multiple cell surface glycoproteins and activates T cells through both the CD3[12] and CD2[13] molecules, but this activation on its own is not sufficient to induce vigorous proliferation of T cells. The results described here show that DC costimulate $CD4^+$T cell response to PHA and augment PHA-induced responsiveness to IL-2 and IL-4, but do not mediate $CD4^+$T cell interaction with PHA. It implies that DC do not present PHA to $CD4^+$T cells, but presumably provide $CD4^+$T cells with a distinct costimulatory signal. It also indicates that DC can activate T cells in the process, which does not involve antigen processing and antigen presentation by DC.

Our results support a concept that DC functions are not limited to antigen presentation, but DC "sensitize" T cells in addition to antigen presentation[14]. Our results suggest also that DC "sensitize" $CD4^+$T cells to IL-2 and IL-4 in the absence of identifiable exogeneous antigenic or mitogenic stimuli.

DC costimulatory function probably involves adhesion molecules expressed on DC surface. The differential costimulatory effect of four different adhesion molecules on the response of naive and memory $CD4^+$T lymphocytes has ben shown recently[15] with the use of soluble fusion chimeras of these molecules co-immobilized with moAb directed at TCR or CD3.

In our study we have used PHA in submitogenic doses to examine $CD4^+$T cell requirements for costimulatory signals in the response to PHA. Differential activation requirements have been reported for naive and memory $CD4^+$T lymphocytes stimulated by moAb to CD3 complex[16], whereas both naive and memory subsets responded well to PHA used at optimal doses in the presence of accessory cells[7,16]. Our results suggest that costimulatory effects of DC and monocytes on the response of naive and memory $CD4^+$T lymphocytes to PHA in submitogenic doses are different, but the observed differences are not clear-cut.

REFERENCES

1. J.W.Young and R.M.Steinman, Accessory cell requirements for the mixed-leucocyte reaction and polyclonal mitogens , as studied with a new technique for enriching blood dendritic cells, *Cell.Immunol.* 111:167(1988).
2. T.Hunig,M.Loos,and A.Schlimpl.The role of accessory cells in polyclonal T cell activation. I.Both induction of interleukin 2 production and interleukin 2 responsiveness by Concanavalin A are accessory cell dependent, *Eur.J.Immunol.* 13:1 (1983).
3. R.J.Armitage,A.E.Namen,H.M.Sassenfeld,and K.H.Grabstein, Regulation of human T cell proliferation by IL-7, *J.Immunol.*144:938(1990).
4. J.Lifson,A.Raubitschek,C.Benike,K.Koths,A.Ammann,P.Sondel,and E.G.Engleman, Purified interleukin-2 induces proliferation of fresh human lymphocytes in the absence of exogeneous stimuli, *J.Biol.Response Mod.*5:61 (1986).
5. B.K.Mookerje and J.L.Pauly, Human recombinant interleukin-2 is mitogenic to human lymphocytes, *J.Leuk.Biol.* 38:553 (1985).
6. H.Spits,H.Yssel,Y.Takebe,N.Arai,T.Yokota,F.Lee,K.Arai,J.Banchereau,and J.E. de Vries, Recombinant interleukin 4 promotes the growth of human T cells, *J.Immunol.*139:1142 (1987).
7. S.H.Smith,M.H.Brown,D.Rowe,R.E.Callard,and P.C.L.Beverley,Functional subsets of human helper inducer cells defined by a new monoclonal antibody,UCHL-1, *Immunology* 58:63 (1986).
8. C.Huber,M.Merkenschlager,C.Gattringer,I.Royston,U.Fink,and H.Braunsteiner, Human autologous mixed reactivity is primarily specific for xenoprotein determinants absorbed to antigen-presenting cells during rosette formation with sheep erythrocytes, *J.Exp.Med.*155:1222 (1982).

9. V.Tsai and N.J.Zvaifler,Dendritic cell-lymphocyte clusters that form spontaneously in rheumatoid arthritis synovial effusions differ from clusters formed in human mixed leucocyte reactions, *J.Clin.Inv.* 82:1731 (1988).

10. S.Markowicz and E.G.Engleman,Granulocyte-macrophage colony-stimulating factor promotes differentiation and survival of human peripheral blood dendritic cells in vitro, *J.Clin.Inv.* 85:955 (1990).

11. E.G.Engleman,C.J.Benike,F.C.Grumet,and R.L.Evans,Activation of human T lymphocyte subsets: helper and suppressor/cytotoxic cells recognize and respond to distinct histocompatibility antigens, *J.Immunol.*127:2124 (1981).

12. M.A.Valentine,C.D.Tsoukas,G.Rhodes,J.H.Vaughan,and D.A.Carson, Phytohemagglutinin binds to the 20-kDa molecules of the T3 complex, *Eur.J.Immunol.*15:851 (1985).

13. K.O'Flynn,A.M.Krensky,P.C.L.Beverley,S.J.Burakoff,and D.C.Linch, Phytohemagglutinin activation of T cells through the sheep red blood cell receptor, *Nature* 313:686 (1985).

14. R.M.Steinman, Cytokines amplify the function of accessory cells, *Immunology Letters* 17:197 (1988).

15. N.K.Damle,K.Klussman,P.S.Linsley,and A.Aruffo, Differential costimulatory effects of adhesion molecules B7, ICAM-1, LFA-3, and VCAM-1 on resting and antigen-primed CD4+T lymphocytes, *J.Immunol.* 148:1985 (1992).

16 .J.A.Byrne,J.L.Butler,and M.D.Cooper,Differential activation requirements for virgin and memory T cells, *J.Immunol.* 141:3249 (1988).

ANALYSIS OF CYTOKINE AND CYTOKINE RECEPTOR PRODUCTION BY HUMAN DENDRITIC CELLS

Victoria L. Calder, Timothy C.R. Prickett,
Judith L. McKenzie and Derek N.J. Hart

Haematology Department
Christchurch Hospital
Christchurch, New Zealand

INTRODUCTION

Dendritic cells (DC) have a unique ability to stimulate a primary T lymphocyte response. The presence of adhesion molecules on DC[1] promotes the association of DC with T lymphocytes and the MHC antigen-peptide complex on DC triggers the responding T lymphocyte TCR (signal I). It is clear that additional cellular properties of the DC (signal II), as well as signalling via the adhesion molecules, are crucial to its function as a specialist antigen presenting cell (APC).[2] The nature of the DC co-stimulatory signal II remains unexplained but the possibility that DC produce cytokines such as IL-1[3] or IL-6,[4] which activate T lymphocytes, has been suggested. Convincing evidence that DC produce cytokines has not been forthcoming[5,6] but it is clear that exogenous cytokines such as IL-1[7] and GM-CSF[8] do impact on DC. These effects may be considered as a third signal involved in T cell activation (regulated by CD45) which primes DC to maximise their APC function.[9]

It has been difficult to purify sufficient DC to analyse cytokine production by biological or immunological assays. Likewise, only limited analysis of cytokine mRNA levels has been possible by Northern blotting. The polymerase chain reaction (PCR) now makes it possible to analyse cytokine and cytokine receptor mRNA[10] in highly purified preparations of DC. This report describes the PCR analysis of the IL-1 cytokines,[3] receptors[11,12] and antagonist.[13] It also complements our previous conventional IL-1 analysis[5] and provides preliminary data on other biologically important cytokines and their receptors.

MATERIALS AND METHODS

Dendritic Cell Preparations

Highly purified tonsil DC preparations were obtained using immunomagnetic bead separations as described previously.[1]

Blood DC were isolated using a similar procedure but this included additional panning steps using anti-Fc and anti-Ig coated dishes as well as a second density gradient separation. Separation of contaminating cells was carried out on a FACS IV using additional CD16 and CD57 antibodies to remove large granular lymphocytes.

Both tonsil and blood preparations were shown to be greater than 90% pure by repeat antibody mix[1] immunolabelling.

Peripheral Blood Mononuclear (PBMC) and B cell preparations

Tonsil B cells were isolated after T cell depletion and panning using CD3 antibody. PBMC were isolated over a 1.077 Ficoll/Hypaque gradient and the cells cultured for three days in the presence of 10 ug LPS/ml or 5 ug PHA/ml.

Preparation of RNA and PCR Analysis

Total mRNA was prepared from approximately 5 x 10^4 DC using the Fast Track mRNA isolation kit (Invitrogen, USA) and cDNA prepared using oligo dT priming and AMV reverse transcriptase. Approximately 0.5 ul (of 20 ul final reaction) was then used as template for PCR reactions designed to detect the different cytokines and their receptors. Primer sequences (available on request) were designed to amplify across spliced exons excluding problems with amplication of genomic DNA. Conditions for each PCR were optimised using appropriate positive controls. The cDNA was shown in each case to amplify a 499 bp product of the dihydrofolate reductase gene. Each PCR amplified a band of the predicted size and amplified products were separated on agarose gel, blotted onto Hybond N^+ (Amersham, UK) filters and probed with a ^{32}P-gamma ATP labelled internal oligonucleotide. After exposure to the filters autoradiographs were interpreted as negative (-), weak (+/-), moderate (+) or strong (++) signal.

RESULTS

IL-1 Analysis

Tonsil DC do not produce IL-1 alpha or IL-1 beta as assessed by immunolabelling, bioassays and ELISA analysis.[5] A similar analysis shows that blood DC do not produce IL-1 (McKenzie and Hart - unpublished). Furthermore, polyclonal antisera capable of blocking the functional effects of IL-1 alpha and IL-1 beta failed to influence the ability of tonsil or blood DC to stimulate allogeneic T cell responses.

Analysis by PCR suggests that tonsil DC produce small amounts of either IL-1 alpha or IL-1 beta mRNA but in each case the signal on Southern blotting was weak compared to the stimulated PBMC control (Table 1). Blood DC appear to produce very small amounts of IL-1 beta mRNA and no detectable IL-1 alpha mRNA. Thus although in certain circumstances some IL-1 mRNA may be detectable in DC this does not result in the production of functionally relevant IL-1 protein.

TABLE 1. Interleukin 1 cytokine, receptor and antagonist expression in DC preparations.

	Tonsil DC 1	Tonsil DC 2	Blood DC 1	Blood DC 2	B cells	PBMC (stimulated)
IL-1alpha	+/-	+	-	-	+/-	++
IL-1beta	+/-	+	-	+/-	+/-	++
IL-RI	+	+	+/-	+	+	++
IL-RII	-	-	-	-	-	++
IL-Ra	+	+	-	-	-	+

IL-1 Receptor Expression

Although IL-1 has been reported to stimulate mouse[7] and human tonsil DC[5] function the presence of IL-1 receptors on DC has not been demonstrated previously. Significant amounts of IL-1RI mRNA (645 bp product) was produced by both tonsil and blood DC (Figure 1), indeed this appeared to be more than produced by tonsil B cells. No IL-1RII mRNA was detected except a strong band (600 bp product) in the PBMC positive control (Figure 1 and Table 1).

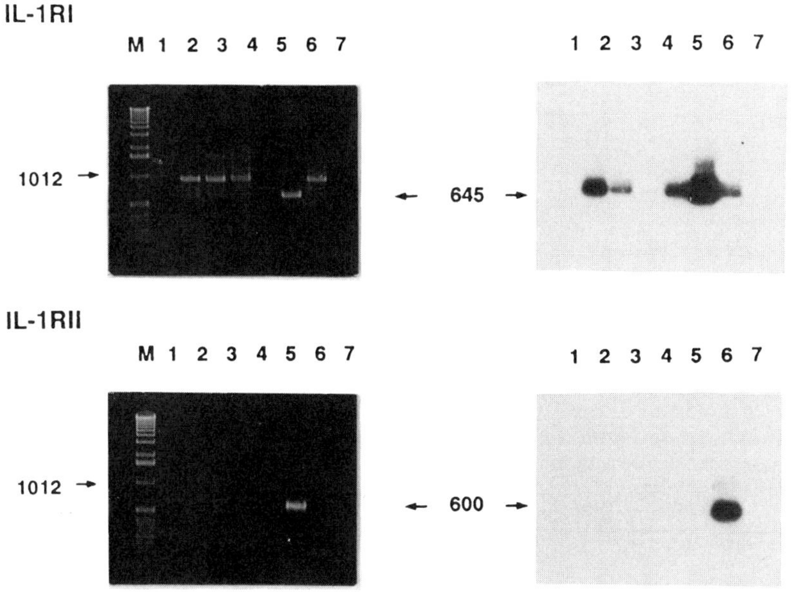

Figure 1. PCR products for IL-1RI (top) and IL-1RII (bottom) were electrophoresed through an agarose gel, ethidium stained (left) and after Southern transfer, membranes were probed with specific internal ^{32}P labelled oligonucleotides and the resulting autoradiographs exposed for two hours (right). The IL-1RI 625 bp product and the IL-1RI 600 bp product are arrowed. Lane 1: negative control; Lane 2 - 3: tonsil DC preps; Lane 4 - 5: blood DC preps 1 & 2; Lane 6: stimulated PBMC; Lane 7: tonsil B cells; M = BRL 1Kb ladder 1 & 2.

IL-1 Receptor Antogonist

The regulation of cytokines may involve the production of soluble cytokine antagonists. The IL-1 - IL-1R interaction is unusual in that a specific nonfunctional IL-1 antagonist, IL-1Ra, has been identified that binds to but does not stimulate either the IL-1RI or the IL-1RII.[11] Significant amounts of IL-1Ra mRNA (424 bp product) were found in tonsil DC but not in blood DC (Table 1).

DISCUSSION

This study indicates the potential of PCR for analysing the transcription of gene products within a limited cellular sample of highly purified DC. High quality cDNA was obtained from small numbers of cells and it was possible to analyse several cytokine and cytokine receptor transcripts from the same preparations. The major issue as to the applicability of this technique to DC analysis is the purity of the cellular preparations available. It is therefore notable that although B cells are the most likely contaminating cell type in our tonsil DC preparations, we isolated the T cell form of the IL-1 receptor (type 1) from the tonsil DC. Semi-qualitative PCR analysis will allow comparison of mRNA levels with other cell types and we are developing PCRs to identify lineage specific molecules to provide molecular as well as morphological data to support the purity of our preparations.

The data suggests that DC do produce low levels of IL-1 transcripts. However, it appears that these transcripts (they are probably full length with a poly A tail) undergo some form of post-transcriptional regulation as no significant IL-1 alpha or IL-1 beta protein release has been documented.[5] We have evidence that IL-1 post transcriptional regulation also occurs in the Hodgkins derived cell lines KM-H2 and HDLM-2 (J.L. McKenzie, W. Egner, V. Calder, D.N.J. Hart - in press). It will be interesting to test DC for the presence of the novel cysteine protease recently described to be necessary for IL-1 beta processing in monocytes.[12] Furthermore, IL-1Ra mRNA was readily detectable on tonsil DC preparations (note again that no IL-1Ra mRNA was detected in tonsil B cells). Differential transcription rates for IL-1 beta and IL-1Ra are recognised in monocytes.[13] The apparent predominance of IL-1Ra again argues against a contribution of IL-1 to DC signal II.

This is the first direct demonstration that DC express IL-1 receptors and at the mRNA level these are determined to be type I (T cells, endothelial cells, keratinocytes, fibroblasts and hepatocytes) and not type II (B cells, monocytes, macrophages and granulocytes). It appears from our data that the type II form of the IL-1R described to date on activated B cells[12] is not expressed on resting tonsil B cells. The presence of IL-1Ra mRNA in human DC concurs with mouse data suggesting that the type I form of the IL-1 receptor is important in IL-1 (DC) mediated thymocyte proliferation.[7,15] Other cytokine receptors have been analysed (data not shown) and it appears both tonsil and blood DC produce mRNA for both TNF RI and II (II > I). It is of interest that GM-CSF R are detectable on human tonsil DC but not in preliminary experiments on blood DC. These cytokine receptors on DC may therefore constitute part of our proposed signal III which activates the full APC capabilities of DC.

Preliminary experiments suggest that DC may express low levels of mRNA for several other cytokines including TNF alpha and IL-6. It is hoped that this approach will allow a full evaluation of DC transcripts for immunologically significant molecules.

ACKNOWLEDGEMENTS

This work was supported by grants from the Canterbury Medical Research Foundation and the Health Research Council of New Zealand. D.N.J. Hart is also supported by the McClelland Trust.

REFERENCES

1. T.C.R. Prickett, J.L. McKenzie and D.N.J. Hart, Adhesion molecules on human tonsil dendritic cells, *Transplantation*, 53:4831(1992).
2. D.L. Mueller, M.K. Jenkins, and R.H. Schwartz, Clonal expansion versus functional clonal inactivation: a costimulatory signalling pathway determines the outcome of T cell antigen receptor occupancy, *Ann.Rev. Immunol.*.7:445(1989).
3. S.B. Mizel, Interleukin 1 and T cell activation, *Immunol. Today*.8:330(1987).
4. T. Kishimoto, The biology of interleukin-6, *Blood*.74:1(1989)
5. J.L. McKenzie, T.C.R. Prickett and D.N.J. Hart, Human dendritic cells stimulate allogeneic T cells in the absence of interleukin 1, *Immunology*.67:290(1989).
6. N. Bhardwaj, U. Santhanam, L.L. Lau, S.B. Tatter, J. Ghrayeb et al, IL-1/IFN-beta2 in synovial effusions of patients with rheumatoid arthritis and other arthritides, *J. Immunol.* 143:2153(1989).
7. S.L. Koide, K. Inaba and R.M. Steinman, Interleukin 1 enhances T-dependent immune responses by amplifying the function of dendritic cells, *J. Exp. Med.* 165:515(1987).
8. K. Inaba, R.M. Steinman, M. Witmer Pack, H. Aya, M. Inaba, T. Sudo, S. Wolpe and G. Schuler, Identification of prolierating dendritic cell precursors in mouse blood, *J. Exp. Med.*. 175:1175 (1992).

9. T.C.R. Prickett and D.N.J. Hart, Anti-leucocyte common (CD45) antibodies inhibit dendritic cell stimulation of CD4 and CD8 T lymphocyte proliferation, *Immunology*. 69:250(1990).
10. M.J. Dallman, R.A. Montgomery, C.P. Larsen, A. Wanders and A.F. Wells, Cytokine gene expression analysis using Northern blotting, polymerase chain reaction and in-situ hybridization, *Immunol. Rev.* 119:163(1991).
11. J.E. Sims, R.B. Acres, C.E. Grubin, C.J. McMahan, J.M. Wignall et al, Cloning the interleukin 1 receptor from human T cells, *Proc. Natl. Acad. Sci. USA*. 86:8946(1989).
12. C.J. McMahan, J.L. Slack, B. Mosley, D. Cosman, S.D. Lupton et al, A novel IL-1 receptor, cloned from B cells by mammalian expression, is expressed in many cell types, *EMBO.J.* 10:2821(1991).
13. C.A. Dinarello, R.C. Thompson, and R.C. Brocking, IL-1 interleukin 1 receptor antagonist in vivo and in vitro, *Immunol. Today*. 12:404(1991).
14. N.A. Thornberry, H.G. Bull, J.R. Calaycay, R.T. Chapman, A.D. Howard et al, A novel heterodimeric cysteine protease is required for interleukin 1B processing in monocytes, *Nature*. 356:768(1992).
15. K.W. McIntyre, G.J. Stepan, K.D. Kolinsky, W.R. Benjamin, J.M. Plozinski et al, Inhibition of interleukin (IL-1) binding and bioactivity in vitro and modulation of acute inflammation in vivo by IL-1 receptor antagonist and anti-IL-1 receptor monoclonal antibody, *J. Exp. Med.* 173:931(1991).

COSTIMULATING FACTORS AND SIGNALS RELEVANT FOR ANTIGEN PRESENTING CELL FUNCTION[1]

Marc K. Jenkins, Dimuthu R. DeSilva, Julia G. Johnson, and Steven D. Norton

Department of Microbiology
University of Minnesota Medical School
420 Delaware Street, S.E.
Minneapolis, MN 55455

INTRODUCTION

The antigen-specific activation of CD4+ T lymphocytes is absolutely dependent on ligands expressed on the surface of specialized hematopoietic cells[1] known as antigen-presenting cells (APC). Recent results from many laboratories suggest that these ligands include: major histocompatibility complex (MHC) -encoded class II molecules that are important for presenting antigenic peptides to the T cell antigen receptor (TCR); adhesion molecules that facilitate the formation of T cell/APC conjugates; and molecules that transduce nonspecific costimulatory signals upon binding their complementary T cell receptors[1-3]. The efficiency of various APC types in their capacity to stimulate T cells can therefore be influenced by qualitative and/or quantitative differences in the expression or function of any of these molecules. Here we describe our analysis of the contribution of antigen-presentation, adhesion, and costimulation to the functional APC potency of dendritic cells and B cells.

THE TWO SIGNAL MODEL OF T CELL ACTIVATION

Although initial experiments with transformed T cell lines indicated that T cells could be activated to produce IL-2 solely by stimulation through the TCR, it is now clear that nontransformed T cells have more stringent activation requirements[1,3]. In a purified state, that is in the absence of viable APC, freshly isolated T cells or resting T cell clones do not produce maximal amounts of IL-2 in response to mitogenic lectins, immobilized anti-TCR antibodies, or peptide/MHC complexes in spite of the fact that these agents all induce TCR signal transduction as measured by inositol phosphate and diacylglycerol production and increases in intracellular calcium and protein kinase C activation. In contrast, if viable APC are present these agents all stimulate maximal amounts of IL-2 production even though the quantity of TCR-associated second messengers does not change. This costimulatory function of APC is clearly separate from antigen-presentation in that class II MHC molecules are not involved. APC must, however, physically interact with the responding T cells for

[1]This work was supported by NIH grant AI-27998 and by a Pew Scholars Award (M.K.J.). S. D. N. was supported by an NIH training grant (AI-07313); J.G.J. by an NRSA (AI-08523).

costimulation to be delivered. Because antigen-presentation and costimulation both depend on T cell/APC interaction, it is not surprising that disruption of adhesion molecules interactions such as LFA-1/ICAM-1 and CD2/LFA-3 inhibits T cell activation[2].

Although many soluble and membrane-bound APC-derived molecules have been reported to enhance T cell activation (see below), the best candidates for the T cell receptor/APC ligand pair responsible for the costimulatory activity of APC are CD28 and B7. CD28, a 44 kilodalton homodimeric integral membrane protein, is a member of the immunoglobulin superfamily that is expressed exclusively on T cells[4]. A ligand for CD28 is the B7 molecule[5]. B7 is a 60 kilodalton integral membrane protein that is also a member of the immunoglobulin superfamily[6]. B7 is expressed on lipopolysaccharide-, Epstein Barr virus-, anti-immunoglobulin-, or anti-class II MHC-activated B cells; but not resting T cells, B cells, or macrophages, or nonhematopoietic tissues[7]. This expression pattern correlates reasonably well with the capacities of these APCs to stimulate T cells. In addition, agonistic anti-CD28 antibodies or B7 transfectants provide costimulation to purified T cells stimulated with immobilized anti-TCR or anti-CD3 antibodies[8-11]. As in the case of viable APC, costimulation by anti-CD28 antibody does not result in increases in intracellular calcium or protein kinase C activity, biochemical events associated with inositol phospholipid hydrolysis[12], although it does appear to activate a tyrosine kinase(s)[13].

Table 1. Relative functional potency of various APCs.

APC Type	Number of APC Required for 25% Maximal Response ($x10^{-3}$)[1]: Antigen-Stimulated Inositol Phosphate Production[2]		Anti-CD3-Stimulated Proliferation[3]		Antigen-Stimulated Proliferation[4]	
Dendritic Cells	21	(1) [5]	12	(1)	3	(1)
Large B Cells	110	(5)	100	(8)	60	(20)
Resting B Cells	165	(8)	1,300	(108)	1,800	(600)

[1]Derived by plotting the number of APC on the X-axis and response on the Y-axis and graphically determining the point on the X-axis at which the APC dose response curve intersected the 25% response point.
[2]The A.E7 murine T cell clone was labeled with tritiated myoinositol and stimulated ($1.8x10^5$/tube) for 60 minutes with varying numbers of the indicated B10.A APC and 1 μM pigeon cytochrome *c* fragment 81-104. Labeled water soluble inositol phosphates were measured as previously described[14]. The maximal response was 2,326 CPM; the response in the absence of APC was 81 CPM.
[3]A.E7 cells (10^4/well) were stimulated for 3 days with the indicated irradiated (1,000 rads) B10.A APC in wells containing immobilized (10 μg/ml coating) anti-CD3 monoclonal antibody as described[16]. The ensuing T cell proliferative response was quantitated by tritiated thymidine incorporation. The maximal response was 14,177 CPM, the response in the absence of APC was <1,000 CPM.
[4]As in 3 except that 1 μM pigeon cytochrome *c* fragment 81-104 was used.
[5]Parenthetical values represent the relative fold reduction in potency compared to dendritic cells.

DETERMINANTS OF APC POTENCY

Based on the model of T cell activation presented above, APC could differ in their ability to stimulate T cells because of differences in antigen presentation or costimulation. In addition, both antigen presentation and costimulation would be expected to be affected by differences in T cell/APC adhesiveness. APC do in fact differ with respect to their functional potency. For example, we reported that splenic low density APC (a heterogeneous mixture of activated B cells, macrophages, and dendritic cells) and resting B cells, populations that express similar numbers of class II MHC molecules, differ greatly in their capacity to stimulate T cell proliferation and lymphokine secretion[14]. We have extended this observation by studying the T cell stimulatory capacities of purified dendritic cells and activated B cells. Splenic dendritic cells were enriched as described by Crowley et al.[15] by taking advantage of the facts that these cells are transiently adherent to plastic and are less dense than most lymphocytes. Dendritic cells were purified from the enriched population by flow cytometric sorting of cells that expressed the dendritic cell-specific molecule recognized by the 33D1 monoclonal antibody. Splenic B cells were purified (to ~95% homogeneity as assessed by surface Ig expression) by treating plastic-nonadherent cells with anti-Thy 1, anti-CD4, and anti-CD8 monoclonal antibodies plus complement. B cells were then separated into activated and resting cells on a Percoll density gradient. As shown in Table 1, dendritic cells were by far the most potent APC at inducing antigen-specific proliferation by a pigeon cytochrome *c* peptide/I-E^k-specific IL-2-producing murine T cell clone. Only 3×10^3 dendritic cells were required to stimulate 25% of the maximal response whereas 6×10^4 activated B cells and 1.8×10^6 resting B cells were required. Therefore, dendritic cells were 20-fold better than activated B cells which were 30-fold better than resting B cells at stimulating antigen-specific T cell proliferation.

When APC are used to stimulate antigen-specific T cell proliferation, the APC must both present antigen and provide costimulation to the responding T cells. We therefore, determined whether APCs differed with respect to antigen-specific T cell stimulation because of differential antigen-presentation or differential provision of costimulation. The generation of inositol phosphates in the responding T cell was used to assess the antigen-presenting capacity of the various APCs. Inositol phosphate production provides a direct measure of the amount of antigen presentation because it occurs purely as a consequence of TCR occupancy and is unaffected by costimulatory signals[1,14]. The costimulatory function of the APCs was measured by assessing their ability to enhance the proliferation of T cells stimulated with immobilized anti-CD3 antibody. Because the TCR is directly crosslinked by the anti-CD3 antibody, the antigen-presenting capacity of the APC is irrelevant and, therefore, only the costimulatory capacity is measured[16].

Dendritic cells were again more efficient than activated or resting B cells at stimulating antigen-specific inositol phosphate production by a T cell clone (Table 1). The relative difference was much less in this case, however, than that detected in the antigen-specific proliferation assay (compare the left and right columns in Table 1). Dendritic cells were only 5-times more potent than activated B cells at stimulating inositol phosphate production and activated B cells were only 1.6-times more potent than resting B cells. These differences correlated well with the expression of class II MHC molecules (Table 2): dendritic cells expressed 5-times more class II MHC than activated B cells which expressed 2-times more than resting B cells. Therefore, as inferred from other experimental approaches[17], it is likely that the amount of class II MHC expression determines how much peptide is presented to the TCR. Adhesion molecules like ICAM-1 may indirectly improve antigen-presenting capacity by facilitating conjugate formation between the APCs and T cells. In fact we observed a reasonable correlation between the ability to stimulate inositol phosphate production and ICAM-1 expression: dendritic cells expressed 12-times more ICAM-1 than activated B cells which expressed 1.6-times more than resting B cells. However, the antigen presentation differences between dendritic cells, activated B cells, and resting B cells as measured by their abilities to induce inositol phosphate production in the T cells could not account for their great differences in stimulating antigen-specific T cell proliferation. Dendritic cells were also the most efficient cell type at directly providing costimulation to T

Table 2. Flow cytometric quantitation of surface molecule expression on APC.

	Relative Fluorescence Intensity[1]					
APC Type	Class II MHC		ICAM-1		B7	
Dendritic Cells	1,870	(1) [2]	1,320	(1)	2,350	(1)
Large B Cells	350	(5)	110	(12)	120	(20)
Resting B Cells	190	(10)	70	(19)	20	(118)

[1]The indicated populations were stained with anti-class II MHC (14.4.4S, mouse IgG2a), anti-ICAM-1 (YN174, rat IgG), or CTLA-4Ig (a fusion protein between the extracellular domains of CTLA-4 and the Fc of human IgG1) followed by an appropriate FITC-labeled goat anti-Ig reagent. CTLA-4 is a second membrane receptor for B7[18]. The fluorescence intensities (shown in arbitrary units) of 10,000 cells/group were measured on a Becton Dickinson FACStar[PLUS].
[2]Parenthetical values represent the relative fold reduction in surface expression compared to dendritic cells.

cells cultured with immobilized anti-CD3 antibody. Dendritic cells were 8-times more potent than activated B cells which were 14-times more potent than resting B cells (Table 1). Therefore differences in the provision of costimulation more accurately reflect the differences in the capacities of APC to stimulate antigen-specific T cell proliferation than do differences in antigen presentation. In addition, costimulatory capacity correlated well with expression of B7: dendritic cells expressed 20-times more B7 than activated B cells which expressed 6-times more than resting B cells (Table 2). Costimulation alone, however, could not fully account for the great differences between these APC types at stimulating antigen-specific T cell proliferation. For example, dendritic cells were 108-times more potent than resting B cells in the costimulation assay but were 600-times more potent in the antigen-stimulated proliferation assay.

ANTIGEN PRESENTATION, COSTIMULATION, AND ADHESION

Our results on the different capacities of dendritic cells, activated B cells, and resting B cells to stimulate antigen-specific T cell proliferation are best explained by a model in which costimulation and antigen-presentation both contribute to APC potency. For example, the ~600-fold difference between dendritic cells and resting B cells in stimulating antigen-specific proliferation can be explained by the combination of an ~8-fold difference in antigen-presentation and a ~100-fold difference in costimulation.

Interplay between antigen presentation and costimulation could be explained by the induction of distinct but required DNA-binding transcription factors. At least five critical sets of sequences in the 5' IL-2 enhancer are bound by distinct transcription factors: NFAT-1, Oct-1/2, CD28RC, NF-κB, and AP-1[19,20]. Some of these factors, for example NFAT-1, appear to be responsive to TCR-mediated signals, whereas AP-1 and CD28RC appear to be responsive to APC-derived signals, such as the costimulatory signal transduced by CD28. Because the IL-2 enhancer is dependent on each individual binding site, all of the factors must bind before transcription can be initiated. Therefore the T cell must receive both TCR and costimulatory signals before IL-2 can be made.

The close correlation between B7 expression and provision of costimulation suggests that B7 is the major costimulatory ligand, at least for dendritic cells and B cells. How can this finding be reconciled with the numerous reports that suggest that CD2/LFA-3, VLA-

4/VCAM-1, and LFA-1/ICAM-1 are costimulatory receptor/ligand pairs[2] and that cytokines such as IL-1 and IL-6 play a role in costimulation[21,22]? The conclusion that adhesion molecules can deliver costimulatory signals is often based on the finding that co-immobilization of the adhesion molecule, or an antibody specific for its receptor, with anti-CD3 antibody leads to an enhancement of T cell proliferation[2]. In addition, transfection of adhesion molecules into APC often improves their capacity to stimulate T cells[2]. Although it is possible that this type of enhancement is due to costimulation it is equally likely that the adhesion molecule allows the responding T cell to interact more avidly with the surface containing the TCR ligand, thereby indirectly enhancing TCR signaling. This effect can be ruled out by immobilizing the TCR ligand and adhesion molecule on different surfaces[23], for example in the anti-CD3 costimulation assay used in this paper. ICAM-1[23,24] and LFA-3[25] function poorly or not at all when they are not on the same surface as the TCR ligand suggesting that they do not deliver independent costimulatory signals and function mainly as adhesion molecules. VCAM-1 has not been tested in this type of assay.

IL-1 and IL-6 have also been shown to enhance T cell activation[21,22]. However, although IL-1 is a potent costimulus for some IL-4-producing murine T cell clones[3], many studies indicate that IL-1 is not a costimulatory molecule for IL-2-producing T cell clones or purified peripheral T cells. Furthermore, potent APC such as dendritic cells are poor IL-1 producers[26]. IL-6 has also been reported to be a costimulus for purified T cells particularly in combination with IL-1[22]. Our studies, however, indicate that IL-6 either alone or together with IL-1 is not a potent costimulus for purified IL-2-producing T cells[1,9]. It should be noted, however, that IL-1 and IL-6 have been shown to enhance the effects of anti-CD2, purified ICAM-1, or fixed APC on T cells stimulated through their TCRs[23,27]. It is possible that two costimulation systems exist, one that depends on signals transduced by the CD28/B7 interaction, and one that depends on signals transduced by adhesion molecules and that is amplified by IL-1 and IL-6.

REFERENCES

1. D.L. Mueller, M. K. Jenkins, and R. H. Schwartz, Clonal expansion versus functional clonal inactivation: a costimulatory signalling pathway determines the outcome of T cell antigen receptor occupancy. *Annu. Rev. Immunol.* 7:445 (1989).
2. G.A. van Seventer, Y. Shimizu, and S. Shaw, Roles of multiple accessory molecules in T-cell activation: bilateral interplay of adhesion and costimulation. *Curr. Opin. Immunol.* 3:294 (1991).
3. C.T. Weaver, and E.R. Unanue, The costimulatory function of antigen-presenting cells. *Immunol. Today* 11:49 (1990).
4. C.H. June, J.A. Ledbetter, P.S. Linsley, and C.B. Thompson, Role of the CD28 receptor in T-cell activation. *Immunol. Today* 11:211 (1990).
5. P.S. Linsley, E.A. Clark, and J.A. Ledbetter, T-cell antigen CD28 mediates adhesion with B cells by interacting with activation antigen B7/BB-1. *Proc. Natl. Acad. Sci. USA* 87:5031 (1990).
6. G.J. Freeman, A.S. Freedman, J.M. Segil, G. Lee, J.F. Whitman, and L.M. Nadler, B7, a new member of the Ig superfamily with unique expression on activated and neoplastic B cells. *J. Immunol.* 143:2714 (1989).
7. T.R. Yokochi, D. Holly, and E.A. Clark, B lymphoblastoid antigen (BB-1) expressed on EBV-activated B cell blasts, B lymphoblastoid cell lines, and Burkitt's lymphomas. *J. Immunol.* 128:823 (1982).
8. C.H. June, J.A. Ledbetter, M.M. Gillespie, T. Lindsten, and C.B. Thompson, T-cell proliferation involving the CD28 pathway is associated with cyclosporine-resistant IL-2 gene expression. *Mol. Cell. Biol.* 7:4472 (1987).
9. M.K. Jenkins, P.S. Taylor, S.D. Norton, and K.B. Urdahl, CD28 delivers a costimulatory signal involved in antigen-specific IL-2 production by human T cells. *J. Immunol.* 147:2461 (1991).
10. P.S. Linsley, W. Brady, L. Grosmaire, A. Aruffo, N.K. Damle, and J.A. Ledbetter, Binding of the B cell activation antigen B7 to CD28 costimulates T cell proliferation and IL-2 mRNA accumulation. *J. Exp. Med.* 173:721 (1991).

11. C.D. Gimmi, G.J. Freeman, J.G. Gribben, K. Sugita, A.S. Freedman, C. Morimoto, and L.M. Nadler, B-cell surface antigen B7 provides a costimulatory signal that induces T cells to proliferate and secrete IL-2. *Proc. Natl. Acad. Sci. USA* 88:6575 (1991).
12. A. Weiss, B. Manger, and J. Imboden. 1986. Synergy between the T3/antigen receptor complex and Tp44 in the activation of human T cells. *J. Immunol.* 137:819 (1986).
13. P. Vandenberghe, G.J. Freeman, L.M. Nadler, M.C. Fletcher, M. Kamoun, L.A. Turka, J.A. Ledbetter, C.B. Thompson, Antibody and B7/BB-1-mediated ligation of the CD28 receptor induces tyrosine phosphorylation in human T cells. *J. Exp. Med.* 175:951 (1992).
14. M.K. Jenkins, E. Burrell, and J.D. Ashwell, Antigen presentation by resting B cells. Effectiveness at inducing T cell proliferation is determined by costimulatory signals, not T cell receptor occupancy. *J. Immunol.* 144:1585 (1990).
15. M. Crowley, K. Inaba, M. Witmer-Pack, and R.M. Steinman, The cell surface of mouse dendritic cells: FACS analyses of dendritic cells from different tissues including thymus. *Cell. Immunol.* 118:108 (1989).
16. M.K. Jenkins, C. Chen, G. Jung, D.L. Mueller, and R.H. Schwartz, Inhibition of antigen-specific proliferation of type 1 murine T cell clones after stimulation with immobilized anti-CD3 monoclonal antibody. *J. Immunol.* 144:16 (1990).
17. L. Matis, L. Glimcher, W. Paul, and R. Schwartz, Magnitude of response of histocompatibility-restricted T -cell clones is a function of the product of the concentrations of antigen and Ia molecules. *Proc. Natl. Acad. Sci. USA* 80:6019 (1983).
18. P.S. Linsley, W. Brady, M. Urnes, L.S. Grosmaire, N.K. Damle, and J.A. Ledbetter, CTLA-4 is a second receptor for the B cell activation antigen B7. *J. Exp. Med.* 174:561 (1991).
19. G.R. Crabtree, Contigent genetic regulatory events in T lymphocyte activation. *Science* 243:355 (1989).
20. C. Go, and J. Miller, Differential induction of transcription factors that regulate the IL-2 gene during anergy induction and restimulation. *J. Exp. Med.* 175:1327 (1992).
21. S.K. Durum, J.A. Schmidt, and J.J. Oppenheim, Interleukin 1: an immunological perspective. *Annu. Rev. Immunol.* 3:263 (1985).
22. J. Van Snick, Interleukin-6: an overview. *Annu. Rev. Immunol.* 8:253 (1990).
23. G.A. van Seventer, Y. Shimizu, K.J. Horgan, G.E. Ginther Luce, D. Webb, and S. Shaw, Remote T cell co-stimulation via LFA-1/ICAM-1 and CD2/LFA-3: demonstration with immobilized ligand/mAb and implication in monocyte-mediated co-stimulation. *Eur. J. Immunol.* 21:1711 (1991).
24. P. Kuhlman, V.T. Moy, B.A. Lollo, and A.A. Brian, The accessory function of murine ICAM-1 in T lymphocyte activation: contributions of adhesion and co-activation. *J. Immunol.* 146:1773 (1991).
25. P. Moingeon, H. Chang, B.P. Wallner, C. Stebbins, A.Z. Frey, and E.L. Reinherz. CD2-mediated adhesion facilitates T lymphocyte recognition function. *Nature* 339:312 (1989).
26. S. Koide, and R.M. Steinman, Induction of murine IL-1: stimuli and responsive cells. *Proc. Natl. Acad. Sci. USA* 84:3802 (1987).
27. K. Kawakami, Y. Yamamoto, K. Kakimoto, and K. Onoue, Requirement of delivery of signals by physical interaction and soluble factors from accessory cells in the induction of receptor-mediated T cell proliferation. Effectiveness of IFN-γ modulation of accessory cells for physical interaction with T cells. *J. Immunol.* 142:1818 (1989).

DISTINCT T CELL STIMULATION MECHANISM AND PHENOTYPE OF HUMAN BLOOD DENDRITIC CELLS

Hui Xu,[1] Ulrike Friedrichs,[1] Robert K. H. Gieseler,[1] Jörg Ruppert,[1] Göran Ocklind,[2] and J. Hinrich Peters[1]

[1]Department of Immunology, University of Goettingen, Kreuzbergring 57 D-3400 Goettingen, Germany
[2]University of Uppsala, Zoophysiological Institute S-75122 Uppsala, Sweden

INTRODUCTION

Dendritic cells (DC) have been found to exist as antigen-presenting cells in non-lymphoid and lymphoid tissues.[1,2] They play an important role in modulation of the immune reaction.[3,4] Blood dendritic cells (BDC), like other types of DC, exhibit distinct cellular characteristics and a strong potency to stimulate lymphocyte proliferation.[5,6] Cell interaction via adhesion molecules has been demonstrated to function as a co-stimulatory signal in the T cell stimulation.[7,8] The blockage of adhesion molecules or their ligands by antibodies could reduce cell interaction and function.[9,10]

In the present report, we have identified human BDC by means of double staining in flow cytometry, and studied mechanisms underlying stimulation of T cell proliferation in the allogeneic mixed lymphocyte reaction. Our data provide further insights into the cytological and functional properties of human BDC.

RESULTS AND DISCUSSION

Identification of human blood dendritic cells

BDC investigated in our experiments were isolated according to the methods of Young and Steinman[11] and identified by their phagocytosis activity and phenotype. By using E.coli opsonized with FITC-labelled antibodies in combination with FACS-analysis, BDC were demonstrated not to phagocytize the bacteria. Double staining of BDC in FACS-analysis has shown that these cells did not express specific markers for T cells (CD3), B cells (CD20 and CD21) and myeloid cells (CD11b, CD13 and CD14), but abundant HLA-DR and CD11a molecules (Fig. 1). The detailed phenotypical analysis (not presented) revealed similar results to those reported by Steinman.[12]

Dendritic Cells in Fundamental and Clinical Immunology,
Edited by Kamperdijk *et al.*, Plenum Press, New York, 1993

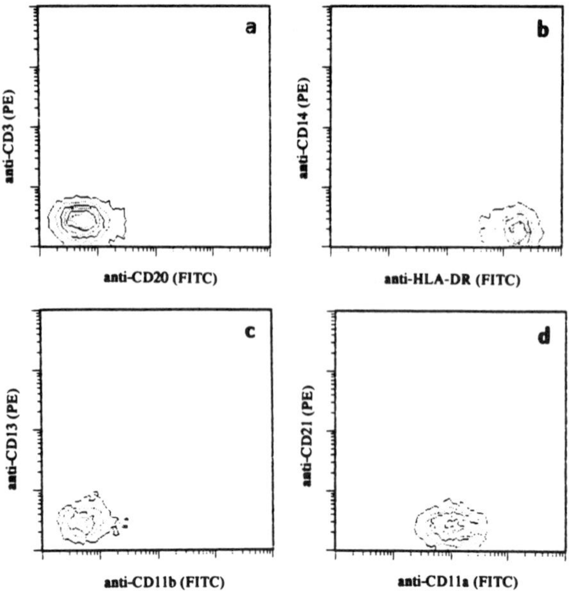

Figure 1. Phenotypical identification of BDC were carried out by FACS-analysis and BDC fraction was gated according to their forward and side light scatter. Double staining of BDC shows that BDC do not express specific T and B lymphocyte markers CD3 and CD20 (a); CD21 (d) as well as myeloid cell surface antigens CD14 (b); CD13 and CD11b (c). These cells, however, were strongly positive for HLA-DR and CD11a molecules (b and d). PE: Phycoerythrin (red); FITC: Fluorescine isothiocyanate (green).

Effects of the mAbs on the stimulation of T cell proliferation induced by BDC

Functional studies on mechanisms of T cell proliferation induced by DC have been reported in mouse as well as in man with tissue DC or Langerhans' cells,[13] but not yet with human BDC. When using BDC as stimulator in a one-way MLR, addition of mAbs directed against adhesion and MHC class II molecules has been demonstrated to affect T cell proliferation (Fig. 2). The mAbs against CD11a and CD11c reduced T cell proliferation, whereas the mAb against CD18 (the ß chain of CD11a, b and c) had no effect. Conflicting findings (King and Katz[14]) reported that mAbs against CD11a and CD18, but not CD11c, could inhibit T cell proliferation induced by human tonsil DC in an oxidative mitogenesis assay. This may not only result from different cell types and concentrations of mAbs used, but may also indicate different mechanisms in mitogenesis assays and MLR. Otherwise, the mAbs used in the studies may react with different epitopes.

The LFA-3 molecule has been demonstrated to be involved in cluster formation and initial events of human tonsil DC-induced T cell activation.[14] Our results indicate that the anti-LFA-3 mAb could not inhibit cluster formation of BDC with T cells, which is in agreement with a recent report in which human BDC have been used as antigen presenting

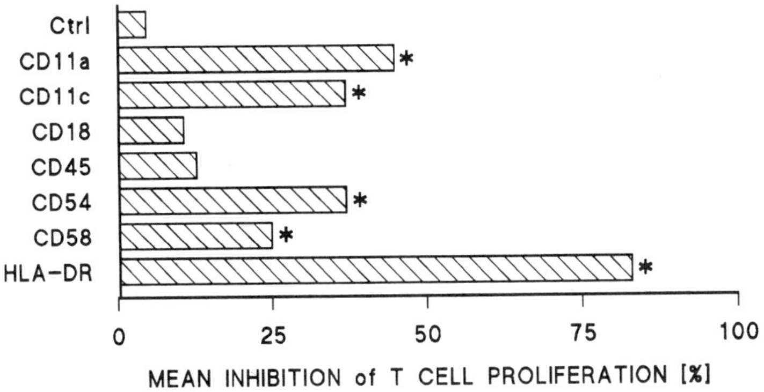

Figure 2. Effect of mAbs on the activity of BDC to stimulate T lymphocyte proliferation in a one-way allogeneic MLR. The mAbs were added in the beginning of the MLR at 5 μg/ml. After 5 days, the culture was pulsed with ^{3}H thymidine (0.2 μCi/well) for 18 hours and then harvested to determine ^{3}H thymidine incorporation (triplicates). CD11a (25.3.1. IgG1), CD11c (BU15. IgG1), CD18 (BL5. IgG1), CD45 (ALB12. IgG2b), CD54 (84H10. IgG1) and HLA-DR (monomorphic) mAbs were purchased from Dianova, Hamburg, Germany. The CD58 mAb (HB205. anti-LFA-3) were obtained from the ATCC. The mean inhibition (%) of ^{3}H thymidine incorporation by mAbs is calculated as follows: Mean inhibition (%)=mean cpm of sample (mAb)-mean cpm of control (mouse IgG)/mean cpm of control (mouse IgG). * means significant difference ($P<0.05$).

cells.[15] Moreover, the ability of LFA-3 mAb to suppress T cell proliferation induced by BDC has been confirmed by our tests, indicating the role of a CD2-LFA-3 pathway in allogeneic T cell responses.

The interaction of LFA-1 and ICAM-1 molecules has been confirmed to provide a co-stimulatory signal for T cell activation.[16] The inhibitory effect of the CD54 mAb in the MLR indicated a critical role of ICAM-1 molecules in the induction of T cell proliferation by BDC.

The strong suppressive effect of anti-MHC class II mAbs on the stimulatory function of DC has been reported by many others and confirmed by our results.

The role of adhesion molecules in cluster formation of T cells and BDC

Intriguingly, though mAbs against CD11a, CD11c, LFA-3 and ICAM-1 suppressed T cell proliferation in MLR, they were unable to disrupt cluster formation between BDC and T cells (Table 1). This does not coincide with the findings by Scheeren et al.[15] who demonstrated a strong inhibition of anti-CD11a, CD18 and ICAM-1 mAbs on clustering of BDC-T cells. It may result from a different mAb concentration and incubation time used in the report. The possibility that the mAbs react with different epitopes may also not be ruled out. In the mouse system, the LFA-1 molecule has been reported to be involved in T cell activation, but not to be implicated in initial cluster formation of DC-T cells.[17] The interaction of adhesion molecules might contribute not only to favouring antigen and co-stimulatory factor presentation,[18,19] but, more importantly, to triggering signals necessary for T cell activation. It was evident that ICAM-1 deposited on plastic with the nonmitogenic immobilized CD3 mAb resulted in a potent activating stimulus for T cells. In this case, the costimulation may not be readily explained as an adhesion, but

Table 1. The effect of mAbs (5 μg/ml) on cluster formation of BDC with T cells in the MLR after one or 18 hours. Cells (2x10^4 BDC to 2x10^5 T cells/well) were cultured in 10% human serum with mAbs at the beginning, and cluster formation was monitored under an inverted microscope.

mAbs	effect of mAbs on cluster formation	
	one hour	18 hours
CD11a	+	++
CD11c	+	++
CD18	+	+++
CD45	+	++++
CD54	+	++
CD58	+	++
HLA-DR	+	++
Control (mouse IgG)	+	++

Note. + : small; ++ : moderate; +++ to ++++ : large.

Table 2. Comparison of the antigen expression on BDC before and after 8 days of culture in the presence of GM-CSF. Cells were cultured on hydrophobic Teflon in standard medium with 10% normal human serum and 100 U/ml GM-CSF. The analysis of antigen expression was performed by immune fluorescence staining in FACS.

antibodies	fresh	cultured
CD11a	14.3*	16.4
CD11c	14.0	16.0
CD18	19.1	14.9
CD58	32.7	19.9
CD54	45.7	143.1
HLA-DR	237.3	206
HLA-DQ	15	46
HLA-DP	73.6	119.5
CD14	-	-
CD16	-	-

*: Relative fluorescence intensity (rFI) calculated according to the formula mean fluorescence of sample/mean fluorescence of negative control.

rather as a signal.[20] Given that the interaction of adhesion molecules acts as a co-stimulatory factor for T cell activation, we have confirmed that a group of molecules (LFA-3, ICAM-1, CD11a and CD11c) is involved in this process. Blocking any of these molecules can interfere with T cell activation to a certain degree.

Phenotyping of cultured BDC

The ontogeny of BDC remains to be delineated, albeit in vitro differentiation of BDC was reported with respect to adherence ability and decreased expression of MHC class II antigens.[21] Our efforts to define differentiation of BDC have demonstrated that these cells revealed no dramatic changes in phenotype after 8 days of culture in the presence of GM-CSF (100 U/ml) (Table 2). Though the expression of ICAM-1 and HLA-DQ molecules increased about three times as indicated by rFI, the surface antigens specific for myeloid cells (CD14 and CD16) remained undetectable on the cultured BDC. The phenotype pattern of BDC is clearly different from that of follicular dendritic cells[22] but almost identical to that of interdigitating dendritic cells (IDC) (2), strongly suggesting a relationship of BDC with IDC. BDC may represent the migratory form of DC from non-lymphoid through the blood stream into lymphoid tissues. This migratory way of DC has been proven by Larsen et al. in rodent cardiac transplantation.[23]

In summary, human BDC exhibit distinct mechanisms in clustering with T cells and stimulating T cell proliferation as analyzed by addition of a panel of mAbs in an allogeneic MLR. The freshly isolated as well as cultured BDC present a phenotypic profile which is different from that of any defined cell types, but almost identical to that of IDC, suggesting a close relationship between BDC and IDC.

Acknowledgments

This work was supported by the Deutsche Forschungsgemeinschaft through SFB 330, grant C6, and grant Pe 192/6-1. J. R. received a grant from the Boehringer Ingelheim Fonds.

We thank Mrs Dorothea Ostermeier, Ms Rita Kuhn, and Mr. Detlef Friedrichs for technical assistance, and Mrs Ingrid Teuteberg for typing the manuscript.

REFERENCES

1. D.N.J. Hart, J.W. Fabre, Demonstration and characterization of Ia-positive dendritic cells in the interstitial conective tissues of rat heart and other tissues, but not brain, *J Exp Med* 154:347 (1981).
2. P.D. King, D.R. Katz, Mechanisms of dendritic cell function, *Immunol Today* 11:206 (1990).
3. J.M. Austyn, Lymphoid dendritic cells, *Immunology* 62:161 (1987).
4. J.P. Metlay, E. Pure, and R.M. Steinman, Control of the immune response at the level of antigen-presenting cells: a comparison of the function of dendritic cells and B lymphocytes, *Adv Immunol* 47:45 (1989).
5. W.C. van Voorhis, L.S. Hair, R.M. Steinman, and G. Kaplan, Human dendritic cells: enrichment and characterization from peripheral blood, *J Exp Med* 155:1172 (1982).

6. W.C. van Voorhis, J. Valinsky, E. Hoffman, J. Luban, L.S. Hair, and R.M. Steinman, Relative efficacy of human monocytes and dendritic cells as accessory cells for T cell replication, *J Exp Med* 158:174 (1983).
7. T.A. Springer, M.L. Dustin, T.K. Kishimoto, and S.D. Marlin, The lymphocyte function-associated LFA-1, CD2, and LFA-3 molecules: cell adhesion receptors of immune system, *Ann Rev Immunol* 5:223 (1987).
8. B.E. Bierer, B.P. Sleckman, S.E. Ratnofsky, and S.J. Burakoff, The biologic roles of CD2, CD4, and CD8 in T-cell activation, *Ann Rev Immunol* 7:579 (1989).
9. M.W. Makgoba, M.E. Sanders, and S. Shaw, The CD2-LFA-3 and LFA-1-ICAM-1 pathways: relevance to T-cell recognition, *Immunol Today* 10:417 (1989).
10. L.H. Dang, M.T. Michalek, F. Takei, B. Benaceraff, and K.L. Rock, Role of ICAM-1 in antigen presentation demonstrated by ICAM-1 defective mutants, *J Immunol* 144:4082 (1990).
11. J.W. Young, R.M. Steinman, Accessory cell requirments for the mixed-leukocyte reaction and polyclonal mitogens, as studied with a new technique for enriching blood dendritic cells, *Cell Immunol* 111:167 (1988).
12. P.S. Freudenthal, R.M. Steinman, The distinct surface of human dendritic cells, as observed after an improved isolation method, *Proc Natl Acad Sci USA* 87:7698 (1990).
13. R.M. Steinman, The dendritic cell system and its role in immunogenicity, *Ann Rev Immunol* 9:271 (1991).
14. P.D. King, D.R. Katz, Human tonsillar dendritic cell-induced T cell responses: analysis of molecular mechanisms using monoclonal antibodies, *Eur J Immunol* 19:581 (1989).
15. R.A. Scheeren, G. Koopman, S. van der Baan, C.J.L.M. Meijer, and S.T. Pals, Adhesion receptors involved in clustering of blood dendritic cells and T lymphocytes, *Eur J Immunol* 21:1101 (1991).
16. D.M. Altmann, N. Hogg, J. Trowsdale, and D. Wilkinson, Cotransfection of ICAM-1 and HLA-DR reconstitutes human antigen-presenting cell function in mouse L cells, *Nature* 338:512 (1989).
17. K. Inaba, R.M. Steinman, Monoclonal antibodies to LFA-1 and to CD4 inhibit the leukocyte mixed reaction after the antigen-dependent clutering of dendritic cells and T lymphocytes, *J Exp Med* 165:1403 (1987).
18. E.R. Flechner, P.S. Freudenthal, G. Kaplan, and R.M. Steinman, Antigen-specific T lymphocytes efficiently cluster with dendritic cells in the human primary mixed-leukocyte reaction, *Cell Immunol* 111:183 (1988).
19. K. Inaba, N. Romani, and R.M. Steinman, An antigen-independent contact mechanism as an early step in T cell-proliferative responses to dendritic cells, *J Exp Med* 170:527 (1989).
20. G.A. van Seventer, Y. Shimizu, K.J. Horgan, and S. Shaw, The LFA-1 ligand ICAM-1 provides an improtant costimulatory signal for T cell receptor-mediated activation of resting T cells, *J Immunol* 144:4579 (1990).
21. S. Markowicz, E.G. Engleman, Granulocyte-macrophage colony-stimulating factor promotes differentiation and survival of human peripheral blood dendritic cells in vitro, *J Clin Invest* 85:955 (1990).
22. F. Schriever, A.S. Freedman, G. Freeman et al., Isolated human follicular dendritic cells display a unique antigenic phenotype, *J Exp Med* 169:2043 (1989).
23. C.P. Larsen, P.J. Peter, and J.M. Austyn, Migration of dendritic leukocytes from cardiac allografts into host spleens, *J Exp Med* 171:307 (1990).

THE ROLE OF DENDRITIC CELLS IN THE REGULATION OF T CELL CYTOKINE SYNTHESIS

Joanna Ellis, Mohammed A A Ibrahim, Benjamin M Chain, and David R Katz

Departments of Histopathology, Immunology and Biology
University College London
48, Riding House St., London W1P 7PN

INTRODUCTION

Antigen presenting cells (APC) have been implicated frequently in the numerous cytokine network models, chiefly as a hypothetical source of a critical mediator that is important in induction of an immune response. The example usually quoted is interleukin 1 (IL-1) (1), even though it has been clear for some time that the most potent form of APC, the dendritic cell (DC), is an IL-1 responder rather than producer (2). Other cytokines, such as IL-6, have also been proposed as important in induction (3); but IL-6 effects are so pleiotropic that it is difficult to ascribe a specific APC related role to this cytokine (4).

Recently another way that cytokines may be involved in DC pathways, and DC - APC function, has been suggested. The differentiation of Langerhans cells into migratory veiled cells and interdigitating DC has been shown to be due to granulocyte - macrophage colony stimulating factor (GM-CSF) (5). Thus GM-CSF, too, may be regarded as a "pro-presentation" cytokine.

The third way that the role of cytokines has been documented during DC - T cell interaction is seen in many assays of APC function, which frequently make use of a cytokine, IL-2, as a "read-out" system. DC are the most potent known inducers of supernatants which contain IL-2, and monitoring DC - induced IL-2 levels is a useful way to show the different phases of T cell activation that lead to proliferation and clonal expansion (6).

IL-2 is not, however, the only cytokine effector molecule produced by T cells. It has not been clear, for example, what role (if any) the DC play in induction of IL-4 synthesis, or of gamma interferon synthesis, and hence in the control of the so- called TH2 population (those T cells that make IL-4 and gamma interferon rather than IL-2) or indeed of the so-called promiscuous THOs (7), which are the postulated precursors for both TH1 and TH2 populations.

In the studies reported here we have investigated the hypothesis that APC, and in particular DC, play a pivotal role in determining how these patterns of T cell cytokine synthesis are controlled. One possibility that we tried to address is the superficially attractive notion that different APC induce synthesis of different cytokines. This has been proposed frequently in the literature (8) and implies that theoretically DC should be inducers of TH1 (IL-2 producing) cells but not TH2 (IL-4 producing) cells. There is, however, little experimental evidence for this view at present; and that little evidence relies chiefly upon pre-selected long term T cell clones, and has not taken into consideration the possible role of selected potent APC, such as DC.

MATERIALS AND METHODS

Animals

Balb/c ($H-2^d$) and CBA/Ca($H-2^k$) mice were from the Imperial Cancer Research Fund Animal Breeding Unit, Clare Hall, Potters Bar, Hertfordshire.

Cell isolation and purification

APC were from CBA/Ca mice. Spleen cell suspensions were used as a source for DC, macrophages and B cells. The DC and splenic FcR+ macrophages (MO) were isolated as described previously (9) except that Iscove's Fortified medium (I/F) was used throughout. Antibodies used were NLDC 145 rat anti-mouse dendritic cell antibody (gift of Dr G Kraal) (10) HB32 and HB3 anti - class II major histocompatibility complex (MHC) antibodies (ATCC); and F4.80 anti - macrophage (gift of Dr S Gordon) (11). DC were >95% NLDC 145 +; >95% class II MHC +; splenic FcR + MO were >70% NLDC 145 +; 40-60% class II MHC +; and >70% F4.80+ by flow cytometry.

B cells were isolated from the high density (HD) cells by T cell depletion using rat anti-Thy 1 monoclonal antibody (YTS-154, gift of Dr S Cobbold, Dept. of Pathology, Cambridge) plus rabbit complement (Buxted). B cells were activated by treatment of T - depleted HD cells with lipopolysaccharide (LPS) (0.010 mg/ml) (LPS W S typhosa, Difco Labs, Detroit, Michigan) for 48 h., followed by repeat T cell lysis. Purified B cells were washed and resuspended in I/F - 5% foetal calf serum (FCS). Surface immunoglobulin expression was monitored by indirect immunofluorescence. B cells were >95% purified with no residual T cells present.

Peritoneal MO were lavaged with 5ml Modified Eagle's Medium (MEM) supplemented with 5% FCS. Activated MO were from mice pre-treated with 0.1ml complete Freund's adjuvant (Difco Labs., Detroit, Michigan) intraperitoneally 7 days before harvesting. The peritoneal lavage cells were washed x 2 and enriched by rosetting with rabbit anti-sheep red blood cell (SRBC) - SRBC complexes. The MO phenotype was confirmed using anti-class II MHC and F4.80 antibodies.

Responder Balb/c splenic T cells were prepared from HD cells, followed by nylon wool passage, Class II MHC + depletion (HB32 / HB3 and complement), passage through a Sephadex G10 column (Pharmacia), and adherent cell removal by incubation for 2h in I/F - 5%FCS. Non- adherent cells were harvested, washed x 2 and re-suspended in I/F with 5% FCS. T cell phenotype was confirmed using the 2C11 anti-CD3 (gift of Dr O Leo) (12).

In vitro proliferation assays

Responder T cells (10^6/ml) were cultured with different numbers of APC (10^3 - 3 x 10^5/ml) in 0.2ml I/F with 5%FCS volume in flatbottom microtitre plates for 72 hours at 37°C in 5% CO_2. During the last 6h each well was pulsed with 0.001mCi [^{3}H]- thymidine, harvested on to fibreglass filters, and incorporation of thymidine measured in a liquid scintillation counter.

In vitro cytokine production

For cytokine assays T cell / APC cultures were performed in 10ml I/F with 5% FCS in 50ml tissue culture flasks (Nunc). In primary cultures 10^6 / ml T cells were cultured with 2 x 10^4 / ml APC and 0.5ml culture supernatant was harvested on various days after stimulation and stored at -20°C until assayed.

For secondary MLR cultures 10^6 / ml T cells were cultured with 2 x 10^4 / ml DC for 7 days as above. Viable T cells were recovered (ficoll hypaque) and recultured at 2 x 10^5/ml with 2 x 10^4 / ml different allogeneic APC.

For tertiary MLR cultures viable T cells were recovered from secondary MLR cultures stimulated by 2 cycles of allo - DC. 2 x 10^5 / ml T cells were re-incubated with 2 x 10^4 / ml allogeneic APC.

For eight cycles of allogeneic stimulation 2 x 10^6/ml T cells were cultured with 2.5 x 10^4/ml DC for 7 days. T cells were recovered and recultured with fresh DC each week for 6 weeks. For the final cycle 2 x 10^5/ml T cells were cultured with 2 x 10^4/ml APC.

Cytokine assays

IL-2 was detected by maintenance of growth of CTLL cells (13). (CTLL cells are sensitive to IL-4, but require a concentration of > 100 units IL-4 per ml). Control assays were not inhibited in the presence of anti-IL-4 (11.B11, gift of Dr T Mossman, DNAX, Palo Alto, California); and were inhibited by anti-IL-2 receptor (TIB 222, from the ATCC). CTLL cells do not respond to IL-3 or IL-6.

IL-3 was detected using IL-3 dependent cells, FDCP-2 (14) which do not respond to IL-2, IL-4, and IL-6, but which do respond to GM-CSF, and which cannot therefore be distinguished from IL-3 in this study.

IL-6 was detected using IL-6 dependent B9 cells, with inhibitory anti - IL-6 as a control (gift of Prof M Feldmann, Charing Cross - Sunley Research Centre) (15).

For IL-2, IL-3 and IL-6 assays triplicate supernatants were titrated over a wide range of concentrations. Cell survival after 24 hours (IL-2 and IL-3) or 48 hours (IL-6) was measured by metabolism of MTT (3-4, 5-dimethylthiazole - 2-y/-2,5 - diphenyl tetrazolium bromide, 5mg / ml) assessed as optimal density at 590nm (16).

The data has been converted to units cytokine / ml by fitting the linear portion of the titration curves obtained to standard curves using recombinant cytokines (Human IL-2, first International standard, NIBSC, Potters Bar, Hertfordshire, UK; murine recombinant IL-3 and human recombinant IL-6, Genzyme, Hatfield, Hertfordshire). The experimental standard deviation was calculated for the raw data (i.e. optical densities) but is not shown as the data is expressed as units.

IL-4 was detected by supernatant induction of expression of Class II on splenic B cells (17). T cell depleted B10.Sc.Sn (H-2^b) splenocytes (pre-treated with rat anti-Thy 1 and complement as above) were cultured at 3 x 10^5 cells per well in 0.1ml I/F with 5% FCS and 0.1ml supernatant in round bottom microtitre plates for 16 hours at 37°C and 5% CO_2. Controls included anti - IL-4 (as above). Class II expression was examined directly with fluoresceinated anti-Ia^b antibody, 3TP (gift of Drs R Germain and R Lechler) in a FACScan (Becton Dickinson, Mountain View, California). Mean fluorescence channel number on a linear scale in the presence of antibody was subtracted from the value obtained without antibody and units of IL-4 calculated by comparison to a standard curve generated using murine rIL-4 (Genzyme, Hatfield, Hertfordshire).

γifn was detected by a capture ELISA on Immunolon plates (Nunc) were coated for 16 hours at 4°C with purified R46A2 monoclonal antibody at 0.005mg/ml (ATCC HB170) in pH 9.6 carbonate buffer. After blocking with 2.5% BSA coating buffer (1 hr, 37°C), supernatants were added to wells containing 0.050ml PBS and 0.05% Tween 20. A standard curve was generated using recombinant murine γifn (10^7 U/mg specific activity, Lot 4408-41, Genentech) (gift of Boehringer Ingelheim, Germany). After 2 hours at 37°C, plates were washed with PBST and polyclonal rabbit anti-γifn added at 1:1000. After 1 hour peroxidase conjugated goat anti-rabbit IgG (Kirkegaard and Perry Labs. Inc., MD, USA) was added for 30 minutes followed by ABTS substrate. Plates were read at 410nm using a Dynatech automatic plate reader. Levels of γifn in each sample were determined from the linear portion ($r>0.98$) of the standard curve. The sensitivity range was from 100 - 6000 pg/ml and data are presented as γifn production for each series of samples. Variation between the samples was less than 10%.

RESULTS AND DISCUSSION

The functional characterisation of the DC as the most potent inducing APC was confirmed in alloproliferative responses (Table 1) which showed a typical hierarchy of APC with DC as the most potent type, with FcR+ splenic M0 as the only other significant APC. Activated peritoneal M0, and B cells, both activated and resting, did not stimulate proliferation (data not shown).

In parallel with this proliferative response, cytokine release into the supernatant fluid was also monitored at 48 and 72 hours (Table 2). DC induced higher levels of IL-2, IL-3, and γifn than any other APC. IL-4 was not detectable even in the presence of DC. IL-6 was synthesised constitutively by T cells in the absence of APC (not shown) and there was no detectable difference when APC were added.

Table 1 APC in primary alloproliferative response induction

Antigen presenting cell		^{3}[H] - thymidine incorporation
Number	Type	(cpm x 10^{-3})
10^3	DC	39.8
	FcR+ MO	6.2
	Peritoneal MO	5.8
5 x 10^3	DC	79.7
	MO	37.5
	Peritoneal MO	5.8
10^4	DC	118.0
	MO	47.5
	Peritoneal MO	5.8

Table 2 The role of different APC in the control of cytokine release (units / ml) into supernatant during primary alloproliferative response induction at 48 hrs.

Cytokine	Antigen presenting cell type					
	DC	FcR+	PMO	A.PMO	B	A.B
IL-2	1	0.5	0	0	0	0
IL-3	53	25	0	9	0	0
IL-4	0	0	0	0	0	0
IL-6	64	73	60	60	62	64
γifn	64	15	0	0	0	0

In secondary allostimulation the pattern of cytokine levels was similar, except that for γifn the differences between DC (100U/ml) and FcR+ tissue MO (80U/ml) were less obvious, and both activated PMO (50U/ml) and activated B cells (50u/ml) were able to induce some detectable activity.

In tertiary allostimulation the pattern of cytokine levels was different, in that for the first time there were detectable IL-4 levels (90U/ml) present, but only in the cultures where DC had been used as APC. Likewise after eight cycles of restimulation, the cultures which incorporated DC as APC had higher IL-4 levels (55U/ml) but at this stage the response was very similar to that seen with gifn, with raised levels in the FcR+ (58U/ml) and activated B cell

(24U/ml) stimulated cultures. Another feature of the later time point was that there was now no detectable IL-2 in the supernatants, irrespective of APC type. Proliferation at these later time pints was also diminished (data not shown).

CONCLUSION

The major effects of DC on cytokine release are summarised in Table 3:

Table 3: The effects of DC on supernatant cytokine levels in allogeneic responses:

1. IL-2: DC induce IL-2 in primary, secondary and tertiary but not octernary cultures, and are the most potent APC inducer at all stages of responsiveness where IL-2 is detectable.
2. IL-3 (and / or GM-CSF): DC induce in primary, secondary, tertiary, and octernary cultures, and are the most potent APC inducer at all stages of responsiveness.
3. IL-4: DC are the only APC inducer in tertiary cultures, and at later timepoints; and they are the most potent APC inducer at all stages of responsiveness
4. γifn: DC are inducers in primary, secondary, tertiary, and octernary cultures; they are not the only inducer at any time point, but are the most potent quantitative inducer at all stages.
5. IL-6: DC have no detectable effect due to high background T cell levels.

Thus DC are the most important inducers of cytokine release in early responses, irrespective of which cytokine is measured. DC are capable of activating release of both TH1 and TH2 associated cytokines. The differential effect between APC with respect to cytokine production decreases with time. DC may drive differentiation of T cells not only for proliferation but also towards an effector phenotype.

Acknowledgements

This research is supported by the Arthritis and Rheumatism Council and the Medical Research Council.

REFERENCES

1. Unanue E.R. Ann. Rev. Immunol. 1984. 2: 395.
2. Koide S.L., Inaba K. and Steinman R.M. J. Exp. Med. 1987. 165: 515.
3. Holsti M.A. and Raulet D.H. J. Immunol. 143: 2514.
4. van Snick J. Ann. Rev. Immunol. 1990 8: 253.
5. Heufler C., Koch F. and Schuler G. J. Exp. Med. 1988 167: 700.
6. Austyn J.M., et al J. Exp. Med. 1983. 157: 1101.
7. Firestein G.S., et al J.Immunol. 1989 143: 518.
8. Mosmann T.R. and Coffman R.L. Ann. Rev. Immunol. 1989. 7: 145.
9. Sunshine G.H., Katz D.R. and Czitrom A.A. Eur. J. Immunol. 1982 12: 9.
10. Breel M., Mebius R.E. and Kraal G. Eur. J. Immunol. 1987. 17: 1555.
11. Austyn J.M. and Gordon S. Eur. J. Immunol. 1981. 10: 805.
12. Leo O., et al Proc. Nat'l. Acad. Sci. (USA). 1988. 84: 1374.
13. Baker P.E., Gillis S. and Smith K.A. J. Exp. Med. 1979. 149: 273.
14. Marcinkiewicz J. and Chain B.M. Immunology. 1989. 68: 185.
15. Helle M., Boeijie L. and Aarden L.A. Eur. J. Immunol. 1988. 18: 1535.
16. Tada H., et al J. Immunol. Methods. 1986. 93: 157.
17. Weinberg A.D., English M. and Swain S. J. Immunol. 1990. 144: 1800.

THE ROLE OF DENDRITIC CELLS AS CO-STIMULATORS IN TOLERANCE INDUCTION

Michal Dabrowski, Mohammad A A Ibrahim, Benjamin M Chain and David R Katz

Depts. of Histopathology, Immunology and Biology, University College London, Riding House St., London W1P 7PN UK

INTRODUCTION

In previous studies we have used an *in vitro* murine primary system to demonstrate that chemically modified antigen presenting cells (APC) were able to induce haplotype specific T cell hypo-responsiveness (1). This results confirmed a potentially important role of the APC in generation of peripheral T cell tolerance.

Under physiological conditions, during an APC-T cell interaction, APC surface accessory molecules are capable of changing from inactive to active state (and vice versa) (2). When the APC are modified, there does not seem to be any quantitative change in cell surface molecule expression. However, a shift from one state of surface molecule to another may be prevented, either by direct inhibition of conformational change, or by abolishing a signalling mechanism within the APC. Both of these changes may affect the dynamic interaction between the APC and the T cell that is required for antigen presentation.

Recent studies in our laboratory have used APC and T cells isolated from human tissues. We demonstrated that modification of human APC abrogates their ability to stimulate proliferation of allogeneic T cells. Notably, this effect was also seen when dendritic cells (DC) were used as APC. The T cells that had been exposed to modified APC showed reduced proliferation when re-challenged with unmodified APC. This reduction was only observed when T cells were negatively purified, which suggests that activation of the T cell may influence the response to the APC tolerogenic signal.

MATERIALS AND METHODS

Media and reagents

The cell culture medium (CM) used was RPMI 1640

supplemented with 10% fetal calf serum (FCS), 10mM Hepes, 2mM L-glutamine, 50μM 2-mercaptoethanol, 100U/ml penicillin, 100μg/ml streptomycin, 2,5μg/ml amphotericin B (Gibco BRL). Hank's balanced salt solution (HBSS) was also from Gibco BRL; collagenase type II, 1-ethyl-3- (3-dimethyloaminopropyl)-carbodiimide (ECDI), rabbit anti-mouse IgG serum (RbαMIgG), leucine methyl ester (LME), were purchased from Sigma; sheep erythrocytes (SRBC) from TCS; Percoll from Pharmacia; and ^{125}5-iodo-2-deoxyuridine (^{125}I-Urd) from Amersham. Tissue culture plastics were from Nunclon.

Antibodies

Monoclonal antibodies: αCD3 (UCHT1), αCD19 (BU12); (0.22μm filtered hybridoma supernatants) were kindly provided by Dr. Diana Wallace, ICRF Human Tumor Immunology Unit, University College London.

Cell preparation

A single cell suspension of human tonsils was prepared by collagenase type II digestion (1mg/ml, 1h, 37°C) and passing through a nylon mesh. Cells were washed twice in HBSS, and spun at 600g on a discontinuous 10% step 30-70% Percoll gradient. Pooled cells from the two lower interfaces formed a high density (HD) fraction and cells from the two upper interfaces are referred to as low density (LD) cells. Residual RBC were lysed, cells were washed and cultured overnight in CM, and non-adherent cells harvested.

To prepare T cells, the HD population was fractionated further by either a positive selection, using rosetting with SRBC, or by a negative selection. For the negative selection HD cells were treated for 1h at 20°C with 5mM LME, washed twice, incubated for 1h at 4°C in 1:20 dilution of αCD19 MoAb in 10% FCS HBSS, washed twice and panned for 1h at 4°C on RbαMIgG coated 10cm Petri dishes. Non-adherent cells were harvested and washed twice before use.

To prepare APC, LD non-adherent cells were depleted of T cells by incubation with 1:20 dilution of αCD3 MoAb followed by panning as above, and rewashed before use. A DC enriched population was prepared by incubating the LD fraction with LME and depletion of CD3 and CD19 positive cells by panning as above. The APC were irradiated (5000R, ^{60}Co) before adding them to T cells.

ECDI modification

APC were modified with ECDI as described previously. Briefly, the cells were incubated at 4°C for 1h in 75mM ECDI / HBSS at 10^7 cells/ml. The ECDI modified cells were washed four times before addition to the T cells.

Proliferative responses

APC from one donor were added at varying cell concentrations to $2x10^5$ T cells from another donor, in a total volume of 0.2 ml of CM per well in 96 well plate. Cultures were incubated for 72h at 37°C in humidified incubator containing 5% CO_2. 0.7μCi of ^{125}I-Urd was added to each well for the final 6h of the culture. Cells were harvested using a Titerek cell harvester (Skatron) and the amount of radiolabel incorporated was measured on NE 1600 γ-counter. Results are expressed as mean cpm +/-SD of three wells.

Induction of hypo-responsiveness

T cells at final concentration $1x10^6$cells/ml were incubated for 18h either in CM alone (control) or with ECDI modified APC, at a final concentration $2x10^6$/ml. T cells were re-isolated by either SRBC rosetting or LME treatment and depletion of CD19 positive cells by panning.

RESULTS

ECDI modification of APC abrogated their ability to stimulate proliferation of allogenic T cells. Results shown were obtained with DC used as APC. Unmodified DC (Figure 1, open bars) were potent as APC. The number of DC that induced maximal T cell proliferation was four to eight times lower than that of LD (data not shown). When the DC had been ECDI treated (hatched bars), however, there was no proliferative response.

To determine whether the T cells that had been exposed to ECDI modified APC became tolerised, the T cells were re-purified and challenged with unmodified APC of the same origin. These T cells showed markedly reduced proliferation, when compared to control T cells. We confirmed that lower ^{125}I-Urd uptake was not due to reduced T cell content during secondary challenge (data not shown). The reduction was only observed in the cultures where the T cells had been negatively selected (Figure 2A). Positively selected T cells (Figure 2B) could not be tolerised.

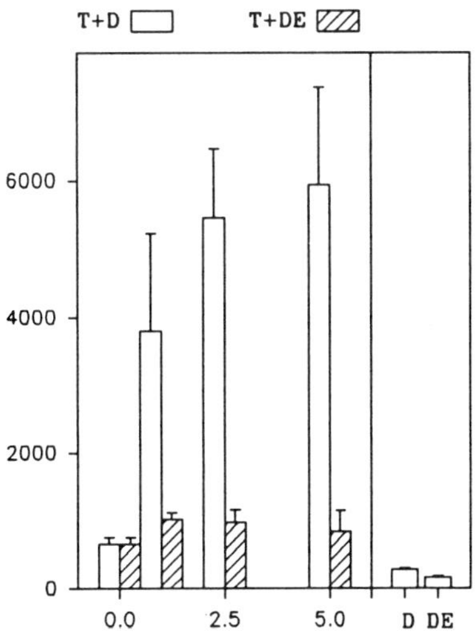

Figure 1. ECDI modification of DC abrogated their allo-stimulatory effect. DC numbers x 10^{-4}.

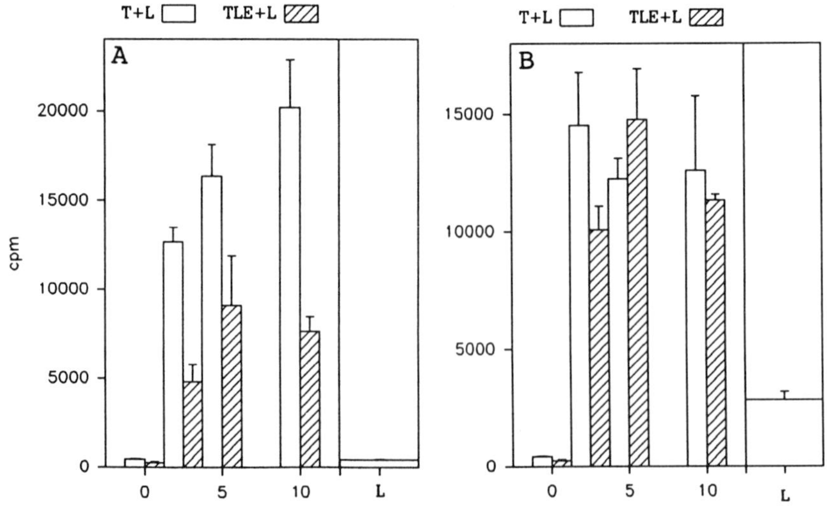

Figure 2. Negatively purified T cells exposed to ECDI modified APC became hypo-responsive. LD cell numbers x 10^{-4}.

DISCUSSION

Several groups have shown that APC may induce either a proliferative response of T cells or a state of anergy, the latter due to reduced level of transcription of interleukin 2 (IL-2)(3). Our studies of the murine tolerance model demonstrated that ECDI modification reduced APC-T cell clustering and led to induction of T cell hypo-responsivness.

This study extends the previous murine findings and shows that ECDI modification of human APC can also induce hypo-responsivness.

The chemical modification abrogates allo-stimulatory capacity of the DC to the same extent as is seen with LD cells, which consist of DC, B cells and monocytes, suggesting that the function(s) affected are common to all APC types. This implies that co-stimulatory activity might reflect preparation of the cell to respond to an encounter with T the cell by up-regulating or modifying ligand molecules for the T cell, rather than constitutive expression of these molecules. This hypothesis is ilustrated in Figure 3.

The activation of the T cell, mediated via SRBC-CD2 signaling appears to prevent the induction of tolerance. A possible mechanism, based on the assumption that binding of positive and negative transcription regulatory factors to the IL-2 promoter is a competitive process, is ilustrated in Figure 4.

This model could also account for a requirement of potent APC such as DC for the induction of primary response of T cells. Furthermore, presentation by the DC, as the most potent co-stimulator, seems most likely to provide the necessary trigger during the phase when tolerance is breached i.e. the critical early stage of auto-immune reactions.

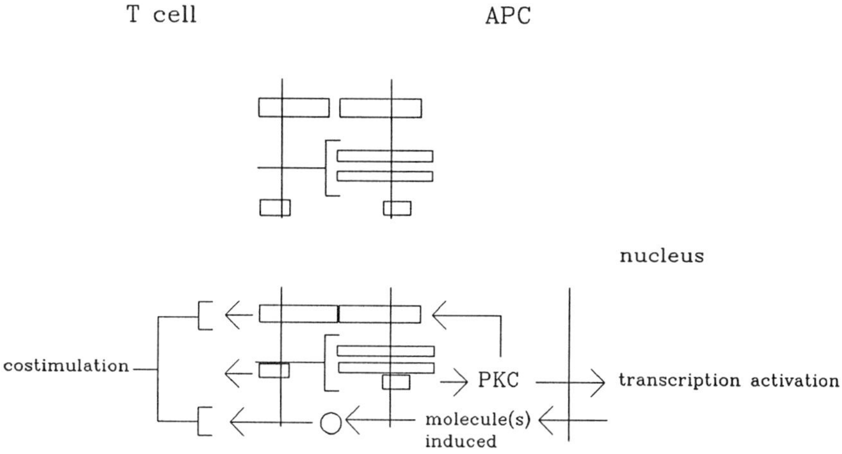

Figure 3. A dynamic model of a co-stimulatory function.

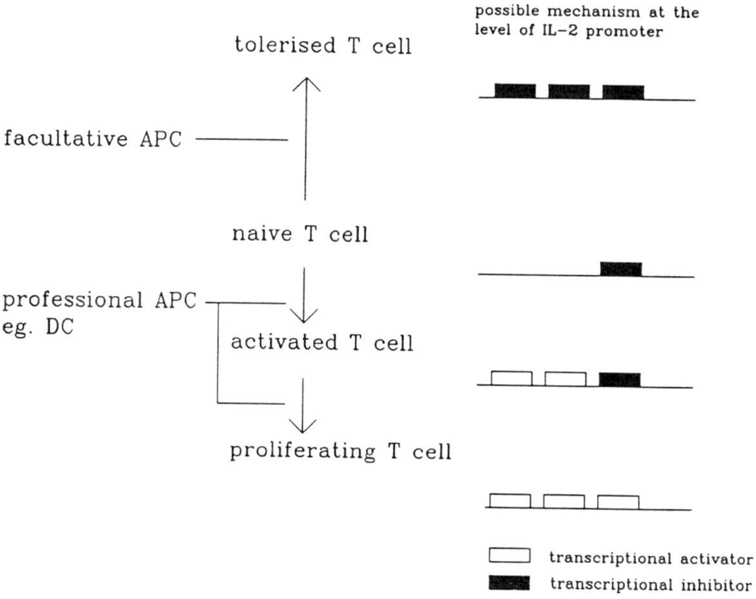

Figure 4. The outcome of APC-T cell interaction depends on both type of APC and state of T cell.

Acknowlegments

This studies were supported by grants from the ECC Tempus Programme and the Arthritis and Rhumatism Council.

REFERENCES

1. Ibrahim M.A., 1991, J. Immunology, 147, 12, 4086-93.
2. Mourad W., Geha R.S., Chatila T., 1990, J. Exp. Med., 172, 5, 1513-6.
3. Jenkins M.K., 1987, Proc. Natl. Acad. Sci. USA, 84, 5409-413.

THE INFLUENCE OF DENDRITIC CELLS ON T-CELL CYTOKINE PRODUCTION

Erna van Wilsem
John Brevé
Ingrid van Hoogstraten[1]
Huub Savelkoul[2]
Georg Kraal

Dept. of Cell Biology, [1]Dept. of Pathology
Vrije Universiteit, Amsterdam
[2]Dept. of Immunology, Erasmus University, Rotterdam
the Netherlands

INTRODUCTION

Pretreatment of mice with picryl chloride via the oral mucosa leads to a T-cell mediated tolerance after sensitization, while mice that do not receive this treatment react with a contact sensitivity (CS) response after topical skin application of picryl chloride(1). It is assumed that both in the oral epithelium and in the skin dendritic Langerhans cells are involved in the uptake of the antigen and subsequent transportation into the draining lymph nodes. Here the Langerhans cells will present antigen to T-cells in the paracortical areas as interdigitating, dendritic cells.

In spite of these similarities in antigen presentation it is obvious that antigen deposition in the two types of epithelia leads to different, opposite, immumological responses. Therefore we wished to determine whether differences would exist in antigen presentation capacity between dendritic cells from lymph nodes draining oral mucosa and skin sites. In addition we looked at the influence of these types of dendritic cells on the stimulation of cytokine profiles.

MATERIALS AND METHODS

Induction of immune tolerance or contact sensitivity

At day 0, BALB/c mice were given a sublingual dose of 50μg picryl chloride (PCl) in 50μl ointment (unguentum hypromellosi) under light anesthesia. After 1 hour the ointment was removed with a small spatula followed by rinsing with warm (37°) water. Ten days later 5% PCl in aceton/etha-

Dendritic Cells in Fundamental and Clinical Immunology,
Edited by Kamperdijk *et al.*, Plenum Press, New York, 1993

nol (1:3) was applied at the shaved abdomen and fore- and hindlegs (150 μl). The mice were challenged 4 days later, with 0.8% PCl in olive oil, 10 μl on each ear. The ear thickness was measured 24 hours later.

For contact sensitivity the same procedure was followed, without pretreatment with a sublingual dose.

Cell separation

Dendritic cells. Draining lymph nodes were cut into small pieces and incubated with a collagenase (type III)/ DNAse solution at 37° for 45 minutes. After pressing through a wire mesh, the cells were spun down, resuspended in medium (RPMI 1640, 1% hepes, 2mM glutamine, 100 U penicillin, 100μg streptomycin (GPS) and 10% FCS (v/v)) and counted. A 2,5 ml cellsuspension of 2.10^7 cells/ml was layered onto 2 ml Nycodenz (14.5 g / 100ml medium) and centrifuged for 20 minutes at 600 g. Cells at the interface were collected, washed once and counted again. This resulted in a dendritic cell fractiom of >70 % purity.

T-cells. Enriched T-cells were obtained from spleen and activated (skin application of PCl at day -10) lymph nodes. Cell suspensions were allowed to adhere for 90 minutes in a culture bottle to remove adhering macrophages. The non-adherent fraction was incubated with monoclonal antibodies (M5/114 and 187/1) against B-cells, after which T- and B-cells were separated using magnetic beads. The purity of the resulting T-cell fraction was over 80%.

Antigen presentation test

Freshly isolated (BALB/c) dendritic cells (DC) were cultured with 4.10^5 (BALB/c) T-lymphocytes in flat bottom 96-wells plates. The cells were cultured in medium supplemented with 5.10^{-5} M β-mercapto-ethanol (2-ME) for 72 hours with or without 100 μg TNP-OVA. After 64 hours 1μCi ^{3}H/well was added and ^{3}H-thymidine incorporation was measured as counts per minute (cpm).

ELISPOT- assay for cytokines

To quantitate the frequency of IFN-γ or IL-5 specific spot forming cells (SFC) the ELISPOT assay was used. For the IL-5-specific ELISPOT assay (1), a 96-well plate with nitrocellulose base was coated overnight at 4° with mo-Ab anti-IL-5 (TRFK 5)(2), washed 3 times with PBS (0,01M) and blocked with PBS/1%BSA for 60 minutes at 37°.
The cells were resuspended in RPMI-1640, 1% BSA, GPS (1.10^3 - 1.10^5 cells/well) and incubated overnight at 37°C with 5% CO2 in air. The plates were washed with PBS/Tween and incubated for 2 hours at 37°C with biotinylated mo-Ab anti-IL-5 (TRFK 4)(2). The plates were washed again with PBS/Tween and incubated with streptavidin-phosphatase diluted 1:1000 in PBS /Tween/1% BSA for 1 hour at 37°. After washing with PBS the plates were incubated with alkaline phosphatase substrate (5BCIP/NBT). The reaction was stopped by running tap water, the plates were dried and the spots were counted by the use of a stereomicroscope (40-60X).

A similar assay, with R4-6A2 (3) as coating mo-Ab and XMG 1.2 (4) as detecting mo-Ab, was performed to determine IFN-γ forming cells.

RESULTS AND DISCUSSION

In this study, contact sensitivity and oral tolerance are used as a model to investigate the role of dendritic cells in the induction of different immune responses. In both cases dendritic Langerhans cells are involved in the uptake and transportation of the antigen to the paracortex of the draining lymph node. There the antigen is presented to the T-cells.

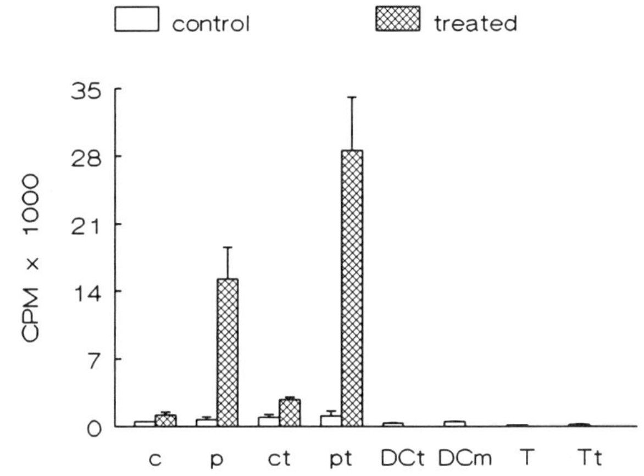

figure 1 Antigen presentation capacity measured as counts per minute (cpm). DC (1.10^4) from oral mucosa draining cervical (c) ln and skin draining peripheral (p) ln were cocultured with activated T-cells (4.10^5), in the presence of 100 μg/ml TNP-OVA (t) as antigen or with a anti-class II monoclonal antibody (m) . Controls, T-cells alone (T) or with TNP-OVA (Tt) and DC alone or with TNP-OVA (Dt), did not stimulate T-cell proliferation.

It would be interesting to know if different signals are given by the DC to the T-cells during antigen presentation, leading to these opposite immune responses. Therefore we compared the antigen presentation capacity of DC isolated from lymph nodes draining oral mucosa and skin sites.

We found that dendritic cells isolated from lymph nodes draining skin sites achieve a better T-cell proliferation compared to DC from oral sites (fig.1).

An important aspect of T-cell stimulation is the induction of cytokine production. We therefore looked at the production of IFN-γ and IL-5 as two major representatives of the Th1 and Th2 cytokines.

Table 1. IFN-γ and IL-5 production in lymph node suspensions of oral mucosa- and skin draining lymph nodes after PCl application.

	interferon-γ	interleukin-5
cervical control	302.0 ± 52.7	134.0 ± 18.5
cervical treated	158.6 ± 74.8	181.0 ± 80.8
peripheral control	265.3 ± 92.5	210.6 ± 59.7
peripheral treated	177.3 ± 11.8	159.6 ± 18.0

The frequency of IFN-γ and Il-5 forming cells in draining lymph node suspensions measured as spots/well by an ELISPOT assay with overnight incubation of 2.5 10^3 cells / well.

Table 2. IFN-γ and IL-5 production after coculturing of isolated DC and naive T-cells.

	interferon-γ	interleukin-5
cervical control	680.7 ± 91.9	281.7 ± 29.5
cervical treated	916.0 ± 152.0	373.7 ± 77.3
peripheral control	796.7 ± 133.5	297.7 ± 17.8
peripheral treated	906.9 ± 318.1	390.0 ± 16.7

Isolated dendritic cells (10/well) were cultured with T-cells (4000/well, spleen) for 48 hours. Cytokine production was measured by the ELISPOT assay. Incubation of DC alone resulted in ± 12 spots/well while after incubation without cells, no spots were formed.

First we investigated the overall cytokine production in the two types of lymph nodes after oral or skin stimulation. The most prominent changes we observed were a reproducible reduction of the number of T-cells producing IFN-γ, coinciding with an increase of IL-5 producing T-cells (table 1).

To look for a role of dendritic cells in these changes, isolated DC from the two types of lymph nodes were cocultured with naive (spleen) T-cells (table 2). Interestingly, no difference in the production of interleukine types was found but it was clear that DC from activated nodes irrespective of their draining sites led to increased interleukin production.

Together this indicates that, although DC isolated from the two types of nodes are different in their T-cell stimulating capacity as measured by proliferation no differences in the induction of cytokine profiles could be found.

A difference in antigen presentation capacity can be explained by possible differences in antigen concentration between the sublingual dose and the peripheral dose. This may result in a different antigen loading of the DC and affect the antigen presentation capacity in vitro. This difference in antigen loading has obviously no effect on the induction of IL-5 or IFN-γ cytokine profiles.

It is possible that a difference in antigen loading represents a mechanism to induce different immune responses. Certain T-helper subsets can preferentially be activated by DC with a low concentration of antigen and start to produce cytokine profiles resulting in oral tolerance. Further experiments are needed to explore these possibilities.

REFERENCES

1. van Hoogstraten I.M.W., D. Boden, B.M.E. von Blomberg, G. Kraal, R.J. Scheper. Persistent immune tolerance to nickel and chromium by oral administration prior to cutaneous sensitization. J.Invest.Dermatol. in press

2. Taguchi T., J.R. McGhee, R.L. Coffman, K.W. Beagley, J.H. Eldridge, K. Takatsu, and H. Kiyono. Detection of individual mouse splenic T-cells producing IFN-γ and IL-5 using the enzyme linked immunospot (ELISPOT) assay. J.Immunol.Methods 128:65. (1990)

3. Schumacher, J.H., A. O'Gara, B. Shrader, A. van Kimmenade, M.W. Bond, T.R. Mosmann, and R.L. Mosmann. The characterization of four monoclonal antibodies specific for mouse IL-5 and development of mouse and human IL-5 enzyme-linked immunosorbent assays. J.Immunol. 141:1567.(1988)

4. Spitalny, G.L., and E.A. Havell. Monoclonal antibody to murine gamma interferon inhibits lymphokine-induced antiviral and macrophage tumoricidal activities. J.Exp.Med. 159:1560. (1984)

5. Cherwinski, H.M., J.H. Schumacher, K.D. Brown, and T.R. Mosmann. Two types of mouse helper T-cell clone. III Further differences in lymphokine sythesis between Th1 and Th2 clones revealed by RNA hybridization, functionally monospecific bioassays, and monoclonal antibodies. J.Exp.Med 166:109. (1987)

PHENOTYPICAL AND FUNCTIONAL CHARACTERIZATION OF DENDRITIC CELLS IN THE HUMAN PERITONEAL CAVITY

Michiel G.H. Betjes, Cees W. Tuk, and Robert H.J. Beelen

Department of Cell Biology
Vrije Universteit
Van der Boechorststraat 7
1081 BT, Amsterdam, The Netherlands

INTRODUCTION

Little is know about antigen presenting cells in the human peritoneal cavity. Mottolese et al.[1] studied the antigen presenting capacity of human peritoneal macrophages (the adherent cell fraction) and concluded that these cells could stimulate allogeneic and syngeneic T cells at a level similar to peripheral blood mononuclear cells. However it can not be excluded that contaminating dendritic cells (DC) biased these results.

Human peritoneal cells (PC) can easily be obtained from patients undergoing chronic ambulatory peritoneal dialysis (CAPD) and by suction of the peritoneal fluid in the cul-du-sac of women undergoing laparoscopic sterilization (PClap)[2]. The CAPD peritoneal cells (PCcapd) are potent stimulators of an allogeneic mixed leukocyte reaction[3] and evidence was obtained for a relatively high number of DC among these peritoneal cells[4].

In the present study the immunophenotype and function of human peritoneal DC is described and compared to peritoneal macrophages and monocytes.

MATERIAL AND METHODS

Culture Medium

Culture medium was RPMI 1640 supplemented with 10 % heat inactivated FCS, 2mM L-glutamine, 50 U/ml penicillin-50μg/ml streptomycin and 50 μM ß-mercaptoethanol.

Peritoneal Cells and T cells

PC were obtained from the overnight dialysis effluents of patients (N=17) undergoing CAPD and from the peritoneal fluid of women undergoing laparoscopic sterilization (N:10), as described previously[2]. The yield of peritoneal cells per effluent varied widely (median:15.0 x 10^6 cells, range:1.0-30.0 x 10^6 cells) and contained less than 5% granulocytes. The median yield of PClap was 3 x 10^6 (range:1.0 x 10^5-10.0 x 10^6). PC were resupended in culture medium after isolation.

DC and Mϕ Enrichment. DC and macrophages were enriched as described previously3. Briefly, PC were cultured overnight (16 h) at 37^0 C in a humidified 5% CO_2 atmosphere. The adherent cells were used as macrophage (Mϕ) enriched cell fraction. The contaminating Mϕ in the non-adherent cell fraction (NAC) were removed by incubating these cells on IgG coated petridishes. The remaining lymphocytes were separated from the DC by layering the Fc receptor (FcR) negative cells on a 14.5 % metrizamide gradient. The interfase (low density cells, LoD) contained the DC-enriched fraction and the pellet (high density cells, HiD) the peritoneal lymphocytes.

Peripheral blood (10-20 ml) was taken from patients and healthy volunteers by venapuncture into sterile heparine containing glass tubes. Peripheral blood mononuclear cells were isolated on a Lymphoprep gradient (Nycomed). A monocyte enriched fraction (>80% purity) was obtained by harvesting the adherent cells after overnight culture on plastic tissue culture petri-dishes.

Isolation of T Cells. T cells for the allogeneic MLR were isolated from buffycoats by E-rosetting of NAC cells after 16 h of culture on plastic tissue culture petri-dishes.

Autologous T cells were isolated from the NAC of the PBMC (see above). NAC were incubated with a mixture of anti-CD22, anti-CD14 and anti-HLA-DR/DQ monoclonal antibodies for 30-45 min on ice. After washing twice in culture medium the cells were incubated with sheep anti-mouse IgG coated magnetic particles (Dynabeads, Dynal AS, Oslo, Norway) for 30-45 min on ice in a ratio of 20 Dynabeads per positive cell (assuming that approximately 20% of the cells were reactive with the used McAb's). The Dynabead negative population contained >92% CD2 positive cells, <1% monocytes, and DC were not detectable.

Cell viability of all cell populations was always above 90% as determined by trypan blue exclusion.

Allogeneic MLR and Antigen Presentation

Allogeneic MLRs were set up as described by Mottolese et al.[1].

Monoclonal Antibodies (mAbs)

The mAbs used were: anti-CD22 (Dakopatts, Copenhagen, Denmark); anti-CD2 (CLB, Amsterdam, The Netherlands); 3C10 (ATTC, Rockville, NY, USA) reactive with CD14[5]; EBM11 (Dakopatts) reactive with CD68, a lysosomal associated antigen with a high specifity for monocytes and Mϕ[4]; 9.3F10 (ATCC) reactive with HLA-DR\DQ and especially used for its bright staining with DC[5]; W6/32 specific for a nonpolymorphic determinant of HLA-ABC[6]; TS2/9 reactive with LFA-3[7]; F10.2 reactive with ICAM-I[8]; anti-CD11a (Dakopatts); anti-CD11b (Dakopatts); anti-CD11c (Dakopatts); anti-CD18 (Dakopatts); OKT6 recognizing CD1a[9].

Immunocytochemistry

Cytocentrifuge preparations of cells were tested for their reactivity with the different mAb using an immunoperoxidase staining as described[4]. Thereafter cells were stained for acid phosphatase activity[4].

DC were identified as large cells with an irregular outline, eccentric nucleus, acid phosphatase staining or EBM11 reactivity in a spot paranuclear[4] and strong HLA Class II expression on their cell surface.

FACS Analysis

For the fluorescence flow cytometry cells were incubated on ice (30 min) with the different mAbs followed by an incubation with FITC conjugated goat anti-mouse IgG (Tago, Burlingame, CA, USA) in PBS+ with 5% pooled human serum. Between succesive incubations cells were washed twice in PBS + 0.5% BSA. Cell bound fluorescence was measured on a FACS (Becton and Dickinson). Incubation of cells with an irrelevant antibody of the same IgG isotype (Dakopatts) showed no significant increase in fluorescence above background.

RESULTS

Enrichment Procedure

DC comprised on average 5.8% ± 1.0 of CAPD and 5.4% ± 1.6 laparoscopy peritoneal cells. The relatively high percentage of DC made it possible to enrich DC up to 20-44%. On average 20-30% of the Mϕ adhered to plastic, irrespective of the source of PC. Panning for FcR was a very efficient method to remove the peritoneal Mϕ (>97%). On average 33.6% ± 12.2 (PClap) and 46.2% ± 8.4 (PCcapd) of DC was recovered from the NAC cells after FcR panning, indicating that a substantial part of the DC in the PC was bearing FcR on their cell surface. A metrizamide gradient was required for further enrichment of the DC from the PCcapd. After this procedure the LoD fraction yielded a similar total number of cells as was obtained for the PClap after FcR panning (on average 3.2 x 10^5 cells, range 1.6-8.0 x 10^5 cells). This reflected a DC recovery from the unfractionated PC of 13.9% ± 3.3 for the PCcapd and 25.0% ± 3.6 for the PClap.

Table 1. FACS analysis of the immunophenotype of peritoneal DC, Mϕ and monocytes.

Determinant	DC	Mϕ capd[1]	Mϕ lap[2]	Mon[3]
HLA-DR\DQ	+++	++	++	+/++
HLA-ABC	+++	++	++	++
CD14	-/+	++	++	+
CD11A	+	+	+	+
CD11B	-/+	+	+	+
CD11C	+	+	+	+
CD18	+	+	+	+
ICAM-1	+	+	+	+
LFA-3	+	+	+	+

[1]CAPD peritoneal macrophages;[2]laparoscopy peritoneal macrophages;[3]monocytes;[4]-/+: <5 x background (bg), +: 5-50 x bg, ++: 50-500 x bg, +++: >500 x bg.

Immunophenotype

DC did not react with anti-CD22 or anti-CD2 and showed no expression of CD14 using a immunoperoxidase staining. CD1 positive cells were not observed among peritoneal cells. For FACS analysis the cell populations were gated on the non-lymphoid cell populations. Results showed no differences between the monocytes of CAPD patients or healthy controls (data not shown). The expression of CD14, the integrins or MHC molecules was the same for monocytes and peritoneal Mϕ irrespective of CAPD treatment. Dendritic cell enriched fractions showed a 5 to 10 fold increased expression of MHC class I and II, a 5-10 times lowered expression of CD11b and a 10-100 times lowered expression of CD14.

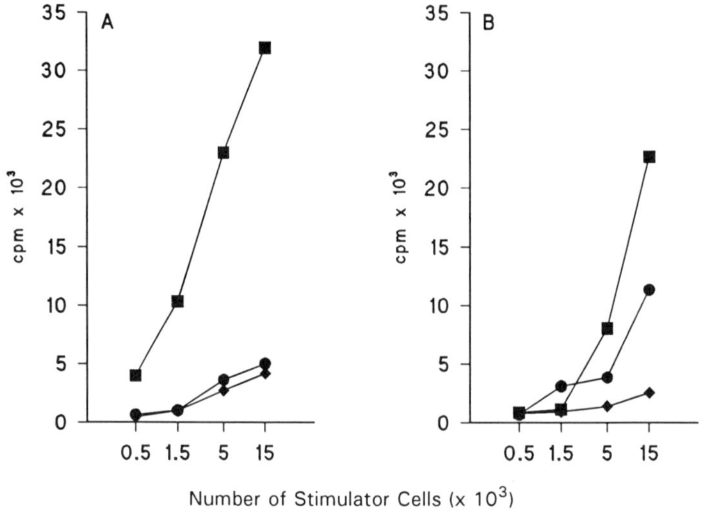

Figure 1. Stimulatory activity of (■) peritoneal DC, (●) macrophages (adherent cell fraction) and (◆) monocytes for allogeneic T cells. Figure 1A shows the peritoneal cells obtained by laparoscopy, figure 1B the peritoneal cells obtained from the CAPD dialysate. Stimulator cells were added to 5.0 x 10^4 T cells. The background response of T cell was always below 200 cpm.

Antigen Presentation

The allo-antigen presentation was most effectively done by DC and CAPD peritoneal Mϕ. Allogeneic T cell responses became marginal when the LoD/T cell ratio was below 1:30 (corresponding to a percentage of DC <1-2%). In contrast to the PClap, peritoneal Mϕ in the PCcapd were potent stimulator cells. Monocytes from CAPD patients and healthy controls were relatively poor stimulator cells when compared to the CAPD peritoneal Mϕ. Taking the stimulatory activity of CAPD Mϕ and the percentage of DC in the LoD into account, it was estimated that peritoneal DC were about 10 times more potent stimulator cells in the allogeneic MLR when compared to CAPD peritoneal Mϕ and 30 times more potent than monocytes.

DISCUSSION

The immunophenotype and functional characteristics have not yet been described for human peritoneal DC and studies sofar undertaken have investigated peripheral blood derived DC[10,11], tonsil derived DC[12] and DC isolated from synovial effusions[13]. The percentage of DC in the peritoneal cavity was surprisingly high when compared to the peripheral blood DC which do not exceed 1 % of the PBMC[14]. In human bronchoalveolar lavages the percentage of DC (using the same criteria for DC) is below 0.5 % (personal communication C.E.G. Havenith). The reason for such high numbers of DC in the peritoneal cavity is not known.

Dendritic cells isolated from the peripheral blood have been reported to express high numbers of MHC class I and II molecules on their cell surface when analysed with the FACS[14]. Furthermore DC from different sources have been shown to be CD14 negative and to express no or low levels of CD11b. The immunophenotype of peritoneal DC also showed a high expression of MHC molecules, but a low expression of CD14 and moderate CD11b expression on the FACS. Differences in the staining methodology[11,12] (for instance, peritoneal DC were CD14 negative using an immunoperoxidase staining), or in culture period[11,12,13] could explain these differences. Also selective depletion of CD11b[13] or Mϕ marker positive cells[12] should be considered.

The DC and Mϕ seem to be at least very closely related members of the same family because there were many overlapping properties of DC and Mϕ. Peritoneal DC showed expression of CD14, CD11b and CD68, and there was a strong indication that part of the DC were FcR positive. Also CD14 and FcR negative Mϕ were co-enriched during the DC enrichment procedure. In a recent study it was shown that without a short culture period peripheral blood DC can (immuno)phenotypically not be distinguished from other monocytes[4].

The low stimulatory capacity of laparoscopy derived peritoneal Mo in the allogeneic MLR is in accordance with the findings on human alveolar Mϕ[15,16]. Also in the animal model Mϕ show a low stimulatory capacity for allogeneic T cells[17]. In contrast, CAPD peritoneal Mϕ showed a suprisingly high stimulatory capacity. This could implicate that the dialysate-activated condition of the CAPD peritoneal Mϕ[18] also upregulates their antigen presenting capacity.

Sofar it is not known why DC have a potent capacity to initiate a primary immune response and induce a strong syngeneic MLR. Furthermore no human DC lineage specific antibody has been raised up to now and DC are usually isolated by selecting for cells that do not express monocyte/macrophage markers or functions and lack T and B cell markers[11,12,13,14]. It is reported that the LFA-1/ICAM pathway is important in the clustering of DC with allogeneic T cells[19] and it is known that the functional status of the integrin molecules can be influenced for instance by a difference of phosphorylation[20]. One might therefore speculate that DC or activated Mϕ have a different functional status of their integrines making them suitable cells for initiating a primary immune response.

In conclusion, human PC are a rich source for isolation of DC with phenotypic and functional characteristics that closely resemble those of peripheral blood DC. The peritoneal Mϕ of CAPD patients showed an increased function as APC which is probably caused by their activated condition.

REFERENCES

1. Mottolese M, P.G. Natali, G. Atlanta, A. Cavallari, F. Difilippo, and S. Ferrone. Antigenic profile and functional characterization of human peritoneal macrophages. *J.Immunol.* 135:200 (1985).
2. Goldstein C.S., J.S. Bomalaski, R.B. Zurier, E.G. Neilson, and S.D. Douglas. Analysis of peritoneal

macrophages in continuous ambulatory peritoneal dialysis patients. *Kidney Int.* 26:733 (1984).
3. Betjes M.G.H., C.W. Tuk, L. Arisz, and R.H.J. Beelen. Peritoneal Mϕ of patients treated with continuous ambulant peritoneal dialysis are very potent stimulator cells of an allogeneic mixed leukocyte reaction, *in*: Lymphatic Tissues and In-Vivo Immune Responses. Marcel Dekker, Inc., New York (1992).
4. Betjes M.G.H., M.C. Haks, C.W. Tuk, and R.H.J. Beelen. Monoclonal antibody EBM11 (anti-CD68) discriminates between dendritic cells and macrophages after short-term culture. Immunobiol. 183:79 (1992).
5. Van Voorhis, W.C., R.M. Steinman, L.S. Hair, J. Luban, M.D. Witmer, S. Koide, and Z.A. Cohn. Specific anti-mononuclear phagocyte monoclonal antibodies: application to the purification of dendritic cells and tissue localization of macrophages. *J. Exp. Med.* 158:126 (1983).
6. Barnstable C.J., W.F. Bodmer, G. Brown, G. Galfre, C. Milstein, A.F. Williams, and A. Ziegler. Production of monoclonal antibodies to group A erythrocytes, HLA and other human cell surface antigens. *Cell.* 14:9 (1978).
7. Sanchez-Madrid, F., A.M. Krensky, C.F. Ware, E. Robbins, J.L. Strominger, S.J. Burakoff, and T.A. Springer. Three distinct antigens associated with human T-lymphocyte mediated cytolysis:LFA-1, LFA-2, and LFA-3. *Proc. Natl. Acad. Sci. USA.* 79:7489 (1982).
8. Rothlein R., M.L. Dustin, S.D. Marlin, and T.A. Springer. A human intercellular adhesion molecule (ICAM-1) distinct from LFA-1. *J. Immunol.* 137:1270 (1986).
9. Fithian, E., S. Kung, G. Goldstein, M. Rubenfeld, C. Fenoglio, R. Edelson. Reactivity of Langerhans cells with hybridoma antibody. *Proc. Natl. Acad. Sci. USA.* 78:2541 (1981).
10. Kuntz Crow M., and H.G. Kunkel. Human dendritic cells: major stimulators of the autologous and allogeneic mixed leucocyte reactions. *Clin. Exp. Immunol.* 49:338 (1982).
11. Young J.W., and R. M. Steinman. Accesory cell requirements for the mixed-leukocyte reaction and polyclonal mitogens as studied with a new technique for enriching blood dendritic cells. *Cell. Immunol.* 111:167 (1988).
12. Hart, D.N.J. and J.L. Mckenzie. Isolation and characterization of human tonsil dendritic cells. *J. Exp. Med.* 168:157 (1988).
13. Zvaifler N.J., R.M. Steinman, G. Kaplan, L.L. Lau, and M. Rivelis. Identification of immunostimulatory dendritic cells in the synovial effusions of patients with rheumatoid arthritis. *J. Clin. Invest.* 76:789 (1985).
14. Freudenthal P.S., R. M. Steinman. The distict surface of human blood dendritic cells, as observed after an improved isolation method. *Proc. Natl. Acad. Sci. USA.* 87:7698 (1991).
15. Toews, G.B., W.C. Vial, M.M. Dunn, P. Guzetta, G. Nunez, P. Stastny, and M.F. Lipscomb. The accessory cell function of human alveolar macrophages in specific T cell proliferation. *J. Immunol.* 132:181 (1984).
16. Lipscomb, M.F., C.R. Lyons, G. Nunez, E.J. Ball, P. Stastny, W. Vial, V. Lem, J. Weisssler, L.M. Miller, and G.B. Toews. Human alvolar macrophages: HLA-DR-positive macrophages that are poor stimulators of a primary mixed leukocyte reaction. *J. Immunol.* 136:497 (1986).
17. Havenith, C.E.G., A.J. Breedijk, and E.C.M. Hoefsmit. Effect of Bacillus Calmette-Guérin on numbers of dendritic cells in bronchoalveolar lavages of rats. *Immunobiol.* 184:336 (1992).
18. Bos H.J., D.G. Struijk, C.W. Tuk, J.C. de Veld, T.J.M Helmerhorst, E.C.M. Hoefsmit, L. Arisz, and R.H.J. Beelen. Peritoneal dialysis induces a local sterile inflammatory state and the mesothelial cells in effluent are related to the bacterial peritonitis incidence. Nephron. 59:508 (1991).
19. Scheeren, R.A., G. Koopman, S. van der Baan, C.J.L.M. Meijer, and S.T. Pals. Adhesion receptors involved in clustering of blood dendritic cells and T lymphocytes. *Eur. J. Immunol.* 21:1101 (1991).
20. Springer, T.A. Adhesion receptors of the immune system. *Nature.* 346:425 (1990).

DENDRITIC CELLS ISOLATED FROM RAT AND HUMAN NON-LYMPHOID TISSUE ARE VERY POTENT ACCESSORY CELLS

Robert H.J. Beelen, Ellen van Vugt, Joke J.E. Steenbergen,
Michiel G.H. Betjes, Carin E.G. Havenith and Eduard W.A. Kamperdijk

Department of Cell Biology, Division of Electron Microscopy
Faculty Medicine, Free University
van der Boechorststraat 7, 1081 BT Amsterdam
The Netherlands

INTRODUCTION

In the past it was shown that dendritic cells (DC) are the in vitro equivalent of interdigitating cells present in T-cell area's of lymphoid organs (1). Recently it has been shown that DC can also be isolated from rat normal steady state peritoneal cells (PC) and that these numbers can be increased after intraperitoneal injection of BCG (2,3). The same can be demonstrated from rat broncho-alveolar lavages after installation of BCG (4). Finally DC can be isolated from the PC population in human, especially from CAPD (continuous ambulant peritoneal dialysis) patients (5), which represent a chronic inflammatory state (6).

In this study we induced a chronic inflammatory state in the rat by thioglycollate broth (TG), which is commonly used in research since a high macrophage (Mϕ) exudate is formed (7). It is shown that TG-DC can be easily isolated, which proved to be very efficient in antigen-presentation and moreover can be distinguished from Mϕ immunophenotypically, using the general rat Mϕ-marker ED1 (8).

MATERIAL AND METHODS

Animals

ACI/MaI rats were used, weighing 180-200g, which are high responders for the copolymer GT (see for details (9)). Peritoneal exudates were induced by intraperitoneal administration of 10 ml TG and harvested 4 days later by peritoneal lavage.

Human CAPD PC were obtained as described in detail elsewhere (10).

Mouse (balb/c) TG-PC were obtained as described elsewhere (11).

Immunochemistry

Staining with the general Mϕ-markers ED1 for the rat (9), EBM11 (CD 68) for human (6), and MOMA-2 for the mouse (11) was done according standard procedures.

DC-isolation and Antigen-presentation Assay

Total peritoneal cells (TPC) were cultured overnight. Non-adherent cells (NAC) were put over a Nycodenz gradient (9) and the low density fraction was used as TG-DC (more than 90% pure). Adherent cells (AC, almost pure Mϕ) were harvested with a rubber policeman. Primed T-cells were isolated as described elsewhere (12) and antigen presentation was done according standard procedures (9) with different numbers of TG-DC.

RESULTS AND DISCUSSION

Rat TPC obtained after TG consisted of about 40 million cells. After overnight culture the NAC were enriched further for DC by nycodenz gradient with a final recovery of about 0.5-1 million TG-DC (purity better than 90% according standard criteria (see 9). TPC, NAC, AC and TG-DC (dose-response curve) were tested for antigen presentation (figure 1).

The data show that NAC show a higher antigen presentating capacity (APC) than TPC, while AC show only a slight response. TG-DC show a very efficient APC, which could be diluted downwards in a dose-dependent way. Most remarkably 20 TG-DC are as potent in APC as 20.000 TG-Mϕ. TG-AC might even partly suppress responses (11), which is in detail discussed elsewhere (9). Clearly our data show that DC have a

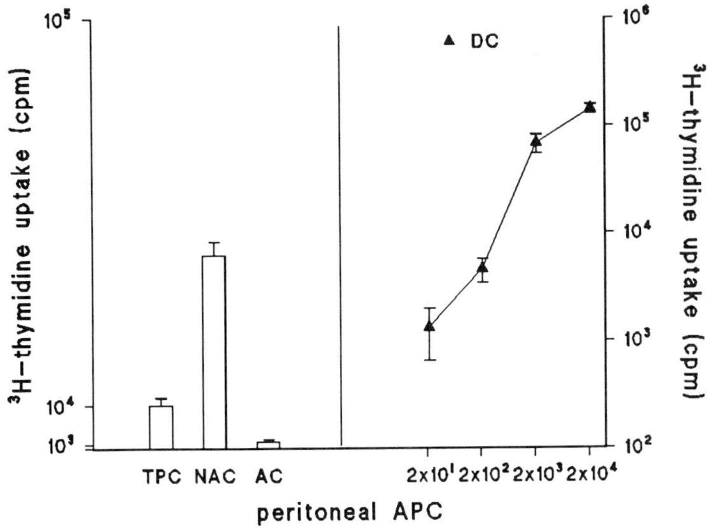

Figure 1: Antigen presentation of the antigen GT by several TG-PC to primed T-cells. For TG-DC a dose response curve is shown.

superior APC, however this does not exclude that under circumstances also Mϕ may clearly APC (as is the case for CAPD-Mϕ (10)). Recently we demonstrated that the general human cytoplasmic MA marker CD 68 located in a discrete juxtanuclear spot (5) in PC-DC (after culture). We therefore studied the expression of the general rat Mϕ-marker ED1 in our TG-DC and moreover studied the expression of the general mouse Mϕ-marker MOMA-2 in mouse TG-DC, isolated according the same procedure with fairly the same results in recovery and purity.

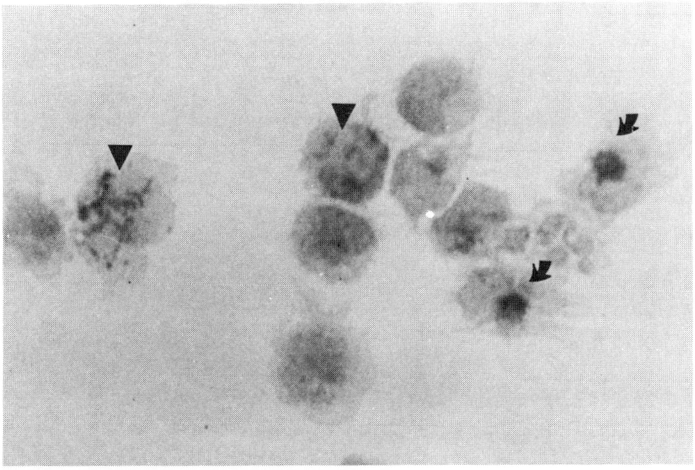

Figure 2: Localization of the general rat Mϕ-marker ED1 in the characteristic distinct spot in TG-DC (➘) and all-over the cytoplasm in Mϕ (▼).

Clearly as is shown in figure 2 also for the rat the general cytoplasmic Mϕ-marker is expressed in a distinct spot localization for DC. We have to state, that this distinct spot localization is only seen after culture and so probably is related with the maturation process of DC. Finally the same was found for the mouse system (MOMA-2), which leads to our scheme as given in table 1.

Table 1. General Mϕ-markers, which locate in a discrete spot in DC

Species	Mϕ-marker
HUMAN	EBM-11 (CD 68)
RAT	ED 1
MOUSE	MOMA-2

In conclusion our results have shown that not only lymphoid organs but also TG-PC are a very convenient source for DC and moreover our data show that the precursor DC and Mϕ must have a very close interrelation, since the discriminating marker pattern (with is a general phenomenon) can only be demonstrated after overnight culture.

References

1. Kamperdijk EWA et al, this symposium.
2. Vugt E van, Arkema JMS, Verdaasdonk MAM, Beelen RHJ, Kamperdijk EWA. Morphological and functional characteristics of rat steady state peritoneal dendritic cells. *Immunobiol.*184, 14-24 (1991).
3. Vugt E van, Verdaasdonk MAM, Beelen RHJ, Kamperdijk EWA. Induction of an increased number of dendritic cells in the peritoneal cavity of rats by i.p. administration of BCG. *Immunobiol.* in press (1992).
4. Havenith CEG, Breedijk AJ, Hoefsmit ECM. Effect of BCG inoculation on numbers of dendritic cells in bronchoalveolar lavages of rats. *Immunobiol.* in press (1992).
5. Betjes MGH, Haks WC, Tuk CW, Beelen RHJ. Monoclonal antibody EBM 11 (CD 68) discriminates between dendritic cells and macrophages. *Immunobiol.* 183, 79-87 (1991).
6. Bos HJ, Struijk D, Tuk CW, Veld JC de, Helmerhorst TJM, Hoefsmit ECM, Arisz L, Beelen RHJ. Letter to the editor: Peritoneal dialysis induces a local sterile inflammation and the mesothelial cells in the effluent are related to the bacterial peritonitis incidence. *Nephron* 59, 508-509 (1991).
7. Beelen RHJ, Walker WS. Dynamics of cytochemically distinct subpopulations of macrophages in elicited peritoneal exudates. *Cell. Immunol.* 82:246-257 (1983).

8. Beelen RHJ, Eestermans IL, Dopp EA and Dijkstra CD. Monoclonal antibodies ED1, ED2 and ED3 against rat macrophages: expression of recognized antigens in different stages of differentiation. *Transplant. Proc.* 19:3166-3170 (1987).
9. Vugt E van et al, this symposium.
10. Betjes MGH et al, this symposium.
11. Wijffels JFAM, Rover Z de, Rooyen N van, Kraal G, Beelen RHJ. Chronic inflammation induces the expression of dendritic cell markers not related to functional antigen presentation on peritoneal exudate macrophages. *Immunobiol.* 184, 83-92 (1991).
12. Havenith CEG, Breedijk AJ, Verdaasdonk MAM, Kamperdijk EWA, Beelen RHJ. An improved and rapid method for the isolation of rat lymph or spleen T lymphocytes for T cell proliferation assays. *J.Immunol. Methods*, in press (1992).

ANTIGEN PRESENTING CAPACITY OF PERITONEAL MACROPHAGES AND DENDRITIC CELLS

Ellen van Vugt, Marina A.M. Verdaasdonk, Eduard W.A. Kamperdijk, and Robert H.J. Beelen

Department of Cell Biology
Division of Electron Microscopy
Medical Faculty
Vrije Universiteit, van der Boechorststraat 7
1081 BT Amsterdam, The Netherlands

INTRODUCTION

Dendritic cells (DC) are very potent antigen presenting cells (1), that can be isolated from many lymphoid and non-lymphoid tissues from different species (2). Recently, we have demonstrated that in the peritoneal cavity of rats approximately 1 % of the cells are DC, based on similar morphological and functional characteristics as DC isolated from lymphoid organs (3). Next to this small number of DC, a large number of Mϕ (macrophages) is present in the peritoneal cavity, which are also capable of presenting antigen although to a lesser extent. Both cell types are bone marrow derived (4), but there are numerous morphological and functional differences (5) that suggest the existence of different cell lineages.

In this article we describe the effect of culture (in vitro) or in vivo stimulation on the phenotype, number and antigen presenting capacity of peritoneal macrophages and dendritic cells. By i.p. administration of Bacillus Calmette-Guérin (BCG) an acute inflammation is induced, whereas i.p. administration of thioglycollate (TG) induces a chronic inflammatory state after 4 days with mainly Mϕ. The effect of a different state of inflammation on the number, phenotype and function of peritoneal DC and Mϕ is studied by comparing them with Mϕ and DC obtained from an unstimulated peritoneal cavity.

MATERIALS AND METHODS

Animals and Immunization Procedures

For the experiments male ACI/Mal rats were used, weighing 160 - 180 g. These rats are high responder for the copolymer GT. They were obtained from Harlan-CPB (Zeist, The Netherlands). To obtain GT-primed T cells, rats were immunized with an equal amount of the copolymer L-glutamic acid : L-tyrosine 1:1 (GT; 500 μg; ICN Immuno-Biochemicals, Lisle, Israel; dissolved in 1 % Na_2CO3 in saline) and Freund's complete adjuvant, which was equally distributed over all four footpads. Peritoneal exudates were induced by the i.p. administration of Bacillus Calmette-Guérin (BCG, 5 x 10^7 cp dissolved in 5 ml PBS) or administration of Thioglycolate Broth (TG, 10 ml).

Cytology and Electron Microscopy

Cytocentrifuge preparations were processed for May-Grünwald/Giemsa (MGG) staining or for indirect immuno- and enzymcytochemistry (3). The following monoclonal antibodies (MoAb) were used: MRC Ox6 (6), specific for MHC class II antigens (Ia); MRC Ox19 and Ox52 (7), specific for T cells. Acid phosphatase was demonstrated as per Burnstone (3), and was double stained with Ia by sequential incubation. Peritoneal DC-fraction were processed for electron microscopy according to standard procedures (3).

Peritoneal Cell Suspensions, Isolated T Cells and DC, and Antigen Presentation Assay

Total peritoneal cell (TPC) suspensions were harvested from steady state rats, 1 day after BCG and 4 days after TG by peritoneal lavage (3). The non-adherent (NAC) and the adherent peritoneal cell fraction (AC) were obtained after 90 min of culture. Dendritic cells were isolated from peritoneal lavages 4 days after i.p. administration using an overnight adhesion culture step followed by a Nycodenz gradient (8).

GT-primed T cells were isolated 2 weeks after immunization from the lymph nodes of rats, using a Nylon wool column and subsequent incubation steps with the anti-Ia MoAb Ox6 and Dynabeads (9). In the antigen presentation assay 2 x 10^4 (TPC, NAC of AC) or 2 x 10^3 (DC) accessory cells were added to 2 x 10^5 T cells with or without GT (20 μg/well) in a total volume of 200 μl in flat-bottomed 96-wells microtiter plates. T cell proliferation was determined by measuring ^{3}H-thymidine incorporation. Results are given as the average counts per minute (cpm) for triplicate cultures $\pm$ sem.

RESULTS

Cellular Composition of Peritoneal Cell Fractions

On cytocentrifuge preparations DC were characterized by an irregular outline with many cell processes, a strongly Ia positive plasma membrane and an acid phosphatase spot of varying intensity close to the eccentrically localized bean-shaped nucleus. Macrophages also had an irregular outline but had less cell processes. Moreover only part of the Mϕ were moderately Ia positive (13 % in a steady state, 45 % 1 day after i.p. BCG and 15% 3 days after i.p. T.G.). Peritoneal Mϕ had a round to oval shaped nucleus which lies eccentrically, and all Mϕ had a strong acid phosphatase activity spread over the whole cytoplasm of the cell. The cellular composition of the TPC in a steady state and after BCG or TG are given in table 1.

Table 1. Absolute number (x 10^6) and percentage of macrophages (Mϕ), dendritic cells (DC), polymorphonuclear cells (PMN) and lymphocytes (Lym) present in the different peritoneal cell suspensions.

	Mϕ		DC		PMN		Lym	
TPC type	nr	%	nr	%	nr	%	nr	%
steady state	6.9	73	0.1	1	1.2	13	0.5	5
1 day after i.p. BCG	16.2	50	1.1	3	13.5	42	1.3	4
3 days after i.p. TG	43.1	77	1.0	2	11.2	20	1.7	3

Table 2 shows the total cell number present in the different peritoneal lavages and the number of DC that were isolated from these lavages. The DC fractions isolated from the peritoneal lavages were all more than 85 % pure. The number of DC isolated from the peritoneal cavity had slightly increased after culture of the TPC, but had largely increased after in vivo stimulation with BCG or TG.

Table 2. Total peritoneal cell number and number of isolated DC from a peritoneal lavage from steady state rats, 24 hr after culture, 1 day after i.p. Bacillus Calmette-Guérin (BCG), and 3 days after i.p. thioglycollate (TG).

peritoneal cells	total cell number (x 10^6)	number of isolated DC (x 10^4)
steady state	11.2	4.2
24 hr culture (in vitro)	-	6.7
1 day after i.p. BCG	32.4	33.7
3 days after i.p. TG	56.0	31.0

On ultrastructural level peritoneal DC had an irregular outline with many cell processes. Near the bean shaped nucleus a complex of cell organelles was present including a Golgi apparatus, lysosomes and other vesicles. At the periphery of the cell some strands of rough endoplasmic reticulum (RER) and a small number of mitochondria were present. The ultrastructural characteristics of peritoneal DC had not changed after i.p. administration of BCG or TG (Figure 1).

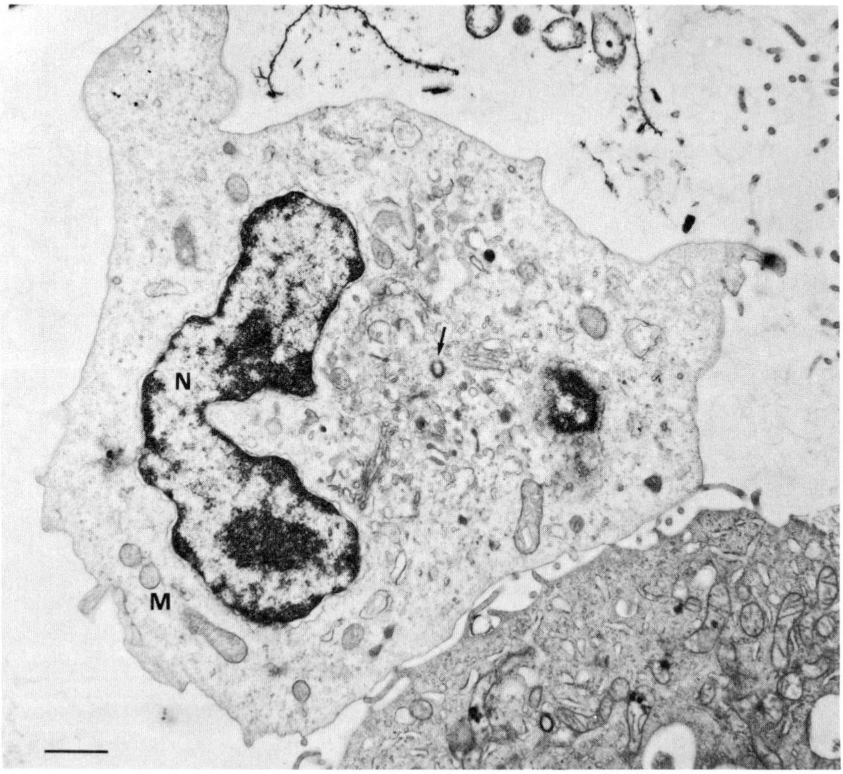

Figure 1. Electron microscopical photograph of a DC isolated from the peritoneal cavity of a rat 1 day after i.p. administration of BCG. The cell has an irregular outline with many cell processes, an eccentrically localized nucleus (N) and small mitochondria (M). Around the cytocentre (arrow) small lysosomes and some Golgi apparatusses were present. Bar = 1 μm.

Antigen Presentation

In the T cell fraction antigen presenting cells (APC) were no longer present because this fraction did not respond to the antigen GT and on cytocentrifuge preparations it appeared to be more than 98 % pure (data not shown).

Figure 2A shows the antigen presenting capacity of TPC, NAC (this fractions contained the DC) and AC (mainly Mϕ). In comparison with a steady state, the antigen presenting capacity of cultured TPC is slightly increased, whereas after TG and especially BCG it is decreased. After dividing the TPC in NAC and AC it appeared that

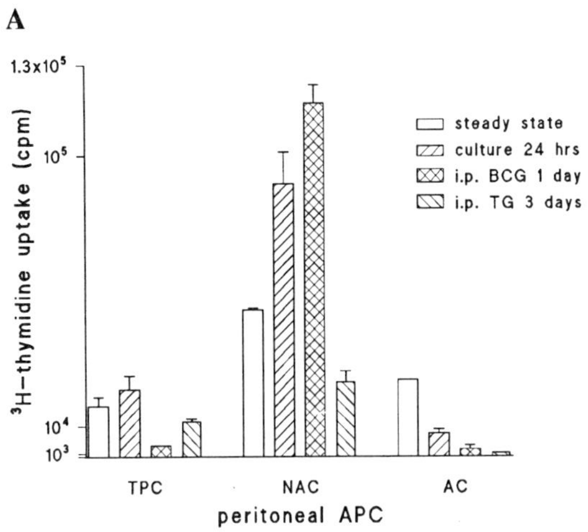

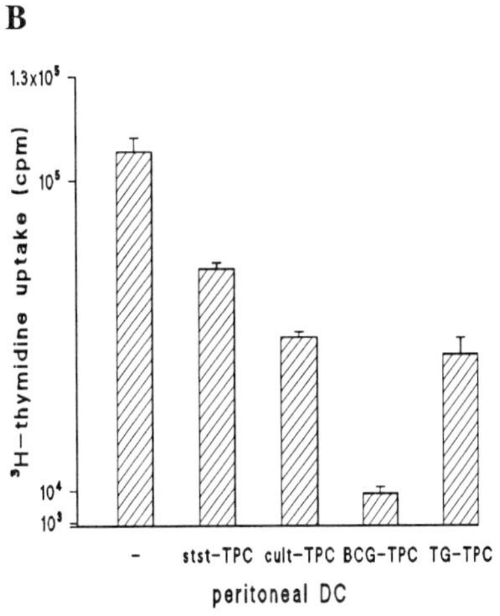

Figure 2. Capacity of different APC to present the copolymer GT to GT-primed T cells. A: APC were steady state, cultured, BCG- or TG-TPC, NAC or AC. B: Effect of addition of the different TPC on the T cell proliferation induced by peritoneal DC (isolated 4 days after i.p. administration of TG).

antigen presentation by the Mϕ always decreased, only slightly in the cultured cells and more in the BCG and TG-TPC. In contrast to the AC, the antigen presenting capacity of the NAC is increased after culture and BCG and only decreased after TG.

For all types of TPC tested it is obvious that the antigen presenting capacity of the TPC is not the sum of the NAC and AC, but is always lower. This might be explained by a suppressive effect of the AC, the Mϕ, on the antigen presentation by DC. This is shown in figure 2B, where addition of TPC on peritoneal DC always decreased the T cell proliferation induced by DC only. BCG-TPC appeared to have a very high suppressive effect in comparison with TG-TPC.

DISCUSSION

On grounds of morphological and functional similarities with lymphoid DC, we recently described the presence of a small number (approximately 1 %) in the peritoneal cavity of unstimulated rats (3). We also described an isolation method for DC, based on the method described by Knight et al. (8), resulting in a very pure DC fraction from either lymphoid organs (80 % pure) or the peritoneal cavity (85 % pure).

For lymphoid organs it had been demonstrated that the number of DC is increased after antigenic stimulation (10,11). It appeared that also in the non-lymphoid micro-environment of the peritoneal cavity the number of peritoneal DC is increased 24 hr after i.p. administration of BCG and 3 days after i.p. administration of TG. This increase is the result of an influx of cells, or a more rapid maturation of 'precursor DC', or both. The slightly increased number of isolated DC 24 after culture of peritoneal cells is indicative for the presence of a 'precursor DC', which does not yet has the typical DC phenotype.

Peritoneal DC are very potent antigen presenting cells, far more potent than peritoneal Mϕ, which have a suppressive rather than an antigen presenting function, especially after i.p. BCG. A suppressive function of Mϕ has also been described by others (12). The function of peritoneal Mϕ appears to differ and is related to the conditions in the peritoneal cavity. After BCG the Mϕ are hardly capable to present antigen to primed T cells and they are highly suppressive on T cell proliferation induced by the peritoneal DC. In contrast, TG-Mϕ are also poor in antigen presentation but are much less suppressive. The mechanism of suppression by peritoneal Mϕ is now in detail studied in our group.

We conclude that, as also described by others (13), in the peritoneal cavity the DC is the antigen presenting cells whereas the Mϕ has a more modulating function.

References

1. Inaba, K., J.P. Metlay, M.T. Crowley, and R.M. Steinman. Dendritic cells pulsed with protein antigens in vitro can prime antigen-specific, MHC-restricted T cells in situ. *J.Exp.Med.*172: 631 (1990).
2. Austyn, J.M. Lymphoid dendritic cells. *Immunol.*62: 161 (1987).
3. van Vugt, E., J.M.S. Arkema, M.A.M. Verdaasdonk, R.H.J. Beelen, and E.W.A. Kamperdijk. Morphological and functional characteristics of rat steady state peritoneal dendritic cells. *Immunobiol.*, in press (1992).
4. Bowers, W.E. and M.R. Berkowitz. Differentiation of dendritic cells in cultures of rat bone marrow cells. *J.Exp.Med.*163: 872 (1986).
5. Van Voorhis, W.C., M.D. Witmer, and R.M. Steinman. The phenotype of dendritic cells and macrophages. *Fed.Proc.*42: 3114 (1983).
6. McMaster, W.R., and A.F. Williams. Identification of Ia glycoproteins in rat thymus and purification from rat spleen. *Eur.J.Immunol.*9: 426 (1979).
7. Lawetzky, A., G. Tiefenthaler, R. Kubo and T. Hünig. Identification and characterization of rat T-cell subpopulations expressing T cell receptors α/β and γ/δ. *Eur.J.Immunol.*20: 343 (1990).
8. Knight, S.C., J. Farrant, A. Bryant, A.J. Edwards, S. Burman, A. Lever, J. Clarke, and A.D.B. Webster. Non-adherent, low-density cells from human peripheral blood contain dendritic cells and monocytes, both with veiled morphology. *Immunology* 57: 595 (1986).
9. Havenith C.E.G., A.J. Breedijk, M.A.M. Verdaasdonk, E.W.A. Kamperdijk and R.H.J. Beelen. An improved and rapid method for the isolation of rat lymph node or spleen T lymphocytes for T cell proliferation assays. *J.Immunol.Methods*, in press (1992).
10. Kamperdijk, E.W.A., M.L. Kapsenberg, M. van den Berg, and E.C.M. Hoefsmit. Characterization of dendritic cells, isolated from normal and stimulated lymph nodes of the rat.*Cell Tissue Res.*242: 469 (1985).
11. Kinnaird, A., S.W. Peters, J.R. Foster and I. Kimber. Dendritic cell accumulation in draining lymph nodes during the induction phase of contact allergy in mice. *Int.Arch. Allergy.Appl.Immunol.* 89: 202 (1989).
12. Travniczek, E., G. Boltz-Nitulescu, C. Holzinger, and O. Förster. Macrophages as regulatory cells in mitogen induced spleen cell proliferation. 1.Macrophages as suppressor cells. *Immunobiol.*168: 260 (1984).
13. Solbach, W., H. Moll and M. Röllinghoff. Lymphocytes play the music but the macrophage calls the tune. *Immunol. Today* 12: 4 (1991).

T-CELL REPERTOIRE DEVELOPMENT IN MHC CLASS II DEFICIENT HUMANS

Marja van Eggermond[1)], Marijke Lambert[1)], Françoise Mascart[2)], Etienne Dupont[2)] and Peter van den Elsen[1)]

[1)]Department of Immunohaematology and Bloodbank
University Hospital Leiden
P.O.Box 9600
2300 RC Leiden, The Netherlands

[2)]Department of Immunology, Haematology and Transfusion
Hôpital Erasme
Route de Lennik 808
B-1070 Brussels, Belgium

INTRODUCTION

We have investigated the impact of MHC class II deficiency on the development of the peripheral T-cell repertoire. During T-cell development, interactions between T-cell receptor (TcR) and MHC class I and class II molecules play a pivotal role in the positive and negative selection events that occur within the thymic micro-environment[1,2,3,4,5]. These MHC dependent selection processes influence in greater part the composition of the peripheral T-cell compartment, where only T-cells with moderate affinity for "self" MHC class I or class II molecules can be found[6]. Various studies in the mouse have shown that interference in these TcR/MHC interactions has a dramatic effect on the composition of the peripheral T-cell compartment with respect to development of T-cells co-expressing the accessory molecules CD4 or CD8 respectively[7,8,9,10].

In humans, the impact of MHC class II antigen expression on the selection of the peripheral T-cell repertoire can be studied in patients with the Bare Lymphocyte Syndrome (BLS, currently also referred to as MHC class II deficiency syndrome). In particular type III BLS patients lack detectable levels of MHC class II expression on cells which normally express these antigens including those involved in thymic differentiation[11,12].

We have analyzed the composition of the peripheral T-cell compartment in a family with type III BLS children with respect to the expression of the accessory molecules CD4 and CD8 and TcR V-gene segment usage within these distinct T-cell subsets. The results emerging from our studies have revealed that in a human

Dendritic Cells in Fundamental and Clinical Immunology,
Edited by Kamperdijk *et al.*, Plenum Press, New York, 1993

environment devoid of MHC class II antigen expression the numbers of peripheral $CD4^+CD8^-$ T-cells are greatly reduced. In addition, these $CD4^+CD8^-$ T-cells have an in vitro growth disadvantage compared to $CD4^+CD8^-$ T-cells derived from MHC class II expressing family controls[14]. In the MHC class II deficient children of this family most of the TcR V-gene segments tested for are employed within the $CD4^+CD8^-$ T-cell subset. However, skewing in the usage frequencies of some of the $V\alpha^-$ gene segments towards the $CD8^+CD4^-$ T-cell was noticeable in the MHC class II deficient patients that differed from those observed in MHC class II expressing family controls. On the basis of these findings it is concluded that in humans the lack of classical MHC class II antigen expression does not result in complete abrogation of the acquisition of the $CD4^+CD8^-$ T-cell phenotype nor to restriction in the development of the T-cell receptor repertoire with respect to TcR V-gene usage.

MATERIALS AND METHODS

$CD4^+CD8^-$ and $CD4^-CD8^+$ T-cell lines were derived from PBMC following FACS sorting. Isolated cell populations were stimulated by poly-clonal activation with 0.5% PHA (Welcome) and 100 u rIL-2 (Cetus) in the presence of irradiated allogenic feeder cells[12].

Cells were collected following Ficoll/Isopaque gradient centrifugation and total RNA was isolated by the RNAzol method (Cinna/Biotecx). An aliquot of the RNA preparation was converted into cDNA with reverse transcriptase and subjected to PCR amplification to determine the frequency of TcR $V\alpha$ and $V\beta$ gene segment usage as described[12]. Samples were drawn in the logarithmic phase of the PCR and analyzed by Southern blotting using ^{32}P-labelled $C\alpha$ and $C\beta$ probes. The membranes were exposed to Kodak XAR-5 films that were subsequently analyzed by densitometry to determine the relative usage frequencies of the various TcR $V\alpha$ and $V\beta$ gene segments as described[13].

RESULTS

In figure 1. the pedigree of family BI is shown. As can be seen in this family both type III Bare Lymphocyte Syndrome children (MBI and ABI) were HLA-identical, as determined by standard serological typing for MHC class I and by oligonucleotide typing for MHC class II, whereas the healthy siblings had a different HLA-haplotype.

Table 1. Expression of CD4 and CD8 antigens determined by staining of PBMC with a mixture PE-conjugated anti-CD8 and FITC-conjungated anti-CD4 antibody. Results are expressed as percentage of positive cells.

	KBI	ZBI	MBI	ABI
$CD4^+CD8^-$	52.8	47.9	22.6	37.0
$CD4^-CD8^+$	27.0	12.8	24.7	33.4
$CD4^+/CD8^+$ratio	2.0	3.7	0.9	1.1

Double staining of PBMC with anti-CD4 and anti-CD8 revealed that in both type III BLS siblings the numbers of peripheral $CD4^{+}CD8^{-}$ T-lymphocytes were reduced compared to the MHC class II expressing parental controls (Table 1).

Using the PCR and oligonucleotides specific for most of the to date known TcR V-gene segment families we have investigated in this type III BLS family whether the MHC class II deficient environment would have an impact in the use of the various TcR V-gene families at the transcriptional level within this greatly diminished peripheral $CD4^{+}CD8^{-}$ T-cell subset. The results of these PCR analyses are shown in figures 2 and

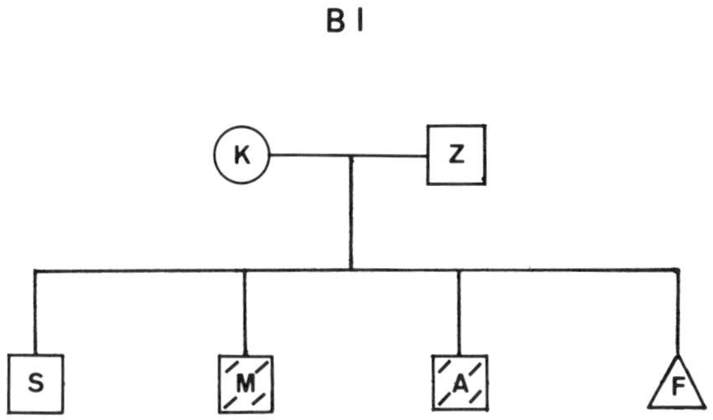

Figure 1.

Father ZBI	;	A2,11;B39,52;CW-;DR2(Dw21,Dw12);DQw1.
Mother KBI	;	A28,A26;B13,44;Cw6;DR7,4;DQw2,w3.
Sibling S	;	A28,11;B13,52;Cw6,w-;DR7,2;DQw2,w1.
BLS si.MBI	;	A28,2;B13,39;CW6,w-;DR7,2(Dw21);DQw2,w1.
BLS si.ABI	;	A28,2;B13,39;CW6,w-;DR7,2(Dw21);DQw2,w1.
Fetal FBI	;	A26,11;B44,52;Cw-;DR4,2;DQw2,w1.

3. As can be seen from figure 2 all $V\alpha$-gene family segments tested for are employed in the $CD4^{+}CD8^{-}$ as well as the $CD4^{-}CD8^{+}$ T-cell subsets of the MHC class II expressing family controls (ZBI, KBI and FBI). When compared, the relative usage frequencies of the various TcR $V\alpha$ gene families between the T-cell subsets differs whereas an individual specific pattern of TcR $V\alpha$ use can be discerned in both of the T-cell subsets. The type III BLS siblings (MBI and ABI) in this family do use the various TcR $V\alpha$ gene segments within the $CD4^{+}CD8^{-}$ and $CD4^{-}CD8^{+}$ T-cell subsets with the exception of $V\alpha$-12 that was used at extreme low frequencies in MBI but at normal frequencies in ABI. Despite of the fact that both type III BLS siblings in this familiy had an identical HLA background, the overall patterns of TcR $V\alpha$ usage were different. In particular the observed skewing of several TcR $V\alpha$-gene segments ($V\alpha$ 1,2,3,7,17 and 19) to the $CD4^{-}CD8^{+}$ T-cell subset in MBI, that was not noticeable in the MHC class II expressing controls nor in ABI was striking. This might be related to the clinical status of this patient when PBMC samples were taken for analysis.

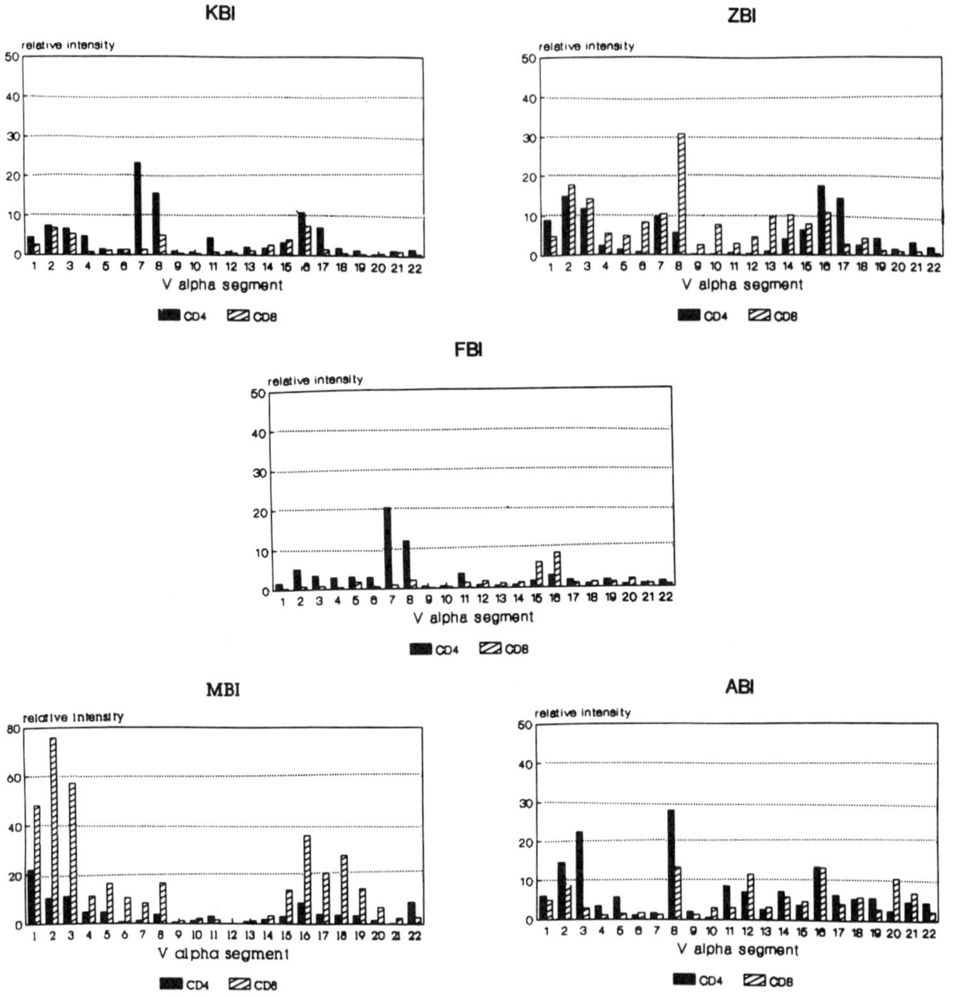

Figure 2. TcR Vα-gene segment usage in MHC class II expressing controls ZBI, KBI and FBI in comparison to TcR Vα-gene segment usage in type III BLS siblings MBI and ABI.

Likewise, for the frequencies of TcR Vβ-gene segment use by both type III Bare Lymphocyte Syndrome siblings in the $CD4^+CD8^-$ as well as $CD4^-CD8^+$ T-cell subsets, the TcR V-gene segments tested for were employed at similar frequencies compared to the MHC class II expressing controls[12]. In general, usage of the TcR Vβ gene segments by both MBI (not shown) and ABI (shown in figure 3) was less influenced by the MHC class II deficient environment compared to the usage frequencies of the various TcR Vα gene segments.

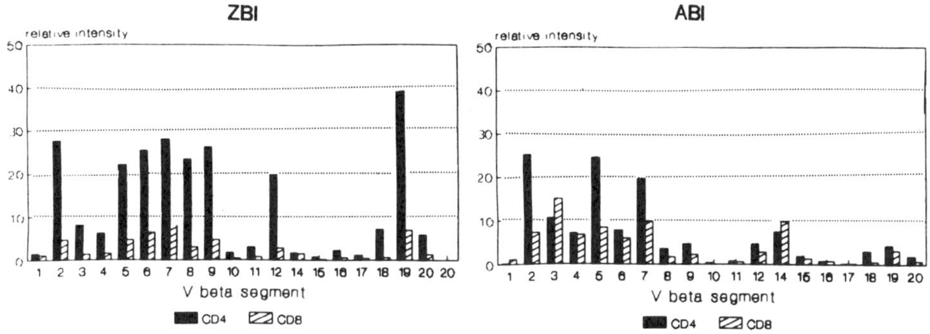

Figure 3. TcR Vβ gene segment usage in ZBI and BLS sibling ABI. The bars in the graph represent the relative quantity of the TcR Vβ-gene segments in the $CD4^+CD8^-$ and $CD4^-CD8^+$ subsets of T-lymphocytes as measured by densitometry.

DISCUSSION

We have analyzed the impact of an environment devoid of MHC class II antigen expression on the composition of the peripheral T-cell repertoire. We have shown by PCR analysis that these patients lack detectable levels of MHC class II transcription[12,14]. Noteworthy was the observation that $CD4^+CD8^-$ T-lymphocytes can be observed in the periphery albeit their numbers are reduced (see table I). However the various TcR V-gene segments within this impaired $CD4^+CD8^-$ T-cell subset are employed at the family level in normal frequencies in most cases indicating that the presumably MHC class II negative thymic micro-environment does not result in aberrant selection processes that have an impact on the use of the various TcR V-gene segments.

These observations were for a long time in seemingly contradiction with studies in the mouse that were aimed to define the role of the mouse MHC class II antigens on the peripheral T-cell compartment. Using monoclonal antibodies in neonates, it was demonstrated that these treatments resulted in a rapid and almost complete depletion of peripheral $CD4^+CD8^-$ T-lymphocytes[7]. However, recent experiments with MHC class II deficient mice as a result of disruptive homologous recombination, have indicated that despite the lack of the to-date known mouse MHC class II structures $CD4^+CD8^-$ T-cells can be found in the periphery, albeit at reduced numbers[9,10]. The TcR V-gene usage within this reduced $CD4^+CD8^-$ T-cell subset in these MHC class II deficient mice is not restricted as has been determined by the use of various TcR V-region specific monoclonal antibodies[9].

In conclusion, we have shown that in MHC class II deficient humans reduced numbers of $CD4^+CD8^-$ T-cells can be found in the periphery. These $CD4^+CD8^-$ T-lymphocytes are not restricted in their use of the various TcR V-gene segments indicating that the MHC class II deficient environment does not have a major impact on the generation of TcR-specificity with respect to TcR V-gene employment. Whether or not these $CD4^+CD8^-$ T-cells are selected via non-classical MHC class II structures or MHC class I expressed in the thymic micro-environment remains to be elucidated. Alternatively, these $CD4^+CD8^-$ T-cells might have escaped from the MHC class II mediated negative thymic selection processes.

ACKNOWLEDGEMENTS

The authors would like to thank mr.W.Verduyn for oligonucleotide typing and Prof.Dr.J.J.van Rood for his support. This research was supported in part by the Netherlands Organization for Research (NWO, grant H92-90 to P.v.d.E.), the Macropa Foundation and the J.A.Cohen Institute for Radiopathology and Radiation Protection (IRS).

REFERENCES

1. P.J.Fink and M.J.Bevan, H-2 antigens of the thymus determine lymphocyte specificity, J.Exp.Med. 148:766 (1978).
2. R.M.Zinkernagel, G.N.Callahan, A.Althage, S.Cooper, P.A.Klein, and J.Klein, On the thymus in differentiation of "H-2 self-recognition" by T-cells: evidence for dual recognition?, J.Exp.Med. 147:882 (1978).
3. J.Sprent, D.Lo, E.K.Gao, and Y.Ron, T cell selection in the thymus, Immunol.Rev. 101:173 (1988).
4. J.Nicolic-Zugic, and M.J.Bevan, Role of self peptides in positively selecting the T-cell repertoire, Nature 344:65 (1990).
5. H.S.Teh, A.M.Garvin, K.A.Forbush, D.A.Carlow, and R.M.Perlmutter, Participation of CD4 coreceptor molecules in T-cell repertoire selection, Nature. 349:241 (1991).
6. P.Marrack, D.Lo, R.Brinster, R.Palmiter, L.Burkly, R.H.Flavell, and J.Kappler, The effect of thymus environment on T cell development and tolerance, Cell 53:627 (1988).
7. A.M.Kruisbeek, J.J.Mond, B.J.Foulkes, J.A.Carmen, S.Bridges, and D.L.Long, Absence of the Lyt-2, L3T4+ lineage of T cells in mice treated neonatally with nti-I-A correlates with the absence of intrathymic I-A bearing antigen presenting cell function, J.Exp.Med.161:1029 (1985).
8. M.Zijlstra, M.Bix, N.E.Simister, J.M.Loring, D.H.Raulet, and R.Jaenisch, β2-microglobulin deficient mice lack CD4-8+ cytolytic T-cells, Nature 344:742 (1990).
9. D.Cosgrove, D.Gray, A.Dierich, J.Kaufman, M.Lemeur, C.Benoist, and D.Mathis, Mice lacking MHC class II molecules, Cell 66:1051 (1991).
10. M.J.Grusby, R.S.Johnson, V.E.Papaioannou, and L.H.Glimscher, Depletion of CD4+ T cells in major histocompatibility complex class II-deficient mice, Science 253:1417 (1991).
11. H.J.Schuurman, F.P.van den Wijgaert, J.Huber, R.K.B.Schuurman, B.J.M.Zegers, J.J.Roord, and L.Kater, The thymus in "Bare Lymphocyte" syndrome: significance of expression of major histocompatibility complex antigens on thymic epithelial cells in intrathymic T-cell maturation, Human Immunol. 13:69 (1985).
12. M.Lambert, M.van Eggermond, F.Mascart, E.Dupont, and P.van den Elsen, TcR Vα and Vβ-gene segment use in T-cell subcultutes derived from a type III bare lymphocyte patient deficient in MHC class II expression, Dev.Immunol. in press (1992).
13. G.Hawes, L.Struyk, and P.van den Elsen, T-cell receptor V-gene usage in T-cell subsets of monozygotic twins, submitted (1992).
14. M.Lambert, M.van Eggermond, M.Andrien, F.Mascart, E.Vamos, E.Dupont, and P.van den Elsen, Analysis of the peripheral T-cell compartment in the MHC class II deficiency syndrome, Res.Immunol.142:789 (1991).

RAT THYMIC DENDRITIC CELLS

Ed W.A. Kamperdijk, Joanne M.S. Arkema, Marina A.M. Verdaasdonk, Robert H.J. Beelen and Ellen van Vugt

Dept. Cell Biology, Medical Faculty, Vrije Universiteit, Van der Boechorststraat 7, 1081 BT Amsterdam, The Netherlands

INTRODUCTION

The aim of this study was to compare rat thymic dendritic cells in situ (i.e. interdigitating cells, IDC) with isolated dendritic cells (1) using enzyme cytochemical, immunocytochemical and electron-microscopical methods. Moreover the phenotypical and morphological changes of these isolated dendritic cells were studied after culture. To get more information about the influence of size (volume) and/or micro-environment on antigen presentation, we also compared the capacity to present GT to primed T cells by freshly isolated dendritic cells from thymus, spleen and peritoneal cavity.

MATERIALS AND METHODS

Male Wistar rats, weighing 130-150 gram were used for the characterization of the thymic dendritic cells in situ (IDC) and in vitro (DC). For antigen presentation experiments male ACI/Mal rats were used. These animals are high responder for the copolymer GT (glutamine-tyrosine) on a MHC class II (Ia) dependent manner (2).

Enzyme and immunocytochemistry IDC in situ. Frozen thymi were cut into cryosections and fixed in 100% acetone for 10 min. at 20°C. After overnight drying a combined Ia and acid phosphatase (APh) staining was performed.

Ultrastructure IDC in situ. Fixation of the thymus was performed using retrograde perfusion with 1.5% glutaraldehyde in 0.15 M phosphate buffer (pH 7.3) after a short prewash with Meyler's salt solution. Ultrathin sections were stained with uranylacetate and leadcitrate.

Enrichment for DC was performed as described earlier (3) using ficoll gradient, irradiation and rozetting. DC used for antigen presentation experiments were obtained by using a modified procedure originally developed by Knight et al (4) using Nycodenz gradients (5). Iscove's medium was used as culture medium.

Dendritic Cells in Fundamental and Clinical Immunology,
Edited by Kamperdijk *et al.*, Plenum Press, New York, 1993

Enzyme- and immunocytochemistry of DC for light microscopy

Acid phosphatase (APh) activity on frozen section (for INC) cytocentrifuge preparations (for DC) was demonstrated using naphthol-AS-BI-phosphate (Sigma) as substrate with hexazotized pararosaniline as diazonium salt. For Ia labelling an indirect immunoperoxidase method was used with OX-6 as primary antibody and rat anti mouse IgG/peroxidase as conjugate (DAKO).

Ultrastructure DC. Cells from enriched fractions were fixed in 1.5 % Glutaraldehyde in phosphate buffer overnight, followed by a postfixation in 1% OsO_4 in phosphate buffer for 1 hr. at 4°C. Thereafter the cells were pelleted in agar and embedded in an epon/araldite mixture. Ultrathin sections were stained with uranylacetate and leadcitrate.

GT primed proliferation assay. Antigen presentation by DC from thymus, spleen and peritoneal cavity was measured using ^{3}H-thymidine incorporation in co-cultures of the antigen presenting cell (DC), GT primed T-cell enriched lymph node cells and GT (6). To test the role of Ia antigens in antigen presentation blocking experiments were performed using anti Ia (OX6).

RESULTS AND DISCUSSION

Thymic IDC in situ

In the thymic cortico-medullary region (CMR) and medulla characteristic IDC were present. These Ia positive cells had an abundant cytoplasm with eccentrically localized nucleus. In contrast to macrophages which had APh activity throughout the whole

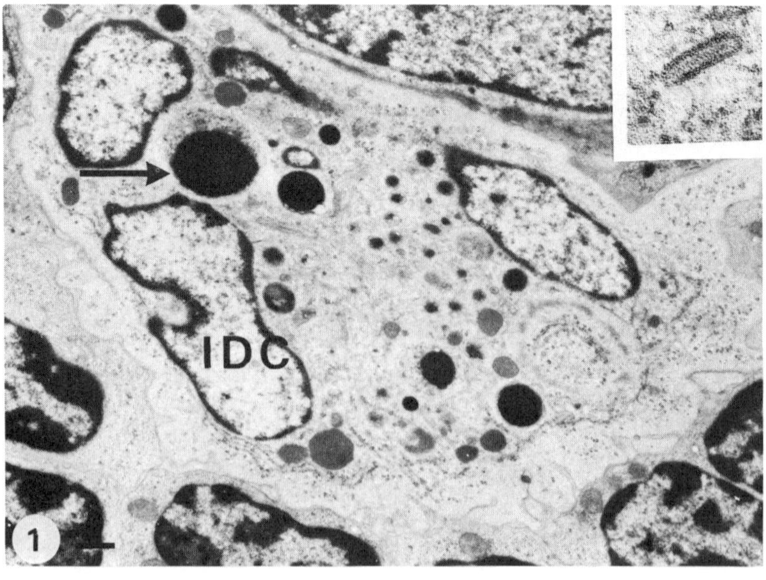

Fig.1 Ultrastructure phagocytic IDC. The electron lucent cytoplasm contains cellular debris (arrow). Inset Birbeck granulum. Bar: 1.8 μm

cytoplasm (7), they only showed this enzyme in a central spot of the cell. Ultrastructural studies revealed that they had a varying number of small and medium sized (phago)lysosomes in a central area. Occasionally cellular debris was present in these phagolysosomes (fig. 1). Only this type of cell (mean 60% of the cells) demonstrated Birbeck granules, the characteristic organelles of the epidermal Langerhans cells, suggesting a relationship between these different types of cell.

Freshly isolated dendritic cells

After separation of the low density cells, which were irradiated and EA rosetted to remove the contaminating lymphocytes and FcR positive macrophages, a highly enriched DC fraction (about 90%) was obtained. Cytocentrifuge preparations showed that they not only had dendritic cell processes but also persistently expressed MHC class II (Ia) antigens on the plasmamembrane with an APh spot of varying intensity and size in a juxtanuclear position (fig. 2). On ultrastructural level these cells had an irregular shaped outline with a nucleus with many indentations eccentrically localized. The cytoplasm contained mitochondria and strands of smooth and rough endoplasmic reticulum. Vacuoles were sometimes observed, especially near the cell membrane. The cells did not phagocytose, but frequently contained irregular shaped (phago)lysosomes in the centre of the cell (fig. 3). In many cases Birbeck granules were present (about 50% of the cells). Very occasionally in the rosetted fraction (i.e. the FcR positive cells) cells were shown containing Birbeck granules.

Thymic dendritic cells after culture

After culture in Iscove's medium for 5 days thymic DC were still recognizable by their irregular outline, APh activity in a juxtanuclear position and MHC class II (Ia) antigen expression. However the cells were larger and some of them showed phagocytic activity (mainly necrotic cells, fig. 4,5) as IDC in situ do. Large numbers of myelin figures were also present in the cytoplasm. In many cases these cells seemed to be multinucleated. Birbeck granules were not present any more. Addition of ^{3}H-thymidine in the culture medium never resulted in incorporation of this label in the nucleus of these cells as demonstrated by autoradiography. Moreover never mitotic activity was observed. Table 1 summarizes these results.

Table 1. Characteristics of IDC in situ, freshly isolated DC and DC after culture.

IDC (in situ)	DC freshly isolated	DC after culture
Ia pos	Ia pos	Ia pos
APh + (spot)	APh + (spot)	APh + (spot)
Bg ++ (60%)	Bg ++ (50%)	Bg −
Phagocytic activity ±	Phagocytic activity −	Phagocytic activity ±

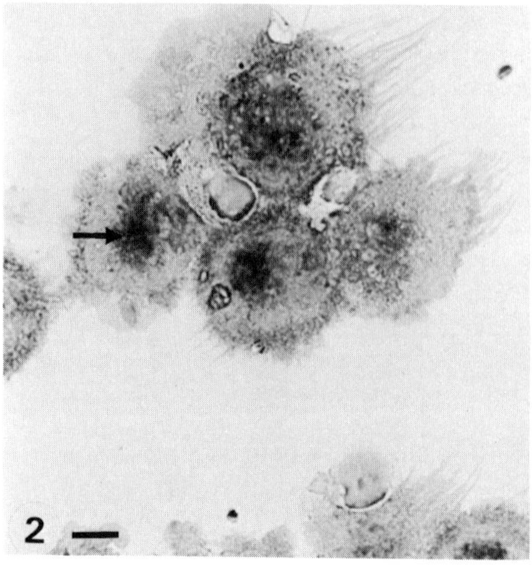

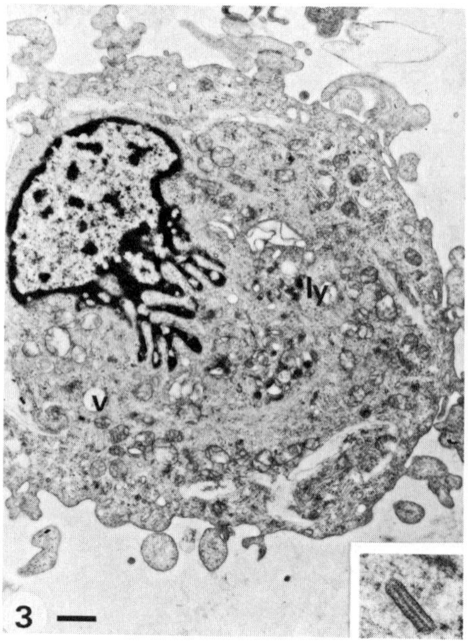

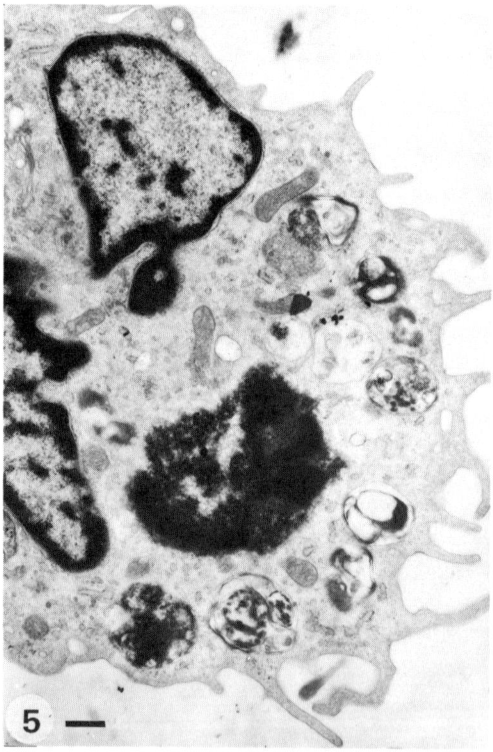

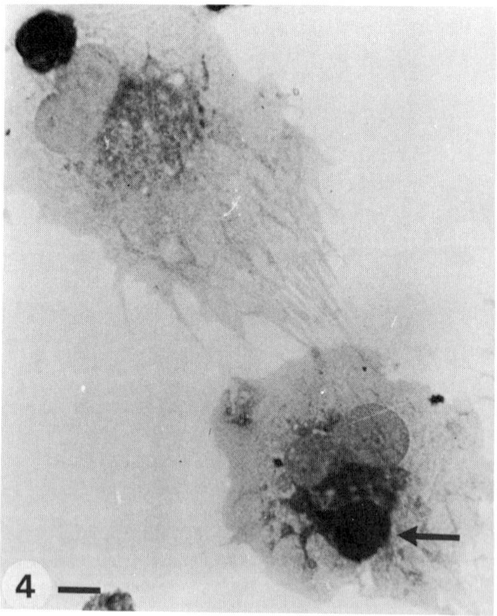

Fig.2 Cytocentrifuge preparation of freshly isolated DC. APh activity is present in a central spot (arrow). Bar: 8 μm.

Fig.3 Ultrastructure of freshly isolated DC. The cytoplasm contains relatively small lysosomes (ly) v: vacuole. Inset: Birbeck granule. Bar: 1.7 μm

Fig.4 Cytocentrifuge preparation DC, 5 days after culture. In the central area phagocytic activity (probably cellular debris) is present (arrow). Bar: 12.5 μm.

Fig.5 Ultrastructure phagocytic DC, 5 days after culture. Birbeck granules are absent. Bar: 2.9 μm.

Antigen presenting capacity of thymic DC, spleen DC and DC from the peritoneal cavity

As shown in fig.6 no distinct difference in the capacity to present GT to GT primed T cells was observed between thymic DC, spleen DC and peritoneal DC. In all cases even 20 DC gave a distinct response.
After adding anti Ia (Ox6) a mean 40% reduction in antigen presentation was observed.

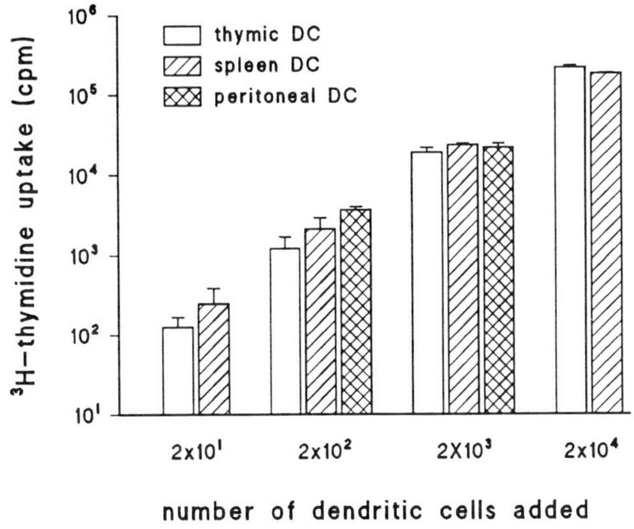

Fig.6 The capacity of thymic, spleen and peritoneal DC to present GT to GT primed lymph node cells.

From the results it can be concluded that:

a) there exists a relationship between thymic IDC and (isolated) DC as indicated by their MHC class II (Ia) antigen expression, characteristic APh activity localization and presence of Birbeck granules. However, IDC may show phagocytic activity, especially after irradiation of the thymus (8) which was not seen in freshly isolated DC. Probably phagocytic IDC are not selected during our enrichment procedure. For lymph nodes the relation between DC and a subpopulation IDC (the non actively phagocytozing cells) has already been described (9).
b) At least part of the thymic IDC are FcR positive similar to the epidermal Langerhans cells, since a (very small) number of rozetted cells in the high density fraction contained Birbeck granules.
c) During culture thymic DC may develop phagocytic activity, this indicates a relationship between these cells and cells from the monocyte/macrophage lineage.
d) The presence of Birbeck granules depends on the microenvironment in which the cells are present, since disappearance of those cell organelles is observed during culture.
e) During culture DC never showed mitotic activity.
f) Although there exists a difference in volume (size) of the DC (thymic DC are larger than spleen DC, Arkema et al, to be published) or microenvironment (the peritoneal cavity is a non lymphoid microenvironment) there is no significant difference in GT presenting capacity.

Acknowledgements: the authors wish to thank Mr. Paniry for preparing the photographs.

REFERENCES

1. R.M. Steinman and Z.A. Cohn. Identification of a novel celltype in peripheral lymphoid organs of mice. *J. Exp. Med.* 137, 1142-1162(1973)
2. D. Armerding, D.H. Katz and B. Benacerrof. Immune respons genes in inbred rats Analysis of responder status of the synthetic polypeptides and low doses of bovine serum albumin. *Immunogenetics,* 1 329-339(1974)
3. E.W.A. Kamperdijk, M.L. Kapsenberg, M. van den Berg, and E.C.M. Hoefsmit. Characterization of dendritic cells isolated from normal and stimulated lymph nodes of the rat. *Cell Tissue Res.* 242 469-474(1985)
4. S.C. Knight, J. Farrant, A. Bryant, A.J. Edwards, S. Burman, A. Lever, J.K. Clarke and A.D.B. Webster. Non-adherent, low density cells from human peripheral blood contain dendritic cells and monocytes, both with veiled morphology. *Immunology* 57, 595-603(1986)
5. E. van Vugt, J.M.S. Arkema, M.A.M. Verdaasdonk, R.J.H. Beelen and E.W.A. Kamperdijk. Morphological and functional characteristics of rat steady state peritoneal dendritic cells. *Immunobiol.* 184 14-24(1991)
6. E.W.A. Kamperdijk, M.A.M. Verdaasdonk and R.H.J. Beelen. Visual and functional expression of Ia antigens on macrophages and dendrtitic cells in ACI/Ma1 rats. *Transplant. Proc.* XIX 3024-3028(1987)
7. E.W.A. Kamperdijk, E.B.J. van Nieuwkerk, J.M.S. Arkema, A.M. Duijvestijn and E.C.M. Hoefsmit. Macrophages and dendritic cells. "Accessory Cells in HIV and other Retroviral infections." pp 136-144. Procs, Dijkstra C.D. and Gluckman J.C. (eds) Basel, Karger (1991)
8. A.M. Duijvestijn, Y.G. Köhler and E.C.M. Hoefsmit. Interdigitating cells and macrophages in the acute involuting rat thymus. An electronmicroscopic study on phagocytic activity on population development. *Cell Tissue Res.* 224 291-301(1982)
9. E.W.A. Kamperdijk, E.B.J. van Nieuwkerk, M.A.M. Verdaasdonk and E.C.M. Hoefsmit. Macrophages in different compartments of the non-neoplastic lymph node. *Curr. Topics in Path.* 84 219-245(1990)

RAT THYMIC DENDRITIC CELLS: FLOW CYTOMETRY ANALYSIS

María Purificacion Bañuls, Alberto Alvarez, Isabel Ferrero, Agustín Zapata and Carlos Ardavin

Department of Cell Biology
Faculty of Biology
Complutense University
28040 Madrid, Spain

INTRODUCTION

Thymic dendritic cells (DC) are considered to be involved in negative selection of T cells leading to the acquisition of self-tolerance[1,2], and many lines of evidence suggest that they may participate in the maturational process of thymocyte development[3], in the clonal amplification of mature medullary thymocytes[4], and in intrathymic presentation of non-MHC antigens[5,6]. Besides, recent reports analyzing the cell surface markers of murine thymic DC[7,8] have demonstrated that these cells express some interesting molecules, particularly the CD8 antigen. The expression of CD8 by thymic DC may promote a mechanism for elimination of self-reactive thymocytes, as Sambhara and Miller[9] have demonstrated that clonal deletion can be triggered if T cells that have been signaled through their TCR, also receive a second signal through the α_3 domain of their class I MHC molecules, which can be delivered by a CD8 molecule.

As a comparative study with regard to our previous studies on mouse thymic DC[7,8], we have performed a cell surface marker analysis of rat thymic DC by flow cytometry using a large panel of monoclonal antibodies.

MATERIALS AND METHODS

Five to six-week-old Wistar rats thymuses were cut into small fragments, digested in RPMI 1640 medium (Sigma Chemical Co., St Louis, MO) containing 2mg/ml collagenase (Boehringer-Mannheim, West Germany) for 15 min at 37°C, and the resulting cell suspension washed twice in RPMI 1640 medium. The cells were resuspended at $1x10^8$ cells/ml in 5% FCS-BSS containing 5mM EDTA (EDTA-BSS-FCS) to dissociate DC-thymocyte complexes. The cell suspension was distributed into round-bottomed 12 ml-centrifuge tubes, carefully layering 4 ml of cells above 4ml of metrizamide-HBSS solution (Nyegaard, Oslo, density= 1075 g/cm^3), and centrifuged at 1800g for 15 min at 4°C. The low density cell fraction was collected, resuspended in RPMI 1640 medium with 10% FCS at $5x10^7$ cells/ml, and incubated for 60 min at 37°C in a 5% CO_2-in-air gas phase. The nonadherent cells were removed by gentle washing with warm RPMI 1640 medium, and the adherent cells, i.e. DC and macrophages, cultured overnight in the same medium at 37°C with 5% CO_2. After this culture period, DC became nonadherent and can be collected from the culture supernatant, whereas most macrophages remained adherent.

Cells were stained using the mouse mAb listed in Table 1 followed by rabbit serum immunoglobulins, and using a phycoerythrin (PE)-conjugated rabbit F(ab')2 anti-mouse IgG (RAM-PE) as second stage (Southern Biotecnology, Birmingam, AL). All the staining steps were performed at 4°C, in EDTA-BSS-FCS, and propidium iodide (Sigma Chemicals Co., St Louis, MO) was included at 10μg/ml in the final wash to selectively stain dead cells. Flow cytometry analyses were performed using a FACScan flow cytometer (Becton Dickinson, Mountain View, CA). Dead cells were gated out on the basis of low forward scatter and high propidium iodide staining.

RESULTS

DC were isolated after collagenase digestion, on the basis of their low buoyant density, and their differential adherence characteristics. The DC-enriched preparation had a purity of >90% as assessed by phase contrast microscopy and flow cytometry. The contaminating population was essentially composed of thymocytes, macrophages constituting <1% of the final preparation. The forward (FS) versus side (SS) scatter profile of the DC-enriched preparation (data not shown) showed two clearly defined cell subpopulations: the majority subpopulation, corresponding to DC, displayed high FS and low-intermediate SS, whereas thymocytes had both low FS and SS. Gating for cells with light scatter characteristics of DC produced an increase in MHC class II expression from 90% to 98% of positive cells, indicating that the gated DC population had a purity of >98%, and that the contaminating population corresponded mainly to MHC class II-negative thymocytes. In all subsequent analyses for surface antigen expression, rat thymic DC were analyzed for phycoerythrin (PE) fluorescence by gating for the high FS, low-intermediate SS population; results are summarized in Table 1 and Fig. 1. The level of expression by rat thymic DC of a particular cell marker is sometimes difficult to determine, particularly when the staining profiles do not allow a clear definition of a positive and/or negative cell population. Therefore, an arbitrary cut-off of five times the background was used to define positive staining. PE-fluorescence intensity levels are indicated in Table 1.

Rat thymic DC expressed high-intermediate levels of MHC class I and class II molecules, of the adhesion molecules Mac-1, LFA-1 and ICAM-1, of the rat leukocyte antigen defined by the mAb OX44, and of the leukocyte common antigen CD45. They were negative for the T and B cell-specific forms of CD45, CD45R and CD45RA (B220), and for the B cell marker OX12. DC were positive for the rat macrophage and interdigitating cell (IDC) marker ED1, expressed low levels of ED2, and were negative for ED3.

Concerning T cell marker expression, rat thymic DC were negative for the mAb R73 specific for the rat TCR, but according to the criterion previously established, they expressed intermediate levels of CD2, CD4 and CD8. Moreover, about 50% of the cells expressed intermediate levels of the interleukin-2 receptor α chain, 80% expressed the T cell activation antigen recognized by the mAb OX48, and about 50% of DC were positive for CD5 with intermediate levels. Interestingly, the staining pattern for the Thy-1 molecule allowed a clear definition of two DC subsets according to their level of Thy-1 expression: about 60% displayed high levels of Thy-1, whereas 40% expressed intermediate levels of this T cell marker.

Table 1. Cell surface markers of rat thymic dendritic cells

Clone	Ig class	Specificity	% positive DC	PE-fluorescence[1]	Source[2]
OX18	IgG1	MHC class I	95-100	++	A
OX17	IgG1	MHC class II	95-100	+++	A
OX42	IgG2a	Mac-1	95-100	++	A
WT1	IgG2a	LFA-1	95-100	++	B
1A29	IgG1	ICAM-1	95-100	++	B
OX44	IgG1	OX44 antigen	95-100	++	A
OX1	IgG1	CD45	95-100	++	A
OX22	IgG1	CD45R	<5	-	A
OX33	IgG1	CD45RA (B220)	<5	-	A
OX12	IgG2a	Ig kappa chain	<5	-	A
ED1	IgG1	rat macrophages & IDC	70-90	++	C
ED2	IgG2a	rat macrophage subsets	30-50	+	C
ED3	IgG2a	rat macrophage subsets	<5	-	C
R73	IgG1	TCR	<5	-	D
OX55	IgG1	CD2	45-75	++	A
OX38	IgG2a	CD4	25-30	+/++	A
OX8	IgG1	CD8	30-40	++	A
OX39	IgG1	IL-2Rα	45-55	++	A
OX48	IgG1	OX48 antigen	75-85	++	E
OX19	IgG1	CD5	45-55	++	A
OX7	IgG1	Thy-1	30-45	++	A
			55-70	+++	

[1]Levels of PE-fluorescence are indicated as follows: -: negative, <5x background; +: low levels, 5-10x background; ++: intermediate levels, 10-100x background; +++: high levels, >100x background.
[2]A: Serotec, Oxford, UK; B: Products for Life Science, Tokyo, Japan; C: Dr. C. Dijkstra, Free University, Amsterdam, The Netherlands; D: Dr. T. Hunig, Ludwig-Maximilians University, Munich, FRG; E: Dr. A. Williams, University of Oxford, UK.

DISCUSSION

Rat thymic DC have been isolated after collagenase digestion and separation of the low density cell fraction, on the basis of their adherence properties, following essentially our previous work on murine thymic DC[7].

In agreement with previous reports on human and murine thymic and peripheral DC[10], rat thymic DC express MHC class I and class II molecules, the leukocyte common antigen CD45, and are negative for B cell markers as OX33 (rat CD45RA, B220) and OX12 (rat Ig kappa chain).

Concerning adhesion molecule expression, rat thymic DC express Mac-1, LFA-1 and ICAM-1. Mac-1 expression by DC is controversial: whereas human thymic DC have been described as Mac-1-negative[11], DC from human peripheral blood[12], murine thymus, spleen and lung[7,13,14] and rat lymph[15] have been considered to express this integrin molecule. These different results are likely related with the origin of the cells, but also with the methods used for analyzing their cell surface antigens, as previously suggested[7]. The adhesion molecules LFA-1 and ICAM-1, implicated in cluster formation involving heterotypic interactions between accessory cells and T lymphocytes[16], are also expressed by murine thymic DC[7,8], and peripheral DC[10].

According to previous results obtained after inmunoperoxidase staining[17], rat thymic DC express the rat macrophage and interdigitating cell marker ED1, and are low positive for the rat macrophage ED2 antigen, and do not express the ED3 macrophage marker.

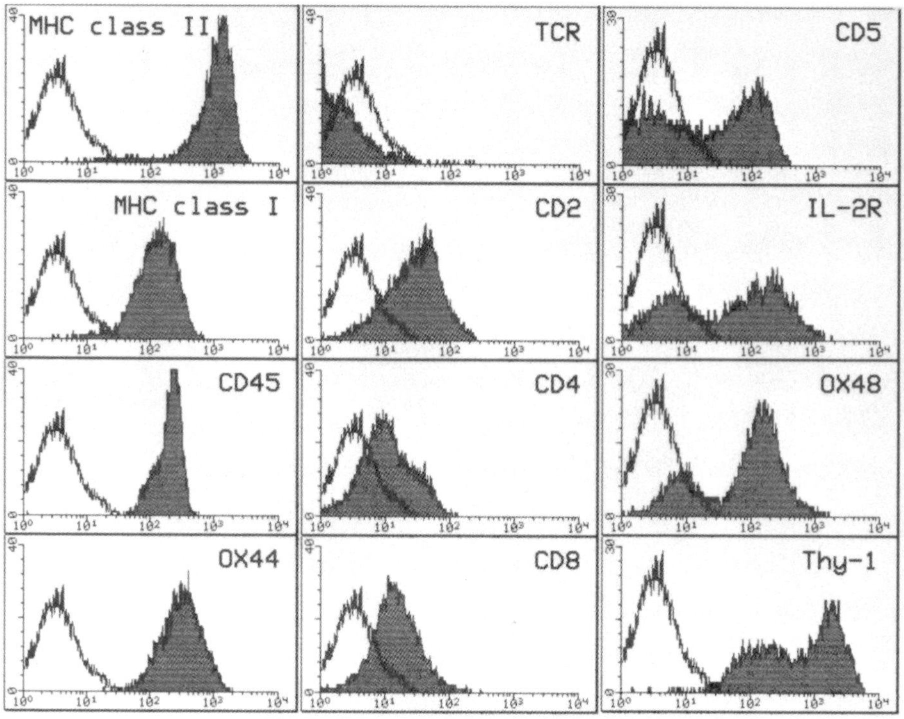

Figure 1. Cell surface phenotype of rat thymic DC. Grey profiles show the expression of the indicated marker by thymic DC; the black line indicates the background staining of rat thymic DC with RAM-PE. Frequency vs PE-fluorescence intensity are representative of 4 experiments with similar results; fluorescence intensity is represented in a 4-decade log_{10}-scale.

The expression by rat thymic DC of the OX44 antigen, also expressed by rat monocytes, B cells, and T cells in the late stages of differentiation, may be related to the suggested role of thymic DC in T cell maturation[3,4], as the OX44 molecule is a member of a family of proteins which appear to be involved in the regulation of cell proliferation[18].

Regarding T cell marker expression, rat thymic DC are negative for the rat TCR recognized by the mAb R73, but they express cell surface levels of CD2, CD4 and CD8 comparable to those of thymocytes. CD4 expression has been previously reported for human thymic DC and murine splenic DC[19,20], but this T cell marker does not appear to be expressed by murine thymic DC[7,8,13]. As previously suggested, CD8 expression by thymic DC could be involved in the elimination of autoreactive T cell clones and hence, in the induction of tolerance. In this sense, it has been demonstrated[9] that the CD8 molecule can deliver a signal to the α_3 domain of the class I MHC molecule, leading to clonal deletion of T cells, first signaled through their TCR.

IL-2R expression by rat thymic DC is in agreement with our previous results on murine DC using a similar isolation protocol[7], but not with other data obtained on murine splenic and thymic DC isolated whithout including a 37°C incubation[8,20], suggesting that this antigen can be upregulated in culture. MacPherson et al.[21] have shown that lymph DC bind recombinant IL-2 in a dose dependent manner, suggesting that IL-2R may be implicated in the primary activation of T cells, and therefore, in T cell development.

Although the physiologic function of the CD5 is still unclear, this antigen appears to be involved in T and B cell activation. In this regard, CD5 expression by DC, which has not been reported previously, could be related with an activated state of these cells, as recent studies on the cell cycle status of rat thymic DC, performed in our laboratory, (unpublished results) show that a higher proportion of thymic DC in G2/M phase is found within the CD5+ DC subset, than within the CD5-.

Concerning Thy-1 expression, our results are in agreement with previous results on rat lymph and splenic DC[15,22], and interestingly, thymic macrophages[23] and epithelial cells[24] appear to be Thy-1+. The expression of the Thy-1 antigen on thymic stromal cells may be involved in T cell differentiation by promoting a mechanism of interaction with thymocytes, as this molecule appear to be implicated in T cell activation and is known to support the adhesion of mouse thymocytes to thymic epithelial cells[25].

Acknowledgments

We are extremely grateful to Dr. A. Silva (CSIC, Madrid, Spain), Dr. T. Hünig (Ludwig-Maximilians University, Munich, FRG) and Dr. C. Dijkstra (Free University, Amsterdam, The Netherlands) for providing some of the mAb used in this study.

REFERENCES

1. P. Matzinger and S. Guerder, Does T-cell tolerance require a dedicated antigen presenting cell?, *Nature* 338:74 (1989).
2. F. Ramsdell and B.J. Fowlkes, Clonal deletion versus clonal anergy: the role of the thymus in inducing self tolerance, *Science* 248:1342 (1990).
3. M. Lafontaine, D. Landry, N. Blanc-Brunât, M. Pelletier, and S. Montplaisir, IL-1 production by human thymic dendritic cells: studies on the interrelation with accessory function, *Cell.Immunol.* 135:431 (1991).
4. D. Landry, L. Doyon, J. Poudrier, M. Lafontaine, M. Pelletier, and S. Montplaisir, Accessory function of human thymic dendritic cells in ConA induced proliferation of autologous thymocyte subsets, *J. Immunol.* 1990. 144: 836.
5. B.A. Kyewski, C.G. Fathman, and R.V. Rouse, Intrathymic presentation of circulating non-MHC antigens by medullary dendritic cells, *J. Exp. Med.* 163:231 (1986).
6. M. Zöller, Intrathymic presentation by dendritic cells and macrophages: their role in selecting T cells with specificity for internal and external nominal antigen, *Immunology* 74:407 (1991).
7. C. Ardavin and K. Shortman, Cell surface marker analysis of mouse thymic dendritic cells, *Eur. J. Immunol.* 22:859 (1992).
8. D. Vremec, M. Zorbas, R. Scollay, D. Saunders, C. Ardavin, L. Wu, and K. Shortman, The surface phenotype of dendritic cells purified from mouse thymus and spleen: investigation of the CD8 expression by a subpopulation of dendritic cells, *J. Exp. Med.* In press.
9. S.R. Sambhara and R.G. Miller, Programmed cell death of T cells signaled by the T cell receptor and the $\alpha3$ domain of class I MHC, *Science* 252:1424 (1991).
10. R.M. Steinman, The dendritic cell system and its role in immunogenicity, *Annu. Rev. Immunol.* 9:271 (1991).
11. D. Landry, M. Lafontaine, M. Cossette, H. Barthelemy, C. Chartrand, S. Montplaisir, and Pelletier, Human thymic dendritic cells. Characterization, isolation and functional assays, *Immunology* 65:135 (1988).
12. P.S. Freudenthal and R.M. Steinman, The distinct surface of human blood dendritic cells, as observed after an improved isolation method, *Proc. Natl. Acad. Sci. USA* 87:7698 (1990).
13. M. Crowley, K. Inaba, M. Witmer-Pack, and R.M. Steinman, The cell surface of mouse dendritic cells: FACS analysis of dendritic cells from different tissues including thymus, *Cell. Immunol.* 118:108 (1989).
14. A.M. Pollard and M.F. Lipscomb, Characterization of murine lung dendritic cells: similarities to Langerhans cells and thymic dendritic cells, *J. Exp. Med.* 172:159 (1990).
15. G.C. MacPherson, Properties of lymph-borne (veiled) dendritic cells in culture I. modulation of phenotype, survival and function: partial dependence on GM-CSF, *Immunology* 68:102 (1989).
16. G.A. van Seventer, Y. Shimizu, and S. Shaws, Roles of multiple accessory molecules in T-cell activation, *Curr. Opin. Immunol.* 3:294 (1991).
17. J.G. Damoiseaux, E.A. Dopp, J.J. Neefjes, R.H. Beelen, and C.D. Dijkstra, *J. Leucocyte Biol.* 46:556 (1989).

18. A. Bellacosa and P.A. Lazo, The rat leukocyte antigen MRC OX-44 is a member of a new family of cell surface proteins which appear to be involved in growth regulation, *Mol. Cell. Biol.* 11:2864 (1191).
19. D. Landry, M. Lafontaine, H. Barthelemy, N. Paquette, C. Chartrand, M. Pelletier, and S. Montplaisir, Human thymic dendritic cell-thymocyte association: ultrastructural cell phenotype analysis, *Eur. J. Immnunol.* 19:1855 (1989).
20. M.T. Crowley, K. Inaba, M.D. Witmer-Pack, S. Gezelter, and R.M. Steinman, Use of the fluorescence activated cell sorter to enrich dendritic cells from mouse spleen, *J. Immunol. Methods* 133:55 (1990).
21. G.G. MacPherson, S. Fossum, and B. Harrison, Properties of lymph-borne (veiled) dendritic cells in culture II. Expression of the IL-2 receptor: role of GM-CSF, *Immunology* 68:108 (1989).
22. S.M. Setum, J.R. Serie, and O.D. Hegre, Comparative analysis of potency of splenic dendritic and adherent cell (macrophages) as alloantigen presenters in vivo, *Diabetes* 40:1719 (1991).
23. R. Navarro, C. Ardavin, A.M. Fontecha, A. Alvarez, and A. Zapata, In vitro characterization of rat thymic macrophages, *Immunology* 73:114 (1991).
24. C.L. Tucek and R.L. Boyd, Surface expression of CD4 and Thy-1 on mouse thymic stromal cells, *Int. Immunol.* 2:593 (1990).
25. H. He, P. Naquet, D. Caillot, and M.J. Pierres, Thy-1 supports adhesion of mouse thymocytes to thymic epithelial cells through a Ca^{2+}-independent mechanism, *J. Exp. Med.* 1991. 173: 515.

ULTRASTRUCTURE OF INTERDIGITATING CELLS IN THE RAT THYMUS DURING CYCLOSPORIN A TREATMENT

Eric J. de Waal,[1] Louk H.P.M. Rademakers,[2] Henk-Jan Schuurman,[1,2] and Henk van Loveren[2]

[1] National Institute of Public Health and Environmental Protection, P.O. Box 1, 3720 BA Bilthoven
[2] Departments of Pathology and Internal Medicine, University Hospital, P.O. Box 85.500, 3508 GA Utrecht, The Netherlands

INTRODUCTION

Cyclosporin A (CsA) is well-known for its immunosuppressive characteristics.[1] A peculiar activity ascribed to the compound is the induction of auto-immune phenomena resembling graft-versus-host disease. This emerges in rodents after a short-course CsA treatment during the recovery phase after lethal irradiation and syngeneic bone marrow transplantation.[2] A role for the thymus has been implicated in this so-called syngeneic graft-versus-host disease.[3] Histology of the thymus after CsA treatment shows the almost absence of medullary areas.[4,5] The normal medulla is characterized by the presence of medium-sized lymphocytes with the medullary T-cell immunophenotype, and by the microenvironment consisting of medulla-type epithelium and interdigitating cells (IDC). Besides some small remnants, these components of the medulla are lacking in animals after a two-week period of CsA treatment. The disappearance of medullary epithelium and IDC has been related to a disturbance of negative selection of lymphocytes, resulting in the export of potentially autoreactive cells, that give auto-immune phenomena in the periphery.

The disappearance of IDC has been documented by the absence of immunoreactivity for MHC class II antigen. It is not known whether this loss of MHC class II immunolabeling on tissue sections is due to the actual disappearance of IDC, or due to the down-regulation of MHC class II on persistent cells. This question prompted us to investigate the presence of IDC in the thymic medulla of rats after CsA treatment at the ultrastructural level.

MATERIAL AND METHODS

This study included nine male WAG/CPB ($RT1^u$) rats, at 4-8 weeks of age, that were divided in three groups. The first two groups (6 rats) were treated with daily subcutaneous injections of CsA (15 mg/kg body weight) that was prepared from Sandimmun stock solution (Sandoz, Basel, Switzerland). Animals in the first group (n=3) were sacrificed

Dendritic Cells in Fundamental and Clinical Immunology,
Edited by Kamperdijk *et al.*, Plenum Press, New York, 1993

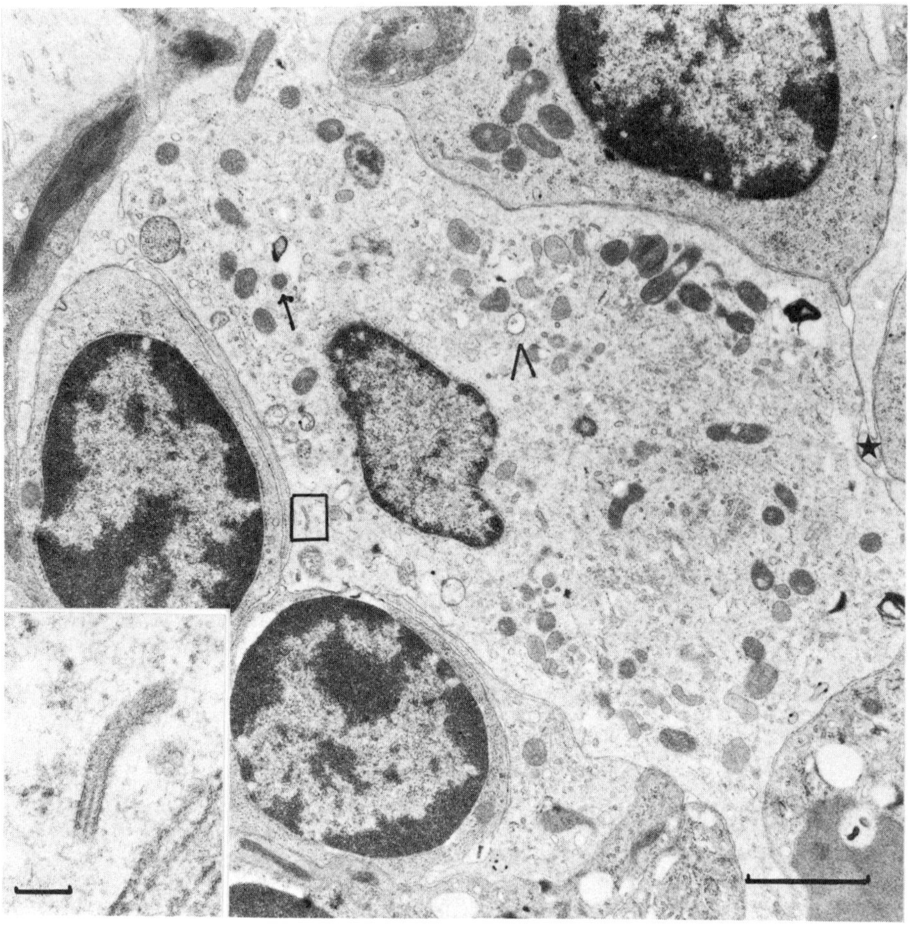

Figure 1. Electron micrograph of the normal rat thymic medulla. An IDC is presented showing abundant electron-lucent cytoplasm, inconspicuous membrane intrerdigitations (asterisk), electron-lucent vesicles with an internal coating containing amorphous material (arrowhead), and Birbeck granules (arrow). Tiny cytoplasmic projections interdigitate with surrounding cells. Bar, 1.5 μm.
Inset. Detail showing a Birbeck granule near the plasma membrane. Bar, 100 nm.

at day 7 after treatment, and those in the second group (n=3) at day 14 after treatment. Animals in the third group (n=3) served as control. They were treated with daily subcutaneous injections of the Sandimmun vehicle (1 ml/100 g body weight, consisting of 650 g/l polyoxyethylated ricinus oil [Cremophor] and 33% ethanol) and sacrificed at day 14.

At sacrifice, thymus was sampled. Small pieces (about 1 mm^3) of one lobe were fixed for at least 2 hours at 4°C, by immersion in 0.1 M Na-cacodylate buffer, pH7.4, containing 2.5% (v/v) glutaraldehyde, 2% (w/v) paraformaldehyde, 250 μM $CaCl_2$, and 500 μM $MgCl_2$. Postfixation was done in 1% (w/v) osmiumtetroxide in 0.1 M Na-cacodylate buffer, pH7.4, supplemented with 500 μM $CaCl_2$ and 1000 μM $MgCl_2$. Then, the specimens were dehydrated and embedded in Epon 812. For two specimens from each individual thymus, we selected medullary areas in light microscopy of 1 μm-thick sections stained with pararosaniline. Ultrathin sections (50 nm) were contrasted with uranyl acetate and lead citrate, and read using a Philips 201c electron microscope.

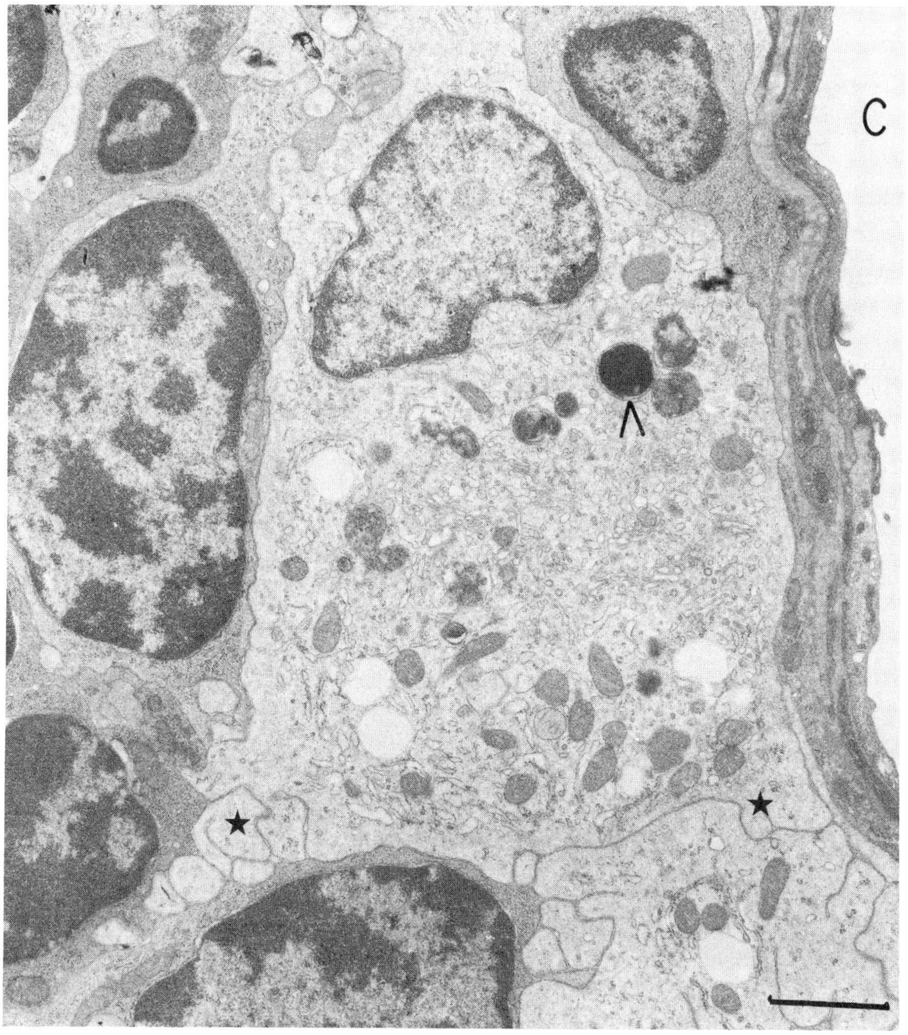

Figure 2. Thymus at day 7 of CsA treatment. The IDC depicted shows abundant electron-lucent cytoplasm and inconspicuous membrane interdigitations (asterisk). Arrowhead: dense bodies. The cell is located near a blood capillary (C). Bar, 1.5 μm.

RESULTS AND DISCUSSION

In the normal thymus medulla, IDC are present in high density (Fig. 1). The nucleus is located excentrically and often is indented. The large amount of cytoplasm is electron-lucent, and manifests electron-lucent vesicles with a flocculent internal coating. In some cells electron-dense bodies are observed. Birbeck granules (inset of Fig. 1), representing a general characteristic of IDC, are present in moderate numbers, in the cytocentre and in distal cell projections. These observations are in accord with descriptions in the literature.[6]

Histology of the thymus at day 7 and 14 after the start of CsA treatment showed an almost total reduction of the medulla. Only small medullary remnants were still present.

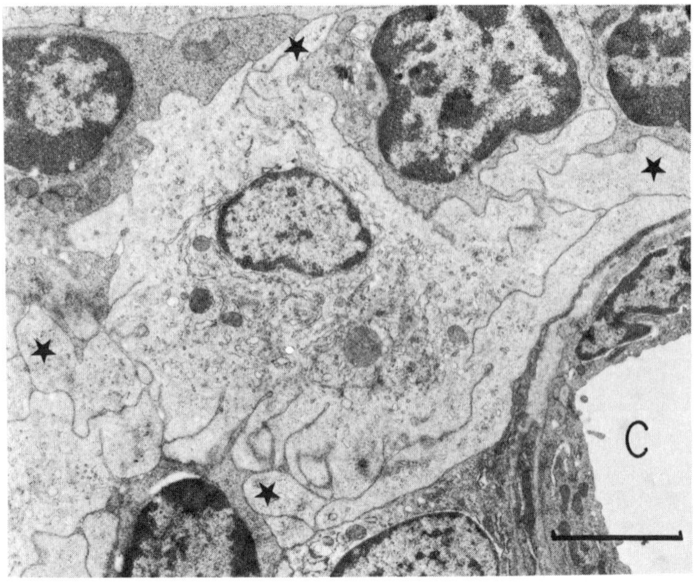

Figure 3. Thymus at day 14 of CsA treatment. The IDC shown is located near a blood capillary (C). The cell contains only a few cytoplasmic organelles, which contrasts to IDC in normal control thymus. Membrane interdigitations are well developed (asterisks). There are no Birbeck granules detectable. Bar, 2.5 μm.

In electron microscopy, sections of thymus at day 7 manifested in these areas a few IDC, mainly around capillaries (Fig. 2). The cells had inconspicuous membrane interdigitations. We were unable to identify Birbeck granules in these cells. Also coated vesicles containing amorphous or membranous material were not present. Only incidentally we observed dense bodies in IDC. This ultrastructural appearance is indicative of an immature stage of differentiation. Together with the tissue location, this suggests that the cells had just migrated into the thymus.

In sections of thymus at day 14 (Fig. 3), only small numbers of IDC were observed, located in foci around blood capillaries and solitary at scattered location. Their numbers were somewhat higher than those in the thymus at day 7. The ultrastructural morphology of these cells was similar to that of IDC in the normal thymus, except for the absence of Birbeck granules and coated vesicles containing amorphous or membranous material. The cells showed well developed interdigitations, and some contained numerous dense bodies. These ultrastructural observations indicate the normalization of the IDC population with regard to the situation at day 7 of CsA treatment.

From this data we conclude that CsA actually induces the disappearance of IDC from the thymus, and does not down-regulate marker substances on persisting cells. This conclusion fits with hypotheses in the literature.[7] It is tempting to hypothesize on the toxic effect of CsA on the thymus in the following way.[8] At day 7 of treatment, IDC have completely disappeared from the medulla. The very few IDC found are immature and may have just migrated into the organ. At day 14 of treatment, the recovery is more prominent, as indicated by the somewhat larger (but still very low) numbers of cells, and by the higher development of cytoplasmic organisation. We elsewhere have documented that IDC at other locations of the body such as spleen, i.e. in another local microenvironment, apparently are not vulnerable to CsA action.[5] Following this data, a direct action of CsA on IDC seems unlikely. Rather, the cells may be affected by changes in the local microenvironment.

REFERENCES

1. B.D. Kahan, Cyclosporine, *N. Engl. J. Med.* 321:1725 (1989).
2. A. Glazier, P.J. Tutschka, E.R. Farmer, and G.W. Santos, Graft-versus-host disease in Cyclosporin A-treated rats after syngeneic and autologous bone marrow reconstitution, *J. Exp. Med.* 158:1 (1983).
3. S. Sakaguchi, and N. Sakaguchi, Thymus and autoimmunity. Transplantation of the thymus from cyclosporin A-treated mice causes organ-specific autoimmune disease in athymic nude mice, *J. Exp. Med.* 167:1479 (1988).
4. W.E. Beschorner, J.D. Namnoun, A.D. Hess, C.A. Shinn, and G.W. Santos, Cyclosporin A and the thymus. Immunopathology, *Am. J. Pathol.* 126:487 (1987).
5. H.-J. Schuurman, H. van Loveren, J. Rozing, A. van Dijk, J.G. Loeber, and J.G. Vos, Cyclosporin and the rat thymus. An immunohistochemical study, *Thymus* 16:235 (1990).
6. A.M. Duyvestijn, Y.G. Köhler, and E.C.M. Hoefsmit, Interdigitating cells and macrophages in the acute involuting rat thymus. An electron-microscopic study on phagocytic activity and population development, *Cell Tissue Res.* 224:291 (1982).
7. W.E. Beschorner, and A.O. Armas, Loss of medullary dendritic cells in the thymus after cyclosporine and irradiation, *Cell. Immunol.* 132:505 (1991).
8. E.J. de Waal, L.H.P.M. Rademakers, H.-J. Schuurman, and H. van Loveren, Interdigitating cells in the rat thymus during Cyclosporin A treatment. Ultrastructural observations, *Thymus* in press (1992).

T CELL TOLERANCE AND ANTIGEN PRESENTING CELL FUNCTION IN THE THYMUS

David J. Izon, John D. Nieland, Lori A. Jones and
Ada M. Kruisbeek

Division of Immunology
The Netherlands Cancer Institute
Plesmanlaan 121
1066 CX Amsterdam
The Netherlands

INTRODUCTION

The repertoire of T cell receptor specificities is shaped by two processes that occur in the thymus, positive and negative selection (1-5). Positive selection is responsible for generating a T cell repertoire that has the ability to recognize antigenic peptides in association with self-major histocompatibility complex (MHC) molecules. The other process is called negative selection, and assures that tolerance for self-antigens is achieved. During negative selection, potentially autoreactive T cells (i.e. those with high affinity for "self" antigens presented on self-MHC molecules) are actually deleted from the T cell repertoire (2, 6, 7) or are clonally inactivated (1, 8). "Self" is defined here as those self or foreign antigens which are present in the thymus at the moment of selection, a prime moment for which appears to be the early neonatal period: neonatally thymectomized mice develop a variety of tissue-specific autoimmune diseases later in life (9-11). Although factors controlling these autoimmune diseases are poorly understood, defects in clonal inactivation and clonal deletion may induce autoimmune diseases (9, 11, 12).

While it is likely that multiple known and unknown adhesion molecules and perhaps tissue-specific receptors are involved in the cellular interactions that dictate intrathymic selection steps, known cell surface molecules demonstrated to participate include CD2, CD4, CD8, the CD3-TCR complex and LFA-1 (on the T cell side), and class I-and II MHC antigens, LFA-3 and ICAM (on the side of the antigen presenting cells) (3, 4, 13). As these molecules are involved in antigen-specific activation of T cells as well, one of the most puzzling questions immunologists have to face is how such different outcomes can result from seemingly similar interactions: peripheral

Dendritic Cells in Fundamental and Clinical Immunology,
Edited by Kamperdijk *et al.*, Plenum Press, New York, 1993

interactions between mature T cells and foreign antigenic peptides/self MHC complexes result in functional activation of T cells, while intrathymic interactions between immature T cells and self-peptide/self-MHC complexes result in either positive selection or negative selection. While one could postulate that the difference lies at the level of the T cells (i.e., thymocytes may respond differently than peripheral T cells to TCR-mediated activation), that still leaves the question of how positive and negative selection are distinguished. The current view holds that positive selection is mainly mediated through interactions with thymic epithelial cells (TE), while negative selection would be a consequence of interactions with cells of hemopoietic origin, such as dendritic cells (1, 4, 8, 13, 14). Yet, expression of certain self-antigens on TE cells may result in immunological tolerance, albeit not through the induction of clonal deletion, but through induction of specific clonal nonresponsiveness (1, 4, 14). Predictions regarding the consequences of expression of self-antigens on B cells cannot easily be made, either: both clonal deletion as well as clonal anergy have been suggested to occur, and may be dependent in part on whether the presenting B cells are activated or not (15, 16). Activation of B cells induces costimulatory activity, including expression of B7 (see below), while resting B cells have no costimulatory activity. The present review summarizes how differences in the outcome of interactions between developing T cells and antigen presenting cells (APC) in the thymus may be dictated by differences in APC function, most notably by those properties that result in differences in costimulatory ability.

NONRESPONSIVENESS THROUGH ABSENCE OF COSTIMULATORY ACTIVITY

As mentioned above, T cells with self-reactive antigen receptors are usually removed through intrathymic clonal deletion (1, 2). If, however, that process fails for any number of reasons (i.e., a self-antigen is not expressed in the thymus, or expressed on non-hemopoietic antigen presenting cells, APC), other safeguards are built in: *in vivo* clonal non-responsiveness of T cells (anergy) with self-reactive receptors was demonstrated in a variety of models in which clonal deletion was precluded (1, 8, 12). Distinctions between clonal anergy and clonal deletion are described elsewhere in this volume (Jenkins et al, "Costimulatory factors and signals relevant for antigen presenting cell function "). Briefly, anergy is the result of absence of so-called "costimulatory signals" in situations in which normal TCR/MHC-peptide interactions occur. This rather ill-defined costimulatory activity is derived from APC: class II-MHC restricted T cells stimulated with antigenic peptides and metabolically inactive or non-bone marrow derived APC become anergic, rather than activated, and this function can be reconstituted by nonfixed APC expressing irrelevant class II MHC (17-19). Thus, one of the ways in which *in vivo* tolerance for self-peptides may be maintained is through recognition of self-peptides on APC incapable of providing costimulatory signals.

The best candidate costimulatory signal that determines whether TCR-occupancy leads to a productive immune response is that generated by the interaction of CD28 (on T cells) and its ligand B7 (on APC) (20-24): anti-CD28 crosslinking is the only intervention identified today that can replace APC, and bypass anergy induction. Furthermore, mAb against human B7 blocks generation of co-stimulatory signals in human B cells. Obviously, blocking of CD28-B7 interactions might present a powerful immunosuppressive intervention, with the potential to control undesirable immune responses against self-antigens or transplantation antigens.

T cell repertoire selection can also be anticipated to be powerfully affected by presence of B7 and other costimulatory molecules on thymic selecting elements: interactions of developing T cells with thymic elements that do not express B7 may induce anergy, and while the activation requirements leading to clonal deletion are unknown, the recent description of participation of LFA-1/ICAM interactions in induction of clonal deletion (25) suggests that other molecules, such as B7, may be involved as well. In this context, we therefore analyzed the expression and function of mouse B7 and several other lineage specific markers on thymic stroma cells possibly

Table 1. Reagents used to subset thymic stromal cells by surface staining.

mAb	reactivity
CDR1	Thymic cortical epithelium
G8.8a	Thymic medullary epithelium
33D1	Spenic dendritic cells
NLDC 145	Thymic and splenic dendritic cells Thymic cortical epithelium
Mac-1	Macrophages
B220	B cells
CD5	Subset of B cells

involved in T cell repertoire selection, with a special focus on possible consequences for T cell tolerance acquisition. Recently developed chimeric proteins (21) consisting of mouse CD28 or another counter receptor to B7 (mouse CTLA4), coupled to human Ig, were used as mock-B7 reagents. These reagents were made available by Dr. Peter L. Linsley and have an affinity for B7 of Kd=12-100 nM, i.e., comparable to the affinity of most mAbs. Additonal reagents used in this study are listed in Table 1, and can distinquish between the main candidate thymic stromal cell types, i.e., dendritic cells, B cells, cortical epithelial cells, and medullary epithelial cells. Other adhesion molecules to which costimulatory activity has been ascribed include ICAM-1 and-2, VCAM-1, and LFA-3 (25), and their possible contributions to T cell repertoire development are just beginning to be explored (13, 26); space limitations preclude these studies form being included in the present review. Overall, the findings presented provide the framework within which future studies into the interaction structures involved in T cell development will occur.

RESULTS

Thymic Stromal Cells Possibly Involved in T Cell Development

We first investigated which thymic stromal cells (TSC) were represented in a population of non-T cells from the thymus, obtained through collagenase digestion and after thymocyte removal through selective gating. Both CDR1 and G8.8a (staining thymic cortical and medullary cells, respectively) could be detected on separate but small subsets of this TSC mixture (<10%), demonstrating that both cortical and medullary epithelial cells are present. NLDC 145 staining of TSC detected two populations of TSC: a major peak of dull cells is followed by a brighter minor peak; we contend that these correspond to cortical epithelium and TDC, respectively, as TDC express NLDC 145 at a higher level; earlier studies reported that NLDC 145, besides staining DC, also stained cortical epithelium (28). Another surprising result was that B220 stained a small but significant population of TSC; data to be published elsewhere reveal that this population of cells represents a subset of activated B cells (Izon and Kruisbeek, submitted for publication). Mac-1 stained a bright population (~10%) of TSC, and B7 is weakly expressed on this mixed population. Two-and three-parameter analysis of these cell populations is currently in progress, somewhat hampered by the tendency of TSC to form multi-cellular clusters, such that assignment of B7 expression in this mixed population is difficult. However, analysis of B7 expression on a large panel of non-lymphoid, non-T cell thymic stromal cell lines has never identified a single thymic epithelial line that expressed B7; we therefore presume that the B7 expression in the mixed TSC preparation reflects expression on TDC, macrophages and B cells (see below).

Splenic vs Thymic Dendritic Cells and Thymic B Cells

Thymic dendritic cells (TDC) have been implicated in negative selection (1, 8, 12, 29, 30), as have splenic dendritic cells (SDC) (29-31). However, some studies claim that only a combination of thymic B cells and DC can induce clonal deletion (30). It was therefore considered worthwhile to see if any phenotypic differences between these different DC populations could be detected. Strikingy, the antibody 33D1 stained all SDC but not TDC; this is at odds with a recent publication (32). In contrast, 100% of the DC populations from both thymus and spleen is stained by NLDC 145. The low level staining revealed on both populations by CDR1 and G8.8a did not appear to be specific when compared to other mAb. Staining with T cell markers such as CD3, CD4 and CD8 revealed that both SDC and TDC expressed CD8, although SDC demonstrated less reactivity; these results confirm the recent identification of CD8 as one of the T cell markers expressed on DC (32, 33). It was found that about 50-60% of thymic B cells expressed CD5, consistent with previous results (34). This is in contrast to the B cells found in the peritoneum which all express CD5 (35). Anti-class II staining showed that most but not all B220 cells expressed class II, and most of the B220-expressing cells were negative for Mac-1, thus revealing a conflict with a previous study that revealed significant Mac-1 staining on thymic B cells (34).

SUMMARY

B7 expression appears much more extensive than previously recognized with anti-human B7 reagents on human leucocyte populations: it is extremely high on splenic and thymic DC, moderate on macrophages and activated B cells, and low on resting B cells. Additionally, B7 is entirely undetectable on any thymic epithelial cells belonging to a panel of transformed cell lines and T cells, but its expression on epithelial cells in situ is still under investigation. This expression pattern is consistent with the hierarchy of costimulatory signal activity among these cell types, with DC being the most effective, and epithelial cells (like other cells of non-hemopoietic origin) not at all. Future studies will investigate to which extent B7 is involved in clonal deletion, i.e., the selection process dependent on self-antigen presentation by DC and B cells.

REFERENCES

1. Ramsdell F, Fowlkes BJ (1990) Clonal deletion versus clonal anergy: The role of the thymus in inducing self tolerance. Science 248, 1343-1348.
2. Kappler JW, Roehm N, Marrack P (1987) T cell tolerance by clonal elimination. Cell 49, 273-280.
3. Zuniga-Pflucker JC, Jones LA, Chin LT, Kruisbeek AM (1991) CD4 and CD8 act as co-receptors during thymic selection of the T cell repertoire. Sem Immunol 3, 167-175.
4. Kisielow P, Teh HS, Blüthmann H, Von Boehmer H (1988) Positive selection of antigen-specific T cells in thymus by restricting MHC molecules. Nature 335, 730-733.
5. Blackman M, Kappler J, Marrack P (1990) The role of the T cell receptor in positive and negative selection of developing T cells. Science 248, 1335-1341.
6. Kappler JW, Staerz U, White J, Marrack PC (1988) Self-tolerance eliminates T cells specific for Mls-modified products of the major histocompatibility complex. Nature 332, 35-41.
7. MacDonald HR, Schneider R, Lees RK, Howe RC, Acha-Orbea H, Festenstein H, Zinkernagel RM, Hengartner H (1988) T cell receptor Vb use predicts reactivity and tolerance to Mls^a-encoded antigens. Nature 332, 40-44.
8. Roberts JL, Sharrow SO, Singer A (1990) Clonal deletion and clonal anergy in the thymus induced by cellular elements with different radiation sensitivities. J Exp Med 171, 935-940.
9. Kojima A, Prehn RT (1981) Genetic susceptibility to post-thymectomy autoimmune diseases in mice. Immunogenetics 14, 15-27.
10. Smith H, Chen I-M, Kubo R, Tung KSK (1989) Neonatal thymectomy results in a repertoire enriched in T cell deleted in adult thymus. Science 245, 749-752.
11. Yunis EJ, Hong R, Grewe MA, Martinez C, Cornelius E, Good RA (1967) Postthymectomy wasting associated with autoimmune phenomena. Antiglobulin-positive anemia in A and C57BL/6 Ks mice. J Exp Med 125, 947-956.
12. Frontiers in Research: Immunological Tolerance. (1990) Science 248, 1335-1393.
13. Carlow DA, Van Oers NSC, Teh S-J, Teh H-S (1992) Deletion of antigen-specific immature thymocytes by dendritic cells rquires LFA-1/ICAM interactions. J Immunol 148, 1595-1603.
14. Carlow DA, Teh S-J, Teh H-S (1992) Altered thymocyte development resulting from expressing a deleting ligand on selecting thymic epithelium. J Immunol 148, 2988-2995.
15. Eynon EE, Parker DC (1992) Small B cells as antigen-presenting cells in the induction of tolerance to soluble protein antigens. J Exp Med 175, 131-138.
16. Lin R-H, Mamula MJ, Hardin JA, Janeway Jr CA (1991) Induction of autoreactive B cells allows priming of autoreactive T cells. J Exp Med 173, 1433-1439
17. Jenkins MK, Schwartz RH (1987) Antigen presentation by chemically modified splenocytes induces antigen-specific T cell unresponsiveness *in vitro* and *in vivo*. J Exp Med 165, 302-319.

18. Jenkins MK, Taylor PS, Norton SD, Urdahl KB (1991) CD28 delivers a costimulatory signal involved in antigen-specific IL-2 production by human T cells. J Immunol 147, 2461-2466.
19. Mueller DL, Jenkins MK, Schwartz RH (1989) Clonal expansion versus functional clonal inactivation. A costimulatory signalling pathway determines the outcome of T cell antigen receptor occupancy. Ann Rev Immunol 7, 445-480.
20. June CH, Ledbetter JA, Linsley PS, Thompson CB (1990) Role of the CD28 receptor in the T cell activation. Immunol Today 11, 211-216.
21. Linsley PS, Brady W, Grosmaire L, Aruffo A, Damle NK, Ledbetter JA (1991) Binding of the B cell activation antigen B7 to CD28 costimulates T cell proliferation and interleukin 2 mRNA accumulation. J Exp Med 173, 721-730.
22. Azuma M, Cayabyab M, Buck D, Phillips JH, Lanier LL (1992) CD28 interaction with B7 costimulates primary allogeneic proliferative responses and cytotoxicity mediated by small, resting T lymphocytes. J Exp Med 175, 353-360.
23. Koulova L, Clark EA, Shu G, Dupont B (1991) The CD28 ligand B7/BB1 provides costimulatory signal for alloreactivation of the $CD4^+$ T cells. J Exp Med 173, 759-762.
24. Harding FA, McArthur JG, Gross JA, Raulet DH, Allison JP (1992) CD28-mediated signalling co-stimulates murine T cells and prevents induction of anergy in T-cell clones. Nature 356, 607-609.
25. Damle NK, Klussman K, Linsley PS, Aruffo A (1992) Differential costimulatory effects of adhesion molecules B7, ICAM-1, LFA-3, and VCAM-1 on resting and antigen-primed $CD4^+$ T lymphocytes. J Immunol 148, 1985-1992.
26. Fine JS, Kruisbeek AM (1991) The role of LFA-ICAM-1 interactions during murine T lymphocyte development. J Immunol 147, 2852-2859.
27. Crowley M, Inaba K, Witmer-Pack S, Steinman RM (1989) The cell surface of mouse dendritic cells: FACS analyses of dendritic cells from different tissues including thymus. Cell Immunol 118, 108-125.
28. Kraal G, Breel M, Janse M, Bruin J (1986) Langerhans' cells, veiled cells and interdigitating cells in the mouse recognized by a monoclonal antibody. J Exp Med 163, 981-990.
29. Inaba M, Inaba K, Hosono M, Kumamoto T, Ishida T, Muramatsu S, Masuda T, Ikehara S (1991) Distinct mechanisms of neonatal tolerance induced by dendritic cells and thymic B cells. J Exp Med 173, 549-559.
30. Mazda O, Watanabe Y, Gyotoku J-I, Katsura Y (1991) Requirement of dendritic cells and B cells in the clonal deletion of Mls-reactive T cells in the thymus. J Exp Med 173, 539-547.
31. Matzinger P, Guerder S (1989) Does T cell tolerance require a dedicated antigen presenting cell? Nature 338, 74-77.
32. Vremec D, Zorbas M, Scollay R, Saunders DJ, Ardavin CF, Wu L, Shortman K (1992) The surface phenotype of dendritic cells purified from mouse thymus and spleen: Investigation of the CD8 expression by a subpopulation of dendritic cells. J Exp Med 176, 47-58.
33. Ardavin C, Shortman K (1992) Cell surface marker analysis of mouse thymic dendritic cells. Eur J Immunol 22, 859-865.
34. Miyama-Inaba M, Kuma S-I, Inaba K, Ogata H, Iwai H, Yasumizu R, Murumatsu S, Steinman RM, Ikehara S (1988) Unusual phenotype of B cells in the thymus of normal mice. J Exp Med 168, 811-820.
35. Hayakawa K, Hardy RR, Parks DR, Herzenberg LA (1983) The "Ly-1B" cell population in normal, immunodefective and autoimmune mice. J Exp Med 157, 202-211.

TOLERIZING MICE TO HUMAN LEUKOCYTES:

A STEP TOWARD THE PRODUCTION OF MONOCLONAL ANTIBODIES

SPECIFIC FOR HUMAN DENDRITIC CELLS

Una O'Doherty, William J. Swiggard, Kayo Inaba[1], Yasunori Yamaguchi, Iris Kopeloff, Nina Bhardwaj, and Ralph M. Steinman

Laboratory of Cellular Physiology and Immunology
The Rockefeller University
New York, NY 10021 USA, and
[1]Department of Zoology, Faculty of Science
Kyoto University
Kyoto 606, Japan

ABSTRACT

Despite several attempts to isolate a mAb specific for human dendritic cells, none currently exists. Recent attempts have utilized an improved dendritic cell purification method to prepare immunogens[1] and a rapid two-color flow cytometric screening procedure that allows large numbers of hybridoma supernatants to be examined in each fusion[2]. Yet these improvements have also failed, yielding only hybridomas that bind "shared" antigens expressed by both dendritic cells and other leukocytes. Dendritic cells express many shared antigens, including CD45 [leukocyte common antigen], CD40, leukocyte [ß2] integrins CD11a and CD11c, CD54 [ICAM-1], CD44 [Pgp-1], CD58 [LFA-3], and the B7/BB1 antigen. Therefore, we are attempting to bias the immune response toward rarer, dendritic cell-specific clones by tolerizing or immunosuppressing our animals to shared antigens.

In one approach, adult mice held in barrier cages are injected with "nondendritic" cells and cyclophosphamide [CP], in order to ablate responding "nonspecific" B cell clones. Fifteen days after the last dose of CP, they are challenged with nondendritic cells. A week later they are bled, and serum antibody titers against nondendritic cells are determined by FACS, in order to demonstrate tolerance compared to controls injected with CP alone.

In the second approach, neonatal mice are injected with human T lymphoblasts at birth, followed by boosting at 1 week. In adulthood, they are challenged sequentially with sheep erythrocytes [sRBC], then with T blasts, to demonstrate that they can respond to unrelated cells but not to tolerogenic cells. One week after each kind of challenge, mice are bled and serum antibody levels are determined for treated and sham-injected mice.

Dendritic Cells in Fundamental and Clinical Immunology,
Edited by Kamperdijk *et al.*, Plenum Press, New York, 1993

When these two approaches were compared, CP led only to nonspecific immunosuppression, while neonatal injections produced selective, antigen-specific nonresponsiveness to the tolerizing T blasts.

INTRODUCTION

After a standard dendritic cell immunization [3 doses of ≥ 2 x 10^6 blood DC] and fusion, one-third of the hybridoma supernatants stain leukocytes. Thus far, one hundred percent of these supernatants have proven to stain other types of leukocytes as well as dendritic cells. If mice could be made tolerant to shared, immunodominant leukocyte antigens, it might be possible to isolate hybridomas specific for rarer antigens expressed only by human blood dendritic cells. Two methods have been described to prevent antibody production to one set of antigens while preserving the capacity to respond to others. In the first method, CP, a cytotoxic antiproliferative agent, is administered along with the undesired antigens in order to preferentially kill B cells that respond to unwanted epitopes[3]. In the second method, tolerance is induced by injecting the undesired antigens during the neonatal period[4,5]. Neither method has been used previously to try to raise mAbs to the cell surfaces of subsets of leukocytes.

In this study, we compare the efficacies of these two methods by using serum titers to assay for tolerance. This enables us to treat many more mice per group than in prior studies. To assess nonspecific immunosuppression, we include groups of control mice that were treated with CP alone or sham-injected as neonates.

MATERIALS AND METHODS

CP immunosuppression. Three groups of 5 male BALB/c x DBA/2 [CD2] F1 mice [Trudeau Institute] were 7-9 weeks old at day 0 of the study. One-gram CP Isopacks [Sigma] were dissolved with 100 mL of sterile PBS, and the unused portion was discarded after 7 days. The human EBV-transformed B cell line KET was kindly provided by Dr. Janet Lee of the Sloan-Kettering Institute. The injection schedule is shown in Fig. 1, and was adapted from Ou et al[6].

Neonatal tolerization. Within 24 hours of birth and one week later, CD2 F2 pups [Trudeau] were injected with human T blasts. T cells were enriched by rosetting human blood mononuclear cells with neuraminidase [Calbiochem] treated sheep erythrocytes [Cocalico]. This T cell-enriched fraction at 1-2 x 10^6 cells/mL was stimulated with concanavalin A [Boehringer Mannheim] at 3μg/mL and phorbol myristic acetate [Sigma] at 5 ng/mL. After 3 days, clustered cells were treated with α-methyl mannoside [Sigma] at 500μM for 1 h to release T blasts. Blasts were washed twice in RPMI, and resuspended at 1.5 x10^8/mL. 100μL of resuspended cells was injected ip, or 50 μL was injected iv into the orbital branch of the facial vein. Tolerization schedules are shown in Fig. 3.

Serum antibody assay for specific vs. nonspecific immunosuppression. Mice were bled retroorbitally one week after challenge with cells [either B cell line, sRBC or T blasts]. Dilutions of sera were used to stain tolerogenic or irrelevant cells by indirect immunofluorescence [fluorescein-conjugated goat anti-mouse IgG and IgM, Pierce]. Stained cells were analyzed on a FACScan [Becton Dickinson], and fluorescence histograms were plotted to determine mode fluorescence values at each serum dilution. Data is reported as means of mice in a group ± SEM.

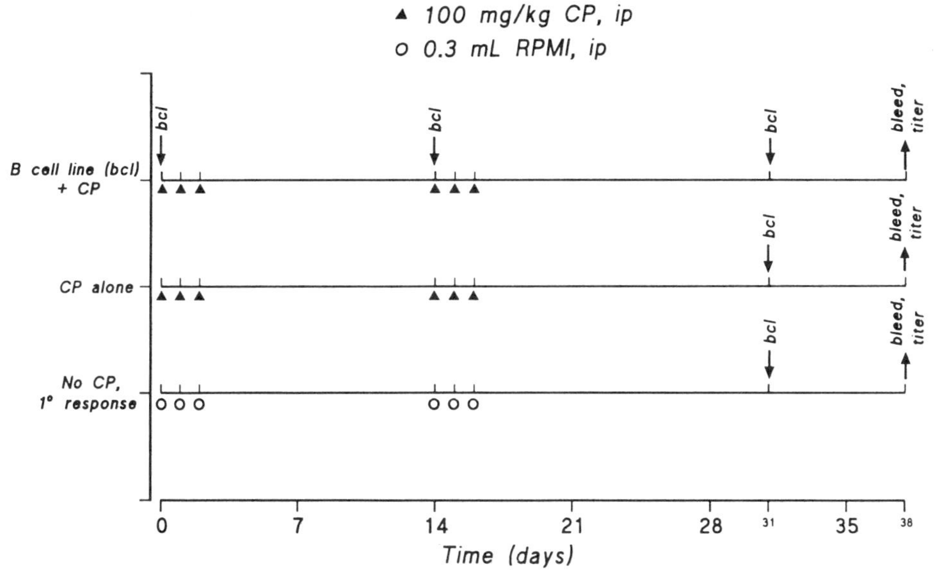

Figure 1. CP injection schedule. One group received 5 x 10^6 EBV-transformed B cells ip, followed 30 min later by 100 mg/kg CP [▲] ip. CP alone was injected 24 and 48 hours later. On days 14, 15 and 16 this procedure was repeated. A second group [CP alone] received the same treatment except no B cell line was injected. Sham-injected controls received neither cells nor CP, but were injected with 300 μL RPMI [○] ip. On day 29, mice in all 3 groups were challenged with 5 x 10^6 EBV-transformed B cells ip. They were bled one week later, and dilutions of sera were used to stain the B cell line, as described in Materials and Methods.

RESULTS

With the dose and schedule chosen [Fig. 1], CP did not induce antigen-specific nonresponsiveness [Fig. 2]. On challenge with tolerogenic cells, mice that had received the B cell line with CP and control mice that received CP alone made indistinguishable low-level antibody responses. If injection of tolerogenic cells with CP had led to selective ablation of tolerogen-responsive clones, the titers on tolerogen should have been lower in mice that received cells with CP than in those that received only CP. Responses of both groups of CP-treated mice were significantly less than a normal primary response [$p<0.05$], demonstrating that CP produces general, nonspecific immunosuppression.

In contrast, neonatal injections of human T blasts at the doses and schedules indicated [Fig. 3] did produce antigen-specific nonresponsiveness at the humoral level [Fig. 4]. Mice that had been injected with either T blasts or RPMI as neonates were challenged

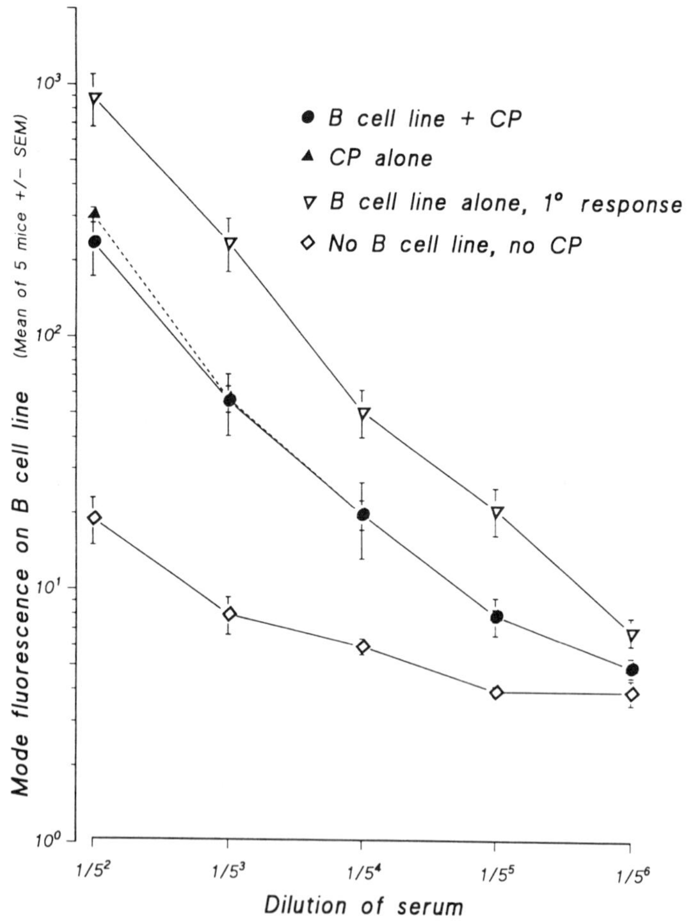

Figure 2. Immunosuppression with CP is not antigen-specific. Serum titers of mice treated with B cell line [bcl] + CP, CP alone and sham-injected controls are compared by flow cytometry. On the y axis, mode fluorescence is proportional to the concentration of antibody bound by the bcl. Responses of mice treated with bcl + CP are identical to those of mice treated with CP alone. Both are between a normal primary response [sham-injected, challenged] and naive mouse serum [no challenge].

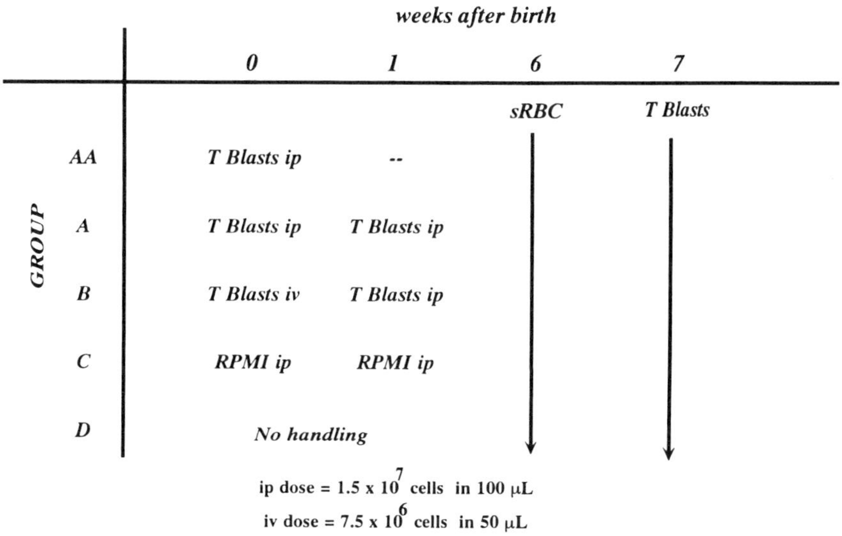

Figure 3. Tolerizing schedule. Within 24 hours of birth, mice received ip (A and AA) or iv (B) injections of human T blasts, RPMI ip (C), or were not handled (D). Seven days later, mice received a booster ip injection, except group AA. At 42 days, they were challenged with 10^8 sRBC. One week later, they were bled, and sera were then assayed for antibodies to sRBC. At 51 days, those mice that made detectable responses to sRBC were challenged with 5 x 10^6 human T blasts ip. They were bled again one week later, and sera were assayed for antibodies that bound T blasts.

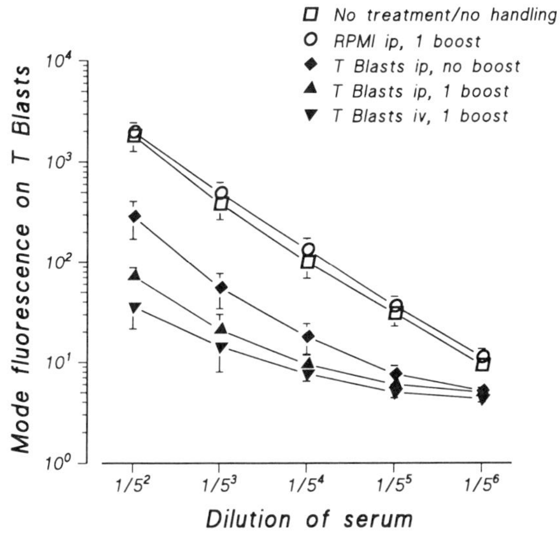

Figure 4. Neonatal injections of human T blasts produce tolerance. To determine if the treated mice were immunocompetent, they were first challenged with irrelevant cells [sRBC]. One week later, they were bled, and their sera were used to stain sRBC. Mice that responded to sRBC were then challenged with T blasts. One week later, they were bled again, and dilutions of sera used to stain T blasts. Boosted mice are specifically nonresponsive to T blasts. Unboosted mice make blunted responses.

Group	Treatment	6 weeks +	6 weeks −	9 weeks +	9 weeks −
		Number of mice responding (+) or not responding (-) to sRBC at age:			
A	*T blasts ip + boost*	2	1	1	0
AA	*T blasts ip, no boost*	5	1	6	0
B	*T blasts iv + boost*	3	1	4	0
C	*RPMI ip + boost*	6	1	6	0
D	*No handling*	5	0	5	0

Figure 5. Responses of neonatally-injected animals to challenge with sRBC at 6 and 9 weeks. Mice were challenged at 6 weeks with sRBC in order to identify and exclude from the study animals that could not respond to irrelevant cells. A week after challenge, they were bled, and sRBC were stained with dilutions of sera. Sera giving mode fluorescences equal to those from unchallenged mice were considered negative responses.

with antigenically-unrelated sRBC at 6 weeks to assess whether they were immunocompetent. On average, immunocompetent mice that had been neonatally injected with T blasts failed to respond to T blasts. Sham-injected controls made full primary responses to T blasts. Mice that received only one injection of T blasts at birth made intermediate responses to T blasts, suggesting that a booster injection early after birth is necessary for complete tolerance. Mice that received ip and iv injections at birth made equivalent responses.

Although antigen-specific nonresponsiveness was demonstrated in most of the animals injected neonatally with T blasts, a small number of T blast-injected animals [23% in Fig. 5] were unable to respond to sRBC at 6 weeks, suggesting that they were not immunocompetent. These mice were excluded from the rest of the study. One of the sham-injected mice was similarly unable to respond to sRBC. In contrast, mice that were not handled during the neonatal period all made anti-sRBC responses. This suggests that sRBC unresponsiveness occurred because of neonatal handling. It was clear that some of the handled mice were runted, and it is likely that their immunosuppression is related to this.

DISCUSSION

In 1959, Schwartz and Dameshek[7] showed that animals could be prevented from making antibody responses to a specific protein by administering that protein followed by the antimetabolite 6-mercaptopurine. Matthew and Sandrock[8] and others[6,9] have exploited this observation as the basis of a method for enhancing production of mAbs to developmentally-restricted antigens in brain extracts. Mice are injected with shared antigens followed by the DNA-alkylating agent cyclophosphamide [CP]. Theoretically, CP would preferentiallly kill dividing cells, including those B cell clones that proliferate in response to shared antigens. After two or more rounds of this treatment, mice are immunized with a mixture of shared and target antigens. In these studies, the assay for tolerance is to perform a fusion and determine the percentage of hybridomas secreting mAbs that bind the target antigen, but not the shared antigens. Because screening large numbers of hybridomas is very laborious, relatively few animals were included in these studies. Furthermore, control mice were spared the CP injections, and so were not exposed to the general immunosuppressive effects and toxicity of CP alone.

In our experiments, only a single dose of CP [100 mg/kg] and a single treatment schedule [Fig. 1] were studied. These conditions have been reported to be effective by Ou et al[6] and others[8,9], but these groups failed to report controls for nonspecific immunosuppression. Using these published conditions, mice treated with CP alone and mice treated with a B cell line plus CP made the same weak antibody response after challenge with the B cell line. There was no evidence that treatment with cells plus CP biased the humoral response away from the B cell line. Antigen selectivity might be achievable with [1] a different CP dose; or [2] a modified treatment schedule; or [3] injection with adjuvant-treated leukocyte preparations instead of viable cells. In addition, it is possible that the original regimen described by Schwartz and Dameshek[7], involving a longer treatment period and a different class of drug [6-mercaptopurine] might lead to drug-induced tolerance. Although drug-only controls were not included in that study, it was shown that treated animals were able to respond to antigens unrelated to the tolerogen. Subsequent studies switched to CP [a less-hazardous, bioactivated prodrug] after it was shown to be just as effective as 6-mercaptopurine in *general* immunosuppression[10]. A perfect treatment regimen would ablate all B cell clones that respond to tolerogen, and permit immunization with the target antigen before they are replenished. However, we have not attempted to optimize pharmaceutical immunosuppression here.

Experiments in the 1950's[11-13] based on the classic work of Medawar's group on tolerance to transplantation antigens[14] showed that animals could be prevented from making antibody responses to protein antigens by adminstering those proteins during the fetal or neonatal periods. Hockfield[4] and others[5,6] have incorporated these ideas into methods for increasing the probability of obtaining mAbs to selected nervous system antigens in mixtures. Mice are injected neonatally with shared antigens, then challenged with the target antigen in adulthood. The assay for tolerance is a fusion. Again, the numbers of mice studied are very small. Furthermore, the controls are unhandled littermates, not animals exposed to the stresses of sham injection.

In this experiment, we utilized larger numbers of mice, and included sham-injected controls. We found that neonatal mice can be made tolerant to antigens on the surfaces of intact human leukocytes. This tolerance is specific, and does not interfere with the ability of immunocompetent animals to mount humoral responses to unrelated cells.

CONCLUSIONS

Injection of intact human cells into neonatal mice leads to a state of antigen-specific nonresponsiveness, where mice are unable to mount an antibody response to tolerogenic cells, but can respond to sheep erythrocytes.

In contrast, injection of intact human cells into adult mice followed by treatment with CP using the dose and schedule described induces a nonspecific state of immunosuppression.

REFERENCES

1. P.S. Freudenthal and R.M. Steinman, The distinct surface of human blood dendritic cells, as observed after an improved isolation method, *Proc.Natl.Acad.Sci.USA* 87:7698 (1990).

2. J.P. Metlay, M.D. Witmer-Pack, R. Agger, M.T. Crowley, D. Lawless, and R.M. Steinman, The distinct leukocyte integrins of mouse spleen dendritic cells as identified with new hamster monoclonal antibodies, *J.Exp.Med.* 171:1753 (1990).
3. W.D. Matthew and P.H. Patterson, The production of a monoclonal antibody that blocks the action of a neurite outgrowth-promoting factor, *Cold Spring Harbor Symp.Quant.Biol.* 48:625 (1983).
4. S. Hockfield, A mAb to a unique cerebellar neuron generated by immunosuppression and rapid immunization, *Science* 237:67 (1987).
5. G.S. Golumbeski Jr. and R.L. Dimond, The use of tolerization in the production of monoclonal antibodies against minor antigenic determinants, *Anal.Biochem.* 154:373 (1986).
6. S.K. Ou, C. McDonald, and P.H. Patterson, Comparison of two techniques for targeting the production of monoclonal antibodies against particular antigens, *J.Immunological Methods* 145:111 (1991).
7. R. Schwartz and W. Dameshek, Drug-induced immunological tolerance, *Nature* 183:1682 (1959).
8. W.D. Matthew and A.W. Sandrock Jr., Cyclophosphamide treatment used to manipulate the immune response for the production of monoclonal antibodies, *J.Immunological Methods* 100:73 (1987).
9. C.V. Williams, C.L. Stechmann, and S.C. McLoon, Subtractive immunization techniques for the production of monoclonal antibodies to rare antigens, *Biotechniques* 12:842 (1992).
10. M.C. Berenbaum, Immunology. A screen for agents inhibiting the immune response and the growth of tumours, *Nature* 196:384 (1962).
11. R. Hanan and J. Oyama, Inhibition of antibody formation in mature rabbits by contact with the antigen at an early age, *J.Immunol.* 73:49 (1954).
12. F.J. Dixon and P.H. Maurer, Immunologic unresponsiveness induced by protein antigens, *J.Exp.Med.* 101:245 (1954).
13. R.T. Smith and R.A. Bridges, Immunological unresponsiveness in rabbits produced by neonatal injection of defined antigens, *J.Exp.Med.* 108:227 (1958).
14. R.E. Billingham, L. Brent, and P.B. Medawar, "Actively acquired tolerance" of foreign cells, *Nature* 172:603 (1953).

MORPHOLOGICAL AND FUNCTIONAL DIFFERENCES BETWEEN HLA-DR$^+$ PERIPHERAL BLOOD DENDRITIC CELLS AND HLA-DR$^+$ IFN-ALPHA PRODUCING CELLS

Stuart E Starr, Santu Bandyopadhyay, Vedapuri Shanmugam, Nassef Hassan, Steven Douglas, Stephanie J. Jackson, Giorgio Trinchieri*, and Jihed Chehimi

The Children's Hospital of Philadelphia, Division of Infectious Diseases and Immunology, and *The Wistar Institute of Anatomy and Biology Philadelphia, Pennsylvania 19104

INTRODUCTION

Several cell types, including B cells, NK cells, and monocytes have been proposed to be the cells responsible for IFN-α production in-vitro in response to herpesviruses and other viruses. When Perussia et al[1] described IFN-α producing cells as HLA-DR+ nonadherent, non-monocyte, non-B peripheral blood cells, dendritic cells (DC) became an attractive candidate cells for this function. Subsequently, Bandyopadhyay et al[2,3] showed that the major leukocyte subset in human peripheral blood responsible for IFN-α production upon culture with herpesvirus-infected target cells consists of nonadherent HLA-DR+ cells, which bear no surface markers of B, or T cells, monocytes or NK cells and that these cells also serve as accessory cells for NK cell-mediated lysis of the herpesvirus and HIV-infected targets. Later, Chehimi et al[4] reported that human peripheral blood mononuclear cells contain at least two distinct HLA-DR+ cell subsets, both of which lack any B and T cell, monocyte and NK cell surface markers. One HLA-DR+ cell subset is plastic nonadherent, produces IFN-α in response to herpesvirus and HIV and provides accessory help to NK cells for lysis of virus-infected targets, but does not induce mixed leukocyte reaction (MLR). The other population has the morphology of dendritic cells (DC), is loosely adherent to plastic, stimulates T cells in the MLR, but neither produces IFN-α when exposed to virus-infected targets nor provides accessory help to NK cells. In this report, we show for the first time morphological, as well as new functional differences, between HLA-DR+ IFN-α producing cells (IPC) and HLA-DR+ DC.

MATERIALS AND METHODS

Enrichment of DC and IPC. DC were enriched as previously described[4]. Briefly, peripheral blood mononuclear cells (PBMC) from healthy adult donors (seronegative for hepatitis B and HIV), were obtained by Ficoll-Hypaque density gradient centrifugation of leukocyte concentrates. After 90 min incubation (37°C, 5% CO_2) in plastic flasks non-adherent cells were carefully removed and plastic adherent cells were further incubated for 18h at 37°C. The loosely adherent cells that detached during this 18h incubation were treated with MAbs to T and B cells, NK cells and monocytes in the presence of baby rabbit complement,(C) for 10 min. Cells were washed twice and then fractionated by centrifugation on a continuous density gradient of colloidal silica particle (Sepracell-MN). This gradient separation was repeated 2 to 3 times. Low-density cells which appeared as a compact band just below the meniscus were removed, washed with RPMI 1640, 10% FBS and residual Fc receptor positive monocytes were removed by panning over human IgG-coated plastic dishes. The enrichment of IPC was done following our modified method previously described[4]. Nonadherent PBMC were subjected to a second adherence cycle on plastic flasks for 60 min at 37°C. The nonadherent cells (50 to 60 x 10^6 cells in 3 ml RPMI 1640, 10% fetal bovine serum) were layered onto 2 ml of metrizamide solution as described by Knight et al[5]. Low density interface cells were collected, washed twice and nonadherent HLA-DR+ cells were further enriched by depleting T and B cells, monocytes and NK cells using MAbs and baby rabbit complement. The following antibodies were used: Leu 11b (anti-CD16) for NK cell depletion, B36.1 (anti-CD5) for T cells, B52.1 for monocytes and B1 (anti CD-20) for B cells. Viable cells were collected by centrifugation over Ficoll-Hypaque and remaining monocytes were removed by panning over human IgG coated plastic petri dishes. The cells were further incubated with goat anti-mouse FITC and negatively selected (FITC negative) cells were sorted using a florescence-activated cell sorter. Both subset of HLA-DR positive cells (DC and IPC) were analyzed by indirect immunofluorescence. Morphological analysis was done after May-Grunwald Giemsa staining of cytocentrifuged cells.

Interferon induction and cytotoxicity assays. Nonadherent HLA-DR+ cells were cocultured with CMV-infected FS4 cells or with cell-free NDV for 18 h at 37°C in a total volume of 0.2 ml. Cell-free supernatant was collected after centrifugation (100 x g for 10 min) and quantitation of IFN-α was done as described previously[3]. Titers were calculated in International Units (IU) based on results obtained with the National Institutes of Health reference standard for IFN-α (G-023-901-527). For IFN-γ induction, cell suspensions were incubated with IL-2 (100 U/ml) for 24 h, or with cell free CMV-AD169 for 72 h. Supernatants were then harvested and tested for IFN-γ production by radioimmunoassay (RIA) as described[6,7]. International Units (IU) based on results obtained with the NIH reference standard for IFN-γ (Gg-23-901-530). NK assays were done as described previously[8] using CMV-FS4 as target cells.

RESULTS AND DISCUSSION

Starting from 0.3-0.5 x 10^9 PBMC, we recovered 0.5-2 x 10^6 DC. The enriched DC preparations contain ~80% HLA-DR+ cells, <2% T cells, <2% B cells, <2% NK cells, <2% monocytes and ~ 10% small cells lacking any known lineage cell markers as judged by FACS. We recovered 2-4 x 10^6 partially enriched (30-40%) IPC from individual buffy coat. Using such partially enriched preparations as the starting populations, we recovered 0.5-1.0 x 10^6 cells, ≥80% HLA-DR+ non-T, non-B, non-NK, non-monocytes after sorting. These high yields of enrichment allowed us to compare the morphology of purified DC and IPC. DC had typical indented nuclei and clear abundant cytoplasm without granules and an irregular ruffled membrane (Figure 1A). In contrast IPC (Figure 1B) had scant cytoplasm containing very few if any granules and the membrane was smoother compared to DC. Both IPC and DC showed different morphology when compared to monocytes (Figure 1D) or NK cells (not shown). Cultured DC (Figure 1C, and E) also showed a morphology different than that of cultured monocytes (figure 1F) or NK blasts (not shown).

TABLE 1. IFN-α production after viral induction.

Responder cells[1]	IFN-α (U/ml) after induction with:	
	CMV-FS4	NDV
PBMC	125.0 ± 51	224.0 ± 190
HLA-DR⁻ PBMC	<4	4.5 ± 5
IPC	1295.0 ± 79	1428.0 ± 220
DC	11.0 ± 4	6.5 ± 3

[1]Responders (1-2 x 10^6/0.1 ml) were mixed with CMV-FS4 cells (5 x 10^4/0.1 ml) or with cell-free NDV and incubated at 37°C in 5% CO2 for 24 h. Cell free supernatants were harvested and assayed for IFN-α.

We compared the ability of the two HLA-DR+ cell subsets to produce IFN-α upon viral stimulation and to provide accessory help for NK cell-mediated lysis of CMV-infected target cells. When different viruses, such as CMV and NDV, were used, the amount of IFN-α

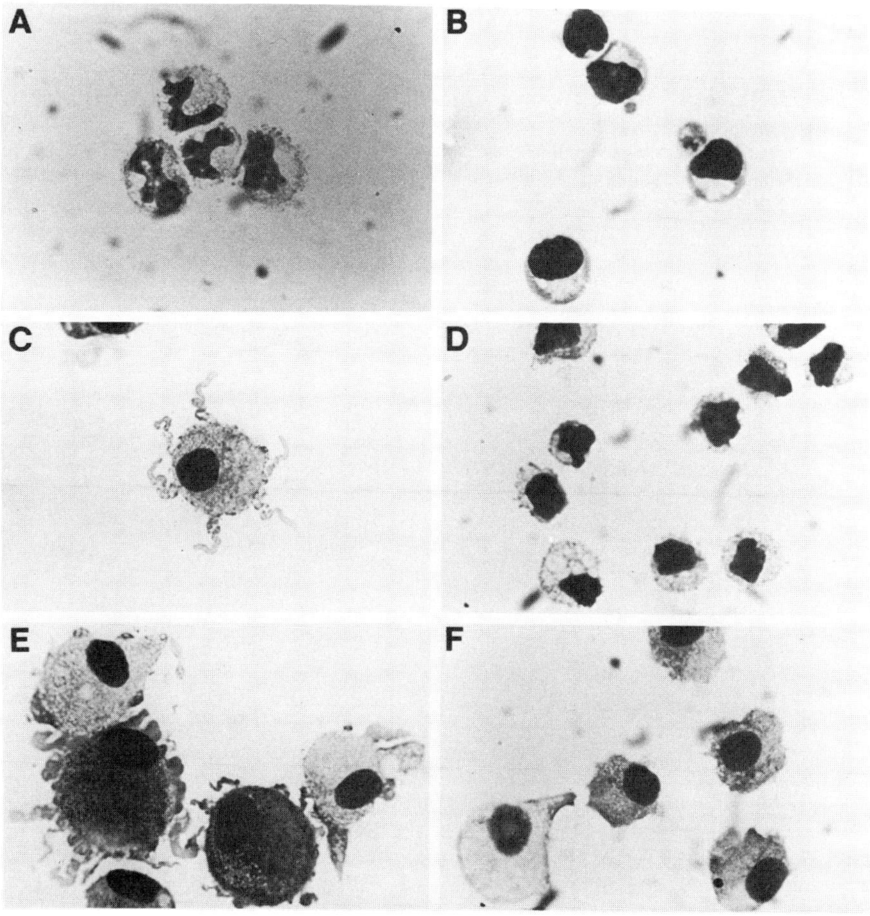

Figure 1. Morphology of the different populations of HLA-DR+ cells. Cells were cytocentrifuged and stained with May-Grunwald Giemsa, (A), DC; (B), IPC; (C and E), cultured DC; (D), fresh monocytes; (F), cultured monocytes.

produced by IPC was 5 to 10-fold higher than that produced by PBMC. DC, on the other hand, produced 10 to 30-fold lower amounts of IFN-α (Table 1). Production of IFN-α by PBMC in response to CMV or NDV was markedly reduced when HLA-DR+ cells were depleted. These results confirm and extend our previous observations where we showed the ability of nonadherent HLA-DR+ cells to produce IFN-α in response to HSV, CMV, VZV and HIV. As shown in Figure 2, PBMC depleted of HLA-DR+ cells or Leu11b+ (CD16+) cells, lysed the CMV-infected target cells poorly. Similarly IPC or DC alone failed to lyse CMV-FS4 target cells. Addition of IPC but not of DC to the HLA-DR- cells, restored lysis to the PBMC levels indicating that IPC and not DC are required for lysis of CMV-FS4 cells.

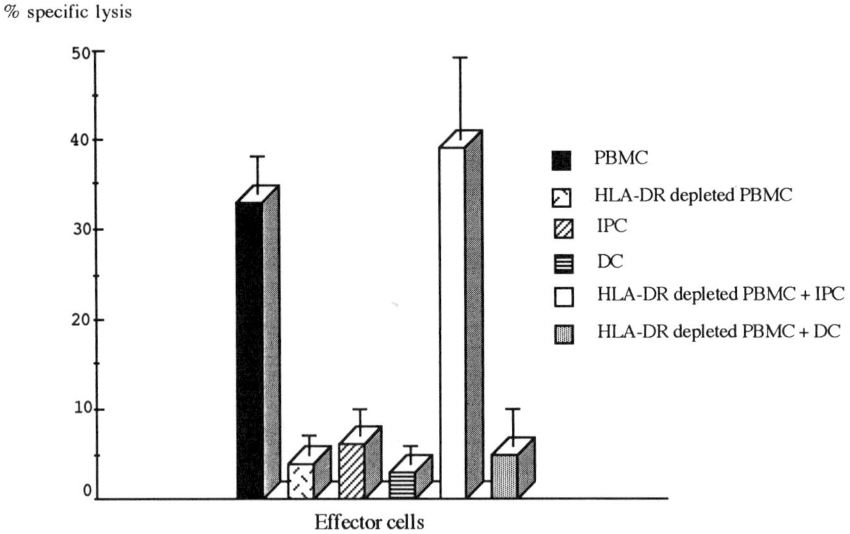

Figure 2. Cytotoxicity mediated against CMV-infected fibroblasts.

Several conflicting reports exist in the literature regarding the nature of accessory cells needed for IFN-γ induction. Monocytes as well as DC were claimed to be required. Lee et al described the requirement of thymic and tonsillar DC for IFN-γ production upon induction with IL-2. In their study, Lee et al enriched DC after one step (18 h) adherence. We believe that the nonadherent population used to enrich for HLA-DR+ DC was a mixture of IPC and DC. To further emphasize the difference in accessory function of these two HLA-DR+ cells, the requirement for IPC and DC for IFN-γ production was compared using cell-free CMV and IL-2. In preliminary experiments, we have found that CMV-induced IFN-γ production by PBMC from seropositive donors were maximal at 72 h. When CMV-seronegative donors were used, CMV-induced IFN-γ was barely detectable. In-vitro production of IFN-γ by normal PBMC in response to IL-2 was markedly reduced when PBMC were depleted of HLA-DR+ cells or NK cells. Depletion of T cells, to levels undetectable by indirect immunofluorescence (flow cytometry), reduced IFN-γ titers only by 50%. Depletion of monocytes did not change the IFN-γ production. When NK-depleted PBMC or T-depleted PBMC of individual donors were mixed at a ratio of 1:1 with HLA-DR depleted PBMC and stimulated with IL-2, IFN-γ production was restored, indicating the requirement for HLA-DR+ accessory cells for IL-2 -induced IFN-γ production by NK and T cells (Table 2). When CMV was used as an induce, IFN-γ production was reduced by 30% and 99% when NK cells and T cells were removed by incubation of PBMC with appropriate MAbs plus C. When T- or NK- PBMC were mixed with HLA-DR- PBMC and stimulated with CMV, IFN-γ production was reconstituted, suggesting confirming the requirement for HLA-DR+ cells (Table 2).

TABLE 2. HLA-DR+ cells are the accessory cells for IFN-γ production after induction with IL-2 or with cell-free CMV

Responder cells[1]	IFN-γ (U/ml) after stimulation with:	
	IL-2[2]	CMV
PBMC	98.0	200.0
HLA-DR- PBMC	<1.0	8.0
NK- PBMC	<1.0	135.0
Monocyte- PBMC	157.0	192.0
T- PBMC	45.0	<1.0
HLA-DR-PBMC + NK-	85.0	182.0
HLA-DR- PBMC +T-	89.0	170.0

The results presented are representative of two similar experiments.
[1]Depletion of NK cells, T cells, HLA-DR+ cells and monocytes was done using different MAbs and baby rabbit C. Incubating PBMC with these MAbs without complement had no effect on IFN-γ production in response to IL-2 or CMV.
[2]Cells (1-2 x 10^6/well) were incubated with 100 U/ml of rIL-2 in a total volume of 0.2 ml, for 24 h at 37°C in 5% CO2. The supernatant fluids were harvested and assayed for IFN-γ by radioimmunoassay.
[3]CMV-seropositive donor cells (1 x 10^6) were incubated with cell-free CMV-AD169 (1 x 10^6 PFU/ml) at 37°C in for 72 h. Supernatant fluids were harvested for IFN-γ titration.
[4]NK- and T- cells were added to HLA-DR- PBMC at a ratio of 1:1.

We then investigated whether enriched IPC or DC could reconstitute IFN-γ production. As shown in Table 3, DC and IPC alone were unable to produce IFN-γ after stimulation with IL-2 or CMV. When stimulated with IL-2, IPC were efficient accessory cells for IFN-γ production by HLA-DR- PBMC. In contrast, addition of large number of DC or monocytes to HLA-DR- failed to induce IFN-γ production (Table 3). These data strongly indicate that the HLA-DR+ accessory cell subset required for IFN-γ production after IL-2 stimulation consists of neither DC nor monocytes, but of non-adherent IPC. Similarly, addition of IPC or monocytes to HLA-DR- cells did not induce IFN-γ production after stimulation with CMV (Table 3). Only addition of DC could restore IFN-γ production, suggesting that the DC are required for CMV-induced IFN-γ production.

TABLE 3. Requirement of IPC and DC for IL-2 or virus induced-IFN-γ production.

Responder Cells[1]	IFN-γ (U/ml) production after induction with	
	IL-2	CMV
PBMC	217 ± 59	89 ± 52
HLA-DR- PBMC	6 ± 4	<4
IPC	5 ± 2	<4
DC	<4	<4
Monocytes	<4	<4
HLA-DR-PBMC + DC (1:10)	21 ± 15	101 ± 15
HLA-DR- PBMC + IPC (1:10)	381 ± 79	7 ± 3
HLA-DR- PBMC + Monocytes (1:10)	<4	<4

[1]Responder cells (1-2 10^6/well) were incubated with 100 U/ml of rIL-2 or cell-free CMV (1 x 10^6 PFU/ml) at 37°C in 5% CO2. Supernatant fluids were harvested and assayed for IFN-γ.
[2]Mean ± SD of 6 experiments are shown

Several investigators reported deficiency in IFN-α as well as IFN-γ production by peripheral blood mononuclear cells of patients with AIDS. This decline was found to be predictive of opportunistic infections. The IPC subset may play a key role in natural resistance by producing IFN-α in response to many viruses. Of interest, administration of IFN-α has been shown to be of benefit for the treatment of AIDS-associated Kaposi's sarcoma, and several investigators have described the suppression of HIV replication by IFN-α. The possible involvement of DC in AIDS pathogenesis has been suggested by several recent studies[9,10]. Therefore, one can speculate that functional defects of these two HLA-DR+ cell populations, DC and IPC, as a result of HIV-infection along with that of T helper cells and monocytes may contribute to part of the immunosuppression observed in AIDS patients.

Acknowledgments: This work was supported by Public Health Service grant AI-31368 and CA-20833 from the National Institutes of Health. We thank Jeffrey Faust and John Gibas for fluorescence-activated cell sorter analysis.

References

1. B. Perussia, V. Fanning, and G. Trinchieri, A leukocyte subset bearing HLA-DR antigen is responsible for in vitro interferon-alpha production in response to viruses, *Nat Immun Cell Growth Regulation.* 4:120 (1985).
2. S. Bandyopadhyay, B. Perussia, G. Trinchieri, D.S. Miller, and S.E. Starr, Requirement for HLA-DR+ accessory cells in natural killing of cytomegalovirus-infected fibroblasts, *J Exp Med.* 164:180 (1986).
3. S. Bandyopadhyay, S. H. Ho, J. Chehimi, U. Ziegner, D.S. Miller, J.A. Hoxie, and S.E. Starr, Natural killer cell-mediated lysis of target cells infected with cytomegalovirus and human immunodeficiency virus, *in* "Natural killer cells and host defense", E.W Ades, C. Lopez, eds., S. Karger, Basel (1989).
4. J. Chehimi, S.E. Starr, H. Kawashima, D.S. Miller, G. Trinchieri, B. Perussia, and S. Bandyopadhyay, Dendritic cells and IFN-alpha-producing cells are two functionally distinct non-B, non-monocytic HLA-DR+ cell subsets in human peripheral blood, *Immunology.* 68:486 (1989).
5. S.C. Knight, J. Farrant, A. Bryant, A.J. Edwards, S. Burman, A. Lever, J. Clark, and A.D.B. Webster, Non-adherent, low-density cells from human peripheral blood contain dendritic cells and monocytes, both with veiled morphology, *Immunology.* 57:595 (1986).
6. M. Murphy, R. Loudon, M. Kobayashi, and G. Trinchieri, Gamma interferon and lymphotoxin, released by activated T cells, synergyze to inhibit granulocyte-monocyte colony formation, *J Exp Med.* 164:263 (1986).
7. J. Chehimi, S.E. Starr, I. Frank, M. Rengaraju, S.J. Jackson, C. Llanes, M. Kobayashi, B. Perussia, D. Young, E. Nickbarg, S.F. Wolf, and G. Trinchieri, Natural killer (NK) cell stimulatory factor increases the cytotoxic activity of NK cells from both healthy donors and human immunodeficiency virus-infected patients, *J Exp Med.* 175:789 (1992).
8. J. Chehimi, S. Bandyopadhyay, K. Prakash, B. Perussia, N.F. Hassan, H. Kawashima, D. Campbell, J. Kornbluth, and S.E. Starr, In-vitro infection of natural killer cells with different immunodeficiency virus type 1 isolates, *J Virol.* 65:1812 (1991).
9. L.J. Eales, J. Farrant, M. Helbert, and A.J. Pinching, Peripheral blood dendritic cells in persons with AIDS and AIDS-related complex: loss of high intensity class II antigen expression and function, *Clin Exp Immunol.* 71:425 (1988).
10. S.E. Macatonia, R. Lau, S. Patterson, A.J. Pinching, and S.C. Knight, Dendritic cell infection, depletion and dysfunction in HIV-infected individuals, *Immunology.* 71:38 (1990).

THREE MONOCLONAL ANTIBODIES TO ANTIGEN PRESENTING CELLS IN THE RAT WITH DIFFERENTIAL INFLUENCE ON CELLULAR INTERACTIONS

Jan Damoiseaux, Ed Döpp, Marina Verdaasdonk,
Ed Kamperdijk, and Christine Dijkstra

Department of Cell Biology
Medical Faculty
Vrije Universiteit
Amsterdam, The Netherlands

INTRODUCTION

A physical interaction between antigen presenting cells (APC) and T cells is essential for evoking an immune response[1]. It has been proposed that T cells first bind to APC by an antigen-independent mechanism. Antigen specific lymphocytes are then selected and activated within the APC-T cell cluster resulting in T cell proliferation[2]. In the first stage several accessory molecules, like LFA-1, ICAM-1, CD2, and LFA-3, are involved in the antigen-independent interaction[1,2,3]. Thereafter the antigen-dependent interaction, mediated by the T cell receptor (TCR), CD3, CD4, and the MHC Class II molecule, becomes more important and induces cytokine production triggering T cell proliferation[1,3,4,5]. In the third stage, which can occur in the absence of APC, IL-2 alone mediates proliferation of the responsive T cells[6].

Recently we described three monoclonal antibodies (mAb), ED7, ED8, and ED9, recognizing mononuclear phagocytes, dendritic cells and neutrophils[7]. By preclearing experiments we showed that ED7 and ED8 recognize complement receptor type 3 (CR3), like the mAb OX-42[8], whereas ED9 recognizes an epitope on a different as yet unidentified molecule[9].

The expression of these antigens on interdigitating dendritic cells (IDC) raised the question whether they are involved in the induction of cell-cell contact between APC and T cells during the initiation of an immune response. The influence of the recognized antigens are studied in *in vitro* clustering assays as well as in an antigen presentation assay.

MATERIALS AND METHODS

Cell Populations

Cell suspensions of rat spleens were enriched for DC using a Nycodenz sugar gradient (analytical grade, Nyegaard, Norway)[7]. Contaminating B-cells were removed by incubation with sheep-anti-rat IgG coated DYNA Beads (DYNAL, Oslo, Norway). Peritoneal macrophages were obtained by peritoneal lavage with 10 ml RPMI-1640 medium.

Dendritic Cells in Fundamental and Clinical Immunology,
Edited by Kamperdijk *et al.*, Plenum Press, New York, 1993

T cells were prepared from rat lymph node suspensions. DC and B cells were removed as described above, macrophages were removed by adherence. Wistar rat T cells were activated by 2.4 mM $NaIO_4$ in case of antigen independent clustering[6]. T cells for the antigen-dependent clustering and the antigen presentation assay were isolated from lymph nodes of ACI/Mal rats 2 weeks after priming with the copolymer glutamine-tyrosine (GT) in the hind footpads[10].

Cluster Assay

The cluster assay was performed according to Austyn et al.[3]. APC were incubated differentially with mAb for 15 min at 4^0C. Antibody excess was removed by extensive washing. APC ($2x10^5$) and T cells ($6x10^5$) were mixed in 0.5 ml ice cold RPMI-medium containing 0.5% BSA and 0.1% DNase I (2000 U/mg). In antigen (GT)-dependent clustering the polymer GT was added in a concentration of 100 µg/ml. For preincubation, the cells were left on ice for 45 min. The cells were then sedimented at 500*g* for 5 min, and either transferred to a 37^0C water bath for 10 min or placed on ice. After incubation the cells were gently resuspended twice, and non-clustered T cells were counted in a hemocytometer. The morphological difference between T cells and APC could clearly be observed. The percentage of clustering was determined by the formula:

$$\% \text{ clustering} = 100 \text{ x } \frac{\text{Total T cells - T cells unbound}}{\text{Total T cells}}$$

Antigen Presentation Assay

As APC 50 µl of $4x10^4$ DC/ml or $4x10^5$ peritoneal macrophages/ml, both obtained from ACI/Mal rats, were used and added to 100 µl $2x10^6$ T cells/ml. Cells were incubated in a total volume of 200 µl in flatbottomed microtitre plates differentially supplemented with 20 µl GT as antigen. Inhibition studies were performed with 30 µl ProteinA purified mAb, applied in two concentrations (0.5 µg/µl and 0.1 µg/µl). Tritiated thymidine (6.7 Ci/mmol, Amersham International, Amersham, UK) was added at 72 hr after the start of culture. The amount of isotope incorporation was determined 18 hr later using an automatic cell harvesting device and liquid scintillation counting. Results are given as the average counts per minute (cpm) ± one standard deviation for triplicate cultures.

RESULTS

Cluster Assay

The obtained DC population was enriched to 70-90% purity. DC were defined as cells with irregular shaped, eccentric nuclei, extensive cell processes, acid phosphatase activity restricted to a distinct area in the cytocenter and strongly Ia positive as determined by mAb OX-6 by the immunoperoxidase technique. The obtained T cell population was enriched to 95% purity as defined with mAb OX-19. In case periodate activated T cells were incubated with syngeneic DC, clusters were formed at 37^0C in a temperature dependent way. The aggregation at 4^0C was strongly increased in case the cells were incubated at 37^0C. These clusters comprised T cells and DC as determined by morphology in an hemocytometer. About 30% of the T cells could cluster specifically, i.e. clustering at 37^0C minus clustering at 4^0C, in this assay (Fig. 1).

To measure the effects of mAb on the cluster formation the DC were preincubated with 10 µl ascites of the appropriate mAb. The three ED-mAb had different effects on the cluster formation. The mAb ED7 strongly inhibited clustering in all experiments. The mAb OX-42 and ED8 did not cause any inhibition of clustering. Like ED7, the mAb ED9 also inhibited the cluster formation although to a lesser extend (Fig. 1). No clustering between DC only was noticed, indicating that crosslinking by the mAb did not occur.

In case GT-primed T cells were used for clustering, no temperature dependent clustering could be observed in the absence of GT. This was the same in case DC or macrophages were used as APC. After addition of GT during clustering a strong increase in clustering at 37^0C

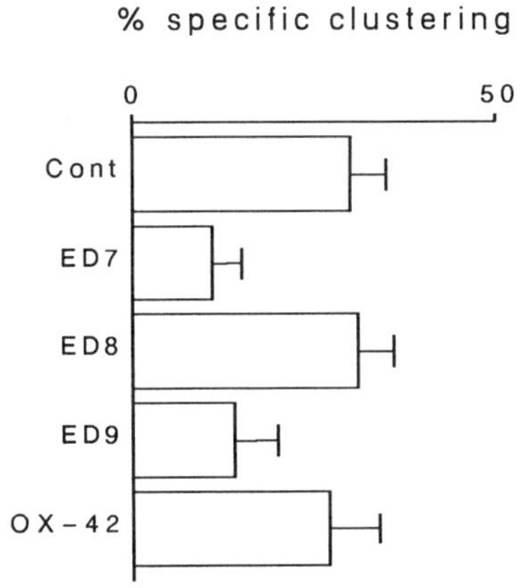

Figure 1. Effects of mAb on syngeneic antigen independent clustering. Clustering was performed with dendritic cells and periodate activated T cells in a ratio of 1:3. The mAb were added as 10 µl ascites to purified dendritic cells and incubated 15 min at 4^0C prior to the clustering assay. Inhibition of clustering by the mAb was performed at 37^0C. Results are presented as the mean ± SD percentage specific clustering, i.e. clustering at 37^0C minus 4^0C clustering, of five experiments.

could be seen with DC as well as with macrophages. This *in vitro* clustering appeared completely independent of the studied ED-antigens (Table 1).

Antigen Presentation

The influence of the ED-antigens expressed by APC on the induction of T cell proliferation was tested in an antigen presentation assay. One representative experiment is shown in figure 2 & 3. DC induced proliferation was only inhibited by the mAb OX-6 and OX-42. The ED-mAb had no distinct effect on the proliferation (Fig. 2). When macrophages were used as APC, T cell proliferation was influenced in a concentration dependent manner by OX-6 as well as by the three mAb reacting with CR3, but not by mAb ED9 (Fig. 2).

DISCUSSION

Like mAb OX-42, the mAb ED7 and ED8 both recognize CR3, whereas mAb ED9 reacts with a different as yet unidentified antigen[7,8]. Although the tissue distribution of ED7, ED8, and OX-42 is very alike, ED7 exclusively recognizes cilia in the bronchus epithelium[7].

Table 1. Effects of mAb on antigen-dependent APC-T cell clustering.

Temp	mAb	%Clustering			
		DC		Mph	
		-GT	+GT	-GT	+GT
4^0C	-	32	34	25	23
37^0C	-	28	60	33	50
37^0C	ED7	27	63	33	48
37^0C	ED8	28	63	27	52
37^0C	ED9	27	66	24	55

Clustering was performed with APC from normal ACI/Mal rats and T cells from antigen (GT) primed ACI/Mal rats in an APC-T cell ratio of 1:3.

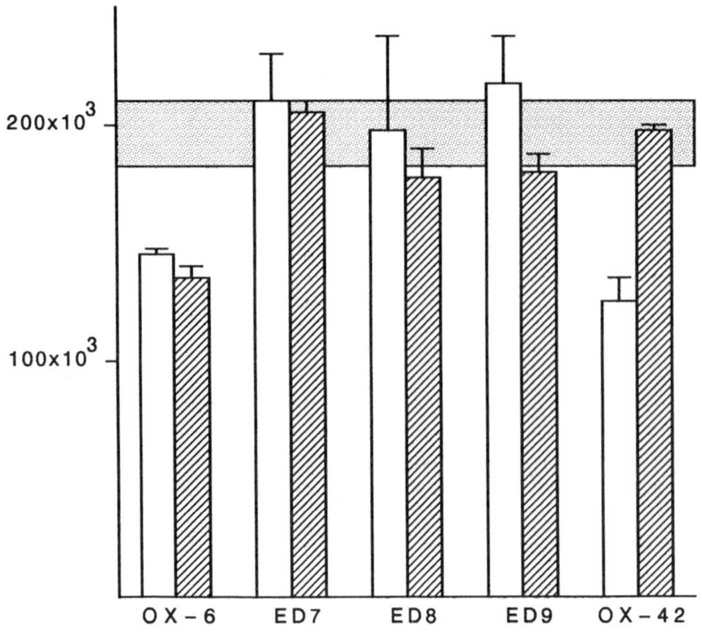

Figure 2. Effects of mAb on antigen presentation by dendritic cells. Proliferation was assayed after 3 days incubation of APC with T cells isolated from ACI/Ma1 rats 2 weeks after priming with GT. Antibodies were purified using Protein A-Sepharose and applied in two concentrations: 5 μg/μl (open bars) and 1 μg/μl (shadowed bars). Tritiated thymidine uptake was measured after an 18 hr pulse before the end of the incubation period. Results are given as the average counts per minute ± SD for triplicate cultures. The horizontal bar represents the mean of the control ± SD.

Furthermore, upon chronic inflammation the OX-42 expression on airway epithelial DC is strongly increased in contrast to the ED8 expression which is only slightly increased[11]. CR3 is a member of the LFA-1 family of adhesion molecules (CD11/CD18), consisting of heterodimers which share the same 95 kD ß-subunit, but structurally and antigenically distinct α-subunits of 175, 165, and 150 kD respectively for CD11a, CD11b, and CD11c[12]. Both the α and the ß chain, being non-covalently linked, are transmembrane proteins with a small cytoplasmic portion and a single membrane-spanning domain[13]. In mouse and human CR3 is involved in cell-matrix interaction like adhesion to plastic[14]. In the rat we were unable to inhibit binding of peritoneal macrophages to plastic by the mAb ED7, ED8, or ED9[9]. As far as cellular interactions are concerned, murine CR3 mediates rosette formation between phagocytic cells of the thymic reticulum and cortical type thymocytes[15]. Furthermore, the involvement of CR3 in binding to endothelium has been described[14]. In this study we showed that in the rat CR3 also is involved in clustering between DC and periodate activated T cells and thereby in induction of T cell proliferation.

The interaction between APC and T cells is essential in the initiation of an immune response. This interaction starts in an antigen independent manner. The antigen-independent clustering between DC and periodate activated T cells was only performed with DC, because macrophages are reported to be virtually inactive in this type of clustering[2]. There is evidence that antigen-independent clustering does not involve MHC class II molecules on DC or CD4 molecules on the T cells[4]. However, this clustering can be partially inhibited by blocking the CR3 epitope recognized by mAb ED7 as well as by blocking the ED9 antigen. Untill now several other membrane molecules have been described to play an essential role in clustering between APC and T cells[1,2,3,4], but CR3 has not been described to be involved in DC-T cell interaction before.

In contrast to the antigen-independent clustering, all types of APC can interact with T cells in an antigen dependent fashion. This type of clustering is primarily based on an interaction between antigen/MHC on the APC and the specific TCR on the T cells[2]. Our results show that indeed the antigen GT is needed for clustering. In this assay blocking of the accessory molecules recognized by the ED-mAb did not result in dissociation of the established clusters.

After clustering between APC and T cells has occurred, the immunostimulatory capacities of the APC can take place. This results in IL-2 production and responsiveness by T-cells. Inaba and Steinman showed that mAb to CD11a and CD4 did not influence clustering in the mixed leukocyte reaction, but the mAb blocked the function of clusters by interfering with the stability and the IL-2 release respectively[5]. This is conformable to our findings for we showed that the ED-antigens are involved as accessory molecules in the antigen-independent clustering, but they are superfluous in the antigen-dependent clustering. However, the mAb reactive with CR3 differentially inhibited T cell proliferation in the

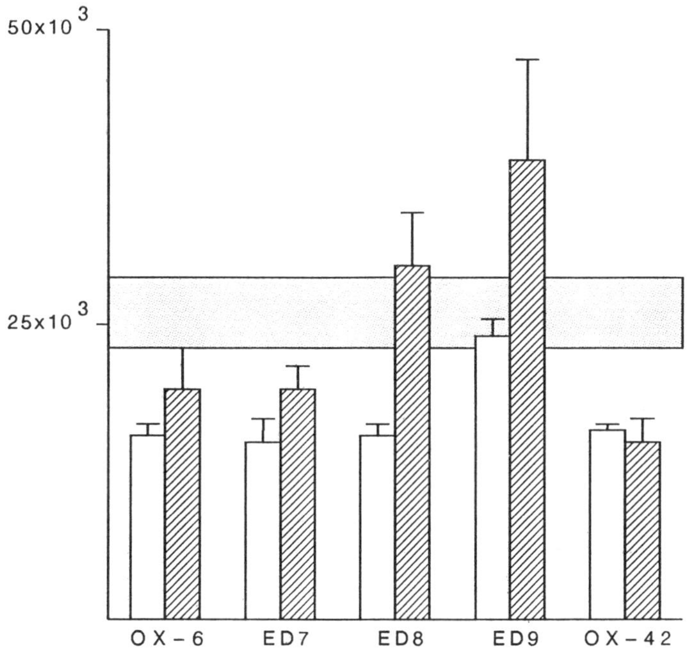

Figure 3. Effects of mAb on antigen presentation by peritoneal macrophages. Proliferation was assayed after 3 days incubation of APC with T cells isolated from ACI/Ma1 rats 2 weeks after priming with GT. Antibodies were purified using Protein A-Sepharose and applied in two concentrations: 5 μg/μl (open bars) and 1 μg/μl (shadowed bars). Tritiated thymidine uptake was measured after an 18 hr pulse before the end of the incubation period. Results are given as the average counts per minute ± SD for triplicate cultures. The horizontal bar represents the mean of the control ± SD.

antigen presentation assay depending on the type of APC used as accessory cell. It appears that the stability of the macrophage-T cell clusters is the weakest, for all the anti-CR3 mAb inhibit T cell proliferation in case macrophages function as APC, whereas only OX-42 blocks T cell proliferation when DC are used as APC. This difference might be explained by the difference in morphology and MHC clas II expression between these cell types. DC have more elongated cell processes as well as higher numbers and different types of MHC class II molecules enabling more intense contact with T cells[16], and making the accessory molecules partially superfluous.

In conclusion, we have demonstrated that three different mAb, recognizing rat APC, have differential effects on APC-T cell interaction. We also showed that at least in the rat CR3 is involved in APC-T cell clustering and antigen presentation.

REFERENCES

1. P.D. King and D.R. Katz. Mechanisms of dendritic cell function. *Immunol Today* 11:206 (1990).
2. K. Inaba and R.M. Steinman. Accessory cell-T lymphocyte interactions. Antigen-dependent and independent clustering. *J. Exp. Med.* 163:247 (1986).
3. J.M. Austyn and P.J. Morris. T-cell activation by dendritic cells: CD18-dependent clustering is not sufficient for mitogenesis. *Immunology* 63:537 (1988).
4. J. Green and R. Jotte. Interactions between T helper cells and dendritic cells during the rat mixed lymphocyte reaction. *J. Exp. Med.* 162:1546 (1985).
5. K. Inaba and R.M. Steinman. Monoclonal antibodies to LFA-1 and to CD4 inhibit the mixed leukocyte reaction after the antigen-dependent clustering of dendritic cells and T lymphocytes. *J. Exp. Med.* 165:1403 (1987).
6. J.M. Austyn, R.M. Steinman, D.E. Weinstein, A. Granelli-Piperno and M.A. Palladino. Dendritic cells initiate a two-stage mechanism for T lymphocyte proliferation. *J. Exp. Med.* 157:1101 (1983).
7. J.G.M.C. Damoiseaux, E.A. Döpp, J.J. Neefjes, R.H.J. Beelen and C.D. Dijkstra. Heterogeneity of macrophages in the rat evidenced by variability in determinants: two new anti-rat macrophage antibodies against a heterodimer of 160 and 95 kd (CD11/CD18). *J. Leukocyte Biol.* 46:556 (1989).
8. A.P. Robinson, T.M. White and D.W. Mason. Macrophage heterogeneity in the rat as delineated by two monoclonal antibodies MRC OX-41 and MRC OX-42, the latter recognizing complement receptor type 3. *Immunology* 57:239 (1986).
9. J.G.M.C. Damoiseaux. Macrophage Heterogeneity in the rat. PhD thesis, Vrije Universiteit Amsterdam (1991).
10. E.W.A. Kamperdijk, M.A.M. Verdaasdonk and R.H.J. Beelen. Visual and functional expression of Ia-antigen on macrophages and dendritic cells in ACI/Mal rats. *Transplant. Proc.* 19:3024 (1987).
11. M.A. Schon-Hegrad, J. Oliver, P.G. McMenamin and P.G. Holt. Studies on the density, distribution, and surface phenotype of intraepithelial class II major histocompatability complex antigen (Ia)-bearing dendritic cells (DC) in the conducting airways. *J. Exp. Med.* 173:1345 (1991).
12. F. Sanchez-Madrid, J.A. Nagy, E. Robbins, P. Simon and T.A. Springer. A human leukocyte differentiation antigen family with distinct α-subunits and a common β-cubunit. *J. Exp. Med.* 158:1785 (1983).
13. P.A. Detmers and S.D. Wright. Adhesion-promoting receptors on leukocytes. *Curr. Opin. Immunol.* 1:10 (1988).
14. H. Rosen and S. Gordon. Monoclonal antibody to the murine type 3 complement receptor inhibits adhesion of myelomonocytic cells in vitro an inflammatory cell recruitment in vivo. *J. Exp. Med.* 166:1685 (1987).
15. S. El Rouby, F. Praz, L. Halbwachts-Mecarelli and M. Papiernik. Thymic reticulum in mice IV. The rosette formation between phagocytic cells of the thymic reticulum and cortical type thymocytes is mediated by complement receptor type three. *J. Immunol.* 134:3625 (1985).
16. C.J.M. Melief. Dendritic cells as specialized antigen-presenting cells. *Res. Immunol.* 140:902 (1989).

THE MHC EXPRESSION OF DENDRITIC CELLS FROM MOUSE SPLEEN ISOLATED BY CENTRIFUGAL ELUTRIATION IS UPREGULATED DURING SHORT TERM CULTURE

Miriam A. Ossevoort[1], René E. M. Toes[1], Marloes L. H. De Bruijn[2], Cornelis J. M. Melief[1], Carl G. Figdor[2] and W. Martin Kast[1]

[1] Department of Immunohematology and Bloodbank University Hospital, Building 1, E3-Q, P.O.BOX 9600, 2300 RC, Leiden, The Netherlands
[2] Division of Immunology, The Netherlands Cancer Institute Amsterdam, The Netherlands

Dendritic cells (DC) are extremely potent antigen presenting cells (APC), which can even induce an anti-viral specific primary T cell response in vitro (Macatonia et al,1989). Both the quantitatively and qualitatively (fewer sialic acids) superior MHC expression and the large surface area of DC with long dendritic projections possibly allow a better interaction of the MHC-peptide complex on the DC with the T cell receptor (Boog et al., 1989). Generally, fewer DC than other types of APC are required to induce a specific T cell response (Kast et al., 1988).

DC used in these previous studies are isolated by the standard method (Steinman et al, 1979). This method takes two days and includes selective adherence to tissue culture plastic surfaces. Adherence may induce biological and biochemical changes in these cells which might result in an increase in MHC expression as has been described for human monocytes (Kelley et al., 1987; Figdor et al, 1986). A very pure fraction of non-activated monocytes could be isolated by centrifugal elutriation (CE) (Figdor et al, 1984). This cell separation procedure avoids cell adhesion. Because DC belong to the same cell lineage as monocytes it was of interest to know if non-activated DC with a low MHC expression could be isolated by CE.

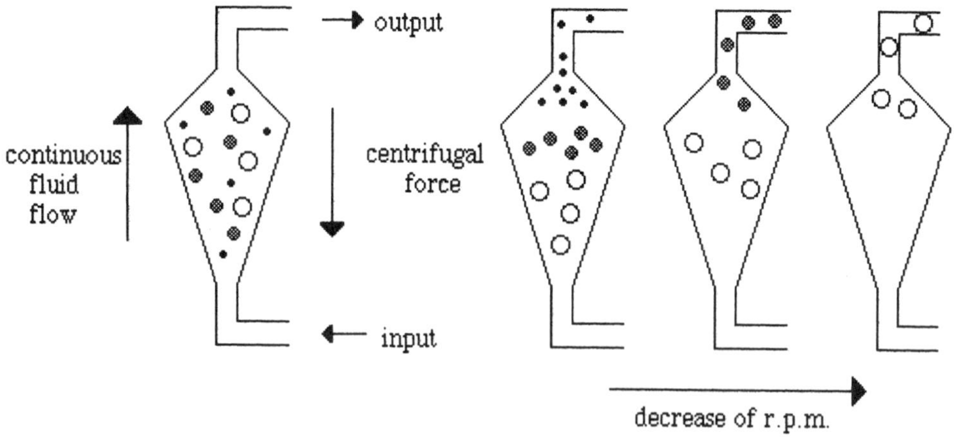

Figure 1. Schematic representation of centrifugal elutriation.

CE is used widely for the separation of different cell types (Pretlow and Pretlow, 1979). Cells, present in a chamber placed excentrically on a rotor are subjected to two opposing forces, the centrifugal force and a force caused by a continuous fluid flow through the chamber that is directed towards the rotor axis. At equilibrium all cells are positioned according to size and density. By a step wise decrease of the rotor speed, this equilibrium shifts and cells with the smallest volume are elutriated (Figure 1).

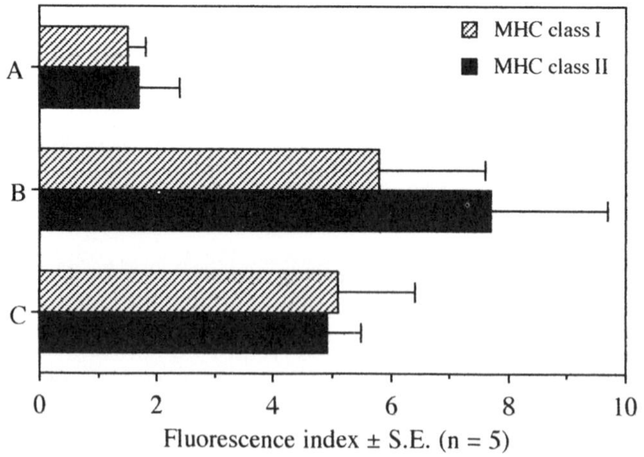

Figure 2. MHC expression of 33D1-positive cells of freshly isolated (A) or cultured (B) DC isolated by CE or DC isolated by the standard method (C) measured by two-color FACS analysis. Cell populations were stained with 33D1-biotin and PE-streptadivin plus a FITC-labeled antibody against MHC class I (H2-D^b, H141.31.10) or MHC class II (I-A, 17-227).

We isolated DC from mouse spleen by CE followed by discontinuous density centrifugation (Ossevoort et al., 1992). This method takes 5 hours. The isolation protocol is performed at 4°C and therefore does not allow DC to adhere to tissue culture plastic surfaces. DC isolated by CE contain 35 - 55% 33D1-positive cells (33D1 = DC-marker; Nussenzweig et al, 1982). Although DC isolated by the standard method contain 70 - 90% 33D1-positive cells, the absolute number of 33D1-positive cells isolated by both methods is comparable (Ossevoort et al., 1992).

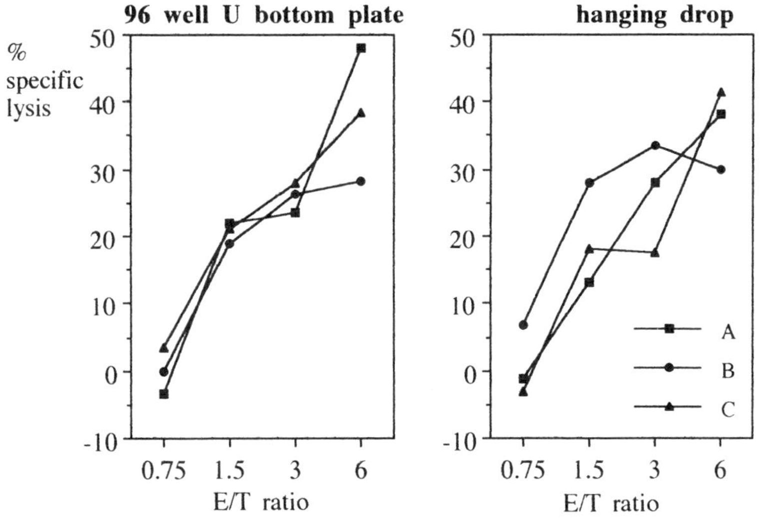

Figure 3. Stimulatory activity of DC isolated by CE, cultured for 18 hours in Teflon bags (no adherence)(A) or cultured for 18 hours in glass Petri dishes (adherence)(B) in comparison with DC isolated by the standard method(C). 10^5 nylonwool passed spleen cell of mice immunized with non-virulent Sendai virus were cocultured for 5 days with 10^3 irradiated (2500 rad) Sendai virus infected DC isolated by the various methods in a hanging drop culture or in 96 U-bottomed plate culture. Effector cells were tested in a ^{51}CR release assay on Sendai virus infected LPS induced B cell blast target cells in different E/T ratio's. Data are presented as mean of 3 determinations. Non-specific lysis of LPS induced B cell blast target cells was less than 5%.

Adherence during an 18 hour culture period which is a part of the standard method results in an increase of MHC class I and II expression on 33D1-positive cells isolated by CE to the same extent as the MHC expression of 33D1-positive cells isolated by the standard method (Figure 2). During this same period their morphology changed from a round shaped to a dendritic shaped appearance.

Despite the difference in MHC expression DC isolated by both methods have a comparable stimulatory capacity to induce secondary virus specific and primary viral peptide specific T cell responses (Ossevoort et al.,

1992). Because DC isolated by CE can adhere to tissue culture plastic surfaces during the induction of the T cell response, their MHC expression probably rises during culture. This could result in an increase in stimulatory capacity. However, when DC isolated by CE are precultured for 18 hours in Teflon bags (no adherence, so no increase in MHC expression) or glass Petri dishes (adherence) and the induction of a secondary virus specific cytotoxic T cell response is performed in a hanging drop culture, where cells can not adhere to the tissue culture plastic surfaces, their APC activity is comparable (Figure 3). Thus adherence to tissue culture plastic surfaces alone does not influence the stimulatory activity of DC isolated by CE. Possibly cell-cell contact or release of cytokines during the induction of the antigen specific T cell response might also induce an increase in MHC expression resulting in an excellent antigen presenting capacity.

Since DC isolated by CE have a low MHC expression they probably have a low MHC expression in vivo. DC isolated by the standard method which have a high MHC expression are inherently activated as a result of adherence of these cells to glass Petri dishes during the 18 hour culture period. Girolomoni et al. (1990) have shown that the MHC expression on DC isolated by cell adhesion during 90 minutes was lower compared to DC isolated after 18 hours indicating that a longer period of cell adherence results in a high MHC expression.

In conclusion, DC can be isolated by centrifugal elutriation more rapidly and without activation through cell adherence which inherently results in activation of these cells. Therefore this new isolation procedure makes it possible to perform further studies on the activation process of DC.

Acknowledgments

This work was supported by grants from the Netherlands Organization for Scientific Research (NWO), grant 900-509-151 and 900-509-092 and by the Dutch Cancer Society grant IKW 90-23. W. M. Kast is a fellow of the Royal Netherlands Academy of Arts and Sciences (KNAW).

References

Boog, C.J.P, Neefjes, J.J., Boes, J., Ploegh, H.L. and Melief, C.J.M., 1989, Specific immune responses restored by alteration in carbohydrate chains of surface molecules on antigen-presenting cells, *Eur. J. Immunol.* 19: 537.

Figdor, C.G., Van Es, W.L., Leemans, J.M.M. and Bont, W.S., 1984, A centrifugal elutriation system of separating small numbers of cells, *J. Immunol. Meth.* 68:37.

Figdor, C.G., Te Velde, A.A., Leemans, J. and Bont, W.S., 1986, Differences in functional, phenotypical and physical properties of human peripheral blood monocytes(Mo) reflect their various maturation stages, in Leukocytes and Host Defense, J.J. Oppenheim and D.M. Jacobs, Alan R. Liss ed., New York, p. 283.

Girolomoni, G., Simon, J.C., Bergstresser, P.R. and Ponciano Jr, D.C., 1990, Freshly isolated spleen dendritic cells and epidermal langerhans cells undergo similar phenotypic and functional changes during short term culture, *J. Immunol.* 145: 2820.

Kast, W.M., Boog, C.J.P., Roep, B.O., Voordouw, A.C. and Melief, C.J.M., 1988, Failure or succes in the restoration of virus-specific cytotoxic T lymphocyte response defects by dendritic cells, *J. Immunol.* 140:3186.

Macatonia, S.E., Taylor, P.M., Knight, S.C. and Askonas, B.A., 1989, Primary stimulation by dendritic cells induces antiviral proliferative and cytotoxic T cell responses in vitro, *J. Exp. Med.* 169:1255.
Nussenzweig, M.C., Steinman, R.M., Witmer, M.D. and Gutchinov, B., 1982, A monoclonal antibody specific for mouse dendritic cells, *Proc. Natl. Acad. Sci. U.S.A.* 79: 161.
Ossevoort, M.A., Toes, R.E.M., De Bruijn, M.L.H., Melief, C.J.M., Figdor, C.G. and Kast W.M., 1992, Rapid isolation procedure for dendritic cells from mouse spleen by centrifugal elutriation. *J. Immunol. Meth.in press.*
Pretlow, T.G. and Pretlow, T.P., 1979, Centrifugal elutriation (Counterstreaming centrifugation) of cells, *Cell Biophysics*, 1:195.
Steinman, R.M., Kaplan, G., Witmer, M.D. and Cohn, Z., 1979, Identification of a novel cell-type in peripheral lymphoid organs of mice. V. Purification of spleen dendritic cells, new surface markers and maintenance in vitro, *J. Exp. Med.* 149:1.

IL-6 AND ITS HIGH AFFINITY RECEPTOR DURING DIFFERENTIATION OF MONOCYTES INTO LANGERHANS CELLS

Gertrud Rossi,[1] Markus Schmitt,[2] Martin Owsianowski,[3]
Bernhard Thiele,[4] and Harald Gollnick[3]

[1] Institut für Molekularbiologie, Freie Universität Berlin
[2] Krankenhaus Zehlendorf, Bereich Heckeshorn, Berlin
[3] Hautklinik, Universitätsklinikum Steglitz, Berlin
[4] Charite der Humboldt-Universität Berlin

INTRODUCTION

In the skin variable concentrations of IL-6 are a result of local production by Langerhans cells (LC), keratinocytes and endothelial cells (1), as well as of posthepatic accumulation (2). Although LC precursors leaving the capillaries must migrate through the dermis, it remains unclear how far IL-6 influences the development and function of LC.

Recently, a Langerhans cell phenotype could be differentiated from peripheral blood monocytes under conditions of low serum concentrations. These monocyte derived LC (MoLC) were potent stimulators of mannan-specific T-cell proliferation, characterized by high HLA-DR and CD1a expression, constant mannose receptors and decreasing CD14 (3). MoLC were strictly dependent on their on conditioned medium which contained high concentrations of IL-6.

In the present work we studied the kinetics of IL-6 release and the development of high affinity receptors and we tried to derive the moment of maximal induction via IL-6 receptors and calculate its time relationship to the first appearence of CD1a and IL-4 receptors.

MATERIALS AND METHODS

Buffy coat monocytes were isolated by centrifugation on Ficoll-Paque and adherence on polystyrene surfaces. They were cultured in Iscove`s modified Dulbecco medium (IMDM) (Sigma) with 2% FCS (Gibco), in a 7% CO_2 in air atmosphere. The concentration of IL-6 in the supernatants was measured every 2 hours, later every 4 hours in an ELISA test established with mAbs and pAbs from Janssen and British Biotecnology (1). Flow

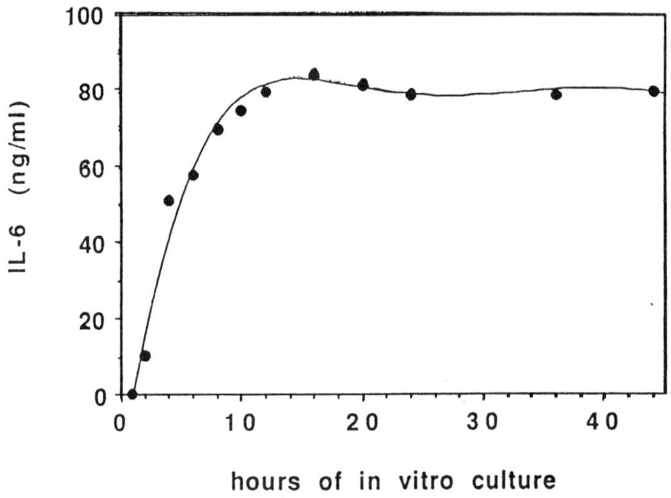

Figure 1. IL-6 concentration in supernatant of monocytes

cytometric analysis of IL-6 receptors was performed with the phycoerythrin-labelled ligand, IL-4 receptors were detected with biotin-labelled IL-4 which was stained with streptavidin-FITC (Biermann). CD1a was detected by FITC-labelled OKT6 (Ortho). High affinity receptors for IL-6 were deduced from scatchard plot analysis of ^{125}J-IL-6 binding. 10^6 cells were labelled in 0.1ml volumes on top of a silicon oil layer for one hour on ice. The cells were separated from unbound ligand by centrifugation through the oil layer (4). Treatments with GM-CSF (10 ng/ml) and IL-4 (100 U/ml) for 48 hours were used in order to study their influence on the development of MoLC.

RESULTS

The release of IL-6 by the monocyte layers was spontaneous and was detectable within one hour after adherence. The strongest increase was observed between 2 and 4 hours, and the maximal concentration in the supernatant was reached 12 hours after adherence. This concentration remained constant during the next 2 days, indicating that the release slowed down, and that an equilibrium was established between release and binding of IL-6 (Figure 1). Depending on the cell number which varied between 5 and $9x10^5$/ml, concentrations of 40 to 80 ng/ml were measured.

IL-6 receptors were detectable by flow cytometry in all the phases of MoLC development, from the PBM to 4 weeks old MoLC. Dot plot analysis of IL-6-PE binding showed the gated monocytes as a single positive population (Figure 2).

However, schatchard plot analysis of 125J-IL-6 binding revealed great variations in the number of high affinity receptors, which increased between one and twelf hours in vitro together with the concentration of released IL-6 (Figure 3). Since high affinity receptors are the result of association between the receptor and its signal transmitter in the presence of

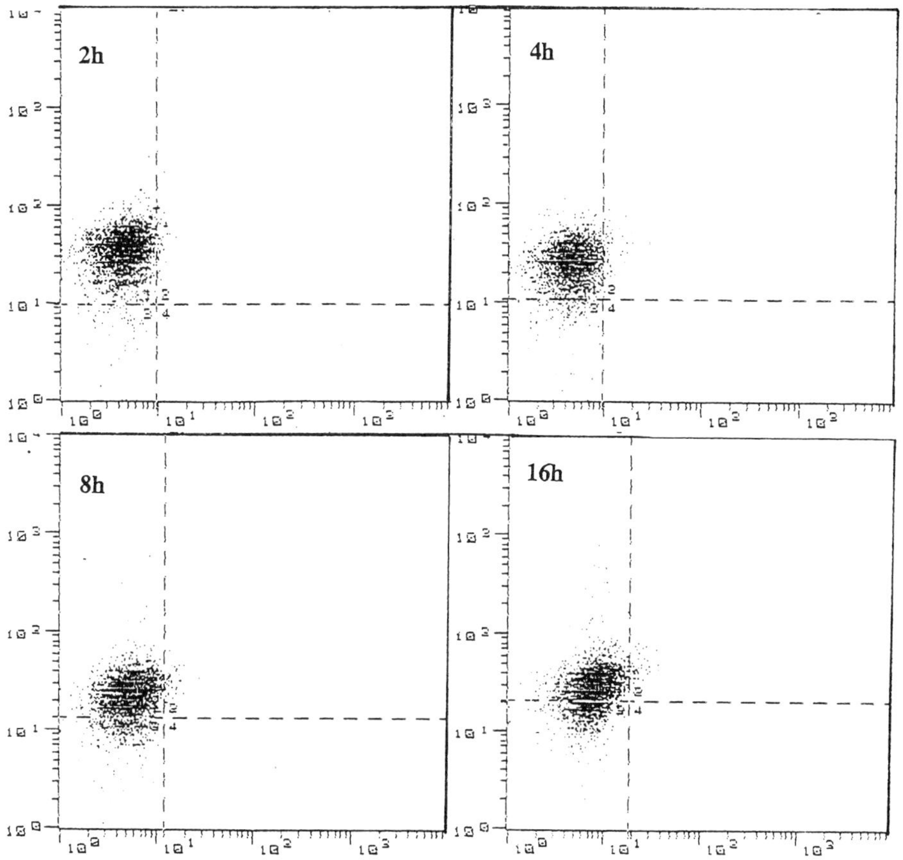

Figure 2. Binding of phycoerythrin-labelled IL-6

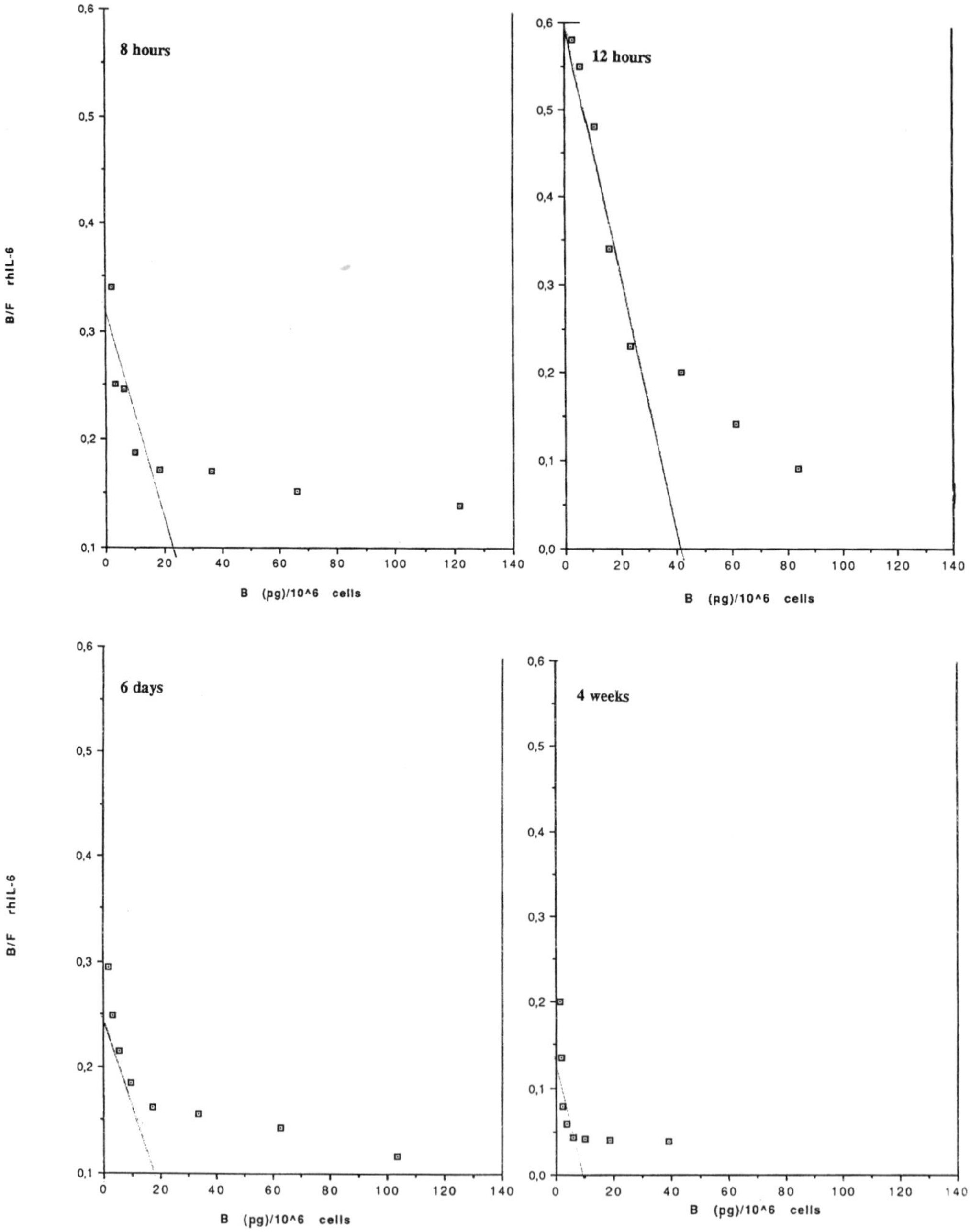

Figure 3. Development of high affinity receptors for IL-6

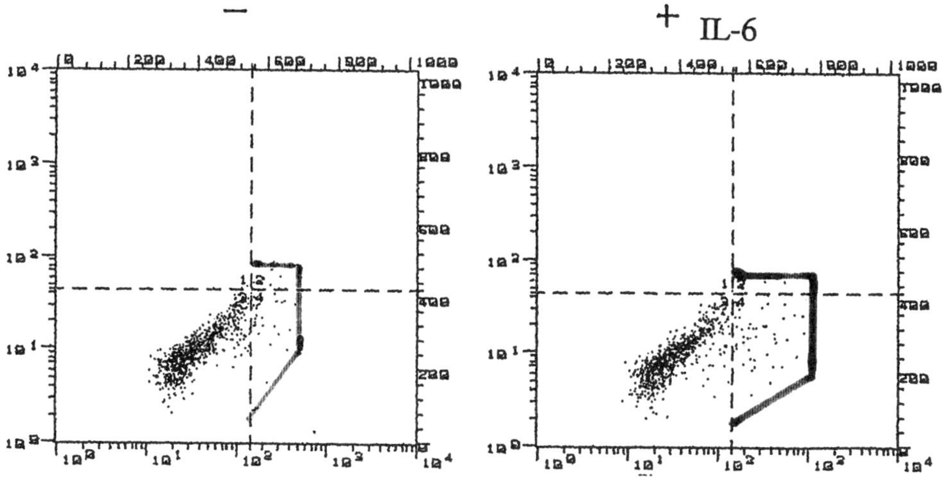

Figure 4. CDla expression, 20 hours in vitro. Effect of the addition of IL-6 (40ng/ml) immediately after adherence

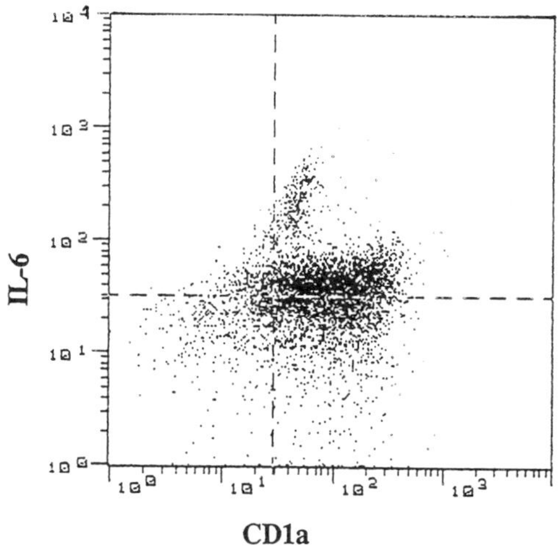

Figure 5. MoLC, 48 hours in vitro. Double labelling of CDla and IL-6 receptors

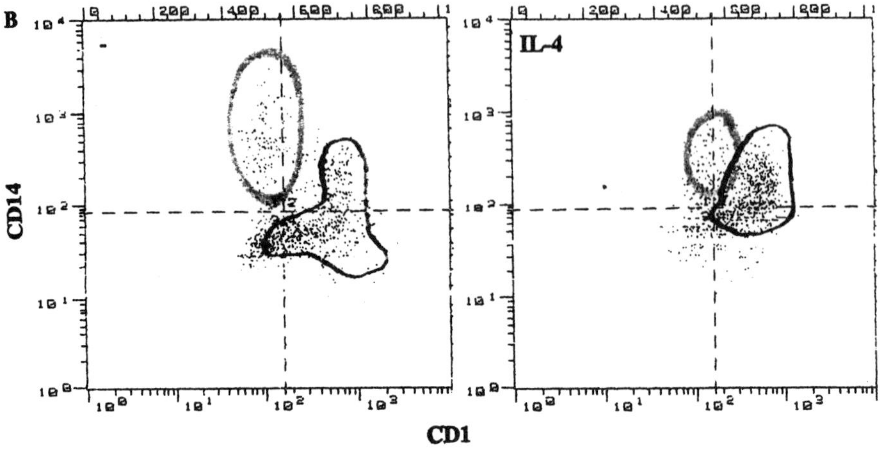

Figure 6. Enhancement of CD1a and decrease of CD14 expression after treatment with IL-4

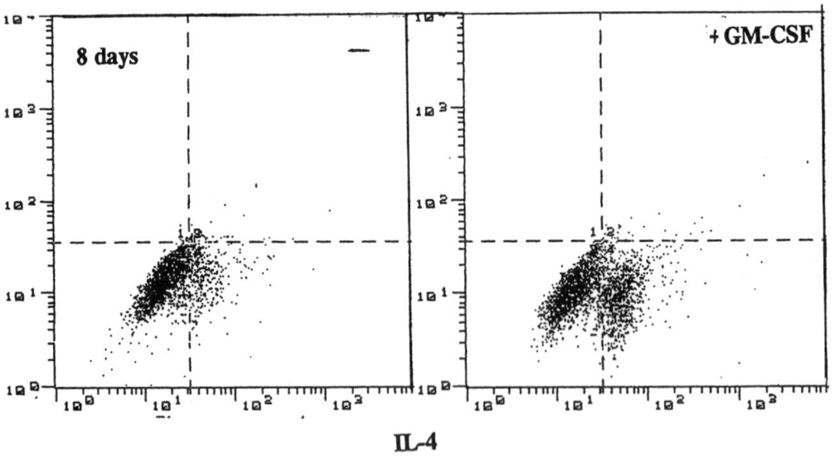

Figure 7. Enhancement of IL-4 receptor expression by treatment of MoLC with GM-CSF

ligand (4), we assume that around 12 hours in vitro the cells receive their maximal stimulation via the IL-6 receptor.

Between this moment and the first detection of CD1a positive cells a time interval of 8-12 hours occurred. The first outgrowth of CD1a positive cells was normally seen at 20-24 hours in vitro. We obtained a slight increase of the CD1a+ population in 20 hours old cells when 40ng/ml rhIL-6 were added immediately after adherence (Figure 4).

After 48 hours of in vitro culture, when CD1a was fully expressed, the IL-6 receptor bearing young MoLC were equally distributed from low to high CD1a expressing cells (Figure 5).

From day 4 in vitro MoLC were divided in one low density CD1 population which expressed CD14 and a second high density CD1 population with weak or no CD14. These two phenotypes remained stable for 4 weeks in the presence of their own conditioned medium or rhIL-6. A further enhancement of CD1a and decrease of CD14 expression was obtained only by IL-4 (Figure 6).

IL-4 receptors appeared as soon as 2 hours after adherence, too early for an induction by the released IL-6. They decreased from day 2 to day 8 and could then be induced again by GM-CSF (Figure 7), which itself did not increase the density of CD1a labelling.

Conclusions

The development of MoLC is essentially complished within 2 days (3). Like cord blood LC (5) they coexpress CD1a, the first marker of LC differentiation, and CD14 which decreases with increasing CD1 expression.

This way initiates with an enormous release of IL-6. From the contemporaneous increase of high affinity receptors we deduce an autocrine stimulation via the IL-6 R signal transmitter with its maximum at 12 hours in vitro.

The time interval of 8-12 hours between this stimulation and the first detection of CD1a would be conform either with direct or indirect induction of CD1 by IL-6. This is subject of next investigations.

MoLC remain non adherent, clustering and veiled cells phenotipically young LC if cultured in their own conditioned medium or rhIL-6. A further differentiation into mature LC needs lymphokines. It involves GM-CSF which enhances IL-4 receptor expression and IL-4 which increses CD1a expression together with a down regulation of CD14.

References

1. Detmar, M., Hettmannsperger, U., Owsianowski, M. et al. (1991)
Int. Archives Dermatol. 96, Abstracts, 999

2. Castell, J., Klapproth, J., gross, V., Walter, E., Andus, T., Snyers, L., Content, J., Heinrich, P.C. (1990) Eur. J. Biochem 189, 113-118
3. Rossi, G., Heveker, N., Thiele, B., Gelderblom, H., Steinbach, F. (1992) Immunology letters, 31, 189-198
4. Taga, T., Hibi, M., Hirata, Y., Yamasaki, H., Yasukawa, K., Matsuda, T., Hirano, T., Kishimoto, T. (1989) Cell 58, 573-581
5. Hanau, D., Gothelf, Y., Schmitt, D.A., Fabre, M., Garaud, J.C. Gazit, E., Cazenave, J.P. (1987) J. Invest. Dermatol. 89, 172-177

PHAGOCYTOSIS OF ANTIGENS BY LANGERHANS CELLS

Caetano Reis e Sousa and Jonathan M. Austyn

Nuffield Department of Surgery
University of Oxford
John Radcliffe Hospital
Headington
Oxford OX3 9DU
United Kingdom

INTRODUCTION

Unlike B cells, T cells do not recognise antigen in its native form. Instead, T cell receptors have evolved to recognise peptide fragments in association with MHC molecules on cell surfaces. MHC molecules of the class I type normally associate with peptides derived from endogenous antigens, such as cytosolic proteins, and are ideally placed for presenting viral antigens in infected cells[1]. MHC class II molecules (Ia) are primarily involved in presenting peptides derived from exogenous antigens and are restricted to a more limited number of cell types such as the dendritic cell[2].

Accessory cells must have four properties in order to function as efficient presenters for exogenous antigen. They 1) have to internalise antigen from the external milieu by endocytosis (soluble antigens) and phagocytosis (particulate antigens); 2) degrade internalised antigen to peptides suitable for binding to Ia. 3) synthesise class II molecules and target them to the compartment along the endocytic pathway where peptide association occurs; and 4) express the class II–peptide complexes on the cell surface and deliver appropriate costimulatory signals to T cells recognising those complexes. Lymphoid dendritic cells (DC) are potent accessory cells with high surface class II expression[3] and the unique capacity to activate naive T cells in an antigen–specific MHC–restricted manner[4]. However, splenic DC have been characterised as non–phagocytic and poorly endocytic *in vitro*[5]. Although a limited endocytic capacity may be sufficient to generate peptides for efficient presentation of soluble antigens (see below), the discovery that lymphoid DC originate from immature non-lymphoid dendritic leukocytes (DL) in peripheral tissues (see A.S Rao *et al*) has led to the idea that antigen acquisition is primarily a function of these precursors. Here we review evidence for uptake of different antigens by DL and discuss data from our laboratory suggesting that Langerhans cells (LC) are capable of handling particulate antigens.

LYMPHOID DC CAN ACQUIRE ANTIGENS FOR PRESENTATION

Until recently, it was thought that murine splenic DC were unable to present native antigens[6]. However, DC are purified in a lengthy procedure which includes an overnight step and leads to phenotypic changes[7, 8]. Recent data shows that freshly isolated DC can be pulsed with enough antigen, during but not after this step, to prime T cells *in vivo*[4] and generate antibody responses[9]. Experiments where proteins injected *in vivo* are presented by DC isolated shortly thereafter[10] also suggest that lymphoid DC can acquire antigens since the time course of activity makes it unlikely that antigen was acquired by non-lymphoid DC that subsequently migrated to spleen. These observations imply that lymphoid DC have the cellular machinery to generate peptides but do not formally rule out the hypothesis that processing is extracellular[11]. However, the visualisation of small amounts of antigen inside DC pulsed overnight[4] argues in favour of conventional intracellular processing. Recently, Nair *et al*[12] have reported introduction of exogenous protein into the class I presentation pathway of DC through pH–sensitive liposomes, again implying these were internalised prior to releasing their contents. In addition, we have observed that a fraction of fresh N418$^+$ low density splenocytes are phagocytic and down regulate this function during overnight culture (C. Reis e Sousa *et al*, manuscript in preparation). These and other observations[10] also argue against DC passively acquiring peptides regurgitated by a small number of MØ that may contaminate DC preparations.

It is not clear whether the loss of ability to process native antigen during overnight culture is due to down regulation of endocytic functions and/or a loss of processing machinery. Endocytosis can be a limiting factor in antigen presentation[13] and our unpublished observations suggest that changes in antigen handling observed for DC may be related to changes in endocytic capacity.

IMMATURE DL IN THE PERIPHERY AS SENTINELS OF THE IMMUNE SYSTEM

Many recent findings support the tentative theory that LC function to carry antigen penetrating epithelial barriers to the lymphoid tissues where it has maximal chance of encountering antigen-specific T cells. *In vitro*, LC mature into potent immunostimulatory cells which are virtually indistinguishable in phenotype from lymphoid DC[14, 15]. *In vivo*, this process is thought to be concomitant with migration to the lymphoid organs via the lymph and blood[16]. Cultured LC have lost the capacity to present native proteins although they present processed antigen very efficiently[6]. This functional change appears to correlate with loss of certain acidic vesicles from the endocytic pathway[17] but other studies have shown that there is no difference in accumulation of acridine orange, a lysosomotropic fluorochrome, in fresh and cultured LC[18]. The loss of antigen–processing ability and shutdown of class II synthesis[19] may have evolved to ensure that LC retain antigens encountered *in situ* and that they do not become loaded with irrelevant peptides *en route* to lymphoid tissues. This model explains how antigen encountered peripherally might be efficiently presented centrally and has important clinical implications, namely in the field of vaccine development.

Further support for it is provided by *in vivo* observations on LC and their migratory form in the lymph, the veiled cell (VC). Early reports implicated skin LC[20] and afferent lymph cells[21, 22] in the induction of contact sensitivity and LC have also been postulated to play a role in atopic disease[23]. Silberberg-Sinakin *et al*[24] injected ferritin intradermally into guinea pigs and found ferritin inside cells with Birbeck granules in draining lymph nodes (LN). Whole LN cell suspensions from immunised animals could transfer delayed-type hypersensitivity to ferritin when injected into naive syngeneic guinea pigs. Similarly, experiments using mice skin-painted with contact sensitising fluorochromes have demonstrated an influx of labelled dendritic cells into draining lymph nodes which can stimulate syngeneic T cells[25, 26] and can adoptively sensitise normal animals[25]. Although it is possible that in these experiments antigen drained to the LN and was picked up locally by LN DC, a study in sheep[27] shows that VC isolated from pseudo–afferent lymph do carry proteins injected intradermally, as assessed by antigen–specific proliferation assays.

None of the studies discussed above present direct evidence for uptake of antigen and intracellular processing by DL. Puré *et al*[19] have shown that antigen–processing by LC is sensitive to chloroquine and that LC endocytose rhodamine–ovalbumin. But other such evidence is controversial and, often, circumstantial. Endocytosis *in situ* by LC has been reported for peroxidase injected intradermally in humans[28] or administered intravaginally in mice[29] but no evidence for uptake of ^{125}I-flagellin was found in similar experiments on guinea pigs[30]. VC from mice internalise little antigen *in vitro*[31] and VC from sheep do not endocytose immune complexes *in vitro* but may do so *in vivo*[32]. Uptake of viruses has also been reported[33] but it is unclear whether it represents a physiological process or a viral mechanism leading to infection. Some reports of Fc receptor–mediated uptake by sheep VC fail to differentiate between true internalisation and binding to the cell surface[34].

In the light of such conflicting reports it is interesting to note that the unique structural marker of LC in many species, the Birbeck granule (BG), has an endocytic origin[35, 36]. Recently, it has been reported that BG are lost as part of the maturational change undergone by LC in culture[17]. In addition, BG are accessible to surface MHC molecules cross–linked by antibodies[37, 38] and they may recycle antigens back to the cell membrane[39]. These observations could have important implications for antigen processing and presentation.

PHAGOCYTOSIS OF ANTIGENS BY LC

Phagocytosis of particulate antigens is a process fundamentally distinct from the uptake of soluble substances discussed above. Most cells are capable of endocytosing solutes from the external milieu but uptake of particles is believed to be restricted to specialised cells, often called professional phagocytes. Studies on DL have emphasised their non–phagocytic nature[5, 40-44] without addressing the paradox that a cell that initiates immune responses to exogenous antigen must be able to take up particulate, often pathogenic, organisms. Reports of phagocytosis by immature DL are sparse and, largely, inferential; many fail to distinguish between endocytosis and phagocytosis. Veiled cells from rat contain inclusions reminiscent of a phagocytic past but do not phagocytose *in vitro*[40]. Poulter *et al*[45] describe rare CD1a$^+$ cells in the dermis of leprous skin which contain intracellular *Mycobacterium leprae*, but none in the epidermis. VC from sheep may take up latex beads[33].and one report mentions that murine LC have some phagocytic capacity *in vitro*[46]. Other reports of phagocytosis by DL suggest they can play a non–immunological role in homeostasis. LC phagocytose apoptotic epithelia cells in vaginal epithelium[47] and VC often show vacuoles containing erythrocytes[32]. Also, interdigitating cells in LN ingest allogeneic lymphocytes injected intravenously into rats[48].

We have made use of a combination of flow cytometry and light and electron microscopy to examine the phagocytic ability of LC *in vitro*. Our results (C. Reis e Sousa *et al*, manuscript in preparation) indicate that freshly isolated LC are phagocytic for a wide variety of markers such as latex beads, bacteria and yeast–derived particles. Unlike MØ, LC do not appear to phagocytose IgG–coated sheep erythrocytes or colloidal carbon. The extent of phagocytosis for other particles is also limited compared to that of MØ, supporting the notion that it is mainly concerned with acquisition of antigen for presentation rather than with immune clearance. Further support for this hypothesis comes from the observation that LC down regulate their phagocytic potential during culture and come to resemble lymphoid DC in that respect. We have also identified mouse strain differences in the uptake of particles with complex surfaces such as zymosan which may be due to differences in expression of specific phagocytic receptors. Inhibition studies suggest there are at least three receptors mediating phagocytosis of zymosan by LC, two of which, the mannose[49] and glucan[50] receptors, have not previously been described in LC. The third one, CR3, has not been attributed a functional role in LC although it is a well characterised phagocytic receptor in MØ. It is tentative to speculate that down regulation of phagocytic activity is yet another functional manifestation of the maturation process that LC undergo in culture. Given that phagocytic activity may be a property conferred by specific receptors rather than a property of a given cell type[51], we have tried to monitor the expression of mannose receptor (MR) during culture of LC. Specific uptake of mannosylated bovine serum albumin, a ligand for MR, decreases in culture with similar kinetics to those observed for the uptake of zymosan suggesting the same receptor is involved in both cases.

Thus, our data support the notion that uptake of antigen is primarily a function of non–lymphoid DC. Putative kidney and heart DC are also phagocytic *in vitro* (A.S. Rao, J.A. Roake, C.R.S. AND J.M.A., unpublished observations) and become less so after maturation. Cultured LC have not totally lost phagocytic potential and resemble fresh splenic DC in that respect. Phagocytosis of parasites by murine LC could explain the recent observation that they can present antigens from whole *Leishmania major*[52], opening up a new line of investigation into the potential role of DC in parasitic diseases. Our results support the idea that antigen is carried into central lymphoid tissues associated with dendritic cells.

REFERENCES

1. J.J. Monaco, A molecular model of MHC class-I-restricted antigen processing, *Immunol. Today* 13:173 (1992).
2. J.J. Neefjes and H.L. Ploegh, Intracellular transport of MHC class II molecules, *Immunol. Today* 13:179 (1992).
3. R.M. Steinman, The dendritic cell system and its role in immunogenicity, *Annu. Rev. Immunol.* 9:271 (1991).
4. K. Inaba *et al*, Dendritic cells pulsed with protein antigens in vitro can prime antigen-specific, MHC-restricted T cells in situ, *J. Exp. Med.* 172:631 (1990).
5. R.M. Steinman and Z. Cohn, Identification of a novel cell type in peripheral lymphoid organs of mice: II. Functional properties in vitro, *J. Exp. Med.* 139:380 (1974).
6. N. Romani *et al*, Presentation of exogenous protein antigens by dendritic cells to T cell clones. Intact protein is presented best by immature, epidermal Langerhans cells, *J. Exp. Med.* 169:1169 (1989).
7. M.T. Crowley *et al*, Use of the fluorescence activated cell sorter to enrich dendritic cells from mouse spleen, *J. Immunol. Methods* 133:55 (1990).
8. G. Girolomoni *et al*, Freshly isolated spleen dendritic cells and epidermal Langerhans cells undergo similar phenotypic and functional changes during short-term culture, *J. Immunol.* 145:2820 (1990).
9. T. Sornasse *et al*, Antigen-pulsed dendritic cells can efficiently induce an antibody response in vivo, *J. Exp. Med.* 175:15 (1992).
10. M. Crowley, K. Inaba and R.M. Steinman, Dendritic cells are the principal cells in mouse spleen bearing immunogenic fragments of foreign proteins, *J. Exp. Med.* 172:383 (1990).
11. P.M. Kaye, B.M. Chain and M. Feldmann, Non-phagocytic dendritic cells are effective accessory cells for anti-mycobacterial responses in vitro, *J. Immunol.* 134:1930 (1985).
12. S. Nair *et al*, Soluble proteins delivered to dendritic cells via pH-sensitive liposomes induce primary cytotoxic T lymphocyte responses in vitro, *J. Exp. Med.* 175:609 (1992).
13. B. Stockinger, Capacity of antigen uptake by B cells, fibroblasts or macrophages determines efficiency of presentation of a soluble self antigen (C5) to T lymphocytes, *Eur. J. Immunol.* 22:1271 (1992).
14. G. Schuler and R.M. Steinman, Murine epidermal Langerhans cells mature into potent immunostimulatory dendritic cells in vitro, *J. Exp. Med.* 161:526 (1985).

15. N. Romani *et al*, Cultured human Langerhans cells resemble lymphoid dendritic cells in phenotype and function, *J. Invest. Dermatol.* 93:600 (1989).
16. J.M. Austyn and C.P. Larsen, Migration patterns of dendritic leukocytes: implications for transplantation, *Transplantation* 48:1 (1990).
17. H. Stössel *et al*, Disappearance of certain acidic organelles (endosomes and Langerhans cell granules) accompanies loss of antigen processing capacity upon culture of epidermal Langerhans cells, *J. Exp. Med.* 172:1471 (1990).
18. G. Girolomoni *et al*, Vacuolar acidification and bafilomycin-sensitive proton translocating ATPase in human epidermal Langerhans cells, *J. Invest. Dermatol.* 96:735 (1991).
19. E. Puré *et al*, Antigen processing by epidermal Langerhans cells correlates with the level of biosynthesis of major histocompatibility complex class II molecules and expression of invariant chain, *J. Exp. Med.* 172:1459 (1990).
20. W.B. Shelley and L. Juhlin, Langerhans cells form a reticuloendothelial trap for external contact antigen, *Nature* 261:46 (1976).
21. B. Søeberg, T. Sumerska and B.M. Balfour, The role of the afferent lymph in the induction of contact sensitivty, *Adv. Exp. Med. Biol.* 66:191 (1976).
22. B. Søeberg *et al*, Contact sensitivity in the pig, *Int. Arch. Allergy Appl. Immunol.* 57:114 (1978).
23. G.C. Mudde *et al*, IgE: an immunoglobulin specialized in antigen capture?, *Immunol. Today* 11:440 (1990).
24. I. Silberberg-Sinakin *et al*, Antigen-bearing Langerhans cells in skin, dermal lymphatics and in lymph nodes, *Cell. Immunol.* 25:137 (1976).
25. S.E. Macatonia, A.J. Edwards and S.C. Knight, Dendritic cells and the initiation of contact sensitivity to fluorescein isothiocyanate, *Immunology* 59:509 (1986).
26. S.E. Macatonia *et al*, Localization of antigen on lymph node dendritic cells after exposure to the contact sensitizer fluorescein isothiocyanate. Functional and morphological studies, *J. Exp. Med.* 166:1654 (1987).
27. R. Bujdoso *et al*, Characterization of sheep afferent lymph dendritic cells and their role in antigen carriage, *J. Exp. Med.* 170:1285 (1989).
28. K. Wolff and E. Schreiner, Uptake, intracellular transport and degradation of exogenous protein by Langerhans cells, *J. Invest. Dermatol.* 54:37 (1970).
29. M.B. Parr, L. Kepple and E.L. Parr, Antigen recognition in the female reproductive tract. II. Endocytosis of horseradish peroxidase by Langerhans cells in murine vaginal epithelium, *Biol. Reprod.* 45:261 (1991).
30. S. Barbey *et al*, Skin Langerhans cells fail to trap bacterial antigen in non-sensitized guinea-pig, *Ann. Immunol. (Paris)* 191:111 (1981).
31. J.M. Rhodes *et al*, Comparison of antigen uptake by peritoneal macrophages and veiled cells from the thoracic duct using isotope-, FITC-, or gold-labelled antigen, *Immunology* 68:403 (1989).
32. J.G. Hall and D. Robertson, Phagocytosis, in vivo, of immune complexes by dendritic cells in the lymph of sheep, *Int. Arch. Allergy Appl. Immunol.* 73:155 (1984).
33. R. Barfoot *et al*, Some properties of dendritic macrophages from peripheral lymph, *Immunol.* 68:233 (1989).
34. G.D. Harkiss, J. Hopkins and I. McConnell, Uptake of antigen by afferent lymph dendritic cells mediated by antibody, *Eur. J. Immunol.* 20:2367 (1990).
35. M. Ishii *et al*, Sequential production of Birbeck granules through adsorptive pinocytosis, *J. Invest. Dermatol.* 82:28 (1984).
36. M. Takigawa *et al*, The Langerhans cell granule is an adsorptive endocytic organelle, *J. Invest. Dermatol.* 85:12 (1985).
37. D. Hanau *et al*, Human epidermal Langerhans cells internalize by receptor-mediated endocytosis T6 (CD1 "NA1/34") surface antigen. Birbeck granules are involved in the intracellular traffic of the T6 antigen, *J. Invest. Dermatol.* 89:172 (1987).
38. D. Hanau *et al*, Human epidermal Langerhans cells cointernalize by receptor-mediated endocytosis "nonclassical" major histocompatibility complex class I molecules (T6 antigens) and class II molecules (HLA-DR antigens), *Proc. Natl. Acad. Sci. U.S.A.* 84:2901 (1987).
39. A. Ray *et al*, Reappearance of CD1a antigenic sites after endocytosis on human Langerhans cells evidenced by immunogoldrelabeling, *J. Invest. Dermatol.* 92:217 (1989).
40. C.W. Pugh, G.G. MacPherson and H.W. Steer, Characterization of non-lymphoid cells derived from rat peripheral lymph, *J. Exp. Med.* 157:(1983).
41. D. Landry *et al*, Human thymic dendritic cells. Characterization, isolation and functional assays, *Immunology* 65:135 (1988).
42. D.N. Hart and J.L. McKenzie, Isolation and characterization of human tonsil dendritic cells, *J. Exp. Med.* 168:157 (1988).

43. A.M. Pollard and M.F. Lipscomb, Characterization of murine lung dendritic cells: similarities to Langerhans cells and thymic dendritic cells, *J. Exp. Med.* 172:159 (1990).
44. W.J. Xia *et al*, Accessory cells of the lung. II. Ia+ pulmonary dendritic cells display cell surface antigen heterogeneity, *Am. J. Respir. Cell Mol. Biol.* 5:276 (1991).
45. L.W. Poulter *et al*, Parasitism of antigen presenting cells in hyperbacillary leprosy, *Clin. Exp. Immunol.* 55:611 (1984).
46. D.R. Katz and G.H. Sunshine, Comparative accessory cell function of Langerhans cells isolated from mouse skin, *J. Exp. Pathol.* 67:157 (1986).
47. M.B. Parr, L. Kepple and E.L. Parr, Langerhans cells phagocytose vaginal epithelial cells undergoing apoptosis during the murine estrous cycle, *Biol. Reprod.* 45:252 (1991).
48. S. Fossum and B. Rolstad, The roles of interdigitating cells and natural killer cells in the rapid rejection of allogeneic lymphocytes, *Eur. J. Immunol.* 16:440 (1986).
49. S.J. Sung, R.S. Nelson and S.M. Silverstein, Yeast mannans inhibit binding and phagocytosis of zymosan by mouse peritoneal macrophages, *J. Cell Biol.* 96:160 (1983).
50. J.K. Czop and K.F. Austen, A ß-glucan inhibitable receptor on human monocytes: its identity with the phagocytic receptor for particulate activators of the alternative complement pathway, *J. Immunol* 134:2588 (1985).
51. R. Ezekowitz et al, Molecular characterization of the human macrophage mannose receptor: demonstration of multiple carbohydrate recognition-like domains and phagocytosis of yeasts in Cos-1 cells, *J. Exp. Med.* 172:1785 (1990).
52. A. Will *et al*, Murine epidermal Langerhans cells are potent stimulators of an antigen–specific T cell response to *Leishmania major*, the cause of cutaneous leishmaniasis, *Eur. J. Immunol.* 22:1341 (1992).

DISSECTION OF HUMAN LANGERHANS CELL ALLOSTIMULATORY FUNCTION. MODULATION BY INTERFERON-γ.

J. Péguet- Navarro, C. Dalbiez-Gauthier, C. Dezutter-Dambuyant, D. Schmitt.

INSERM U346, Hôpital E. Herriot
Pav. R, Lyon, France.

INTRODUCTION

It has now become evident that dendritic cells (DC) are specialyzed in initiating primary T cell-dependent immune responses[1]. A particular large pool of DC, namely Langerhans cells (LC) are present within epidermis where they are the only cells which express MHC class II antigens constitutively. LC play a key role in the initiation of T cell responses to cutaneous antigens by picking up the antigen and migrating to the draining lymph nodes where they trigger specific T cell activation.

Recent studies revealed that several LC characteristic features change during a 2-3 day in vitro incubation, giving to LC an interdigitating cell appearance[2,3]. Cultured LC become more dendritic, lose or markedly reduced cytoplasmic Birbeck granules and expression of CD1 antigens. Concurrently the cells express higher levels of MHC antigens and adhesion molecules such as ICAM-1 and LFA-3. Furthermore, it was recently reported in the murine system that cultured LC become more potent stimulators of resting T cells and the mediator of this maturation process was identified as GM/CSF[4]. The mechanism leading to enhanced stimulatory function remains, however, to be clarified.

In the present study we dissected the stimulating activity of both freshly isolated and 3-day cultured human LC in the induction of primary allogeneic T cell responses. In order to dissociate membrane-bound stimuli from soluble activating factors, LC were fixed with paraformaldehyde (PF) before their addition to the T cells.

MATERIAL AND METHODS

Culture medium: RPMI-1640 medium (Gibco Laboratories, USA) supplemented with 10% human AB serum, 1mM L-Glutamine (Gibco), 1μg/ml indomethacine (Sigma, USA) and antibiotics was used and referred to as complete culture medium.

Epidermal cell suspensions: Epidermal cell suspensions were obtained from normal skin removed during plastic surgery through the action of 0.05% trypsin for 18h at 4°C. Enriched-LC suspensions (eLC, 10-20% LC) were prepared by density centrifugation.

eLC incubation: eLC were incubated for 1 to 3 days in RPMI complete medium supplemented or not with interferon-γ (IFN-γ, Genzyme Corporation, Boston, USA, 250u/ml). Then the cells were washed, layered on a Lymphoprep gradient to remove dead cells, and enumerated.

PF fixation: Freshly prepared or 3 day-incubated eLC were fixed in 2% PF, for 30 min at 20°C. Cells were then extensively washed before their addition to the T cells.

T cells: Peripheral blood mononuclear cell suspensions were obtained from the peripheral blood of allogeneic donors. Monocytes were depleted by adherence on plastic dishes and purified T cells were prepared by rosetting with sheep erythrocytes. The resulting cell suspension was unable to proliferate in response to Concanavalin A.

Proliferation assays: Mixed epidermal cell lymphocyte reaction (MELR) were performed in 96-well microtiter plates by adding 10^5 T lymphocytes to 10^3 stimulator cells. In some cases, recombinant IL-1 (Genzyme Corporation) or/and IL-6 (Genzyme Corporation) was added to the wells at the beginning of culture. After 5 day incubation at 37°C, the cells were pulsed with 1μCi of ^{3}H-methyl thymidine and harvested 18hrs later.

Immunogold labeling and electron microscopy analysis: LC suspensions were labeled with anti HLA-DR monoclonal antibody coupled to gold granules for 30 min at 4°C. After washing the cells were fixed and processed for electron microscopy analysis.

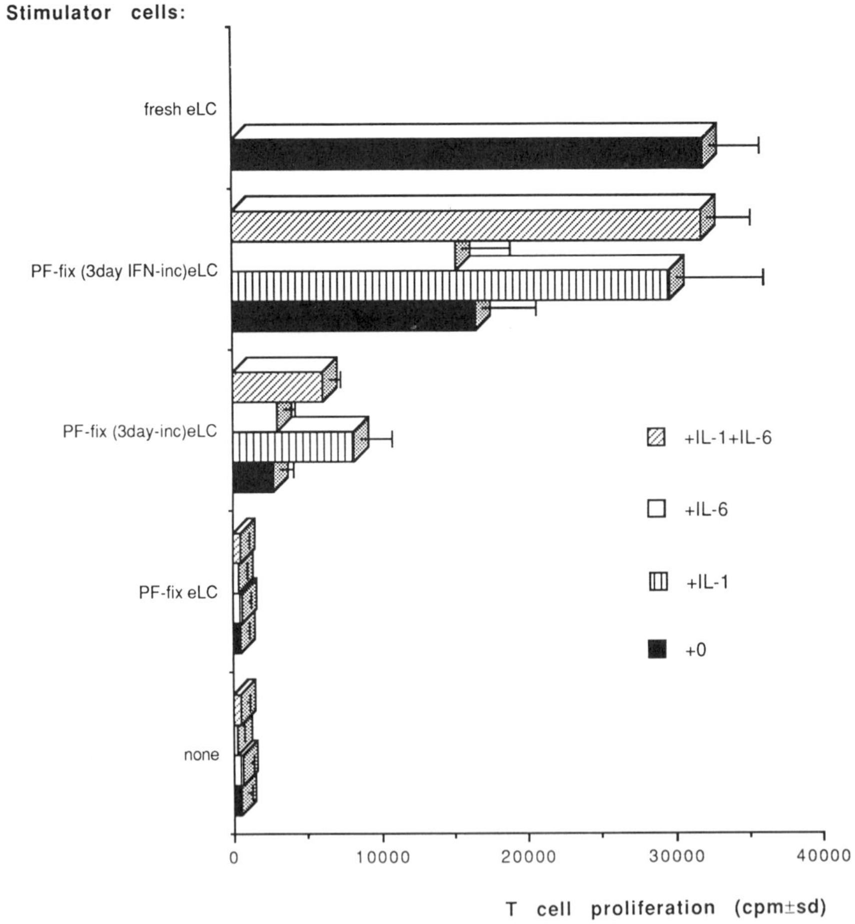

Figure 1. Allostimulatory function by PF-fixed human eLC. eLC were fixed either after isolation from skin samples or after a 3 day-incubation in medium supplemented or not with IFN-γ (250u/ml). Fixed LC (10^3 cells) were then added to allogeneic T cells (10^5 cells) in the presence, or not, of IL-1 (10u/ml) or/and IL-6 (100u/ml). Controls were performed using fresh LC as stimulators. T cell proliferation was assessed by ^{3}H-thymidine incorporation during the last 18hrs of culture. Results were the mean±sd of triplicate cultures.

RESULTS

PF-Fixed eLC Failed to Initiate Allogeneic T Cell Responses

As shown in Fig 1, freshly prepared eLC were good stimulators of primary allogeneic T cell responses and as few as 10^3 LC were sufficient to induce potent T cell proliferation. By contrast, eLC fixation with PF completely abrogated this property, and addition of either IL-1 (10-50u/ml) or/and IL-6 (10-100u/ml) failed to restore the T cell response. The absence of T cell proliferation was not related to a suppressive influence, such as leaching of the fixator, since PF-fixed eLC did not supress T cell response to non-fixed eLC (data not shown). Furthermore, as evidenced by electron microscopy, PF fixation had no apparent effect on the expression of MHC class II encoded determinants (data not shown).

Allostimulatory Function of PF-Fixed 3 Day-Incubated eLC

In the following experiments, eLC were first incubated for 3 days in complete RPMI medium before fixation. As reported in Fig 1, these cells exhibit a low but significant allostimulatory property. Moreover, T cell response was consistantly enhanced by the addition of IL-1 (10u/ml), but not IL-6 (10-100u/ml) during the MELR. We did not observe any potentiating effect when the two cytokines were added simultaneously.

Since trypsin treatment was necessary to isolate epidermal cells from human skin samples, allostimulatory function of incubated LC might result from a mere restoration of cell membrane from enzyme treatment. To test this possibility, incubated eLC were submitted to 0,05% trypsin for 30 min at 37°C before fixation. No effect of the enzyme was found on human LC function (data not shown).

Interferon-γ (IFN-γ) is expected to be released during LC-T cell interactions. Therefore, this cytokine was added to eLC during the 3 day incubation. IFN-γ pretreatment mediated a potent allostimulatory function by PF-fixed eLC (Fig 1). Furthermore, T cell proliferation was increased by the addition of IL-1, but not IL-6, during the MELR, leading in some experiments to a complete recovery of the control response induced by a similar number of fresh LC (Fig1).

Allostimulatory Function by PF-fixed 3 Day-Incubated eLC did not Correlate with Increased HLA-DR Expression

Because 3 day-incubated LC were known to increase HLA class II antigen expression and because IFN-γ is known to upregulate these antigens at the surface of most cells, it was important to consider whether the enhanced expression of MHC products could account for the above reported results. This assumption was unlikely, however, since we found, using electron microscopy, that IFN-γ did not significantly potentiate the increase of HLA class II molecules on 3 day-incubated human LC (Table I).

Table 1. Effect of IFN-γ on HLA-DR expression by incubated human LC.

Cells	Addition of IFN-γ	mean gold particle count / 100μm membrane[1]
fresh LC	-	1601±787
18hr inc-LC	-	2543±688
	+	2508±676
3day inc-LC	-	2529±633
	+	2562±643

[1]The number of gold particles along the cell membrane was counted on 30 LC in each population

DISCUSSION

Interaction of T cell-receptor with antigen MHC-class II complex on antigen presenting cells (APC) is, in most cases insufficient to trigger T lymphocyte proliferation and APC must convey additionnal signals that enable the T cells to be fully activated and to proliferate. These accessory signals involved both membrane-bound ligands which react with specific receptors on T cells and soluble cytokines[5]. DC are major APC since they are unique in initiating primary T cell responses, but little is known about the mechanism of this superior APC function[6].

In the present study, we analyzed the signals required for induction of a primary allogeneic T cell response using PF-fixed human eLC as APC. We found that freshly prepared eLC lost their allostimulatory activity after PF fixation and that addition of IL-1, or/and IL-6, failed to restore the reaction. It appears unlikely that PF fixation prevented recognition of allogeneic determinants. Indeed, we found that when LC were first incubated in vitro and especially in the presence of IFN-γ before fixation, they retained significant allostimulatory property. The resulting T cell proliferation was inhibited by anti-class II monoclonal antibody (data not shown), providing evidence that allodeterminants are still effective on the fixed cells. Based on these results, it is likely that freshly isolated human LC are not equipped to trigger allogeneic T cell response and that in vitro incubation allowed expression of membrane-bound costimulatory molecules that were required in this process. Indeed profound morphologic and phenotypic changes have been demonstrated on human LC after a short in vitro culture[2,3]. The present study using PF-fixed LC therefore provides evidence that these cell-membrane modifications are associated with increased accessory function. Whether upregulated adhesion molecules such as ICAM-1 and LFA-3 play a major role in this process is currently under investigation.

An interesting finding is that major LC activation developped after in vitro incubation with IFN-γ, suggesting that fully development of LC accessory function requires LC-T cell interaction. Furthermore, the results showed that together with membrane-bound stimuli, IL-1 can serve as a costimulatory factor in the primary T cell response induced by allogeneic LC.

REFERENCES

1. K. Inaba, R.M. Steinman, Resting and sensitized T lymphocytes exhibit distinct stimulatory (antigen-presenting cell) requirements for growth and lymphokine release, *J. Exp. Med.* 160:1717 (1984).

2. G. Schuler, R.M. Steinman, Murine epidermal Langerhans cells mature into potent immunostimulatory dendritic cells in vitro, *J. Exp. Med.* 161:526 (1985).

3.N. Romani, A. Lenz, H. Glassel, H. Stossel, U. Stanzl, O. Majdic, P. Fritsch, G. Schuler, Cultured human Langerhans cells resemble lymphoid dendritic cells in phenotype and function, *J. Invest. Dermatol.* 93:600 (1989).

4. M.D. Witmer-Pack, W. Olivier, J. Valinsky, G. Schuler, R.M. Steinman, Granulocyte-macrophage colony-stimulating-factor is essential for the viability and function of cultured murine epidermal Langerhans cells, *J. Exp. Med.* 166:1484 (1987).

5. T.D. Geppert, L.S. Davis, H. Gur, M.C. Wacholtz, P.E. Lipsky, Accessory cell signals involved in T cell activation. Immunol. Rev. 117:5 (1990).

6. R.M. Steinman, The dendritic cell system and its role in immunogenicity, *Annu. Rev. Immunol.* 9:271 (1991).

HUMAN *IN VITRO* T CELL SENSITIZATION USING HAPTEN-MODIFIED EPIDERMAL LANGERHANS CELLS

Corinne Moulon,[1] Josette Péguet-Navarro,[1] Pascal Courtellemont, [2]
Gérard Redziniak,[2] and Daniel Schmitt [1]

[1] INSERM U 346, Hôpital E. Herriot, Pav. R, Lyon, France
[2] Centre de Recherches PCD, Saint-Jean de Braye, France

INTRODUCTION

Epidermal Langerhans cells (LC), a population of non-lymphoid dendritic cells which constitutively express MHC class II molecules, have been demonstrated to play a key role in the development of contact hypersensitivity reactions by picking up the haptens within epidermis and migrating to draining lymph nodes where antigen presentation to specific T cells occurs[1]. Recent studies reported that after 2-3 day *in vitro* incubation, LC undergo profound phenotypic changes and acquire an interdigitating cell appearance[2,3]. In the murine system, at least in some strains of mice, the incubated LC become substantially more potent accessory cells than fresh LC, while they are relatively inefficient in processing of protein antigens[4]. It has been thus suggested that cultured LC may represent the *in vitro* counterparts of antigen-bearing LC that have migrated to regional lymph nodes[5].

Although previous papers have reported *in vitro* sensitization of naive human T cells to haptens, these studies used peripheral blood cells as antigen presenting cells (APC)[6,7]. In the present study, we analysed the capacity of both freshly isolated and 2-day incubated human LC to elicit *in vitro* primary T cell sensitization to the hapten TNP.

MATERIALS AND METHODS

Medium

The culture medium was RPMI 1640 medium supplemented with 10% heat-inactivated human AB serum, 100μg/ml gentamicin, 2mM L-glutamine and 1μg/ml indomethacin: hereafter designated complete medium.

Langerhans Cell Enriched Epidermal Cell Suspensions

Epidermal cell (EC) suspensions were obtained from normal human skin by trypsinization (0.05% trypsin, 18hrs at 4°C). The epidermal cells were either frozen or enriched in LC by density gradient centrifugation. The cells from the interface were used as fresh Langerhans cell-enriched EC (fLC= 8 to 30% LC) or incubated (10^6/ml) for 48 h at 37°C in culture medium containing 200 UI/ml human recombinant GM-CSF (cLC= 15 to 30% LC). Enrichment for LC was quantified by flow cytometry analysis using FITC coupled anti HLA-DR mAb.

Chemical Modification Of Antigen Presenting Cells

Modification of APC with the trinitrophenyl hapten (TNP) was performed according to the method of Shearer[8]. Briefly, cell pellets were resuspended in HBSS (pH 7.2) containing 5mM 2,4,6 trinitrobenzenesulfonic acid (TNBS). For control of hapten specificity, APC were as well conjugated to 100μg/ml of fluorescein isothiocyanate (FITC). Cells were incubated with these haptens for 10min at 37°C, and then washed extensively before use as modified-APC in cell culture.

T Cells

Autologous peripheral blood mononuclear cells (PBMC) were obtained by density centrifugation. T lymphocytes from the non adherent cell population were isolated by the technique of rosetting with sheep erythrocytes. The T cell population contained 95% or more CD3-positive cells as assessed by cytofluorimetry.

Sensitizing Cultures

Hapten modified fresh LC (TNP-fLC) or 2-day incubated LC (TNP-cLC) were cultured with autologous T cells in complete medium in 5 ml tube for 9-11 days : the purified T cells were incubated with LC at 37°C at a ratio of about 100:1. Viable T cells were then recovered and used for restimulation assays. To determine primary proliferation, this sensitizing culture was also set up in 96-well U-bottomed microtitre plates : the autologous T cells were cultured at 10^5 cells per well with 10^3 fresh or incubated hapten-coupled-LC. A kinetic study of T cell proliferation was performed by pulsing the cells with 1μCi of [^{3}H]-thymidine for the final 18 hrs of culture.

Restimulation Assays

After 9-11 days of primary cultures, secondary cultures of *in vitro* primed T cells were carried out in 96-well microtitre plates. Viable T cells were recovered by density gradient centrifugation and restimulated (10^5 LT) for 3 days with thawed and TNP-modified autologous fresh or 2-day incubated LC (5.10^3 LC). T cell proliferation was determined by addition of [^{3}H]-thymidine for the final 18hrs of culture. The specificity of T cell proliferation was assessed by the use of non-modified LC, or LC treated with another hapten (FITC).

RESULTS

Primary *In Vitro* T Cell Response To TNP

As shown in figure 1, addition of TNP-modified fresh LC to autologous naive T cells did not induce any detectable T lymphocyte proliferation. By contrast, upon stimulation with TNP-modified 2 day-incubated LC, the T cell population was able to proliferate. This *in vitro* primary proliferative response was evident on day 3 of culture and became maximal 2 days later.

In Vitro Secondary Response To TNP-Modified Fresh LC

In order to determine whether the T cells recovered after a 9 day primary culture with TNP-modified fresh LC could have been primed without proliferating, these lymphocytes were restimulated *in vitro* with thawed and TNP-modified fresh LC. Figure 2 shows data of a triplicate experiment, representative of regularly obtained results. We thus observed that after a primary culture with fresh TNP-LC, a low but significant secondary T cell proliferation was obtained by restimulation with hapten-conjugated fresh LC. This response was TNP specific as shown by the absence of T cell proliferation in the presence of fresh LC either non-modified or treated with an irrelevant hapten such as FITC.

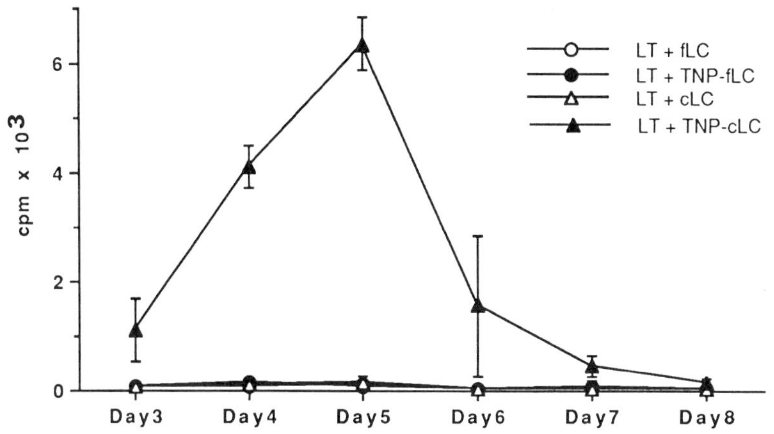

Figure 1. T cell primary response to TNP-modified fresh or 2 day-incubated Langerhans cells. T lymphocytes (10^5) were incubated with either non- or TNP-modified fresh LC, or non- or TNP- modified 2 day-incubated LC (10^3 LC). T cell proliferation was assessed by a [^{3}H] thymidine pulse for the final 18 hrs of culture. Results are expressed in cpm ± SD of triplicate cultures.

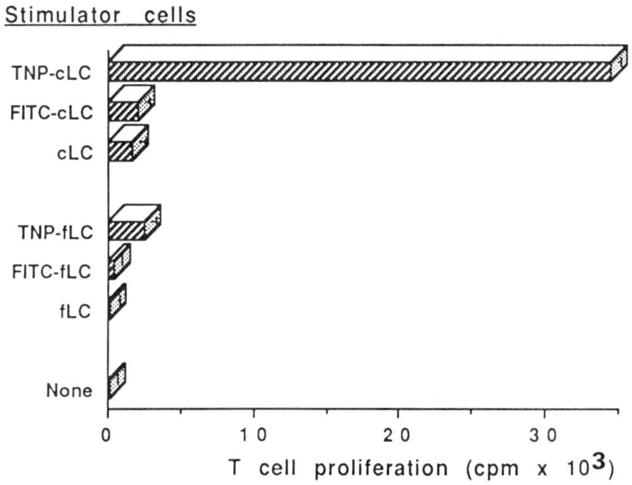

Figure 2. Revelation of *in vitro* T cell sensitization to the TNP hapten using either fresh or cultured TNP-modified Langerhans cells. Viable T cells were recovered from a 9 day primary culture with TNP-modified fLC. They were then restimulated for 3 days with fresh or 2 day-incubated LC which have been non- or TNP- or FITC-coupled. T cell proliferation was assessed by [^{3}H] thymidine incorporation for the final 18 hrs of culture. Results are expressed in cpm ± SD of triplicate cultures.

cLC Are More Potent APC Than fLC For *In Vitro* Primed Lymphocytes

We have compared the ability of fresh and 2 day incubated LC to present TNP to *in vitro* primed T cells. As shown in figure 2, although it was possible to reveal the sensitization to TNP with fresh LC, the magnitude of the hapten-specific proliferative reaction was considerably enhanced by the use of TNP-cLC. These cLC enable us to develop T cell lines which were then expanded by non specific mitogenic stimulations.

FACS analysis of T cell surface antigens revealed a cell population mainly composed of CD4+, TcR αß+ cells (data not shown). These cell lines kept their specificity towards the sensitizing TNP hapten and were able to proliferate with TNP-fLC. However, as reported in secondary responses, T cells exhibited stronger proliferative reactions when stimulated by TNP-cultured LC.

DISCUSSION

Primary contact of T cells with an antigen on APC results in the development of antigen specific T lymphocytes, which upon secondary exposure to this immunogen, will initiate a more rapid and vigourous immune response. It has now become evident that in hypersensitivity skin reactions, the APC function is achieved by LC, the dendritic cells from epidermis[1].

In the present study, we developed an *in vitro* model of T cell sensitization to haptens using TNP-modified fresh or cultured human LC. Results showed that both freshly prepared and cultured human LC are able to sensitize naive T lymphocytes to TNP *in vitro*. This was evidenced by the capacity of T cells, recovered from an *in vitro* primary culture, to proliferate in secondary responses when restimulated with TNP-modified LC. We showed, however, that cultured LC are more efficient APC than fresh LC in eliciting TNP-specific T cell responses. Indeed, naive T lymphocyte primary proliferative response to TNP was only obtained using TNP-cLC. Moreover, under secondary stimulation, TNP-cLC supported a greatly enhanced T cell proliferation, as compared to TNP-fLC. *In vitro* cultured LC have been shown[2,3] to upregulate MHC class II antigens, as well as cell-adhesion molecules such as ICAM-1 and LFA3, which might explain their enhanced APC function. Unexpectedly, secondary T cell proliferative responses to TNP-cultured LC were similar whether TNP-modified fresh or cultured LC had been used during T cell sensitization. This might suggest that maximal proliferative level was reached in both cases.

The present findings extend previous studies in the mouse[9], showing that cLC are potent APC in primary hapten-dependent proliferation assays. Furthermore, they emphasized the role of human epidermal Langerhans cells in the induction of contact sensitivity.

Using TNP-cLC as APC, we showed that *in vitro* sensitized T cells can be expanded without loosing hapten-specificity. Attemps to develop specific T cell clones are currently underway. These cell lines and clones should be very useful to analyse the as yet undefined process of hapten uptake and presentation by LC. Furthermore, this *in vitro* priming assay might be used as a predictive test for detection of sensitizing compounds.

REFERENCES

1. M.L. Kripke, C.G. Munn, A. Jeevan, J.M. Tang and C. Bucana, Evidence that cutaneous antigen-presenting cells migrate to regional lymph nodes during contact sensitization, *J. Immunol.* 145:2833 (1990).
2. N. Romani, A. Lenz, H. Glassel, H. Stössel, U. Stanzl, O. Majdic, P. Fritsch and G. Schuler, Cultured human Langerhans cells ressemble lymphoid dendritic cells in phenotype and function, *J. Invest Dermatol.* 93:600 (1989).
3. M.B.M. Teunissen, J. Wormmeester, S.R. Krieg, P.J. Peters, I. M.C. Vogels, M.L. Kapsenberg and J.D. Bos, Human epidermal Langerhans cells undergo profound morphologic and phenotypical changes during in vitro culture, *J. Invest. Dermatol.* 94:166 (1990).
4. S. Aiba and S.I. Katz, The ability of cultured Langerhans cells to process and present protein antigens is MHC-dependent, *J. Immunol.* 146:2479 (1991).
5. J.W. Streilein and S.F. Grammer, In vitro evidence that Langerhans cells can adopt two functionally distinct forms capable of antigen presentation to T lymphocytes, *J. Immunol.* 143:3925 (1989).
6 W. Newman, G.L. Stoner, Primary in vitro sensitisation of human T cells, Nature 269:151 (1977).
7 D. Charmot and C. Mawas, The in vitro cellular response of human lymphocytes to trinitrophenylated autologous cells: HLA-D restriction of proliferation but apparent absence of HLA restriction of cytolysis, Eur. J. Immunol. 9:723 (1979).
8 G.M. Shearer, Cell-mediated cytotoxicity to trinitrophenyl-modified syngeneic lymphocytes, *Eur. J. Immunol.* 4:527 (1974).
9 C. Hauser and S.I. Katz, Generation and characterization of T-helper cells by primary in vitro sensitization using Langerhans cells, Immunol. Rev. 117:67 (1990).

MONOCYTE-DERIVED LANGERHANS CELLS FROM DIFFERENT SPECIES - MORPHOLOGICAL AND FUNCTIONAL CHARACTERIZATION

Falko Steinbach[1] and Bernhard Thiele[2]

[1] Dept. of Virology, FU Berlin
Nordufer 20
1000 Berlin 65 and
[2]Charité, Humboldt University of Berlin
Tucholskystr.4
1140 Berlin

INTRODUCTION

It has been demonstrated earlier that peripheral blood monocytes can be differentiated into antigen presenting accessory cells[e.g.1,2], and also that mouse blood contains precursors of dendritic cells[3]. In the human system we have recently demonstrated the in-vitro differentiation of a Langerhans cells phenotype from peripheral blood monocytes for the first time,[4] indicating the possibility to study the suggested pathway from monocytes via Langerhans cells (LC) to dendritic cells (DC)[5] in-vitro.

As we were interested in comparative studies, we worked with human and horse monocytes. Both were differentiated into accessory cells as well as into macrophages (MΦ). As it has been our aim to study the functional properties of our cells too, we concentrated not only on the surface markers like CD1a, which LC share with interdigitating dendritic cells (IDC) and CD14, which is mostly lost by epidermal LC (eLC), but characterized them also by their phagocytic and non-specific esterase activity, as well as their capacity to stimulate lymphocyte proliferation in a mixed lymphocyte reaction (MLR).

MATERIALS AND METHODS

Purification of Monocytes

Isolation of the monocytes was performed as described before.[6] Briefly, buffy coats were diluted with 1/5 Vol. PBS, containing 5mM EDTA before they were layered on top of Ficoll-Paque®. PBMC were collected and washed with PBS/EDTA to get rid of platelets. Monocytes were purified by adherence to plastic dishes. So obtained monocytes were purified more than 95% as demonstrated by May-Grünwald-Giemsa staining and had a viability >98% as checked by trypan-blue dye exclusion. Iscove's modified Dulbeccos Medium (IMDM) was supplemented either with 3% FCS to obtain the accessory cell type, or with 20% autologous serum to obtain macrophages.

Immunofluorescence

$5x10^4$ cells were spun down for 10 min at 400g. They were resuspended in 100μl PBS and monoclonal antibody was added according to the distributor. If indirect immunofluorescence was performed, the cells were afterwards washed with ice-cold PBS and labelling with the second antibody was performed like with the first. Monoclonal antibodies (mAbs) used for this contribution have been: α-CD1a (NA1/34), kindly provided from Dr. Milstein (Cambridge, UK); α-CD14 (IOM 2), Immunotech, α-HLA-DR (HHLA 06), Silenius; α-MHC II, horse (Bo 139) kindly provided from Dr. Schubert, Hannover. In the case of indirect immunofluorescence, the following polyclonal antisera were used: goat α-Mouse-PE (R 9670), Sigma and goat α-Mouse-FITC (115-095-062), Immunotech.

Flow cytometry was performed, using a FACScan® (Becton Dickinson) with FACScan research®, Consort 30® and DNA Cell-cycle Analysis® software. Data acquisition and analysis were performed as described before.

Functional assays

Cell-cycle analysis of human monocytic cells was performed in order to evaluate the proliferation capacity of the monocyte-derived Langerhans cells. Assay was performed in accordance to the method described by Ormerod.[7] Briefly, cells were pelleted and fixed with ice-cold ethanol, before treatment with DNase-free RNase and staining with propidiumiodide. Finally, the cells were analysed by flow cytometry.

Phagocytosis has been measured by internalization of Candida albicans cells in serum free IMDM during an incubation period of 45min. Internalized cells can easily be counted after May-Grünwald-Giemsa staining.

An allogeneic primary mixed lymphocyte reaction was performed in order to investigate the lymphocyte stimulation capacity. Therefore, peripheral blood T-lymphocytes had been isolated by passaging lymphocytes over a nylon wool collumn after selective adherence of the PBMC to plasma-coated petri dishes in order to get rid of contaminating monocytes. Afterwards, 10^6 of the freshly isolated T-lymphocytes were co-cultivated with $5x10^4$ monocytic cells, in-vitro cultured for three days as described. After 72h, DNA synthesis was measured by a 24h pulse of 4μCi [^{3}H] thymidine.

RESULTS

Figure 1 represents a FACS-analysis of three day cultured human monocytes. They express CD1a, whereas their CD14 expression varies from 25-50%. $CD14^+$ Langerhans cells are widely accepted as being a precursor form of mature Langerhans cells. As for their origin, they represent a monocyte-derived Langerhans cell phenotype (MoLC), before they loose the rest of their CD14 (data not shown). Fig.2 illustrates that analogous results have been obtained with equine monocytes by the use of cross-reactive human antibodies.

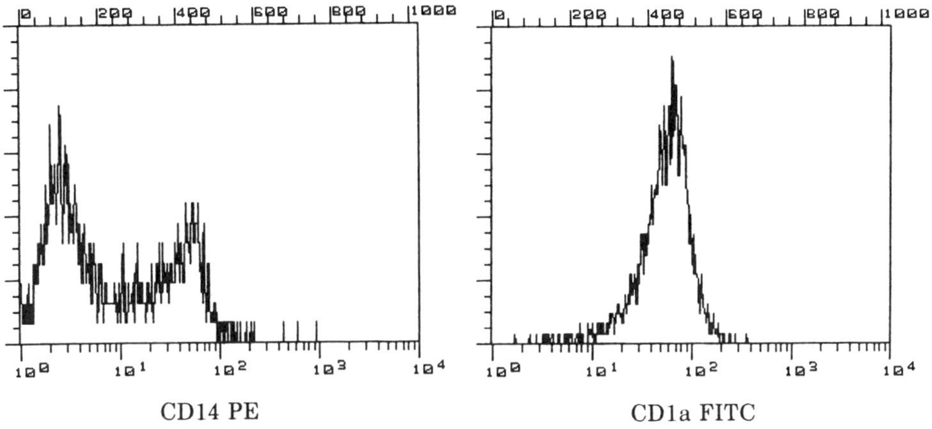

Figure 1. FACS-analysis of in-vitro cultured human monocytes, which have already lost most of their CD14 (left), while expressing CD1a (right). As for their origin, they will be referred to as monocyte-derived Langerhans cells (MoLC).

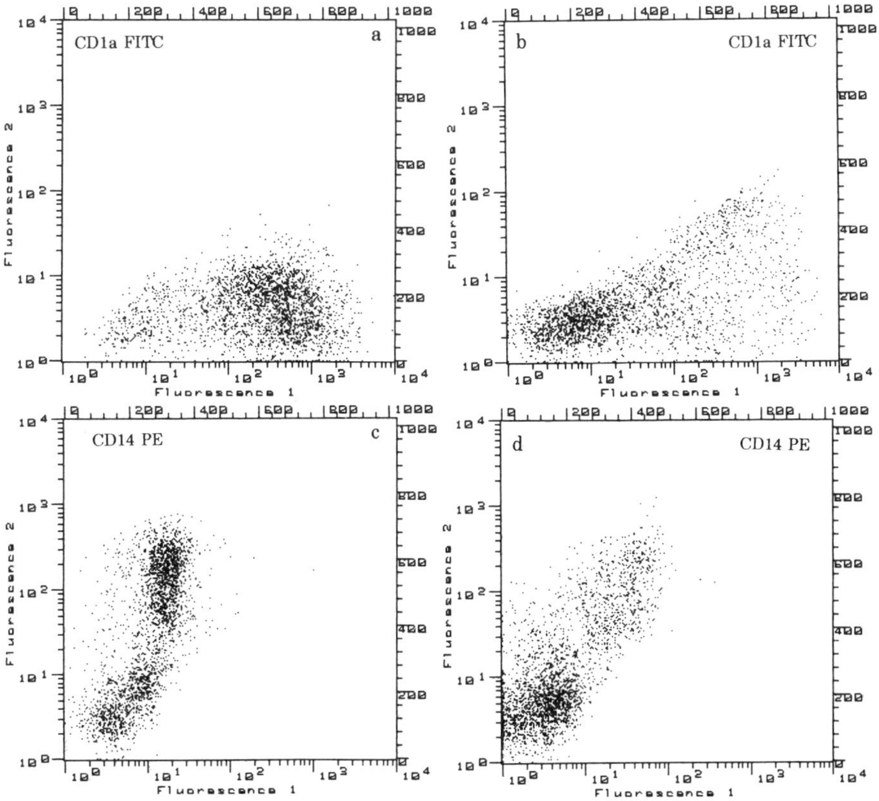

Figure 2. Flow cytometry of horse monocytes after 2 (a,c) or 5 (b,d) days of in-vitro culturing, using cross-reactive human antibodies to detect the highly conserved structures CD1a and CD14.

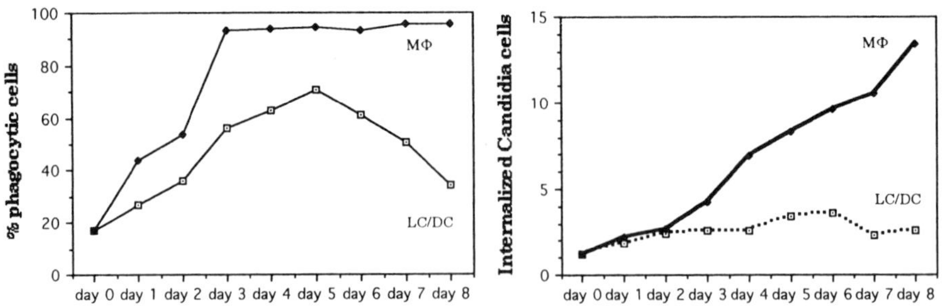

Figure 3. Phagocytosis of Candida albicans cells by in-vitro cultured horse monocytes. As to be seen, monocytes being cultured for becoming macrophages (MΦ) rapidly increase their phagocytic capacity as a population (left), as well as individually (right). In contrast to these results, monocytes, which were cultured the accessory pathway (LC/DC), never become really phagocytic; neither as a population, nor as the individual cell.

Figure 3 compares horse monocytic accessory cells and macrophages due to their phagocytic activity, which has been measured by the internalization of Candida albicans cells. It should be taken into account that at least differentiated macrophages express the mannosyl-fucosyl receptor (MFR), which does accelerate internalization of Candida cells. It is clearly visible that MoLC and LC have a much lower phagocytic capacity than macrophages. Non-specific esterase activity is expressed in all MoLC and decreases in LC (data not shown), while they mature into dendritic cells (DC) loosing their CD1a. Analogous data were obtained in the human system.

As dense cultures of human MoLC form non-adhering clusters between day 2 and day 4 of in-vitro culture, we were interested in their proliferating capacity, which is demonstrated in Figure 4 by cell-cycle analysis. As shown with three days cultured MoLC, there are cells in the S-phase as well as in the G_2/M-phase.

Stimulating capacity of human MoLC in a MLR is shown in the last figure. MoLC were compared to cells cultured with 20% autologous serum, freshly isolated monocytes or ConA. The highest proliferation rate was found with

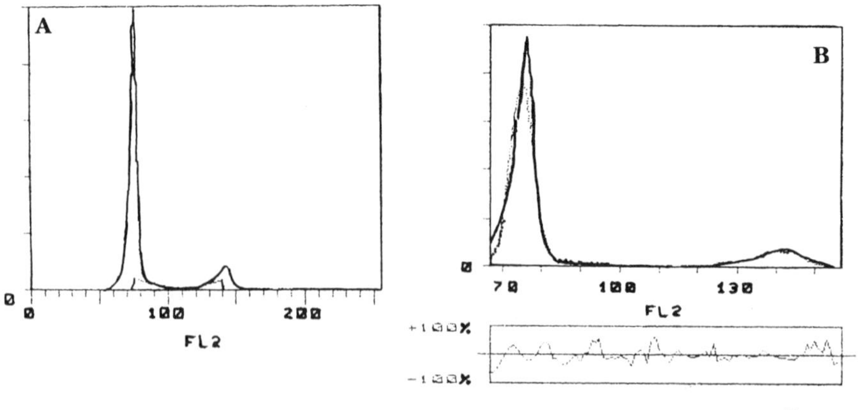

Figure 4. Cell cycle analysis of three days cultured human MoLC. The method[7] excludes measurement of contaminating lymphocytes. The amount of proliferating cells is variable between different donors. In the typical case presented here, 12% (+/-2) of the cells are in the S-phase and 12% (+/-1) in the G_2/M-phase. A) and B) represent different statistical programs used.

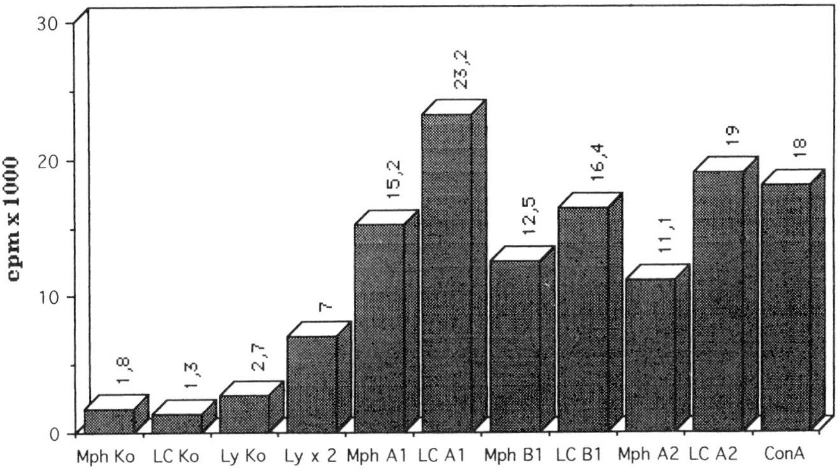

Figure 5. Stimulating capacity of three day old huMoLC in an autologous MLR. Stimulating capacity of MoLC (LC) is distinctively higher than the stimulating capacity of three day old monocytes cultured with 20% autologous serum (Mph). Individual differences are due to different donor combinations (A1,A2,B1).

found with MoLC, reaching the level of stimulation by ConA. Lymphocyte controls were probably influenced by contaminating monocytic cells. It should be emphasized that the responding cells have been freshly isolated peripheral blood lymphocytes, which have a lower responder intensity than spleen lymphocytes, depending upon their state of activation.

DISCUSSION

In order to study monocyte-derived Langerhans cells, we used an in-vitro differentiating system, which was described earlier.[6] These MoLC proofed to be LC precursors due to both their phenotype and functional characteristics. Comparative studies with human and horse cells demonstrated that the system is not species restricted. CD1a expression in the human system is consistently reproducible and stable during the period of constant culture conditions, whereas horse MoLC spontaneously differentiate into a $CD1a^-$ $CD14^-$ phenotype of dendritic morphology (data not shown). It should be mentioned that CD1a expression is not to be seen in each case of horse monocyte differentiation. The MHC II expression is constantly high on MoLC and distinctively higher than on MΦ (data not shown).

Phagocytic activity of human and horse MoLC is comparable to the low phagocytic activity of dendritic cells described by others.[1] The phagocytosis of Candida cells may be influenced by the existence of surface lectins, able to recognize the surface mannan of C. albicans. Spectrofluorimetric inhibition experiments, using the ligand mannosyl-BSA-FITC, indicate the existence of a mannose-specific lectin on MoLC (data not shown), but the phagocytic capacity of MoLC is allways lower than on comparable MΦ, which express the mannosyl-fucosyl receptor. This 175kD receptor is not expressed by freshly isolated monocytes. Nevertheless, the ligand mannosyl-BSA-FITC is a very good marker to detect all mononuclear phagocytes in flow cytometry,[6] which may partly be due to a cross-reactivity with the mannose-6-phosphate receptor.[8]

Proliferation of human MoLC has been underlined by the detection of Ki-67 (data not shown). The data correspond with the postulation of proliferating dendritic cell precursors in mouse blood[3] and provide further evidence for the proliferation capacity of eLC.[9] It is to be mentioned that proliferation has not yet been found in the horse system.

Immunostimulation of freshly isolated peripheral blood lymphocytes by MoLC in a primary mixed lymphocyte reaction is comparable with other dendritic cells in the human system described so far[10] and as effective as ConA. It has to be stressed that the monocytes cultured just in 20% serum after three days do not represent macrophages. On the other hand, we did on purpose not purify the T-cells further than by nylon-woll collumn separation, in order to define a high comparative background level for our cells. Therefore, it can finally be summarized that the data presented for the phenotype and the function of MoLC correspond with the data obtained by studying eLC. Studies are under way to further define the accessory function of the MoLC at different stages of differentiation.

Acknowledgements

We thank Dr. H. Keller and the Klinik für Pferde, FU Berlin for blood samples of clinically healthy horses and Dr. Kerstin Borchers, Dept of Virol., FU Berlin for helpful discussions.

REFERENCES

1. H.M. Najar, A.C. Bru-Capdeville, R.K.H. Gieseler and J.H. Peters, Differentiation of human monocytes into accessory cells at serum-free conditions, *Eur. J.Cell Biol.*, 51:339 (1990)

2. G. Rossi, N. Heveker, B. Thiele and H. Gelderblom, Mannan antigen presentation by monocyte-derived dendritic cells, *in*: Progress in AIDS research in the Fed. Rep. of Germany M. Schauzu, ed., MMV Medizin Verlag München (1990)

3. K. Inaba, R.Steinman, M. Witmer Pack, H. Aya, M. Inaba, T. Sudo, S. Wolpe and G. Schuler, Identification of proliferating dendritic cell precursors in mouse blood, *J .Exp. Med.*, 175:1157 (1992)

4. G. Rossi, N. Heveker, B. Thiele, H. Gelderblom and F. Steinbach, Development of a Langerhans cell phenotype from peripheral blood monocytes, *Imm. Lett.* 32:189 (1992)

5. G. Schuler and R.M. Steinman, Murine epidermal Langerhans cells mature into potent immunostimulatory dendritic cells in vitro, *J. Exp. Med.*, 161:526 (1985)

6. F. Steinbach and B. Thiele, Phenotypical investigation of mononuclear phagocytes by FACS analysis, *in*: N.N., P. Poidron, D. Wachsman eds., Intercept, Andover, G.B.(in press)

7. M. Ormerod, Analysis of DNA, *in*: Flow Cytometry, M. Ormerod ed., IRL Press, Oxford (1990)

8. A.C Roche, P. Midoux, P. Bouchard and M. Monsigny, Membrane lectins on human monocytes, *FEBS Lett.*, 193:63 (1985)

9. S. Miyauchi and K. Hashimoto, Mitotic activities of normal epidermal Langerhans cells, *J.Invest.Dermatol.*, 92:120 (1989)

10. E.R. Flechner, P.S. Freudenthal, G. Kaplan and R.M. Steinman, Antigen-specific T-lymphocytes efficiently cluster with dendritic cells in the human primary mixed lymphocyte reaction, *Cell. Imm.*, 111:183 (1988)

A SERIAL SECTION STUDY ON MICE LANGERHANS CELL GRANULES AFTER DNFB PAINTING

Akira Senoo, Nobuo Imazeki,
Yoshifusa Matsuura and Yusuke Fuse

Second Department of Pathology
National Defense Medical College
3-2 Namiki, Tokorozawa
Saitama, 359 Japan

INTRODUCTION

There are numerous ultrastructural studies on the cells containing Langerhans cell granules. And general features of Langerhans cell granules have been studied and reported with detailed description.[1] The profiles of Langerhans cell granules were identified as discoid bodies with a focal vesicular expansion at the margin of the disc from three dimensional reconstruction.[2,3] Our previous report[4] also mentioned a three dimensional morphology of normal mice Langerhans cell granules.

On the other hand, it was proposed that the epidermal Langerhans cells and lymphoid dendritic cells etc., are specialized cell line to be referred as "dendritic cell" which act important role for cellular immune response.[5,6] However, the significance of the Langerhans cell granule remain uncertain, in present paper we describe the ultrastructural changes of the granules in the cells in antigen challenged skin sites and regional lymph nodes.

MATERIALS AND METHODS

Eight weeks old mice (BALB/c) were used for present study. The animals were sensitized by 3 daily applications of 1 drop (0.02 ml) of 0.5% 2,4-dinitrofluorobenzene (DNFB) in 4:1 acetone olive-oil to foot pad. Applied skin sites and popliteal lymph nodes were rejected on 1,2,3,4 days after the last DNFB painting.

Freshly collected tissue samples were fixed in 2% paraformaldehyde-2.5% glutaraldehyde mixture in 0.1 M phosphate buffer(pH 7.3) for 3 hours, post fixed in 2% osmium tetroxide in the same buffer for 2 hours. The fixed specimens were dehydrated in increasing concentration of ethanol, cleared in propylene oxide and embedded in epoxy resin.

Serial ultrathin sections cut with diamond knife were mounted on a formvar coated single hole copper grid. They were stained with uranyl acetate and lead citrate and examined with a JEM 100C electron microscope.

Three dimensional morphology of Langerhans cell granules was examined from series of electron micrographs(x.30000) taken from complete serial sections. The mean maximum diameter (± standard deviation) of disc portion of the granules was calculated from 100 granules in epidermal Langerhans cells and each 30 granules in the cells in subcapsular sinus and paracortical area (T area) of the lymph nodes.

Dendritic Cells in Fundamental and Clinical Immunology,
Edited by Kamperdijk *et al.*, Plenum Press, New York, 1993

RESULTS

On 1 and 2 days after DNFB painting, it was noted as striking features in epidermal Langerhans cells that the large part of the granules had expanded vesicular portion at the margin of disc in one plane of sectioned materials. And serial section study revealed that the almost all granules showed obvious vesicular expansions (Fig. 1).

The maximum diameter of the disc portion of the granules was measured from series of micrographs of serial sections, and the mean maximum diameter of the disc portion of 100 granules (230.26±35.36 nm, Fig. 5) was clearly reduced compared with normal counterpart (388.49±49.58 nm, Fig .5). In addition to this finding, increased number of cup–shaped granules were noted, and from serial section study it was evident that the disc portion of these granules were larger than the ordinary granules shown in Fig. 1(Fig. 2). On 3 and 4 days after DNFB painting, some of the

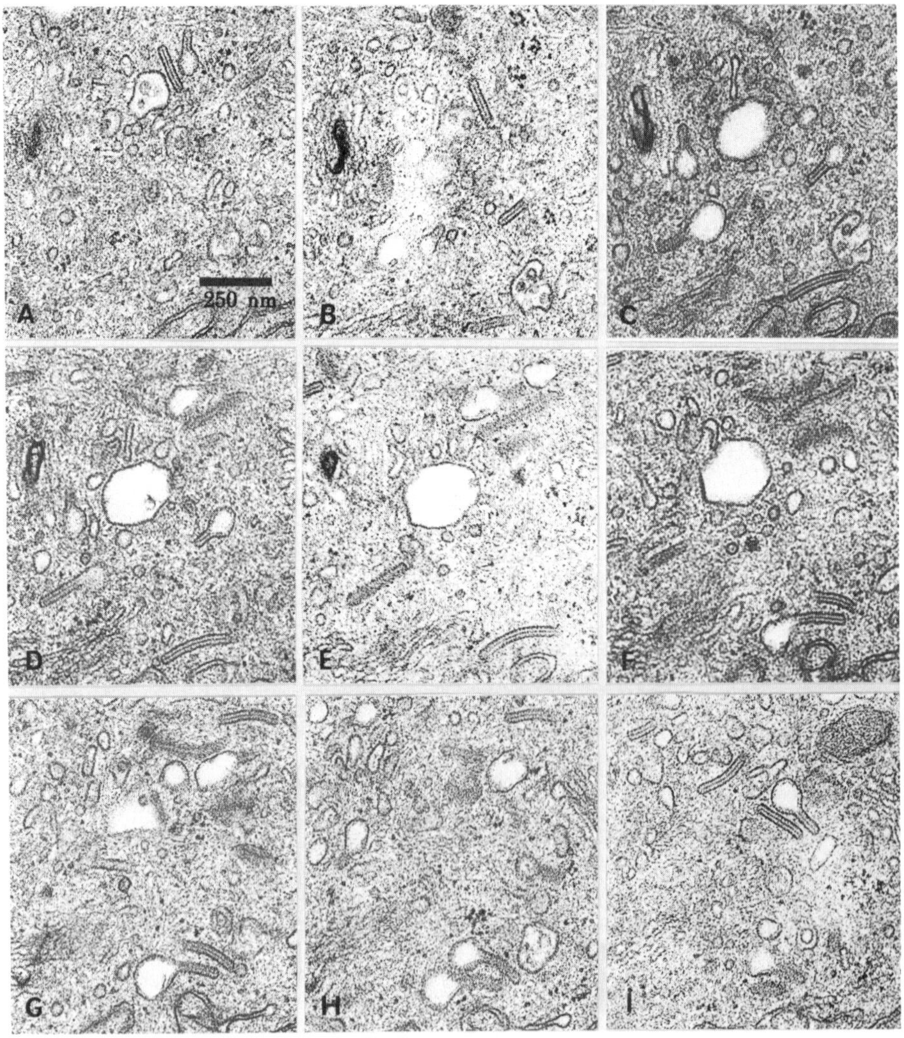

Figure 1. Serial sections of epidermal Langerhans cell after 3 daily DNFB painting. The granules have expanded vesicular portions at the margin of the discs.

granules showed above mentioned ultrastructural changes, but not conspicuous.

The Langerhans cell granules in the cells in subcapsular sinus and paracortical area of popliteal lymph nodes showed no remarkable changes in size and structure compared with that in normal mice. However, on 1 and 2 days after the painting, the Langerhans cell granule-containing cells were increased in number in the subcapsular sinus and paracortical area of the lymph nodes. At the periods of 3 and 4 days after the painting, Langerhans cell granule-containing cell was characteristically reduced in number.

The vesicular complex near Langerhans cell granules was noted in the cells in subcapsular sinus and paracortex in the lymph node(Figs. 3,4), but never in the epidermal Langerhans cells.

Schematic illustration of Langerhans cell granules of normal and DNFB painted mice in epidermis and lymph node was shown in Fig. 6.

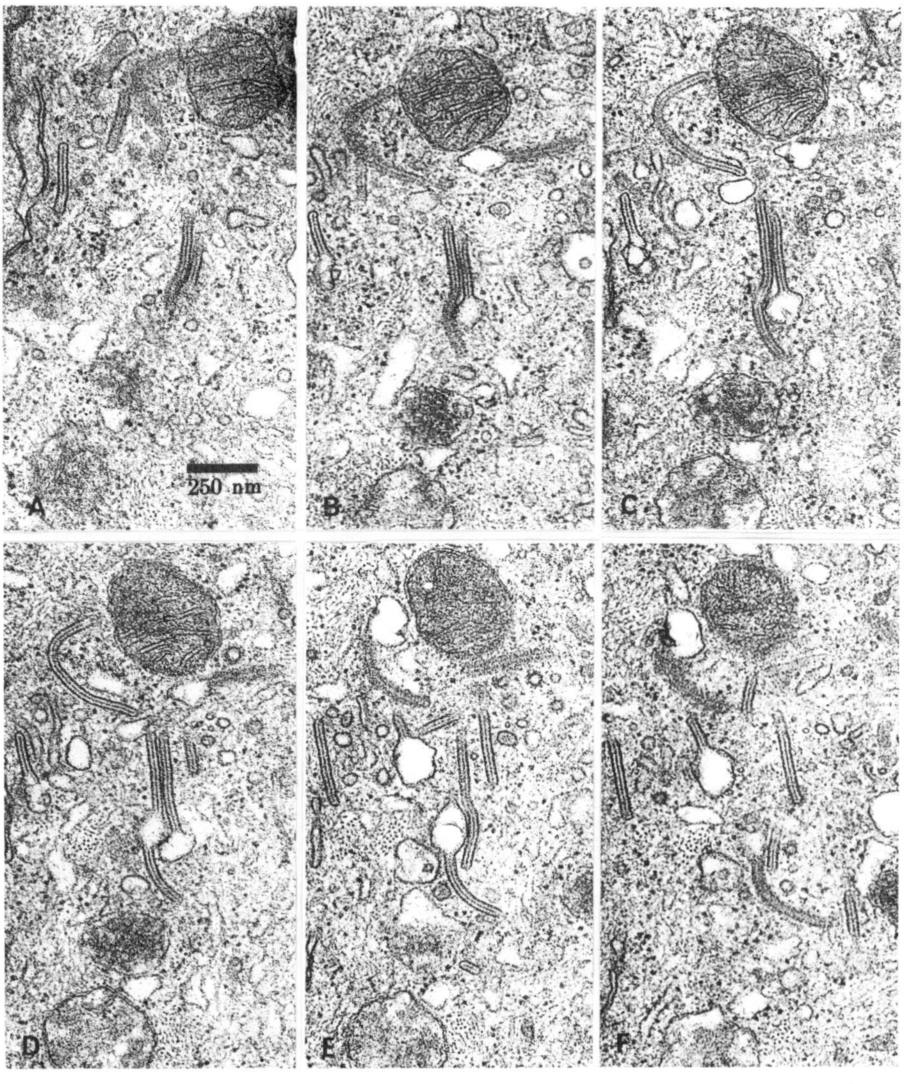

Figure 2. Serial section of epidermal Langerhans cell after 3 daily DNFB painting. Increased number of cup-shaped and irregular-shaped granules are noted.

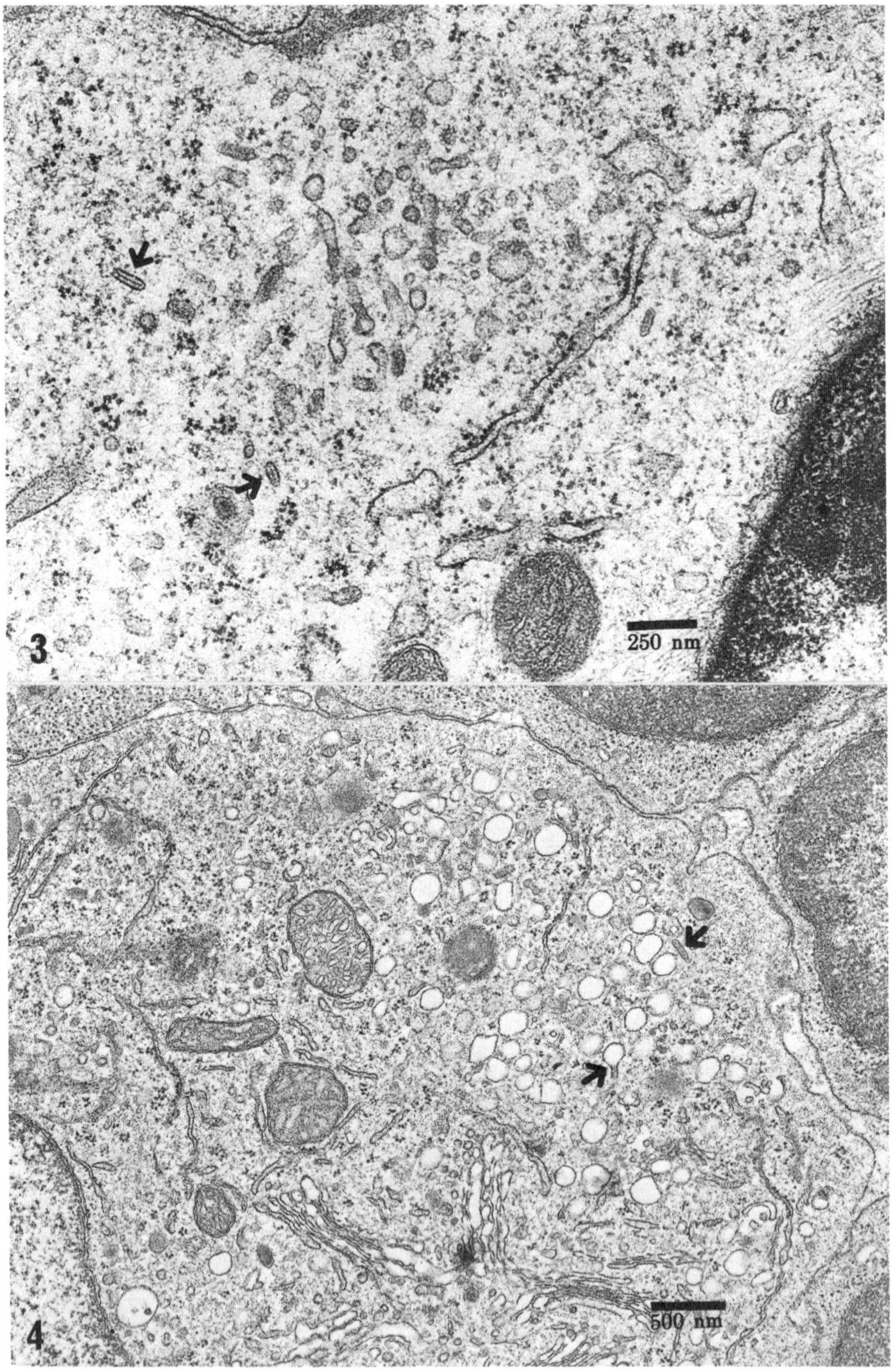

Figure 3,4. An intimate localization of vesicular complex and Langerhans cell granules(arrows) in the cells in subcapsular sinus(Fig. 3) and paracortex(Fig. 4) of lymph node.

DISCUSSION

There are numerous reports dealt with Langerhans cell and its granules, but no report is found about an experimentally induced ultrastructural change of the granule. In the present paper, we described the first an ultrastructural changes of Langerhans cell granules after antigenic stimulation using DNFB applied mice skin. It was reported that the Langerhans cell has an ability to present an antigen to T lymphocytes,[7,8] therefore the ultrastructural changes of Langerhans cell granule in DNFB applied skin site may related to an antigen presenting function of the epidermal Langerhans cell.

In previous our serial section study[4] on Langerhans cell granules in normal mice, we proposed an idea concerning with the exhaustive process of Langerhans cell granules, that to say the epidermal Langerhans cells mainly form disc shaped–granules which acquire the vesicular portion as the cells leave the epidermis and migrate into the regional lymph node. Within the lymph node, the disc portion progressively transform to the vesicular

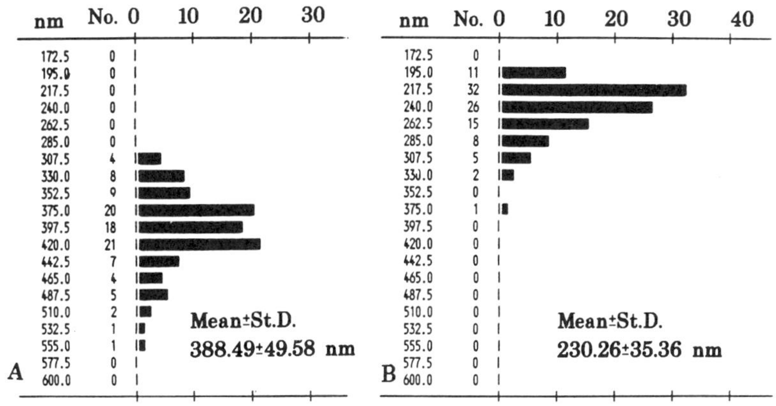

Figure 5. Histograms of the maximum diameter of disc portion of the granules in epidermal Langerhans cells in normal (A) and DNFB painted mice (B).
The mean maximum diameter ± standard deviation of the disc is also showed.

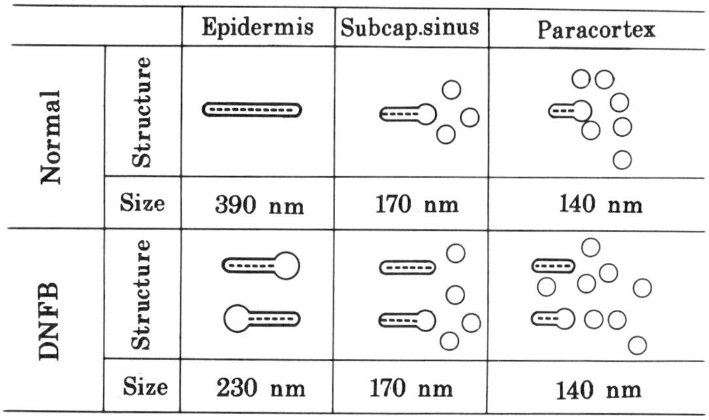

Figure 6. Schematic illustration of mice Langerhans cell granules in epidermis and lymph nodes.

portion of the granules, and the vesicles detach from the disc, finally the granules change to the vesicular complex in the cytoplasm. The present result that the expanded vesicular portion of the granules in epidermal cells after DNFB painting may suggest the accelerated sequence of the exhaustive process of the granules by antigenic stimulation. The intimate localization of Langerhans cell granules and vesicular complex in the cells of subcapsular sinus and paracortex of the lymph node may also support this idea.

It has been reported that fresh Langerhans cells are weak stimulators of primary T cell proliferative response in vitro, and during culture Langerhans cells become more active and these cells rapidly lost the granules.[9] Therefore, it may appear that reduced size and number of the granules in the cells in lymph nodes after DNFB application as well as normal condition may represent to immunological maturation of Langerhans cells during immigration from the skin.

Considerable evidences support the immigration of Langerhans cells into regional lymph nodes from the skin.[10,11,12] However, another possibility that Langerhans cells in the lymph node arise from transformation of the resident interdigitating cells was suggested.[13] An increased number of Langerhans cell granule-containing cells in the lymph nodes were noted at early periods in time course study after DNFB painting to the foot pads, this finding support the possibility of the immigration of epidermal Langerhans cells to the regional lymph nodes.

Finally, we are speculating that the vesicular complex in the cells in lymph node may originate from Langerhans cell granules, but it is not clear the precise meaning of the vesicular complex and also the functional significance of Langerhans cell granules, in these points more study is needed.

REFERENCES

1. Wolff,K., 1972, The Langerhans cell. Curr. Probl. Dermatol., 4: 79.
2. Wolff,K., 1967, The fine structure of the Langerhans cell granule. J. Cell Biol., 35: 468.
3. Sagebiel,R.W, and Reed,T.H., 1968, Serial reconstruction of the characteristic granules of the Langerhans cell. J. Cell Biol., 36: 595.
4. Senoo,A.,Imazeki,N.,Matsuura,Y.,and Fuse,Y., 1990, A serial section study on Langerhans cell granules in the cells in epidermis and lymph nodes of mice. Dendritic cell, in press.
5. Steinman,R.M. 1981, Dendritic cells. Transplantation, 31: 151.
6. Tew,J.G.,Thorbecke,G.J.,and Steinman,R.M., 1982, Dendritic cells in the immune response: Characteristics and recommended nomenclature(A report from the Reticuloendothelial Society Committee on nomenclature). J. Reticuloendothel. Society, 31:371.
7. Stingel,G.,Stingel-Gazze,L.A.,Aberer,W., and Wolff,K., 1981, Antigen presentation by murine epidermal Langerhans cells and its alteration by ultraviolet B light. J. Immunol., 127: 1707.
8. Aberer,W.,Stingel-Gazze,L.A. and Wolff,K., 1982, Langerhans cell as stimulator cells in the murine primary epidermal cell-lymphocyte reaction: alteration by UV-B irradiation. J. Invest. Dermatol., 79: 129.
9. Schuler,G., and Steinmam,R.M., 1985, Murine epidermal Langerhans cells mature into potent immunostimulatory dendritic cells in vitro. J. Exp. Med., 161: 526.
10. Silberberg-Sinakin,I.,Thorbecke,G.J.,Baer,R.L.,Rosenthal,S.A.,and Berezowsky,V., 1976, Antigen-bearing Langerhans cells in skin, dermal lymphatics and in lymph nodes. Cell. Immunol., 25: 137.
11. Raush,E.,Kaiserling,E.,and Goos,M., 1977, Langerhans cells and interdigitating reticulum cells in the thymus-dependent region in human dermatopathic lymphadenitis. Virchow(Cell Pathol.)., 25: 327.
12. Kamperdijk,E.W.A.,and Hoefsmit,E.C.M.m 1978, Birbeck granules in lymph node macrophages. Ultramicroscopy, 3: 137.
13. Sonoda,Y.,Asano,S.,Miyazaki,T.,and Sagami,S., 1985, Electron microscopic study on Langerhans cells and related cells in lymph nodes of DNCB-sensitive mice, Arch. Dermatol. Res., 277: 44.

SKIN DENDRITIC CELL - LYMPHOCYTE INTERACTIONS IN AUTOLOGOUS SYSTEM

Hanna Galkowska and Waldemar L. Olszewski

Department of Surgery Research & Transplantation
Medical Research Centre, Polish Academy of Sciences
02004 Warsaw, Poland

INTRODUCTION

Skin tissue fluid and lymph contain migrating immune cells, among them large dendritic (veiled) cells (VC) and lymphocytes [1]. These cells are continuously transported with lymph stream to the regional lymph nodes. This way the information about the presence of antigen can be transferred within minutes to the lymph node [2].

Clustering of dendritic cells with lymphocytes is the first phase of antigen presentation to lymphocytes. These cells bind lymphocytes in antigen-independent pathway. Little is known about the mechanism involved in such clustering, including expression of adhesion molecules [3].

In this study we tried to elucidate the mechanisms of "spontaneous" binding of VC from the canine skin afferent lymph with autologous lymphocytes in their own environment, which is the lymph, and in the absence of a known antigen. Controlling this process may be helpful in mitigation of skin immune reactions.

MATERIALS AND METHODS

Dogs

Outbred dogs with chronic lymphedema, after surgical interruption of afferent lymphatics, were used as lymph donors [4].

Collection of lymph

Lymph was obtained by direct percutaneous puncture of dilated lymphatics and collected into heparin (10 U/ml). The average concentration of cells was 2.2± 1.3 x 10^6/ml, the percent of cells with VC morphology was 3.2± 1.7 and 3.3± 2.8 % of all VC have been found in clusters with lymphocytes immediately after collection.

Dendritic Cells in Fundamental and Clinical Immunology,
Edited by Kamperdijk *et al.*, Plenum Press, New York, 1993

Veiled cell-lymphocyte binding assay

Binding of cells was quantitated after incubation in lymph mixed 1:1 (v/v) with appriopriate reagent solutions. The number of VC with two or more lymphocytes attached per 100 of VC was counted under light microscope. The period of cell incubation with applied agents was chosen depending on the peak of their effect in vitro.

Reagent solutions

EDTA, verapamil(Isoptin R), xylocaine, deoxy-D-glucose and NaN_3 were dissolved in 0.15 M NaCl. The following enzymes were prepared in 0.15 M NaCl: trypsin, collagenase type II, DN-ase type II, neuraminidase (Cl.perfringens type X). Retinoic acid was dissolved in ethanol to 10^{-2} M and next in 0.15 M NaCl to concentration of $5x10^{-3}$ M as a stock solution. The following sugars were used and stored frozen at -20^{0}C at 200 mM in 0.15 M NaCl:galactose, mannose, methyl-α-mannoside. Glucose-6-phosphate and mannose-6-phosphate were dissolved and stored in H_2O. Lymph cells were mixed with reagents to the final concentrations: sugars- 100 mM (osmolarity 300 mOsm), heparin- 250 μg/ml, and incubated for 4 h at 37^{0}C. All steroid immunosuppressants (hydrocortisone, dexamethasone, methylprednisolone) were dissolved in ethanol to 10^{-2} M and stock solutions were mixed with 0.15 M NaCl. Cyclosporine A and azathioprine were dissolved in ethanol to 5 mg/ml and FK 506 to 10 mM stock solution before mixing with 0.15 M NaCl. Indomethacin and acetylsalicylic acid were dissolved in ethanol to 10 and 20 mg/ml, respectively, before mixing with NaCl.

Effect of N-ase induced cluster formation on lymphocyte response

Lymph cells, 9 $x10^6$ cells in 0.15 M NaCl with 2 U/ml of N-ase were incubated for 1 h at 37^{0}C. Next, cells were washed 4 times with RPMI 1640 medium with 10% fetal calf serum and cultured in 0.2 ml of culture medium in microcultures at density 2 $x10^5$ cells, with or without 3 doses of PHA (HA15,Wellcome). Culture medium consisted of RPMI 1640 medium supplemented with 20% FCS and 10 mM HEPES buffer, 100 U/ml penicillin and 100 μg/ml streptomycin, 2 mM L-glutamine. Cultures were incubated for 6 days (auto-MLR) and 3 days (PHA) in a humidified atmosphere of 5% CO_2 in air, with 0.4 μCi of ^{3}H-thymidine for last twenty hours.

To compare the results the Student's *t* test was used.

RESULTS

Time, medium and temperature dependence of VC-lymphocyte binding

The interaction of lymph cells increased as a function of time during cells were incubated together, from 9.5% after 1 h to 20.7% after 4 h incubation ($p<0.05$) (Table 1). After 20 h incubation at 37^{0}C the rate of clustering was the same (data not shown). Table 2 shows data of cluster formation at different temperatures in saline and in lymph. Clustering in 0.15 M NaCl was similar at 37 and 22^{0}C and significantly lower at 4 and 39^{0}C. The percentage of clusters formed in the presence of lymph was reduced at temperatures 4 to 37^{0}C as compared to clustering in saline ($p<0.05$).

Table 1. Effect of different agents on spontaneous binding of lymph cells after incubation at 37^{0}C (mean ± SD).

Time	Cell treatment		Doses	% of clusters	
0	Directly from lymph			3.3 ± 2.8	
1 h	NaCl	(M)	0.15	9.5 ± 6.2	#
	EDTA	(mM)	20	16.5 ± 4.9	
	Xylocaine	(%)	0.2	0.7 ± 0.5	*
			0.05	3.7 ± 1.8	
	Trypsin	(mg/ml)	1	13.8 ± 8.2	
	Collagenase	(mg/ml)	0.1	12.0 ± 1.4	
	DN-ase	(μg/ml)	100	18.0 ± 12.2	
	N-ase	(U/ml)	2	69.0 ± 24.5	*
			1	48.0 ± 30.5	
4 h	NaCl	(M)	0.15	20.7 ± 2.8	#
	Verapamil	(M)	10^{-5}	23.0 ± 1.4	
	Retinoic acid	(M)	10^{-4}	10.0 ± 2.1	*
			10^{-6}	13.0 ± 1.4	*
	Deoxy-D-glucose	(mM)	50+10	13.6 ± 3.3	*
	+ azide		25+5	22.8 ± 3.3	

* $p<0.05$ vs saline, # $p<0.05$ 4h vs 0 or 1h

Treatment with EDTA and Verapamil

Treatment of lymph cells with EDTA and verapamil, a Ca^{++} channels blocker, did not affect the binding rate (Table 1).

Treatment with enzymes

Cells subjected to proteolysis or hydrolysis with DN-ase for 1 h were not changed in their ability to interact. Treatment of cells with N-ase resulted in the increased cell clustering (Table 1).

Effect of retinoic acid and xylocaine

Cell binding was significantly decreased in the dose dependent manner when cells were treated with xylocaine or retinoic acid, the drugs influencing cell membrane fluidity (Table 1).

Effect of metabolic blockers

Spontaneous cell binding in vitro was energy dependent, since both oxidative and glycolytic energy production inhibitors, azide and deoxy-D-glucose were inhibitory for cell binding (Table 1).

Effect of sugars and heparin

In order to investigate the involvement of carbohydrate structures in cluster formation the effect of sugars and heparin was studied. Neither heparin nor sugars hindered the clustering of lymph cells (data not shown).

Effect of immunosuppressants and anti-inflammatory drugs

Among steroids only treatment with methylprednisolone resulted in the decreased percentage of formed clusters (Table 3) ($p<0.05$).

Other immunosuppressants, Cyclosporin A and azathioprine had no effect on cell binding. In contrast, FK 506 was a potent inhibitor of lymph cells clustering.

Effect of N-ase induced clusters on lymphocyte responsiveness

As presented in Table 4 N-ase treated lymph cells cultured for 3 days showed higher responsiveness of lymphocytes to suboptimal doses of PHA than the control cells, however, the same level of lymphocyte stimulation ratio was observed in auto-MLR, irrespective of desialation.

Table 2. Effect of temperature and medium on cell interaction after 4 h incubation (mean±SD). *p<0.05 vs NaCl, #p<0.05 vs 37°C

Temperature (°C)	% of clusters	
	Lymph	Saline
4	5.0±1.1 *	12.1±4.4 #
22	11.5±1.9 *	16.2±3.3 #
37	13.9±5.2 *	20.7±2.8
39	8.3±1.5	11.8±3.8 #

Table 4. Effect of N-ase treatment on cell responsiveness to PHA or in auto-MLR (mean ± SD). * p< 0.05 vs control

Test	cpm x 10^{-3}	
	control	N-ase
auto-MLR	1.3±0.2	1.5±0.2
PHA 90μg	12.4±1.2	11.5±0.4
PHA 18μg	3.5±0.2	4.9±0.4 *
PHA4.5μg	0.7±0.1	1.2±0.1 *

DISCUSSION

Spontaneous binding of VC with lymphocytes proceeded as a function of time, during which cells were incubated together, in absence of any antigen, reaching the maximal level at 4 h. Cell attachment occurred even at 4°C, and this suggests the invovement of CD2 and LFA3 adhesion pathway, since LFA1-ICAM1 pathway functions only at 37°C [5]. It was found that sheep VC express a high level of LFA3 [6]. Experiments performed at 39°C revealed the reduced cell attachment in the presence both of lymph and saline. It could be explained by the effect of hyperthermia, since the canine skin temperature is 33-34°C (unpublished observation). Lower cell binding was observed in the presence of lymph as compared with saline. It can not be ruled out that some humoral factors present in lymph are involved, and the further studies remain to be carried out.

The present study indicates that the spontaneous binding of lymphocytes to autologous VC is mediated by divalent cation independent pathway. Both EDTA and verapamil, Ca^{++} chanells blocker did not interfere with cell binding in our system. Experiments performed in the presence of metabolic blockers, deoxy-D-glucose and azide, revealed the energy dependence of lymph cells binding.

Our data are consistent with notion that treatment of DC or Langerhans cells with proteolytic enzymes does not alter their attachment functions. We showed that the ability of VC to bind lymphocytes was inhibited by xylocaine, the drug which interferes with lymphocyte immune functions. Also retinoic acid abolished cell adhesion. This can explain the observed by others influence of retinoic acid on efficiency of antigen presentation by spleen DC or lymphocyte response to epidermal cells in MLR [7,8].

Table 3. Effect of immunosuppressants and anti-inflammatory drugs on lymph cell binding after 4 h incubation at 37^{0}C. (mean ± SD) * $p < 0.05$ vs saline

Drugs	Doses		% of clusters
Control - NaCl	(M)	0.15	20.1 ± 3.5
Hydrocortisone	(M)	10^{-5}	21.1 ± 7.1
Dexamethasone	(M)	10^{-5}	18.0 ±10.7
Methylprednisolone	(M)	10^{-5}	14.1 ± 5.5 *
		10^{-6}	13.5 ± 4.4 *
Cyclosporine A	(µg/ml)	5	19.3 ± 5.8
Azathioprine	(µg/ml)	5	20.3 ± 5.5
FK 506	(µg/ml)	4	9.1 ± 2.4 *
		0.4	11.5 ± 2.8 *
		0.04	12.0 ± 3.2 *
Indomethacin	(µg/ml)	5	16.3 ± 7.8
Acetylsalicylic acid	(µg/ml)	10	17.8 ± 4.8

In several systems of cell-cell interaction carbohydrates were identified as acceptor structures. Our studies showed that VC-lymphocyte binding is not a lectin-dependent process.

We attempted to analyse the effect of widely clinically used corticosteroids on cluster formation. Only methylprednisolone used in both physiological and pharmacological doses caused a statistically significant decrease in the percentage of in vitro spontaneously formed clusters. Other immunosuppressants, Cyclosporin A and azathioprine did not prevent spontaneous cluster formation. In contrast, FK 506 had a potent inhibitory effect on clustering in vitro in our system. The immunosuppressive effect of FK 506 on the prolonged skin allo- and xeno- grafts survival observed by others indicates that this agent is superior to Cyclosporin A [9]. We found that the anti-inflammatory drugs, indomethacin and acetylsalicylic acid had no effect on cluster formation.

Spontaneous binding in our system was enhanced by neuraminidase and this nonspecific interaction did not affect the level of autologous MLR, however lymphocytes responsiveness to suboptimal doses of PHA was higher than in control. Lack of enhanced response in auto-MLR after N-ase treatment could indicate that the specific antigenic signal for clustering had been acquired in vivo and no additional response should be expected in culture, although more clusters were formed due to desialation. It has been demonstrated[10] that sialic acid plays the restrictive role in antigen presentation by dendritic cells.

Antigen-independent clustering of lymph cells in skin tissue fluid in vivo could provide a temporary state of antigen-presenting cell-T-cell contact, which may be essential for the interaction to occur. A large array of environmental antigens penetrate the skin following microtrauma. Self-antigens of dying tissue cells may also contribute to stimulation of VC. This is the subject of our further studies.

REFERENCES

1. W.L.Olszewski. "In Vivo Migration of Immune Cells," CRC Press, Boca Raton (1987).
2. S.Knight, Veiled cells-dendritic cells of the peripheral lymph, *Immunobiology* 168:349(1984).
3. R.M.Steinman, The dendritic cell system and its role in immunogenicity, *Annu.Rev.Immunol.* 9:271(1991).
4. H.Galkowska, and W.L.Olszewski, Cellular composition of lymph in experimental lymphedema, *Lymphology*.19:139(1986).
5. M.W.Makgoba, M.E.Sanders, and S.Shaw, The CD2-LFA3 and LFA1-ICAM pathways: relevance to T-cell recognition, *Immunology Today*. 10:417(1989).
6. R.Bujdoso, J.Hopkins, B.M.Dutia, P.Young, and I.McConnell, Characterization of sheep afferent lymph dendritic cells and their role in antigen carriage, *J.Exp.Med*.170: 1285(1989).
7. P.Dupuy, M.Bagot, M.Heslan, and L.Dubertret, Synthetic retinoids inhibit the antigen presenting properties of epidermal cells in vitro, *J.Invest.Dermatol.* 93:455(1989).
8. P.A.Bedford, and S.C.Knight, The effect of retinoids on dendritic cell function, *Clin.Exp.Immunol.* 75:481(1989).
9. N.Inamura, K.Nakahara, T.Kino,T.Goto, et al,Prolongation of skin allograft survival in rats by a novel immunosuppressive agent FK 506, *Transplant.* 45:206(1988).
10. M.L.Kapsenberg, F.E.M.Stiekema, A.Kallan, J.D.Bos, and R.C. Roozemond, The restrictive role of sialic acid in antigen presentation to a subset of human peripheral $CD4^+$ T lymphocytes that requires antigen presenting dendritic cells, *Eur.J.Immunol.* 19:1829(1989).

INDUCTION OF THE LOW AFFINITY RECEPTOR FOR IgE (FcεRII/CD23) ON HUMAN BLOOD DENDRITIC CELLS BY INTERLEUKIN-4

Susanne Krauss, Elfriede Mayer, Gerti Rank and E. Peter Rieber

Institute for Immunology
University of Munich
Munich, FRG

INTRODUCTION

The human low affinity receptor for IgE (FcεRII/CD23) is expressed on a wide range of cells many of which have the capacity to process and present antigen. Thus, CD23 expression is constitutive on B cells and EBV-transformed B cell lines and can be induced on monocytes, macrophages and epidermal Langerhans cells (LHC) upon incubation with interleukin-4 (IL-4)[1,2,3].

In previous studies we have shown that FcεRII/CD23 on antigen presenting cells (APC) can focus IgE-complexed antigen for presentation to human T cells with increased efficacy when compared with conventional routes of antigen uptake[4]. This FcεRII/CD23-mediated antigen uptake might play a pivotal role in maintaining and enhancing allergen specific T cell activation[4]. The demonstration of in vivo binding of preformed IgE to epidermal LHC and dermal dendritic cells (DC) in patients with atopic eczema[5,6] lends support to this view.

These findings led us to the question as to whether FcεRII/CD23 can also be induced on DC which represent professional APC and whether CD23 could mediate uptake of IgE-complexed antigen by these cells.

LHC and mouse spleen cells have been shown to loose rapidly their capacity to process antigen when cultivated in vitro[7,8]. Therefore, functional tests on DC require a rapid isolation procedure. We developed a method for enrichment of DC which is based on the complete removal of T cells, B cells, monocytes and NK cells by the direct monoclonal antibody rosetting technique (DART) which has been shown to be a very efficient method for large scale separation of mononuclear cells[9]. By this method enrichment of DC is obtained within 6-8 hours. We show that DC isolated from peripheral blood express CD23 when cultivated in the presence of IL-4 whereas IFN-γ failed to induce CD23.

RESULTS

Enrichment of Dendritic Cells

Enrichment of dendritic cells from peripheral blood was performed as outlined in table 1.

Dendritic Cells in Fundamental and Clinical Immunology,
Edited by Kamperdijk *et al.*, Plenum Press, New York, 1993

Table 1. Enrichment of human blood dendritic cells.

1. Isolate mononuclear blood cells by Ficoll Paque centrifugation from buffy coat cells obtained from 400ml blood.
2. Remove T cells by a direct monoclonal antibody rosetting technique (DART) using CD2 mAb M-T910 as described[9]. Very briefly, mix cells with a twentyfold excess of bovine erythrocytes coated with CD2 mAb and spin down at 230 g for 10 min. Incubate for 40 min on ice. Separate rosetted and non rosetted cells by centrifugation over Ficoll Paque (δ=1,077g/ml).
3. Remove remaining T cells and monocytes by a second round of DART using a mixture of erythrocytes coated with CD2 mAb and with CD14 mAb (M-M42), respectively.
4. Deplete B cells and remaining monocytes by a third round of DART using a mixture of erythrocytes coated with CD37 mAb (M-B371) and with CD14 mAb.
5. Deplete remaining B cells and NK cells by a fourth DART step using a mixture of erythrocytes coated with CD16 mAb (B73.1) and CD37 mAb.
6. Remove remaining CD16 positive cells by rosetting with CD16 coated erythrocytes.

The final cell population contained < 1% T cells, B cells, monocytes and CD16 positive cells as evidenced by FACS analysis. NaF-sensitive naphthol-AS esterase containig cells could not be detected by cytochemistry. On the basis of strong MHC class II antigen expression and simultaneous presence of dendritic protrusions 20 - 30% of the cells were considered to be DC. Immunofluorescence staining followed by FACS analysis revealed the following reactivity: $CD1^-$, $CD2^-$, $CD3^-$, $CD11b^-$, $CD14^-$, $CD16^-$, $CD19^-$, $CD21^-$, $CD23^-$, $CD32^-$, $CD64^-$, $HLA\text{-}Dr^+$.

Autologous Mixed Leukocyte Reaction

In a second approach DC were tested for their reacitivty towards autologous T cells. For this purpose enriched DC were added to purified autologous T cells (>99% purity) from the same donor which were prepared as $CD2^+$ cells using DART. T cell depleted PBL ($PBL\text{-}T^-$) were used as control APC. As shown in Fig.2. T cells are stimulated by autologous DC, whereas purified autologous blood monocytes failed to induce an autologous reaction.

Comparison of Blood Dendritic Cells and Monocytes as Stimulatory Cells in Heterologous Mixed Leukocyte Reaction

As a further indication for the enrichment of DC their stimulatory capacity in a mixed leukocyte reaction was evaluated and compared with purified peripheral blood monocytes. For this purpose monocytes from the same donor were purified as $CD14^+$ cells using DART. For positive selection the rosetted cells were centrifuged over Percoll (δ =1,082) and rescued from the pellet[9]. Purity of monocytes was > 95 % as revealed by light microscopy of rosettes counterstained with crystal violet and of Giemsa stained smears of rosetted cells. As shown in Fig. 1 enriched DC were by far more effective in allogeneic stimulation when compared with blood monocytes. Significant proliferation of allogeneic PBL was observed at cell concentrations as low as 600 DC per 2 x 10^4 responder cells.

IL4- induced Expression of FcεRII/CD23 on Blood Dendritic Cells

In view of the antigen focusing capacity of CD23 on APC, it was of interest to see

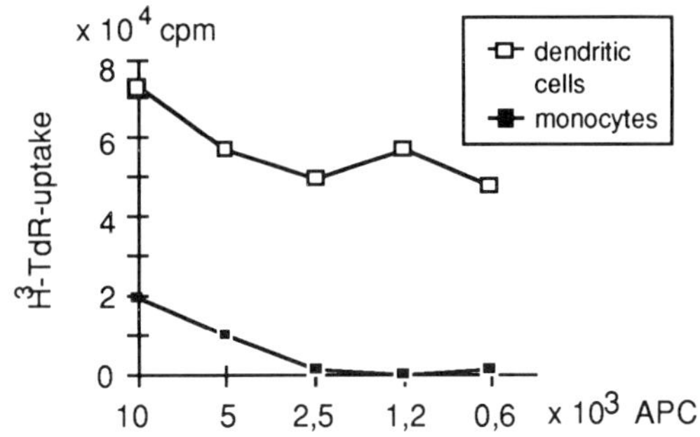

Figure 1. APC of an HLA-typed donor were irradiated with 2500 rad and mixed with PBL from an HLA-different donor at various cell ratios. The assay was performed in RPMI medium containing 10% human AB serum in U-bottom 96 well plates. After 5 days the cells were pulsed with 2 μCi/well ^{3}H-TdR for 12h. Results are expressed as mean cpm of triplicates.

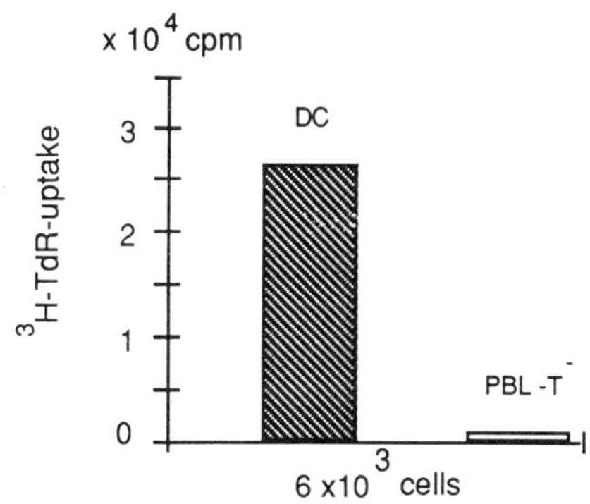

Figure 2. Reactivity of DC in autologous mixed lymphocyte reaction (AMLR). 1 x10^4 APC were irradiated with 2500 rad and mixed with 2 x 10^4 purified T cells of the same donor. The assay was performed in RPMI medium containing 10% human AB serum in U-bottom 96 well plates. After 5 days the cells were pulsed with 2 μCi/well ^{3}H-TdR for 12h. Results are expressed as mean cpm of triplicates.

whether DC can be induced to express CD23. For these experiments presence of monocytes had to be excluded. Therefore the protocol for enrichment of DC was modified. Blood mononuclear cells were first depleted of monocytes and NK cells by two rounds of DART using erythrocytes coated with CD14 mAb and CD16 mAb. The cells were then cultivated in the presence of rIL-4 and rIFN-γ. After 2 days T cells and B lymphocytes were removed by two rounds of rosetting with a mixture of erythrocytes coated either with CD2 mAb or with CD37 mAb.

Freshly isolated DC did not express CD23 as evaluated by FACS analysis or rosetting assays (data not shown). Culture in medium for 2 days lead only to a weak CD23 expression on DC (Fig 3a). However, after stimulation with 300U/ml rIL-4 for 2 days a significant CD23 expression was observed. The mean channel of CD23-fluorescence on the strongly MHC class II positive cells increased from 537 to 708 (Fig. 3b). No induction of CD23 could be observed upon incubation with varying doses of IFN-γ (data not shown). Control staining before IL-4 stimulation with CD19 mAb excluded contaminating B cells. Absence of monocytes was evidenced by negative staining with CD14 mAb before culture.

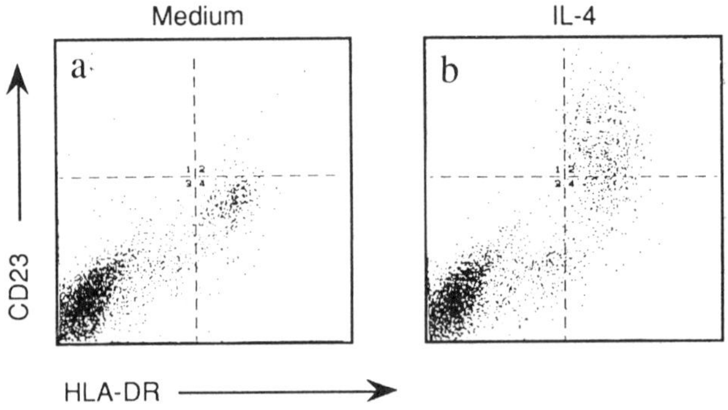

Figure 3. FcεRII/CD23 expression on blood dendritic cells. DC enriched from peripheral blood were cultivated with medium alone (a) or with 300U/ml IL-4 (b). After two days cells were doublestained in immunofluorescence with biotinylated CD23 mAb M-L25, detected by Avidin-Phycoerythrin and FITC labelled HLA-DR-specific mAb L243.

DISCUSSION

It is widely accepted that DC take up antigen in the periphery of the organism and transport it to the T cell dependent regions of lymphatic organs. During the period of antigen uptake DC seem to be weak in sensitizing T cells[10]. LHC have been shown to loose the ability to process antigen on their way to lymphatic organs via the afferent lymphatics and to develop into potent stimulatory APC[11]. The functional state of DC circulating in the blood remains to be determined.

Interestingly, human dendritic cells do not possess Fc receptors and with the exeption of freshly isolated LHC[12] do not phagocytose[13]. On the other hand, LHC have been shown to express the low affinity receptor for IgE upon stimulation with IL-4 and IFN-γ[3] and to bind preformed IgE in atopic patients[5,6]. We had shown previously that CD23 on monocytes and B cells can focus IgE-complexed antigen for presentation to human T cells[4]. It was therefore of interest to know whether also DC can utilize this way of antigen uptake when low amounts of IgE-antigen complexes are present.

In a first approach the ability of DC to express FcεRII/CD23 was tested. A novel protocol was developed which allows enrichment of peripheral blood DC within 6-8 hrs. This method resulted in DC of about 30% purity as evidenced by morphology, strong MHC II expression and absence of T cell, B cell, monocyte and NK cell surface marker expression on MHC class II positive cells. Of the remaining MHC class II negative lymphoid cells <1% expressed lineage specific surface marker. In addition, functional tests underlined the effective enrichment of DC from peripheral blood: first, they displayed an outstanding stimulatory capacity in heterologous MLR. Even at a concentration as low as 600 DC per 2 x 10^4 allogeneic PBL a strong stimulation was observed.This stimulatory capacity of DC became particularly evident when compared with purified blood monocytes which were by far less effective in activating allogeneic T cells. Second, DC were quite effective in stimulating autologous T cells, whereas T cell depleted PBL failed to activate autologous T cells. This is also in line with reported observations on DC[13].

Since monocytes strongly express CD23 upon induction with IL-4, absence of monocytes in the DC preparation had to be ascertained: 1. The discriminative expression of CD14 on monocytes and CD16 on NK cells is partially lost upon incubation with IL-4[14]. Therefore, CD14-positive and CD16-positive cells were removed prior to cultivation with IL-4. The cultures, however, still contained T cells since it had been observed that T cells supported the IL-4 induced CD23 expression on monocytes in some donors (own unpublished observation). 2. Absence of cells positive for the monocyte specific NaF-sensitive a-naphthol-AS esterase in the final cell population was proved by cytochemical analysis of cytospins.

Based on the observation that LHC can be induced to express FcεRII/CD23 by stimulation with IL-4 and IFN-γ DC were cultivated in the presence of these two lymphokines. Only IL-4 was able to induce CD23 on DC.

This observation seems particularly important in view of the possible role of CD23 expressing cells which can bind IgE-antigen complexes at low concentrations and present it more efficiently to T lymphocytes. This preferential activation of T helper cells by IgE-complexed antigen might represent a pivotal amplification mechanism of the allergic immune response.

Acknowledgements

This work was supported by the Deutsche Forschungsgemeinschaft SFB 217 and by the Wilhelm-Sander Stiftung.

REFERENCES

1. T. Defrance, J.P. Aubry, F. Rousset, B. Vandervliet, J.Y. Bonnefoy, N. Arai, Y. Takebe, T. Yokota, F. Lee, K. Arai, J. De Vries and J.J. Banchereau,Human recombinant interleukin-4 induces Fcε receptors (CD23) on normal human B lymphocytes, *J.Exp. Med.* 165:1459 1987).
2. D. Vercelli, H.H. Jabara, B.W. Lee, N. Woodland, R.S. Geha and D.Y.M. Leung, Human recombinant interleukin-4 induces FcεR2/CD23 on normal human monocytes, *J.Exp.Med.* 167:1406 (1988).
3. T. Bieber, A. Rieger, C. Neuchrist, J.C. Prinz, E.P. Rieber, G. Boltz-Nitulescu, O. Scheiner, D. Kraft, J. Ring and G. Stingl, Induction of Fc epsilon receptor RII/CD23 on human epidermal Langerhans cells by human recombinant interleukin 4 and gamma interferon, *J.Exp.Med.* 170:309 (1989).

4. U. Pirron, T. Schlunck, J.C. Prinz and E.P. Rieber, IgE-dependent focusing by human B lymphocytes is mediated by the low affinity receptor for IgE, *Eur.J.Imunol.* 20:1547 (1990).
5. C. Bruynzeel-Koomen, D.F. van Wiche, J. Toonstr, L. Berrens and P.L.B. Bruynzeel, *Arch.Dermatol.Res.* 278:199 (1986).
6. T. Bieber, B. Dannenberg, J.C. Prinz, E.P. Rieber, W. Stolz, O. Braun-Falco and J. Ring, Occurence of IgE-bearing epidermal Langerhans cells in atopic eczema: a study of the time course of the lesions and with regard to the serum IgE levels, *J.Invest.Dermatol.* 92:215 (1989).
7. E. Puré, K. Inaba, M.T. Crowley, L. Tardelli, M.D. Witmer-Pack, G. Ruberti, G. Fathman, and R.M. Steinman, Antigen processing by epidermal Langerhans cells correlates with the level of biosynthesis of MHC class II molecules and expression of invariant chain, *J.Exp.Med.* 172:1459 (1990).
8. K. Inaba, J.P. Metlay, M.T. Crowley and R.M. Steinman, Dendritic cells pulsed with protein antigens in vitro can prime antigen-specific, MHC-restricted T cells in situ, *J.Exp.Med.* 172: 631 (1990).
9. M. Wilhelm, H. Pechumer, G. Rank, E. Kopp, G. Riethmüller and E.P. Rieber, Direct monoclonal antibody rosetting. An effective method for weak antigen detection and large scale separation of human mononuclear cells, *J.Immunol. Methods.* 90, 89, (1986).
10. H. Stössel, F. Koch, E. Kämpgen, P. Stoger, A. Lenz, C. Heufer, N. Romani and G. Schuler, Disappearance of certain acidic organelles (endosomes and Langerhans cell granules) accompagnies loss of antigen processing capacity upon culture of epidermal Langerhans cells, *J.Exp.Med.* 172:1472, (1990).
11. N. Romani, S. Koide, M. Crowley, M. Witmer-Pack, A.M. Livingstone, C.G. Fathman, K. Inaba and R.M. Steinman, Presentation of exogenous protein antigen by dendritic cells to T cell clones. Intact protein is presented best by immature, epidermal Langerhans cells, *J.Exp.Med.* 169:1169 (1989).
12. C.R. Sousa, V. Kery, P.D. Stahl, P.J. Morris and J.M. Austyn, Phagocytosis of antigens by Langerhans cells, 2nd International Symposium on dendritic cells in fundamental and clinical immunology, Amsterdam, (1992).
13. R.M. Steinman, The dendritic cell system and its role in immunogenicity, *Annu.Rev.Immunol.* 9:271 (1991).
14. R.P. Lauener, S.M. Goyet, R.S. Geha, and D. Vercelli, Interleukin 4 down-regulates the expression of CD14 in normal human monocytes, *Eur.J.Immunol.* 20:2375 (1990).

FcϵRI MEDIATES IGE-BINDING TO HUMAN EPIDERMAL LANGERHANS CELLS

Oliver Kilgus*, Binghe Wang*, Armin Rieger*, Birgit Osterhoff*, Kenichi Ochiai+, Dieter Maurer*, Dagmar Födinger*, Jean-Pierre Kinet+, Georg Stingl*.

From the *Department of Dermatology I, Division of Cutaneous Immunobiology, University of Vienna Medical School, Vienna, Austria; and from the +Molecular Allergy and Immunology Section, National Institute of Allergy and Infectious Diseases, National Institutes of Health, Rockville, M.D., U.S.A.

In 1986, it was first recognized that epidermal Langerhans cells (LC) in lesional and non-lesional skin of atopic dermatitis (AD) patients have IgE-binding capacity (1,2). Furthermore, it was shown that, after acid-stripping of cell-bound IgE, the anti-IgE reactivity of LC can be restored by incubation of tissue substrates with native, but not with heated autologous IgE-rich serum (1). In view of the heat lability of the Fcϵ fragment, this observation led to the concept that IgE binds to LC via its Fcϵ fragment (1). Since exposure of LC-enriched epidermal cells from non-atopics to IL-4 and/or IFN-γ leads to the surface expression of anti-FcϵRII/CD23-reactive moieties on LC (3), it was originally thought that LC in AD skin bind IgE via FcϵRII/CD23 induced by cytokines present in the AD skin microenvironment. However, IgE$^+$ LC were subsequently also detected in other diseases, provided that the serum IgE level exceeded 100kU/L (4). It was therefore reasonable to assume that the presence of IgE-binding sites on human

Dendritic Cells in Fundamental and Clinical Immunology,
Edited by Kamperdijk *et al.*, Plenum Press, New York, 1993

epidermal LC is not a consequence of the respective disease process, but rather represents a constitutive property of these cells.

In order to test this hypothesis, we exposed normal-appearing skin sections of non-atopic individuals to various IgE preparations and then visualized cell-bound IgE in an indirect immunoperoxidase and immunofluorescence method. We found that this immunolabeling procedure regularly reveals IgE/anti-IgE reactive cells in both the epidermis and dermis (Fig. 1A,C). Incubation of IgE-labelled sections with anti-IgG antibodies did not result in detectable staining (Fig. 1B). IgE^+ epidermal cells were dendritic in shape, located at a suprabasal position, reacted with mAb against CD1a, and, thus, qualified as LC (5; Figure 1 C,D). Flow cytometric studies confirmed that $CD1a^+$ cells are the only IgE-binding epidermal cells and, conversely, that most of $CD1a^+$ epidermal cells (i.e. LC) are capable of IgE-binding. In the dermis, a portion of IgE^+ cells reacted with the mast-cell label FITC/avidin (5; data not shown). Other dermal IgE^+ cells were FITC/$avidin^-$, $CD45^+$, dendritic in shape, and were preferentially located around the dermal microvasculature. Preliminary data suggest that the latter represent a heterogeneous group of cells with respect to factor XIIIa, RFD1, and CD1a/c antigen expression (B.O. et al, in preparation). We next carried out inhibition experiments aimed at defining the IgE-binding structure on epidermal LC by preincubating the tissue with reagents capable of blocking the IgE-binding sites of the known IgE-binding structures (5), i.e., a) the tetrameric FcϵRI consisting of one α-, one ß- and two γ-chains, which binds homologous and heterologous monomeric IgE with high affinity (6,7); b) FcϵRII/CD23, a single chain molecule which binds IgE with an approximately 100 times lower affinity than FcϵRI (8,9); c) the ϵ-binding protein (ϵBP), a galactoside-specific lectin (10); d) the single-chained FcγRII/CD32, which binds monomeric IgG with very low affinity (7), and which, in the murine system, is capable of binding monomeric IgE (J.-P. Kinet, unpublished observation). We found that IgE-binding to LC could not be prevented by preincubation of cryostat sections with reagents known to block Ig binding to FcϵRII/CD23 (i.e. mAb MHM6), to ϵBP (i.e. lactose) and to FcγRII (i.e. mAb IV-3), but could be entirely abrogated by preincubation with the anti-FcϵRIα mAb 15-1 (5). A direct testing of the anti-FcϵRIα mAb 15-1 and 19-1 on cryostat sections in an indirect immunodouble-labeling technique showed that, in contrast to a panel of different anti-FcϵRII/CD23 mAb, these mAb react with the majority of CD1a-bearing epidermal cells (i.e. LC). At an ultrastructural level, 15-1 immunogold-labeling was confined to the surface

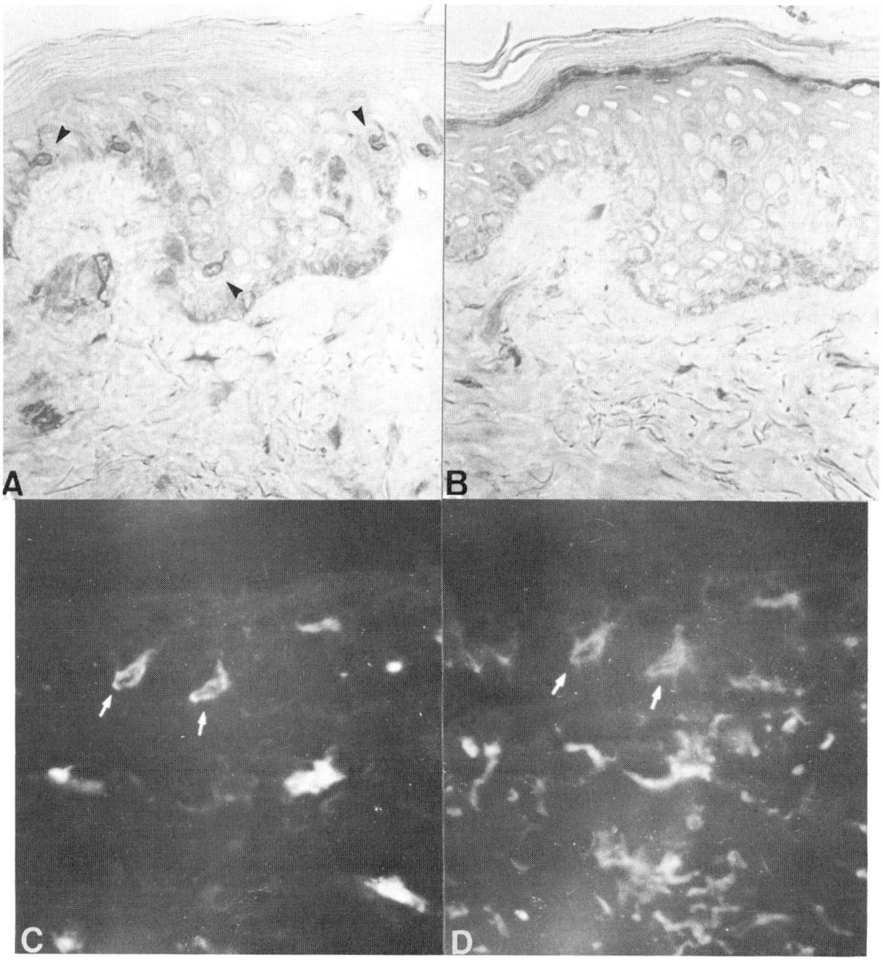

Figure 1. Demonstration of IgE bound to LC in normal human skin
After preincubation with IgE, tissue sections were reacted with biotinylated anti-IgE (A and C, arrows) or anti-IgG (B) serum from goat, developed with streptavidin/biotin-peroxidase complex (A,B) or texas-red/streptavidin (C) and counterstained with anti- CD1a/FITC (D, arrows).

of cells exhibiting Birbeck granules. In the dermis, FcϵRI$^+$ cells were frequently clustered around the dermal microvasculature, included FITC/avidin$^+$ as well as FITC/avidin$^-$ (dendritic) cells, and exhibited the same phenotypic heterogeneity as observed for IgE-binding dermal cells. In order to determine whether FcϵRI in LC have the same molecular configuration as in mast cells, we searched for the respective gene transcripts in LC by PCR technology.

In four independent experiments, we were able to amplify transcripts of comparable intensity from LC-enriched epidermal cells for the α- and γ-chains. Note that, within a given donor, these signals were stronger than those generated from dermal cells. In contrast, amplification using tryptase-specific primers yielded dermal but no epidermal signals. Thus, the observed FcεRI gene expression was not due to mast cells contaminating the epidermal cell preparation. When we searched for ß-chain transcripts, we obtained only weak epidermal signals; this low level of FcεRI ß-chain mRNA in LC-enriched epidermal cell suspensions may simply be attributable to differences in amplification efficiency. Alternatively, the ß-chain may be absent in LC or may be substituted for by another protein. A heterodimeric configuration of this receptor consisting only of α- and γ-chains has been shown to suffice for surface expression of high-affinity IgE-binding sites in transfectants (11). Collectively, these data clearly demonstrate that epidermal Langerhans cells express the high affinity Fc receptor for IgE. In addition, similar findings reported by an independent group of investigators confirm these results (12).

While a role for FcεRI in the IgE-binding to LC in AD skin has yet to be shown, one is tempted to speculate that in AD skin antigenetically cross-linked IgE on LC may stimulate events in these cells similar to those occurring in mast cells resulting in the production and release of soluble mediators (13) and cytokines (14,15,16). The recent observation that IgE-bearing LC derived from AD skin are superior to IgE^- LC in their capacity to present house dust antigen to sensitized T cells (17) indicates that FcεRI on LC may also play a crucial role in the uptake and/or processing of allergens by these cells. Thus, our findings provide a possible link between elevated IgE serum levels and atopic eczema, an association long recognized but hitherto not understood. The additional demonstration of FcεRI reactive, hematopoetic, dermal non-mast cells indicates that high-affinity IgE binding may have a widespread role in antigen presentation. Should further studies establish and substantiate the role of this receptor in diseased skin, the development of a topical agent capable of interfering with FcεRI function may be a useful strategy for the therapy of allergic skin diseases.

References

1. Bruynzeel-Koomen C, Van Wichen DF, Toonstra J, Berrens L, Bruynzeel PLB: The presence of IgE molecules on epidermal Langerhans cells in patients with atopic dermatitis. Arch Dermatol Res 278: 199-205, 1986

2. Bruynzeel-Koomen C, Van der Donk EMM, Bruynzeel PLB, Capron M, De Gast GC, Mudde GC: Associated expression of CD1 antigen and Fc receptor for IgE on epidermal Langerhans cells from patients with atopic dermatitis. Clin Exp Immunol 74: 137-142, 1988
3. Bieber T, Rieger A, Neuchrist C, Prinz JC, Rieber EP, Boltz-Nitulescu G, Scheiner O, Kraft D, Ring J, Stingl G: Induction of FcϵR2/CD23 on human epidermal Langerhans cells by human recombinant interleukin 4 and γ interferon. J Exp med 170: 309-314, 1989
4. Bieber T, Braun-Falco O: IgE-bearing Langerhans cells are not specific to atopic eczema but are found in inflammatory skin diseases. J Amer Acad Dermatol 24: 658-659, 1991
5. Wang B, Rieger A, Kilgus O, Ochiai K, Maurer D, Födinger D, Kinet JP, Stingl G: Epidermal Langerhans cells from normal human skin bind monomeric IgE via FcϵRI. J Exp Med 175; 1353-1365, 1992
6. Metzger H, Alcaraz G, Hohman R, Kinet JP Pribluda V, Quarto R: The receptor with affinity for immunoglobulin E. Ann Rev Immunol 4: 419-470, 1986
7. Ravetch JV, Kinet JP: Fc receptors. Ann Rev Immunol 9: 457-492, 1991
8. Spiegelberg HL: Structure and function of Fc receptors for IgE on lymphocytes, monocytes and macrophages. Adv Immunol 35: 61-88, 1984
9. Kikutani H, Inui S, Sato R, Barsumian EL, Owaki H, Yamasaki K, Kaisho T, Uchibayashi N, Hardy RR, Hirano T, Tsunasawa S, Sakiyama F, Suemura M, Kishimoto T: Molecular structure of human lymphocyte receptor for immunoglobulin E. Cell 47: 657-665, 1986
10. Robertson MW, Albrandt K, Keller D, Liu FT: Human IgE-binding protein: a soluble lectin exhibiting a highly conserved interspecies sequence and differential recognition of IgE glycoforms. Biochemistry 29: 8093-8100, 1990
11. Chisei R, Jouvin MHE, Kinet JP: Complete structure of the mouse mast cell receptor for IgE (FcϵRI) and surface expression of chimeric receptors (rat-mouse-human) on transfected cells. J Biol Chem 264: 15323-15327, 1989
12. Bieber T, de la Salle H, Wollenberg A, Hakimi J, Chizzonite R, Ring J, Hanau D, de la Salle C: Human epidermal Langerhans cells express the high affinity receptor for Immunoglobulin E. J Exp Med 175, 1282-1290, 1992
13. Parker CW: Lipid mediators produced through the lipoxygenase pathway. Ann Rev Immunol 5: 65-84, 1987
14. Plaut M, Pierce JH, Watson CJ, Hanley-Hyde J, Nordan RP, Paul WE: Mast cell lines produce lymphokines in response to cross-linkage of FcϵRI or to calcium ionophores. Nature 339: 64-67, 1989
15. Wodnar-Filipowicz A, Heusser CH, Moroni C: Production of the haemopoietic growth factors GM-CSF and interleukin-3 by mast cells in response to IgE receptor mediated activation. Nature 339: 150-152, 1989
16. Burd PR, Rogers HW, Gordon JR, Martin CA, Jayaraman S, Wilson SD, Dvorak AM, Galli SJ, Dorf ME: Interleukin 3-dependent and -independent mast cells stimulated with IgE and antigen express multiple cytokines. J Exp Med 170: 245-257, 1989
17. Mudde GC, Van Reijsen FC, Boland GJ, De Gast GC, Bruijnzeel PLB, Bruijnzeel-Koomen CAFM: Allergen presentation by epidermal Langerhans cells from patients with atopic dermatitis is mediated by IgE. Immunology 69: 335-341, 1990

MURINE EPIDERMAL LANGERHANS CELLS AS A MODEL TO STUDY TISSUE DENDRITIC CELLS

Gerold Schuler, Franz Koch, Christine Heufler, Eckhart Kämpgen, Gerda Topar, and Nikolaus Romani

Department of Dermatology
University of Innsbruck
Anichstr. 35
A-6020 Innsbruck, Austria

INTRODUCTION

Dendritic cells (DC) form a system of widely distributed antigen presenting cells that seems essential to initiate immmune responses [1]. The study of epidermal Langerhans cells (LC) was important to unravel how the individual components of this system interact [2]. It became evident that DC accomplish their task as "nature's adjuvant" in the afferent limb of the immune system in three discrete steps: immature DC are located in peripheral non-lymphoid tissues ("tissue DC") and are specialized to capture and process antigen ("sentinel function"). Following rapid downregulation of their antigen processing capacity they begin to develop into mature DC by acquiring the capacity to stimulate *resting* T cells. These cells then migrate as "veiled cells" via afferent lymph (or blood) to the T areas of the lymph node (or spleen) ("migratory function"), where they appear as "lymphoid DC", select antigen-specific T cells from the circulating pool and stimulate them ("sensitizing function"). Here we will review the studies of *murine* LC that have helped to establish this concept of DC function. We will also outline recent data about the modulation of LC function by cytokines, the evidence that DC and LC constitute a distinct hematopoietic lineage, and hints for a novel role of LC as regulators of keratinocyte proliferation.

LANGERHANS CELLS ARE IMMATURE PRECURSORS OF LYMPHOID DENDRITIC CELLS

Dendritic Cells (DC) were first identified in peripheral lymphoid organs and later on in afferent lymph and blood [1]. DC from all these sites exhibited similar features and were characterized by a typical morphology (such as "veils"), characteristic surface phenotype (such as constitutive and high surface expression of MHC class II molecules), and most importantly a unique stimulatory capacity for *resting* T cells. Immunohistochemical studies revealed the presence of leukocytes with a dendritic shape and constitutive expression of MHC class II molecules in the interstitial tissues of various non-lymphoid organs, suggesting the presence of DC at these sites as well. Due to difficulties in isolating these cells their exact relationship to DC remained, however, unclear. Several years ago we chose, therefore, to study another candidate non-lymphoid DC that was located in an epithelial compartment rather than in the interstitium, namely the epidermal Langerhans cell (LC). We found that freshly isolated LC stimulated *preactivated* T cells perfectly well, yet were only very weak stimulators of *resting* T cells [3,4]. Freshly isolated LC (as in vitro equivalents of resident LC) thus clearly did not qualify as typical DC. Quite surprisingly, however, LC upon culture acquired potent T cell sensitizing capacity and became virtually indistinguishable from DC isolated from lymphoid organs with respect to morphology and phenotype as well [3]. Cultured LC thus appeared to represent in vitro equivalents of DC isolated from lymphoid organs. These findings suggested, that resident LC were actually direct, yet immature precursors of the DC in the draining lymph nodes (*concept of LC maturation*). Further in vitro studies tried to disclose the mechanisms responsible for the potent T cell sensitizing capacity of *cultured* LC. Studies using the anti-CD3 model of T cell activation clearly showed that the increase in stimulatory capacity was not simply due to an increased number of ligands for the T cell receptor [5]. An interesting observation was that LC during culture became capable of binding T cells antigen-independently [4]. This antigen-independent binding of resting T cells was shown to occur as a first recognizable step of T cell activation in the anti-CD3 model [6] suggesting that this mechanism might indeed be critical for the unique capacity of mature DC to stimulate *resting* T cells. Although it has been shown that adhesion molecules such as ICAM-1, LFA-3 [1], and B7/BB1 (our unpublished observations) are upregulated during culture and maturation of LC, it is not yet known whether these or rather as yet undefined molecules cooperate to mediate the antigen-independent clustering of T cells [1]. Experiments are in progress to identify such putative cluster molecules by using monoclonal antibody and molecular biology approaches.

Studies addressing the antigen processing capacity of LC led to the unexpected observation that freshly isolated LC readily process antigen and generate MHC class II - peptide complexes, but rapidly downregulate this capacity during culture and thus become similar to lymphoid DC in this regard as well [7,8]. In contrast to a recent report [9] we were unable to find significant differences in downregulation of antigen processing capacity amongst various strains of mice (manuscript in preparation). There is now some understanding how and why this downregulation may occur. It was found that freshly isolated LC synthesize and express on their cell surfaces large amounts of MHC class II molecules [10,11,12] and possess many endosomal structures [13]. Interestingly, class II synthesis was rapidly shut down upon culture [10,11] and acidic vacuoles disappeared [13]. Presentation on MHC class II molecules is thought to be most effective if exogenous proteins are digested in acidic vacuoles and the resulting peptides gain access to newly synthesized class II molecules. The above findings thus readily explain why cultured LC and mature DC even though still capable of some endocytosis 10 do not present exogenous proteins antigens well. The turnover of MHC class II molecules on LC turned out to be very slow [11] which explains how LC can retain immunogenic peptides on the cell surface over several days.

Taken together, the in vitro studies demonstrate that antigen processing capacity and T cell sensitizing capacity are inversely expressed in fresh versus cultured LC. Evidence is emerging that other "tissue DC" behave similarly as LC [14] (see also contribution by Abul S. Rao at this meeting), and several studies suggest that LC / DC behave *in vivo* [15,16] as observed *in vitro*. These findings helped to establish the current concept of LC / DC function.

CYTOKINES REGULATE THE MATURATION OF LANGERHANS CELLS INTO POTENT IMMUNOSTIMULATORY DENDRITIC CELLS

The maturation of LC into potent immunostimulatory DC was discovered when unfractionated epidermal cells which contain 1-3 % LC were cultured for 2 to 3 days. To study LC maturation in more detail techniques had to be developed to highly enrich LC [2]. Purified LC when cultured in medium alone rapidly died, but addition of keratinocyte supernatant was sufficient to promote viability and maturation of LC. GM-CSF was identified as the critical, soluble keratinocyte-derived factor [17,18]. Recombinant GM-CSF when added to purified LC kept LC alive and allowed all the various components of the maturation process (see above) to occur normally. Although there is thus little doubt that GM-CSF is the principal mediator of LC maturation, there is evidence that others are involved as well. MHC class II synthesis, for example, increases even in the presence of blocking doses of anti-GM-CSF antibody [19] , and might thus be regulated otherwise. We had also observed that Interleukin-1 (IL-1) although incapable of keeping LC viable enhanced their stimulatory function when combined with GM-CSF [18]. We are currently testing the hypothesis that IL-1 acts as a first signal to induce LC maturation by induction of GM-CSF receptors, since we recently found that fresh LC express high-affinity IL-1 receptors (see contribution by E. Kämpgen at this meeting), whereas GM-CSF receptors are probably expressed only following culture and maturation of LC.

Recently, Enk et al., using reverse transcriptase PCR analysis, investigated early molecular events in the induction phase of contact hypersensitivity [20]. They observed that following epicutaneous application of various haptens (contact sensitizers, tolerogens, and non-sensitizers) as well as irritants GM-CSF mRNA was uniformly upregulated. Interestingly, however, IL-1 alpha and beta mRNA, as well as class II mRNA were upregulated only after application of contact sensitizers. These observations provide evidence that during acquisition of contact hypersenitivity a) the maturation of LC does indeed occur *in situ* (as class II and IL-1 beta mRNA were expressed by LC but not other epidermal cells [20], and are upregulated upon culture and maturation of LC, see [10,11] and below), and b) cytokines might indeed regulate LC maturation *in vivo* as suggested by the in vitro studies.

TNF alpha has been found to keep LC alive but in contrast to GM-CSF without inducing their functional maturation [19]. We recently found that TNF alpha disturbs the antigen presenting function of LC by causing loss of immunogenic peptides from their MHC class II molecules. This appears to be due to the induction of isolated and excessive synthesis of invariant chain in LC (manuscript in preparation). While these findings demonstrate that GM-CSF exerts a maturation effect that is distinct from maintaining LC in a viable, MHC class II-rich state the significance of the observed effects of TNF alpha escapes us at present. One might speculate that TNF alpha keeps LC *in situ* in a viable but immature state and / or that excessive TNF alpha production by keratinocytes mediates UVB-mediated immunosuppression as has been suggested by others [21].

EVIDENCE THAT LANGERHANS CELLS AND DENDRITIC CELLS CONSTITUTE A DISTINCT HEMATOPOIETIC LINEAGE

We have recently finished a study to determine the lineage of LC / DC. First we demonstrated that DC / LC are not lymphocytes as T cell and Ig receptor genes are in germ line. We then analyzed cytokine receptors on LC and DC by using ^{125}I-labeled cytokines and equilibrium binding. Spleen DC and cultured LC expressed high numbers of a single class of high-affinity GM-CSF receptors (2-4 x 10^3 / cell versus 1 x 10^3 / J774 macrophage tumor line), but unlike macrophages they lacked M-CSF receptors. These findings suggest that LC / DC belong to the myeloid lineage but are distinct from granulocytes and monocytes (manuscript in preparation). Given our data on the cytokine receptor profile of LC / DC it is also no longer surprising that M-CSF unlike GM-CSF does not support the viability of LC in vitro [17,18,19], and numbers of LC in M-CSF-deficient osteopetrotic mice are essentially normal whereas macrophage numbers are depressed (unpublished observations). Based on our findings and conclusions regarding the lineage of LC / DC it was possible to identify proliferating DC progenitors [22].

THE CYTOKINE GENE EXPRESSION PROFILE OF LANGERHANS CELLS SUGGESTS A NOVEL ROLE OF LANGERHANS CELLS

By using reverse transcriptase PCR analysis for screening and Northern blotting for further examination we found that out of 18 cytokines tested only 4 were expressed by LC, and fell into two differentially regulated groups. IL-1 beta and IL-6 were upregulated upon culture, whereas MIP-1 alpha and MIP-2 were downregulated. IL-1 beta and MIP-1 alpha were expressed by LC but not other epidermal cells [23]. The cytokine gene expression pattern of LC is clearly different from that of typical macrophages. Macrophages would not differentially regulate the above cytokine genes, and would express additional cytokines such as IL-1 alpha, TNF alpha, INF alpha, and GM-CSF. While the cytokine gene expression profile of LC supports the notion that LC / DC are a distinct member of the myeloid lineage its physiological role remains to be determined. The massive upregulation of IL-1 beta during culture is certainly notable, but its relevance is unclear. It is, for example, very unlikely that IL-1 beta is a critical signal for the activation of *resting* T cells as DC can sensitize T cells in the apparent absence of this cytokine [1]. MIP-1 alpha, which acts as a stem cell inhibitor in the bone marrow [24], was recently found to inhibit keratinocyte stem cell proliferation as well [25]. This points to a novel, non-immunological role of LC in epidermal homoeostasis.

CONCLUSION

Epidermal Langerhans cells have proven a rewarding model to study tissue dendritic cells. We have learned that "tissue DC" might be direct, though immature precursors of "lymphoid DC". It was thus not until LC were studied in more detail that it became evident that DC populations in peripheral tissues, in the circulation, and lymphoid organs are not functioning separately but seem actually connected in series. The finding that the maturation of LC into potent immunostimulatory DC is regulated by cytokines, notably GM-CSF, suggests novel approaches to immunotherapy. It should be possible to up- as well as downregulate immunogenicity by influencing the maturation of tissue DC *in situ*. As proliferating DC precursors have recently been identified it is also likely that DC will be used for inducing T cells capable of adoptive immunity. It might even be possible to modify DC and to use them for specific downregulation of antigen-specific T cells. It is likely that further *in vitro* studies of LC will be important to achieve these goals.

ACKNOWLEDGMENTS

We appreciate the continuous support of Dr. P. Fritsch and Dr. R.M. Steinman. Our work on Langerhans Cells is supported by a grant from the Austrian Science Foundation (FWF-P 4985 MED).

REFERENCES

1. R.M. Steinman, The dendritic cell system and its role in immunogenicity, *Ann. Rev. Immunol.* 7:371 (1989).
2. G. Schuler. " Epidermal Langerhans Cells", CRC Press,Inc., Boca Raton (1991).
3. G. Schuler and R.M. Steinman, Murine epidermal Langerhans cells mature into potent immunostimulatory dendritic cells in vitro, *J. Exp. Med.* 161:526 (1985).
4. K. Inaba, G. Schuler, M.D. Witmer, J. Valinski, B. Atassi, and R.M. Steinman, Immunologic properties of purified epidermal Langerhans cells. Distinct requirements for stimulation of unprimed and sensitized T lymphocytes, *J. Exp. Med.* 164:605-613 (1986).
5. N. Romani, K. Inaba, E. Pure, M. Crowley, M. Witmer-Pack, and R.M. Steinman, A small number of anti-CD3 molecules on dendritic cells stimulates DNA synthesis in mouse T lymphocytes, *J. Exp. Med.* 169:1153 (1989).
6. K. Inaba, N. Romani, and R.M. Steinman, An antigen-independent contact mechanism as an early step in T cell-proliferative reponses to dendritic cells, *J. Exp. Med.* 170:527 (1989).
7. N. Romani, S. Koide, M. Crowley, M. Witmer-Pack, A.M. Livingstone, C.G. Fathman, K. Inaba, and R.M. Steinman, Presentation of exogenous protein antigens by dendritic cells to T cell clones: intact protein is presented best by immature epidermal Langerhans cells, *J. Exp. Med.* 169:1169 (1989).
8. J. W. Streilein, and S.F. Grammer, In vitro evidence that Langerhans cells can adopt two functionally distinct forms capable of antigen presentation to T lymphocytes, *J. Immunol.* 143:3925 (1989).
9. S. Aiba, and S. I. Katz, The ability of cultured Langerhans cells to process and present antigens is MHC-dependent, *J. Immunol.* 146:2479 (1991).
10. E. Puré, K. Inaba, M.T. Crowley, L. Tardelli, M.D. Witmer-Pack, G. Ruberti, G. Fathman, and R.M. Steinman, Antigen processing by epidermal Langerhans cells correlates with the level of biosynthesis of major histocompatibility complex class II molecules and expression of invariant chain, *J. Exp. Med.* 172:1459 (1990).
11. E. Kämpgen, N. Koch, F. Koch, P. Stöger, C. Heufler, G. Schuler, and N. Romani, Class II major histocompatibility complex molecules of murine dendritic cells: Synthesis, sialylation of invariant chain, and antigen processing capacity are down-regulated upon culture, *Proc. Natl. Acad. Sci. USA* 88:3014 (1991).
12. D. Becker, A.B. Reske-Kunz, J. Knop, and K. Reske, Biochemical properties of MHC class II molecules endogenously synthesized and expressed by mouse Langerhans cells, *Eur. J. Immunol.* 21:1213 (1991).
13. H. Stössel, F. Koch, E. Kämpgen, P. Stöger, A. Lenz, C. Heufler, N. Romani, and G. Schuler, Disappearance of certain acidic organelles (endosomes and Langerhans cell granules) accompanies loss of antigen processing capacity upon culture of epidermal Langerhans cells, *J. Exp. Med.* 172:1471 (1990).

14. P. G. Holt, J. Oliver, C McMenamin, and M.A. Schon-Hegrad, Studies on the surface phenotype and functions of dendritic cells in parenchymal lung tissue of the rat, *Immunology* 75:582 (1992).
15. C.P. Larsen, R.M. Steinman, M. Witmer-Pack, D.F. Hankins, P.J. Morris, and J.M. Austyn, Migration and maturation of Langerhans cells in skin transplants and explants, *J. Exp. Med.* 172:1483 (1990).
16. K. Inaba, J.P. Metlay, M.T. Crowley, and R.M. Steinman, Dendritic cells pulsed with protein antigens in vitro can prime antigen-specific, MHC-restricted T cells in situ, *J. Exp. Med.* 172:631 (1990).
17. M.D. Witmer-Pack, W. Olivier, J. Valinsky, G. Schuler, and R.M. Steinman, Granulocyte/macrophage colony-stimulating factor is essential for the viability and function of cultured murine epidermal Langerhans cells, *J. Exp. Med.* 166:1484 (1987).
18. C. Heufler, F. Koch, and G. Schuler, Granulocyte-macrophage colony-stimulating factor and interleukin-1 mediate the maturation of murine epidermal Langerhans cells into potent immunostimulatory dendritic cells, *J. Exp. Med.* 167:700 (1988).
19. F. Koch, C. Heufler, E. Kämpgen, D. Schneeweiss, G. Böck, and G. Schuler, Tumor necrosis factor alpha maintains the viability of murine epidermal Langerhans cells in culture but in contrast to granulocyte/macrophage colony-stimulating factor without inducing their functional maturation, *J. Exp. Med.* 171:159 (1990).
20. A.H. Enk and S.I. Katz, Early molecular events in the induction phase of contact sensitivity, *Proc. Natl. Acad. Sci. USA* 89:1398 (1992).
21. M. Vermeer and J.W. Streilein, Ultraviolet B light-induced alterations in epidermal Langerhans cells are mediated in part by tumor necrosis factor-alpha, *Photodermatol. Photoimmunol. Photomed.* 7:258 (1990).
22. K. Inaba, R.M. Steinman, M. Witmer-Pack, H. Aya, M. Inaba, T. Sudo, S. Wolpe, and G. Schuler, Identification of proliferating dendritic cell precursors in mouse blood, *J.Exp. Med.* 175:1157 (1992).
23. C. Heufler, G. Topar, F. Koch, B. Trockenbacher, E. Kaempgen, N. Romani, and G. Schuler, Cytokine gene expression in murine epidermal cell suspensions: Interleukin-1 beta and Macrophage Inflammatory Protein-1 alpha are selectively expressed in Langerhans cells but are differentially regulated in culture, *submitted.*
24. G.J. Graham, E.G. Wright, R. Hewick, S.D. Wolpe, N.M. Wilkie, D. Donaldson, S. Lorimore, and I.B. Pragnell, Identification and characterization of an inhibitor of haemopoietic stem cell proliferation, *Nature* 344:442 (1990).
25. E.K. Parkinson, G.J. Graham, P. Daubersies, J.E. Burns, C. Heufler, M. Plumb, G. Schuler, and I.B. Pragnell, A haematopoietic stem cell inhibitor (SCI/MIP-1alpha) also inhibits clonogenic keratinocyte proliferation, *submitted.*

DIFFERENTIATION OF DENDRITIC CELLS IN CULTURES OF RAT AND MOUSE BONE MARROW CELLS

William E. Bowers, Shengqiang Yu, and Syeedur Khandkar

Department of Microbiology and Immunology
University of South Carolina School of Medicine
Columbia, SC 29208

Although dendritic cells originate from bone marrow,[1,2,3,4] they are not detectable in fresh preparations of bone marrow cells.[5] We previously developed an *in vitro* system in which dendritic cells are produced when rat bone marrow cells are cultured for 4-5 days.[6] We recently established an *in vitro* bone marrow culture system for mouse that enables us to use the greater number of recombinant cytokines, monoclonal antibodies, cell lines, and reagents available for this species. Similar results have been obtained in both species. This paper summarizes the major findings.

Freshly isolated bone marrow cells do not have accessory activity. A sensitive assay for one dendritic cell function - the ability of a single dendritic cell to provide accessory activity enabling as many as 300 periodate-treated T lymphocytes to respond - is barely detectable in freshly isolated bone marrow cells. However, readily detectable dendritic cells and measurable accessory activity are found in 4-day cultures of bone marrow cells.

In order to eliminate interfering effects of serum factors on dendritic cell differentiation and to facilitate studies on the role of recombinant cytokines on the production of dendritic cells, we culture bone marrow cells in a serum-free medium. Without additional supplement(s), dendritic cell production is low.

Supplementing serum-free medium with a 20% supernatant from Con A-treated spleen cells leads to a large increase (5-20-fold) in the production of dendritic cells (Table I). This increase is considerably greater than the overall increase in the number of viable cells (1.5-3 times) that also occurs. The number of dendritic cells produced in the presence of a Con A supernatant is proportional to the number of

Dendritic Cells in Fundamental and Clinical Immunology,
Edited by Kamperdijk *et al.*, Plenum Press, New York, 1993

TABLE I. Production of dendritic cells from low density rat bone marrow cells cultured for four days at different cell concentrations ± Con A spleen cell supernatant.

	bone marrow cells cultured (X 10^{-6})			
	1.25	2.5	5	10
Medium	2*	6	19	33
+ Con A Supernatant	32	52	93	160

*number of dendritic cells produced (X 10^{-3})

bone marrow cells cultured. The Con A supernatants are produced overnight in serum-free medium with mouse or rat spleen cells; both are effective.

Dendritic cells function as accessory cells for periodate-treated T lymphocytes in a dose-dependent manner. In rat, no other Ia^+ cell type studied (macrophages, B cells) has accessory activity[8]; in mouse, macrophages and B cells exhibit accessory activity that is lower than that of dendritic cells by an order of magnitude or more. Dendritic cells produced in bone marrow cultures containing a Con A supernatant have a low buoyant density and are 7-10-fold more active on a per cell basis than those recovered in the low density fraction from control cultures.

Even though dendritic cells have a characteristic morphology, an independent means of identification is required. A number of monoclonal antibodies (mAb) are available for this purpose. In rat we have developed three mAb (A3C, B6G, C11B) showing a high degree of specificity for lymphoid dendritic cells. Along with G489, which detects class II MHC, these mAb stain strongly the dendritic cells produced in cultures of bone marrow cells. The same cells are negative for macrophages (ED2), T cells (0X19), and B cells (0X33).

In mouse, B cells stain strongly with B220, but dendritic cells are negative. Macrophages stain positively with F4/80, but dendritic cells are weakly stained or negative. B21-2, a mAb specific for Class II MHC, stains most strongly for dendritic cells and B cells, less so for macrophages. Weak to no staining occurs with two anti-dendritic cell mAb, 33D1 and J11D, which was anticipated because it is known that dendritic cells express few copies of the molecules these mAb detect.

The ability of these and other mAb to produce lysis with complement provided another approach. Mouse bone marrow cells were cultured for four days, incubated with mAb plus complement, and tested for accessory activity with periodate-treated lymphocytes as responders. The results are presented in Table II. B21-2 virtually abolishes the response, confirming that T lymphocyte responses require an accessory cell bearing class II MHC molecules. The anti-dendritic cell mAb, 33D1 and J11D, also significantly reduce the response. The macrophage and Langerhans cell marker

TABLE II. Accessory activity of mouse bone marrow cells cultured for four days and treated with mAb and complement.

mAb	detects	complement	% reduction of response
--	--	-	0
--	--	+	0
B220	B cells	+	8 ± 4
F4/80	macrophages, Langerhans cells	+	24 ± 7
33D1	dendritic cells	+	75 ± 5
J11D	dendritic cells, B cells, others	+	82 ± 8
B21-2	class II MHC	+	95 ± 4

Dose-response curves were established with periodate-treated T lymphocytes as responders. The values are the mean and standard deviation for four experiments.

detected by F4/80 has a small effect. Finally, elimination of B cells by B220 and complement only slightly reduces the responses. Taken together, these findings provide strong evidence that dendritic cells are produced in cultures of mouse bone marrow cells.

Most dendritic cells in bone marrow cultures are produced from dividing precursors. The results from two different methods indicate that precursors divide to give rise to dendritic cells: 1) irradiation, and 2) ^{3}H-thymidine incorporation and autoradiography. Irradiation of bone marrow cells with 1000R prior to culture abolishes most of the dendritic cells and accessory activity. (In contrast, irradiation does not alter either cell number or accessory activity of dendritic cells produced after 4-5 days in cultures of bone marrow cells or of those isolated from tissues.) Autoradiography of bone marrow cells cultured for four days in the presence of low levels of ^{3}H-thymidine reveals grains over the nuclei of most dendritic cells.

The phenotype of the precursor differs from that of the dendritic cell produced in bone marrow cultures. Dendritic cells produced in cultures of bone marrow cells phenotypically resemble mature dendritic cells isolated from tissues. For each species, different means were used to determine whether precursors also express the same phenotypic markers. For rat, one or more of the following methods was used with each mAb: 1) panning, 2) rosetting, and/or 3) sorting with a flow cytometer. Freshly isolated bone marrow cells were exposed to the mAb under appropriate conditions, the mAb$^+$ and mAb$^-$ cells separated, and each population cultured for four days. The production of dendritic cells in each population was then compared to that of unfractionated cells. For mouse, advantage was taken of the ability of the mAb to lyse cells in the presence of complement; treated and untreated preparations of bone marrow cells were cultured for four days and the production of dendritic cells determined. A summary of the findings is presented in Table III.

TABLE III. Phenotype of dendritic cell precursors in bone marrow.

Species	mAb	detects	precursors	dendritic cells
rat	0X7	Thy 1	+	+
	0X22	leukocyte common antigen	+	+
	A3C	dendritic cells	-	+
	C11B	dendritic cells	-	+
	G489	class II MHC	-	+
mouse	B220	B cells	-	-
	F4/80	macrophages, Langerhans cells	-	±
	33D1	dendritic cells	-	+
	J11D	B cells, dendritic cells	-	+
	B21-2	class II MHC	-	+

In both species, antigens expressed on the dendritic cells produced in cultures of bone marrow cells (and on mature, tissue-derived dendritic cells) are not present on precursors. Class II MHC, which is constitutively expressed on dendritic cells, also is not expressed on the precursor.

GM-CSF is the major cytokine controlling differentiation of dendritic cells. GM-CSF was found to increase significantly the production of dendritic cells in cultures of bone marrow cells. Table IV presents the results of three experiments which compare G-CSF, M-CSF, and GM-CSF. Accessory activity per dendritic cell is also greatly enhanced by GM-CSF, but not by G-CSF or M-CSF. Combinations of these three cytokines do not increase the number of dendritic cells or accessory activity above that produced by GM-CSF alone.

TABLE IV. Effect of cytokines on the production of dendritic cells in cultures of mouse bone marrow cells.

	Number of dendritic cells ($x10^{-3}$)		
Addition	Expt. 1	Expt. 2	Expt. 3
--	30	50	110
G-CSF	75	80	130
M-CSF	85	120	160
GM-CSF	300	420	360

Other recombinant cytokines (IL-1, IL-3, IL-4, IL-7 and TNFα) have no or little effect on the production of dendritic cells; in no case was accessory activity elevated above that of the control. Combinations of cytokines likewise have no effect unless GM-CSF is also present, in which case the number of dendritic cells produced and accessory activity are identical to those of cultures containing only GM-CSF.

In conclusion, we have demonstrated that dendritic cells that are morphologically, phenotypically, and functionally indistinguishable from those isolated from tissues can be generated in cultures of bone marrow cells from rat and mouse. The fact that mature dendritic cells are not present in fresh preparations of bone marrow cells suggests that an immature form of dendritic cell is released from the bone marrow and matures elsewhere in the body, presumably under the control of GM-CSF. The lack of an effect of other cytokines defines these events as occurring late in the differentiative pathway. Our findings support a myeloid lineage for dendritic cells, in agreement with Reid et al.[9] and Gieseler et al.[10]

REFERENCES

1. J.G. Frelinger, L. Hood, S. Hill, and J.A. Frelinger, Mouse epidermal Ia molecules have a bone marrow origin, *Nature (Lond.)* 282:321 (1979).
2. S.I. Katz, K. Tamaki, and D.H. Sachs, Epidermal Langerhans cells are derived from cells originating in bone marrow, *Nature (Lond.)* 282:324 (1979).
3. A.N. Barclay and G. Mayrhofer, Bone marrow origin of Ia-positive cells in the medulla of rat thymus, *J. Exp. Med.* 153:1666 (1981).
4. G. Mayrhofer, C.W. Pugh, and A.N. Barclay, The distribution, ontogeny and origin in the rat of Ia-positive cells with dendritic morphology and of Ia antigen in epithelia, with special reference to the intestine, *Eur. J. Immunol.* 13:112 (1983).
5. R.M. Steinman and Z.A. Cohn, Identification of a novel cell type in peripheral lymphoid organs of mice. I. Morphology, quantitation, tissue distribution, *J. Exp. Med.* 137:1142 (1973).
6. W.E. Bowers and M.R. Berkowitz, Differentiation of dendritic cells in cultures of rat bone marrow cells, *J. Exp. Med.* 163:872 (1986).
7. S. Yu, S. Khandkar, and W.E. Bowers, Dendritic cells in cultures of mouse bone marrow cells, *FASEB J.* 5:A1349 (1991).
8. W.E.F. Klinkert, J.H. LaBadie, and W.E. Bowers, Accessory and stimulating properties of dendritic cells and macrophages isolated from various rat tissues, *J. Exp. Med.* 156:1 (1982).
9. C.D.L. Reid, P.R. Fryer, C. Clifford, A. Kirk, J. Tikerpae, and S.C. Knight, Identification of hematopoietic progenitors of macrophages and dendritic Langerhans cells (DL-CFU) in human bone marrow and peripheral blood, *Blood* 76:1139 (1990).
10. R.K.H. Gieseler, R.-A. Röber, R. Kohn, K. Weber, M. Osborn, and J.H. Peters, Dendritic accessory cells derived from rat bone marrow precursors under chemically defined conditions *in vitro* belong to the myeloid lineage, *Eur. J. Cell Biol.* 54:171 (1991).

TNF AND GM-CSF DEPENDENT GROWTH OF AN EARLY PROGENITOR OF DENDRITIC LANGERHANS CELLS IN HUMAN BONE MARROW

Cecil DL Reid[1], Arthur Stackpoole[2], and Jaak Tikerpae[1]

Department of Haematology[1]
Department of Transplantation Biology[2]
Northwick Park Hospital and Clinical Research Centre
Harrow, Middlesex HA1 3UJ UK

INTRODUCTION

It has been demonstrated recently that colonies of cells with dendritic morphology can be generated in-vitro in semi-solid cultures of both human bone marrow and peripheral blood mononuclear cells[1]. The cells express antigens typically found on skin Langerhans cells (CD1a, HLA-DR/DQ, CD4) and are functionally active in primary mixed leucocyte reactions (MLR). The colony forming cells are the earliest progenitors of dendritic/Langerhans cells so far identified and have been termed dendritic Langerhans colony forming units - DL-CFU. This bone marrow culture technique has now been applied in order to characterise further these progenitors of dendritic cells and to determine their regulation by growth factors in-vitro.

Studies of hematopoiesis in vitro may be complicated by the impurity of cultured cell populations and by the undetected presence in cultures of molecules in serum or conditioned medium or released by accessory cells that have activity on progenitors. Therefore in this study the more stringent conditions of a serum deprived and chemically defined culture system has been used to investigate cytokine regulatory activity. These studies have shown that TNF and GM-CSF are essential for the growth of dendritic cells from their early progenitor in-vitro[2].

METHODS

Bone marrow mononuclear cells (BMC) from normal donors were depleted of adherent cells by a one hour incubation in 20% fetal calf serum (FCS) and cultured for 14 days in 35 mm Petri dishes at between 0.05 and 3.0 x 10^4 cells/ml as previously described[1]. Serum replete cultures contained Iscoves modified Dulbecco medium (IMDM) with 30% FCS, 1% bovine serum albumin, 10^{-4} M 2-beta mercaptoethanol, 1% methylcellulose and PHA leucocyte conditioned medium (PHA-LCM - 5%). The technique of culture without serum was adapted from that described previously by Migliaccio et al[3]. The growth medium was

Dendritic Cells in Fundamental and Clinical Immunology,
Edited by Kamperdijk *et al.*, Plenum Press, New York, 1993

IMDM containing adenosine, cytidine, guanosine, uridine, 2' deoxy- adenosine, 2' deoxycytidine, 2' deoxyguanosine and thymidine (Sigma Cell Culture Reagents) all at 0.1 mg/ml. Added inorganic salts (Sigma culture tested) were manganous sulphate and ammonium molybdate at 10^{-7} M, ammonium metavanadate, nickel chloride and stannous chloride all at 5 x 10^{-7} M and ferrous sulphate at 4 x 10^{-6} M. Also included were insulin 2.5 ug/ml,1 cholesterol 72 ug/ml, 0.7 mg/ml transferrin saturated with $FeCl_3$, BSA 2%, 10^{-4} M 2-beta mercaptoethanol and 1% methyl cellulose. All cultures were scored for colonies on the inverted microscope at 14 days after incubation in 5% CO_2 in air. Granulocyte and macrophage colonies were readily recognised by conventional criteria. Colonies containing dendritic cells were easily recognised because of their highly distinctive morphology (Figure 1a) and in some instances this was confirmed either by positive immuno-enzymatic labelling (APAAP - Figure 1c) of cytospin preparations or by electron microscopy (immunogold) with the antibody NA1/34 for CD1a membrane expression. Many of these colonies also contain macrophages but were scored together with the pure dendritic cell colonies as DL-CFU in this study.

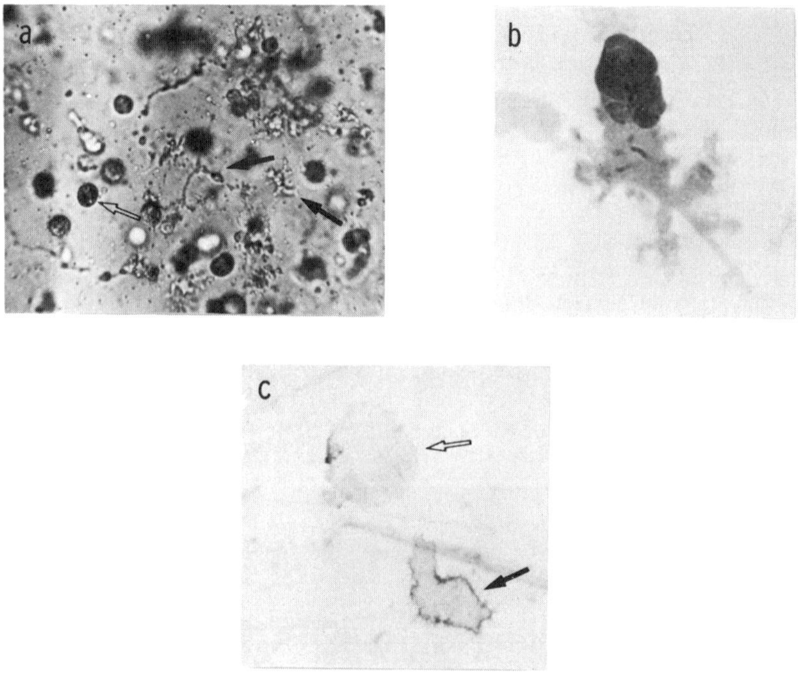

Figure 1. Microscopic appearance of 14 day colony cells in methylcellulose cultures of human bone marrow cells : a) mixed dendritic macrophage colony (x 58). b) Giemsa stained dendritic cell from colony (x 825). c) CD1a+ve dendritic cell and CD1a-ve macrophage (APAAP x 330). ← = dendritic cell ⇐ = macrophage.

RESULTS

DL-CFU Proliferation in Serum Replete and Serum Deprived Cultures

The growth of DL-CFU had previously been found to be optimal when both fetal bovine serum (FCS 30%) and PHA-LCM(5%) were present in the cultures[1]. In order to determine the relative dependence of DL-CFU as well as of granulocyte (CFU-G) and macrophage progenitors (CFU-M) on the presence of serum, colony yields were studied in chemically defined medium with the addition of serum at concentrations of 1 - 30%. The negligible granulocyte growth in serum confirmed CFU-G dependence on external sources of cytokines however both macrophage (CFU-M) and dendritic (DL-CFU) growth were close to optimal levels at 30% FCS. It was clear that further study of cytokine regulation of dendritic colonies would require the exclusion of serum from culture.

Table 1. Colony growth* in serum replete or serum deprived cultures + cytokines

serum replete	CFU-G	CFU-M	DL-CFU
1%	0	0	0
10%	0	45	0
30%	4	86	79
serum-deprived + additions			
Nil	0	0	0
GM-CSF	57 (19)	20 (0.3)	0
GM-CSF+IL-3	85 (20)	14 (6)	0
GM-CSF+IL-3+G-CSF	200 (42)	43 (8)	0
GM-CSF+IL-3+M-CSF	116 (28)	51 (16)	1

* colony numbers are percentage (SE) of optimal values in 30% FCS & 5% PHA-LCM

Serum deprived cultures were studied with various combinations of GM-CSF (4.5×10^{-10} Molar), G-CSF (2.3×10^{-9} Molar), and IL-3 (2×10^{-9} Molar), concentrations that had previously been found to be optimal for colony growth in serum replete cultures (data not shown). M-CSF where present was at a concentration of 20 U/ml. The lower portion of Table 1 shows that virtually no dendritic cell colonies were observed under any of these conditions. CFU-G numbers, although stimulated suboptimally by either GM-CSF alone or by GM-CSF and IL-3, were markedly enhanced when G-CSF was also added to the cultures. CFU-M numbers were suboptimal under all these culture conditions and rose to half optimal values when M-CSF (20 U/ml) was added.

Table 2. Serum deprived cultures - effects of IL-1α*

serum-deprived	CFU-M	DL-CFU
GM-CSF+IL-3+M-CSF (20u)	43	1
GM-CSF+IL-3+M-CSF+IL-1 (5u)	51	4
GM-CSF+IL-3+M-CSF+IL-1 (10u)	60	1
GM-CSF+IL-3+M-CSF (50u)	49	4

* colonies expressed as numbers per 1×10^5 cells cultured

Interactions of IL-1α or TNF with Growth Factors in Serum Deprived Cultures

Table 1 showed that GM-CSF alone could stimulate CFU-M growth and that the addition of IL-3 and M-CSF increased this response. Virtually no DL-CFU were observed

Table 3. Serum deprived cultures - effects of TNFα*

cytokine additions		CFU-G	CFU-M	DL-CFU
GM-CSF+IL-3+G-CSF		158	51	1
GM-CSF+IL-3+G-CSF+TNF	0.1 U	183	27	0
GM-CSF+IL-3+G-CSF+TNF	2 U	243	0	13
GM-CSF+IL-3+G-CSF+TNF	5 U	129	4	33
GM-CSF+IL-3+G-CSF+TNF	50 U	43	0	17
GM-CSF+IL-3+G-CSF+TNF	100 U	20	0	14

* colonies are expressed as numbers per 1 x 10^5 cells cultured. Mean (SE) numbers of CFU-G, CFU-M and DL-CFU in serum replete cultures with 5% PHA-LCM were 120(11), 134(16) and 48(8) respectively.

even when the M-CSF concentration was increased to 2000 U/ml (not shown). Table 2 shows that the addition of IL-1α (10 U/ml) resulted in a further increase in CFU-M and this combination of factors also stimulated a very small number of dendritic cell colonies.

TNFα or TNFβ alone had no effect when added to serum deprived cultures, Table 3 shows that the addition of TNFα at low concentrations to cultures stimulated by GM-CSF + IL-3 + G-CSF resulted in the appearance of numerous DL-CFU colonies at 14 days.

Similar results were observed when G-CSF was omitted from the cultures (not shown). Numbers of colonies fell at higher TNFα concentrations. TNFβ had a very similar effect at 5 U/ml but this stimulation persisted at 100 U/ml (not shown). CFU-G colonies were also markedly stimulated by both TNFα and β. However TNFα appeared to suppress CFU-M growth at all concentrations tested.

To see whether TNF was acting either early or late in progenitor development TNFα addition to the cultures was delayed in two experiments to day 3, 7 or 10. DL-CFU numbers fell to 85% and 57% of day 0 numbers respectively at day 3 addition and to zero when addition was delayed to the 7th or 10th day in culture. Since the colonies are scored at 14 days and are not recognisable as DL-CFU prior to one week in culture, it appears that TNF must be acting at an early proliferative phase of progenitor development.

DISCUSSION

We have recently identified primitive cells in both human bone marrow and peripheral blood that are capable of proliferation in-vitro into colonies of dendritic cells and have termed these early progenitors DL-CFU[1]. The presence of macrophages within many of these colonies suggested a close relationship between macrophage and dendritic cell development. We therefore first examined whether in-vitro conditions optimal for macrophage colonies (CFU-M) might also be sufficient stimulus for DL-CFU.

The initial experiments (Table 1) showed that, since fetal calf serum alone could stimulate both CFU-M and DL-CFU in the absence of any additional source of growth factors, it would be necessary to employ a serum free culture assay in order to fully characterise their growth factor responses. CFU-G growth under these stringent conditions was found to be optimal when early acting factors (GM-CSF and IL-3) were acting together with G-CSF (Table 1). However sub-optimal growth of CFU-M was observed with IL-3 and GM-CSF with or without M-CSF and this accords with other recent studies[4] that have found human rM-CSF to be a poor stimulus for human progenitors though more effective in mice. IL-1α appeared to augment the CFU-M colony growth induced by GM-CSF, IL-3 and M-CSF and a very small number of DL-CFU (less than 25% of optimal), was observed in the presence of 5 to 50 U/ml of IL-1 (Table 2).

The most striking observations were those made in serum deprived cultures with TNF (Table 3). Both TNFα and TNFβ had marked effects on dendritic colonies and TNFα stimulated up to 82% of optimal DL-CFU growth provided that GM-CSF and IL-3 were also present in the cultures. It is noteworthy that a similar striking enhancement of CFU-G growth, up to 203% of optimal, was also observed. However CFU-M were inhibited by TNFα and zero growth of these cells occurred at concentrations above 0.1 U/ml of this cytokine.

Previous studies have indicated a purely suppressive role for TNF in hematopoiesis[5,6] however in some cases these studies were with serum and CM which may contain unspecified cytokines including TNF itself. Where these confounding factors were excluded TNF was suppressive at concentrations of greater than 25 U/ml[5] and thus greater than levels we have found to stimulate granulocyte and dendritic cell growth in this study. More recent reports[7,8] have shown that both TNFα and β may stimulate myeloid proliferation in-vitro from GM-CSF stimulated CFU-GM and in one case this was confirmed with a purified $CD34^+$ cell population. It has been suggested that TNF may act in these cases by up or down regulating GM-CSF or G-CSF receptors respectively[9]. It is not possible to conclude from our study what the nature of the synergy between TNF and GM-CSF or IL-3 might be. Since we were unable to induce any DL-CFU growth with GM-CSF/IL-3 combinations at any concentration tested (data not shown) it seems unlikely that TNF is acting on these cells by simple regulation of receptor expression.

Both TNF and GM-CSF have been shown to have activity on mature cells of the dendritic-Langerhans lineage. GM-CSF promotes survival and may enhance the antigen presenting functions of mature dendritic cells[10,11,12] whereas TNF appeared to enhance survival in culture without promoting functional maturation[12]. Although these findings demonstrated an effect on mature cell function and structure our studies failed to show that the delayed addition of TNFα to bone marrow cultures could influence DL-CFU proliferation or maturation. The observation that TNF effectively synergised with GM-CSF only during the first 3 days of a 14 day culture implies an effect on very early maturational events at progenitor level.

These findings suggest a novel biological role for TNF. Since TNFα appears to suppress monocyte-macrophage growth even at the low concentrations that were found to stimulate dendritic cell progenitors it is possible that this cytokine may act by directing early bone marrow progenitors along the pathway of dendritic rather than macrophage development. It is possible that TNF, like GM-CSF, M-CSF and G-CSF, is constitutively produced by bone marrow stromal cells and perhaps, like these other growth factors, acts in close association with proteoglycans at the stromal cell surface. It will be of interest to determine whether DL-CFU like some other myeloid progenitors express receptors for the newly described stem cell factor (c-kit ligand) and what part, if any, this growth factor may have in regulating its development.

REFERENCES

1. Reid, C.D.L., P.R. Fryer, C. Clifford, A. Kirk, J. Tikerpae, and S.C. Knight. 1990. Identification of hematopoietic progenitors of macrophages and dendritic Langerhans cells (DL-CFU) in human bone marrow and peripheral blood. Blood 76:1139

2. Reid, C.D.L., A. Stackpoole, A. Meager, and J. Tikerpae. 1992. Interactions of tumour necrosis factor with granulocyte-macrophage colony-stimulating factor and other cytokines in the regulation of dendritic cell growth in vitro from early bipotent CD34+ progenitors in human bone marrow. J. Immunol. 149:2681

3. Migliaccio, G., A.R. Migliaccio, and J. Adamson. 1988. In-vitro differentiation of human granulocyte/macrophage and erythroid progenitors: Comparitive analysis of the influence of recombinant human erythropoietin, G-CSF, GM-CSF, and IL-3 in serum-supplemented and serum deprived cultures. Blood 72:248

4. Sonada, Y., Y.C. Yang, G.G. Wong, S.C. Clarke, and M. Ogawa. 1988. Analysis in serum free culture of the targets of recombinant human hemopoietic growth factors: Interleukin 3 and granulocyte/macrophage-colony-stimulating factor are specific for early developmental stages. Proc. Natl. Acad. Sci. USA. 85:4360

5. Li Lu, K. Welte, J.L. Gabrilove, G. Hangoc, E. Bruno, R. Hoffman, and H.E. Broxmeyer. 1986. Effects of recombinant human tumour necrosis factor, recombinant human interferon, and prostaglandin E on colony formation of human hematopoietic progenitor cells stimulated by natural human pluripotent colony-stimulating factor, pluripoietin, and recombinant erythropoietin in serum-free cultures. Cancer Res. 46:4357

6. Broxmeyer, H.E., D.E. Williams, Li Lu, S. Cooper, S.L. Anderson, G.S. Beyer, R. Hoffman, and B.Y. Rubin. 1986. The suppressive influences of human tumour necrosis factor on bone marrow hematopoietic progenitor cells from normal donors and patients with leukaemia: synergism of tumour necrosis factor and interferon gamma. J. Immunol. 136:4487

7. Caux, C., S. Saeland, C. Favre, V. Duvert, P. Mannoni, and J. Banchereau. 1990. Tumour necrosis factor strongly potentiates interleukin-3 and granulocyte-macrophage colony stimulating factor-induced proliferation of human CD34+ hematopoietic progenitor cells. Blood 75:2292

8. Piacibello, W., F. Sanavio, A. Severino, S. Morelli, A.M. Vaira, A. Stacchini, and M. Aglietta. 1990. Opposite effect of tumour necrosis factor on granulocyte colony-stimulating factor and granulocyte-macrophage colony-stimulating factor-dependent growth of normal and leukaemic hemopoietic progenitors. Cancer Res. 50:5065

9. Shieh, J.H., R. Peterson, D.J. Warren, and M.A.S. Moore. 1989. Modulation of colony stimulating factor-1 receptors on macrophages by tumor necrosis factor. J. Immunol. 143:2534

10. Witmer-Pack, M.D., W. Olivier, J. Valinsky, G. Schuler, and R.M. Steinman. 1987. Granulocyte/macrophage colony-stimulating factor is essential for the viability and function of cultured murine epidermal Langerhans cells. J. Exp. Med. 166:1484

11. Heufler, C., F. Koch, and G. Schuler. 1988. Granulocyte/macrophage colony-stimulating factor and inter-leukin 1 mediate the maturation of murine epidermal Langerhans cells into potent immunostimulatory dendritic cells. J. Exp. Med. 167:700

12. Koch, F., C. Heufler, E. Kampgen, D. Schneeweiss, G. Bock, and G. Schuler. 1990. Tumour necrosis factor maintains the viability of murine epidermal Langerhans cells in culture, but in contrast to granulocyte/macrophage colony-stimulating factor, without inducing their functional maturation. J. Exp. Med. 171:159

HUMAN BONE MARROW CONTAINS POTENT STIMULATORY CELLS FOR THE ALLOGENEIC MLR WITH THE PHENOTYPE OF DENDRITIC CELLS

William Egner, Judith L. McKenzie,
Stewart M. Smith, Michael E.J. Beard,
and Derek N.J. Hart

Haematology/Immunology Research Group
Christchurch School of Medicine
Christchurch Hospital
Christchurch, NZ

INTRODUCTION

Lymphoid and interstitial dendritic cells (DC) originate from the bone marrow (BM) in rodents,[1,2] as does the human Langerhans cell,[3] but cells with appropriate morphological features and stimulatory properties have yet to be isolated from fresh, uncultured marrow in these species. We have shown that human BM, unlike that of rodents, is capable of stimulating resting T cells in the allogeneic mixed leucocyte reaction (MLR)[4,5] without a period of in-vitro differentiation.

Identification and characterisation of the most potent MLR stimulatory cell in human BM has potentially important applications for the understanding and treatment of human transplant failure[6] and graft-versus-host disease. We therefore undertook a novel approach to the identification and purification of the MLR stimulatory cells in human BM, by FACS cell sorting with a combination of antibodies to other haemopoietic cell lineages. In this way we have isolated a potent MLR stimulatory fraction in which putative DC precursors could be identified.

MATERIALS AND METHODS

Donors

Peripheral blood and human BM was obtained with informed consent from healthy volunteer donors, with Christchurch Hospital Ethics Committee approval.

Monoclonal Antibodies

The CMRF monoclonal antibodies CMRF-7 (CD15), CMRF-14 (anti-glycophorin A), CMRF-12 (CD45) and CMRF-31 (CD14) were produced in this laboratory. OKM1 (CD11b), OKT3 (CD3), HB55 (anti-HLA-DR), HB96 (anti-HLA-DQ) and HNK1 (CD57) were obtained from the American Type Culture Collection (Rockville, USA). Other antibodies were obtained as gifts from the following sources: FMC63 (CD19, Dr Zola, Adelaide), HUNK2 (CD16, Prof McKenzie, Melbourne), B7/21 (anti-HLA-DP, Prof Brodsky, USA).

FACS Cell Sorting

BM mononuclear cells were prepared over a Ficoll/Hypaque gradient (density 1.077), washed, and labelled with a saturating mixture of antibodies to the erythroid, myeloid and lymphoid lineages (CD3, CD11b, CD15, CD16, CD19, CD57 and Glycophorin A) for 30 minutes at 4°C. After a single wash, and a second incubation with FITC-conjugated polyclonal sheep anti-mouse antibodies (Silenus, Australia) (FITC-SAM), cells were washed and resuspended in 10% heat-inactivated AB serum in RPMI supplemented with 2mM glutamine, 100 U/ml penicillin and 100 ug/ml streptomycin. Cell preparations at approximately 2 - 5 x 10^5 cells/ml were then sorted on a FACS IV fluourescence activated cell sorter (Becton Dickinson, USA) into positive and negative populations. Purity was assessed by reanalysis on the FACS. Negative controls incubated with FITC-SAM alone or 16.4.4. (mouse-anti-rat class I) and FITC-SAM were analysed simultaneously.

Mixed Leucocyte Reaction

The stimulatory activity of positive and negative BM cell fractions were compared to starting and labelled BM in an allogeneic MLR, as described elsewhere.[7] Briefly, cells were added in graded doses to 1 - 1.2 x 10^6/ml allogeneic peripheral blood mononuclear cells (PBMC) prepared over Ficoll Hypaque, and cultured in 96 well microtest plates (Nunc, Denmark) for five days at 37°C, in 5% CO_2. Proliferation was measured with a pulse of 0.5 uCi of ^3H thymidine (Amersham, UK) for the final eight hours of culture. Cells for MLR stimulation were incubated with mitomycin C for 30 minutes at 30°C and washed three times before labelling and FACS sorting as described above.

Immunoperoxidase Labelling

Immunoperoxidase labelling was performed as described elsewhere.[7,8] Briefly, FACS separated, mix-negative cells were incubated with saturating primary antibodies to HLA-DR, HLA-DP, HLA-DQ or CD45 for 30 minutes at 4°C, washed once, cytospun onto glass slides, and fixed with paraformaldehyde and acetone for 45 seconds, prior to blocking of endogenous peroxidase for 30 minutes at room temperature, in 5:1 methanol/hydrogen peroxide (H_2O_2). Cells were subsequently labelled with polyclonal rabbit anti-mouse antibody (Dakopatts, Denmark), followed by polyclonal peroxidase-conjugated goat anti-rabbit antibody (Tago, USA) for 30 minutes at room temperature in a humidified tray. The cells were developed for 10 minutes in diaminobenzidine/H_2O_2 developer.[7,8] Developed slides were mounted under coverslips and examined by light microscopy, compared to nil, CMRF-3 (mouse anti-mycoplasma) and 16.4.4 incubated negative controls. Positive controls for all antisera consisted of unseparated BM cells.

RESULTS

MLR Stimulation

Human BM, like blood, stimulates a strong allogeneic MLR response. BM stimulatory activity was generally greater than that of peripheral blood (data not shown) from the same donor. Studies using the individual antibodies CMRF-7, CMRF-14, OKM1, OKT3, HNK1, HUNK2 and FMC63 revealed that the most stimulatory cell fraction lacked each of these markers of myeloid (CMRF-7, OKM1, HUNK2), erythroid (CMRF-14) and lymphoid lineages (HNK1, HUNK2, OKT3 and FMC63) (data not shown, Egner et al, in preparation). We then selected each of the antibodies for inclusion in an antibody mix to negatively select for the highly stimulatory cell population. Labelling with a mixture containing saturating amounts of all the antibodies resulted in a mix-negative population which was markedly (> 10 times) more stimulatory than the mix-positive BM population. Both were of high purity (82 - 97% for the negative population, 98 - 99% for the positive) (Fig 1).

Immunoperoxidase Phenotype and MGG Morphology

The mix-negative population (Fig 2A) was enriched for cells with the morphology of putative dendritic cells, having irregular nuclei with prominant nucleoli, and basophilic cytoplasm lacking apparent granules or vacuoles. Other large cells were also present which appear to consist of immature haemopoietic progenitors, which had rounded or kidney-shaped nuclei and a smaller nuclear/cytoplasmic ratio. The mix-negative population also contained many small lymphoid cells.

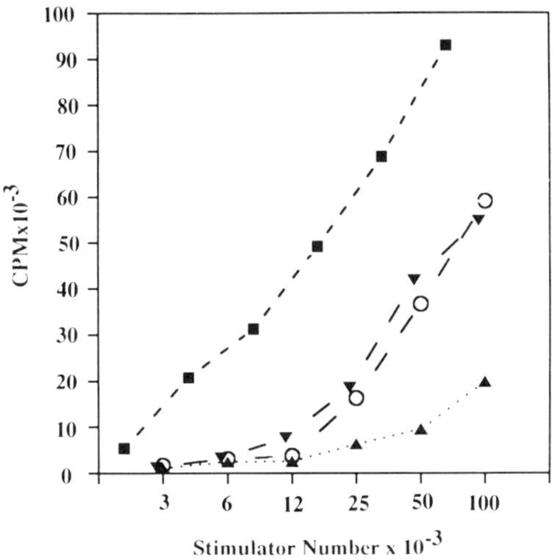

Figure 1. The most potent allostimulatory cell resides in the mix-negative population of human BM. BM cells were labelled with a saturating mixture of the antibodies CMRF-7, CMRF-14, OKM1, OKT3, FMC63, HNK1 and HUNK2 prior to sorting into positive (97% positive) and negative (82% negative) fractions on a FACS IV cell sorter, and use as stimulators of PBMC in the allogeneic MLR.
○ = starting unlabelled BM cells, ▼ = mix labelled BM cells,
■ = mix-negative sort, ▲ = mix-positive sort.

Immunophenotyping (Fig 2B - E) revealed that the majority of the large mix-negative cells were HLA-DR positive (Fig 2B), but that the putative DC precursors were strongly stained. HLA-DP (Fig 2C) and HLA-DQ (Fig 2D) reactivity was more restricted but once again the putative DC precursors stained strongly for both alleles, as did a subpopulation of the large progenitor cells. This is a notable feature of human blood DC,[9] and interstitial DC[10] which appears to differentiate them from monocytes and cultured macrophages. Only a minority of the lymphocytes stained positive for HLA-DP or HLA-DQ, and to a lesser extent than the putative DC precursors or the larger cells. The CD45 antibody CMRF-12 stained most of the cells in the negative preparation, including all three types of cells, but most strongly for the small lymphocytes (Fig 2E).

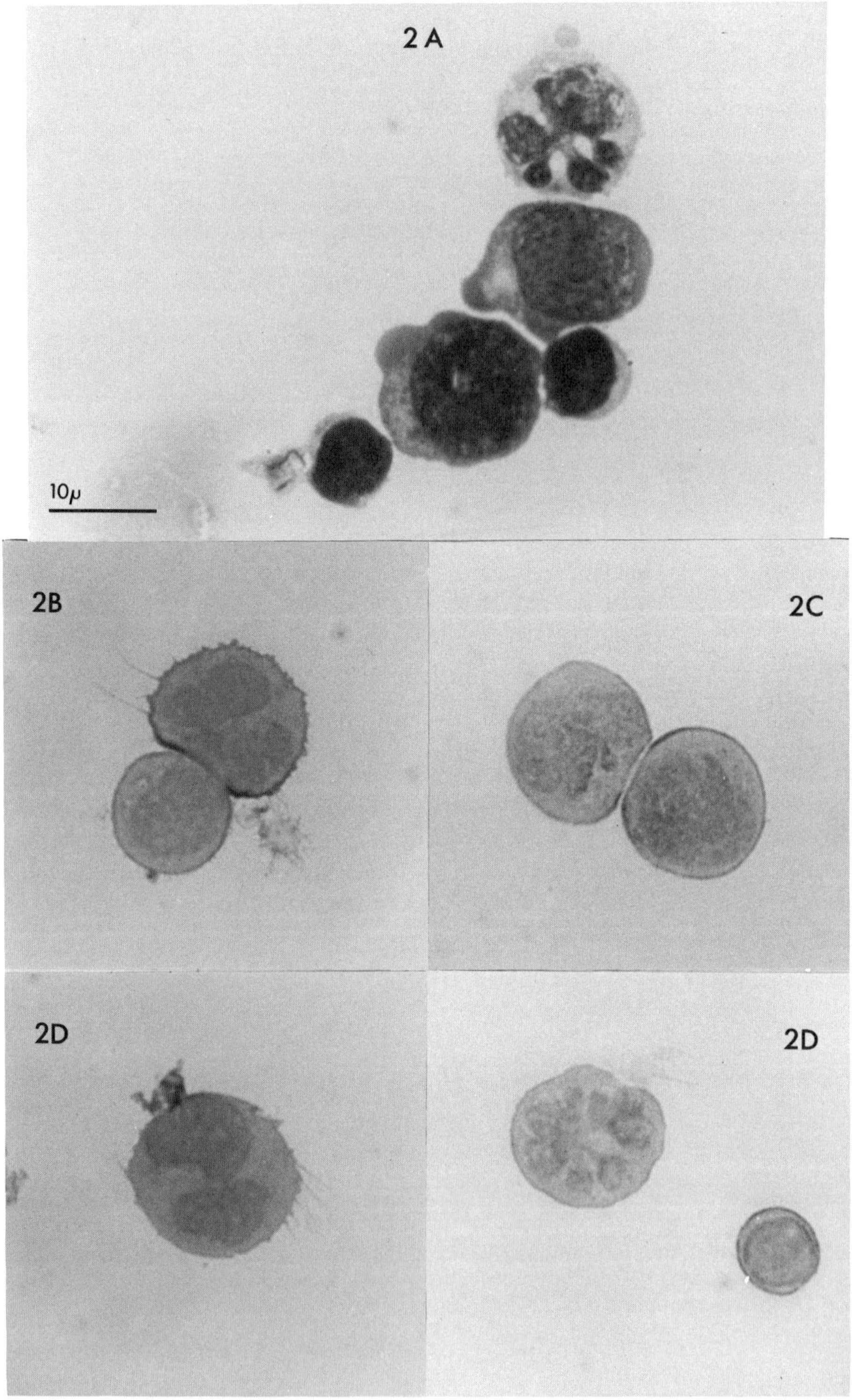

Figure 2. Mix-negative cells were cytospun for MGG staining (2A) or immunoperoxidase staining for HLA DR (2B), DP (2C), DQ (2D) or CD45 (2E). Note the cytoplasmic processes on the strongly HLA DR, HLA-DP and HLA-DQ positive cells, which share some morphological features with human dendritic cells. CD45 expression is relatively weak in contrast to the strongly positive small lymphoid cells.

DISCUSSION

Our initial attempts to identify the stimulatory cell populations within human BM have identified a minor population of large, strongly HLA class II positive cells which express all three HLA-class II loci products and the CD45 antigen (Fig 2A - E). This population also has morphological and immunophenotypic similarities to the human dendritic cells from other tissue[9,10] and is an excellent candidate for the DC precursor or an early differentiation stage of the DC in human BM which has been inferred from clinical studies.[3] This cell lacks the CD3, CD11b, CD15, CD16, CD19, CD57 and glycophorin A antigens, demonstrating that it lacks markers of the mature lymphoid, myeloid and erythroid lineages, including those of early committed granulocytic (CD15), monocytic (CD15, CD11b) and B-lymphoid (CD19) precursors. Further phenotypic studies are underway to further define phenotypic differences between this cell and the presumed haemopoietic progenitor cells which copurify with it, however lack of CD15 and CD11b suggest that the DC precursor is not of later myeloid origin, and in contrast to the report of Czitrom et al[11] is not a granulocytic precursor.

This is the first study to define human allostimulatory cells on a functional basis, unlike previous attempts to phenotypically define DC-like cells[12,13] and precursors. Further characterisation of these cells, and in-vitro culture, may clarify their relationship, if any, to the previously reported myeloid precursors of the Langerhans cell.[12,13]

ACKNOWLEDGEMENT

This work was supported by grants from the Canterbury Medical Research Foundation and the Health Research Council of New Zealand.

REFERENCES

1. R.M. Steinman, D.S. Lustig, and Z.A. Cohn, Identification of a novel cell type in the peripheral lymphoid organs of mice III. Functional properties in-vivo, *J. Exp. Med.* 139:1431 (1974).
2. D.N.J. Hart, and J.W. Fabre, Demonstration and characterisation of Ia positive dendritic cells in the interstitial connective tissues of rat heart and other tissues but not brain,*J. Exp. Med.* 153:347 (1981).
3. B. Volc-Platzer, G. Stingl, K. Wolff, W. Hinterberg, and W. Schredl, Cytogenetic identification of allogeneic epidermal Langerhans cells in a BM graft recipient, *N. Engl. J. Med.*. 310:1123 (1984).
4. W.E.F. Klinkert, Rat bone marrow precursors develop into dendritic cells under the influence of conditioned medium, *Immunology*. 168:414 (1984).
5. J. Sprent and M. Schaefer, Antigen-presenting cells for Lyt2$^+$ cells: I. Stimulation of umprimed Lyt2$^+$ cells by H-2 different Thy 1$^-$ Ia$^-$ cells prepared from spleen and bone marrow, *J. Immunol.* 140:3745 (1988).
6. J.L. McKenzie, M.E.J. Beard, and D.N.J. Hart, The effect of donor pretreatment on interstitial dendritic cell content and rat cardiac allograft survival, *Transplantation*. 38:371 (1984).
7. T.C.R. Prickett, J.L. McKenzie, and D.N.J. Hart, Adhesion molecules on human tonsil dendritic cells, *Transplantation.* 53:483 (1992).
8. D.N.J. Hart, and J.L. McKenzie, Isolation and characterization of human tonsil dendritic cells, *J. Exp. Med.* 168:157 (1988).
9. W. Egner, T.C.R. Prickett, and D.N.J. Hart, Adhesion molecule expressions: a comparison of human blood dendritic cells, monocytes and macrophages. *Transpl. Proc,* (in press) (1992).
10. T.C.R. Prickett, J.L. McKenzie, and D.N.J. Hart, Characterization of interstitial dendritic cells in human liver, *Transplantation*.46:754 (1988).
11. A.A. Czitrom, T.S. Axelrod, and B. Fernandes, Granulocytic precursors are the principle cells in bone marrow that stimulate allospecific cytolytic T lymphocyte responses, *Immunology*. 64:655 (1988).

12. A. de Fraissinette, D. Schmitt, C. Dezutter-Dambuyant, D. Guyolat, M. Zabot, and T. Thivott, Culture of putative Langerhans cell bone marrow precursors: characterization of their phenotype, *Exp. Haematol.* 16:764 (1988).
13. C.D.L. Reid, P.R. Fryer, C. Clifford, A. Kirk, J. Tikerpal, and S.C. Knight, Identification of haemopoietic progenitors of macrophages and dendritic Langerhans cell (DL-CFU) in human bone marrow and peripheral blood. *Blood*. 76:6 (1990).

RECOMBINANT GM-CSF INDUCES IN VITRO DIFFERENTIATION OF DENDRITIC CELLS FROM MOUSE BONE MARROW

Christoph Scheicher,[1] Maria Mehlig,[1] Reinhard Zecher,[1] Friedrich Seiler,[2] Petra Hintz-Obertreis,[2] and Konrad Reske[1]

[1]Institut für Immunologie, Johannes Gutenberg-Universität, 6500 Mainz, Germany
[2]Behringwerke Research Laboratories, 3550 Marburg, Germany

INTRODUCTION

The unprecedented functional capacity of dendritic cells (DC) in sensitizing resting T cells and their role in triggering T dependent immune responses attract increasing interest in this unique accessory cell population. Like macrophages (Mph) DC have been described to originate in the bone marrow (BM) (1). While the cytokine-promoted *in vitro* differentiation of Mph from BM-cells is well established, a convincing *in vitro* culture system for propagating mouse DC from BM-cells has not yet been reported. This work demonstrates the differentiation of DC from mouse bone marrow cells by a short term *in vitro* culture system supplemented with rGM-CSF.

In line with observations by Metcalf et al. (2) who observed that the development *in vitro* of discrete hematopoietic cell populations *i.e.* Mph, granulocytes, eosinophils, and GM colonies from BM-cells was dependent of the GM-CSF dose, differentiation of DC in our culture system was also found to be dose dependent. While low doses of GM-CSF gave rise to substantial quantities of DC their appearance in the presence of high doses of the cytokine became rather difficult to detect due to outgrowing BM-Mph and granulocytes.

In contrast to coinduced BM-Mph BM-DC exhibited constitutive synthesis of MHC class II molecules. This feature was employed to establish an efficient one-step purification procedure that allows to obtain DC in elevated quantities (3). On the basis of morphological, cytochemical and functional characteristics (1), BM-DC closely resemble DC purified from various tissues.

RESULTS AND CONCLUSION

Earlier work from our laboratory on influences of GM-CSF on the antigen presenting function of BM derived Mph revealed a higher antigen presenting potential of these cells as compared to IFN-γ stimulated BM-Mph, whereas class II expression exhibited the opposite correlation (4). We therefore explored the possibility that trace amounts of DC contained within the GM-CSF-treated BM-Mph population might be responsible for their superior antigen presentation function. When BM-cells were cultured in the presence of low doses of GM-CSF, cells with dendritic morphology became detectable floating in the culture medium as early as day 2 following culture onset. The dendritic appearance of these cells became progressively pronounced and reached its optimal development on day 6-7 of culture (see Fig. 1A, arrow). Cells showing this morphology were clearly distinct from coinduced Mph and granulocytes. As indicated in Fig. 1B, strong staining was observed by immunofluorescence microscopy using the I-A^k specific mAb 10-2.16.

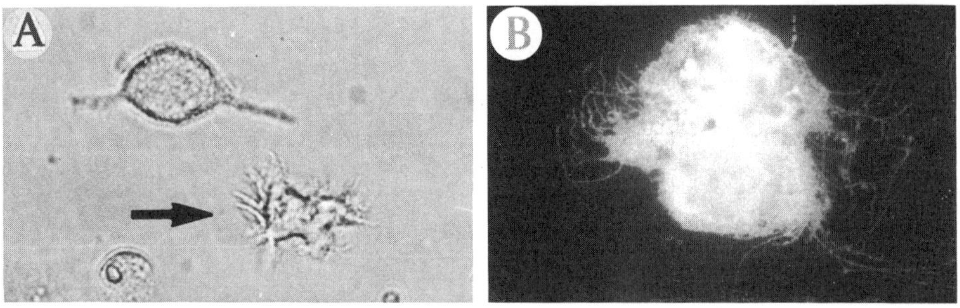

Figure 1. A: BM-cell culture in the presence of 20 ng rGM-CSF/mL on day 7 showing a DC (arrow) and a Mph. B: MAb 10-2.16 (anti-I-A^k) staining of a single DC taken from the culture depicted.

The induction of BM-DC is GM-CSF dose dependent

We investigated the effect of GM-CSF on mouse bone marrow more rigorously and found a strict dose dependence of the DC inducing capacity of GM-CSF. Since in the presence of the low doses of GM-CSF required to observe optimal BM-DC induction, DC, in contrast to BM-Mph, were found to be strongly Ia^+ we used this phenotypic trait to follow DC development during the standard 7 day culture period applied. As indicated in Figure 2, maximal DC numbers were detectable in the presence of 10-20 ng GM-CSF. Lower concentrations proved suboptimal, whereas higher concentrations resulted in overgrowth by Mph and granulocytes. Thus, in contrast to the M-CSF driven differentiation of BM-Mph from whole BM-cells which results in a rather homogeneous population of hematopoietic cells counting approximately 95% BM-Mph, GM-CSF-supplemented cultures always yielded heterogeneous populations of hematopoietic cells. However, since DC are the main cell type that expresses class II determinants constitutively under the culture conditions applied, a convenient one-step enrichment procedure using immunomagnetic bead-selection could be developed to obtain DC in sizeable numbers for experimentation. DC enriched by this technique were extensively

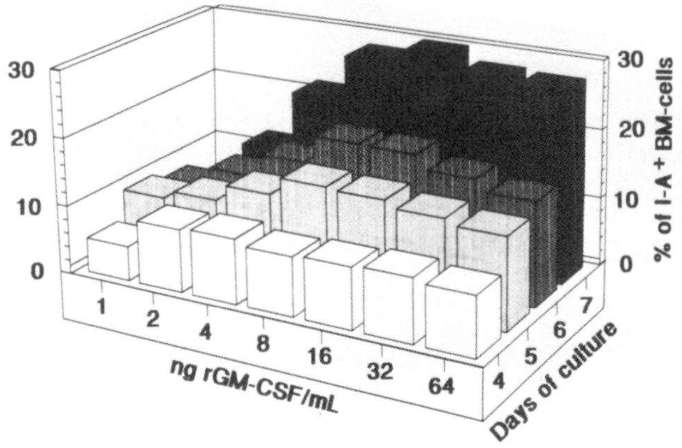

Figure 2. Dose and time dependence of the development of Ia$^+$ BM-DC in GM-CSF supplemented BM-cell cultures as monitored by FACScan analysis. 5x10^5 BM-cells/ mL were cultured in the presence of different doses of GM-CSF in petri dishes, harvested at the time points indicated and stained for I-A expression with mAb 10-2.16, showing an optimum at day 7 with 20 ng GM-CSF/mL.

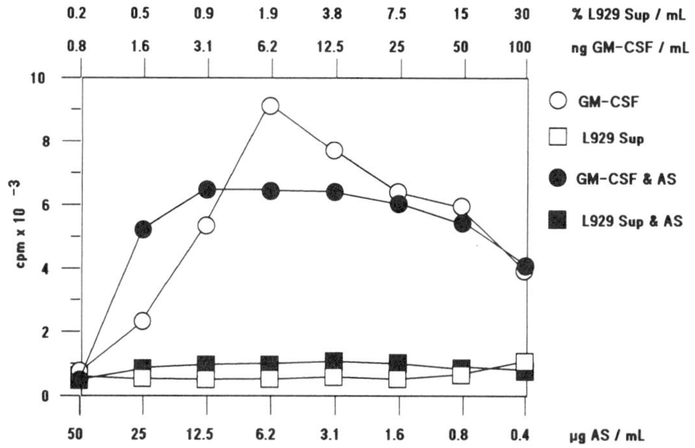

Figure 3. Inhibition of MLR response of GM-CSF grown BM-DC in the presence of anti-GM-CSF-antiserum. BM-cells (5x10^5/mL) were cultured with serial dilutions of L929 supernatant (squares) or GM-CSF (circles) for 7 days in the absence of anti-GM-CSF-antiserum (open symbols) or were cultured with serial dilutions of anti-GM-CSF-antiserum in addition to 100 ng GM-CSF/mL or 30% L929 supernatant (closed symbols), respectively. The cultures were then washed and irradiated with 20 Gy *in situ* and naive nylon wool nonadherent BALB/c T cells (5x10^5/mL) were added. The cultures were further incubated for 90 h and were labelled for the last 6 h with 25 kBq [^{3}H]-TdR/mL. Cpm are means for triplicate cultures.

characterized. They were nonphagocytic, specific esterase negative and functionally competent (2).

Inhibition of DC Differentiation by an anti-GM-CSF-Antiserum

Because DC stimulate the primary allo-MLR very effectively (5), this function was used to evaluate the presence of DC within BM-cultures. In parallel cultures BM-cells were grown for 7 days with graded doses of GM-CSF or L929 supernatant as a source of M-CSF, or with a constant dose of GM-CSF or L929 supernatants combined with graded doses of anti-GM-CSF-antiserum, respectively. The cultures were washed on day 7, irradiated and T cells were added for primary MLR. The results are depicted in Figure 3. High doses of GM-CSF in the absence of antiserum resulted in diminished MLR responses due to a suboptimal DC development and an increase in Ia^- cellss

Efficient DC Development requires 96 h of GM-CSF-supported Culture of BM-Cells

In an attempt to determine the culture period required for BM-cells to generate DC we monitored the MHC class II expression of developing DC including their capacity to stimulate a primary MLR. BM-cells were grown in multiwell-plates in the presence of an optimal dose of 20 ng/mL GM-CSF. A GM-CSF neutralizing concentration of anti-GM-CSF-antiserum was added to discrete culture aliquots at distinct time points beginning 24 h after culture onset for the first culture aliquot and proceeding in 24 h intervals for the remaining aliquots. All cell aliquots were harvested on day 6 and MHC class II expression was assessed by a cell-ELISA. Cells of a parallel experimental set-up were harvested on day 6, irradiated, and subjected to a MLR by adding naive spleen T cells enriched by nylon wool fractionation. After

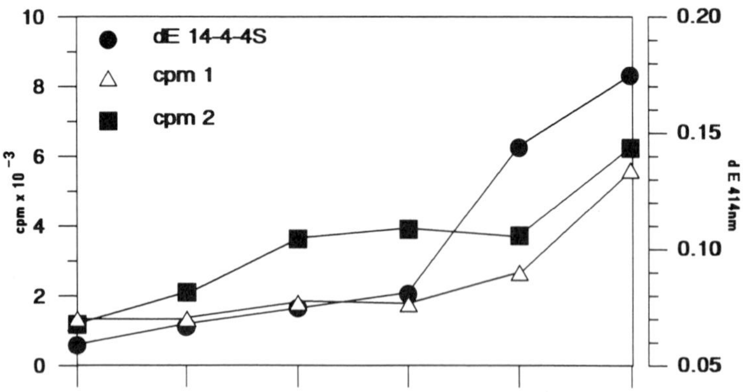

Figure 4. Kinetics of DC development in GM-CSF supplemented BM-cell cultures. BM-cells ($5x10^5$/mL) were cultured with an optimal dose of 20 ng/mL GM-CSF. Discrete parallel cultures contained a blocking concentration of anti GM-CSF antiserum for various time periods (24 h up to 144 h). After a total culture time of 144 h I-E expression, MLR activity and cell proliferation were assessed.

another 90 h DNA synthesis was determined by incubation with [^{3}H]-TdR for an additional 6 h. In a third parallel culture set cell proliferation was determined by pipetting [^{3}H]-TdR to the cultures 6 h before the day 6 cell harvest. The presence of anti-GM-CSF-antiserum at 50 μg/mL abrogated the response completely. The results are shown in Figure 4.

It is apparent that very low numbers of DC appear to be present already on day 2 consistent with a small increase in primary MLR stimulation. However, while a prominent increase in class II expression requires the continued presence of GM-CSF for at least 96 to 120 h, efficient MLR capacity developed exclusively in unblocked BM-cultures (144 h). These findings are in accordance with earlier observations (2) showing an increase in I-A expression on day 4-5 of GM-CSF supplemented BM-cells.

We conclude from these experiments that the surface expression of class II molecules appears to be an early event in the functional maturation of DC towards potent primary MLR stimulators. Furthermore, these findings correlate with those of other groups who noticed the development of DC from Ia$^-$ precursor cells by *in vitro* culture with GM-CSF (6).

In summary, we demonstrated the development of DC from mouse bone marrow in the presence of GM-CSF during a 7 day primary culture. The sustained presence of GM-CSF for at least 4 days is mandatory for prominent DC induction. Bone marrow culture derived DC exhibit key properties of cells of the dendritic lineage like high levels of constitutive class II expression and predominant stimulatory potential in the primary allo-MLR. Our data are in accordance with a recent report that demonstrates the GM-CSF-mediated growth and differentiation of DC in cultures of mouse blood (6).

REFERENCES

1. R.M. Steinman, The dendritic cell system and its role in immunogenicity, *Annu. Rev. Immunol.* 9:271 (1991).
2. D. Metcalf, A.W. Burgess, G.R. Johnson, N.A. Nicola, E.C. Nice, J. DeLamarter, D.R. Thatcher, and J.-J. Mermod, In vitro actions on haemopoietic cells of recombinant murine GM-CSF purified after production in Escherichia Coli: comparison with purified native GM-CSF, *J. Cell. Physiol.* 128:421 (1986).
3. C. Scheicher, M. Mehlig, R. Zecher, and K. Reske, Dendritic cells from mouse bone marrow: In vitro differentiation using low doses of recombinant granulocyte-macrophage colony-stimulating factor, *J. Immunol. Meth.* in press.
4. H.-G. Fischer, S. Frosch, K. Reske, and A.B. Reske-Kunz, Granulocyte-macrophage colony-stimulating factor activates macrophages derived from bone marrow cultures to synthesis of MHC class II molecules and to augmented antigen presentation function, *J. Immunol.* 141:3882 (1988).
5. R.M. Steinman, B. Gutchinov, M.D. Witmer, and M.C. Nussenzweig, Dendritic cells are the principal stimulators of the primary mixed leukocyte reaction in mice, *J. Exp. Med.* 157:613 (1983).
6. K. Inaba, R.M. Steinman, M. Witmer-Pack, H. Aya, M. Inaba, T. Sudo, S. Wolpe, and G. Schuler, Identification of proliferating dendritic cell precursors in mouse blood, *J. Exp. Med.* 175:1157 (1992).

SIGNALS REQUIRED FOR DIFFERENTIATING DENDRITIC CELLS FROM HUMAN MONOCYTES IN VITRO

J. Hinrich Peters,[1] Hui Xu,[1] Jörg Ruppert,[2]
Dorothea Ostermeier,[1] Detlef Friedrichs,[1]
Robert K.H. Gieseler[1]

[1]Department of Immunology, University of Göttingen
[2]Cytel Corp., San Diego, CA, U.S.A.

ABSTRACT

Human peripheral blood monocytes (Mo) can quantitatively be differentiated into potent accessory cells which exhibit dendritic cell (DC) function and phenotype. This alternative differentiation of Mo into DC rather than into macrophages (Mφ) will be triggered when signals leading to Mφ differentiation are omitted from the culture. Serum contains such stimulatory signals and was therefore omitted from the cultures. The cells were cultured on solid agarose surfaces. This newly developed technique allows for the attachment-free differentiation of DC. In the absence of signals, Mo do not survive in culture. IL-1 and IL-6 are endogenously produced by Mo and create an autokrine stimulatory milieu which increases the accessory function. However, also mature Mph will respond by an increased accessory activity upon stimulation by these cytokines. Cyclic AMP is the most likely second messenger to trigger an increase in accessory activity. IL-4 plus GM-CSF further act to upregulate dendritic cell properties and function. By action of these mediators, virtually all markers and functions of Mo/Mφ are lost, and the cells convert to the phenotype and function of dendritic cells.

INTRODUCTION

Knowing the life history of dendritic cells (DC) may enable us to grow these cells from their precursors as well as to manipulate their development in vivo. We have critically evaluated two conflicting hypotheses regarding DC development. Since DC lack all characteristic signs of monocytes/macrophages (Mo/Mφ), they have been regarded as a line distinct from phagocytes[1]. Alternatively, inasmuch as Mφ express the property of antigen presentation, it has been proposed that DC may derive from the Mo/Mφ lineage[2].

We have found that differentiation of Mo into Mφ is an inducible process. Such inducers of Mφ development are present in serum. In an appropriate serum-free medium Mo therefore cannot develop into Mφ[3]. Instead, they lose adhesion ability, round up, and

Dendritic Cells in Fundamental and Clinical Immunology,
Edited by Kamperdijk *et al.*, Plenum Press, New York, 1993

develop veils within two days of culture (Fig. 1). Their capacity to act as antigen-presenting cells and to stimulate T lymphocyte proliferative response increases, concomitant to the increased expression of class II molecules and the loss of markers typical of Mφ[4].

We have studied signals which are responsible for inducing the novel way of Mo differentiation towards DC. In the absence of inducers, Mo in culture will die (unpublished). We have found that differentiation of Mo into Mφ or into accessory cells

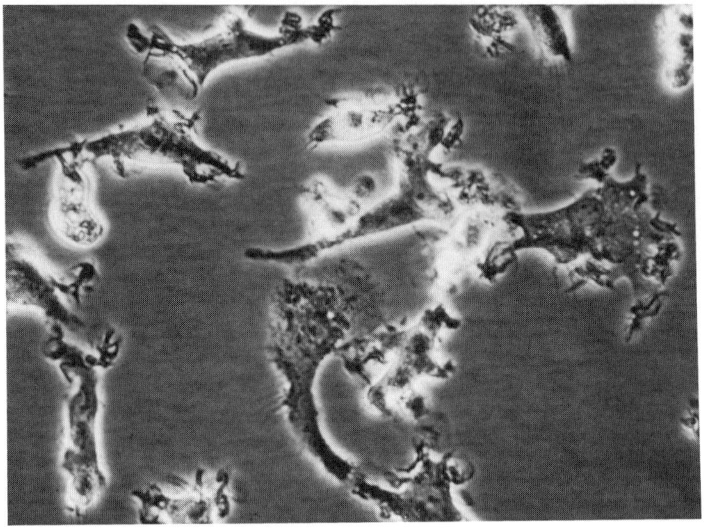

Fig. 1. Cells of a dendritic phenotype differentiated from human monocytes after 12 days of serum free culture.

depends on the factors present in culture. Serum, IFNγ and attachment enhance Mφ development, as characterized by typical morphology, Mφ markers, and a decreased accessory activity. For contrast, IL-1, IL-6[5], IL-4[6], GM-CSF, and c-AMP[7] act as factors controlling DC development, as identified by dendritic morphology, a typical phenotype and a high accessory activity.

RESULTS

Culture on solid agarose. When human Mo were prepared by selective attachment and cultured on tissue culture plastic in the absence of serum, they developed a phenotype close to DC, paralleled by an increase of MHC class II expression and accessory potency[3]. However, we have hypothesized that the process of attachment could already provide a differentiation signal for Mo. We therefore prepared Mo by a hypertonic density gradient centrifugation, as modified from Recalde[8]. These cells were then cultured under attachment-free conditions. We found that a solid agar or agarose (2%) surface most efficiently prevents spreading of endothelial cells, fibroblasts and also Mo/Mφ (not demonstrated). We therefore observed monocytic development on solid agarose under serum-free conditions.

As we have shown recently, in serum-containing culture IL-4 induces a reduction of CD14 expression on the surface of Mo[6]. We show in an accompanying paper (Ruppert et al., this volume) that in serum-free culture GM-CSF as a co-factor is required for IL-4

Table I. Expression of CD14 as influenced by various concentrations of IL-4 and the culture substrate. Human peripheral blood monocytes were prepared without attachment[8] and plated either on a solid (2%) agarose layer or on TC-plastic. They were cultured in CG medium (Camon, Wiesbaden, Germany) and treated with various doses of IL-4. After 6 days the expression of CD14 was measured in the FACS. The relative fluorescence intensity (rFI) was calculated as follows: rFI=mean fluorescence intensity of sample/mean fluorescence intensity of negative control.

substrate	IL-4 added to the culture [U/ml]					
	0	0.3	1	3	10	30
TC-pastic	41.45	51.82	50.19	33.17	38.24	37.75
Agarose	34.91	36.82	33.71	19.25	18.16	11.45

Table. II. Regulation of surface antigens on Mo by IL-4 (30 U/ml) and IFNγ (100 U/ml) after 6 days in serum- and attachment-free culture. The expression of antigens was analyzed by immune fluorescence staining using a FACS.

	CD14	CD16	CD32	HLA-DR
Control	30	5.5	10	41
IL-4	10	0.4	3.9	95
IFNγ	35.5	7.2	12.8	59
IL-4/IFNγ	23.8	3.6	8.7	78

induced downregulation of CD14. In this paper we demonstrate that absence of attachment also is able to act as a co-factor for IL-4 induced downregulation of CD14. Cells cultured under such conditions revealed a significantly reduced expression of CD14 (Table I).

Effects of cytokines on the Mo phenotype. As shown in Table II, expression of other Mo/Mφ markers such as CD16 and CD32 is also reduced by IL-4 treatment at attachment- and serum-free culture. For contrast, expression of HLA-DR is strongly enhanced. IFNγ antagonizes this regulation either when given alone or in combination with IL-4.

Increase of accessory activity. As shown in Fig. 2, Mo following the "dendritic pathway" of differentiation exhibit an increased expression of HLA-DR molecules on their surface. This finding argues for an increase in their functional activity as accessory cells. IL-4 treated cells increased in their accessory activity when they were used as stimulator cells of an allogeneic mixed lymphocyte reaction (Fig. 2). For contrast, IFNγ acts in an antagonistic way, either alone or in combination with IL-4.

DISCUSSION

The Mo/Mφ lineage is known to give rise to a number of differentiated cells apart from macrophages, such as osteoclasts and microglia. Although it is known that Langerhans cells derive from the bone marrow which is also most likely for other cells of the "dendritic cell family", such as interdigitating reticulum cells and dendritic cells, it has been unclear at which stage of differentiation they may branch off.
Recently we have shown that in rat bone marrow cultures development of DC can be triggered under controlled conditions. These results suggest that DC may develop from monocytes[9] and they support our contention that also in the human system monocytes can give rise to DC[4].

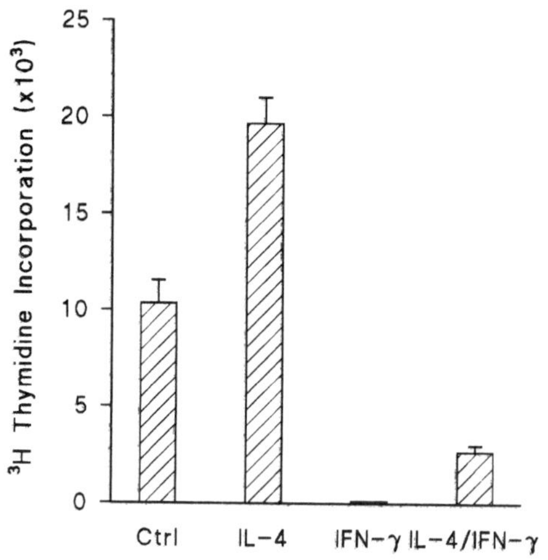

Figure 2. Effect of IL-4 or/and INFγ on the accessory activity of Mo. The cells were cultured for 6 days under serum- and attachment-free conditions with IL-4 (30 U/ml) or/andIFNγ (100 U/ml) and used as stimulator cells in an allogeneic mixed lymphocyte culture to quantify their T cell-stimulating activity. The values are indicated as means+SEM of triplicates.

Findings of this kind may imply that monocytes still carry the property of pluripotency which otherwise is a property of embryonic cells. Whereas the bone marrow has for long time regarded as a remnant of embryonicity even in the adult organism, a clear cut marker for this property has been missing in the past. However, we have shown that nuclear lamins which in mammals appear in mainly three types, may reflect the state of differentiation of a cell in a rather general way. By means of these markers, most bone marrow cells lack the lamins A/C of differentiated cells, and thus appear as embryonic cells. Like other peripheral tissues of the adult organism, also most of the nucleated cells in the peripheral blood express the differentiation markers A/C. However, peripheral blood Mo are still negative for this marker which suggests to attribute them to the compartment of incompletely differentiated cells. When culturing Mo under conditions to develop either into Mφ or into DC we found that both developments were accompanied with the acquisition of lamins A/C[10]. We therefore conclude that both alternative developmental pathways fulfill the criteria of a regular differentiation (see also the accompanying paper by Gieseler et al.).

Furthermore, it could be argued that development of DC-like cells from Mo in vitro may be artificial in the sense that the stimuli required for DC development are unphysiological. In fact, use of serum-free conditions has always to be questioned for its relevance. Serum-free conditions have been employed by us to supply controlled conditions, inasmuch as serum contains manifold triggers which may act as hidden inducers or inhibitors. One example has been shown in the accompanying report by Ruppert et al. in this volume.

We have, therefore, studied inducers in serum-free conditions. In a second instance, the same inducers were added again to serum-containing cultures in order to demonstrate their principal effectiveness also in this condition which is closer to the in vivo situation.

Judging from our data, signals required for DC differentiation from monocytes can be grouped into three categories:

1. Factors present in the serum which induce Mφ development should, ideally, left away. If, however, serum is present, it can be shown that the two antagonistic principles are present in a balanced form. Therefore, each of them can be dominated by the other, dependent on their concentrations. In addition to lymphokines and growth factors which act in this way, physical factors such as attachment, as demonstrated in this paper, or pH, as shown before[3], can fulfill this requirement.
2. Autocrine factors mainly act during the first phase of monocyte development. IL-1 plus IL-6 have been found to support survival of monocytes as well as development into potent accessory cells[5]. It is quite feasible to assume that monocytes, after leaving the blood stream and entering the tissues, may be entering a microenvironment which favors the short range accumulation of these lymphokines in order to reach the critical concentration for lymphokine action.
3. In a second stage IL-4 and GM-CSF act synergistically to further support DC development, as shown by Ruppert et al. in an accompanying paper. We have found that cAMP is likely to be the main intracellular second messenger mediating DC development, whereas an increase in intracellular cGMP is known to correlate with Mφ differentiation[7].

In conclusion, we have applied lymphokines and other physiological triggers and demonstrated a network of inducers acting in either a cooperative or a sequential manner. They antagonize the well-known development of Mo into Mφ and instead favor the quantitative shift of monocytes into cells which carry markers and function of dendritic cells. These cells, which can then be obtained in high numbers and at a high purity, may help to study the cooperation between AC and lymphocytes and may as well be useful in influencing immune reactions in vivo.

ACKNOWLEDGMENTS

This work was supported by the Deutsche Forschungsgemeinschaft (Pe 192/6-1, SFB 330, grant D9, and by the Boehringer Ingelheim Fonds through a grant to J.R.. We are indebted to Mrs. R. Döhne for typing the manuscript.

REFERENCES

1. R.M. Steinman, Gutchinov, B., M.D. Witmar, and M.C. Nussenzweig, Dendritic cells are the principal stimulator of the primary mixed leukocyte reaction in mice. *J. Exp. Med.* 157:613 (1983)
2. E.R. Unanue, The regulation of lymphocyte functions by the macrophage. *Immunol Rev.* 40: 227 (1978)
3. H.M. Najar, A.C. Bru-Capdeville, R.K.H. Gieseler and J.H. Peters, Differentiation of human monocytes into accessory cells at serum-free conditions. *Eur. J. Cell Biol.* 51:339 (1990)
4. J.H. Peters, S. Ruhl, D. Friedrichs, Veiled accessory cells deduced from monocytes. *Immunobiol.* 176:154 (1987)
5. J. Ruppert, and J.H. Peters, Interleukin-6 (IL-6) and interleukin-1 (IL-1) enhance the accessory activity of human blood monocytes during differentiation to macrophages. *J. Immunol.* 146:144 (1991)
6. J. Ruppert, D. Friedrichs, H. Xu, and J.H. Peters, IL-4 decreases the expression of the monocyte differentiation marker CD14, paralleled by an increasing accessory potency. *Immunobiol.* 182:449 (1991)
7. J.H. Peters, T. Börner, J. Ruppert, Accessory phenotype and function of macrophages induced by cyclic adenosine monophosphat. *Int. Immunol.* 2:1195 (1990)
8. R.H. Recalde, A simple method of obtaining monocytes in suspension. *J. Immunol. Meth.* 69:71 (1984)
9. R.K.H. Gieseler, R.-A. Röber, R. Kuhn, W. Weber, M. Osborn, and J.H. Peters, Dendritic cells (DC) derived from rat bone marrow precursors under defined conditions in vitro belong to the myeloid lineage. *Eur. J. Cell Biol.* 54:171 (1991)
10. R.-A. Röber, R.K.H. Gieseler, M. Osborn, K. Weber und J.H. Peters: Induction of nuclear lamins A/C in macrophages in in vitro cultures of rat bone marrow precursor cells and human blood monocytes, and in macrophages elicited in vivo by thioglycollate stimulation. *Exptl. Cell Res.* 190:185 (1990)

DOWN-REGULATION AND RELEASE OF CD14 ON HUMAN MONOCYTES BY IL-4 DEPENDS ON THE PRESENCE OF SERUM OR GM-CSF

Jörg Ruppert[1], Christiane Schütt[2], Dorothea Ostermeier[1], J. Hinrich Peters[1],

[1]Department of Immunology, University of Göttingen, Germany
[2]Department of Immunology, University of Greifswald, Germany

ABSTRACT

IL-4 induces down-regulation of CD14 expression on human monocytes only when the cells are cultured with serum. In serum-free cultures we failed to down-regulate CD14 by IL-4. Instead of serum, GM-CSF was required as a co-factor to restore the regulatory effect of IL-4 on CD14-expression. After 4 days of culture human monocytes were quantitatively CD14-negative as determined by flow-cytometry. On day 6, high amounts of CD14 molecules were detected in the SUP of these cultures, whereas intracellular immunofluorescence staining revealed no detectable CD14 in cytokine-treated monocytes. Thus, CD14 is lost by down-regulation (as shown by others) as well as by delivery into the medium. We previously hypothesized that dendritic cells may originate from monocytes. Our present finding support that one of the key markers, distinguishing monocytes/macrophages from dendritic cells, can be lost upon physiological stimuli.

INTRODUCTION

Monocytes have been found to represent a population of cells of a high developmental flexibility. Beside of their long-known potency of developing into macrophages, it becomes increasingly evident that along their path of maturation other differentiations, namely osteoclasts, Langerhans cells and others may branch off.

Normal monocyte/macrophage maturation is known to be strictly dependent on the presence of human serum[1]. As we could show repeatedly, monocytes cultured in serum-free medium differentiate into highly accessory cells that differ from mature macrophages in morphology, phenotype and function[2] and approach a rather dendritic phenotype. Therefore, human serum, because of its complex composition, still represents an enigma regarding the factors that actually lead to macrophage maturation[3] and renders it difficult to interprete effects obtained with single exogenously added mediators.

The CD14 surface molecule is an indirect LPS receptor[4]. The expression of these surface antigen has recently been shown to be regulated by the T-cell derived cytokine IL-4[4,5]. Interestingly, IL-4 exerted opposite effects on the two functionally distinct groups of molecules: The expression of MHC-antigens on human monocytes was enhanced, while the receptor molecules CD14 and CD16 disappeared after treatment with IL-4[5]. These findings suggested a major role for IL-4 as a regulatory factor in the development of accessory cells from the monocytic lineage during an immune response. Here we demonstrate that the effect of IL-4 on CD14 expression is almost abrogated in serum-free culture.

The purpose of this study was to define the role of serum in the regulation of the IL-4 mediated shift of surface antigen expression on human monocytes. The IL-4 mediated decrease of CD14 expression could be reestablished only by addition of GM-CSF. Our results indicate that at least these two factors are involved in regulation of the monocyte differentiation marker and indirect LPS receptor molecule CD14.

RESULTS

Decrease of CD14 expression by IL-4 is dependent on the presence of serum. In order to examine whether the down-regulation of the CD14 expression on human monocytes is solely mediated by IL-4 we tested the effect of this cytokine in serum-free versus serum-containing cultures. Monocytes were cultured for 7 days in medium containing 20% of human serum or in serum-free medium. Both cultures were subjected to 20 units of recombinant IL-4. As shown in Figure 1, the expression of CD14 decreased on monocytes that were allowed to differentiate in serum-containing medium. Contour plot (not demonstrated) revealed that the entire population of monocytes underwent this change.

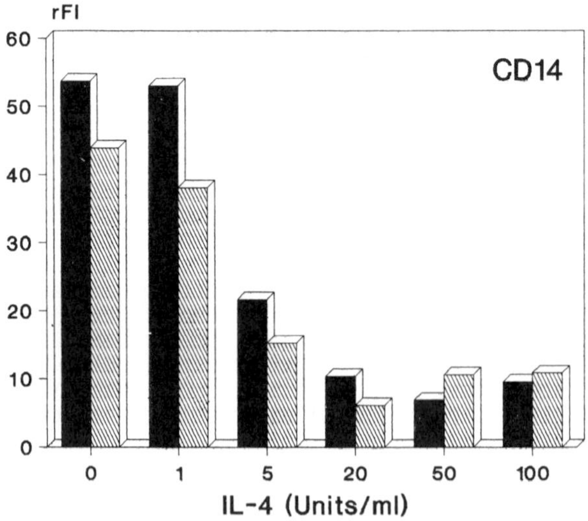

Fig. 1. Effect of various doses of IL-4 on CD14 expression in human monocytes in serum-containing culture. Cells from two donors (black, dashed) were cultured for 6 days in CG medium (Camon, Wiesbaden, Germany) plus 10% human serum and tested for CD14 expression in the FACS. The results are expressed as relative fluorescence intensity (rFI), calculated as follows: rFI = (mean fluorescence intensity of sample - mean fluorescence of negative control) / mean fluorescence intensity of negative control.

In contrast, virtually no effect of IL-4 was observed when monocytes were cultured in the absence of serum. The failure of reducing the expression in serum-free culture again confers to the whole population and no CD14 negative subpopulation persisted.

Inasmuch as HS plays a crucial role in differentiation and maturation of human monocytes to macrophages, we examined the effect of different mediators that are present in serum and are involved in monocyte activation, differentiation or long time survival in culture. These experiments revealed that in the absence of HS a combination of IL-4 and GM-CSF synergistically reduced the expression of CD14. Different other cytokines, such as IL-1, IL-6 as well as the colony stimulating factors G-CSF and IL-3 (multi CSF) had no effect on CD14 expression when added to serum-free cultures of human monocytes in combination with IL-4 (Table I). When added solely, these mediators also showed no effect (not shown).

Table I. Effect of lymphokine combinations on CD14 expression. Human monocytes were cultured for 6 days in serum-free CG medium supplemented with serum or various combinations of lymphokines as indicated. The expression of CD14 was determined by FACS analysis (see Fig. 1).

medium supplemented with	CD14 expression [rFI]
None	181,5
20 % human AB serum	185,8
IL-4 (20 U/ml)	102,1
GM-CSF (100 U/ml)	131,5
IL-4 plus GM-CSF	24,9
IL-4 plus IL-1 (20 U/ml)	161,9
IL-4 plus IL-3 (20 U/ml)	147,8
IL-4 plus IL-6 (20 U/ml)	167,3
IL-4 plus G-CSF (100U/ml)	164,1

Dose variations using two different doses of GM-CSF (50 or 100 Units/ml) combined with increasing amounts of IL-4 led to a decreased expression of CD14. Addition of 20 units/ml of IL-4 in combination with 100 units/ml of GM-CSF revealed a 90% reduction of the expression of the antigen compared to the non-treated control. In this experiment, the addition of higher concentrations of IL-4 or GM-CSF did not reveal a further decrease of CD14 expression.

Loss of a surface molecule may be due to a down-regulation of specific mRNA, or a turnover of the molecule as well as loss by shedding. We have therefore measured the concentration of soluble CD14 in the culture supernatants. As shown in Fig. 2, the culture fluid contained high amounts of the molecule with and also without lymphokine treatment. GM-CS appeared to enhance the release of sCD14. In addition, immunofluorescence staining no detectable amounts of intracellular CD14.

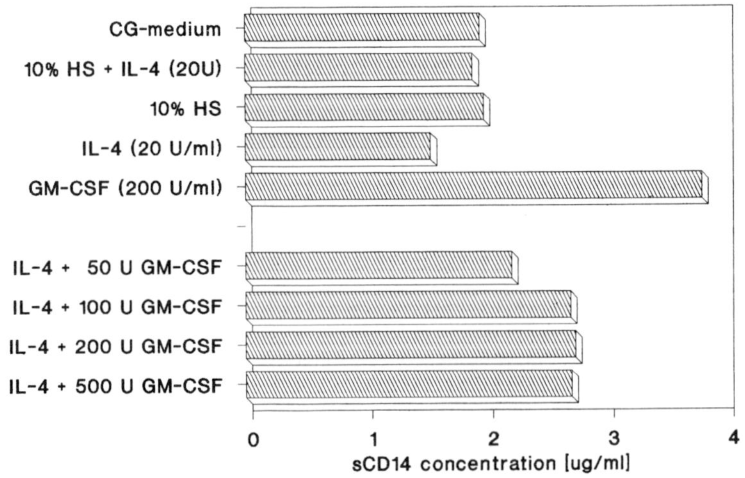

Fig. 2. Human monocytes were cultured on hydrophobic teflon plates (Peteriperm, Bachofer, Nürtingen, Germany) in CG medium containing lymphokines as indicated. Soluble CD14 was measured by a specific ELISA and exppressed as μg / ml / 7.5x105 cells.

Table II. Effect of IL-4 and GM-CSF on monocyte accessory function. Human monocytes were cultured on hydrophobic teflon plates in serum-free CG-medium containing 20 U/ml IL-4 and 100 U/ml GM-CSF. Monocytes were cultured for 6 days, harvested and used as stimulator cells (5%) of an allogeneic mixed lymphocyte culture over 6 days. Tritiated thymidine incorporation of lymphocytes was measured after another 24 hrs labelling period.

Additives	Mixed Lymphocyte Reaction Tritium thymidine incorporation [dpm +-SD]
none	273 +-115
IL-4	6,859 +-312
IL-4 plus GM-CSF	13,768 +-115

Monocyte function after treatment with IL-4 and GM-CSF. The T-cell stimulatory capacity of accessory cells depends on the expression of MHC molecules. Therefore, mediators affecting the expression of these surface molecules are likely to also enhance the accessory activity. As shown earlier IL-4 when used in serum-containing cultures strongly increased the expression of HLA-DR and HLA-DQ and the accessory activity of the cells[5]. Here we show (Table 2) in a functional assay that the T-cell proliferative response to monocytes cultured in serum-free medium and treated with IL-4 alone was already increased even at the low level of 5% stimulator cells. IL-4 when combined with GM-CSF revealed a further increase in the potency of monocytes to act as stimulator cells for lymphocytes. GM-CSF alone had no effect on T-cell proliferation (data not included). These data support the synergistic effect of GM-CSF on the IL-4 induced acquisition of accessory capacity.

DISCUSSION

The CD14 molecule is likely to function as an indirect receptor for LPS. Bacterial endotoxin when bound to an LPS-binding protein present in human serum will be recognized by CD14 on monocytes[6] or by soluble CD14 molecules in the serum[7]. Because of its universal expression it also serves as a rather general marker for monocytes and macrophages.

Down-regulation of the expression of CD14-antigen has been documented by others and ourselves recently[4,5], using conventional culture conditions, IL-4 as inducer and serum as an additive. We have attempted to establish well defined culture systems by using serum-free media. In various serum-free media we, surprisingly, failed to reproduce the above result. This led us to the assumption that serum might contain additional inducer(s) which contribute to the IL-4 induced down-reduction of CD14 expression. According to our recent results serum contains the main inducing activity for monocytes to develop into macrophages. Omitting serum from monocyte cultures prevents macrophage maturation; instead, veiled accessory cells develop which are devoid of typical macrophage markers and function. These cells increase in their antigen presenting activity, by far exceeding that of macrophages[2]. Thus, serum contains inducer(s) with a decisive function on the process of cell differentiation.

This notion is strongly supported by the finding described here. Among the manifold serum constituents we speculated on GM-CSF as a possible candidate as a co-factor, because of its profound action on macrophage growth and differentiation[8-12]. Also, it is known that GM-CSF is present in human serum as a normal constituent[13]. Here we show that GM-CSF synergizes with IL-4 in reducing the expression of CD14 on monocyte surfaces.

Reduced expression is most likely due to a down-regulated synthesis of this molecule. However, other alternatives are feasible. CD14 has been shown to be shed from the macrophage surface and subsequently detected in high amounts in the culture fluid. Our results demonstrate a high concentration of soluble CD14 in serum-free monocyte cultures especially after GM-CSF treatment. We therefore conclude that CD14 is not a stable marker of monocytes/macrophages and rather is influenced by physiological stimuli. Reduction of surface CD14 appears therefore to be due to either a reduced synthesis, a higher rate of intracellular degradation or an extracellular release of the molecule.

ACKNOWLEDGMENTS

This work was supported by the Deutsche Forschungsgemeinschaft, SFB 330 (Grant No. C 6), and grant Pe 192/6-1. J. Ruppert received a scholarship from the Boehringer Ingelheim Fonds.

REFERENCES

1. R.A. Musson, Human serum induces maturation of human monocytes in vitro, *Am. J. Pathol.* III:331 (1983).

2. H.M. Najar, A.C. Bru-Capdeville, R.K.H. Gieseler, and J.H. Peters, Differentiation of human monocytes into accessory cells at serum-free conditions, *Europ. J. Cell Biol.* 51:339 (1990).
3. Y. Akiyama, R. Griffith, P. Miller, G.W. Stevenson, S. Lund, D.J. Kanapa, Effects of adherence, activation and distinct serum proteins on the in vitro human monocyte maturation process. *J. Leukoc. Biol.* 43:224 (1988).
4. R.P. Lauener, S.M. Goyert, R.S. Geha, and D. Vercelli, Interleukin 4 down-regulates the expression of CD14 in normal human monocytes, *Eur. J. Immunol.* 20:2375 (1990).
5. J. Ruppert, D. Friedrichs, H. Xu, and J.H. Peters, IL-4 decreases the expression of the monocyte differentiation marker CD14, paralleled by an increasing accessory potency, *Immunobiol.* 182:449 (1991).
6. S.D. Wright, R.A. Ramos, P.S. Tobias, R.J. Ulevitch, and J.C. Mathison, CD14, a receptor for complexes of lipopolysaccharide (LPS) and LPS binding protein, *Science* 249:1431 (1990).
7. V. Bazil, M. Baudys, I. Hilgert, I. Stevanova, M.G. Low, J. Zbrozek, and V. Horejsi, Structural relationship between the soluble and membrane surface glycoprotein CD14, *Molec. Immunol.* 26:657 (1989).
8. P.J. Morissey, K.H. Grabstein, S.F. Reed, and P.J. Conlon, Granulocyte/macrophage colony stimulating factor: a potent activation signal for mature macrophages and monocytes, *Int. Arch. Allergy Appl. Immunol.* 88:40 (1989).
9. J.M. Alvaro-Garcia, N.J. Zvaifler, and G.S. Firestein, Cytokines in chronic inflammatory arthritis. VI. Granulocyte/macrophage colony stimulating factor-mediated induction class II MHC antigen on human monocytes: a possible role in rheumatoid arthritis, *J. Exp. Med.* 170:865 (1989).
10. D. Metcalf, The granulocyte-macrophage colony stimulatory factors. *Science* 229:16 (1985).
11. M.J. Elliot, M.A. Vadas, J.M. Eglington, L.S. Park, L. Bik-To, G. Cleland, S.C. Clark, and A.F. Lopez, Recombinant human interleukin 3 (IL-3) and granulocyte-macrophage colony-stimulating factor (GM-CSF) show common biological effects and binding characteristics on human monocytes, *Blood* 74:2049 (1989).
12. J.R. Gamble, M.J. Elliott, E. Jaipargas, A.F. Lopez, and M.A. Vadas, Regulation of human monocyte adherence by granulocyte-macrophage colony-stimulating factor, *Proc. Natl. Acad. Sci. USA* 86:7169 (1989).
13. N.A. Nicola and M. Vadas, *in*: "Oncogenes and Growth Factors," R.A. Bradshaw and S. Prentis, eds., Elsevier Science Publishers, New York (1987).

SERUM-FREE DIFFERENTIATION OF RAT AND HUMAN DENDRITIC CELLS, ACCOMPANIED BY ACQUISITION OF THE NUCLEAR LAMINS A/C AS DIFFERENTIATION MARKERS

Robert K.H. Gieseler,[1] Hui Xu,[1] Rolf Schlemminger,[2]
and J. Hinrich Peters[1]

[1]Department of Immunology, Kreuzbergring 57
[2]Department of General Surgery, Robert-Koch-Strasse 40
University of Göttingen
D-3400 Göttingen, Germany

INTRODUCTION

It is generally accepted that tissue macrophages (Mφ) derive from a common bone marrow (BM) precursor. Clearly, it is the myeloid lineage that branches into the development of Mφ and granulocytes. For contrast, the exact origin of organ-specific dendritic cells (DC) is still doubtful. In vitro studies may help to solve this controversy. In this report we therefore present results gained from such investigations in our laboratory.

Our data obtained with cells from rat and man strongly suggest that DC also descend from a myeloid precursor, and that DC and Mφ are intimately related. As could be demonstrated in the human system, monocytes (Mo) may be differentiated into either DC or Mφ. Using chemically defined media, we could identify signals necessary for DC development of both rat and man. Interestingly, while in the presence of serum such DC are transformed into Mφ, an elevation in intracellular cAMP conversely turns Mφ into potent DC. These cell types strongly resemble their tissue-resident counterparts. Additionally, DC and Mφ cultured from rat or human precursors were positive for the nuclear lamins A/C, which are thought to distinguish differentiated cells from their immature precursors. Here, we mainly focus on the comparison of rat and human DC and Mφ obtained in vitro.

MATERIALS AND METHODS

Rat (LEW.1U or 1A) bone marrow cells (BMC) of the non-adherent fraction were cultured in liquid media. After pretesting colony-stimulating factors (CSF)[2] according to Metcalf,[1] we developed a serum-free medium for DC differentiation consisting of 80% RPMI 1640/20% 199 Earle's salts (medium 80/20), multi-CSF, M-CSF, linoleic acid, α-tocopherol, and cholecalciferol.[2] Mφ were cultured in medium 80/20 plus 10% bovine serum.[2]

Dendritic Cells in Fundamental and Clinical Immunology,
Edited by Kamperdijk *et al.*, Plenum Press, New York, 1993

Human monocytes were from buffy coat preparations (University Clinics, Göttingen). Differentiation into DC was obtained with the serum-free medium MCDB 302 (Sigma, Deisenhofen, FRG), and macrophages were gained with medium 80/20 in the presence of 10% human AB serum.[3] Additional experiments were done with serum-free medium 80/20.

We tested the differentiated cells for the expression of antigens characterizing either DC or Mφ. Differentiation antigens were ED2 (rat)[2,4] and CD14 (human).[5,6] Furthermore, we detected lamin B and the lamins A/C with antibodies recognizing the rat and human variants.[7] For functional antigens we chose MHC Class II, i.e. a monomorphic determinant (rat)[2] and HLA-DR, -DP, and -DQ (human),[5,6] as well as CD16/IgG FcR III (human).[5] α-Naphtylbutyrate esterase was determined[2,8] as an enzyme characteristic of Mo/Mφ which is only weakly present in DC.[9] Functional assays were for FcR-dependent phagocytosis, as well as for the capacity of mitogen-triggered (JO^{4-}) Mφ or DC to stimulate a T cell proliferation, as measured by ^3H thymidine incorporation.[2,3,8] We also evaluated the presence of discrete morphological criteria, which are typical of DC and Mφ found in the tissues.[2,3,10]

As an extention of the experiments leading to a chemically defined medium for differentiation of rat BMC into DC[2,3] we furthermore looked for the effect of essential fatty acids, vitamin E, and the vitamin hormone D_3. Also, the interleukins IL-1 and IL-6 as well as dibutyric cAMP (db-cAMP) were tested as activation signals.

RESULTS

The intention of this work is to compare dendritic cells and macrophages obtained from rat BMC precursors or from human monocytic precursors. DC were differentiated in serum-free media, while Mφ were grown in the presence of serum. These cells were tested for a variety of parameters, as summarized in Table 1.

A central function of accessory cells is their ablility to stimulate lymphocyte proliferation. T cells were strongly stimulated by rat and human DC, whereas Mφ had only a weak accessory potency. Essential fatty acids remarkably enhanced the lymphocyte-stimulating potency of rat DC and human DC and Mφ. Linoleic acid (LA) was the most potent one (4 μg/ml for rat cells,[2] and 0.5 μg/ml for human cells). Accessory activity of human Mφ and DC is also enhanced by IL-1 and IL-6,[12] and we could demonstrate that a rise in cytosolic cAMP causes Mφ to convert into DC.[8] The capacity to phagocytose opsonized latex particles was not present in DC, as opposed to Mφ.

Myeloperoxidase was only tested for in human Mφ treated with db-cAMP. These cells reactively transformed into DC. Such DC were only weakly positive for this enzyme, as opposed to Mφ without db-cAMP that displayed a strong activity.[8] In vivo, Mo and Mφ are positive for α-naphtylbutyrate esterase. The Mφ cultured from rat BM or human Mo were strongly stained with the converted substrate dye. For contrast, DC populations are only weakly positive,[9] which was also the case for DC of rat or human origin.

Class II was strongly expressed on rat and human DC, while serum-cultured Mφ were only weakly positive. ED2[4] and CD14 positively identified rat and human Mφ, whereas the DC were virtually negative. IL-4 strongly down-regulated CD14 on Mφ. CD16 was strongly present on human Mφ, but only weakly on IL-4-treated cells.[5] The nuclear A/C lamins mark cells that are completely differentiated. In both rat and human cell cultures, all precursor types were lamin B^+A/C^-, but all DC and Mφ clearly displayed these markers.[7]

Using rat BMC of the non-adherent BM fraction,[11] myeloid precursors only developed in the presence of multi- and M-CSF. These CSF are therefore obligatory for the

Table 1. Comparison between bone marrow-derived cells of the rat and monocyte-derived cells of man. The features chosen are characteristic of DC and Mφ found in lymph or solid tissues.

	RAT		HUMAN	
	DC	Mφ	DC	Mφ
FUNCTIONS				
- FcR-dependent phagocytosis	-[1]	++	-	++
- T cell stimulation	+++	+	+++	+
- Enhancers of DC functions	EFA[2]	N.D.	EFA	EFA
	N.D.[3]	N.D.	IL-1/IL-6	IL-1/IL-6
	N.D.	N.D.	N.D.	IL-4
ENZYMES				
- Myleoperoxidase	N.D.	N.D.	+ (cAMP)	+++
- α-Naphtylbutyrate esterase	±	+++	±	+++
ANTIGENS				
- MHC class II	+++	+	+++	+
- ED2	-	+		
- CD14			± (IL-4)	+++
- CD16 (FcγR III)	N.D.	N.D.	± (IL-4)	+++
- Lamin B	+++	+++	+++	+++
- Lamins A/C	+++	+++	+++	+++
INDUCERS OF DIFFERENTIATION				
- Multi-CSF (IL-3) + M-CSF	+	+	+	+
- Tocopherols and cholecalciferol	+	?[4]	±	?
- Unknown serum factors	-	+	-	+
MORPHOLOGICAL CRITERIA				
- Veils	±	-	++	-
- Dendrites	+++	-	+++	-
- Phagolysosomes	-	+++	-	+++
- Stretched, irregular nucleus	+++	-	+++	-
- Round, regular nucleus	-	+++	-	+++

[1]Parameters are - (negative) to +++ (strongly positive).
In Inducers of Differentiation + and - are "yes" and "no".
[2]Essential fatty acids (primarily *cis*-18:2).
[3]Not determined.
[4]Not known.

early phase of both DC and Mφ differentiation. However, complete differentiation of DC only occurred with both α-tocopherol and cholecalciferol. At any point in time, rat DC were turned into Mφ when 10% serum was added. The differentiation of human Mo, and therefore DC and Mφ, depends on the action of CSF. We have controverse results, in that Mo-derived DC grow in MCDB 302 without further additives, while depending on the addition of 0.025% α-tocopherol, 0.25 μM cholecalciferol, and 0.5 μg/ml LA in medium 80/20. None of the factors alone could support cell survival. When cultures (MCDB 302) of veiled DC were added 10% serum, the cells again turned into Mφ.

Morphologically, the DC from rat and human origin somehow differed in their abundance of veils, while both DC types were strongly dendritic. These cells had a stretched and irregular nucleus. Macrophages, on the other hand, had an almost round nucleus and did not exhibit veils or dendrites. Also, they were packed with phagolysosomes, which were not seen in accessory cells.

DISCUSSION

This article combines data obtained with myeloid descendants of rat and man. In vitro, differentiation of these cells at least depends on multi-CSF and M-CSF. The exact role of vitamin E and the vitamin hormone D_3 is not fully understood, although their profound influence on the differentiation of DC remains undoubted. However, all cells cultured with these molecules present, expressed the lamins A/C in cells of DC or Mφ phenotype. Such cells are completely differentiated, as far as knowledge about the nuclear lamins tells us.[7]

We could successfully culture potent DC without the dendritic cell-activating factor (DACIF) described by Klinkert.[11] Possibly, however, DACIF could be one out of several signals, that seem to significantly and specifically activate functions typical of DC. We could show that IL-1/IL-6, IL-4, and the second messenger (db)cAMP as well as essential fatty acids (which may be transformed into prostanoid mediators) activate such functions. As far as tested, they act on both DC and Mφ.

The features compared between the respective rat and human cell phenotypes strikingly resemble each other. We may assume that minor inconsistencies are due to the different culture conditions. In both cases, DC as well as Mφ were derived from myeloid progenitors, and they were interconvertible. This again demonstrates the tremendous plasticity of myeloid progeny. However, although we do not exclude the possibility of certain organ-resident DC to descend from alternate differentiation pools, the myeloid differentiation tree clearly is capable of generating large numbers of DC.

ACKNOWLEDGMENTS

The authors gratefully acknowledge the support of the German Research Council, SFB 330, Grant D9, as well as Grant Pe 192/61.

REFERENCES

1. D. Metcalf. "Clonal Culture of Hemopoietic Cells: Techniques and Applications," Elsevier, Amsterdam, New York, Oxford (1984).
2. R.K.H. Gieseler, R.-A. Röber, R. Kuhn, K. Weber, M. Osborn, and J.H. Peters, Dendritic accessory cells derived from rat bone marrow precursors under chemically defined conditions in vitro belong to the myeloid lineage, *Eur. J. Cell Biol.* 54: 171 (1991).
3. H.M. Najar, A.C. Bru-Capdeville, R.K.H. Gieseler, and J.H. Peters, Differentiation of human monocytes into accessory cells at serum-free conditions, *Eur. J. Cell Biol.* 51: 339 (1990).

4. C.D. Dijkstra, E.A. Döpp, P. Joling, and G. Kraal, The heterogeneity of mononuclear phagocytes in lymphoid organs: Distinct macrophage subpopulations in the rat recognized by monoclonal antibodies ED1, ED2, and ED3, *Immunology* 54: 589 (1985).
5. J. Ruppert, D. Friedrichs, H. Xu, and J.H. Peters, IL-4 decreases the expression of the monocyte differentiation marker CD14, paralleled by an increasing accessory potency, *Immunobiol.* 182: 449 (1991).
6. G. Ocklind, D. Friedrichs, and J.H. Peters, Expression of CD54, CD58, CD14, and HLA-DR on macrophages and macrophage-derived accessory cells and their accessory capacity, *Immunol. Lett.* 31: 253 (1992).
7. R.-A. Röber, R.K.H. Gieseler, J.H. Peters, K. Weber, and M. Osborn, Induction of nuclear lamins A/C in macrophages in in vitro cultures of rat bone marrow precursor cells and human blood monocytes, and in macrophages elicited in vivo by thioglycollate stimulation, *Exp. Cell Res.* 190: 185 (1990).
8. J.H. Peters, T. Börner, and J. Ruppert, Accessory phenotype and function of macrophages induced by cyclic adenosine monophosphate, *Int. Immunol.* 2: 1195 (1990).
9. R.M. Steinman, W.C. van Voorhis, and D.M. Spalding, Dendritic cells, *in:* "Handbook of Immunological Methods," D.M. Weir, ed., L.A. Herzenberg, C. Blackwell, Leonore A. Herzenberg, co-eds., Blackwell, Oxford, London, Edinburgh, Boston, Palo Alto, Melbourne (1986).
10. J.H. Peters, S. Ruhl, and D. Friedrichs, Veiled accessory cells deduced from monocytes, *Immunobiol.* 176: 154 (1987).
11. W.E.F. Klinkert, Rat bone marrow precursors develop into dendritic accessory cells under the influence of a conditioned medium, *Immunobiol.* 168: 414 (1984).
12. J. Ruppert, and J.H. Peters, IL-6 and IL-1 enhance the accessory activity of human blood monocytes during differentiation to macrophages, *J. Immunol.* 146: 144 (1991).

IMMUNOPHENOTYPIC AND ULTRASTRUCTURAL DIFFERENTIATION AND MATURATION OF NONLYMPHOID DENDRITIC CELLS IN OSTEOPETROTIC (*op*) MICE WITH THE TOTAL ABSENCE OF MACROPHAGE COLONY STIMULATING FACTOR ACTIVITY

Kiyoshi Takahashi,[1] Makoto Naito,[1] Yuki Morioka,[1] and Leonard D. Shultz[2]

[1]Second Department of Pathology, Kumamoto University School of Medicine, Kumamoto, Japan

[2]The Jackson Laboratory, Bar Harbor, Maine, USA

INTRODUCTION

To examine the effect of macrophage colony stimulating factor (M-CSF) or CSF-1 on the differentiation and maturation of nonlymphoid dendritic cells (DCs) *in vivo*, the osteopetrotic mouse (*op*/*op* mouse) is a useful tool. This is because the *op* mutation is shown to be a defect in the coding region of the macrophage colony stimulating factor (*Csfm*) gene and because CSF-1 produced is nonfunctional, although this mouse does produce *Csfm* messenger RNA at normal levels.[1] This mutation is transmitted by an autosomal recessive trait and homozygous (*op*/*op*) mice are characterized by the absence of incisors, a distinctly domed skull, a short tail, and a small body size.[2] These phenotypic abnormalities become evident by ten days after birth. In addition to a marked reduction of osteoclasts, deficiencies of monocytes and monocyte-derived macrophages occur in *op*/*op* mice.[3,4] All these result from the lack of CSF-1 activity. In a recent study, we found immature macrophages in various organs and tissues of *op*/*op* mice, suggesting that these CSF-1-independent macrophages are derived from granulocyte/macrophage colony forming cells (GM-CFCs) or earlier hematopoietic progenitors.[5] However, little is known about the DCs of *op*/*op* mice, including interdigitating cells (IDCs) in the thymus or peripheral lymphoid tissues, epidermal Langerhans cells (LCs), or indeterminate dendritic cells (IDDCs).

In the present study, we examined immunohistochemically and electron microscopically DCs in lymphoid tissues and skin, as well as in the other tissues of *op*/*op* mice. In a culture of normal mouse bone marrow cells with granulocyte/macrophage colony stimulating factor (GM-CSF) or CSF-1, we studied immunohistochemically or electron microscopically the differentiation of the bone marrow cells into macrophages and compared these macrophages with DCs or macrophages in the *op/op* mice.

MATERIALS AND METHODS

(57BL/6J X C3HeB/FeJ)F_2-*op/op* and +/? littermate control mice were raised from +/*op* breeders and maintained in the Animal Center of the Kumamoto University School of Medicine. Thymus, spleen, lymph nodes, liver, lungs, stomach, small and large intestines, brain, ovaries, uterus, testis, and other tissues were excised from the animals. Bone marrow tissues were taken from the bilateral femora. Skin specimens were obtained from the ears of the animals, and epidermal sheets were prepared. These tissues and epidermal sheets were stained with the indirect immunoperoxidase method using anti-mouse panmacrophage monoclonal antibodies (mAb) F4/80[6,7] and BM8,[8] mAb against LCs/IDCs, NLDC-145,[9,10] M1-8,[11] and MIDC-8,[10] and mAb against Ia antigens, M5/114.[12] Histochemical stainings of adenosine di- or triphophatase (ADPase or ATPase) localization were performed on the epidermal sheets for a marker of epidermal LCs.[13,14] In the lymphoid tissues, the numbers of NLDC-145- or M5/114-positive DCs were enumerated. In the epidermal sheets, the numbers of NLDC-145-, Ia- or ADPase-positive DCs were counted. The significance of the difference between means was evaluated by the Student's *t*-test. The tissue specimens were also submitted to electron microscopy.

In order to examine CSF-dependent macrophages immunohistochemically or electron microscopically, bone marrow cells were obtained from male BALB/c mice (4 to 6 weeks of age) and passed twice through the Sephadex G10 column. 5×10^4 bone marrow cells/ml were incubated on soft agar (Bacto-Agar; Difco, Detroit, MI, USA) with recombinant mouse GM-CSF (Genzyme, Cambridge, MA, USA) or human CSF-1 (Midori Jyuji, Tokyo, Japan). For liquid culture, 1×10^5 bone marrow cells/ml were incubated in Iscove's modified Dulbecco's medium (Gibco, Grand Island, NY, USA) with GM-CSF or CSF-1. In both cultures, colony forming cells (CFCs) were examined at 2, 3, 4, 5, 6, and 7 days by immunohistochemistry and electron microscopy, as well as by the combined method of ultrastructural peroxidase cytochemistry and immunoelectron microscopy using mAb.[15]

RESULTS

DC Population

DCs in the thymic medulla, periarteriolar lymphatic sheath of the spleen (PALS), paracortical area of lymph nodes, and in the T zone of the other peripheral lymphoid tissues were positive for NLDC-145, MIDC-8, M1-8, and M5/114 in *op/op* mice and normal littermates. In the epidermal sheets prepared from both *op/op* and normal littermate mice, epidermal LCs and IDDCs stained positive for ADPase or ATPase and reacted with NLDC-145, MIDC-8, M1-8, F4/80, BM8, and M5/114. Numbers of NLDC-145-, ADPase- or Ia-positive DCs in the epidermal sheets from *op/op* mice were not reduced. In the *op/op* mice, DCs in the thymic medulla, PALS, and paracortical area of lymph nodes showed a well-developed tubulovesicular system characteristic of IDCs, while epidermal LCs possessed Birbeck granules, well-developed Golgi complexes, numerous vesicles, and electron-dense or lucent, membrane-bound vacuoles, the same as in the normal littermates (Figure 1). However, these cells were mostly round with a smooth cell surface, and their cytoplasmic processes were not prominent.

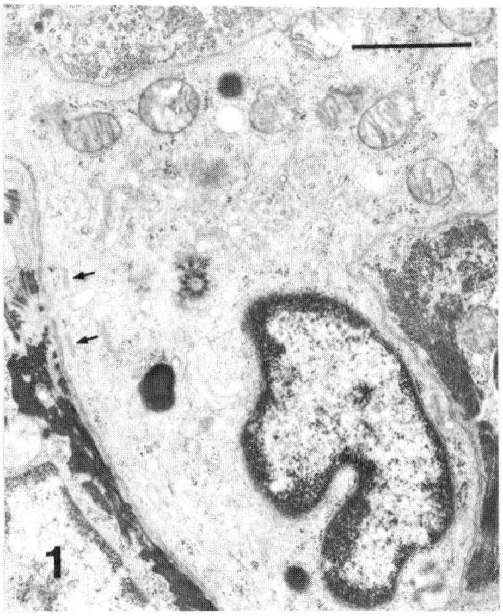

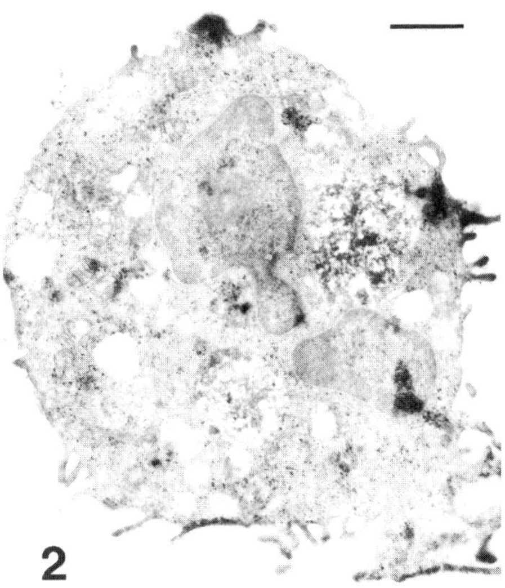

Figure 1. Epidermal LC of an op/op mouse shows the presence of Birbeck granules (arrows) and develops many vesicles, tubules, or vacuoles. The cell surface is smoooth with no cytoplasmic projection. Bar: 2 μ .

Figure 2. GM-CSF-dependent macrophage developed in vitro shows prominent phagocytosis, ultrastructural immaturity of intracellular organelles, and less prominent cell projection. Bar: 2 μ .

Investigation of CSF-dependent Macrophages *in Vitro*

In agar culture, BM8-positive macrophage colonies or clusters developed at two days in the presence of CSF-1 and their colony numbers increased with culture day. In the culture with GM-CSF, BM8-positive macrophage colonies or clusters were detected at five days of culture, and the number of the colonies or clusters increased thereafter. However, no cells with a dendritic morphology were observed in any macrophage colonies or clusters. In the liquid or agar cultures, most of the CSF-1-dependent macrophages showed the ultrastructure of the mature macrophages seen in the normal mice: numerous lysosomal granules, vesicles, or vacuoles, rough endoplasmic reticulum, well-developed Golgi complexes, and abundant cytoplasmic processes. In contrast, GM-CSF-dependent macrophages were small and round with a variable number of vesicles or vacuoles (Figure 2). However, the development of the other organelles was poor and they extended only a few microvilli from the cell surface. These GM-CSF- or CSF-1-dependent macrophages showed no tubulovesicular system or dendritic cytoplasmic projection, both characteristic of DCs.

DISCUSSION

Previous studies have demonstrated that *op/op* mice lack functional CSF-1 activity leading to impaired development of monocytes, differentiation of monocytes into macrophages or osteoclasts, and proliferation of these cells.[1,3-5] In our previous study of a co-culture of normal mouse bone marrow cells on a fibroblast cell line established from a mutant mouse, more than 95% of the bone marrow cells were found to stop differentiating at the stage of monocytes even after two weeks of culture.[5] In the *op/op* mice, we have found immature tissue macrophages in various organs and tissues, particularly in the spleen and brain; these macrophages are called CSF-1-independent macrophages.[5] Since previous studies have substantiated that GM-CSF is a major cytokine normally secreted from *op/op* fibroblasts,[1,3] it appears that GM-CSF plays a major role in the differentiation, maturation, proliferation, and survival of CSF-1-independent macrophages in the mutant mice. Further, it is suggested that the CSF-1-independent macrophages are derived not from monocytes but from GM-CFCs or earlier hematopoietic cells, because the present ultrastructural observation showed that CSF-1-independent macrophages closely resemble GM-CSF-dependent macrophages developed *in vitro*.

Besides the CSF-1-independent macrophages, we demonstrated that DC population exist in the lymphoid tissues and epidermis of the *op/op* mice just as they do as in the normal littermates, thus suggesting that these cells are also derived not from monocytes but from GM-CFCs or earlier hematopoietic cell precursors. Recent *in vitro* studies have provide evidence that GM-CSF is essential for the differentiation, maturation, and viability of DCs,[16-20] and that lineage-specific growth factors, CSF-1 or G-CSF, do not respond to DCs.[14] In this *in vitro* study, however, we could not demonstrate the differentiation of DCs in cultures of normal mouse bone marrow cells in the presence of GM-CSF alone, and the ultrastructure of GM-CSF-dependent macrophages was different from that of DCs. Interestingly, studies noted that besides GM-CSF, other cytokines such as interleukin (IL)-1,[16,19] IL-2,[17] or IL-4[21] are necessary for the differentiation and maturation of DCs. Based on this information, we are trying to investigate *in vitro* differentiation and maturation of DCs using these cytokines. Finally, we found that dendritic projection of the DCs in the *op/op* mice was less prominent than in the normal littermates. This suggests the possibility that the dendritic projection of DCs is influenced by the direct or indirect effect of CSF-1.

SUMMARY

The effect of the *op/op* mutation on the development of DCs including IDCs in the thymus and peripheral lymphoid tissues and epidermal LCs or IDDCs was determined in order to assess the differentiation of such cells *in vivo* in the absence of M-CSF. *op/op* and littermate control mice were examined by immunohistochemistry using F4/80, BM8, NLDC-145, M1-8, MIDC-8, and M5/114. In contrast with the fact that the monocytic cell series, monocyte-derived macrophages and osteoclasts were deficient, DCs in the lymphoid tissues and epidermal LCs from *op/op* mice showed similar immunoreactivities to those of normal littermates and no statistically significant differences in their numbers compared to the normal littermates. Further, the epidermal LCs in the mutant mice stained positively with the histochemical stains for ADPase or ATPase. The development of tubulovesicular system in IDCs and the presence of Birbeck granules in LCs of the *op/op* mice were confirmed by electron microscopy but the cytoplasmic projection of these cells was not prominent. From these results, we concluded that the development and differentiation of DCs are influenced not by M-CSF but by GM-CSF. In our *in vitro* study, however, we found that GM-CSF-dependent macrophages do not resemble DCs ultrastructurally, suggesting that besides GM-CSF, some other cytokines are necessary for the differentiation and maturation of DCs.

REFERENCES

1. H. Yoshida, S-I. Hayashi, T. Kunisada, M. Ogawa, S. Nishikawa, H. Okamura, T. Sudo, L.D. Shultz, and S-I., Nishikawa, The murine mutation "osteopetrosis" (*op*) is a mutation in the coding region of the macrophage colony stimulating factor (*Csfm*) gene, Nature, 345:442 (1990).

2. S.C. Marks. Morphological evidence of reduced bone resorption in osteopetrotic (*op*) mice, Am. J. Anat. 163:157 (1982).
3. W. Wiktor-Jedrzejczak, A. Ahmed, C. Szczylik, and R.R. Skelly, Hematological characterization of congenital osteopetrosis in op/op mouse, Proc. Natl. Acad. Sci. USA, 87:4828 (1990).
4. H. Kodama, A. Yamazaki, M. Nose, S. Niida, Y. Ohgame, M. Abe, M. Kumegawa, and T. Suda, Congenital osteoclast deficiency in osteopetrotic (*op/op*) mice is cured by injections of macrophage colony-stimulating factor. J. Exp. Med. 173:269 (1991).
5. M. Naito, S-I. Hayashi, H. Yoshida, S-I. Nishikawa, L.D. Shultz, and K. Takahashi, Abnormal differentiation of tissue macrophage populations in 'osteopetrosis' (*op*) mice defective in the production of macrophage colony-stimulating factor, Am. J. Pathol. 139:657 (1991).
6. D.A. Hume, A.P. Robinson, G.G. MacPherson, and S. Gordon, The mononuclear phagocyte system of the mouse defined by immunohistochemical localization of antigen F4/80, J. Exp. Med. 158:1522 (1983).
7. D.A. Hume, J.F. Loutit, and S. Gordon, The mononuclear phagocyte system of the mouse defined by immunohistochemical localization of antigen F4/80: Macrophages of bone and associated connective tissue, J. Cell Sci, 66:189 (1984).
8. U. Malorny, E. Michels, and C. Sorg, A monoclonal antibody against an antigen present on mouse macrophages and absent from monocytes, Cell Tissue Res, 243:421 (1986).
9. G. Kraal, M. Breel, M. Janse, and G. Bruin, Langerhans' cells, veiled cells, and interdigitating cells in the mouse recognized by a monoclonal antibody, J. Exp. Med. 163:981 (1986).
10. M. Breel, R.E. Mebius, and G. Kraal, Dendritic cells of the mouse recognized by two monoclonal antibodies, Eur. J. Immunol. 17:1555 (1987).
11. T. Maruyama, S. Tanaka, B. Bozoky, F. Kobayashi, and H. Uda, New monoclonal antibody that specifically recognizes murine interdigitating cells and Langerhans cells, Lab. Invest. 61:98(1989).
12. A. Bhattacharya, M.E. Dorf, and T.A. Springer, A shared alloantigentic determinants on Ia antigens encoded by the I-A and I-E subregions: Evidence for I region gene duplication, J. Immunol. 127:2488 (1981).
13. I.C. Mackenzie, and C.A. Squier, Cytochemical identification of ATPase-positive Langerhans cells in EDTA-separated sheets of mouse epidermis, Br. J. Dermatol. 92:523 (1975).
14. M.B. Chaker, M.D. Tharp, and P.R. Bergstresser, Rodent epidermal Langerhans cells demonstrate greater histochemical specificity for ADP than for ATP and AMP, J. Invest. Dermatol. 82:496 (1984).
15. E.C.M. Hoefsmit and R.H.J. Beelen, The expression of antigen F4/80 and Ia on peritoneal macrophages in normal and BCG-immunized mice, *in*: "Mononuclear Phagocytes, Characterization, Physiology, and Function," R. van Furth, ed., Martinus Nijhoff Publishers, Dordrecht (1985).
16. C. Heufler, F. Koch, and G. Schuler, Granulocyte/macrophage colony-stimulating factor and interleukin-1 mediate the maturation of murine epidermal Langerhans cells into potent immunostimulatory dendritic cells, J. Exp. Med. 167: 700 (1988).
17. G.G. MacPherson, S. Fossum, and B. Harrison, Properties of lymph-borne (veiled) dendritic cells in culture. II. Expression of the IL-2 receptor: Role of GM-CSF, Immunology 68:108 (1989).
18. S. Markowicz and E.G. Engleman, Granulocyte-macrophage colony-stimulating factor promotes differentiation and survival of human peritoneal blood dendritic cells *in vitro*, J. Clin. Invest. 85:955 (1990)
19. M.D. Witmer-Pack, W. Olivier, J. Valinsky, G. Schuler, and R.M. Steinman, Granulocyte/macrophage colony-stimulating factor is essential for the viability and function of cultured murine epidermal Langerhans cells, J. Exp. Med. 166:1484 (1987).
20. R.M. Steinman, S. Koide, M. Witmer, M. Crowley, N. Bhadwaj, P. Freudenthal, J. Young, and K. Inaba, The sensitization phase of T-cell-mediated immunity, Ann. N.Y. Acad. Sci, 546:80 (1988).
21. K. Akagawa. Differentiation of human monocytes and cytokines, J. Jpn. Soc. RES. 32:174 (1992) (*in Japanese*).

LOADING OF DENDRITIC CELLS WITH ANTIGEN IN VITRO OR IN VIVO BY IMMUNOTARGETING CAN REPLACE THE NEED FOR ADJUVANT

T. Sornasse, V. Flamand, G. De Becker, K. Thielemans, J. Urbain, O. Leo and M. Moser

Laboratoire de Physiologie Animale
Université Libre de Bruxelles
rue des chevaux, 67
1640 Rhode-St-Genèse; Belgium

INTRODUCTION

The discovery that T cells are activated upon recognition of antigen presented by MHC products has focused the attention on the major role of antigen-presenting cells in the initiation of immune responses. Although several cell populations can display peptides in association with class II antigens, only dendritic cells appear to efficiently activate specific, naive T cells. Recent studies have shown that antigen presentation and T cell sensitization are independently regulated in dendritic cells, and this specialization of function supports the idea that, among the heterogeneous population of APC, they may be the most efficient cells to present an antigen and to sensitize T cells *in vivo*. In this report, we show that a single injection of DC appropriately pulsed *in vitro* with soluble protein antigens activates a strong specific humoral response *in vivo*. Moreover, antigen targeted on dendritic cells *in vivo* by the use of antibodies induces the secretion of antigen-specific antibodies in the absence of adjuvant.

RESULTS AND DISCUSSION

Fidelity of dendritic cells to the processed antigen. Recent data have shown that fresh but not cultured dendritic cells were able to process native proteins. Thus, in accordance with these data, DC are pulsed with the antigen during the purification steps (1, 2). Downregulation of their capacity to process suggests that DC may retain the initial antigen for a longer period of time, as compared to B cells. Thus, we compared the ability of DC and low density B cells (low density, non-adherent, anti-Thy 1 + complement-treated spleen cells) to induce T cell activation over time. We used a T cell hybridoma, since its

Dendritic Cells in Fundamental and Clinical Immunology,
Edited by Kamperdijk *et al.*, Plenum Press, New York, 1993

activation only requires TCR occupancy and is independent of any costimulatory signal. The APC were pulsed during overnight culture, washed and incubated in antigen-free complete medium until use. Figure 1 shows that, 3 days following Ag pulse, DC retain the ability to stimulate T cell hybridoma, whereas low density B cells have completely lost their capacity to activate T cell hybridoma. Addition of antigen in the culture completely restores

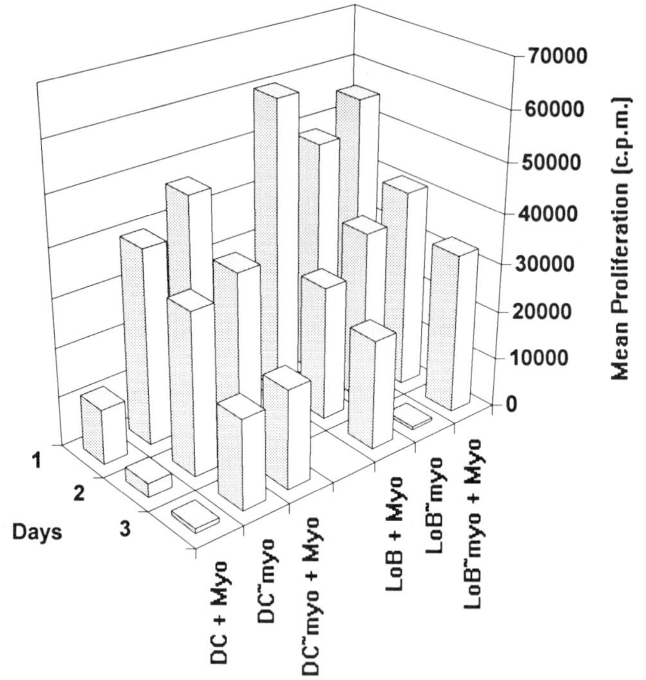

Figure 1: *Dendritic cells retain the original antigen for at least 3 days* . Fresh DC and low-density B cells were cultured overnight in complete media containing myoglobin (DC-Myo, LoB-Myo). Cells were washed and maintained in complete medium before testing. On various days after Ag pulse, cells were counted and $2\ 10^4$ cells were cultured with $3\ 10^4$ T-T hybridoma 13.26.8 (anti Myo /Ia^d) with (+Myo) or without addition of antigen. Control APC (DC, loB) were cultured without Ag during the purification steps. IL-2 was measured by the proliferation of CTL.L clone.

the activation of hybridoma by B cells, but does not significantly enhance the activation by DC. These data show that DC and B cells, isolated from the same low density spleen cells and pulsed in similar conditions with antigen, have distinct antigen-presenting functions: in contrast to DC, B cells retain their capacity of processing over time, as previously described for epidermal Langerhans cells (3). This finding correlates with a recent report showing that, with culture, class II and invariant chain synthesis decreases dramatically, and that synthesized class II molecules remain stable on the surface of DC (4), as compared to B cells (5).

Dendritic cells pulsed in vitro with antigen induce a specific humoral response in vivo. DC were isolated from the spleens of DBA/2 mice and pulsed with HGG (human gamma globulins) during the purification steps. 3 X 10^5 antigen-pulsed DC were injected in syngeneic mice that received 50μg of soluble HGG in saline 5 days later. The data in Figure 2 show that *in vivo* administration of antigen-pulsed DC induces a strong humoral response, whereas control mice, injected with unpulsed DC or with antigen in saline, produce little antibodies. Similar data were obtained with myoglobin (6). It is note that injection of dendritic cells induces a memory response as assessed by the higher concentration of specific antibodies secreted during a secondary response (2). Since the isotype switching is dependent on the Th subset(s) that are activated by the antigen, we characterized the subclasses of the antibodies induced by HGG-pulsed DC. The data show

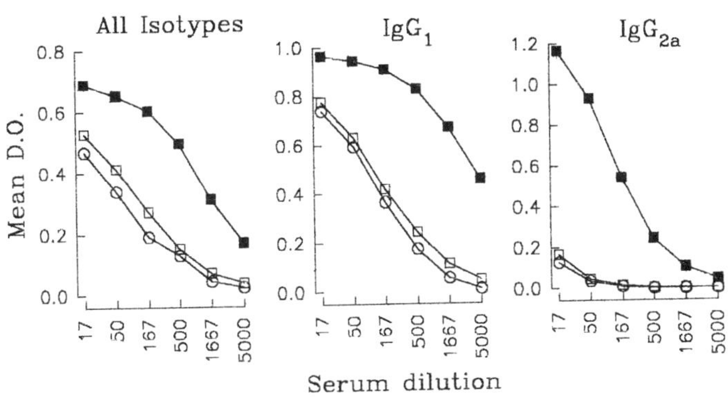

Figure 2: *Dendritic cells pulsed in vitro with antigen induce a specific humoral response in vivo.* DBA/2 mice (7-10 per group) were pretreated with HGG-pulsed DC (filled square), unpulsed DC (open squares) or left untreated (open circle). All mice received 50 μg of Ag in saline 5 days later and bled 21 days after Ag boost. Antigen-specific antibodies were measured in individual sera 21 days after Ag boost. Error bars are omitted for simplicity.

that high concentrations of IgG1 and IgG2a antibodies are synthesized by mice primed with antigen-pulsed DC. These results, although indirect, suggest that DC very efficiently present antigen and activate specific naive T cells, that in turn induce antibody secretion by antigen-specific B cells. The isotype distribution shows that at least Th1 cells are activated since these cells have been shown to secrete interferon-g that is required for isotype switching to IgG2a (6). Recent in vitro data suggest that the accessory cells present during the initial receptor-mediated stimulation of T cells may influence the development of subsets that differ in cytokine production. Based on the present data, it is tempting to speculate that DC may favor the optimal activation of Th1 cells *in vivo,* through the expression of the adequate costimulatory signal. Experiments are in progress in our laboratory to characterize the lymphokine pattern produced by antigen-specific, $CD4^+$ T cells isolated from the mice primed with distinct APC populations.

Immunization by immunotargeting antigens to dendritic cells in vivo. Previous *in vivo* data have shown that immunotargeting avidin to dendritic cells using 33D1 antibodies induced a low but significant secondary humoral response specific for avidin (7). We performed a similar experiment, using the hamster N418 antibody that has been shown to stain the DC *in situ* (8). DBA/2 mice were injected intravenously with 15 μg biotinylated-N418 and received 5μg of streptavidin 6 hours later. 2 hours later, the pretreated mice were injected with 20 μg of biotinylated antigen (HGG-biotin) intravenously. Control mice were injected with biotin-HGG emulsified in complete Freund adjuvant (CFA, positive control), or 20 μg biotin-HGG in saline (negative control). All mice were bled 25 days after immunization and the sera were tested individually.

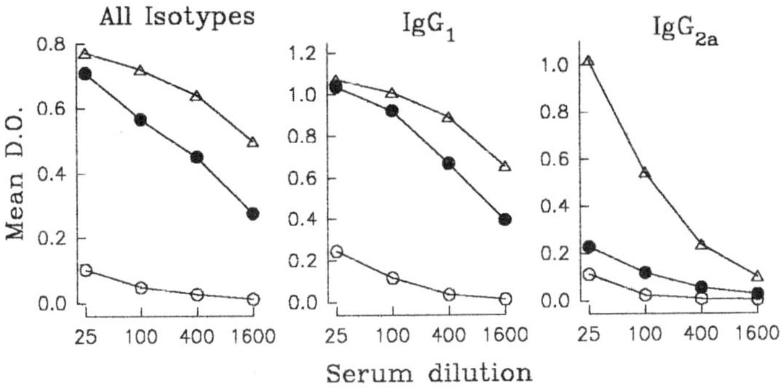

Figure 3: *Immunization by immunotargeting antigens to dendritic cells in vivo.* DBA/2 mice (3 per group) were injected with N418-biot., streptavidin, HGG-biot. (filled circle), with HGG-biot. in CFA (open triangle) or with HGG-biot. in saline (open circle). All mice were bled 25 days later and antigen-specific antibodies were measured in individual sera. Error bars are omitted for simplicity.

The data depicted in Figure 3 show that mice injected sequentially with N418-biotin, streptavidin, HGG biotin and mice injected with the antigen in CFA display a strong humoral response specific for HGG, whereas animals injected with antigen-biotin in saline respond poorly. Surprisingly, however, low levels of specific IgG2a antibodies were detected in the mice primed by immunotargeting, as compared to mice injected with antigen in CFA. Thus, although the delivery of antigen to dendritic cell surface structure provides an efficient way to induce an elevated serological response in the absence of adjuvant, the response appears qualitatively different from the response observed in animals injected with DC that have been pulsed extracorporeally (2). We would hypothesize that the injection of N418 antibodies prevents the maturation of DC *in vivo*, maturation that seems to occur during the overnight culture *in vitro.* Experiments are in progress to characterize the maturation states of dendritic cells isolated from untreated mice or from animals pretreated with N418 antibodies.

Unlike B cells and macrophages, dendritic cells are specialized in antigen presentation (reviewed in 9) and have no other known function. In particular, they have the unique property to activate resting T cells. The induction of primary cellular (1) and humoral (2, and this paper) responses *in vivo* using elements of the immune system avoids the toxic effects often associated with the use of adjuvants and offers new perspectives of immunization that, potentially, could be useful in man.

REFERENCES

1. K. Inaba, J.P. Metlay, M.T. Crowley, and R.M. Steinman, Dendritic cells pulsed with protein antigens in vitro can prime antigen-specific, MHC-restricted T cells in situ, *J.Exp.Med.* 172:631 (1990).

2. T. Sornasse, V. Flamand, G. De Becker, H. Bazin, F. Tielemans, K. Thielemans, J. Urbain, O. Leo, and M. Moser Antigen-pulsed dendritic cells can efficiently induce an antibody response in vivo, *J.Exp.Med.* 175:15 (1992).

3. N. Romani, S. Koide, M. Crowley, M. Witmer-Pack, A.M. Livingstone, C.G. Fathman, K. Inaba, and R.M. Steinman, Presentation of exogenous protein antigens by dendritic cells to T cell clones. Intact protein is presented best by immature, epidermal Langerhans cells, *J.Exp.Med.* 169: 1169 (1989).

4. E. Kampgen, N. Koch, F. Koch, P. Stoger, C. Heufler, G. Schuler, and N. Romani, Class II major histocompatibility complex molecules of murine dendritic cells: synthesis, sialylation of invariant chain, and antigen processing capacity are down-regulated upon culture, *Proc.Natl.Acad.Sci U.S.A.* 88:3014 (1991).

5. P.A. Reid, and C. Watts, Cycling of cell surface MHC glycoproteins through primaquine-sensitive intracellular compartments, *Nature* 346:655 (1990).

6. C.M. Snapper, C. Peschel, and W.E. Paul, IFN-gamma stimulates IgG2a secretion by murine B cells stimulated with bacterial lipopolysaccharide, *J.Immunol.* 140:2121 (1988).

7. G. Carayanniotis, D.L. Skea, M.A. Luscher, and B.H. Barber, Adjuvant-independent immunization by immunotargeting antigens to MHC and non-MHC determinants in vivo. *Mol.Immunol.* 28:261 (1991).

8. J.P. Metlay, M.D. Witmer-Pack, R. Agger, M.T. Crowley, D. Lawless, and R.M. Steinman, The distinct leukocyte integrins of mouse spleen dendritic cells as identified with new hamster monoclonal antibodies, *J.Exp.Med.* 171:1753 (1990).

9. R.M. Steinman, The dendritic cell system and its role in immunogenicity, *Annu. Rev. Immunol.* 9:271 (1991).

MIGRATION OF ALVEOLAR MACROPHAGES FROM ALVEOLAR SPACE TO PARACORTICAL T CELL AREA OF THE DRAINING LYMPH NODE

Theo Thepen[1], Eric Claassen[2], Karin Hoeben[1], John Brevé[1], and Georg Kraal[1]

[1] Department of Histology, Vrije Universiteit, Van de Boechorststraat 7, 1081 BT, Amsterdam, The Netherlands
[2] Department of Immunology and Medical Microbiology, TNO, MBL, P.O. BOX 45, 2280 AA Rijswijk, The Netherlands

SUMMARY

In this report we studied the translocation of fluorescent particulate antigens to the draining lymph node, and the migration of fluorescent labeled alveolar macrophages (AM) and peritoneal macrophages (PM) in mice. The results show that intratracheally (IT) instilled particulate antigens translocate to the paracortical T cell area of the draining lymph node. When labeled AM were injected IT, they were found to migrate from the alveolar space into the paracortical T cell area of the draining lymph node. An indentical localisation was found after IT injection of labeled PM. When either labeled AM, or PM were injected into the peritoneal cavity, a different migration pattern was observed. Via this route the labeled macrophages migrated to the subcapsular sinus and medulla of the draining lymph nodes. It is shown that the migrated cells are not dendritic cells (DC) present in the cellpreparations.

A possible role for the micro-environment of the injection site, and the significance of the specific migration pattern of AM is discussed.

INTRODUCTION

Alveolar macrophages (AM) are the main resident phagocytic cell in lung located in the alveoli at the air-tissue interface. Their anatomical location and phagocytic capacity make the AM one of the first cells to encounter antigens present in ambient air. It has been shown in experiments in vitro that AM are efficient non specific suppressors of T cell activation via secretion of lymphocytostatic mediators (1). In addition we have previously demonstrated that the AM excerts his suppressive action in vivo at the level of local T cells (2-4). These studies show the important role of AM in controlling the local immune response and maintaining immunological homeostasis in the lung.

One of the earliest established functions of the AM is its role as scavenger cell. After non specific phagocytosis of antigen by AM, the antigens are transported up the mucociliary escalator out of the lung and are eventually swallowed (5). Another possible route of clearance of antigen by AM is translocation to the draining lymph node (6-11).

Such transport of antigen to the draining lymph nodes may have important implications. Firstly, transport of potentially pathogenic micro-organisms can lead to systemic dissemination of these organisms and transport of toxic particles can lead to high local tissue doses. Secondly, transport of antigen from the respiratory epithelium to the draining lymph node is considered to be essential in initiating primary immune responses (12).

Dendritic Cells in Fundamental and Clinical Immunology,
Edited by Kamperdijk *et al.*, Plenum Press, New York, 1993

MATERIALS AND METHODS

Animals: BALB/c mice were purchased from Bomholtgård Ltd., Ry, Denmark. The mice were derived from a pathogen-free colony and were maintained in our facilities under clean conditions in isolated rooms with acidified drinking water. The mice were used in the experiments when 6-8 weeks old.

Liposomes: The liposomes were composed of phosphatidylcholine and cholesterol, molar ratio 6:1. They were prepared as described before (13), resulting in a liposome suspension with a constant lipid concentration of 2.5 mg per ml.

Broncho alveolar lavages: The animals were anaesthetized with 40 µl of a 4:3 mixture of Aescoket (Aesculaap N.V. Gent, Belgium) and Rompun (Bayer, Leverkussen, Germany) intramuscularly injected. The lungs were perfused with 10 ml sterile PBS containing 10 U/ml Heparin, and lavaged in situ by 10 consecutive washings with 1 ml sterile RPMI 1640 containing 0.35% lignocaine and 10% FCS at 37°C. after which the cells were washed 3 times. In control experiments broncho alveolar lavage cells were depleted of DC by incubation with cytotoxic concentrations of the monoclonal antibody 33D1 (14), followed by complement lysis.

Peritoneal lavages: The animals were anaesthetized as above, and the periteoneal cavity was washed with 5 ml sterile RPMI 1640 containing 0.35% lignocaine and 10% FCS at 4°C. after which the cells were washed 3 times. The peritoneal lavage cells were enriched for peritoneal macrophages (PM) by depletion of T, B, and dendritic cells, using rat antibodies against mouse T cells, B cells, or dendritic cells (30H12 (15); 187/1 (16); NLDC-145 (17) respectively, conjugated to magnetic Dynabeads. The thus prepared cellsuspension is hereafter referred to as PM.

Labeling of liposomes: PBS liposomes, prepared as described above, were labeled with the hydrophobic fluorescent dye DiI (1,1'-dioctadecyl-3,3,3',3'-tetramethyl-indocarbocyanine perchlorate; D282; molecular Probes, Eugene, USA) as described by Claassen (18) with minor alterations.

Labeling of cells: AM and PM were labeled with the fluorescent hydrophobic dye DiO (3,3'-dioctadecyloxacorbocyanine perchlorate; D-275; Molecular Probes, Eugene, USA), as described before (18) with minor alterations. DiO was dissolved in DMSO/ethanol (1:1) at a concentration of 3 mg/ml and stored at 4°C. Cells were concentrated at $1x10^6$ cells/ml in RPMI 1640 containing 10% FCS, and DiO, final concentration 18 µg/ml, was added and incubated with the cells for 90 min on a rocker platform at 4°C. The cells were spun through serum and washed twice with RPMI and once with PBS to remove free label. The cells were counted and concentrated at $1x10^6$/ ml in saline prior to injection. All solutions were filter sterilized before use.

Administration of liposome suspension or labeled cells: The mice were fixed in an upright position under total anaesthetization. Using a nylon tube, connected to a 1 ml syringe fixed in a micro-manipulator, 100 µl liposome suspension or labeled cells was administered through the glottis into the trachea.

In vivo phagocytosis of DiI labeled liposomes: Animals were given an IT dose of 100 µl DiI labeled liposome suspension, containing 25 µg lipid (1:10 diluted suspension). The lungs were lavaged 12 hours later, and the membrane of the DiI liposome containing AM was subsequently labeled with DiO as described above.

Tissue preparation: At different timepoints spleen, PTLN, and parathymic lymph nodes were removed and snap frozen by immersion in liquid nitrogen. The lungs were inflated in situ with 1 ml of a 1:1 mixture of Tissue Tek (O.C.T. 4583, Miles, USA) and PBS, removed and frozen in liquid nitrogen. Frozen tissues were cut in 8 µm sections on a freezing cryostat (-20°C) and air-dried.

RESULTS

Distribution of DiI labeled liposomes: The labeling with DiI resulted in brightly red fluorescent liposomes, and no free DiI was detected as background fluorescence.

At 2 and 4 hours after IT administration of liposome suspension, containing 25 μg lipid, labeled liposomes were observed in epithelial lining fluid and as individual liposomes inside the majority of the AM throughout the lung. After extensive lavage no label was observed other than in the remaining AM. At 12, 24, and 48 hours the label was observed more dispersed through the AM and incorporated in the cellmembrane.

Table 1. Number of cells (± SEM) retrieved by lavage at different timepoints after IT administration of 25 μg labeled PBS liposomes in 100 μl saline

hours	lavaged cells $x10^6$	% containing label
0	0.41 (± 0.01)	0.0 (± 0.0)
2	0.42 (± 0.03)	62.4 (± 5.5)
4	0.46 (± 0.13)	74.6 (± 7.3)
12	0.48 (± 0.08)	95.6 (± 4.3)
24	0.45 (± 0.06)	93.2 (± 8.6)
48	0.42 (± 0.02)	95.0 (± 6.3)

The number of labeled cells in lavage was quantified at the same timepoints (table 1) and confirmed that the majority of the AM had ingested the labeled liposomes. In PTLN no label could be observed 2 and 4 hours after IT administration of DiI labeled liposomes, but after 12 hours DiI liposomes were found in cells in subcapsular sinus (SCS) and cortical areas (Fig 1).

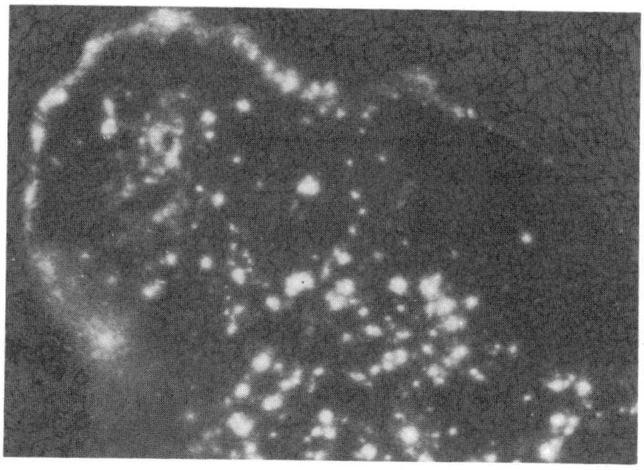

Figure 1. Localisation of DiI labeled fluorescent liposomes in SCS and paracortex of PTLN, 12 hours after IT injection.

At the later timepoints however, numerous cells containing DiI could be observed, mainly located in paracortex. An occasional labeled cell was also found in follicles. The labeled cells in the paracortex possessed long elongated processes. No label could be detected in the spleen.

Distribution of DiO labeled AM: Labeling with DiO resulted in uniformly green fluorescence of the plasma membrane of alveolar macrophages. The viability was greater than 95% in all experiments. The procedure here employed, resulted in less than 10% cell loss.

A dose of $5x10^5$ DiO labeled, viable AM were injected IT and 24 and 48 hours later PTLN, lung and spleen were removed for examination by fluorescence microscopy.

In lung sections, labeled cells could be observed as bright green fluorescent cells throughout the lung at both timepoints. Occassionally, a DiO labeled cell was present in lung interstitium.

In PTLN the distribution of labeled cells was similar to that observed at 24 and 48 hours after IT administration of labeled liposomes, though in the latter experiments higher numbers of cells were found. Again a predominant localization of fluorescent cells in paracortical area was found, with numerous cells showing a dendritic like appearance with elongated processes (Fig 2a and b). No labeled cells were found in the spleen.

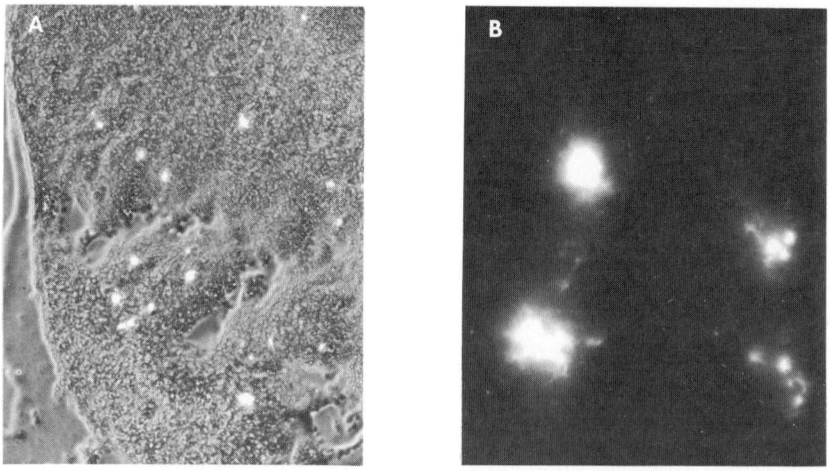

Figure 2. A: Localisation of DiO labeled AM in paracortex of PTLN after IT administration of a dose of $5x10^5$ AM (4x objective, FITC filtersetting)
B: High power fluorescence microphotograph, showing DC like processes of the migrated AM in paracortex (40x objective, FITC filtersetting)

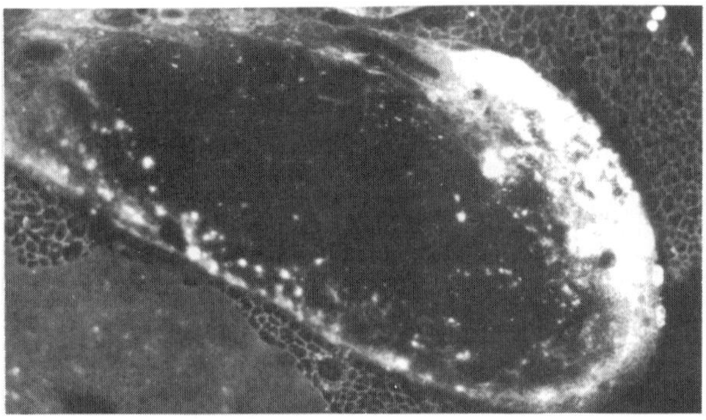

Figure 3. Fluorescence photomicrograph showing localisation of AM in medulla and SCS of PTLN after intraperitoneal injection of $2x10^6$ DiO labelled AM

Distribution of DiO labeled AM containing DiI labeled liposomes: To see whether the uptake of particles and the migration to the draining lymph node was performed by AM, we had AM phagocytose DiI labeled liposomes in vivo as described above. After lavage, the cells were subsequently labeled with DiO and a dose of $5x10^5$ cells, showing both red and green fluorescence, were injected IT in recipient mice. When the lungs of these animals were examined 48 hours later, double labeled cells were found dispersed throughout the lungs, but no single labeled cells were observed. The same observation was made in PTLN where, like in the previous experiments, numerous cells with long elongated processes were found in paracortex containing both labels.

Are the cells that migrate into the lymph nodes Dendritic Cells ? To exclude the possibility that the cells observed in the PTLN were DC present in the cellpreparations, lavaged cells were depleted of DC as described in materials and methods. The cells were subsequently labeled with DiO and transferred to naive recipients. An identical distribution in lung and PTLN, both in number, localization and appearance, was found compared to untreated lavage cells.

Is the specific localization an intrinsic property of AM only ? To determine whether the migration pattern of the AM from lung to paracortical T cell area of the PTLN was an intrinsic property of the AM, or caused by the microenvironment of the lung, we compared this migration pattern with that of IT injected DiO labeled PM and DiO labeled AM or PM injected into the peritoneal cavity. When PM were administered IT, a similar distribution in lung and PTLN was found compared to AM. They migrated mainly to the paracortical area of the PTLN, and in addition to this, they also demonstrated a dendritic like appearance, though not as explicit as seen with migrated AM. After intraperitoneal administration of AM or PM, a completely different migration pattern was observed, labeled cells were found almost exclusively in subcapsular sinus and medulla (Fig 3), which is in contrast to their migrational pattern after IT administration (cf. Figs 2a and b). No fluorescent cells were observed in the spleen.

DISCUSSION

The results of the here presented experiments can be summarised as follows:

- AM, together with antigen, migrate to the T cell area of the PTLN, and AM are responsible for the translocation of particles from the alveoli to the draining lymph node.
- Under normal, non inflammatory conditions particles are translocated to the PTLN incorporated in AM, rather than as free particles.
- The migrational pattern of AM is likely to be determined by micro environmental conditions in the lung.
- The migrating cells are not contaminating DC present in the cell preparations.

The general localisation of macrophages migrating into lymph nodes via afferent lymphatics, is in SCS and medullar region (19).

The abberant localisation of AM in paracortical T cell area is suggestive for a role in antigen presentation or regulation of the initiation of imune responses in the PTLN. Evidence from a number of laboratories indicate that AM generally function poorly as accessory cells for T cell activation, especially compared to macrophages derived from other sources like peritoneal cavity, and monocytes (reviewed in 1). In addition to this we previously demonstrated an increase in response to IT administered antigen after in vivo removal of AM from the lung (2). These observations make it unlikely that AM perform a role as antigen presenting, or accessory cell after migration into the paracortex of the PTLN. It can not be excluded however, that different micro-environmental conditions in the PTLN, compared to the lung can alter the properties of the migrated AM.

The principal antigen presenting cell in respiratory tissue is the DC, present in respiratory epithelium (20-24). The positioning of the DC in repiratory epithelium shows great resemblance with the situation of langerhans' cells in skin, and with a similar DC population in the gut wall. Both DC populations pick up antigen and subsequently migrate to the T cell area of the draining lymph node (25,26). Here they appear as interdigitating cells (IDC) involved in the initiation of immune responses. It is likely that DC in respiratory epithelium play a similar role; sampling antigen in respiratory epithelium and, after migration, initiation of the immune response in the paracortical T cell area of the PTLN. Considering the general suppressive effect of AM, it is therefore more likely that AM, migrating to exactly the same location in the PTLN, provide a down-regulating signal, next to the positive signal of the DC. This would be an additional explanation for the low frequency of pulmonary inflammation under normal circumstances.

REFERENCES

1. P. G. Holt, Down regulation of immune responses in the lower respiratory tract: the role of alveolar macrophages. *Clin. Exp.Immunol.* 63:261 (1986).
2. T. Thepen, N. Van Rooijen, and G. Kraal, Alveolar macrophage elimination in vivo is associated with an increase in pulmonary immune response in mice. *J. Exp. Med.* 170:199 (1989).
3. T. Thepen, C. McMenamin, J. Oliver, G. Kraal, and P .G. Holt, 1991. Regulation of immune responses to inhaled antigen by alveolar macrophages (AM): differential effects of in vivo elimination on the induction of tolerance versus immunity. *Eur. J. Immunol.* 21:2845-2850 (1992).
4. D. Strickland, T. Thepen, U.R. Kees, G. Kraal, and P. G. Holt, Regulation of T cell function by pulmonary alveolar macrophages. *Eur. J. Immunol.* submitted.

5. D.H. Bowden, The Alveolar Macrophage. *Environ. Health Perspect.* 55:327-341 (1984).
6. B.E. Lehnert, Y.E. Valdez, and C.C. Stewart, Translocation of particles to the tracheobronchial lymph nodes after lung deposition: Kinetics and partical-cell relationships. *Exp. Lung Res.* 10:245-266 (1986).
7. D. Corry. P. Kulkarni, and M.F. Lipscomb, The migration of bronchoalveolar macrophages into hilar lymph nodes. *Am. J. Pathol.* 115:321-328 (1984).
8. A.G. Harmsen. B.A. Muggenburg, M.B. Snipes, and D.E. Bice, The role of macrophages in particle translocation from lungs to lymph nodes. *Science* 230:1277-1280 (1985).
9. A.G. Harmsen, M.J. Mason, B.A. Muggenburg, N.A. Gillet, M.A. Jarpe, and D.E. Bice, Migration of neutrophils from lung to tracheobronchial lymph node. *J. Leuk. Biol.* 41:95-103 (1987).
10. J. Ferin,and M.L. Feldstein, Pulmonary clearance and hilar lymph node content in rats after particle exposure. *Environ. Res.* 16:342-352 (1978).
11. I.Y.R. Adamson, and D.H. Bowden, Dose response of the pulmonary macrophagic system to various particulates and its relationship to transepithelial passage of free particles. *Exp. Lung Res.* 2:165-175 (1981).
12. D.E. Bice, and G.E. Shopp, Antibody responses after lung immunization. *Exp. Lung Res.* 14: 133-155 (1988).
13. N. Van Rooijen, The liposome-mediated macrophage 'suicide' technique. *J. Immunol. Methods* 124:1 (1989).
14. M.C. Nussenzweig, R.M. Steinman, M.D. Witmer, and B. Gutchinov, A monoclonal antibody specific for dendritic cells. *Proc. Natl. Acad. Sci. USA.* 84:1659 (1982).
15. J.A. Ledbetter, and L.A. Herzenberg, Xenigneic monoclonal antibodies to mouse lymophoid organs differentiation antigens. *Immunol. Rev.* 47:63 (1979).
16. D.E. Yelton, C. Desaymard, and M.D. Scharff, Use of monoclonal anti-mouse immunoglobulin to detect mouse antibodies *Hybridoma.* 1:5 (1981).
17. G. Kraal, M. Breel, M. Janse, and G. Bruin, Langerhans' cells, veiled cells and interdigitating cells in the mouse recognized by a monoclonal antibody. *J. Exp. Med.* 163:981-997 (1986).
18. E. Claassen, Post formation fluorescent labeling of liposomal membranes: In Vivo detection, localisation and kinetics. *J. Imm. Meth.* 147:231-240 (1992).
19. H. Rosen, and S. Gordon, Adoptive transfer of fluorescence labeled cells shows that resident peritoneal macrophages are able to migrate into specialized lymphoid organs and inflammatory sites in the mouse. *Eur. J. Immunol.* 20:1251-1258 (1990).
20. P. G. Holt, M.A. Schon-Hegrad, and J. Oliver, MHC class II antigen-bearing dendritic cells in pulmonary tissues of the rat; regulation of antigen presentation activity by endogenous macrophage populations. *J. Exp. Med.* 167: 262 (1987).
21. C.L. Rochester, E.M. Goodell, J.K. Stoltenborg, and W.E. Bowers, Dendritic cells from rat lung are potent accessory cells. *Am. Rev. Respir. Dis.* 138: 121 (1988).
22. K. Sertl, T. Takemura, E. Tschachler, V.J. Ferrans, M.A. Kaliner, and E.M. Shevach, Dendritic cells with antigen-presenting capability reside in airway epithelium, lung parenchyma, and visceral pleura. *J. Exp. Med.* 163: 436 (1986).
23. P .G. Holt, M.A. Schon-Hegrad, M.J. Phillips, and C. McMenamin, Ia positive dendriric cells form a tightly meshed network within the human airway eopithelium. *Clin. Exp. Allergy* 19:597-601 (1989).
24. A.M. Pollard, and M.F. Lipscomb, Characterization of murine dendritic cells: Similarities to langerhans cells and thymic dendritic cells. *J. Exp. Med.* 172:159-167 (1990).
25. C.W. Pugh, C.G. MacPherson, and H.W. Steer, Characterisation of non-lymphoid cells derived from rat peripheral lymph. *J. Exp. Med.* 157:1758-1779 (1983).
26. G. Mayrhofer, P. G. Holt, and J.M. Papadimitriou, Functional characteristics of the veiled cells in afferent lymph from the rat intestine. *Immunology* 58:379-387 (1986).

COMPARISON OF LANGERHANS CELLS AND INTERDIGITATING RETICULUM CELLS

Mikihiro Shamoto, Satoru Hosokawa, Masanori Shinzato, and Chiyuki Kaneko

Division of Pathological Cytology, Fujita Health University School of Medicine, Toyoake, Aichi 470-11, Japan

INTRODUCTION

In our previous paper (1, 2) we reported that both OKT-6 and S-100 protein positive cells were observed in superficial and hilar lymph nodes. However, in mesenteric lymph nodes and spleens, only S-100 protein positive and OKT-6 negative cells were found. For immunostaining using OKT-6 and S-100 protein antibodies, mirror sections were used. Immunoelectron microscopy were also undertaken. We concluded that both positive cells were Langerhans cells (LCs), but only S-100 protein positive cells were interdigitating reticulum cells (IDCs). We then tried the double staining method in order to obtain more exact results, namely to prove whether LCs might be positive for both OKT-6 and S-100 protein.

MATERIALS AND METHODS

Lymph nodes and small pieces of the skin taken from ten necropsies, less than 8 hours after death, were used for the experiments. The lymph nodes used were cervical and inguinal (superficial) lymph nodes, and also hilar and mesenteric lymph nodes. These materials were embedded by the AMeX (acetone, methyl benzoate, and xylene) method (3). At first FITC conjugated OKT-6 monoclonal antibody (Ortho Diagnostic Systems Inc.) was used for immunofluorescence staining by the direct method. After taking photographs, sections were washed in glycine-HCl buffer, then immunostained with S-100 protein polyclonal antibody (kindly provided by Prof. H. Hidaka, Nagoya University) using the avidin-biotin-peroxidase complex (ABC: Vectastain ABC kit, Vector Lab. Inc.) method. Sections were incubated in 3,3'-diaminobenzidine (DAB), and then nuclear counterstained with Mayer's hematoxylin solution. Photographs of the same visual fields and the same cells as the photographs of OKT-6 positive cells with the immunofluorescence stain were taken again. Photographs of only S-100 protein positive cells were also taken. The photographs of each pair of OKT-6 and S-100 protein staining cells were compared. And in the sections of the lymph nodes, positive cells were counted.

RESULTS

In the skin both OKT-6 and S-100 protein positive cells were located mainly in suprabasilar portions of the epidermis and the majority of these cells had dendritic processes (Figs. 1a and 1b). These dendritic cells could easily be recognized as LCs. However we

Dendritic Cells in Fundamental and Clinical Immunology,
Edited by Kamperdijk *et al.*, Plenum Press, New York, 1993

Table 1. Comparison between both OKT-6 and S-100 protein positive cells and OKT-6 negative but S-100 protein positive cells in the superficial lymph nodes.

Case No.	Age	Sex	Lymph nodes S-100(+)OKT-6(+):S-100(+)OKT-6(-) Cervical	Inguinal
1	55	M	*1 : 4	2 : 5
2	68	M	1 : 5	2 : 7
3	70	M	0.1 : 3	1 : 4
4	68	F	2 : 6	4 : 8
5	63	M	2 : 7	3 : 7

*Average number of positive cells/ 4-6 visual field, x 400

could not find only OKT-6 positive cells in the epidermis. Both OKT-6 and S-100 protein positive dendritic cells were also observed in the dermis but rarely (Figs.2a and 2b). In the epidermis and dermis a few cells were only positive for S-100 protein and negative for OKT-6.

In the superficial and hilar lymph nodes both OKT-6 and S-100 protein positive cells, and cells which were only positive for S-100 protein were found (Figs.3a and 3b, Figs.4a and 4b). As shown in the Table 1, the numbers of only S-100 protein positive cells were greater than those of both OKT-6 and S-100 protein positive cells. We have already reported that in the mesenteric lymph nodes and the spleens, only S-100 protein positive, and OKT-6 negative cells were found.

DISCUSSION

In this experiment we attempted double staining for OKT-6 and S-100 protein. ABC procedures labeled with peroxidase and/or alkaline phosphatase, and immunofluorescence procedures were carried out by a combination of two of these procedures. Two antibodies were used at the same time, or either OKT-6 or S-100 protein was used prior to one another. However, we could not find double stained LCs in the epidermis or in the superficial lymph nodes. It became apparent that the first antibody was preventing the labeling of the second antibody. So first the sections were stained for OKT-6 using the direct immunofluorescence technique and after taking photographs of the cells and removing OKT-6 reaction products, secondarily, they were stained for S-100 protein by the ABC method

We have immunoelectron microscopically described that LCs with Birbeck granules exist in superficial lymph nodes, but not in mesenteric lymph nodes. Compared with the mirror section method, this experiment has enabled exactly the same cells to be stained with both OKT-6 and S-100 protein. It has become evident from these results that both OKT-6 and S-100 protein positive cells were LCs, and only S-100 protein positive, and OKT-6 negative cells were IDCs. Both OKT-6 and S-100 protein positive cells which rarely existed in the dermis were considered to be LCs. LCs in the dermis were described in 1968 (4). Zelickson et al.(5) reported on indeterminate cells in the skin, and no more needs to be said about IDCs. It has been immunohistochemically ascertained that only S-100 protein positive cells in the skin were indeterminate cells and/or IDCs.

Evidence has been obtained that LCs and IDCs have slightly different distribution in vivo. LCs and IDCs which have some different immunological characteristics may have some different functions. In another paper in this symposium (6) we have reported that LCs have appeared in the tracheal squamous metaplasia of vitamin A deficient rats. We speculate that LCs which belong to the same lineage as IDCs (7, 8) need to make contact with squamous epithelial cells for the formation of Birbeck granules.

It is suggested that LCs in the superficial and hilar lymph nodes are derived from the epidermis (9, 10). LCs do not exist in the mesenteric lymph nodes and in the spleen. The reason for this may be that these tissues do not be drained into squamous epithelia. IDCs which are widely scattered in vivo may migrate from the bone marrow, and some of them may infiltrate into the epidermis and form Birbeck granules.

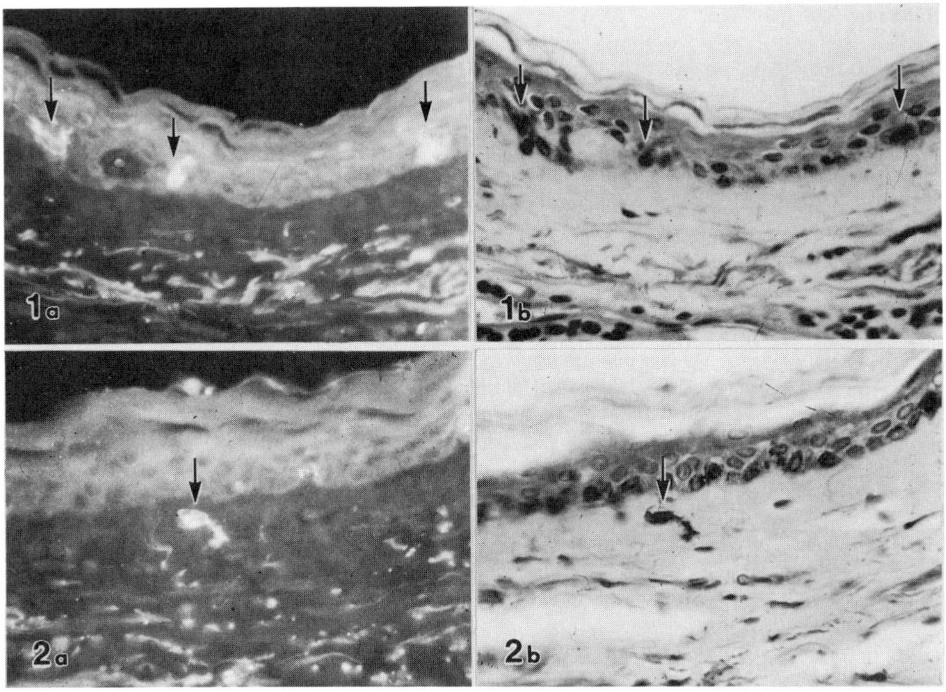

Figs. 1a and 1b: A few LCs in the epidermis (arrows). **a** is stained for OKT-6, and **b** for S-100 protein.

Figs. 2a and 2b: An LC in the dermis (arrow). **a** is stained for OKT-6, **b** for S-100 protein.

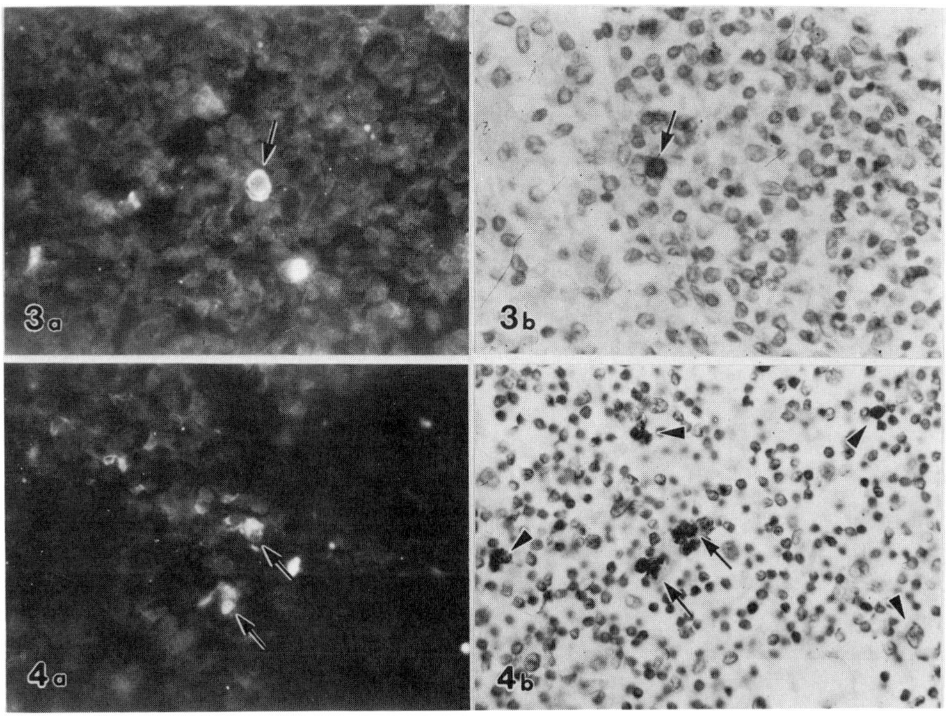

Fig. 3a: A positive cell with OKT-6 (arrow) is LC.

Fig. 3b: An arrow which indicates the same cell as Fig. 3a shows LC.

Figs. 4a and 4b: A different field photograph of Fig 3. Arrows show LCs and arrow heads show IDCs.

SUMMARY

Double immunostaining for OKT-6 and S-100 protein antibodies was carried out. However, it was impossible to stain using two antibodies at the same time, or using one antibody prior to another. We first stained using FITC labeled OKT-6 antibody, and secondary stained for S-100 protein by the ABC method, after taking photographs and removing OKT-6 reaction products. The exact same cells could be stained by both OKT-6 and S-100 protein antibodies. It has been elucidated that both OKT-6 and S-100 protein positive cells are LCs, and only S-100 protein positive, but OKT-6 negative cells are IDCs. LCs only exist in the superficial and hilar lymph nodes. However IDCs exist not only in these lymph nodes, but also in the mesenteric lymph nodes and spleens.

REFERENCES

1. M. Shamoto, M. Shinzato, S. Hosokawa et al., Langerhans cells in the lymph nodes: mirror section and immunoelectron microscopic studies., *Virchows Arch (B)* 561:337 (1992).
2. M. Shamoto, M. Shinzato, The characterization of interdigitating reticulum cells: a comparison between Langerhans cells., *in*: Dendritic Cells in Lymphoid Tissues, Intern Congr Series 926, Excerpta Medica, Y. Imai, J.G. Tew, E.C.M. Hoefsmit, Eds., Amsterdam, pp. 65 (1991).
3. Y. Sato, K. Murai, S. Watanabe et al., The AMeX method. A simplified technique of tissue processing and paraffin embedding with improved preservation of antigens for immunostaining., *Am J Pathol*, 125:431 (1986).
4. K. Hashimoto, W.M. Tarnowski, Some new aspects of the Langerhans cell., *Arch Dermatol*, 97:450 (1968).
5. A.S. Zelickson, J.H. Mottaz, Epidermal dendritic cells, a quantitative study., *Arch Dermatol*, 98:652 (1968).
6. S. Hosokawa, M. Shinzato, C. Kaneko, M. Shamoto, Studies on Langerhans cells in the tracheal squamous metaplasia of vitamin A deficient rats., 2nd Intern Symp on Dendritic Cells in Fundamental and Clinical Immunology, Amsterdam (1992).
7. R.M. Steinman, M.C. Nussenzweig, Dendritic cells: features and functions., *Immunol Rev*, 53:127 (1980).
8. G. Stingl, .Tamaki, S.I. Katz, Origin and function of epidermal Langerhans cells., *Immunol Rev*, 53:149 (1980).
9. B.M. Balfour, H.A. Drexhage, E.W.A. Kamperdijk, E.C.M. Hoefsmit, Antigen presenting cells including Langerhans cells, veiled cells and interdigitating cells., *Ciba Found Symp*, 84:281 (1981).
10. S.M. Breathnach,The Langerhans cells., *Br J Dermatol*, 119: 463 (1988).

MIGRATION OF DENDRITIC CELLS DURING CONTACT SENSITIZATION

S.Hill, +S.Griffiths, *I.Kimber and S.C.Knight

Antigen Presentation Research Group & +Section of Electron Microscopy, Clinical Research Centre, Harrow, Middlesex. U.K.
*Cell & Molecular Biology Section, C.T.L., I.C.I., Alderley Edge, Macclesfield, Cheshire U.K.

INTRODUCTION

Chemicals such as oxazolone, picryl chloride or fluorescein isothiocyanate (FITC) applied to the skin of naive animals can result in the initiation of contact sensitivity. Langerhans' cells (LC) in the skin are thought to form a reticulo-epithelial trap in the epidermis and selectively acquire topically applied antigens[1] and these cells are of major importance in the induction phase of contact sensitivity.

A proportion of dendritic cells (DC) in draining lymph nodes are thought to be derived from skin LC. Guinea-pigs skin painted with ferritin have been shown to have ferritin-bearing cells in draining lymph nodes within 4 hours of skin painting[2]. Further evidence came from studies using an antibody (NLDC-145) which reacted with LC, veiled cells and DC, indicating an origin from the skin[3]. Cultured LC mature into cells *in vitro* that resemble lymphoid DC in both morphology and function[4]. After sensitization with FITC, cells bearing antigen and containing Birbeck granules which are markers for skin LC have been identified in draining lymph nodes[5] indicating origin from the skin.

In a previous study[6], we have shown that following skin painting with FITC, there is a systemic signal that induces the migration of DC to draining, contralateral and distal lymph nodes irrespective of whether these cells carry antigen. This signal is independent of mature T cells as it also occurs in nude mice.

Here, we have examined the origin of cells in draining and contralateral lymph nodes after skin painting with FITC. This was carried out by looking for cells containing Birbeck granules using transmission electron microscopy.

Dendritic Cells in Fundamental and Clinical Immunology,
Edited by Kamperdijk *et al.*, Plenum Press, New York, 1993

MATERIALS AND METHODS

Animals

Female CBA mice were obtained from the specific pathogen-free unit at the Clinical Research Centre.

Sensitization *in vivo*

Fluorescein isothiocyanate (FITC, isomer 1; Sigma, Poole, Dorset U.K.) was dissolved in a 50/50 v/v acetone:dibutylphthalate (Sigma) mixture. Mice were skin painted on the shaved flank with 25μl 0.8% FITC.

Cell suspensions

Two hours after sensitization with FITC, single cell suspensions were prepared from pooled popliteal, inguinal, brachial and axillary lymph nodes from the draining and contralateral lymph nodes (DLN and CLN respectively) of control and sensitized mice.

Lymph nodes were pressed through wire mesh and washed in RPMI-1640 medium (Dutch modification; Flow Labs, Irvine, Ayrshire, U.K.) supplemented with 100IU/ml penicillin, 100 μg/ml streptomycin, 1 x 10^{-5} 2-mercaptoethanol and 10% heat-inactivated foetal calf serum. A cell suspension (5-8ml) at 5 x 10^{6}/ml was layered onto 2ml metrizamide (Nygaard, Oslo, Norway. Analytical grade: 14.5g added to 100ml medium) and centrifuged for 10mins at 600g. Cells at the interface were collected, washed once, resuspended in medium and counted.

Electron microscopy

DC pellets were fixed using a mixture of glutaraldehyde and osmium tetroxide in cacodylate buffer[7] at 4°C for 30 mins. The pellets were block stained in 0.25% uranyl acetate solution for 30 mins and then dehydrated in graded acetone solutions up to 100% acetone. Infiltration and polymerisation were carried out using Spurrs' resin[8] and blocks were sectioned using a diatome diamond knife on a Reichert Ultracut E microtome.

Silver-gold sections were mounted on 200 mesh copper grids and stained in 4% uranyl acetate and Reynolds lead citrate[9].

The stain-mounted sections were given a light carbon coating to maintain stability in the electron beam and specimens were viewed in a Philips EM300 microscope at 60kv.

RESULTS

The origin of some DC in draining and contralateral lymph nodes after skin painting with FITC was examined using transmission electron microscopy. There was an increase in the number of DC in DLN and CLN and a proportion of these contained Birbeck granules indicating an origin from the skin.

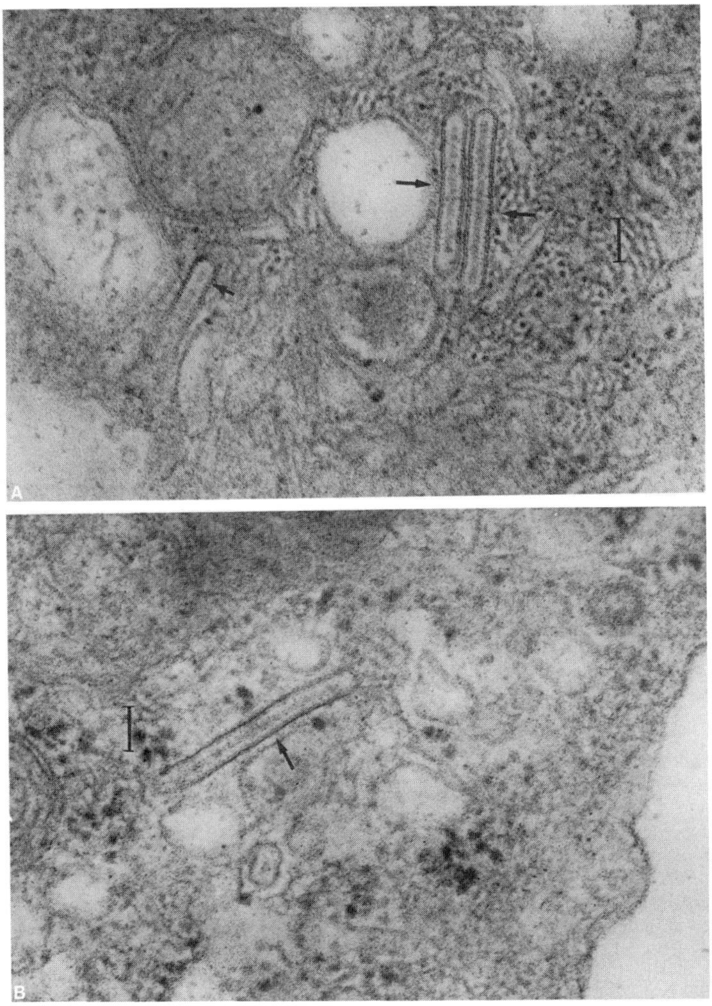

Figure 1. Electron micrographs of dendritic cells isolated from draining (A) and contralateral (B) lymph nodes of mice 2 hours after skin painting with 25μl 0.8% FITC on the flank. Birbeck granules (arrowed) were found in cells from both draining and contralateral lymph nodes (space bar = 0.08μm).

Some DC isolated from DLN, CLN and control nodes (Fig. 1A & 1B) contained Birbeck granules indicating an origin from the skin. In one experiment, Birbeck granules in cells isolated from DLN were less well defined compared with Birbeck granules in CLN or control nodes (data not shown).

In 2 out of 3 experiments there were fewer cells containing Birbeck granules in DLN of sensitized mice compared with CLN or control lymph nodes (Fig.2).

Further experiments to confirm these observations were hampered by viral infection of the mice. This caused an increase in the number of DC in control lymph nodes and also the proportion of cells containing Birbeck granules in both sensitized and control mice (data not shown).

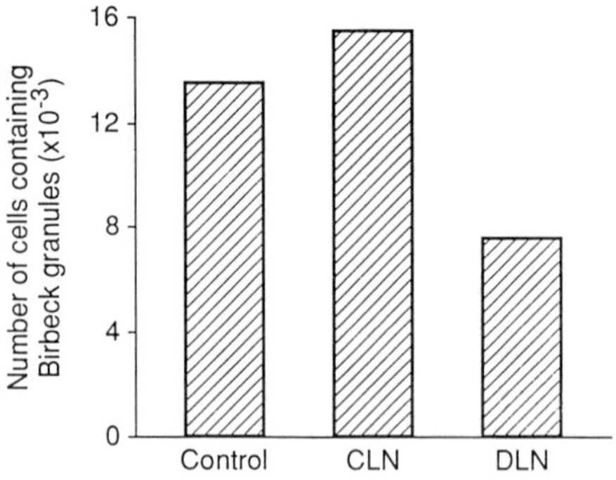

Figure 2. Number of cells containing Birbeck granules isolated from either draining, contralateral or control nodes. Mice were sensitized on the flank with 25μl 0.8% FITC and lymph nodes removed 2 hours later. Birbeck granules were identified using electron microscopy.

DISCUSSION

Previously[6], we have shown that following skin painting with FITC there appears to be a systemic signal that causes the migration of DC to both draining and contralateral lymph nodes and here we show that some of these cells in both DLN and CLN contain Birbeck granules indicating an origin from the skin.

In this study we have examined the origin of some of the DC in lymph nodes following sensitization. There are several pieces of evidence that support the idea that a proportion of DC in DLN are derived from the skin. Hunziker & Winkelmann[10] reported that sensitization caused a loss of LC from the skin. Macatonia *et al*[5] reported that some cells in DLN contained Birbeck granules. Kripke *et al*[11] have shown that if sensitization is carried out through an allogeneic skin graft then antigen-bearing DC of donor MHC are found in the DLN.

The preliminary data presented here is compatible with the published observations[5,10,11] in that cells containing Birbeck granules were found in the DLN and we have also shown that cells containing Birbeck granules are present in CLN irrespective of whether these cells carried antigen.

The appearance of cells containing Birbeck granules in CLN in the absence of detectable antigen is probably due to the systemic signal. Two studies have investigated the nature of the systemic signal[12,13]. These studies have shown that intradermal injections of tumour necrosis factor (TNFα) causes a depletion of LC from the skin[12] and an increase in the number of DC in DLN[13].

A further study[14] has shown that TNFα can act systemically, inhibiting the induction of contact hypersensitivity possibly by causing the loss of LC from the skin.

The data presented here indicates that the systemic signal may cause the loss of LC from the contralateral flank and this is consistent with an increase in the number of DC in CLN. This is compatible with the observations of Yoshikawa & Streilein[14] that show that TNFα can act systemically.

It is known that LC mature into lymphoid DC after culture *in vitro*[4]. The observation that fewer cells containing Birbeck granules are found in DLN may be due to the rapid transformation of LC to lymphoid DC. An alternative to this is that DC may be damaged during sensitization[15] and this may cause the loss of Birbeck granules from these cells.

The preliminary data presented here indicates that skin painting with contact sensitizers may cause the loss of LC from sites distal to the site of sensitization and this may be of importance in the development of antigen-induced non-responsiveness[16].

REFERENCES

1. W.B. Shelley and L. Juhlin. A reticulo-epithelial system: cutaneous trap for antigens.*Trans.Assoc.Am.Physicians*.89: 245 (1976).
2. I. Silberberg-Sinakin, G.J. Thorbecke, R.L. Baer, S.A. Rosenthal and V. Berezowsky. Antigen-bearing Langerhans cells in skin, dermal lymphatics and lymph nodes. *Cell.Immunol.*.25:137 (1976).
3. G. Kraal, M. Breel, M. Janse and J. Bruin. Langerhans cells, veiled cells and interdigitating cells in the mouse recognized by a monoclonal antibody. *J.Exp.Med.*.163:981 (1986).
4. G. Schuler and R.M. Steinman. Murine epidermal Langerhans cells mature into potent immunostimulatory dendritic cells in vitro. *J.Exp.Med.*.161:526 (1985).
5. S.E. Macatonia, S.C. Knight, A.J. Edwards, S. Griffiths and P.R.Fryer. Localization of antigen on lymph node dendritic cells after exposure to the contact sensitizer fluorescein isothiocyanate. *J.Exp.Med.*.166:1654 (1987).
6. S. Hill, I. Kimber, A.J. Edwards and S.C. Knight. System migration of dendritic cells during contact sensitization. *Immunology*.71:277 (1990).
7. J.G. Hirsch and M.E. Fedorko. Ultrastructure of human leukocytes after simultaneous fixation with glutaraldehyde and osmium tetroxide and "postfixation" in uranyl acetate. *J.Cell.Biol.* 38:615 (1968).
8. A.R. Spurr. A low viscosity epoxy resin embedding medium for electron microscopy. *J.Ultrastruct.Res.*.26:31 (1969).
9. E.S. Reynolds. The use of lead citrate at high pH as an electron-opaque stain in electron microscopy. *J.Cell Biol.*17:208 (1963).
10. N. Hunziker and R.K. Winkelmann. Langerhans cells in contact dermatitis of the guinea-pig. *Arch.Dermatol.*.114:1309 (1978).
11. M. Kripke, C.G. Munn, A. Jeevan, J-M. Tang and C. Bucana. Evidence that cutaneous antigen-presenting cells migrate to regional lymph nodes during contact sensitization. *J.Immunol.*.145:2833 (1990).
12. M. Vermeer and J.W. Streilein. Ultraviolet B light induced alterations in epidermal Langerhans cells are mediated in part by tumor necrosis factor alpha. *Photodermatol.Photoimmunol.Photomed.* 7:258 (1990).

13. M. Cumberbatch and I. Kimber. Dermal tumour necrosis factor alpha induces dendritic cell migration to draining lymph nodes and possibly provides one stimulus for Langerhans cell migration. *Immunology*.75:257 (1992).
14. T. Yoshikawa and J.W. Streilein. Tumour necrosis factor alpha and UVB light have similar effects on contact hypersensitivity in mice. *Reg.Immunol.*.3:139 (1990).
15. J.A.A. Hunter. Langerhans cell damage does occur in contact allergic reactions. *Am.J.Dermatopathol.*.8:227 (1986).
16. S. Hill, A. Stackpoole, I. Kimber and S.C. Knight. Function of dendritic cells and changes in T cell proliferation in antigen induced non-responsiveness. *Cell.Immunol.*.139:342 (1992).

PERITONEAL CELL LABELLING: A STUDY ON THE MIGRATION OF MACROPHAGES AND DENDRITIC CELLS TOWARDS THE GUT

Marsetyawan Soesatyo, Theo Thepen, Muhammad Ghufron, Jeike Biewenga, Taede Sminia

Department of Cell Biology
Section of Histology
Medical Faculty, Vrije Universiteit
1081 BT, Amsterdam

INTRODUCTION

Peritoneal macrophages (PM) comprise about 70% of the total peritoneal cell population in the rat, and are mostly resident type cells (1); in addition about 1% dendritic cells (DC) are present in the peritoneal cavity. In microbial infections of the peritoneum, PM efficiently clear bacteria by means of phagocytosis and killing *in situ*, and subsequently remove them into lymphatic channels (2). There are at least two different routes for removal of particulate materials by PM from the peritoneal cavity; firstly, via transdiaphragmatic lymphatic channels that drain into larger intrathoratic lymphatics to the parathymic lymph node (PTLN) before entering into the blood stream (3), and secondly, across the diaphragma directly to the lung interstitium (4).

It has been documented that intraperitoneal (ip) immunization can induce a mucosal as well as a systemic immune response (5). This can be related to PM and DC which are involved as antigen presenting cells in the induction of immune reactions against intraperitoneally-administered antigens (6). Moreover, these reactions are thought to be associated with the sites to which these cells migrate. Besides the above-mentioned migration routes, migration to other sites, including the gut mucosa has not been reported. The present experiments were therefore, designed to follow migration of antigen presenting cells by a labelling technique and especially to study whether these cells migrate to the gut.

MATERIALS AND METHODS

Animals. Adult female Wistar rats purchased from Harlan Dawley-CPB (Zeist, The Netherlands) were used throughout this study. They were kept under routine laboratory conditions.

Peritoneal lavages. Peritoneal cells were collected by lavage with 10 ml sterile cold (4°C) RPMI 1640 (NPBI, The Netherlands) instilled in the peritoneal cavity. After gently shaking the abdomen for about 5 min, the abdominal cavity was opened and the fluid was retrieved using a Pasteur pipette. The cells were washed twice with cold sterile RPMI 1640 and spun down at 1500 rpm, 4°C for 7 min. The final concentration used for cell labelling was 1 x 10^7 cells/ml.

Cell labelling. Peritoneal cells were labelled with a fluorescent cell linker compound, the PKH26 (Zynaxis Cell Science, Inc., Hamburg, Germany). The viable cells (as determined by trypan blue exclusion) concentrated at 1 x 10^7 cells/ml in sterile cold RPMI 1640, were rapidly mixed with PKH26 at a final concentration of 2 x 10^6 molar. During incubation for 5 min at 25°C, the tube was periodically inverted to assure mixing. The staining was stopped by adding an equal volume of sterile fetal calf serum and incubation at 25°C for 1 min. This mixture was then diluted with an equal volume of RPMI 1640 and centrifuged at 1240 rpm, 4°C for 7 min. The cell pellet was washed twice with 10 ml RPMI 1640, centrifuged again, and concentrated at 1 x 10^6 cells/ml.

Administration of labelled cells and tissue preparation. 5 x 10^5 labelled cells (1 x 10^6 cells/ml) were injected into the peritoneal cavity. At different timepoints, *ie.* 24, 48, 72 and 120 h after intraperitoneal injection, the animals were sacrified. Paratracheal lymph node (PTrLN) and parathymic lymph node (PTLN), a piece of small and large intestine, Peyer's patches (PP), proximal colonic lymphoid tissue (PCLT), mesenteric lymph node (MLN) and spleen were sampled and snap frozen in liquid nitrogen.

Fluorescence microscopy. 4 μm cryostat sections were prepared and air-dried at room temperature for at least 1 h prior to examination. The sections were evaluated with a fluorescence microscope (Axioskop, Zeiss, Germany) equipped with a TRITC filter. The labelled cells displayed a bright red fluorescent staining.

RESULTS

The distribution of labelled cells in various tissues at different time intervals is summarized in Table 1.

Table 1. Distribution of labelled cells in various tissues

	Time intervals post-labelling (h)			
Tissue	24	48	72	120
Parathymic lymph node	+	+	+	+
Paratracheal lymph node	+	+	+	+
Intestinal villi	-	-	-	+/-
Peyer's patch	+	+	+	+/-
Proximal colonic lymphoid tissue	-	+/-	+/-	+/-
Mesenteric lymph node	-	-	-	+/-
Spleen	-	-	-	+/-

+: present; -: absent; +/-: occasionally present

24 h post-labelling. Labelled cells were found especially in the PTLN. These brightly red fluorescent cells were round or oval, located in the subcapsular sinus and medullary areas (Fig. 1A). The labelled cells were also seen in the subcapsular sinus of PTrLN. In PP, less brightly fluorescent cells with a granular appearance were detected in the follicle and the dome area (Fig. 1B). No labelled cells were found in the spleen, MLN, intestinal lamina propria or PCLT.

48 h post-labelling

Numerous fluorescent cells were present in the medulla of PTLN and PTrLN. Again hardly any labelled cell was found in other tissues studied, except in the PP where they were located in the dome area and follicle (Fig. 2).

72 h post-labelling. Brightly fluorescent cells were clearly seen in the medulla of PTLN and PTrLN (Fig. 3A). These cells were large, oval shaped and probably represent macrophages. In addition, small labelled cells with dendritic processes were occasionally detected. In PP dome areas and B cell follicle, numerous small labelled cells with cytoplasmic granules and few labelled cells were seen (Fig. 3B). Compared to the fluorescent cells present in the PTLN and PTrLN, these cells were less brightly stained.

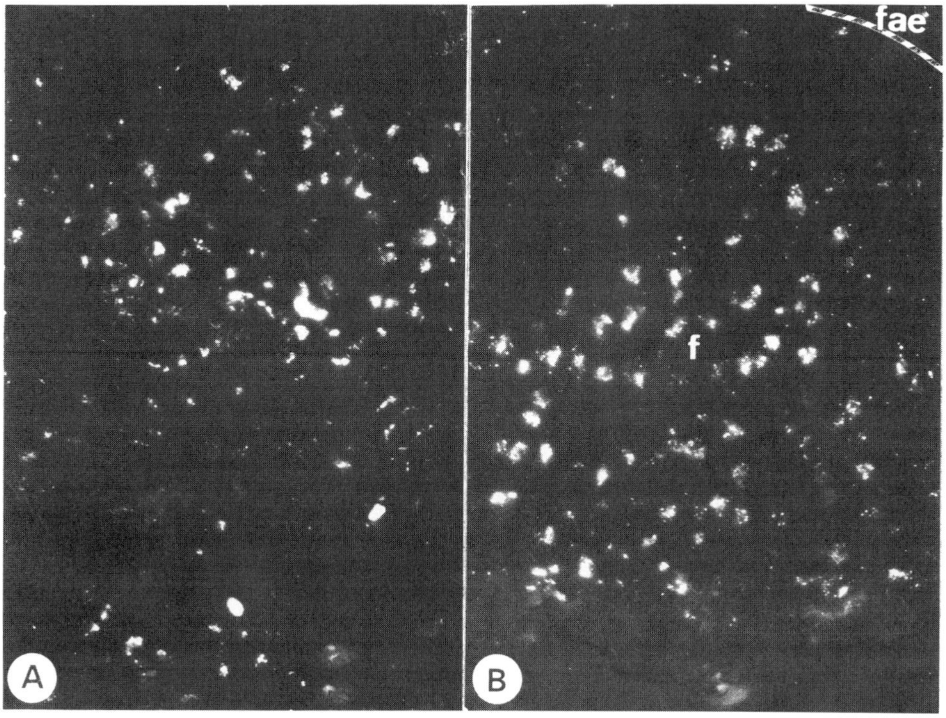

Fig.1. *Fluorescence photomicrographs of tissues 24 h after PKH26-labelling. Fluorescence labelled cells are shown in the medullary area of PTLN (A), and PP follicle (B). f, follicle; fae, follicle-associated epithelium.* A,B: 20x *objective*

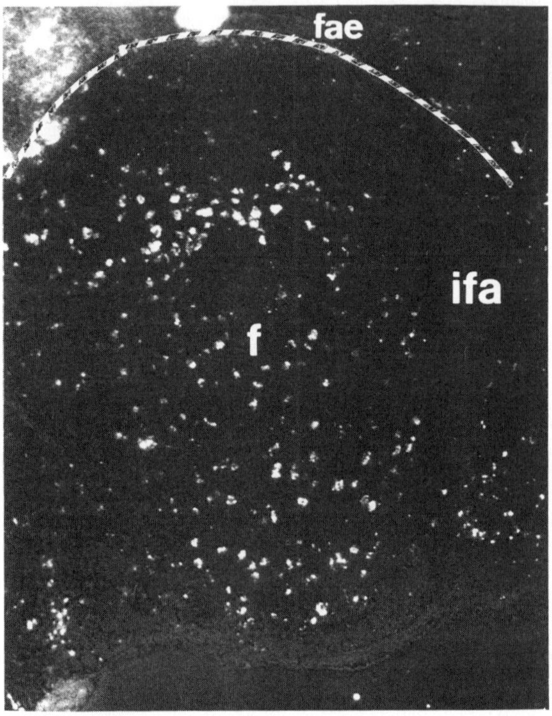

Fig.2. *Numerous labelled cells were seen in the follicle and the dome area of PP 48 h after labelling. f, follicle; ifa, interfollicular area; fae, follicle-associated epithelium.* 10x *objective.*

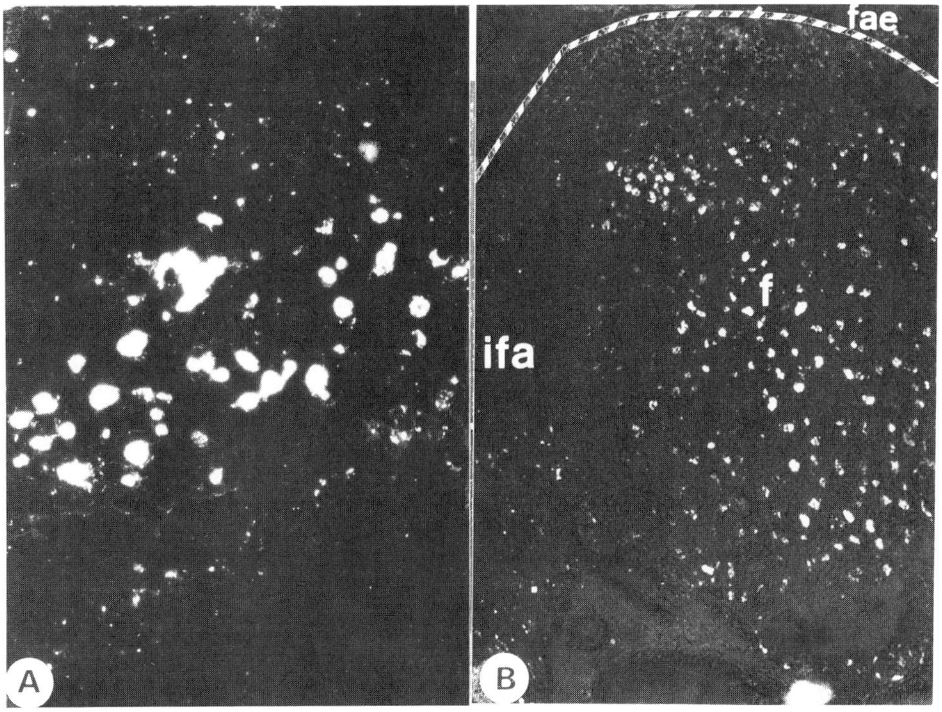

Fig.3. *Fluorescence labelled cells were found in the medulla of PTLN (A) and in PP (B) 72 h after labelling. f, follicle; ifa, interfollicular area; fae, follicle-associated epithelium.* **A,B**: 20x objective

120 h post-labelling. Some fluorescence labelled cells were still detected in the medulla of the PTLN, though their number had decreased. In the gut, labelled cells were occasionally found in PP, PCLT and in the lamina propria of the small intestine. In MLN and spleen, some brightly stained fluorescent cells were occasionally seen.

DISCUSSION

The finding that labelled peritoneal cells, in particular PM migrate to the PTLN was expected and is in agreement with previous observations (3,4). Also bacteria and antigen-laden PM migrate to the PTLN (2,7), but the migration seems to be independent of the presence of antigen. It has been reported that the migration of PM to lymphoid organs is an active process (8). The migration of intraperitoneally (ip)-instilled PM to the subcapsular sinus and medulla of PTLN is remarkable, especially when compared to PM administered intratracheally (it), from where they migrate within 48 h to the T cell area (9). The factors that influence the migration patterns are not known, but again the present findings exemplify the important role of the administration route for the outcome of immunization.

Remarkably, besides their presence in the PTLN, fluorescence labelled cells were detected in PP, though they were less brightly stained. Judged from the morphology of the cells and their granular appearance, many of them possibly are granulocytes (Fig. 1B).

In addition, some labelled cells found in PP might represent macrophages and dendritic cells (DC) based on their morphology. While lymphocytes enter PP through high endothelial venules (HEV), macrophages and DC probably reach this lymphoid tissue through another route. This specific migration may be governed by factor (s), such as cytokines that attract macrophages and dendritic cells. The data presented so far are not sufficient to prove migration of peritoneal cells to the PP.
The present results need further investigation, as for example by following the migration of *in vitro* labelled PM cells through ip injection, and simultaneous visualization of fluorescence (PKH26)-labelled cells as well as cells stained with specific monoclonal antibodies against macrophages and dendritic cells. If however such migration of antigen presenting cells can be demonstrated, this would prime for an IgA immune response.

SUMMARY

In this study the migration of peritoneal cells was investigated by a fluorescence labelling technique. We found that peritoneal cells migrate to the subcapsular sinus and medulla of the parathymic lymph node (PTLN) and paratracheal lymph node (PTrLN). It was also observed that fluorescence labelled cells possibly granulocytes, macrophages and dendritic cells were found in the B cell follicles of Peyer patches and the dome area after intraperitoneal (ip) labelling. The implication of the migration of antigen presenting cells to the gut on the mucosal immune response is discussed.

REFERENCES

1. H.J. Bos, F. Meyer, J.C. de Veld and R.H.J. Beelen. 1989. Peritoneal dialysis fluid induces change of mononuclear phagocyte proportions. *Kidney International* **36**:20.
2. D.L. Dunn, R.A. Barke, N.B. Knight, E.W. Humphrey and R.L. Simmons. 1985. Role of resident macrophages, peripheral neutrophils and translymphatic absorption in bacterial clearance from the peritoneal cavity. *Infect Immun* **49**:257.
3. N.L. Tilney. 1971. Patterns of lymphatic drainage in the adult laboratory rat. *J Anat* **109**:369.
4. M.L.M. Pitt and A.O. Anderson. 1988. Direct transdiaphragmatic traffic of peritoneal macrophages to the lung. *Adv Exp Med Biol* **237**:627.
5. N.F. Pierce and F.T. Koster. 1980. Priming and suppression of the intestinal immune response to cholera toxoid/toxin by parenteral toxoid in rats. *J immunol* **124**:307.
6. M. Soesatyo, J. Biewenga, N. van Rooijen, N. Kors and T. Sminia. 1991. The in situ immune response of the rat after intraperitoneal depletion of macrophages by liposome-encapsulated dichloromethyline diphosphonate. *Res. Immunol.* **142**:142:533.
7. D.L. Dunn, R.A. Barke, D.C. Ewald and R.L. Simmons. 1985. Effects of *Escherichia coli* and *Bacteroides fragilis* on peritoneal host defences. *Infect Immun* **48**:287.
8. H. Rosen and S. Gordon. 1990. Adoptive transfer of fluorescence labelled cells shows that resident peritoneal macrophages are able to migrate into specialized lymphoid organs and inflammatory sites in the mouse. *Eur J Immunol* **20**:1251.
9. T.Thepen. 1992. *Regulation of pulmonary immune responses by alveolar macrophages.* PhD. thesis, Vrije Universiteit, Amsterdam.

DENDRITIC CELLS "IN VIVO": MIGRATION AND ANTIGEN HANDLING

G. Gordon MacPherson and Liming Liu

Sir William Dunn School of Pathology
South Parks Rd.
Oxford OX1 3RE
England

INTRODUCTION

Readers of this volume will be aware that there exist two distinct, and very probably unrelated kinds of dendritic cells (DC) which are involved in immune responses. The follicular DC is present in B cell areas, is probably not bone-marrow derived, is long lived, retains immune complexes on its surface for presentation to B cells and will not be considered further here. The other, T-associated DC, which is the focus of this review, is bone-marrow derived, usually short-lived and is a migratory cell involved in the presentation of antigen to T cells. A primary function of this DC is to acquire antigen in peripheral tissues and to transport it to lymph nodes where the antigen is presented as peptides to T lymphocytes. Research into the DC system has demonstrated many features which underly this specialised "in vivo" function in antigen presentation. This paper will review recent research into "in vivo" DC properties and function, and will include recent data obtained in the model which we have been investigating for several years.

DC can be viewed as a lineage of cells specialised to monitor tissues for the presence of antigen and to maximise the chances of antigen-specific T cells being able to recognise and react to the antigen. A current description of DC function is that they are bone-marrow derived, have a blood-borne precursor which is now being characterised (see this volume), and that this precursor enters tissues in a relatively immature state. At this time it is capable of endocytosing and processing antigen for MHC class II association. After a period of residence in the periphery, the DC enters peripheral lymphatics and migrates to the draining node where it enters the T-dependent paracortical area and is able to present peptides to recirculating CD4+ T cells. After a few days in the node it dies.

Current studies have largely focused on the "in vitro" properties and functions of DC. It is however, essential to point out that in addition to the exploration of DC cell and molecular biology, the complete understanding of DC function demands knowledge of the evolutionary role of DC in the induction of protective immune responses to pathogens "in vivo", and we are still a long way from this position. We are, however, accumulating data which enable us to generate testable some hypotheses.

In this review we will first describe experiments showing that DC can acquire antigen in the periphery. We will then discuss the regulation of DC migration and finally will examine the role of DC in activating naive T cells. We will also point out some critical areas where knowledge is lacking and will try to suggest potential approaches to these problems. We will concentrate on experiments carried out in the rat but will relate our results to those obtained in other systems.

LYMPH-BORNE (VEILED) DC IN THE RAT

DC in peripheral lymph (veiled cells) are normally filtered out in the draining lymph node. Following mesenteric lymphadenectomy, peripheral and central lymphatics join and DC derived from the small intestine can be collected in thoracic duct lymph (1). These DC are a heterogenous, rapidly turning-over population and are potent stimulators of the MLR and other T-dependent responses (1). In contrast to DC from solid tissues, they can be concentrated up to 60-80% by a single centrifugation over Metrizamide and can be collected from TDL for up to four days in as "close to physiological" condition as is at present possible.

ACQUISITION OF ANTIGENS BY DC IN PERIPHERAL TISSUES

One of the first functional properties of DC to be described was their ability to stimulate T cells, both in the allogeneic MLR and by the presentation of protein antigens to sensitized cells (reviewed in 2). It is only more recently that it has been shown that DC can acquire antigen in the periphery. Thus Knight's group have shown that after skin painting with contact-sensitizing agents, DC extracted from draining lymph nodes can stimulate sensitized T cells (3), and Bujdoso et al (4) showed that after the subcutaneous injection of soluble antigen in sheep, lymph draining the site of injection contained DC bearing antigen which could be presented to T cells.

We have recently shown that for a period of about 24h following the injection of ovalbumin (OVA) or horse-radish peroxidase into the intestinal lumen of rats, DC in lymph draining the intestine carry antigen and can present it to sensitized T cells in an antigen-specific, CD4+ and MHC II-dependent manner (5). We also showed that DC were the only cells on this lymph able to present antigen, and that B cells were inert in this assay.

How do DC acquire antigen?

It is very clear that DC in the periphery can acquire contact-sensitizing agents and soluble protein antigens but it is not at all clear how this relates to their role in defence. Most pathogens are particulate and apart from the isolated case of allogeneic lymphocytes (6), mature DC are not, or only very weakly phagocytic. DC in lymph draining the intestine do, however contain inclusions, some of which contain DNA and others peroxidatic material, and electron microscopy shows that there is cellular debris in some inclusions (1). Reis e Sousa (this volume) has shown that Langerhans cells are phagocytic, but much less so than macrophages. Mayrhofer has shown that following a Salmonella infection, rat DC derived from the intestine carry Salmonella antigens but the functional status of these cells was not examined (7). In a recent elegant study, Moll et al (this volume) have shown that epidermal Langerhans cells can be infected "in vivo" by Leishmania, that infected DC migrate to draining nodes and that Leishmania antigens can be presented to sensitized T cells.

Do DC lose the ability to acquire antigens?

Fresh LC are very good presenters of soluble proteins but after "in vitro" culture, this property is lost, although cultured LC retain the ability to present peptides and are potent stimulators of the MLR (8) and see Schuler & Romani, this volume). The implications of these observations are that once DC have acquired antigen in the periphery, peptides derived from this antigen may be retained for presentation for long periods. It has not yet been shown whether this maturation also occurs "in vivo".

Thus there remain several important questions concerning antigen acquisition by DC :-

1. What are the "physiological" antigens acquired by DC. Can they phagocytose bacteria, can they endocytose or be infected by viruses during natural infections?
2. What receptors are involved in antigen uptake?
3. Does antigen acquisition signal changes in DC properties e.g. stimulate their release into lymph?
4. Is the loss of the ability to process and present whole protein antigens a programmed stage of DC differentiation?

MIGRATION OF DC AND ITS REGULATION

Normal Migration

All mammalian peripheral lymph so far examined contains migrating DC (veiled cells) although the actual numbers vary according to the tissue of origin. Under normal conditions the output of DC into lymph remains relatively constant. It is presumed that these migratory DC migrate into lymph nodes and become interdigitating cells (IDC). Fossum (9) has shown that rat intestinally-derived veiled cells injected into rat foot pads develop into IDC in the draining node and he and Austyn (10) have shown that DC injected IV migrate into T cell areas of the spleen. Thus a migratory pathway has been defined but many questions remain :-

1. How is the release of DC determined? LC may spend many weeks in the epidermis before exiting whereas intestinal DC spend only 2-4 days. Is release a programmed maturation step or is it determined by local microenvironmental changes? What changes in the microenvironment stimulate release. Can changes in surface phenotype be correlated with release?
2. How do DC recognise lymphatic endothelium in tissues?
3. What directs DC to appropriate locations in secondary lymphoid tissue?

4. DC spend only a few days in secondary lymphoid tissue and are presumed to die in situ. LC spend a long time in epidermis but die rapidly in culture and "in vivo". Is this death programmed? Is it due to lack of a survival factor present only in peripheral tissues? Is DC death apoptopic?

Stimulated Migration

Kelly (1970) described a marked increase in the output of veiled (frilly) cells in peripheral lymph draining the site of alum-precipitated diphtheria toxoid or bovine RBC injection, detectable at two days and peaking at four to five days. Other groups have shown an increase in the numbers of DC extractable from nodes draining the sites of skin painting with contact sensitizers (3) and also from non-draining nodes (12). We have shown that the output of DC into intestinal lymph is markedly increased after intravenous endotoxin injection (13) and Austyn's group (Roake et al, this volume) have shown that after endotoxin administration to mice, the majority of DC leave hearts and kidneys. The actual mediators of DC migration are not yet defined but it has been shown that following intradermal injection of TNF-α increased numbers of DC can be extracted from draining nodes (14) and we have preliminary evidence that an anti-TNFα monoclonal antibody can block the endotoxin stimulated release of DC into lymph (MacPherson, in preparation). The DC released by endotoxin in rats appear functionally similar those normally found in lymph in their capacity to stimulate a MLR and to present antigen to sensitized T cells.

Thus it is clear that DC release from peripheral tissues can be modulated, and that TNF-α may be involved at some stage in this release, but many questions remain to be answered. Whether TNF is the only mediator and whether TNF acts directly on DC are unknown, as are the changes in DC phenotype and function which may accompany stimulated release.

"IN VIVO" FUNCTIONS OF ANTIGEN-BEARING DC

In rodents, DC stimulate the allogeneic MLR much more efficiently than do other cells expressing MHC II, suggesting that DC may have a particular role in initiating primary immune responses. Knight's group has shown that DC extracted from nodes draining the site of skin painting with FITC can adoptively sensitize naive recipients (3) and Fossum's group (Berg et al, this volume) and Sornasse (this volume) have shown that DC pulsed with antigens "in vitro" can sensitize naive animals for antibody synthesis.

Steinman's group have shown that following the IV injection of antigen, the only cells present in spleen that can present antigen are DC (15). Similarly, we have shown that following antigen injection into the intestine, or after "in vitro" pulsing, DC but not B cells or macrophages are able to present antigen to primed T cells (5) and only DC can prime T cells after injection into naive rats (Liu & MacPherson, in preparation).

In these models it is difficult to exclude a role for host antigen-presenting cells in processing and presenting antigens delivered by the injected DC. To overcome this, Inaba et al (16) used parental strain antigen-pulsed DC to prime a F1 mouse. T cells from the draining node could only be restimulated efficiently by antigen-presenting cells from the priming parental strain, showing that the original priming was restricted to the MHC of the priming DC, and excluding a significant role for host processing and presentation. We have performed similar experiments using DC from animals which had been injected intra-intestinally with OVA and shown a similar restriction of priming (Liu & MacPherson, in preparation).

Thus we can conclude that DC which have acquired antigen "in vivo" in a relatively natural manner from the small intestine can prime T cells in naive recipients. How these observations relate to oral tolerance remains to be established.

CONCLUSIONS

The primary "in vivo" function of DC in the immune response is to acquire antigens in the periphery, to process them, transport them to secondary lymphoid tissue and there to present them to T lymphocytes. The migratory pathway of these cells has been at least partially defined, but the molecular basis of the regulation of migration is still very unclear.

ACKNOWLEDGEMENTS

We are extremely grateful for the expert technical assistance of Chris Jenkins.

REFERENCES

1. Pugh C.W., MacPherson G.G. and Steer H.W. Characterization of non-lymphoid cells derived from rat peripheral lymph. *J.Exp.Med.* 157:1758 (1983) .
2. Steinman R.M. The dendritic cell system and its role in immunogenicity. *Annu. Rev. Immunol.* 9:271 (1991).
3. Macatonia S.E., Edwards A.J. and Knight S. Dendritic cells and contact sensitivity to fluorescein isothiocyanate. *Immunology.* 59:509 (1986).
4. Bujdoso R., Hopkins J., Dutia B.M., Young P. and McConnell I. Characterization of sheep afferent lymph dendritic cells and their role in antigen carriage. *J.Exp.Med.* 170:1285 (1989).
5. Liu L.M. and MacPherson G.G. Lymph-borne (veiled) dendritic cells can acquire and present intestinally administered antigens. *Immunology* 73:281 (1991).
6. Fossum S. and Rolstad B. The roles of interdigitating cells and natural killer cells in the rapid rejection of allogeneic lymphocytes. *Eur.J.Immunol.* 16:440 (1976).
7. Mayrhofer G., Holt P.G. and Papadimitriou J.M. Functional characteristics of the veiled cells in afferent lymph from the rat intestine. *Immunol.* 58:379 (1986).
8. Romani N., Koide S., Crowley M., Witmer-Pack M., Livingstone A.M., Fathman C.G., Inaba-K. and Steinman R.M. Presentation of exogenous protein antigens by dendritic cells to T cell clones. Intact protein is presented best by immature, epidermal Langerhans cells. *J. Exp. Med.* 169:1169 (1989).
9. Fossum S. Lymph-borne dendritic leucocytes do not recirculate, but enter the lymph node paracortex to become interdigitating cells. *Scand. J. Immunol.* 27:97 (1988).
10. Austyn J.M. Migration patterns of dendritic leukocytes. *Res.Immunol.* 140:898, discussion 918 (1989).
11. Kelly R.H. Localization of afferent lymph cells within the draining node during a primary immune response. *Nature* 227:510 (1970).
12. Hill S., Edwards A.J., Kimber I. and Knight S.C. Systemic migration of dendritic cells during contact sensitization. *Immunology* 71:277 (1990) .
13. MacPherson G.G. Properties of lymph-borne (veiled) dendritic cells in culture. II. Expression of the IL-2 receptor: role of GM-CSF. *Immunol.* 68:108 (1989).

14. Cumberbatch M. and Kimber I. Dermal tumour necrosis factor-α induces dendritic cell migration to draining lymph nodes, and possibly provides one stimulus for Langerhans' cell migration. *Immunol.* 75:257 (1992).
15. Crowley M., Inaba K. and Steinman R.M. Dendritic cells are the principal cell in mouse spleen bearing immunogenic fragments of foreign proteins. *J.Exp.Med.* 172:383 (1990).
16. Inaba K., Metlay J.P., Crowley M.T. and Steinman R.M. Dendritic cells pulsed with protein antigens in vitro can prime antigen-specific, MHC-restricted T cells in situ. *J.Exp.Med.* 172:631 (1990).

FOLLICULAR DENDRITIC CELLS:

ISOLATION PROCEDURES, SHORT AND LONG TERM CULTURES

Ernst Heinen, Rikyia Tsunoda, Christian Marcoty, Nadine Antoine, Alain Bosseloir, Nadine Cormann, and Léon Simar

Institute of Human Histology, University of Liège, Belgium, Department of Anatomy, Fukushima Medical College, Japan*

INTRODUCTION

The need to isolate and cultivate follicular dentritic cells (FDC) in vitro is becoming increasingly acute in view of the physiological complexity of the germinal centers and the accumulation of data on their pathology.

Although B lymphoid cells clearly undergo major phases at these sites (activation, proliferation, survival, maturation in B memory cells or precursors of plasma cells), information is scarce concerning mechanisms that control such phenomena. T cells are essential (2), tingible body macrophages discard apoptotic cells, and FDC appear to influence the various evolutionary phases of B cells (8).

Many questions remain without clear answers: what kind of signals do the FDC transmit during their contacts with lymphoid cells ? Do they produce bioactive factors? Why are there different subsets of FDC ? What is the action of lymphoid cells on FDC? Isolation and cultivation of FDC can help unravel these mysteries.

ISOLATION OF FDC

Numerous laboratories have tried to purify FDC, mostly without success. Many have obtained cells which in fact were not FDC but dendritic or lymphoid cells, while others obtained strongly altered FDC. Mostly, FDC have been purified in association with lymphoid cells, in cluster-like formations where FDC envelop other cells with their extensions (7,10,11). Tsunoda et Kojima (17) have shown that these clusters are fragments of the germinal centers, and thus useful for physiological tests. Some rare authors have published convincing results on pure FDC populations devoid of contaminating cells (14, Rieber et al., see this volume), but these have not always clearly demonstrated the functionality or viability of the isolated cells.

In our efforts to develop isolation procedures, priority was given to preserving the physiological properties of FDC, such as their capacity to bind immune complexes or to interact with lymphoid cells. Our usual procedure involves the following steps (see Marcoty et al. in this volume):

1- dissection of the lymph follicles under biomicroscopes
2- mild enzymic digestions to separate the cells
3- gravity fractionation of the cells
4- elimination of macrophages by adherence.

This procedure yields FDC clusters up to 95% pure and retaining their physiological capacities (Fig.1).

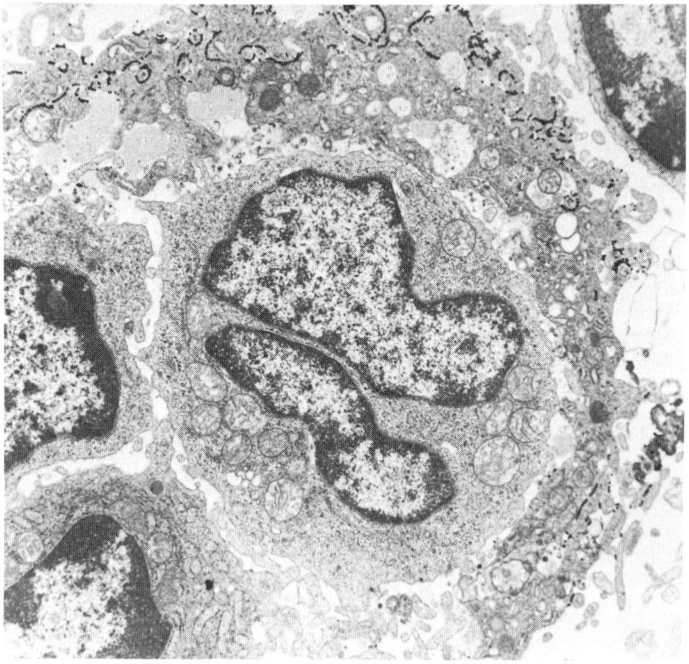

Fig 1 Mouse lymph node FDC, purified and incubated for 30 min at 37°C with gold-labeled TNP-myoglobin-anti-TNP complexes. These FDC remain able to interact with lymphoid cells and retain the complexes on their surface (dark material).

Tsunoda et al. (15) added yet another step: the cells were allowed to sediment and attach to a substrate, and non-attached cells were washed away after a given time. This yielded nearly pure FDC clusters.

Initial trials with Percoll enabled us to concentrate the FDC. Percoll particles remained attached, however, to FDC (Fig.2), so we neglected this medium which has since been used with success by others. Whether these attached Percoll particles affect the behavior of FDC is unknown.

Rieber et al. (this volume) developed another device: they bound specific anti-FDC antibodies to ox red blood cells and, after repeated rosetting, obtained pure lymphoid-cell-free FDC. Schriever et al. (14) first prepared FDC clusters, then dissociated FDC from the lymphoid cells and, using specific antibodies, concentrated the FDC with a cell sorter.

SHORT-TERM EXPERIMENTATION ON FDC

Our isolation procedures yield living FDC which retain most of their surface antigens and are thus useful for short-term experiments. Our group studied receptors for immune complex retention, intercellular adhesion, and production of bioactive factors.

Using antibodies or antigens coupled to fluorescent markers, enzymes, or gold particles, we showed that isolated FDC retain immune complexes involving various different Ig isotypes (except Ig D) as well as do FDC in vivo; complexes containing IgG_{2a} and IgG_{2b} are better retained than other isotypes (6).

After contact at 37°C for 30 min with gold-labeled immune complexes or anti-C3b receptor antibodies, murine as well as human FDC exhibit gold particles in deep membrane infoldings, suggesting that receptors move on their surface (Fig.3). When the cells are then left at 37°C, local focusing and shedding of gold particles occur: membranes, vesicles, and

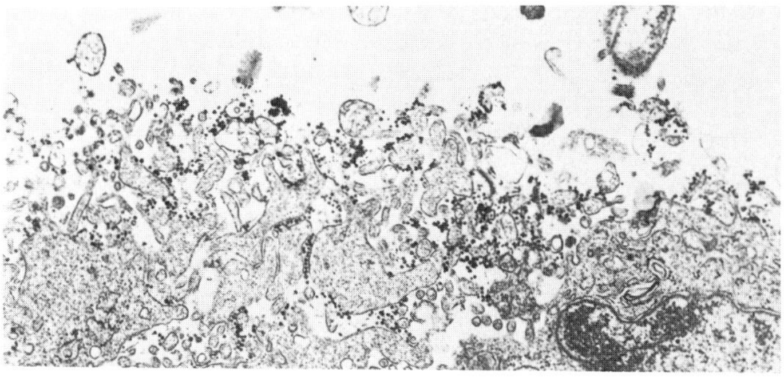

Fig 2. Mouse FDC purified on a Percoll gradient; even after extensive rinsing, Percoll particles remain attached to the FDC.

even dendrites detach from the surface (Fig.4). Sometimes, iccosomes as described by Szakal et al. (15) are observed. However, gold particles mostly appear associated with membrane residues detaching from the FDC and are unfrequently attached to vesicles (Fig.4). Moreover, FDC incubated for several hours after isolation exhibit a lower cell surface density, thus less extensive cytoplasmic evaginations. We hypothesize that FDC can reject cell membranes by a shedding process which allows them to renew their cell surface or receptors.

Using the same procedure, we demonstrated that FDC can perform selective endocytosis: we isolated FDC, incubated them with gold-labeled anti-transferrin receptor antibodies and left them for 1/2 hour before fixation at 37°C. At ultrastructural level, FDC exhibited distinct, gold-containing endocytotic vesicles (see Fig. 5). As under similar conditions FDC do not endocytose gold-labeled immune complexes, we conclude that FDC perform selective endocytosis.

FDC can form close contacts with lymphoid cells. We have shown that this depends on surface receptors (LFA1), on Ca^{++} and Mg^{++}, on the temperature, and on intact actine filaments. Immune complexes bearing or not bearing complement fragments may interfere. This intercellular adherence is species specific and apparently sensitive to anti-DRC1 antibody which reduces it (12). ICAM-1 and CD44 are also determinant for these contacts (9, 10).

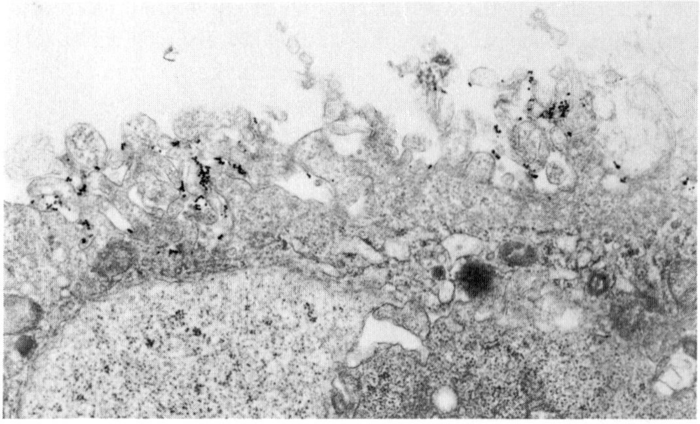

Fig 3. Human FDC after a 30-min incubation at 37°C in presence of gold-labeled tetanus toxoid-anti-tetanus toxoid complexes: most gold particles, at this time, are found imprisoned in deep invaginations.

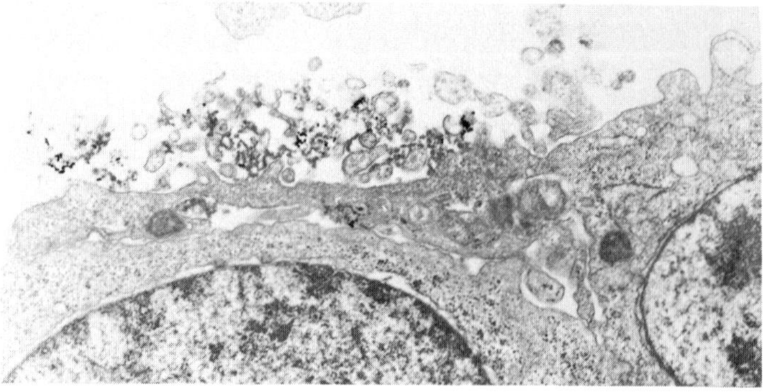

Fig 4. Same conditions as in Fig.3 but with an additional 30-min incubation at 37°C. Under these conditions, gold-labeled material (membranes, vesicles, dendrites) detaching from FDC can be seen.

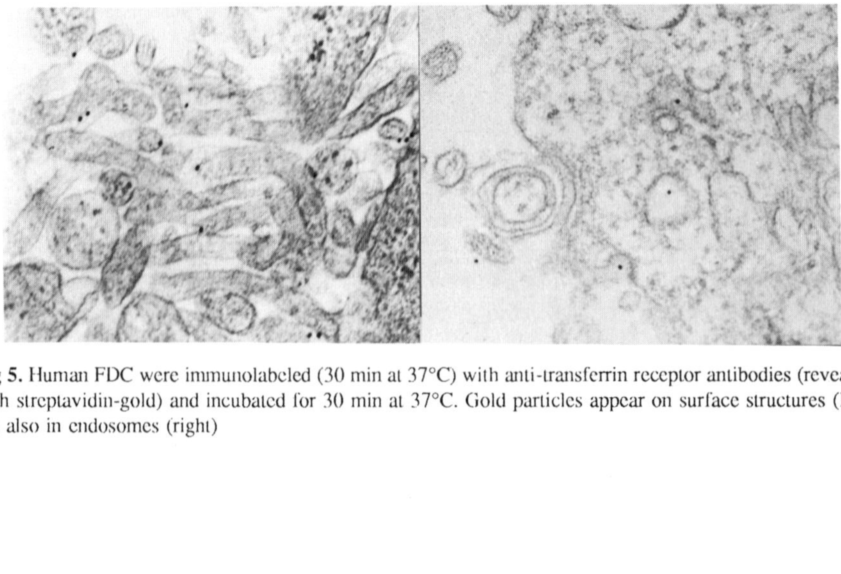

Fig 5. Human FDC were immunolabeled (30 min at 37°C) with anti-transferrin receptor antibodies (revealed with streptavidin-gold) and incubated for 30 min at 37°C. Gold particles appear on surface structures (left) but also in endosomes (right)

CULTURES OF FDC

FDC in culture can be considered under various aspects: phenotype, survival, proliferation, interactions with lymphoid cells, etc...

In culture, FDC lose their typical phenotype: after a few hours, expression of DRC_1, FcR, C_3bR decreases, usually disappearing within 2 to 3 days. Most FDC degenerate, some appear to differentiate into fibroblast-like cells, but without signs of proliferative activity (4). Tsunoda and coworkers were able to maintain living, non- dividing FDC for up to 150 days in culture. These cells adhered to the substrate, spread out, and accumulated stress fibers but remained able to interact with lymphoid cells (Tsunoda et al., this volume).

FDC clusters enlarge during the first hours of culture as lymphoid cells attach to them and proliferate (3). FDC appear to act on lymphoid cells in different ways: they ensure better survival, improve their proliferative capacity, and apparently inhibit their differentiation in Ig-secreting cells (3, 4, 8). Numerous authors have confirmed escape from apoptosis and induction of proliferation. Our observations suggest that FDC exert these effects during close contacts but also at a distance, since supernatants of FDC cultures already contain, after a few hours, products that may improve the proliferative capacity of lymphoid cells (Tables I and II). These products do not appear to be as active in supernatants of B and T cells cultured alone or mixed with macrophages (+/- 3%). In the supernatants of FDC cultures, we found prostaglandins which could be produced by FDC (5). Interestingly, prostagandins can modulate the Ig isotype switch (13). Histamine is absent from such supernatants. Our attempts to identify cytokines gave contradictory results (16): we found IL6, TNFα, very low concentrations of IL4, and no IL2 or IFNgamma. It seems that IL6 arises from T cells contaminating the preparations of germinal center cells, in situ hybridization tests having shown that IL6 is not produced inside the germinal centers but by cells in the interfollicular and sub-epithelial areas (1). Results obtained in such cell mixtures should thus be considered with circumspection.

CONCLUSIONS

Over the last ten years, it has become clear that FDC can be prepared from stimulated lymph organs. Even though they cannot multiply in vitro and are only useful during short-term tests, much information has been gained, mainly on FDC associated in clusters with lymphoid cells. Recent developments have yielded free FDC and should lead to new progress in the study of germinal centers.

Table I. Proliferative capacity (measured by ^{3}H-thymidine incorporation) of tonsillar lymphoid cells cultured in supernatants recovered from 24-hour cultures of lymphoid cells alone (B + T), mixed with macrophages (B + T + M), or with FDC (B + T + FDC).

Origin of the supernatant	B + T	B + T + M	B + T + FDC
^{3}H-thymidine incorporation (cpm) after 3 days of culture	550 +/- 250	1900+/- 250	12,900 +/- 450

Table II. Effect of diluting the supernatant obtained from B + T + FDC cultures on the proliferative capacity of lymphoid tonsillar cells.

Supernatant concentration	100 %	75 %	50 %	25 %
^{3}H-thymidine incorporation (cpm) after 3 days of culture	9320 +/- 600	7400 +/- 400	6700+/- 350	5500+/- 20

REFERENCES

1. Bosseloir, A., E. Dehooghe-Peeters, E. Heinen, N. Cormann, C. Kinet-Denoël, A. Van Haelst, and L. J. Simar. "Production of Interleukin-6 in human tonsils by in situ hybridization". *Europ. J. Immunol.* 19:2379-2382 1989).
2. Bowen, M. B., A. W. Butch, C. A. Parvin, A. Levine, and M. H. Nahm. "Germinal Center T-Cells Are Distinct Helper-Inducer T-Cells". *Hum. Immunol.* 31:67-75 (1991).
3. Cormann N., Heinen E., Kinet-Denoël C., Braun M. et Simar L.J. "Influence de la cellule folliculaire dendritique sur la survie des lymphocytes in vitro". *C.R. Soc. Biol.* 180:218-223 (1986).
4. Cormann, N., F. Lesage, E. Heinen, N. Schaaf-Lafontaine, C. Kinet-Denoël, and L. J. Simar. "Isolation of follicular dendritic cells from human tonsils or adenoids". V. "Effect on lymphocyte proliferation and differentiation". *Immunol. Letters* 14:29-35 (1986).
5. Heinen, E., N. Cormann, M. Braun, C. Kinet-Denoël, J. Vanderschelden, and L. J. Simar."Isolation of follicular dendritic cells from human tonsils and adenoids". VI. "Analysis of prostaglandin secretion". *Annales de l'Inst. Pasteur, Immunol.* 137:369-382 (1986).
6. Heinen, E., P. Coulie, J. Van Snick, M. Braun, N. Cormann, M. Moeremans, C. Kinet-Denoël, and L.J. Simar."Retention of immune complexes by murine lymph node or spleen follicular dendritic cells". Role of antibody isotype. *Scand. J. Immunol.* 24:327-334 (1986).
7. Heinen, E., C. Kinet-Denoël, D. Radoux, and L. J. Simar."Mouse lymph node follicular dendritic cells: Quantitative analysis and isolation". *Adv. Exp. Med. Biol.* 186:171-184 (1985).
8. Heinen, E. and R. Tsunoda."Microenvironments for B cell production and stimulation". *Immunol. Today* 8:142-143 (1987).
9. Koopman G., Parmentier H.K., Schuurman H.J., Newman W., Meijer C.J.L.M., Pals S.T.."Adhesion of human B cells to follicular dendritic cells involves both the lymphocyte function-associated antigen-1 intercellular adhesion molecule-1 and very late antigen-4/vascular cell adhesion molecule-1 pathways". *Journal of Experimental Medecine* 173:6 1297-1304 (1991).
10. Kosco, M. H., Pflugfelder E., Gray D. "Follicular dendritic cell-dependent adhesion and proliferation of B-cells in vitro". *Journal of Immunol.* 148:8 2331-2339 (1992).
11. Lilet-Leclercq C., Radoux C., Heinen E., Kinet-Denoël C., Defraigne J.O., Houben-Defresne M.P. et Simar L.J.. "Isolation of follicular dendritic cells from human tonsils and adenoids". I. "Procedure and morphological characterization". *J. Immunol. Meth.* 66:235-244 (1984).
12. Louis, E., B. Philippet, B. Cardos, E. Heinen, N. Cormann, C. Kinet-Denoël, and L.J. Simar. "Intercellular connections between germinal center cells". "Mechanisms of adhesion between lymphoid cells and follicular dendritic cells". *Acta ORL* 43:297-320 (1990).
13. Phipps R.P., Roper R.L., and Stein S.H.. "Regulation of B cell tolerance and triggering by macrophages and lymphoid dendritic cells". *Immunol Rev* 117:134-150 (1990).
14. Schriever, F. and L. M. Nadler. "Antigenic Phenotype of Isolated Human Follicular Dendritic Cells". *Accessory Cells in HIV and Other Retroviral Infections*. 9-17 (1991).
15. Szakal A., Kosco M.H. and Tew J.G. "A novel in vivo follicular dendritic cell-dependent iccosome-mediated mechanism for delivery of antigen to antigen-processing cells". *J. Immunol.* 140:341-353 (1988).
16. Tsunoda, R., N. Cormann, E. Heinen, K. Onozaki, P. Coulie, Y. Akiyama, K. Yoshizaki, C. Kinet-Denoël, L. J. Simar, and M. Kojima. "Cytokines produced in lymph follicles". *Immunol. Letters* 22:129-134 (1989).
17. Tsunoda, R. and M. Kojima. "A light microscopical study of isolated follicular dendritic cells-clusters in human tonsils". *Acta Pathol. Jpn.* 37:575-585 (1987).

HETEROGENEITY AND CELLULAR ORIGIN OF FOLLICULAR DENDRITIC CELLS

Yutaka Imai, Kunihiko Maeda, Mitsunori Yamakawa, Yasuaki Karube, Mikio Matsuda, Michio Dobashi, Hironobu Sato and Kazuo Terashima

Department of Pathology, Yamagata university School of Medicine
2-2-2 Iida-Nishi, Yamagata 990-23, Japan

INTRODUCTION

Follicular dendritic cells (FDC) are specialized cells which locate within the lymphoid follicles of secondary lymphoid tissues including lymph nodes, spleen, tonsil and mucosal associate lymphoid tissues (Peyer's patches etc). Many studies[1,2,3] have been accumulated to defined the morphological, immuno-phenotypical and functional features of FDC. Nevertheless, the nature of FDC is still unsettled, especially the cellular origin, kinetics, proliferating activity, maturation processes and functional diversity. In this paper, we focused on the two major subjects, the heterogeneity and cellular origin of FDC because these are thought to be most fundamental for FDC research and summarize our recent results.

HETEROGENEITY OF FDC

The human lymph nodes, tonsils, Peyer's patches and mucosal lymphoid apparatus of appendix were carefully examined using immunohistochemical and electron microscopic techniques. Consequently, two types of FDC were distinguishable in a well developed germinal center. One of them are ordinary FDC, which reveals typical morphology including very intricate dendritic processes (Fig.1) and surface phenotypes including not only R4/23, complement receptors (CR) but also CD23 and DF-DRC1. This type of cells also had immunoglobulins, complement components and some complement regulatory factors including properdin, decay accelerating factor (DAF) which suggests the active function to retain immune complexes[4] (Fig.3a, 3b, 3c). On the other hand, another type of FDC were also labeled by R4/23 and anti-CR but they have less intricate processes (Fig.2) and less reactivity with CD23, DF-DRC1. They did not have immunoglobulins, complement components and complement regulatory factors. The former mainly located in the light zone of germinal center and were designated "light zone (LZ) type of FDC", tentatively. The later located in the dark zone and were referred to "dark zone (DZ) type of FDC". Table 1 summaries the features of both types. These diversity is also confirmed by other morphological observation [1,5] but the relationships of both types of FDC or possibility of other types of FDC are still undetermined.

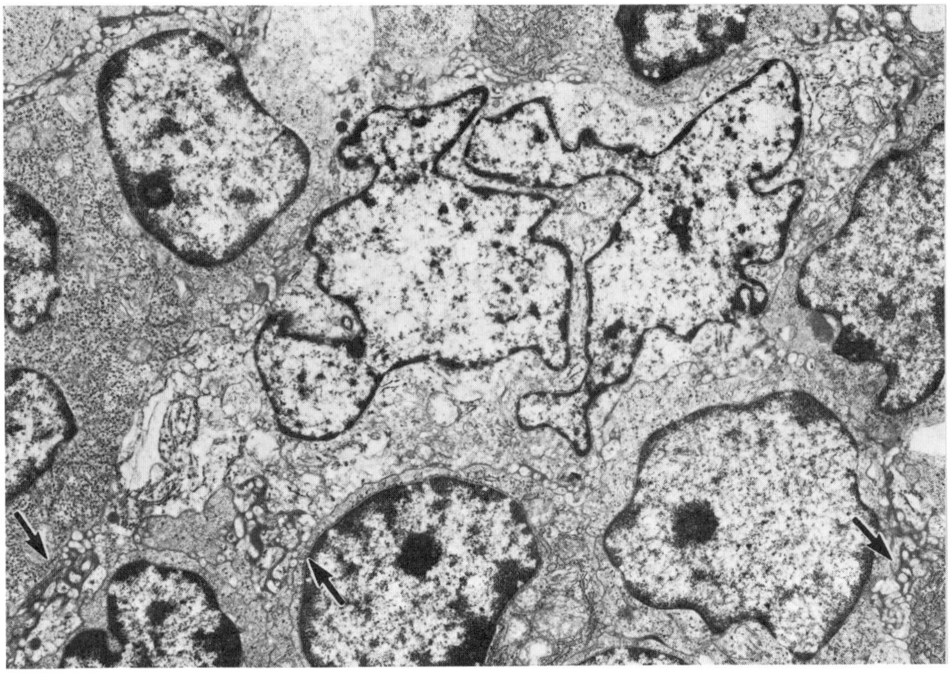

Fig.1 An electron micrograph showing the LZ type of FDC. Note the entangled, intricate cytoplasmic processes emanating among the surrounding lymphocytes (arrows).

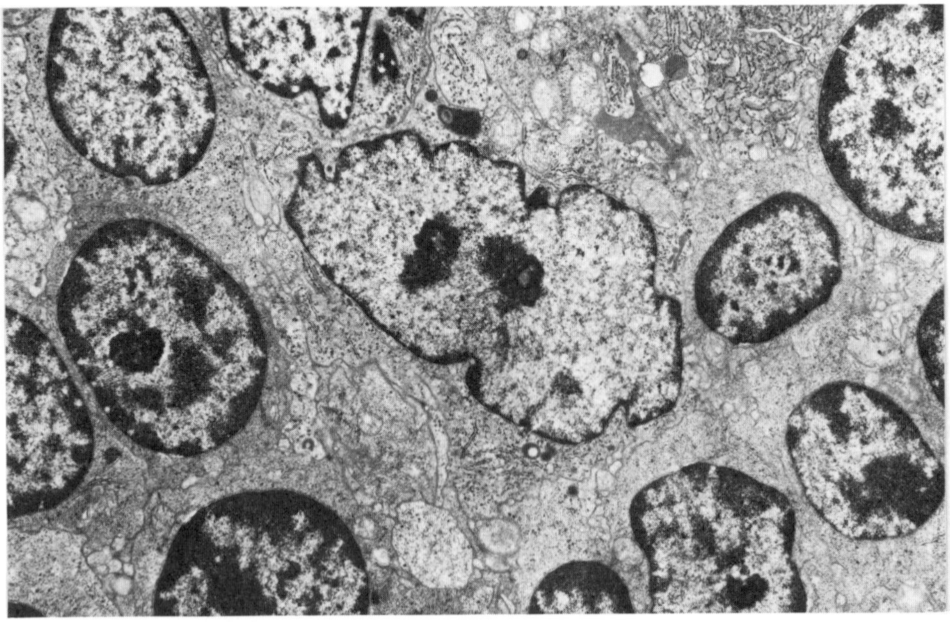

Fig. 2 An electron micrograph showing the DZ type of FDC. Note the less intricate cytoplasmic processes.

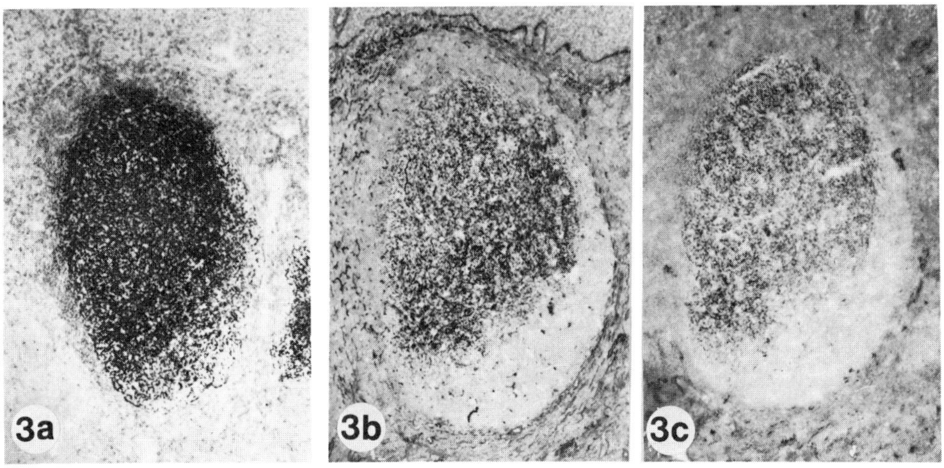

Fig.3 Light micrographs demonstrating the distribution pattern of CD35 (CR1 or C3dR,) (3a), C3d (3b) and properdin (3c) respectively in a germinal center of the reactive tonsillar tissue (serial sections).

Table 1, Comparison of LZ type and DZ type of FDC

Type of FDC	LZ type of FDC (Typical FDC)	DZ type of FDC
Location	light zone of germinal center	dark zone of germinal center
Ultrastructures	intricate, entangled cytoplasmic processes (labyrinth structures) polygonal scanty cytoplasm desmosome-like structures	relatively less intricate cytoplasmic processes desmosome-like structures
Phenotypes		
DRC-1 (R4/23)	++	+/– or –
DF-DRC	++	+/– or –
CD35 (CR1, C3dR)	++	++
CD21 (CR2, C3bR)	++	++
CD23 (FcεRII)	++	+/– or –
Complement components (C1q, C3d, C5, C9)	++	–
Immunoglobulins (IgG, IgM, IgA, IgE)	++ or +	–
Complement regulatory factors (DAF, properdin vitronectin, etc)	++ or +	–

CELLULAR ORIGIN OF FDC

Experimental approach using SCID mice

To definite the cellular origin of FDC experimentally, SCID mice were reconstituted by bone marrow (BM) cells from normal CB17 mice. The BM cells ($1X10^7$ cells) were labeled with BrdU and injected intravenously. After three weeks, the reconstituted SCID mice were preadministrated mouse PAP(peroxidase-antiperoxidase) complex subcutaneously to visualize immunecomplex-retaining cells 24 hours before the extirpation. In the next day, the draining LNs were excised and processed for both of LM-level and EM-level examination. In these experiments, LNs of the reconstituted mice revealed several follicles with detectable retained PAP and some of them had evident germinal centers (Fig.4). In contrast, control (non-reconstituted SCID mice) did not have any follicles (Fig.5) nor retain PAP at all. Ultrastructurally, the retained PAP were observed on the surface of intricate cytoplasmic processes and cell bodies of the cells which were resemble morphologically to typical FDC (Fig.6a, 6b). These cells did not labeled with BrdU, whereas surrounding lymphocytes were clearly labeled in their nuclei. These results suggest that FDC may not be derived from injected donor cells but the BM cells may be necessary to their development .

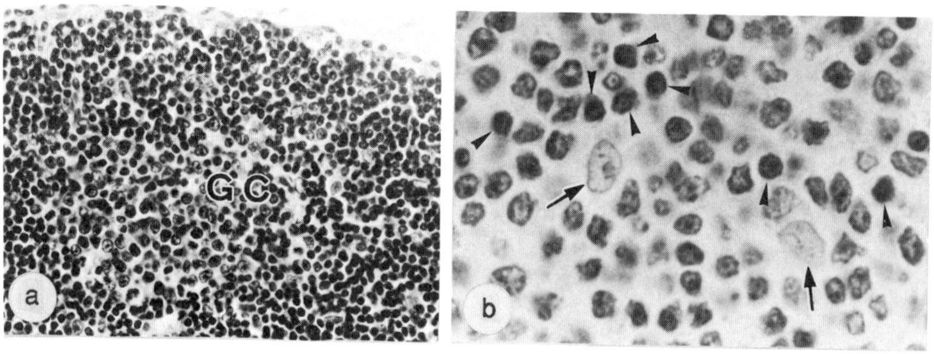

Fig.4 Light micrographs showing the histology (4a) and BrdU labeling (4b) of LN of the reconstituted SCID mouse. Evident germinal centers ("GC") are seen in Fig.4a. In the germinal center, nuclei of FDC (arrows) are not labeled with BrdU whereas nuclei of surrounding germinal center cells are clearly positive (arrow-heads) in Fig.4.

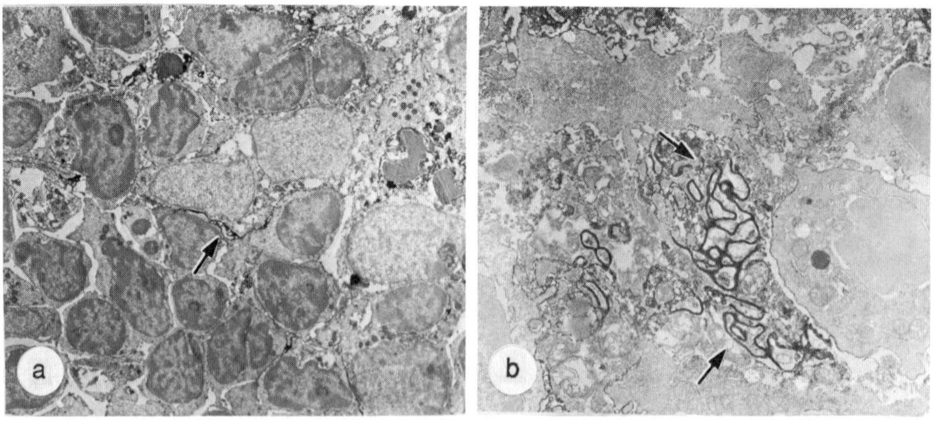

Fig.5 Electron micrographs showing FDC (5a) and their cytoplasmic processes (5b) in the reconstituted SCID mouse. Preadministrated PAP complex were clearly detected as electron-dense products on the surface of the cell body and the cytoplasmic processes (arrows).

Postnatal development of rat and murine FDC

As another approach to determine the origin and developing process of FDC, the postnatal development of rat and murine LNs were observed by electron microscopy[6,7]. In the rat system, PAP complex was preadministrated 24 hours before extirpation of LNs and the distribution and morphology of immune-complex-retaining cells were noticed. In our pilot study, the most of phagocytic cells and a part of lymphoid cells also trapped PAP , but they ingested and processed these complexes promptly and these cells could not retain them over 24 hours. In the mature LNs, representative immune-complex-retaining cells were FDC. The follicular structure with these fully developed FDC were recognized first at 20 days after birth in conventional breeding. But careful observations indicated us a few cells with immune-complex-retaining ability had already appeared in the cortical area, especially beneath the floor of subcapsular sinus of LNs at 10 or 12 days after birth, even before follicular formation . These cells had the morphological features resemble stromal fibroblastic cells and in many cases, a few lymphoid cells closely attached to them. We interpreted these findings suggested the initial developing stage of FDC. Next, we tried to determine the surface phenotypes of these cells using mouse system because a variety of monoclonal antibodies were available in this well defined animal. In murine system, the first follicular appearance were observed at 2 weeks after birth.. Therefore the LNs of neonatal mice, especially from 2 days to 10 days after birth, were focused on. In this period, the cell clusters consisted of several lymphoid cells and one or two stromal cells were noticed in the cortex. The lymphoid cells were labeled with the monoclonal antibody B3B4 (anti-murine CD23) very clearly and the stromal cells also had partial reactivity for this antibody (Fig.7). In addition, this type of stromal cells revealed.distinct reactivity for the monoclonal antibody 7E9 (anti-murine CR-1 and CR-2) (Fig.8). These cells also revealed intense labeling of MK-1 (anti-murine ICAM-1 or related molecules) as well as other stromal fibroblastic cells. Interestingly these cells did not expressed detectable levels of FcγRII (detected by 2.4G2) which were known to express intensely on the surface of mature FDC[8]. These results were interpreted that the small clusters of CD23+ B-lymphocytes and the stromal cells might represent the beginning of follicular development. In other words, these stromal cells might develop to FDC in the microenvironment organized by surrounding CD23+ B-cells. Our observations on the distribution, morphology and surface phenotypes of FDC in malignant lymphoma tissues also suggested B-lymphocytes, especially small cleaved type of cells might associate closely with the development of FDC[9]. Cerny et al[10] also demonstrated the absence of mature FDC in the LNs of B-cell depleted mouse by chronic treatment with anti-IgM antibodies. The relationships between the development or maturation of FDC and surrounding B-cells may be most interesting subject in the fundamentals of in vivo immune responses. The further experimental approaches such as in vitro co-culture of stromal cells and B-cells or induction of in vitro transformation of stromal cells by addition of cytokines are expected to elucidate molecular levels of the mechanism of these relationships.

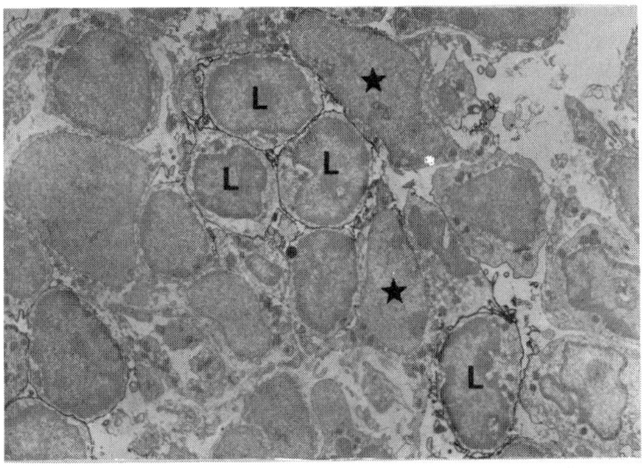

Fig.6 An electron micrograph showing B3B4 (anti-FcεRII) reactivity in the lymph node of 8 days old mouse. Several lymphoid cells (L) with intense labeling and stromal cells (★) are forming the small cellular cluster.

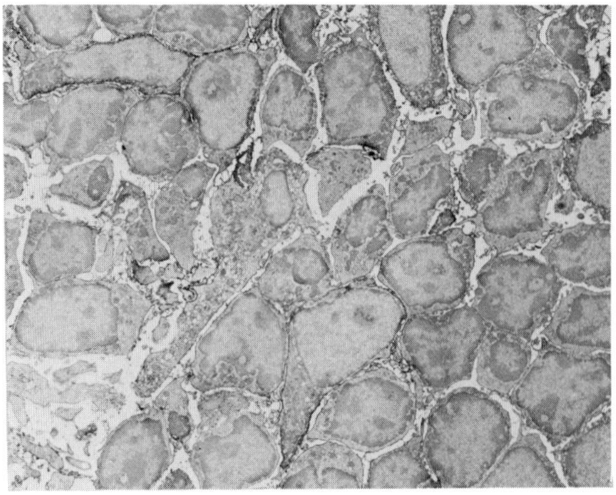

Fig. 7 An electron micrograph showing 7E9 (anti-mouse CR-1 and -2) reactivity in the lymph node of 5 days old mouse. Note the intense labeling on the surface of stromal cells.

REFERENCES

1, Y. Imai, K. Terashima, M. Matsuda, M. Dobashi, K. Maeda and T. Kasajima, Reticulum cell and dendritic reticulum cell –Origin and function–. *Recent Adv. RES Res.* 21: 51 (1983)

2, Y. Imai, M. Yamakawa, A. Masuda, T. Sato, and T. Kasajima, Function of the follicular dendritic cell in the germinal center of lymphoid follicles. *Histol. Histopath.* 1: 341 (1986)

3, J. Tew, M.H. Kosco, G.F. Burton and A.K. Szakal, Follicular dendritic cells as accessory cells. *Immunological Rev.* 117: 185 (1990)

4, M. Yamakawa and Y. Imai Complement activation in the follicular light zone of human lymphoid tissues. *Immunology in press* (1992)

5, L. H. P. M. Rademakers, H.-J. Schuurman, J.F. de Frankrijker and A. Van Ooyen Cellular composition of germinal centers in lymph nodes after HIV-1 infection: Evidence for an inadequate support of germinal center B lymphocytes by follicular dendritic cells. *Clin. Immunol. Immunopathol.* 62: 148 (1992)

6, K. Maeda, M. Matsuda and Y. Imai. Postnatal development of the immune complex retaining cells in the rat lymph nodes. –An immunofluorescent and electron microscopic study–. *in preparation* (1992)

7, K. Maeda, M. Matsuda and Y. Imai Postnatal development of murine follicular dendritic cells. –An immunoelectron microscopic study–. *in preparation* (1992)

8, K. Maeda, G.F. Burton, D.A. Padgett, D.H. Conrad, T.F. Huff, A. Masuda, A.K. Szakal and J.G. Tew Murine follicular dendritic cells and loe affinity Fc receptors for IgE (FceRII). *J. Immunol.* 148: 2340 (1992)

9, K. Maeda, M. Matsuda, R. Nagashima, N. Degawa and Y. Imai Follicular dendritic cells in malignant lymphomas –Distribution, phenotypes & ultrastructures–. in this proceedings (1992)

10, A. Cerny, R.M. Zinkernagel and P. Groscurth Development of follicular dendritic cells in lymph nodes of B-cell-depleted mice. *Cell Tissue Res.* 254:449 (1988)

DIFFERENTIAL UPTAKE AND TRAPPING OF TI-2 ANTIGENS: AN UNEXPECTED ROLE FOR FOLLICULAR DENDRITIC CELLS IN THE INDUCTION OF TI-2 IMMUNE RESPONSES

A.J.M. Van den Eertwegh, M.M. Schellekens, W.J.A. Boersma and E. Claassen

Department of Immunology and Medical Microbiology, TNO Medical Biological Laboratory, P.O. Box 45, 2280 AA, Rijswijk, The Netherlands

INTRODUCTION

The spleen plays an important role in the protection against capsulated micro-organisms e.g. *Streptococcus pneumoniae, Neisseria meningitides and Haemophilus influenzae*[1]. The protective immune response against these micro-organisms is directed mainly against the polysaccharide component of the bacterial capsule, classified as a TI-2 antigen[2]. Following splenectomy patients are at high risk for fulminant infections due to these bacteria. The presence of the spleen seems to be important in the primary encounter of the antigen, whereas secondary humoral immune responses can also be evoked at sites outside the spleen[3]. This has led to the suggestion that in the spleen, specific subsets of B cells may be present or that the splenic micro-environment as such may be crucial for the response of B cells to TI-2 antigens. In particular the marginal zone macrophages of the spleen have received much interest in this respect. These macrophages, which take up and retain carbohydrate macromolecules such as TNP-Ficoll/FITC-Ficoll (model TI-2 antigens), have been suggested to play a role in the processing and presentation of TI-2 antigens[4-7]. However, recent *in vivo* modulation studies demonstrated that marginal zone macrophages are not required in humoral immune responses against TI-2 antigens[8-10].

Evidence for an important role of B cells in TI-2 immune responses was found in experiments with neonatal mice and CBA/N mice, which carry an X-linked immunodeficiency, which were both unable to mount a humoral response against TI-2 antigens[11]. The B cells of these mice have many properties in common with immature B lymphocytes so that their defect is probably due to immaturity, in case of neonatal mice, or due to a maturation arrest in case of the CBA/N mice. Griffioen et al.[12] showed that polysaccharides, TI-2 antigens, were able to activate complement via the alternative pathway. Moreover, they demonstrated that polysaccharides complexed with C3d, a degradation product of complement, were able to bind to CR2 of B cells. In addition, Carter et al.[13,14] demonstrated synergistic signaling of CR2 and membrane immunoglobulin, as measured by intracellular calcium mobilization. These data provide an alternative activation pathway of B cells in TI-2 immune responses. However, the localization studies for FITC-Ficoll indicated the marginal zone macrophages as the principal TI-2 binding cells in the spleen, which is not consistent with the above discussed (direct) B cell activation models. Therefore, we re-evaluated the localization of TI-2 antigens using a highly sensitive immuno-histochemical approach.

Materials and Methods

Animals BALB/c and BCBA.F1 mice were bred at TNO, Rijswijk, The Netherlands and were used at 12-16 weeks of age. CBA/N mice were obtained from Bomholgård, Rye, Denmark.

Chemicals 3-amino-9-ethylcarbazole (A-5754), TNP sulfonic acid (TNBS, grade I) were obtained from Sigma, St. Louis, MO, USA. MHS (maleimidohexanoyl-n-hydroxysuccinimide ester, Pierce, Rockford, IL, USA); β-galactosidase (E. Coli-derived β-D-galactoside galactohydrolase, MW 540 KD) and X-Gal (5-bromo-4-chloro-3-indolyl-β-D-galactopyranoside) were obtained from Boehringer, Mannheim, FRG).

Antigens and bacteria TNP-Ficoll was prepared as previously described[15]. TNP-hydroxyethyl starch (TNP-HES) was a kind gift from the late Prof. Dr. J.H. Humphrey, Royal Post-graduate Medical School, London[7]. Group B streptoccoci (Dr. G. Teti, Istituta di Microbiologia, Messina, Italy) were labeled with 1,1,-dioctadecyl-3,3,3',3'-tetramethylindocarbocyanine perchlorate (DiI), a fluorescent lipophylic dye which diffuses into cell membranes[16].

Reagents The murine monoclonal antibody SH4.1C9, directed against TNP, has been described before[17]. The rat mAb MOMA-2, which recognizes monocytes and macrophages was a kind gift of Dr. G. Kraal, Free University, Amsterdam, The Netherlands[18]. Rabbit-horseradish-peroxidase (HRP) anti-rat Ig, rabbit-horse-radish-peroxidase anti-mouse Ig was obtained from Dakopatts, Copenhagen, Denmark. The affinity-chromotography purified anti-TNP mAb (SH4.1C9) was conjugated to β-galactosidase according to the procedure described by Deelder and De Water with some modifications[19,20].

Experimental design BALB/c mice were immunized i.v. with 100 μg TNP-Ficoll in 200 μl PBS. Mice were anaesthetized and killed at various time intervals: immediately and 5 min, 10 min, 20 min, 40 min, 2 hr, 7 hr, 1, 2, 3, 4, 5, 6, 7, 10, 14, 21 days after injection of the antigen.
CBA/N and BCBA.F1 mice were injected i.v. with 100 μg TNP-Ficoll and killed 7 hr after injection.
BALB/c mice were injected i.p./f.p. with 100 μg TNP-Ficoll and killed after 5 hr.
BALB/c mice were injected i.v. with 500 μg DiI labeled Group B streptococci and killed 7 hr after injection. Spleens and lymph nodes (i.c. of footpad immunization) were removed and immediately frozen in liquid nitrogen and stored at -70°C if not used immediately.

Complement depletion and immunization Complement depletion was performed according to Van de Berg et al.[21]. Complement depleted mice were i.v. injected with 10 μg or 100 μg TNP-Ficoll and were bled and killed 7 hr later. Spleens were removed and immediately frozen in liquid nitrogen and stored at -70°C. Complement depletion of serum was confirmed according to the procedure described by Van Dijk et al.[22].

Macrophage elimination with dichloromethylene-phosphonate (Cl_2MDP) containing liposomes Cl_2MDP containing liposomes were used to eliminate *in vivo* macrophages and were prepared as described earlier[23]. Macrophage depleted mice were i.v. injected with 5 or 50 μg TNP-Ficoll and killed 7 hr later. Macrophage elimination in the spleen was confirmed, as described previously[8].

Immunohistochemistry Immunohistochemistry was performed as earlier described[20,24].

TNP-Ficoll serum levels TNP-Ficoll serum levels were determined by ELISA. 96 well plate were coated with purified SH4.1C9 (5 $\mu g.ml^{-1}$) in PBS; then blocked with 0.1% gelatin. Seriel dilutions of mouse sera were added to wells, followed sequentially by Rabbit anti-TNP, Goat anti-Rabbit alkaline phosphatase, and substrate (p-nitrophenyl phosphate). Standard curves were prepared from TNP-Ficoll diluted in normal mouse serum.

RESULTS AND DISCUSSION

We studied the localization and kinetics of TNP-Ficoll in spleens of BALB/c mice from 5 min until 21 days after injection[20] (Fig. 1,2). Already 5 min after injection of 100 μg TNP-Ficoll there was a small amount of antigen detectable in the marginal zone. The amount of TNP-Ficoll gradually increased in the marginal zone and was slowly spreading to the follicular areas. At 2 hr after TNP-Ficoll injection all the splenic follicles were heavily loaded with TNP-Ficoll and this remained so for the next 5 hr. Thereafter, the amount of follicular TNP-Ficoll decreased gradually, but was still detectable 14 days after injection. During the whole experimental period (0 - 21 days) we observed a large amount of TNP-Ficoll in the macrophages of the marginal zone. From 40 min up to 2 days after TNP-Ficoll injection the amount of TNP-Ficoll in the red pulp gradually increased. Thereafter, the localization of TNP-Ficoll in the red pulp decreased gradually and at 21 days after injection no antigen was observed in the red pulp.

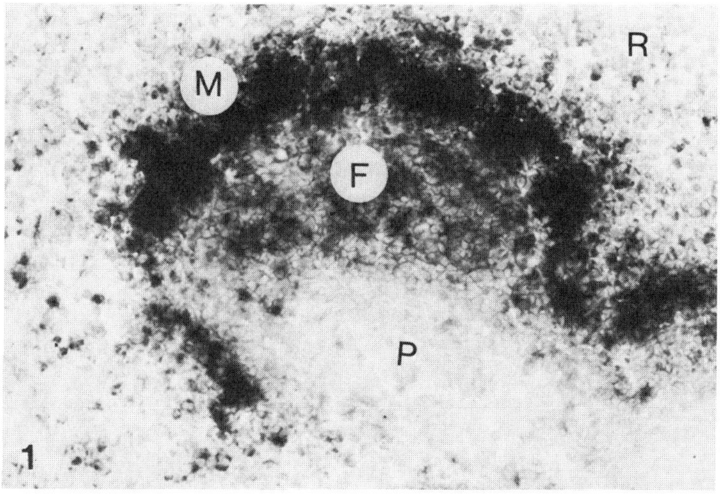

Fig. 1 TNP-Ficoll localization in the spleen 7 hours after i.v. injection. F, follicle; M, marginal zone, P, periarteriolar sheath, R, red pulp.

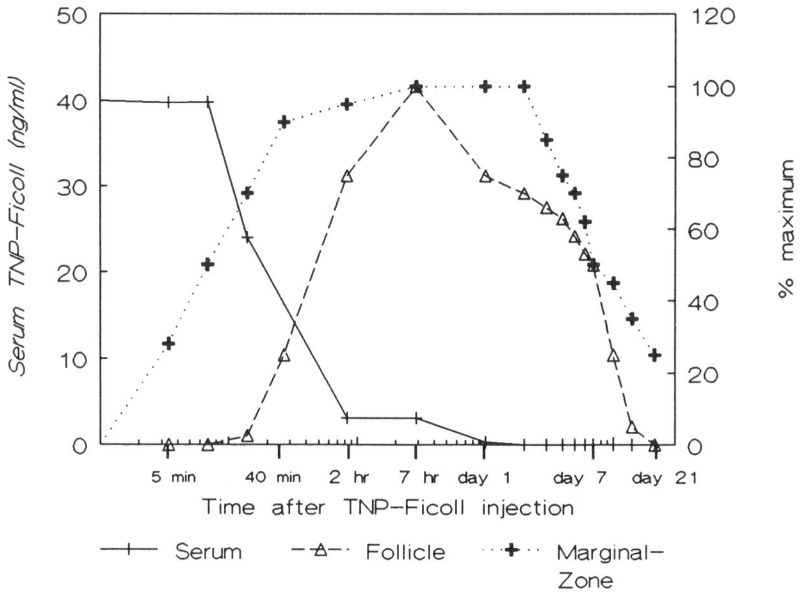

Fig. 2 Kinetics of i.v. injected TNP-Ficoll (100 μg) in sera and in splenic compartments of mice. TNP-Ficoll concentrations of serum were determined by ELISA. The amount of TNP-Ficoll was related to the maximal observed amount of TNP-Ficoll in follicle or marginal zone.

Two hours after injection we observed a striking decrease of TNP-Ficoll serum levels, simultaneously with a marked uptake of TNP-Ficoll by splenic cells (Fig. 2). We found in all tested strains of mice a similar localization pattern as we found in BALB/c mice. In a previous report we ruled out that antibodies were mediating the follicular localization of TNP-Ficoll[20].

Double staining for B cells and TNP-Ficoll revealed that a part of the injected TNP-Ficoll was bound by B cells in the marginal zone, in the follicles and later also in the red pulp of the spleen. However, the follicular localization pattern of TNP-Ficoll could also represent binding of TNP-Ficoll to follicular dendritic cells, as was observed for immune complexes later in the immune response. The close contact of B cells and follicular dendritic cells hampers the immunohistochemical identification of which cells bind TNP-Ficoll in the follicle.

Complement depletion resulted in a total abrogation of follicular TNP-Ficoll after injection of 10 μg of TNP-Ficoll (table 1). After injection of a higher dose of TNP-Ficoll (100 μg) we observed a substantial reduction of follicular TNP-Ficoll. As the population of complement receptor positive B cells in CBA/N mice is markedly lower than in BALB/c mice[25], the similar observed follicular localization pattern after TNP-Ficoll injection together with the complement dependency of this localization are more supportive for follicular dendritic cells than B cells as the TNP-Ficoll binding cells in the follicles.

Double staining revealed that TNP-Ficoll was taken up by the macrophages of marginal zone and red pulp. Macrophage elimination resulted in a drastic decrease of TNP-Ficoll in the marginal zone and red pulp, whereas follicular TNP-Ficoll increased markedly(table 1).

Table 1. Follicular TNP-Ficoll is correlated with humoral TI-2 immune response in case of low dose immunizations

Treatment[a]	CVF		CL_2MDP	
Dose of TNP-Ficoll (μg)	100	10	50	5
Follicular TNP-Ficoll[b]	↓	↓	↑	↑
Humoral immune response	=[26]	↓[26]	=[27]	↑[27]

[a] Treatment of mice: CVF represents complement-depleted mice and CL_2MDP represents macrophage-depleted mice

[b] Follicular localization is related to staining patterns of control

The above described complement depletion and macrophage elimination techniques enabled us to study the role of follicular TNP-Ficoll in the humoral TI-2 immune response (table 1). Matsuda et al.[26] observed in complement depleted mice, corresponding with a decrease in follicular TNP-Ficoll, a significant decrease of the antibody response against TNP-Ficoll in case of immunization with 10 μg of TNP-Ficoll. In addition, we observed in macrophage depleted mice, corresponding with an increase in follicular TNP-Ficoll, a significant increase of the humoral immune response against TNP-Ficoll[27]. These results indicate that the amount of follicular TNP-Ficoll is correlated with the titer of the antibody response against relatively low doses of TNP-Ficoll, suggesting a role for follicular TNP-Ficoll in the induction of TI-2 immune responses.

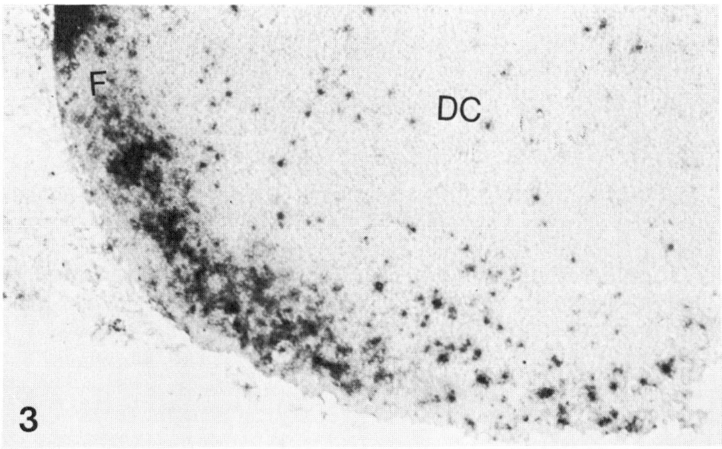

Fig. 3 Localization of TNP-Ficoll (100 μg) in a popliteal lymph node 7 hours after footpad injection. DC, deep cortex; F, follicle.

We also investigated the localization of TNP-Ficoll in popliteal lymph nodes after footpad immunization (Fig. 3). We observed the presence of a large amount of TNP-Ficoll in the subcapsular macrophages and in the macrophages of the medulla. Like in the spleen, we found a high amount of TNP-Ficoll in the follicles of the lymph nodes. Goud et al.[28] demonstrated that antibody responses against TI-2 antigens did not occur in lymph nodes. Since we found that both macrophages and follicular dendritic cells were binding and/or taking up TNP-Ficoll in the lymph node, our results indicate that other factors in the lymph node microenvironment, such as a deficiency of specific cytokine-producing cells and/or specific subsets of B cells/macrophages, may be the cause of this inability.

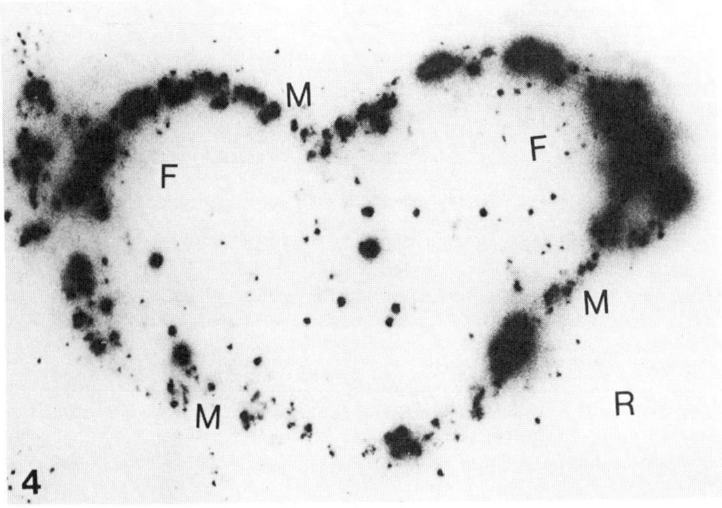

Fig. 4 Localization of Dye I labeled Group B streptococci in the spleen 5 hours after i.v. injection. F, follicle; M, marginal zone; R, red pulp.

After injection of DiI labeled Group B streptococci, a particulate TI-1 antigen, we found that the bacteria localized, like TI-2 antigens, in marginal zone macrophages and to a lower degree in red pulp macrophages (Fig. 4). This is in agreement with experiments of Chao and MacPherson[29], which suggested that the main function of marginal zone macrophages was to clear encapsulated pathogens from the circulation via the TI-2 antigen receptor. We observed hardly any antigen in the follicles, suggesting that only after pre-processing by macrophages fragments of the bacterial capsules may be bound by B cells and follicular dendritic cells[30].

Based on the available data concerning TI-2 immune responses we have postulated a model for TI-2 humoral immune responses[30]. We hypothesize that the role of marginal zone macrophages is not primarily found in antigen-presentation but rather in removal (i.c. of TI-2 antigens), killing and pre-processing of particulate antigen/bacteria, whereas follicular dendritic cells and B cells are most likely involved in the induction of immune responses to TI-2 antigens.

REFERENCES

1. P.L. Amlot, Impaired human antibody response to the thymus-independent antigen, DNP-ficoll, after splenectomy, *Lancet*, 1: 1008(1985).
2. G.T. Rijkers and D.E. Mosier, Pneumococcal polysaccharides induce antibody formation by human B lymphocytes in vitro. *J. Immunol. 135:1(1985).*
3. G. Koch, B.D. Lok, A. Van Oudenaren and R. Benner, The capacity and mechanism of bone marrow antibody formation by thymus independent antigens, *J. Immunol.* 128:1497(1982).
4. J.H. Humphrey, Splenic macrophages. Antigen presenting cells for TI-2 antigens, *Immunol. Lett.* 11: 149(1985).
5. J.H. Humphrey, and D. Grennan, Different macrophage populations distinguished by means of fluorescent polysaccharides. Recognition and properties of marginal zone macrophages, *Eur. J. Immonol.* 11: 221(1981).
6. P.L. Amlot, D. Grennan, and J.H. Humphrey, Splenic dependence of the antibody response to thymus-independent (TI-2) antigens, *Eur. J. Immunol.* 15: 508(1985).
7. J.H. Humphrey, Tolerogenic or immunogenic activity of hapten-conjugated polysaccarides correlated with cellular localization. *Eur. J. Immunol.* 11:212(1981).
8. E. Claassen, Y. Westerhof, B. Versluis, N. Kors, M. Schellekens, and N. Van Rooijen, Effects of chronic injection of sphingomyelin containing liposomes on lymphoid and non-lymphoid cells in the spleen. Transient suppression of marginal zone macrophages. *Br. J. Exp. Pathol.* 96: 865(1988).
9. E. Claassen, E., A. Ott, C.D. Dijkstra, W.J.A. Boersma, C. Deen, N. Kors, M. Schellekens, and N. Van Rooijen, Marginal zone of the murine spleen in autotransplants: Functional and histological observations in the response against a thymus independent type-2 antigen, *Clin. Exp. Immunol.* 77: 445(1989).
10. G. Kraal, H. ter Hart, C. Meelhuizen, G. Venneker and E. Claassen, Marginal zone macrophages and their role in the immune response against T-independent type 2 antigens. Modulation of the cells with specific antibody, *Eur. J. Immunol.* 19: 675(1989).
11. I. Scher, A.D. Steinberg, A.K. Berging and W.E. Paul, X-linked B-lymphocyte immune defect in CBA/N mice. II. Studies of mechanisms underlying the immune defect. *J. Exp. Med.* 142: 637(1975).
12. A.W. Griffioen, G.T. Rijkers, P. Janssens-Korpela, and B.J.M. Zegers, Pneumococcal polysaccharides complexed with C3b bind to human B lymphocytes via complement receptor type 2, *Infection and Immunity* 59: 1839(1991).
13. R.H. Carter, O. Spycher, Y.C. Ng, R. Hoffman and D.T. Fearon, Synergistic interaction between complement receptor type 2 and membrane IgM on B lymphocytes, *J. Immunol.*141: 457(1988).
14. R.H. Carter and D.T. Fearon, Polymeric C3dg primes human B lymphocytes for proliferation induced by anti-IgM, *J. Immunol.* 143:1755(1989).
15. E. Claassen, N. Kors, and N. Van Rooyen, Influence of carriers on the development and localization of anti-trinitrophenyl antibody forming cells in the murine spleen, *Eur. J. Immunol.* 16:271(1986).
16. E. Claassen, Post-formation fluorescent labelling of liposomal membranes, In vivo detection, localisation and kinetics, J. Immunol. Meth., 147:231(1992).
17. I. Claassen, N. Van Rooijen and E. Claassen., A new method for removal of monuclear phagocytes from heterogeneous cell populations in vitro, using liposome-mediated macrophage"suicide" technique, J. Immunol. Methods 134:153(1990).
18. G. Kraal, M. Janse, and E. Claassen, Marginal metallophilic macrophages in the mouse spleen: effects of neonatal injections of MOMA-1 antibody on the humoral immune response, *Immunol. Lett.* 17:139(1988).
19. Deelder, A.M. and De Water, R. J., A comparative study on the preparation of immunoglobulin-galactosidase conjugates, *Histochem. and Cytochem.* 29: 1273(1981).
20. A.J.M. Van den Eertwegh, J.D. Laman, M.M. Schellekens, W.J.A. Boersma and E. Claassen, Complement mediated follicular localization of T-independent type 2 antigens: the role of marginal zone macrophages rivisited, *Eur. J. Immunol.* 22:719(1992).

21. C.W. Van den Berg, P.C. Aerts, and H.J. Van Dijk, In vivo anti-complementary activities of the cobra venom factors from Naja Naja and Naja Haje, *Immunol. Meth.* 136:287(1991).
22. H. Van Dijk, P.M. Rademaker, and J.M.N. Willers, Determination of alternative pathway of complement activity in mouse serum using rabbit erythrocytes, *J. Immunol. Meth.* 36:29(1980).
23. N. Van Rooijen and E. Claassen, In vivo elimination of macrophages in spleen and liver, using liposome-encapsulated drugs: methods and applications. In: Liposomes as Drugs Carriers (ed. G. Gregoradis), John Wiley and Sons, Chichester, U.K. 1988. p131.
24. A.J.M. Van den Eertwegh, M.J. Fasbender, M.M. Schellekens, A. Van Oudenaren, W.J.A. Boersma, and E. Claassen, *In vivo* kinetics and characterization of IFN-γ-producing cells during a thymus independent immune response, *J. of Immunol.* 147: 439(1991).
25. I. Scher, The CBA/N mouse strain: An experimental model illustrating the influence of the X-chromosome on immunity. *Adv. Immunol.* 33: 1(1982).
26. T. Matsuda, G.P. Martinelli, and G. Osler, Studies on immunosuppression of cobra venom factor, II On responses to TNP-Ficoll and DNP-Polycrylamide, *J. Immunol.* 121: 2048(1978).
27. E. Claassen, Q. Vos, A. Ott, M.M. Schellekens, W.J.A. Boersma. Role of the splenic marginal zone in the removal of, and immune response against, neutral polysaccharides. In: Imhof B., Berrih-Aknin S, Ezine S: *Lympathic tissues and in vivo immune responses*, New York: Marcel Dekker 855(1991).
28. S.N. Goud, N. Muthusamy, and B. Subbarao, Differential responses of B cells from the spleen and lymph node to TNP-Ficoll. J. Immunol. 140: 2925(1988).
29. D. Chao, and G.G. MacPherson, Analysis of thymus-independent type 2 antigen uptake by marginal zone macrophages in thin slices of viable lymphoid tissue *in vitro*, *Eur. J. Immunol.* 20: 1451(1990).
30. A.J.M. Van den Eertwegh, W.J.A. Boersma and E. Claassen, Immunological functions and *in vivo* cell-cell interactions of T-lymphocytes in the spleen. Crit. Rev. Immunol. 11(6): (1992)

ULTRASTRUCTURAL HETEROGENEITY OF FOLLICULAR DENDRITIC CELLS IN THE HUMAN TONSIL

Louk H.P.M. Rademakers[1], Henk-Jan Schuurman[1,2]

[1]Department of Pathology
[2]Department of Internal Medicine
University Hospital Utrecht
P.O.Box 85.500, 3508 GA Utrecht, The Netherlands

INTRODUCTION

Follicular dendritic cells (FDC) form the framework of the germinal centre (GC). They act as accessory cells, by providing a favourable microenvironment for the germinal centre reaction.[1-2] Secondary GCs consist of a basal dark zone, mainly composed of immature blast cells and showing extensive mitotic activity, and an apical light zone containing a heterogeneous population of lymphoid cells. FDC are present both in the dark zone and in the light zone of the GC. Only few properties of the FDC present in these zones are known. The FDC density framework shows only minor differences between the dark and the light zone. Cytochemical localization of immune-complexes and the ectoenzyme 5'-nucleotidase indicate differences in the expression of these markers in FDC in these zones.[3] These staining patterns suggest differences in the functional stage of FDC. To date, the ultrastructural morphology of FDC in relation to these functionally different GC compartments is not well established. In human GCs the different types of lymphoid and non-lymphoid cells can be distinguished on the basis of their ultrastructure.[4] We performed a qualitative and quantitative ultrastructural study of the FDC population in the dark and the light zones of GCs. This approach was aimed to define the light and the dark zones of GCs, and to study - on a structure-function basis - the contribution of FDC to these distinct microenvironments.

MATERIAL AND METHODS

Fresh tonsils of children were obtained at tonsillectomy. Small tissue specimen were fixed in a solution of 2.5% glutaraldehyde (v/v), 2% paraformaldehyde (w/v) in 0.1 M Na-cacodylate buffer (pH 7.4), supplemented with 2mM $CaCl_2$ for at least 24 h at 4°C. Post fixation was done in 1% OsO_4 (w/v) in the same buffer and embedding in Epon according to routine procedures. GCs showing distinct dark and light zones of sufficient size for further analysis were selected by light microscopy. Ultrathin sections were contrasted with lead citrate followed by uranyl magnesium acetate treatment and viewed in a Philips 201c electron microscope. The frequency of FDC within the total GC cell population was determined on low power micrographs representing an area of 14,000 μm^2 of both the dark and light zone of ten randomly chosen GCs.

Morphological analysis of FDC was done on micrographs of 420 FDC originating from the dark and light zones of 32 GCs. For this purpose micrographs of all FDC (6-8) present in the dark zone were evaluated. In the light zone a similar number of FDC were randomly chosen in the larger FDC population.

RESULTS AND DISCUSSION

At the ultrastructural level, FDC were present between the lymphoid cells of the GCs as mononucleate or binucleate cells[5]. They form a tight framework with slender cytoplasmic extensions which are interconnected by desmosomes. A marked feature is the presence of fine villous cytoplasmic extensions covered with electron-dense (immune complex) deposits which interdigitate with cell processes of adjacent FDC, centroblasts, and centrocytes. In the dark zone FDC comprised 2.5% of total cell population. In the light zone a significantly higher value of 5.3% was observed.

Within the GC framework the ultrastructural features of FDC varied. Seven FDC subtypes (FDC.1-FDC.7) could be defined on the basis of content of cell organelles, appearance of cellular extensions, presence of electron-dense deposits and

Table 1. Features of FDC subtypes in germinal centres of the human tonsil

FDC subtype	*Ultrastructure*
FDC.1	Very sparse cell organelles; when present, poorly developed. Filamentous cytoplasmic matrix. Fine villous extensions of the plasma membrane absent.
FDC.2	Stellate cells with few organelles; polyribosomes. Submembranous intermediate filaments. Fine villous plasma membrane extensions covered with electron-dense (immune complex deposits absent.
FDC.3	Rounded cells with moderately developed rough endoplasmic reticulum (RER) and inconspicuous Golgi area. Some plumb cytoplasmic extensions covered with electron-dense (immune-complex) deposits. Submembranous filament condensations
FDC.4	Rounded cells with well developed RER and Golgi system. Some submembranous filament condensations. Tiny villous plasma membrane protrusions with dense (immune-complex) deposits.
FDC.5	Like FDC.4, but with a larger amount of cytoplasm. RER, Golgi area and villous plasma membrane web extremely developed. Many Golgi-associated vesicles. Electron-dense (immune-complex) deposits.
FDC.6	Elongated cells with large proportion of cytoplasm. Both RER (dilated) and Golgi area are inconspicuous. Lysosomes. Moderate density of villous extensions with dense (immune complex) deposits which are partly engulfed by broad FDC extensions.
FDC.7	Stellate cells with electron-dense cytoplasmic and nuclear matrix. Dilated RER; inconspicuous Golgi area; clusters of free ribosomes. Lysosomes. Villous extensions and dense deposits not regularly present.

occurrence of intermediate filaments (Table 1, Figs. 1a,b, 2a-d, 3). These FDC subtypes shared general morphological features and showed gradual differences in cell organelle organisation and plasma membrane specializations. All appearances of FDC occurred intermingled in the reticular framework of the germinal centre. Animal studies have shown that mesenchymal cells can transform to FDC and attain comparable morphological features as the GCs develop.[6] From these observation the FDC subtypes may be interpreted in terms of differentation, activation and regression of one particular cell type rather than different cell types having each specific functional properties. Solely from an ultrastructural view on possible function we designated these FDC forms as primitive (FDC.1), undifferentiated (FDC.2), intermediate (FDC.3), differentiated (FDC.4), secretory (FDC.5), regressive, pale (FDC.6), and regressive, dark type (FDC.7). This view is further supported by an analysis of abnormal germinal centres of HIV-1-infected lymph nodes in which the distribution pattern of FDC subtypes particular FDC.4 and FDC.5) was related to the blast and centrocyte content of GCs without alterations in the density of FDC (Rademakers et al.; this issue). In addition, regressive changes at the level of cell organelles in FDC (i.e. swelling of RER and incidence of lysosomes were related with the occurrence of FDC.6 and FDC.7.[7]

The distribution pattern of the FDC subtypes differed. In the dark zone, FDC.1, FDC.2 and FDC.3 with a low degree of differentiation formed the majority of the FDC population (with 21%, 39%, and 31% respectively). In the light zone, the differentiated FDC.4 and FDC.5 predominated in the light zone (28%, and 48% respectively). FDC.6 was mainly located in the apical part of the light zones, whereas FDC.7 was infrequently observed in the light zones. Type FDC.7 is more fequent in abnormal GCs.

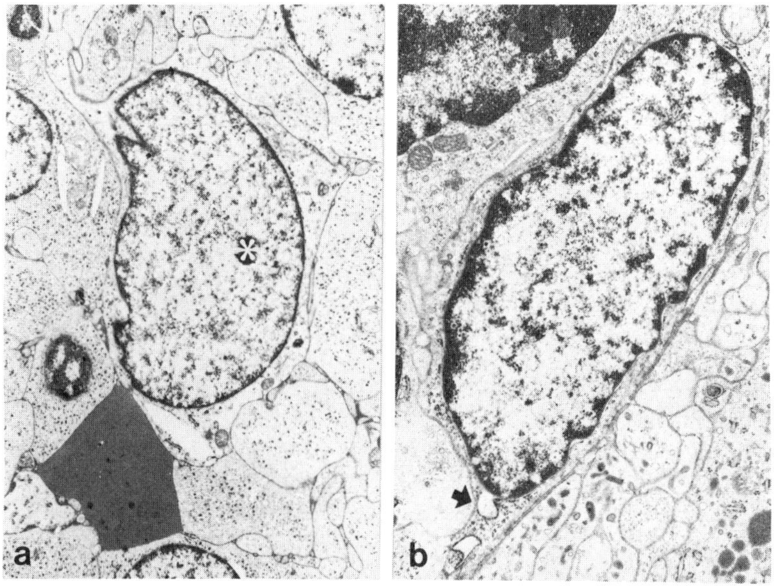

Figure 1a,b. Electron micrographs of FDC subtypes in dark zones. **a** FDC.1 (asterisk), rounded cell with sparce cell organelles, some scattered polyribosomes. x 8,000. **b** FDC.2 showing elongated extensions, poorly developed RER (arrow) and polyribosomes. x 9,000.

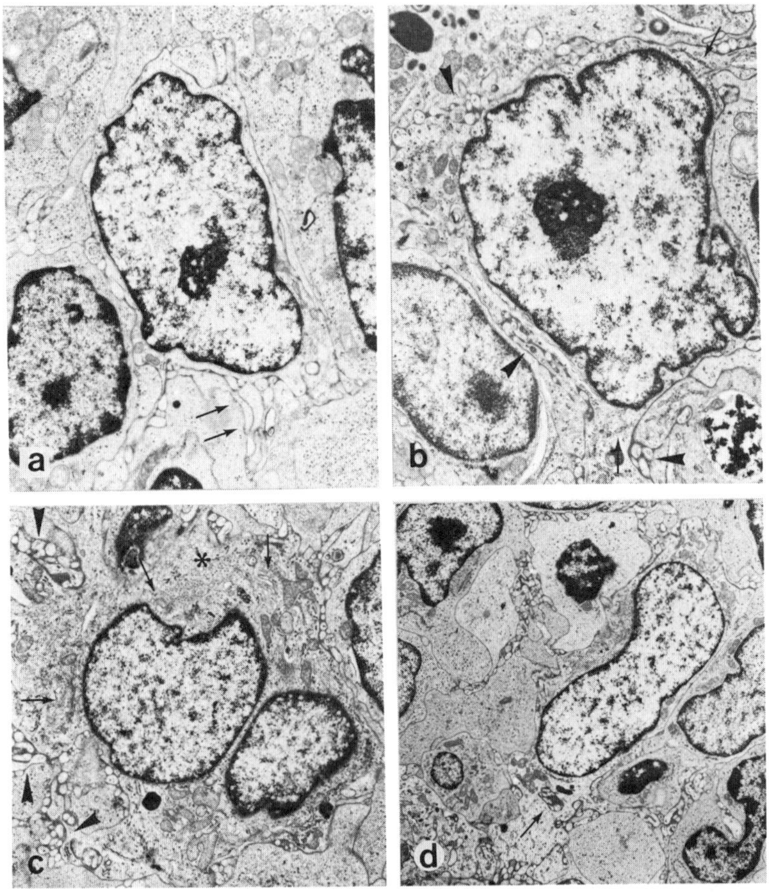

Figure 2a-d. FDC subtypes in dark (a) and light zones (b, c, d). **a** Dark zone, FDC.3, Rounded cell with blunt extension of the cytoplasm (arrows). Dense deposits are not obvious. x 7,000. **b** FDC.4 with numerous villous extensions (arrowheads) covered with electron-dense deposits. Arrows: RER cisterns. x 7,000. **c** FDC.5. Binucleated cell with large cytoplasm. Many villous extensions with dense deposits (arrowheads). Large Golgi area with several Golgi complexes (arrows) and large area with Golgi-associated vesicles (asterisk). x 6,000. **d** FDC.6 with abundant cytoplasm containing relative few organelles. Arrow: villous extensions within the cytoplasm. x 3,500.

The dark and the light zones differ greatly by their content of centroblasts, centrocytes and mitotic activity.[8] The occurence of relative low numbers of undifferentiated, physiologically inactive, FDC in the dark zone suggests that FDC do not promote the proliferation of GC cells. FDC.4 and FDC.5, which exclusively occur in a high density in the light zones, are characterized by immune-complex deposits on the cell surface and by a well developed RER and Golgi complex associated with numerous vesicles. The latter feature indicates that FDC.4 and, even more, FDC.5 can be regarded as secretory active cells. The high incidence FDC with such features may suggest that in the light zone suggest that FDC themselves play a role in the maintenance of the microenvironment which favours B-cell differentiation. This is beside a role in antigen presentation and clonal selection of memory B-cells[9,10]. By modulation of their secretory activity FDC may stimulate differentiation or allow

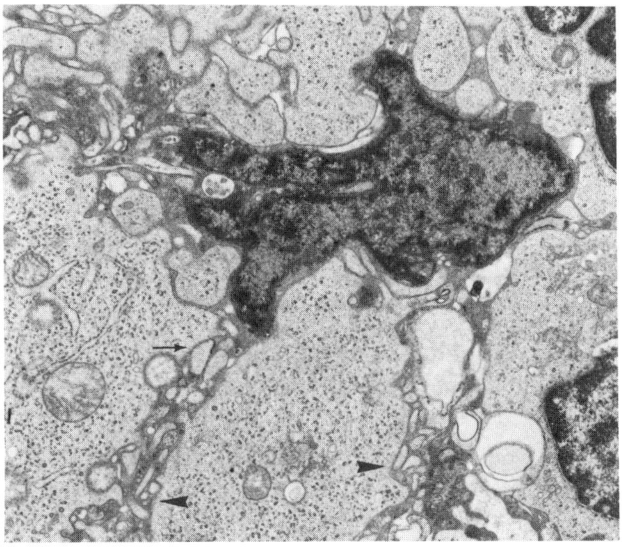

Figure 3. FDC.7 with electron-dense cytoplamic and nuclear matrix. Elongated cytoplasmic extensions elapse between blast cells. RER cisterns (arrow) are extremely dilated. Villous extensions with dense deposits are present (arrowheads). x 8,300.

proliferation of selectively trapped B-cells. A participation of soluble factors in differentiation of GC B-cells has recently been emphasized. Recombinant CD23, a B-cell differentiation antigen which is also expressed on FDC in the light zone, promotes the survival of GC B-cells *in vitro* and induces plasmacytoid differentiation in synergy with interleukin 1α. Rescue of GC B-cells from apoptosis can also be achieved by liganding the CD40 molecule and antigen receptors by monoclonal antibodies.[11]

In conclusion, our ultrastructural observations show that follicular dendritic cells form a heterogeneous cell population having a different distribution pattern of the distinct FDC subtypes in the light and dark zones of GCs. These findings suggest that the dark and light zones have each a specific microenvironment made by FDC forming the basis for the distinct processes of proliferation and differentation of B-lymphocytes within the GC.

REFERENCES

1. E. Heinen, N. Cormann, and C. Kinet-Denoel, The lymph follicle: a hard nut to crack, *Immunol. Today* 9:240 (1989).
2. A.K. Szakal, M.H. Kosco, and J.G. Tew, Microanatomy of lymphoid tissue during humoral immune responses: structure function relationships, *Annu. Rev. Immunol.* 7:91 (1989).
3. H. Stein, J. Gerdes, and D.Y. Mason, The normal and malignant germinal centre, *Clin. Haematol.* 11:531 (1982).
4. J.P.J. Peters, L.H.P.M. Rademakers, RJ de Boer, and J.A.M. van Unnik, Cellular composition of follicles of follicle centre cell lymphomas in relation to germinal centres of reactive lymph nodes. A morphometrical electron microscopical study, *J. Pathol.* 153:233 (1987).
5. J.P.J. Peters, L.H.P.M. Rademakers, J.M.M. Roelofs, D. de Jong and J.A.M. van Unnik, Distribution of dendritic reticulum cells in follicular lymphoma and reactive hyperplasia. Light microscopic identification and general morphology, *Virch .Arch. [Cell Pathology]* 46:215 (1984).
6. U. Heusermann, K.-H. Zurborn, L. Schoeder and H.J. Stutte, The origin of the dendritic reticulum cell, An experimental and enzyme-histochemical study on the rabbit spleen, *Cell Tissue Res.*, 192:1 (1980).
7. L.H.P.M. Rademakers, H.-J. Schuurman, J.F. de Frankrijker, and A. van Ooyen, Cellular composition of germinal centres of lymph nodes after HIV-1 infection: evidence for an inadequate support of germinal centre B lymphocytes by follicular dendritic cells, *Clin. Immunol. Imunopathol.* 62:148 (1992).
8. L.H.P.M. Rademakers, Dark and light zones of germinal centres of the human tonsil: an ultrastructural study with emphasis on the heterogeneity of follicular dendritic cells, *Cell Tissue Res.*, in press.
9. N. van Rooijen, Direct intrafollicular differentiation of memory B cells into plasma cells, *Immunol. Today* 11:154 (1990).
10. J.G Tew, M. H. Kosco, and A.K. Szakal AK, The alternate antigen pathway, *Immunol. Today* 10:229 (1989).
11. Y.-J. Liu, J.A. Cairns, M.J Holder, S.D. Abbot, K.U. Jansen, J-Y Bonnefoy, J. Gordon and I.C.M. McLennan. Recombinant 25kDa CD23 and interleukin 1α promote the survival of germinal centre B-cells: evidence for bifurcation in the development of centrocytes rescued from apoptosis, *Eur. J. Immunol.* 21:1107 (1991).

ULTRASTRUCTURAL ANALYSIS OF HUMAN LYMPH NODE FOLLICLES AFTER HIV-1 INFECTION

Louk H.P.M. Rademakers[1], Henk-Jan Schuurman[1,2]
and Jack F. de Frankrijker[1]

[1]Department of Pathology
[2]Department of Internal Medicine
University Hospital Utrecht
P.O.Box 85.500, 3508 GA Utrecht, The Netherlands

INTRODUCTION

Infection with the human immunodeficiency virus type 1 (HIV-1) may result in persistent generalized lymphadenopathy (PGL). Lymph nodes in PGL most commonly show a florid follicular hyperplasia (FH) in which regressive changes, such as loss of mantle zones and fragmentation of the germinal centre (GC) can be observed.[1,2] Fragmentation of the framework of follicular dendritic cells (FDC) has been demonstrated by immunohistochemistry.[1,3] Intact HIV-1 virus particles are regularly found between the extensions of FDC within GCs of HIV-1 infected individuals.[4,5] These observations have been associated with a cytopathic action of the virus on FDC.[6,7] A proportion of FDC may be infected and produce virus.[8] FDC function as accessory cells in GCs. Impairment of FDC function by cytopathic effects of the virus may influence GC microenvironment. The distinct types of B-cells can be distinguished and quantified at the ultrastructural level.[9,10] Ultrastructural analysis of of the light and dark zones in tonsils shows that FDC subtypes can be recognized representing distinct differentiation stages of FDC (Rademakers and Schuurman, this issue). We applied this approach to GCs in lymph nodes of HIV-1 infected patients to analyze alterations in the GC microenvironment.

METHODS

Twenty-three lymph nodes were studied. Fifteen lymph node biopsies with follicular hyperplasia originated from HIV-1 positive patients. Eight lymph nodes with HIV-1 negative follicular hyperplasia served as a control group. Lymph node specimens were fixed in 2.5% glutaraldehyde (v/v) and 2.5% paraformaldehyde (w/v) in 0.1M Na-cacodylate buffer (pH 7.4) supplemented with 2 mM $CaCl_2$ and processed for electron microscopy according to routine procedures.

Ultrastructural analysis of the cell composition of the lymph node germinal centres was done on low power micrographs originating from three GCs from each lymph node followed by additional data analysis on germinal centre cell types by means of pattern recognition methods. These procedures were performed as earlier described.[9] FDC subtypes (FDC.1-FDC.7) were defined according to ultrastructural features summarized elsewhere in this issue (Rademakers and Schuurman). The frequency distribution of FDC subtypes in lymph nodes was investigated by grading all FDC present in sections of 3-6 germinal centres directly in the electron microscope. This analysis was based on 56-151 FDC per biopsy.

RESULTS AND DISCUSSION

On the basis of the relative frequencies of lymphoid cells types (blast: i.e cleaved blast, immunoblast and centroblast; centrocyte, centroplasmacytoid cell, lymphocyte multilobated cell, and plasma cell; Fig.1) and non-lymphoid cell types (i.e histiocytic reticulum cell or tigible body macrophage, and follicular dendritic cell), two groups of lymph nodes were distinguished by cluster analysis. Six HIV-1 positive lymph nodes as well as all HIV-1 negative lymph nodes were placed in group 1; the remaining 9 HIV-1 positive lymph nodes formed group 2. In contrast to the GCs of lymph nodes of group 1, those of HIV-1 positive lymph nodes of group 2 were characterized (Table 1) by a predominance of lymphoid blasts subtypes with the centroblast as the main component, and by decreased numbers of centrocytes. This resulted in a high centroblast to centrocyte ratio. In GCs of tonsils a light and a dark zone can be distinguished, which each are regarded as different compartments.[10]. Such a zonal structure could not be recognized in the present lymph nodes. However, aspects of these zones were found in the in the HIV-1 positive and HIV-1 negative lymph nodes. The cell composition of GC in group 1 was similar to that of the light zones of tonsillar GCs. This is considered as a compartment of B-cell differentiation. The cell composition of group 2 was almost identical with that of the dark zones of GCs. This is the GC compartment in which proliferation of GC cells is mainly located. However, the high blast counts in GCs of group 2 were not accompanied by a significant increase of mitotic activity when compared with that of group 1. This indicates that in lymph nodes of group 2, the differentiation of blasts into centrocytes is impaired.

Table 1. Mean values of the relative frequencies of the main lymphoid B-cell types and FDCs in GCs of HIV-1 positive (HIV-1^{+}) and HIV-1 negative (HIV-1^{-}) individuals

	Group 1		*Group 2*
Cell type	**HIV-1^{-}** (n=8)	**HIV-1^{+}** (n=6)	**HIV-1^{+}** (n=9)
Lymphoblastoid cells	26.2	26.4	46.8*°
Centrocytes	51.5	46.1	30.9*°
Blast/centrocyte ratio	0.52	0.58	1.6*°
Mitosis in lymphoid cells	1.7	1.8	2.7
FDCs	4.2	6.1	4.2

*: significant difference with HIV-1^{-}; °: significant difference with HIV-1^{+}/group 1 ($P \leq 0.05$).

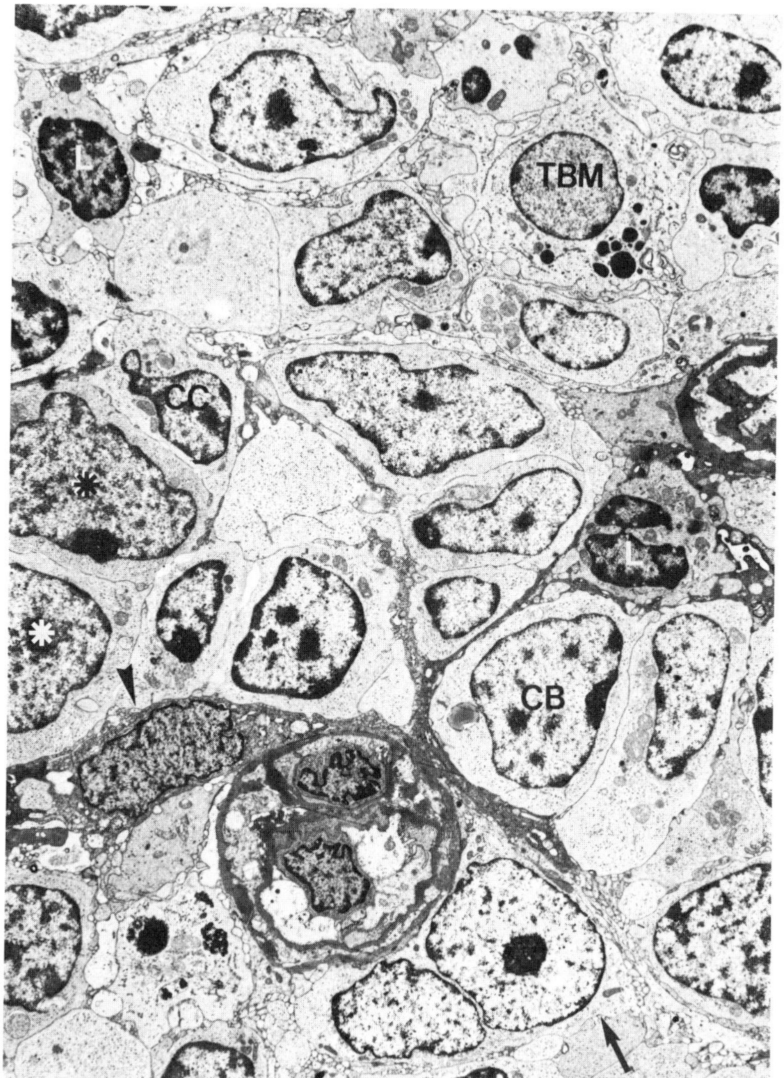

Figure 1. Survey of a GC in a lymph node of a HIV-1$^+$ patient of group 2. The GC is mainly composed of centroblasts (CB) and cleaved blasts (asterisks). Two types of FDC are shown: FDC.6 (arrow) and FDC.7 (arrowhead). CC: Centrocyte; L: lymphocyte; TBM: tigible body macrophage. x 4,000.

The high blast to centrocyte ratio present in the HIV-1 positive GCs of group 2 was not related to differences in FDC numbers. This observation is in contrast to the dark and the light zones of tonsils and is suggestive for an altered participation of FDC in GC B-cell processing.

An intact FDC meshwork was present in all GCs. No signs of fragmentation were observed. FDC showed their usual ultrastructure, bearing numerous fine cytoplasmic extensions covered with electron-dense immune-complex deposits (Fig.1). After HIV-1 infection retroviral particles were regularly observed between the extensions of FDC. These particles were found in all HIV-1 positive lymph nodes of group 1, and in seven of the nine lymph nodes of group 2. As in tonsils, seven FDC subtypes could be recognized on the basis of their ultrastructural characteristics. The distribution pattern of FDC subtypes in GCs of lymph nodes of group 1 and 2 was not identical (Table 2). In group 1, a pattern with the intermediate form FDC.3 and the highly differentiated forms FDC.4 and FDC.5 as the main constituents was observed. In group 2, the intermediate form FDC.3 predominated, and the relative frequencies of the differentiated forms FDC.4 and FDC.5 were lower. There was a relative increase in the frequency of the regressive forms FDC.6 and FDC.7. The FDC subtype distribution patterns of the HIV-1 positive and HIV-1 negative lymph nodes of group 1 were comparable and showed similarities with that of the light zones of GCs. In contrast, the distribution pattern observed in group 2 was less marked and differed from that of both the dark and light zones.[10] These findings suggest that the differentiation stage of FDC is changed. This is in line with the observation of a higher incidence of widened cisterns of the RER and of lysosomal structures in FDC after HIV-1 infection, indicative of an altered cellular physiology.[11]

Table 2. Mean values of the frequencies of FDC subtypes relative to the total FDC population

FDC subtype	*Group 1* HIV-1$^-$ (n=8)	*Group 1* HIV-1$^+$ (n=6)	*Group 2* HIV-1$^+$ (n=9)
FDC.1	0.1	0.2	0.4
FDC.2	5.2	6.4	7.1
FDC.3	21.6	26.5	23.0
FDC.4	32.6	33.5	20.1$^{*\circ}$
FDC.5	20.9	13.7	11.6*
FDC.6	11.6	11.8	21.6°
FDC.7	9.9	7.9	15.8°

*: significant difference with HIV-1-; $^\circ$: significant difference with HIV-1$^+$/group 1 ($P \leq 0.05$).

We conclude that in 9 out of 15 lymph nodes with HIV-1 associated lymphadenopathy early alterations in FDC morphology are observed suggestive of an immature and inactive state, but without alterations in germinal centre architecture. These alterations in FDC coincide with an inversed blast/centrocyte ratio, but without a change in mitotic activity. These results suggest that the after HIV-1 infection, FDC inadequately support the B-cell differentiation in GCs. These changes in the FDC microenvironment may proceed the involution of GC in stages of follicular degeneration and fragmentation.

REFERENCES

1. H.-J. Schuurman, Ph.M. Kluin, F.H.J. Gmelig Meyling and L. Kater, Lymphocyte status of lymph node and blood in acquired immunodefiency syndrome (AIDS) and AIDS-related complex disease, *J. Pathol.* 147:269 (1985).
2. B.F. Burns, G.S. Wood and R.F. Dorfman, The varied histopathology of lymphadenopathy in the homosexual male, *Am. J. Surg. Pathol.* 9:287 (1985).
3. G. Janossy, A.J. Pinching, M. Bofill, J. Weber, J.E. McLaughlin, M. Ornstein, K. Ivory, J.R. Harris, M. Favrot, and D.C. Macdonald-Burns, An immunohistological approach to persistent lymphadenopathy and its relevance to AIDS, *Clin. Exp. Immunol.* 59:257 (1985).
4. K. Tenner-Racz, P. Racz, M. Dietrich and P. Kern, Altered follicular dendritic cells and virus-like particles in AIDS and AIDS-related lymphadenopathy, *Lancet* 1:105 (1985).
5. J.A. Armstrong and R. Horne, Follicular dendritic cells and virus-like particles in AIDS-related lymphadenopathy, *Lancet* 2:370 (1984).
6. P.U. Cameron, R.L. Dawkins, J.A. Armstrong and E. Bonifacio, Western blot profiles, lymph node ultrastructure and viral expression in HIV-infected patients: a correlative study, *Clin. Exp. Immunol.* 68:465 (1987).
7. H.-J. Schuurman, W.J.A. Krone, R. Broekhuizen and J. Goudsmit, Expression of RNA and antigens of human immunodeficiency virus type-1 (HIV-1) in lymph nodes from HIV-1 infected individuals, *Am. J. Pathol.* 133:516 (1988).
8. H.K. Parmentier, D. van Wichen, D.M.D.S. Sie-GO, J. Goudsmit, J.C.C. Borleffs, and H.-J. Schuurman, HIV-1 infection and virus production in follicular dendritic cells in lymph nodes. A case report, with analysis of follicular dendritic cells, *Am. J. Pathol.* 137:247 (1990).
9. J.P.J. Peters, L.H.P.M. Rademakers, R.J. de Boer, and J.A.M. van Unnik, Cellular composition of follicles of follicle centre cell lymphomas in relation to germinal centres of reactive lymph nodes. A morphometrical electron microscopic study, *J. Pathol.* 153:233 (1987).
10. L.H.P.M. Rademakers, Dark and light zones of germinal centres of the human tonsil: an ultrastructural study with emphasis on the heterogeneity of follicular dendritic cells, *Cell Tissue Res.* in press.
11. L.H.P.M. Rademakers, H.-J. Schuurman, J.F. de Frankrijker, and A. van Ooyen, Cellular composition of germinal centres of lymph nodes after HIV-1 infection: evidence for an inadequate support of germinal centre B lymphocytes by follicular dendritic cells, *Clin. Immunol. Immunopathol.* 62:148 (1992).

EMPERIPOLESIS OF LYMPHOID CELLS BY HUMAN FOLLICULAR DENDRITIC CELLS IN VITRO

Rikiya Tsunoda[1], Masayuki Nakayama[2], Ernst Heinen[3], Katsuya Miyake[1], Kazunori Suzuki[1], Hirooki Okamura[1], Naonori Sugai[1] and Mizu Kojima[4]

1. Department of Anatomy Histology, Fukushima Medical College, Fukushima, Japan
2. Japan Rail Road General Hospital, Sapporo
3. Department of Human Histology, University of Liege, Belgium
4. Mito Saiseikai Hospital, Mito

INTRODUCTION

In a previous conference[1], we reported the establishment of a method for the isolation of FDCs (follicular dendritic cells) from human palatine tonsils. In vitro, within 6 days, FDCs rapidly dedifferentiated into flat plastic-adherent cells devoid of surface markers.

It has been reported that thymic epithelial, stromal and nurse cells (TNC), or their cell lines, showed emperipolesis of T-cells and maintained their nurse function with regard to the maturation and the differentiation of T-cells, even after cultivation in vitro. In the present experiments, we attempted to determine the ability of cultured FDC to perform emperipolesis.

MATERIAL AND METHODS

Surgically-removed fresh palatine tonsils were used. The method of enucleating and consequently digesting the lymph follicles, and purification of FDC-clusters has been established in detail elsewhere.

Half of each tonsil was shredded with scissors. Lymphocyte and monocyte fractions (B+T+M) were recovered from floating Lymphoprep (Nycomed, Norway) fractions. T-cell-enriched and B-cell-enriched populations were purified by rosetting (performed twice). Fibroblast cell lines (TIG-7 & TIG-107), gifted from JCRB Cell Bank, Tokyo, Japan., were used for the control experiments.

To examine true emperipolesis, FDCs, at various days of culture, were detached from plastic with trypsin, adjusted to $1x10^4$ cells/ml, combined with thawed tonsillar lymphocytes (5

$x10^5$ cells/ml) recovered from deep-freeze storage, and cultured in hanging-drops in Terasaki plates (S.B. Medical Japan) for 6 to 24h by Nakayama method[2]. After the indicated periods, cells were harvested, washed and cytocentrifuged on glass slides, and then examined to determine the number of FDC-lymphocyte spherical complexes. To examine pseudoemperipolesis (partial envelopment), FDCs adhering on plastic covers were cocultured with lymphocytes ($5x10^5$ cells/ml) for 6 to 24 h. For inhibition test, adherent cultured FDCs and/or lymphocytes (B+T+M) were preincubated with monoclonal antibody against MHC-I & II at saturating binding concentrations of ascites fluid or purifed antibody (1/100 dilution) for 30 min at 4°C, then washed three times with cold PBS, and used for the pseudoemperipolesis study described above.

RESULTS

True emperipolesis

Under the phase-contrast microscope, detached free FDCs in cultures were gradually surrounded by lymphocytes; some of them revealed a typical rosette aspect within 60 min of co-culture. After 6 hours, such FDCs had almost disappeared, and instead, large spherical clusters of lymphocytes covered by FDCs. In 6-day-old cultured FDCs, approximately 60% of the cells formed such complexes. It appeared that FDCs enveloped fewer lymphocytes and formed smaller clusters as the duration of culture increased. The incidence of emperipolesis in co-cultures of FDCs and other allotype lymphocytes is slightly lower. But, there was no clear difference in the size and shape of the clusters in comparision to culture with autologous lymphocytes. TIG-7 and TIG-107 did not show any spherical complex formation with tonsilar lymphocytes (Table 1).

The reconstructed spherical structures of lymphocytes and FDC in 6-day culture were observed by transmission electron microscopy. The fundamental characteristics of these

Table 1. Incidence(%) of True Emperipolesis

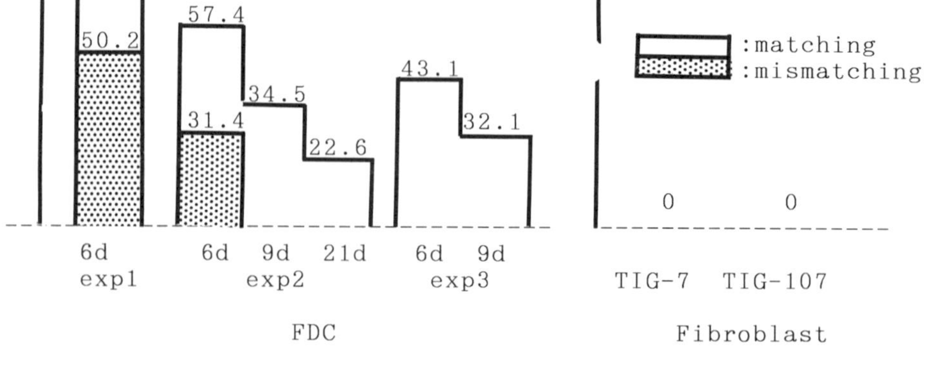

reconstructed clusters were similar to those of the freshly-isolated ones. Two main observations were made regarding the reconstructed FDC-lymphocyte clusters. Firstly, the ultrastructure of the FDCs reflected well the culture period: the developed and ramified dendritic cytoplasmic processes of the FDCs, apparent in-situ and in freshly prepared FDCs, disappeared during culture and were replaced with polypoid pseudopodia; an increase of actin filaments was also noted in these cells. Secondly, some of the enveloped lymphocytes were degenerated and their nuclei showed fragmentation or pycnosis; however, others were found to be in mitosis.

Pseudoemperipolesis

Almost all of the 9-day-cultured FDCs formed clusters with lymphocytes within a few hours in mixed cultures. Simultaneously, some of the lymphocytes began to move under the extended cytoplasmic membranes of the FDCs. The maximum number of trapped lymphocytes per FDC was around 150, although the number of lymphocytes interacting with FDCs varied during the first 6 hours (Table 2). After an additional 18 hours of cultivati[illegible] some lymphocyte populations adhering to the surface of FDCs started to degenerate and to become detached. In contrast, most of the lymphocytes located under the cytoplasmic extensions of the FDCs remained unchanged. About 40 % of TIG-7 and 17 % of TIG-107 showed pseudoemperipolesis for tonsilar lymphocytes, but the number of engulfed lymphocytes was one or two. After fixation and staining, observation of the enveloped lymphocytes was easier. A small percentage of them showed mitotic figures after the first 6 hours of co-culture. Degeneration or fragmentation of the nuclei also appeared in a few of the engulfed lymphocytes during culture periods of 6 to 24h.

Table 2. Pseudoemperipolesis

	FDC(%)(emperip.+)	trapped Ly*	FDC(%)(mitotic Ly)
matching	97.5	21.4	33.3
mismatching		2.5	0
matching	95.5	12.8	15.0
mismatching			4.8
matching	98.5	8.8	14.5
mismatching			2.2
TIG-7	40	1.7	0
TIG-107	17	0.9	0

(open bar): matching
(stippled bar): mismatching
*: number

The phenotype of the enveloped or trapped lymphocytes was independent of the experimental design (culture delays, allotype, etc.). This means that most of the FDCs interacted with both B- and T-cells in the B+T+M fraction. Pure FDC T-cell clusters were not found. In the lymphoid cell population covered by the cytoplasmic membrane of the FDCs. B-cells outnumbered T-cells. Few FDCs showed pseudoemperipolesis with regard to SIgD-positive, CD5-positive B-cells. Among the population of T-cells trapped by FDCs, CD4-positive cells were more numerous than CD8-positive cells. During the co-culture of FDC and purified B-cells, no pseudoemperipolesis with regard to contaminating T-cells was found. During the mixed reaction of FDC and the purified T-cell population, most of the trapped lymphocytes exhibited T-cell surface markers. No monocytes or macrophages reacted with the FDCs. Some mitotic cells exhibited B-cell features, with positive staining for SIgM or CD20. Table 2 also shows the incidence of FDC-pseudoemperipolesis with allotypic lymphocytes. There was no clear difference in form clusters with autologous lymphoytes. However, the mitotic rate among the trapped lymphocytes in matched mixed cultures appeared to be higher than that in mismatched mixed cultures.

Inhibition test by monoclonal antibodies

Table 3 shows the results of experiments related to the inhibition of psuedoemperipolesis by monoclonal antibodies against MHC-I & II. The incidence of FDC-pseudoemperiposis and the number of enveloped lymphocytes did not differ in the different experiments. The mitotic rate, however, was higher in the controls than in experiments conducted in presence of the MoAb.

Table 3. Inhibition test by MoAb against MHC-I & II

FDC	BTM	FDC%(emperip.+)	Trapped Ly	FDC%(mitotic Ly)
NT*	NT	95.5	12.8	15.0
MHCII	NT	95.5	13.8	6.3
NT	MHCII	96.7	9.8	1.3
MHCII	MHCII	97.6	9.8	2.5
MHCI	NT	97.6	10.1	2.5
NT	MHCI	95.5	8.6	0
MHCI	MHCI	98.4	10.2	3.1

* NT: non treated

DISCUSSION

When Lilet-Leclercq et al.[3] first succeeded in isolating FDCs from human tonsils, they found that these cells formed clusters with some germinal center lymphocytes such as thymic nurse cells. This strongly suggested that FDC might also be related to nurse cells, although this was difficult to investigate in the germinal centers in tissue sections. In the present study, we have clearly demonstrated that FDCs separated from other germinal center cells become flat and adherent after culture. At this stage the cells have lost FDC specific markers such as CD23 or DRC-1, but can, nevertheless, engulf newly added germinal center lymphocytes. We can therefore suggest that FDCs are originally destined to differentiate into cells whose function is to engulf such lymphocytes and, in this respect, may be classified as nurse cells.

We can emphasize four characteristics of emperipolesis by FDCs. Firstly, FDCs did not exhibit emperipolesis for monocytes-macrophages. This is in accordance with our previous observations[4] of freshly-isolated FDC-clusters which did not contain any monocytes-macrophages. This indicates that monocytes-macrophages and tingible-body macrophages in situ do not interact in the development or maturation of germinal centers within the FDC microenviroment.

Secondly, the interaction of FDC with lymphocytes was not restricted by the allotype. In mismatching tests of FDCs and lymphocytes, FDCs revealed no significant differences in emperipolesis and monoclonal antibodies against MHC-I or II did not inhibit emperipolesis. Thymic nurse cells were also reported to be unrestricted by allotype.

Thirdly, FDC emperipolesis was not always specific for B-cells. We were unable to characterize the nature of the enveloped lymphocytes in the true emperipolesis experiments, but in pseudoemperipolesis, some T-cells were trapped under FDCs. In addition, large numbers of T-cells were trapped under FDCs in mixed cultures of FDCs and T cell-rich populations. This means that T-cell involvement during emperipolesis by FDC can not be neglected. It must be emphasized that T-cells are always present in the light zone of the germinal centers in situ. Freshly-isolated FDC-clusters contained helper-phenotyped T-cells as 5 % of the engulfed lymphoid cell populations. Braun et al.[5] also showed in an in vitro assay that both B- and T-cells in the tonsils joined the initial adherence with freshly-isolated FDC-clusters and suggested the involvement of T-cells in the FDC-complex phenomenon. These results strongly suggested that T-cells are also important element in the germinal center constitution.

Fourthly, in the psuedoemperipolesis experiment, we found a significant induction of mitosis in entrapped autologous lymphocytes within the 6 initial hours of cultivation. A few mitotic lymphocytes under FDC showed B cell phenotype, but cells with the T-cell phenotype were not observed. This finding strongly suggested that FDCs had a higher affinity for the proliferation of germinal center B-cells than for T-cells. This may also expain why T-cells were scattered in the microenvironment of the FDC without proliferating in situ.

REFERENCES

1 Tsunoda R, Nakayama M, Onozaki K, Heinen E, Cormann N, Kinet-Denoël C, Kojima M (1990) Isolation and long-term cultivation of human tonsil follicular dendritic cells. Virchow Arch[B] 59-105

This was partially cited from our paper: Virchow Archiv [B] Cell Pathology, Springer International (in Press).

2 Nakayama M, Wekerle H (1984) In vitro generation of thymic nurse cells, in: Lymphoid cell function in ageing., Weck AL, ed. EUREG,pp 29-37
3 Lilet-Leclercq C, Radoux D, Heinen E, Kinet-Denoël C, Defraigne JO, Houben-Defresne MP, Simar LJ (1984) Isolation of follicular dendritic cells from human tonsils and adenoids. I. Procedure and morphological characterization. J.Immunol.Meth. 66:235-244
4 Tsunoda R, Kojima M (1987) A light microscopicalstudy of isolated follicular dendritic cell-clusters in human tonsils. Acta Pathol Jpn 37:575-585
5 Braun M, Heinen E, Cormann N, Kinet-Denoël C, Simar LJ (1987) Influence of immunoglobulin isotypes and lymphoid cell phenotype on the transfer of immune complexes to FDC. Cell Immunol. 107:99-106

THE LOCALIZATION OF LYMPHOKINES IN MURINE GERMINAL CENTERS

T. Kasajima, A. Andoh, Y. Takeo, T. Nishikawa

Department of Pathology
Tokyo Women's Medical College, 8-1 Kawada-cho, Shinjuku-ku
Tokyo 162, Japan.

INTRODUCTION

We have already reported that folliclar dendritic cells (FDCs) in the light zone of germinal centers (GCs) reacted positively for CD23, and that there is a close relationship between T cells and CD23-positive ($CD23^+$) FDCs in the pattern of distribution on reactive and neoplastic follicles[1]. It was considered that $CD23^+$ FDCs play an important role in interacting with T cells in follicles, and participate both in the formation of GCs and the preservation of folliclar structure.

However, there are still many unsolved aspects of the interaction between FDCs and T cells, and the function of T cells within GCs *in vivo*. Recently, the existence of many kinds of cytokine has become evident and various immunological functional networks of these substance have been proposed. Among these cytokines, interleukin (IL)-2, IL4, and interferon (IFN)-γ are released by CD4-positive ($CD4^+$) T cells. It would therefore be desirable to confirm that the cytokines produced and released by T cells can effect GC reactions. In this study, we examined the distribution patterns of cytokines (IL2, IL4 and IFN γ), T cells, and antigen-retaining FDCs in murine popliteal lymph nodes.

MATERIALS AND METHODS

Animals

BALB/c and BALB/c nu/nu mice (5-6 weeks old) were purchased from Charles River

Dendritic Cells in Fundamental and Clinical Immunology,
Edited by Kamperdijk *et al.*, Plenum Press, New York, 1993

Japan, Inc. The mice were bred and maintained at the Institute of Laboratory Animals, Tokyo Women's Medical College.

Immunization

Three kinds of immunization were carried out as described below;

Group 1. BALB/c mice were given an injection of 0.5 mg of goat IgG (CAPPEL) into the hind footpads. On day 1, 7, 14 or 21 after primary antigen injection, the popliteal lymph nodes were removed.

Group 2. Fifty days after priming as described above, a secondary antigen challenge was done, and 4 and 7 days after the booster injections, the popliteal lymph nodes were examined.

Group 3. BALB/c nu/nu mice and BALB/c mice were given an injection of 1×10^7/ml sheep red blood cells (SRBC) and alkaloids into the hind footpads . The popliteal lymph nodes were removed 10 days after the SRBC injection.

Immunohistology

Lymph nodes were fixed with periodate-lysine-paraformaldehyde (PLP) solution and frozen in Tissue-Tek OCT compound (Miles Inc., USA). Four-micrometer thick serial sections were used for hematoxylin-eosin staining and the following staining methods described below.

For immunohistochemical staining, rat anti-mouse monoclonal antibodies against IL2 (Pharmigen), IL4 biotinylated (Pharmigen), IFN γ (R46A2; a gift from Dr. Yoshio Kumazawa, School of Pharmaceutical Sciences, Kitasato University, Tokyo, Japan) CD23 (B3B4; Pharmigen), CD32 (2.4G2; Pharmigen), Lytl (CD5; Becton Dickinson), L3T4 (CD4; Becton dickinson), and CR1/CR2 (7G6; a gift from Dr. Taroh Kinoshita, Dept. of Immunoregulation, Research Institute for Microbial Disease, Osaka University, Japan) were used as primary reagents. Anti-rat IgG peroxidase-conjugated goat immunoglobulins (no cross-reaction with mouse immunoglobulins) were used with the indirect immunoperoxidase method as secondary sera. IL4 biotinylated antibody was used with SAB (streptavidin peroxidase; Nichirei Japan). For detection of the antigen distribution, anti-goat IgG peroxidase-conjugated rabbit immunoglobulins (DAKOPATTS) were used for the direct immunoperoxidase method.

For electron microscopy, a pre-embedding method was employed.

RESULTS

Findings after primary antigen injection

In the primary antigen response, the lymph nodes were extremely small, and the GCs were scarcely detectable histologically except for primary follicles (see Fig.1). On day 1 after primary injection, the antigen was detectable in the lumen throughout the entire sinus, the paracortex and in the cytoplasm of scattered macrophages. A small number of cells reactive for IL2, IL4 and IFN γ were scattered within the primary follicles. Electron microscopic examination revealed that the positive reaction was present on the cell membrane of lymphocytes.

On days 7 and 14, the outlines of the GCs appeared, and $CD23^+$ FDCs were distributed in a crescent pattern, which was coincident with the light zones of the GCs. Reticular positivity for IL2, IL4, and IFN γ was detected in the areas corresponding to those positive for CD23 in the GCs. Forthermor, lymphocytes positive for these cytokines were evident in the mantle zone.

On day 21, the GCs were apparent, and $CD23^+$ FDCs were distributed in a crescent pattern. IL4-positive small areas were detected, unlike IL2- and IFN γ -positive areas, which were crescent-shaped. Electron microscopy revealed that these cytokines were expressed on the membrane of FDCs, showing a labyrinth-like structure coincident with the complex cytoplasmic projections of FDC in the light zone.

Findings after secondary antigen injection

Four and 7 days after the booster injections, GCs were developed more predominantly and distinctly than those at 21 days after primary injection, and areas strongly positive for IL4 were demonstrated within them. Simultaneously, small areas weakly positive for IL2 and IFN γ were observed (Fig.1).

GCs in nude mice

In athymic nude mice which had been treated with antigens (SRBC) 10 days earlier, the lymph nodes contained numerous GCs, and $CD23^+$ FDCs were distributed in the GC light zones in a crescent pattern similar to those in stimulated normal BALB/c mice. In addition, IL4-positive areas coincidentally occupied the corresponding territory. IL2- and IFN γ -positive areas were smaller than these positive for IL4 (Fig.1). Considerable numbers of L3T4 (CD4)-positive cells were distributed within the GCs. In BALB/c mice which had been treated with SRBC, the cytokine distribution patterns were the same as these in lymph nodes after secondary antigen challenge.

DISCUSSION

GCs are specialized compartments, occupied mainly by B lymphocytes. Various studies have been done on the characteristics of GCs, and it has been reported that they contain populations of B cells, FDCs, and tingible body macrophages, in addition to selected subsets of T cells.

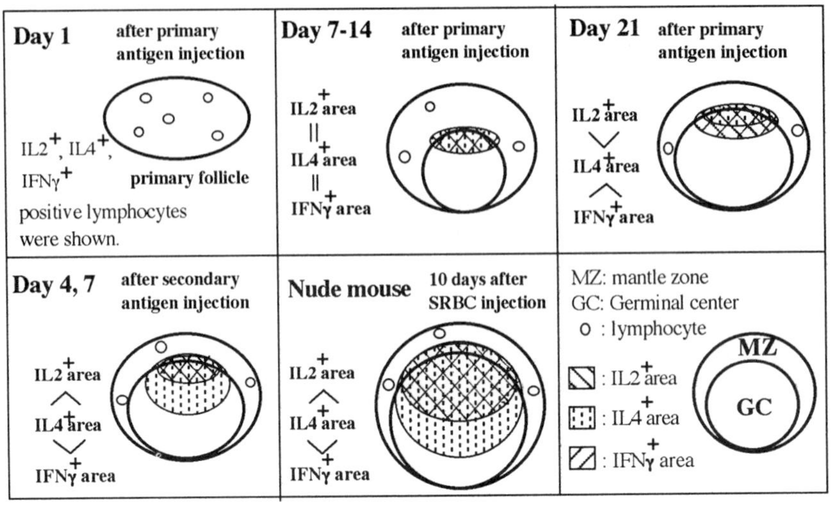

Fig. 1 Cytokine distribution pattern in the murine lymph follicle .

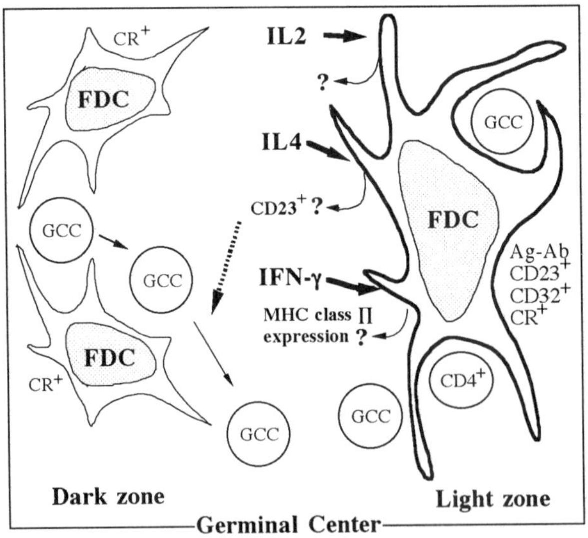

Light zone FDC are expressing much phenotype, retaining Ag-Ab, and having cytokines. The cytokines on the FDC, take a effect for phenotype expressing, i.e. CD23 and MHC class II antigen, in addition, act to rescue indirectly from apoptosis as well as to make effect directly proliferating or differentiating of GCC.

GCC: germinal center cell,
CR: complement receptor,
Ag-Ab: antigen-antibody-complex

Fig. 2 Proposal schema : germinal center reaction and ability of cytokine effects.

In murine lymph follicles, FDCs express complement receptor (CR1/CR2: 7G6)[2], IgG receptor (Fc γ RII; 2.4G2)[3], and low-affinity FcR for IgE (Fc ε RII, CD23; B3B4)[4], as in human FDCs[5]. Furthermore, the function of CD23 and the relationship between CD23 and proliferation of B cells have been described[6,7].

On the other hand, it has been documented that specialized T cells are distributed in GCs[8,9] and primary follicles. The phenotypes of these T cells are $CD3^+$, $CD4^+$. In lymph follicles of human lymph nodes, these T cells are the same as those in murine follicles[10], and express Leu-7[8,9]. Recently, many additional kinds of cytokine released by $CD4^+$ T cells have been documented, along with their phenotypic immunological functions. $CD4^+$ T cells are distributed exclusively in GCs, and may immunologically influence many cell populations through cytokines, although many studies have found neither evidence of cytokine-release from GC T cells[8] *in vitro* nor IL4 ,IL6 messenger ribonucleic acid (mRNA) in GC T cells[11] from human tonsils. Therefore, we attempted to clarify whether cytokines are located in GCs under conditions of stimulation using two kinds of mouse by studying the histological localization of three cytokines using immunocytochemistry. Anti-cytokine antibodies showed various degrees of reaction in lymphoid follicles. IL2, IL4 and IFN γ were positive on FDCs located the light zone of GCs, and on lymphocytes in the mantle zone. These positive reactions were shown to be located on the cell membrane by immunoelectron microscopy. Under stimulated conditions after antigen injection, IL4 was exclusively positive on $CD23^+$FDCs and the areas positive for IL4 varied in size following primary or secondary immunization with antigens. These results suggested that FDCs and a small number of lymphocytes in the mantle zone use these cytokines precisely as immunological regulatory signals.

On the other hand, in nude mice, it has been generally assumed that there are no T cells in thymus-dependent areas of lymphoid tissues. However, $CD4^+$ cells were detectable in primary follicles and GCs in the present study. The numbers of $CD4^+$ T cells distributed in follicles or GCs were not less than those in immunized BALB/c mice (data not shown). These T cells of nude mice were probably produced in bone marrow and might have been generated extrathymically, although further details were not obtained. As we were able to observe the existence of T cells in GCs and cytokines in lymphoid follicles of nude mice, it was considered that T cells in GCs or the mantle zone might have the ability to produce the cytokines *in vivo*.

On the other hand, as it is very difficult to detect cytokines in human lymph nodes, there have been few reports of cytokines in these tissues. We tried to examine the presence of cytokines using numerous samples of lymphoid tissues including lymph nodes, tonsils and various tissues obtained by biopsy, but no unequivocal stainability was achieved .

However, from the present data, we are able to conclude that the formation and preservation of GCs might be regulated by immunological interaction of cytokines and $CD23^+$ FDC (Fig.2).

REFERENCES

1. T. Kasajima, A.Masuda, A.Andoh, T.Nishikawa, M.Kawakami, An immunohistochemical study on the phenotypic immunostaining pattern of follicular dendritic cells in reactive and pathological lymphoid tissues, "Dendritic cells in lymphoid tissues" Y.IMai, J.G.Tew and E.C.M. Hoefsmit, p91-98(1991)
2. Kinoshita T, Lavoie S, Nussenzweig V, *J.Immunol.* 134:2564-(1985)
3. Unkeless,J.C. Characterization of monoclonal antibody directed against mouse macrophage and lymphocyte Fc receptors, *J.Exp.Med* . 150:580(1979)
4. K.Maeda, G.F.Burton, D.A.Padgett, D.H.Conrad, T.F.Huff, A.Masuda, A.K.Szakal, J.G.Tew; murine follicular dendritic cells and low affinity Fc Receptors for IgE (Fc ε RII). *J.Immunol.* 148(8):2340-2347(1992)
5. A.Masuda, T.Kasajima, N.Mori, K.Oka; Immunohistochemical study of low affinity Fc recepter for IgE in reactive and neoplastic follicles. *Clin.Immunol.Immunopathol.* 53:309(1989)
6. Y--J.Liu, J.A.Cairns, M.J.Holder,S.D.Abbot, K.U.Jansen,J-Y.Bonnefoy, J.Gordon, I.C.M.Maclennan; Recombinant 25-kDa CD23 and interleukin 1α promote the survival of germinal center B cells; evidence for bifurcation in the development of centrocyte rescued from apoptosis. *Eur. J. Immunol.* 21:1107-1114(1991)
7. M.J.Holder, Y-J.Liu, T.Defrance, L.F-Romo, I.C.M.Maclennan, J.Gordon; Growth factor requirements for the stimulation of germinal center B cells; evidence for an IL2-dependent pathway of development. *Internatio. Immunol.* 3(12):1243-1251(1991)
8. M.B.Bowen, A.W.Butch, C.A.Parvin, A.Levine, H.Nahm, Germinal center T cells are distinct helper-inducer T cells. *Human Immunol.* 31:67-75(1991)
9. A.Velardi, M.C.Mingari, L.Moretta, C.E.Grossi; Functional analysis of cloned germinal center $CD4^+$ cells with natural killer cell-related features. Divergence from typical T helper cells. *J. Immunol.* 137(9):2808-2813(1986)
10. R.V.Rouse, J.A.Ledbetter, I.L.Weissman; Mouse lymph node germinal centers contain a selected subset of T cells - The helper phenotype. *J.Immunol.* 128(5):2243-2246(1982)
11. A.Bosseloir, E.Hooghe-Perters, E.Heinen, C.Marcoty, N.Cormann, C.Kinet-DeNoel, L.J.Simar, Localization of interleukin-4 and interleukin-6 messenger ribonucleic acid in human tonsils by ISH, p315-(1990)

TWO DIFFERENT MECHANISMS OF IMMUNE-COMPLEX TRAPPING IN THE MOUSE SPLEEN DURING IMMUNE RESPONSES

Kikuyoshi Yoshida, Timo K. van den Berg, Christine D. Dijkstra

Department of Histology, Faculty of Medicine, Vrije University
Van der Boechorststraat 7, 1081 BT Amsterdam, The Netherlands

SUMMARY

The capacity of immune-complex (IC) trapping was examined using purified horse radish peroxidase (HRP)-anti-HRP (PAP) on frozen sections of mouse spleen *in vitro*. We investigated the trapping mechanisms by applying the IC with or without fresh mouse serum added on the spleen sections of naive as well as immunized mice.

When the PAP was applied alone, it mainly located on the macrophages in red pulp. In the splenic white pulp of immunized mice, PAP was trapped on follicular dendritic cells (FDC) in a small area of the germinal center whereas it scarcely bound to the splenic white pulp of non-immunized mice. An antibody against mouse Fc receptor (2.4G2) blocked the trapping but antibodies against mouse complement receptor (8C12, Mac-1 and 7G6) did not.

When the PAP was applied mixed with fresh mouse serum, it bound on FDC in the primary follicles in the spleen of non-immunized mice. The density and area of IC trapping increased in the spleen of immunized mice. IC trapping in the presence of fresh mouse serum was blocked by the antibodies 8C12 and 7G6 but not by 2.4G2 or Mac-1.

INTRODUCTION

Administration of immune complexes (antigen-antibody complex, IC) causes more potent immune responses than that of antigen alone and

Dendritic Cells in Fundamental and Clinical Immunology,
Edited by Kamperdijk *et al.*, Plenum Press, New York, 1993

results in an effective induction of B-cell memory[1]. Complement factors or activation of the complement cascade are necessary for the enhancement of immune responses by the IC administration[2].

Though it is probable that the complement receptor plays a crucial role in the B-cell immune responses, the precise mechanisms of its biological significance are unknown[3]. On the other hand, evidence of IgG-mediated enhancement of humoral immune responses has been reported using mutant mice[4]. It is widely accepted that follicular dendritic cells (FDC) play an important role in the IC retention[5]. FDC may also be involved in capture of IC. However, Humphrey et al.[6] reported that trapping of IC on FDC depends upon the presence of radio- and cyclophosphamide-sensitive population of cells, although the FDC themselves are resistant to such treatments. The site of trapping of IC administered *in vivo*, may change in course of time. By employing an *in vitro* trapping assay using fixed cryostat sections[7], we were able to study the initial steps of IC trapping precisely.

MATERIALS AND METHODS

Animals: Female inbred C3H/Kall mice, purchased from Harlan/Olac/CPB (Amsterdam, the Netherlands) and kept under routine laboratory conditions, were used for the experiments at the age of 8-10 weeks.
Antibodies employed: Rat anti-Mac-1 antibody specific for mouse C3bi receptor (CR3) and rat anti-mouse ICAM-1 (CRL1878) were obtained commercially from ATCC (Rockville, Maryland, USA). 8C12[8], a rat antibody specific for mouse C3b receptor (CR1) and 7G6, a rat antibody against both mouse C3b and C3d receptors (CR1/2) were kindly provided by Dr. Kinoshita (Dept. Immunoregulation, Research Institute for Microbial Disease, Osaka University, Japan). WP-1, a rat monoclonal antibody against lymphoid tissue-specific reticular meshwork was produced in our laboratory[9]. 4C11, a rat antibody against mouse FDC was kindly provided from Dr. Kosco (Basel Institute of Immunology, Switzerland). 2.4G2, a rat antibody against mouse Fc receptor was a kind gift of Dr. Unkeless (The Rockefeller Univ. New York, USA).
Demonstration of IC trapping: Spleen was snap frozen and cryostat sections of 10 μm thickness were picked up on slides, air-dried, fixed in acetone for 30 min and air-dried again. Slides were then incubated overnight either with 1/10 diluted IC (peroxidase-anti-peroxidase, mouse PAP; DAKO Corp. California, USA) alone or with IC plus 1/5 diluted fresh mouse serum (FMS). Control slide were incubated solely with HRP (Wako Pure Chem. Pharm. Co. Ltd. Osaka, Japan) or anti-HRP antibody produced by ourselves. After washing in phosphate-buffered saline, peroxidase activity was demonstrated by incubating the speci-

men in medium consisting 0.2 mg/ml 3,3'-diaminobenzidine tetra-HCl (DAB, Sigma Pharm. Co. Ltd. Missouri, USA) and 0.6 mg/ml sodium azide in 0.05M Tris buffer pH 7.6 in the presence of 0.1% peroxide for 10 minutes at room temperature. Blocking assays were performed by incubating the sections with IC and 10 μg/ml purified antibody.

Immunohistochemistry for light microscopy: Immunostaining was performed by an indirect ABC method as follows. Frozen sections of mouse spleen at 6 μm thickness were fixed in acetone. The specimens were incubated with an optimal concentration of monoclonal antibody at 4 ℃ overnight. After washings, the sections were incubated with 1/50 diluted biotinylated rabbit anti-rat immunoglobulin (Zymed Lab. Inst. California, USA) for 30 min at room temperature. Then, the sections were incubated with 1/200 diluted peroxidase-avidin (Vectastain ABC kit, Vector Lab. Inst., California, USA) and finally visualized by incubating the specimens with DAB solution as described above.

Induction of immune response: Mice were immunized intraperitoneally with sheep red blood cells (SRBC) once a week for three weeks with gradually increasing numbers of SRBC (1×10^7, 2×10^7, 5×10^7 resp.). Ten to 11 days after the last imunization, 10×10^7 SRBC were boosted intraperitoneally. Three days later, the spleens were excised and frozen in liquid nitrogen.

RESULTS

Two patterns in IC trapping on frozen section of mouse spleen

Using the *in vitro* trapping assay, we have demonstrated two different trapping patterns on cryostat sections of mouse spleen by applying IC with or without FMS. First, purified PAP alone was applied as IC source. The IC were mainly detected in the red pulp. Scarce trapping was observed in the primary follicles in the spleens of non-immunized mice. On the other hand, if the PAP was applied together with FMS, the trapping was mainly located in white pulp, whereas binding to the red pulp was only scarce.

Change in trapping capacity during immune responses

The IC trapping capacity of the spleen during an immune response was compared with that in naive state. Trapping of PAP without FMS by red pulp macrophages increased in its density on the cell surface, and significant trapping on stromal cells in a small area of the germinal center became detectable. In addition, in the presence of FMS, the density and area of IC trapping in the germinal center increased simultaneously with the development of the germinal center.

Characterization of IC-binding cells

The IC-binding cells in the follicles after incubation with PAP only, were dendritic shaped cells suggesting that these IC-binding cells belong to the stromal cell population (Fig. 1a). Same observations were gained regarding the cells binding PAP plus FMS (Fig. 1c). These findings make it clear that the IC-binding cells in the germinal center for the two types of IC, with or without FMS, are FDC.

Furthermore, the expression of receptors by FDC, participating in IC trapping, was examined. The trapping pattern of PAP without FMS was very similar to the pattern of expression of Fc receptor in the germinal center (Fig. 1b) and that of PAP plus FMS was nearly identical to the expression of complement receptor types 1/2 (Fig. 1d).

Mechanisms of immune complex trapping

The mechanisms of IC trapping were investigated using blocking studies by various antibodies (Table 1). All antibodies except the anti-CR3 (Mac-1) used in these experiments stained FDC. Trapping of PAP only was blocked by coincubation with antibody anti-Fc receptor (2.4G2), but not with antibodies against complement receptors. On the other hand, trapping of PAP plus FMS was blocked by adding antibodies against CR1/2 (7G6) or CR1 (8C12), but not by adding anti-CR3 (Mac-1) or anti-Fc receptor (2.4G2) antibodies.

DISCUSSION

It is noteworthy that IC bind to different areas if it is applied without or with FMS (Fig. 1). The first type of IC binding is mediated by Fc receptor and the latter by complement receptors (Table 1). Furthermore, these trapping patterns alter according to the immune status. In the absence of FMS, the density of IC bound to the macrophages increase in the spleen after induction of an immune response as compared to the spleen of non-immunized animals. In the white pulp of non-immunized mice, the Fc receptor-mediated trapping was scarcely

Table 1. Mechanisms of IC trapping in the germinal center

Antibodies	PAP only	PAP plus FMS
Anti-CR1(8C12)	Not	Blocked (weak)
Anti-CR1/2(7G6)	Not	Blocked (strong)
Anti-CR3(Mac-1)	Not	Not
4C11	Not	Not
Anti-ICAM-1	Not	Not
WP-1	Not	Not
Anti-FcR(2.4G2)	Blocked	Not

IC were applied mixed with 10μg/ml of a purified antibody.

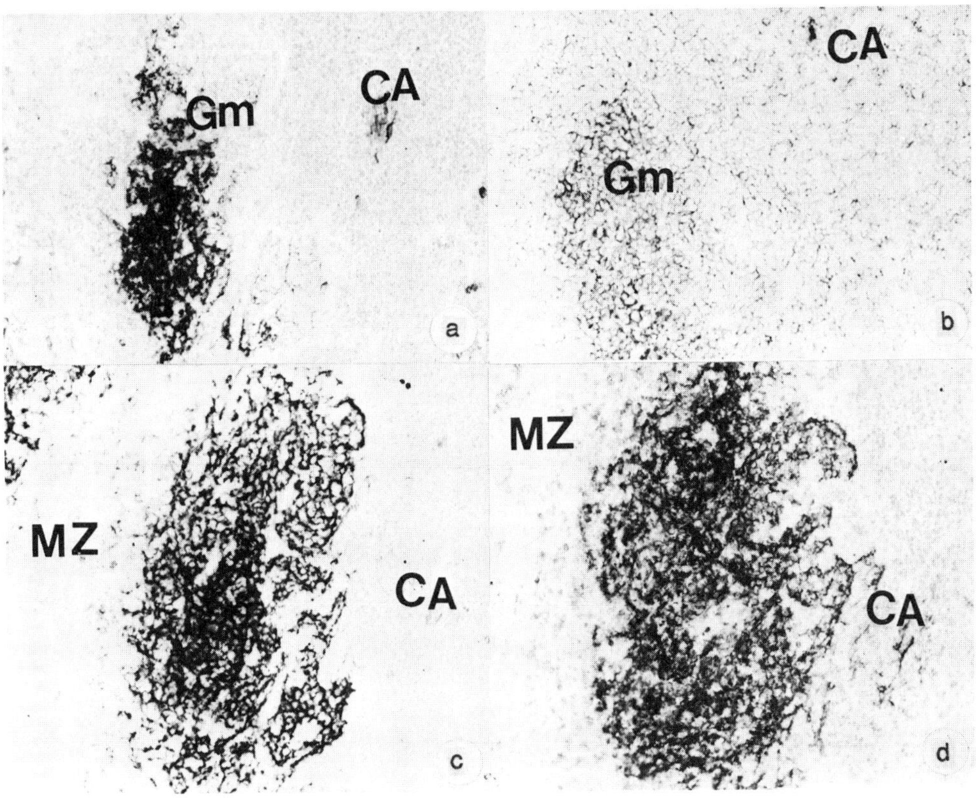

Fig.1 IC trapping and receptor expression on FDC in the germinal center. **a.** Trapping of PAP only. **b.** Expression Fc-R defined by 2.4G2. **c.** Trapping of PAP plus FMS. **d.** Expression CR1/2 by 7G6. CA; central artery, Gm; germinal center, MZ; marginal zone. methyl green counter-staining. x200

observed. However this trapping was clearly detectable in the spleen during an immune response to SRBC. This increased binding of IC coincided with elevated expression of Fc-receptor on FDC which was shown to mediate IC binding in the absence of FMS (Fig. 1a,b). The density and area of IC trapping with FMS also increased during an immune response. This increased trapping of IC coincided with elevated expression of CR1/2 on FDC which was shown to mediate IC binding in the presence of FMS (Fig. 1c,d). Taken together the above data, two functionally different FDC are suggested.

REFERENCES

1. A. Kunkl, G.G.B. Klaus, The generation of memory cells IV. Immunization with antigen-antibody complexes accelerates the development of B-memory cells. The formation of germinal centers and the maturation of antibody affinity in the secondary response. Immunology 43: 371 (1981)
2. T.K. van den Berg , E.A. Döpp, M.P. Daha, G.Kraal, C.D. Dijkstra, Selective inhibition of immune complex trapping by follicular dendritic cells with monoclonal antibodies against rat C3. Eur. J. Immunol. 22: 957 (1991)
3. B. Heyman, E.J. Wiersma, T. Kinoshita, *In vivo* inhibition of the antibody response by complement receptor-specific monoclonal antibody. J. Exp. Med. 172: 665 (1990)
4. E.J. Wiersma, M. Nose, B. Heyman, Evidence of IgG-mediated enhancement of the antibody response *in vivo* without complement activation via the classical pathway. Eur. J. Immunol. 20: 2585 (1990)
5. T.E. Mandel, R.P. Phipps, A. Abbot, A.G. Tew, The follicular dendritic cells: Long term antigen retention during immunity. Immunol. Rev. 53: 29 (1980)
6. J.H. Humphrey, D. Grenann, V. Sundaram, The origin of follicular dendritic cells in the mouse and the mechanism of trapping of immune complexes on them. Eur. J. Immunol. 14: 859 (1984)
7. C.D. Dijkstra, A.A. te Velde, N. van Rooijen, Localization of horseradish peroxidase (HRP)-anti-HRP complexes in cyrostat sections: Influence of endotoxin on trapping of immune complexes in the spleen of the rat. Cell Tissue Res. 232: 1 (1983)
8. T. Kinoshita, J. Takeda, K. Hong, H. Kozono, H. Sakai, K. Inoue, Monoclonal antibodies to mouse complement receptor type 1 (CR1). their use in a distribution study showing that mouse erythrocytes and platelets are CR1-negative. J. Immunol. 140: 3066 (1988)
9. K. Yoshida, N. Tamahashi, N. Matsuura, T. Takahashi, T. Tachibana, Antigenic heterogeneity of the reticular meshwork in the white pulp of mouse spleen. Cell Tissue Res. 266: 223 (1991)

CELLULAR REQUIREMENTS FOR FUNCTIONAL RECONSTITUTION OF FOLLICULAR DENDRITIC CELLS IN SCID MICE

Zoher F. Kapasi[1], Gregory F. Burton[2], Leonard D. Shultz[3], John G. Tew[2] and Andras K. Szakal[1]

Department of Anatomy, Division of Immunobiology[1]
Department of Microbiology and Immunology[2]
Medical College of Virginia, Virginia Commonwealth University
Richmond, Virginia
The Jackson Laboratory, Bar Harbor, Maine [3]

INTRODUCTION

Follicular dendritic cells (FDC) are located in B cell follicles and germinal centers (GC) of secondary lymphoid tissue (Szakal and Hanna, 1968; Nossal et al., 1968). Antigen retention in the form of immune complexes is one of the main properties of FDC (Szakal et al., 1989; Tew et al., 1990). The dendrites of several FDC intertwine to form a three-dimensional network called the antigen-retaining or FDC-reticulum (Szakal et al., 1989). Accumulating evidence indicates that FDC and the retained antigen are important in the production of B-memory cells and the long-term maintenance of secondary antibody response (Tew et al., 1990).

Based on studies with athymic mice (Mitchell et al., 1972; Tew et al., 1979) and rats (Mjaaland and Fossum, 1987) this antigen localization on FDC can occur in the absence of a thymus. However, although antigen trapping by FDC in these thymus dysgenic animals was qualitatively normal, it appeared to be quantitatively reduced (Mitchell et al., 1972; Mjaaland and Fossum, 1987). In addition, one study (Klaus and Kunkl, 1982) reported that nude mice failed to localize immune complexes in lymphoid follicles. Thus, it is difficult to conclude the exact role played by T cells and thymus in immune complex trapping by FDC.

The role of B cells in FDC development has been examined in lymph nodes of nude or normal mice and rats (MacLennan and Gray, 1986; Cerny et al., 1988) depleted of B cells. Both studies reported an absence of FDC in these mice and further demonstrated that the reappearance of B cells preceded the appearance of functional FDC. These results indirectly demonstrate the requirement of B cells for FDC development.

To assess the B and T cell requirements for antigen localization by the FDC-reticulum, we selected the severe combined immunodeficient (SCID) mouse which lacks functional B and T cells (Bosma et al., 1983). The results of transferring either bone marrow cells (BM), B+T cells together, B cells, or T cells alone to SCID mice showed that BM and B+T cells were more effective in FDC development than either B or T cells alone.

Dendritic Cells in Fundamental and Clinical Immunology,
Edited by Kamperdijk *et al.*, Plenum Press, New York, 1993

MATERIALS AND METHODS

Animals

Homozygous mutant C.B.17-scid/scid (SCID) (H-2^d) male mice, aged 15-17 weeks, and normal Balb/cByJ (H-2^d) male mice, aged 6-8 weeks were obtained from Jackson Laboratories, Bar Harbor, ME. Balb/c-nu/nu, nude male mice, aged 6-8 weeks, were purchased from the Medical College of Virginia nude mouse facility. Animals were housed in a virus free environment and given food and water *ad libitum*.. For four weeks after the irradiation and cell transfers, recipient mice were given antibiotic-containing water (0.01% Neomycin sulfate, Sigma).

Cell Transfers for Reconstitution

Prior to cell transfer, SCID mice were irradiated with 300 rad to facilitate reconstitution. Bone marrow cells were obtained from femurs and tibias of Balb/c mice. Unfractionated splenic cells of athymic nude mice were given as the source of B cells. T cells were isolated by passing Balb/c spleen cells through a 10 ml nylon wool column twice with subsequent panning on purified monoclonal rat anti-mouse Ig kappa and lambda (Pharmingen, CA.) coated plates (purity: FACScan data, 97% IgM-negative cells). For the reconstitutions, mice received i.v. 10^7 BM, B or T cells, or 10^7 B plus10^7 T cells.

Immunization

Animals were passively immunized i.p. with 0.75 ml of specific rabbit antiserum 1 day prior to footpad challenge (5µg/0.05 ml in saline) with the histochemically detectable antigen, horseradish peroxidase (HRP, type VI, Sigma).

Histochemistry

For the histochemical localization of HRP, 24 hr after injection of HRP, the tissues were fixed by intracardiac perfusion as described previously (Szakal et al., 1983). The popliteal lymph nodes were serially sectioned at 50µm thickness on a vibratome, and developed for peroxidase activity of HRP in 0.1% diaminobenzidine hydrochloride (DAB)/0.01% hydrogen peroxide substrate (Szakal et al., 1983). The sections were prepared for light microscopy.

Morphometry

The **number** of FDC-reticula per lymph node was determined by following each peroxidase positive FDC-reticulum from beginning to end on serial sections. The three-dimensional size or **volume** of a FDC-reticulum was determined by Bioquant System IV morphometry program (R&M Biometrics Inc., Nashville, TN.). The third parameter calculated was the **compartment size**, defined as the sum of the volumes of all the FDC-reticula per lymph node. This compartment size is a reflection of both the number as well as the volume of FDC-reticula in a given lymph node.

RESULTS

The first objective was to determine if SCID mouse lymph nodes had functional FDC. This was conducted by assessing the capacity to trap and localize immune complexes in situ. No such antigen retaining FDC-reticula could be observed. Secondly, the presence of non-functional FDC in these lymph nodes was evaluated using the anti-mouse FDC monoclonal antibody, FDC-M1 (Kosco et al., 1992). Again, no FDC-reticula could be observed.

We next sought to determine if adoptive transfer of BM would reconstitute functional FDC in SCID lymph nodes. In the popliteal lymph nodes of BM transferred SCID recipients, the number, volume and compartment size of FDC-reticula actually exceeded that of the Balb/c control (Table 1). This data indicates that FDC could be reconstituted in SCID mice.

Table 1. FDC-reticulum parameters in normal and reconstituted SCID mice as compared to Balb/c mice.

	SCID+					Balb/c
	none	B	T	B+T	BM	
NUMBER	0	0.5±0.3	0.5±0.3	4.8±0.4	7.8±1.3	3.8±1.2
VOLUME*	0	0.2±0.1	0.5±0.2	4.4±0.8	5.4±0.7	1.6±0.3
COMP.SIZE*	0	0.2±0.2	0.5±0.4	21.5±3	42.6±7	6.1±2.7

+ SCID mice were reconstituted with various combinations of B and T or bone marrow (BM) cells and 4 weeks later, their popliteal lymph nodes (n=4-8) were evaluated for the number, volume and compartment size of the FDC-reticula. The means of these values was then compared to those of unreconstituted SCID and Balb/c mice.

* values are in μm^3

To determine if B cells or T cells were sufficient for reconstitution of FDC, groups of SCID mice received cell transfers with both B and T cells or B cells or T cells alone. With this approach we expected to assess whether the effects of B and T cells in transfers were additive. Additionally, the effects of T cells could also be evaluated in terms of FDC development. While B cells or T cells alone could induce minimal development of FDC-reticula (Table 1), when B and T cells were transferred together the different parameters of FDC-reticula approximated or even exceeded similar parameters for Balb/c control (Table 1).

Finally, the BM reconstituted SCID mice showed a significant increase in the number (p=0.04) and compartment size (p=0.02) of FDC-reticula as compared to the SCID mice reconstituted with B+T cells.

DISCUSSION AND CONCLUSIONS

The lack of FDC in SCID mice provides a model in which cell requirements for development of FDC-reticulum may be determined. The effects of BM and B+T cell transfers using this model showed that transfer of BM cells, or B+T cells can impart SCID mice with the capacity to rapidly develop FDC-reticula. In contrast, neither B cells or T cells alone were adequate to accomplish this within a 4 week period. Since, B+T cells were more effective in FDC development than B or T cells alone, it indicates that T cells have a definite role in promoting FDC development. The B-cell requirement for the development of FDC-reticula is in agreement with the studies of Cerny et.al. (1988) and MacLennan and Gray (1986) (see introduction).

Bone marrow transplants were found to be more efficient than B+T cell transfers in reconstituting FDC development. This may be only due to the number of B and T cell precursors (Phillips et al., 1977; Cantor et al., 1976) which exist in bone marrow. These cells are known to proliferate rapidly which may result in a more efficient reconstitution of the B and T cell compartments. The number of B and T cells derived from the BM could quickly exceed the 10^7 B and 10^7 T cells transferred to B+T cell recipients after the 4 week period. However, the possibility that antigen transport cell precursors (Szakal et al., 1983) and/or FDC precursors are also supplied by the BM still exist, at least until definitive evidence is available for the nature of the FDC precursor cell.

In conclusion, the B and T cell requirements demonstrated here strongly indicate that B-T cell collaboration and the factors they produce are essential in the development of FDC.

REFERENCES

Bosma G.C., Custer R.P. & Bosma M.J. (1983) A Severe Combined Immunodeficiency Mutation In The Mouse. *Nature,* **301,** 527.

Cantor H. & Weissman I. (1976) Development And Function Of Subpopulations Of Thymocytes And T Lymphocytes. *Prog. Allergy,* **20,** 1.

Cerny A., Zinkernagel R.M. & Groscurth P. (1988) Development Of Follicular Dendritic Cells In Lymph Nodes Of B-Cell-Depleted Mice. *Cell Tissue Res.* **254,** 449.

Klaus G.G.B. & Kunkl A. (1982) The Role Of T Cells In B Cell Priming And Germinal Centre Development. *Adv.Exp.Med.Biol.* **149,** 743.

Kosco M.H., Pflugfelder E. & Gray D. (1992) Follicular Dendritic Cell-Dependent Adhesion And Proliferation Of B Cells In Vitro. *J.Immunol..* **148,** 2331.

Maclennan I.C.M. & Gray D. (1986) Antigen-Driven Selection Of Virgin And Memory B Cells. *Immunol. Rev.* **91,** 61.

Mitchell J., Pye J., Holmes M.C. & Nossal G.J.V. (1972) Antigen Localization In Congenitally Athymic 'nude' Mice. *Aust. J. Exp. Biol. Med. Sci.* **50,** 637.

Mjaaland S. & Fossum S. (1987) The Localization Of Antigen In Lymph Node Follicles Of Congenitally Athymic Nude Rats. *Scand. J. Immunol.* **26,** 141.

Nossal, G.J.V., Abbot, A., Mitchell, J., & Lummus, Z. (1968) Antigens In Immunity XV. Ultrastructural Features Of Antigen Capture In Primary And Secondary Lymphoid Follicles. *J. Exp. Med.* 127:277.

Phillips R.A., Melchers R. & Miller R.G. (1977) Stem Cells And The Ontogeny Of B Lymphocytes. In: *Progress In Immunology* (Eds. T.E. Mandel, C.Cheers, C.G. Hosking, I.F.C. Mckenzie & G.J.V. Nossal), P 155.
Australian Academy Of Science, Canberra City.

Szakal A.K., & Hanna M.G.Jr. (1968) The Ultrastructure Of Antigen Localization And Viruslike Particles In Mouse Spleen Germinal Centers. *Exp.Mol.Pathol.* **8,** 75.

Szakal A.K., Holmes K.L. & Tew J.G. (1983) Transport Of Immune Complexes From The Subcapsular Sinus To Lymph Node Follicles On The Surface Of Nonphagocytic Cells, Including Cells With Dendritic Morphology. *J.Immunol.* **131,** 1714.

Szakal A.K., Kosco M.H. & Tew J.G. (1989) Microanatomy Of Lymphoid Tissue During The Induction And Maintenance Of Humoral Immune Responses: Structure Function Relationships. In: *Annual Reviews Of Immunology* (Eds. W.E. Paul, C.G. Fatham & H. Metzger), Vol. 7, P. 91. Annual Reviews Inc., Palo Alto, California.

Tew J.G., Mandel T.E. & Miller G.A. (1979) Immune Retention: Immunological Requirements For Maintaining An Easily Degradable Antigen In Vivo. *Aust.J.Exp.Biol.Med.Sci.* **57,** 401.

Tew J.G., Kosco M.H., Burton G.F. & Szakal A.K. (1990) Follicular Dendritic Cells As Accessory Cells. *Immunol.Rev.* **117,** 185.

INTERACTION THROUGH THE LFA-1/ICAM-1 PATHWAY PREVENTS PROGRAMMED CELL DEATH OF GERMINAL CENTER B CELLS

Gerrit Koopman, Robert M.J. Keehnen, and Steven T. Pals

Department of Pathology
Academic Medical Center, University of Amsterdam
Meibergdreef 9
1105 AZ Amsterdam
FAX 31-206960389

INTRODUCTION

Germinal centers play a key role in the maturation of the B cell immune response. Although the initial B cell triggering after interaction of the immunoglobulin receptor molecule with the antigen is thought to take place in the paracortical areas of lymph node and MALT, the isotype switch, affinity maturation and differentiation of B cells into memory cells all take place in the germinal center[1-3]. B cells that enter the germinal center divide rapidly. Through the process of somatic hypermutation a great heterogeneity in antigen binding specificity in the proliferating B cell population is generated[4]. However only cells with Ig receptors with high affinity for the antigen are selected for further maturation into memory cells, while the non selected cells die through a process of programmed cell death (PCD)[5,6]. PCD, also called apoptosis constitutes a cell elimination program, that involves generation of DNA strand breaks, chromatin condensation and cell fragmentation.

Follicular dendritic cells (FDC) are thought to play an important role in the B cell selection process. The FDC present the major antigen trapping mechanism of the lymphoid tissues, that bind antigen in the form of immune complexes for long periods of time[7] and present these immune complexes either directly or in the form of immune complex coated vesicles (iccosomes) to the B cells[8]. FDC may also deliver growth factors, like CD23, to the B cells[6]. In a previous paper we have shown that adhesion molecules are important in the interaction between B cells and FDC[9]. Thus the LFA-1 and VLA-4 adhesion receptors on the B cells bind to respectively ICAM-1 and VCAM-1 on the FDC[9,10]. Triggering of adhesion molecules is known to be important in lymphocyte activation[11]. Therefor the adhesive interaction between FDC and B cells may directly contribute to the B cell selection process. In this paper we studied whether adhesion receptors are directly involved in B cell selection, by investigating the effect of these molecules on apoptosis of germinal center B cells. We observed that adhesion of B cells to plastic coated with the LFA-1 ligand ICAM-1 could prevent B cells from entry into apoptosis, while addition of mAb directed against these adhesion receptors had no effect.

Dendritic Cells in Fundamental and Clinical Immunology,
Edited by Kamperdijk *et al.*, Plenum Press, New York, 1993

MATERIALS AND METHODS

Monoclonal Antibodies and Reagents

The mAbs used were: NKI-P2 (IgG1) reactive with CD44; CLB-LFA-1/2 (IgG1), SPV-L7 (IgG1), SPV-L12 (IgG1) and NKI-L16 (IgG2a) all reactive with the α subunit of LFA-1; CLB-LFA-1/1 (IgG1) and IB4 (IgG2a) specific for the ß subunit of LFA-1; F10.2 and RR1/1 (IgG1) specific for ICAM-1; CD38 (Immunotech, Marseille, France); CD39 (Ortho Diagnostic Systems Inc., Raritan, NJ); anti-human IgD mAb (Seralab). Sheep anti-human IgM F(ab')2 antibodies were obtained from ICN Immuno Biologicals (Costa Mesa, CA) and Staphylococcus aureus Cowan strain I was from Calbiochem (La Jolla, CA). Purified ICAM-1 (sICAM-1) was kindly provided by dr. G.A. van Seventer.

B Cell Isolation

Freshly obtained tonsillar tissue was dissected free from surface epithelium, and finely minced into a cell suspension. Mononuclear cells were isolated by Ficoll-Isopaque density gradient centrifugation. Monocytes were removed by plastic adherence (1h. incubation at 37°C in 10 cm petridishes (Costar, Cambridge, MA)). T cells were depleted using 2-aminoethyl-isothiouronium bromide modified SRBC. The purified B cells were then layered on a percoll (Pharmacia, Uppsala, Sweden) gradient, consisting of five density layers (1.085/1.077/1.067/1.056/1.043) and centrifuged for 15 min., 1200g at 4°C. Cells at the 1.043/1.056 interface (low density cells) and the 1.077/1.085 interface (high density cells) were used in the experiments.

FACS Analysis

Cells were sequentially incubated (PBS containing 1% BSA) with appropriate dilutions of the different mAb, PE conjugated Goat anti-Mouse Ig (Southern Biotechnology Associates Inc., Birmingham, AL) and PNA-FITC (Sigma Chemical Co., St. Louis, MO) for 30 min. at 0°C. Fluorescence intensity was measured by FACScan (Becton Dickinson, Mountain View, CA).

Apoptosis Assay

The isolated high and low density B cells were plated in 96 well flat bottomed tissue culture plates (Costar) at 10^5 cells per well. The mAb were added at optimal concentrations as determined in tissue staining and adhesion inhibition assays. For studying the effect of coated substrates, wells were incubated prior to the experiment, for 3 hours at 37 °C with sICAM-1 (10 ng per well) in PBS, which allowed optimal coating as determined in an ELISA.

Cells were harvested after 18 h. incubation. Cytospin preparations were stained with May-Grünwalds/Giemsa and the percentage of apoptotic cells was determined in two different cytospin preparations. Experiments were done in triplo.

RESULTS

B Cell Isolation and Characterization

The composition of the isolated B cell preparations was established by FACS analysis. As is shown in figure 1. the high and low density B cell preparations differed markedly in FSC-SSC pattern. Furthermore 70% of the high density cell population showed a low PNA expression and high IgD expression, while in the low density cell population 70% of the

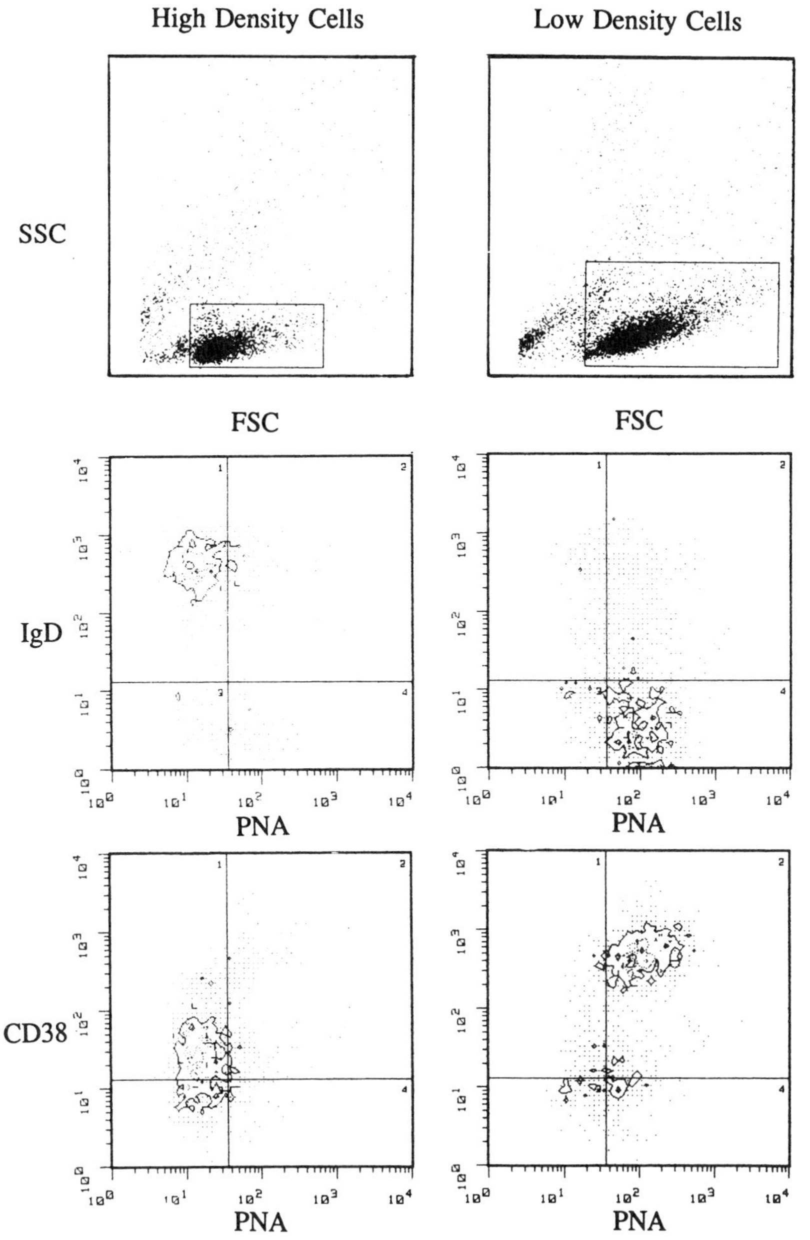

Figure 1. FACS analysis of isolated B cells. High and low density B cells were stained with PNA-FITC in combination with anti-IgD or CD38.

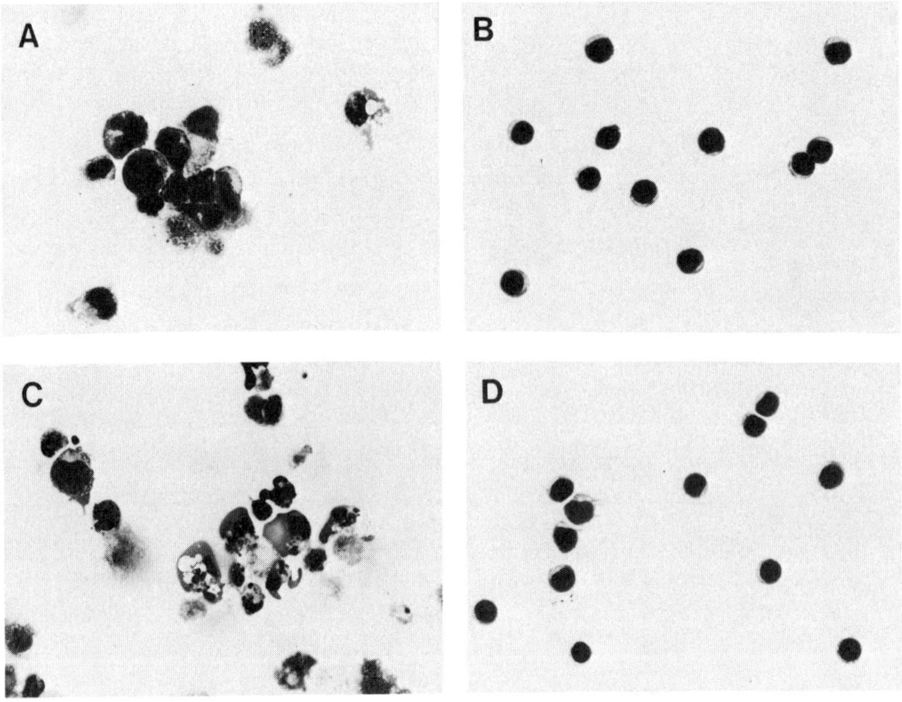

Figure 2. May-Grünwalds/Giemsa stained cytospin preparations of low or high density B cells, that were kept at 4°C (A and B respectively) or incubated for 18 h. at 37°C (C and D respectively). Magnification 460x.

cells had a high PNA expression and low IgD expression (figure 1.). In addition the low density cells showed a high CD38 and low CD39 expression (figure 1. and not shown), which further confirms their germinal center origin, while the high density cells were CD38 negative and CD39 positive.

When the low and high density cells were kept at 4°C they remained viable and no cells with nuclear condensation could be observed in the May-Grünwalds/Giemsa stained cytospin preparations (figure 2A and 2B.). However after culturing the cells for 18 h. at 37°C, part of the low density cell population (50 to 60%) became apoptotic, while almost no apoptotic cells could be detected in the high density population (figure 2C and 2D).

Adhesion and B Cell Apoptosis

For studying the involvement of adhesion receptors in B cell apoptosis, high and low density B cells were plated in 96 well tissue culture plates and anti-adhesion molecule mAb were added or alternatively the cells were plated in wells coated with adhesion molecules. As stated above, almost no apoptotic cells were found in the dense B cell population after 18 h. culture. Furthermore the reagents tested by us, had no effect on apoptosis of these cells (Table 1).

We found that addition of anti-LFA-1α (CLB-LFA-1/2, SPV-L7, SPV-L12, NKI-L16), anti-LFA-1ß (CLB-LFA-1/1, IB4) or anti-ICAM-1 (F10.2, RR1/1) mAb had also no effect on apoptosis of low density B cells (Table 1 and not shown). However when the low density B cells were plated on wells coated with purified ICAM-1 protein, which is the ligand of LFA-1, a strong reduction in the percentage of apoptotic cells was observed. Thus while the anti-adhesion molecule mAb are unable to prevent B cell apoptosis, the immobilized, purified ligand effectively blocks B cell entry into apoptosis. Addition of anti-IgM or Staphylococcus aureus also inhibited B cell apoptosis.

DISCUSSION

In the germinal center, B cells are either selected to become memory cells or plasma cells, or if their Ig receptors do not meet the antigen binding affinity requirements are eliminated through the process of programmed cell death[5,6]. FDC which are intimately associated with the germinal center B cells are thought to be important in this selection process[1,12], to which they can contribute in several ways. First, FDC can trap antigen in the form of immune complexes, and present the antigen to the B cells in a highly immunogenic form, i.e. as immune complex coated vesicles (iccosomes)[8]. Second, FDC may release substances like CD23 that can act as B cell differentiation factors[6]. We now propose a third mechanism in which the adhesive interaction between B cells and FDC affects B cell differentiation.

Table 1. Effect of adhesion molecules on B cell entry into apoptosis.[1]

Antigen	mAb	high density	low density
control		2 ± 1	52 ± 8
LFA-1α	CLB-LFA-1/2	1 ± 1	50 ± 8
LFA-1ß	CLB-LFA-1/1	6 ± 1	51 ± 7
ICAM-1	RR1/1	4 ± 1	45 ± 2
CD44	NKI-P2	2 ± 1	50 ± 5
anti-IgM		2 ± 1	29 ± 3
SAC		6 ± 2	20 ± 8
sICAM-1		5 ± 1	34 ± 2

[1] percentage of apoptotic cells in May-Grünwalds/Giemsa stained cytospin preparations (arithmatic mean ± standart deviation of triplicate wells).

We observed that adhesive interactions can prevent entry into apoptosis of tonsil derived germinal center B cells (defined here as low density cells, predominantly PNA positive, IgD negative and CD38 positive). Plating these cells in 96 well tissue culture plates that were coated with ICAM-1 protein diminished the number of cells becoming apoptotic during 18 h. of culture. Strikingly, mAb against the ICAM-1 ligand LFA-1 had no effect on apoptosis. These findings either implicate that ligand interaction provides a stronger signal to the B cell than interaction with mAb, or alternatively that cross linking of LFA-1, brought about by the immobilized ligand sICAM-1, is required for preventing entry into apoptosis. Interestingly, Liu et al.[13] have recently described that FDC in the basal light zone of the germinal center (in contrast to FDC in the apical light zone) show a strong ICAM-1 expression and that most apoptotic cells are found in this area, which makes it a likely place for B cell selection.

How can adhesive interactions prevent B cell entry into apoptosis? Recent studies have shown that adhesion of T cells, with LFA-1 to immobilized ICAM-1 protein enhances anti-CD3 induced T cell proliferation[11]. Similarly the involvement of adhesion molecules during the interaction between FDC and B cells may provide a signal that results in B cell activation and thereby prevents apoptosis. Selection of B cells appears to be a complex event where antigen, growth factors and adhesion molecules each play a role. However the relative contribution of these factors has yet to be established.

ACKNOWLEDGEMENT

We thank C.G. Figdor for mAb NKI-P2, SPV-L7, SPV-L12 and NKI-L16; R.A.W. van Lier for mAb CLB-LFA-1/2; F. Miedema for mAb CLB-LFA-1/1; A.C. Bloem for mAb F10.2; T.A. Springer for mAb RR1/1; G.A. van Seventer for sICAM-1. This work was supported by grant 88/CR/032/89 from the Dutch Rheumatism Foundation.

REFERENCES

1. Mac Lennan, I.C.M., and Gray, D., Antigen-driven selection of virgin and memory B cells, *Immunol. Rev.* 91:61(1986).
2. Kunkl, A., and Klaus, G.G.B., The generation of memory cells. IV. Immunization with antigen-antibody complexes accelerates the development of B-memory cells, the formation of germinal centers and the maturation of antibody affinity in the secondary response, *Immunology* 43:371(1981).
3. Kocks, C., and Rajewsky, K., Stable expression and somatic hypermutation of antibody V regions in B-cell developmental pathways, *Annu. Rev. Immunol.* 7:537(1989).
4. Griffiths, G.M., Berek, C., Kaartinen, M., and Milstein, C., Somatic mutation and maturation of the immune response to 2-phenyloxazolone, *Nature* 312:271(1984).
5. Liu, J.Y.,. Joshua, D.E., Williams, G.T., Smith, C.A., Gordon, J., and MacLennan, I.C.M., Mechanisms of antigen-driven selection in germinal centres, *Nature* 342:929(1989).
6. Liu, J.Y., Cairns, J.A., Holder, M.J., Abbot, S.D., Jansen, K.U., Bonnefoy, J-Y., Gordon, J., and MacLennan. I.C.M., Recombinant 25 kDa CD23 and interleukin-1α promote the survival of germinal center B cells: evidence for bifurcation in the development of centrocytes rescued from apoptosis, *Eur. J. Immunol.* 21:1107(1991).
7. Tew, J.G., and Mandel, T.E.,Prolonged antigen half-life in the lymphoid follicles of specifically immunized mice, *Immunology* 37:69(1979).
8. Szakal, A.K., Kosco, M.H., and Tew, J.G., Microanatomy of lymphoid tissue during humoral immune responses. Structure function relationships, *Annu. Rev. Immunol.* 7:91(1989).
9. Koopman, G., Parmentier, H.K., Schuurman, H.J., Newman, W., Meijer, C.J.L.M., and Pals, S.T., Adhesion of human B cells to follicular dendritic cells involves both the lymphocyte function-associated antigen 1/intercellular adhesion molecule 1 and very late antigen 4/vascular cell adhesion molecule 1 pathways, *J. Exp. Med.* 173:1297(1991).
10. Koopman, G., and Pals, S.T., Cellular interactions in the germinal center: role of adhesion receptors and significance for the pathogenesis of AIDS and malignant lymphoma, *Immunol. Rev.* 126:21(1992).
11. van Seventer, G.A., Shimizu, Y., Horgan, K.J., and Shaw, S., The LFA-1 ligand ICAM-1 provides an important costimulatory signal for T cell receptor - mediated activation of resting T cells, *J. Immunol.* 144:4579(1990).
12. Tew, J.G., Kosco, M.H., Burton, G.F., and Szakal, A.K., Follicular dendritic cells as accessory cells, *Immun. Rev.* 117:185(1990).
13. Liu, Y.J., Johnson, G.D., Gordon, J., and MacLennan, I.C.M., Germinal centres in T-cell dependent antibody responses, *Immunol. Today* 13:17(1992).

MEMBRANE EXPRESSION OF $Fc\epsilon RII$/CD23 AND RELEASE OF SOLUBLE CD23 BY FOLLICULAR DENDRITIC CELLS

E. Peter Rieber, Gerti Rank, Ingrid Köhler and Susanne Krauss

Institute for Immunology
University of Munich
Munich, FRG

INTRODUCTION

The low affinity receptor for IgE ($Fc_{\epsilon}RII$/CD23) is expressed on a variety of lymphoid and non lymphoid cell types. It is constitutively expressed on μ^{+}/δ^{+} B lymphocytes and can be induced on other cells mainly by IL-4. In addition, soluble CD23 (sCD23) is released by CD23^{+} cells. Various functions have been ascribed to this molecule, its biological role, however, remains elusive[1,2]. In sections of lymphoid tissues a particularly dense expression of CD23 is found on a subpopulation of follicular dendritic cells (FDC) in the light zone of germinal centers. Since detection of CD23 on isolated and in vitro cultivated FDC is controversial[3,4] it is not known whether CD23 found in situ on FDC is produced by these cells or whether it is derived from B cells and adsorbed to the surface of FDC. It seemed important to solve this problem since there is now good evidence that CD23 plays a pivotal role in the interaction of FDC and germinal center B cells[5]. Functional tests on FDC are hampered by the technical problems in preparing these cells in high purity and in sufficient numbers. To this end a modified isolation protocol was developed which is mainly based on the specific enrichment of CD14^{+} cells from tonsillar tissue as described by Schriever et al.[3]. We used the direct monoclonal antibody (mAb) rosetting technique (DART) which has previously been shown to allow sensitive detection of surface antigens and rapid and large scale separation of mononuclear cell subpopulations[6]. When combined with density centrifugation this procedure yields a rather pure population of FDC. Here we show that purified FDC can be induced by IL-4 to express membrane bound CD23 and to release sCD23. In addition, by use of the polymerase chain reaction (PCR) selective expression of the CD23b isoform by FDC is demonstrated.

Table 1. Purification of follicular dendritic cells (FDC) from human tonsils.

1. Cut tonsil into small pieces and mince them through a stainless steel mesh into a petri dish with RPMI 1640 tissue culture medium containing 0.5% BSA, 2 mg/ml collagenase IV and 0.1 mg/ml DNase.
2. Incubate cells for 90 min at 37°C.
3. Collect the supernatant, resuspend the remaining tissue pieces in TC medium by vigorous pipetting and collect the supernatant after sedimentation of tissue clumps. Repeat this treatment three times.
4. Combine the supernatants. After washing isolate the $CD14^+$ cells using a direct monoclonal antibody rosetting technique (DART) as previously described[6]. Very briefly, mix the cells with bovine erythrocytes coated with CD14 mAb M-M42 (5 erythrocytes per leukocyte). Spin down for 5 min at 100 x g and incubate on ice for 1 hr. Resuspend the cell mixture very carefully and centrifuge it through 65% Percoll. Remove supernatant from the pellet and lyse the erythrocytes using a 0.8% (w/v) NH_4Cl buffer.
5. After washing the cells resuspend them in 5 ml of TC medium and layer them on a discontinuous Percoll density gradient consisting of 65%, 46% and 21% Percoll. After centrifugation at 1000 x g for 20 min collect the cells from the first interphase above 21% Percoll (d < 1.030 g/ml) as pure FDC population.
6. Purify FDC contained in the the second interphase above 46% Percoll (d < 1.060) by a second DART using the newly developed FDC-directed mAb M-GC2 which is unreactive with blood monocytes.

MATERIAL AND METHODS

Isolation of FDC from Human Tonsils

FDC were isolated according to the protocol given in table 1.

Phenotyping of Cells

Immunophenotyping of the isolated cell populations was performed with monoclonal antibodies (mAb) using three different methods. For quantitative evaluation cells were stained in direct or indirect immunofluorescence followed by cytofluorographic analysis in a FACScan. When the cell number was limited direct monoclonal antibody rosetting (DART)[6] or immunocytochemistry on cytospin preparations was used. In all staining procedures unspecific binding of antibodies was minimized by preincubation of cells with heat aggregated human IgG.

PCR

The polymerase chain reaction was performed essentially according to Saiki et al.[7]. For the detection of mRNA of the two isoforms CD23a and CD23b specific sense primers were used which span either 21 bp of exon IIa (CD23a) or 21 bp of exon Ib (CD23b) together with a 3' antisense primer out of exon VII resulting in an amplified fragment of 370 bp. IgM message was detected with help of a 5' sense primer out of exon Cμ2 and a 3' antisense primer out of exon Cμ3.

Determination of Soluble CD23

For the quantitative determination of sCD23 in culture supernatants a capture ELISA was established using two CD23 mAbs (M-L25 and M-L234) recognizing different epitopes on the 25 kDa sCD23[8]. Recombinant sCD23 secreted by a transfected CHO cell line served as standard.

RESULTS AND DISCUSSION

Expression of Fc_{ϵ}RII/CD23 by FDC

FDC were isolated from tonsillar tissue as a single cell suspension based on their strong expression of the monocytic differentiation antigen CD14, on their low density and and reactivity with mAb M-GC2. Purity of FDC was ascertained by morphology in Giemsa stained cell smears and by staining with mAbs held to be characteristic for FDC such as DRC1, Ki-M4 and a newly developed mAb M-GC2. For the studies of CD23 expression it was particularly important to exclude the presence of B lymphocytes and monocytes. The very few cells with lymphoid appearance found in the final FDC preparation were not reactive with CD19 or CD20 mAbs. Contaminating monocytes were ruled out by the failure to demonstrate cells positive for the NaF-sensitive naphthol-AS acetate esterase.

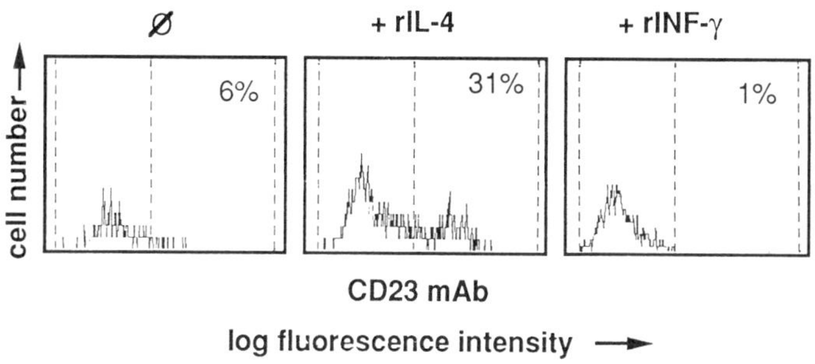

Figure 1. Induction of surface CD23 on FDC. Purified FDC were cultured for 3 days in the presence of 500 U/ml of rIL-4 or 1000 U/ml of rIFN-τ and then stained with the CD23 mAb M-L25. Data obtained by FACScan analysis are displayed as single histograms. Numbers give the percentage of labeled cells; ϕ = medium control.

Despite its strong in situ expression CD23 was detected only on a small subpopulation of freshly isolated FDC after collagenase treatment. When kept in culture for a few days surface expression of CD23 diminished. This finding corresponds to the observations of other authors[9,10]. Since IL-4 has been found to induce or to enhance CD23 expression on monocytes[11], epidermal Langerhans cells[12] and T lymphocytes[13], FDC were cultivated in the presence of 500 U/ml of rIL-4. After 2 days CD23 was found on a certain percentage of FDC ranging from 5% to 42%. Fig.1 shows an example of IL-4-induced CD23 expression by FDC as revealed by cytofluorographic analysis. This finding is in contrast to the observation of Clark et al.[9] who failed to see induction of CD23 on FDC using various cytokines including IL-4. When FDC were incubated with

1000 U/ml of rIFN-τ no CD23 induction was found. In this respect FDC differ from epidermal Langerhans cells where we could show a synergistic effect of IL-4 and IFN-τ in the induction of CD23[12]. On most cells, however, IFN-τ reduces CD23 expression[13,14] and thus seems to be the natural antagonist of IL-4.

Determination of the CD23 Isoform in FDC

CD23 occurs in two isoforms a and b which differ only in the 6 respectively 7 N-terminal intracytoplasmic amino acids[15]. Whether the two isoforms are functionally different is not yet known. Whereas B cells express both isoforms, only CD23b has been found on other cells so far, particularly after induction with IL-4[15,16]. In order to define

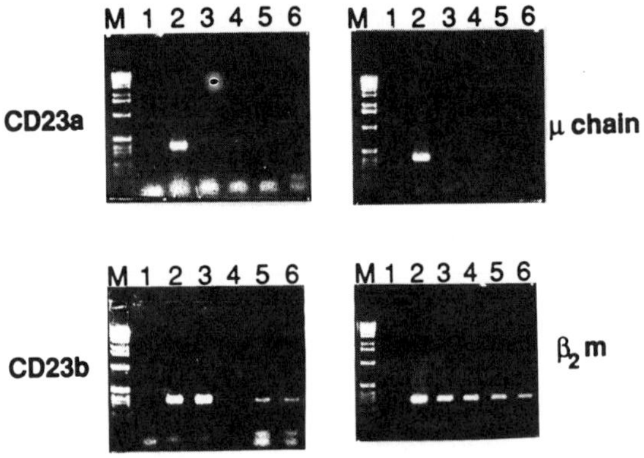

Figure 2. Selective CD23b expression in FDC as revealed by PCR. RNA was extracted from the cells indicated, reverse transcribed into cDNA and amplified by PCR using primers specific for CD23a, CD23b, immunoglobulin μ-chain and β_2-microglobulin. M: marker; lane 1: medium control; lane 2: B cell line WI-L2; lane 3: monocytic cell line U937 cultured with IL-4; lane 4: highly purified FDC cultivated for 3 days without IL-4; lane 5: same FDC cultivated for 3 days in the presence of 500 U/ml rIL-4; lane 6: FDC prepared from another tonsil and cultivated for 3 days in the presence of IL-4.

the isoform expressed by FDC PCRs were performed on mRNA prepared from highly purified FDC using primers specific for either the isoform a or b. Fig. 2 shows the result of a representative experiment where mRNAs from various cell types including highly purified FDC from two different tonsils were analysed. Only message of the b-isoform was detectable in FDC after cultivation with IL-4, whereas no message was seen in the absence of IL-4.

Since IL-4 enhances expression of both CD23a and CD23b in B lymphocytes[15] presence of these cells in the FDC preparation had to be excluded. Therefore, detection of IgM-message was included as a control. β_2m bands indicate the effective reverse transcription of RNA into cDNA.

Release of Soluble CD23 by FDC

Rescue of germinal center B cells from apoptosis and induction of B cell differentiation is considered as an important role of FDC[5]. This activity might at least partially be mediated by CD23, probably in its soluble form. It was therefore of interest to find out whether soluble CD23 is released from FDC. To this end purified FDC were cultivated in the presence of rIL-4 and rIFN-τ. As can be seen from table 2, no sCD23 was found in the supernatant of FDC kept without external lymphokine. When, however,

Table 2. Release of soluble CD23 by purified follicular dendritic cells

	Culture period (days)	Incubation with medium	rIL-4	rIFN-τ
Tonsil 1	2	< 5*	87	< 5
Tonsil 2	2	< 5	70	< 5
Tonsil 3	3	< 5	95	n.d.
Tonsil 4	3	< 5	103	n.d.

* ng / ml / 10^6 cells

Purified FDC were cultivated for 2 or 3 days at a concentration of 1 x 10^6 cells /ml in the presence of 500 U/ml of rIL-4 or of 1000 U/ml or IFN-τ or without lymphokines. sCD23 was determined by an antigen capture ELISA.

FDC were cultured in the presence of rIL-4 a significant amount of sCD23 is released into the supernatant. IFN-τ, on the other hand, was not able to induce sCD23 release. These data clearly show that FDC are able to produce and to release CD23 upon stimulation with rIL-4. However, it cannot be excluded that at least a part of the CD23 found on FDC in situ is B cell derived and binds to a specific ligand on the surface of a subpopulation of FDC. This should be further investigated. It is not yet clear whether it is only IL-4 that is able to induce expression and release of CD23 by FDC or whether other cytokines either alone or in combination have a similar effect. If the expression of CD23 by FDC in vivo depends on cytokines provided by other cells their source in the germinal center remains to be determined. Since we have shown that Fc_{ϵ}RII/CD23 can mediate focusing of IgE-complexed antigen to T cells by antigen presenting cells[17] it would be of interest to know whether this activity is also displayed by FDC.

Acknowledgements

This work was supported by the Deutsche Forschungsgemeinschaft and by the Wilhelm Sander-Stiftung.

We are indebted to Dr.Faas and Dr.Chucholowski, Krankenhaus München-Pasing, and Prof. Kastenbauer and Dr. Grevers, Department of Medicine, University of Munich, for providing us with tonsillar tissue.

REFERENCES

1. G.Delespesse, U.Suter, D.Mossalayi, B.Bettler, M.Sarfati, H.Hofstetter, E.Kilcherr, P.Debre and A.Dalloul, Expression, structure, and function of the CD23 antigen, *Adv. Immunol.* 49: 149 (1991).
2. D.H.Conrad, Fc_{ϵ}RII/CD23: The low affinity receptor for IgE, *Ann.Rev.Immunol.* 8:623 (1990).
3. F.Schriever, A.S.Freedman, G.Freeman, E.Messner, G.Lee, J.Daley, and L.M.Nadler, Isolated human follicular dendritic cells display a unique antigenic phenotype, *J.Exp.Med.* 169: 2043 (1989).
4. K.Sellheyer, R.Schwarting and H.Stein, Isolation and antigenic profile of follicular dendritic cells, *Clin. Exp. Immunol.* 78: 431 (1989).
5. Y.J.Liu, J. Cairns, M.J. Holder. S. Abbot, K.U. Jansen, J.Y. Bonnefoy, J. Gordon and I.C.M. MacLennan, Recombinant 25-kilodalton CD23 and interleukin 1 alpha promote survival and differentiation of germinal center B cells, *Eur.J.Immunol.* 21: 1107 (1991).
6. M.Wilhelm, H.Pechumer, G.Rank, E.Kopp, G.Riethmüller and E.P.Rieber, Direct monoclonal antibody rosetting. An effective method for weak antigen detection and large scale separation of human mononuclear cells, *J. Immunol. Meth.* 90: 89 (1986).
7. R.K.Saiki, S.Scharf, F.Faloona, K.B.Mullis, G.T.Horn, H.A.Erlich and N.Arnheim, Enzymatic amplification of ß-globin genomic sequences and restriction site analysis for diagnosis of sickle cell anemia, *Science* 230: 1350 (1985).
8. J.C.Prinz and E.P.Rieber, Fc-receptors for IgE on human lymphocytes. Detection with a rosetting assay using a recombinant human/mouse IgE antibody and characterization with monoclonal antibodies, *Hybridoma* 6: 1 (1987).
9. E.A.Clark, K.H.Grabstein and G.L.Shu, Cultured human follicular dendritic cells. Growth characteristics and interactions with B lymphocytes, *J.Immunol.* 148: 3327 (1992).
10. R.Tsunoda, M.Nakayama, K.Onozaki, E.Heinen, N.Cormann, C.Kinet-Denoel and M.Kojioma, Isolation and long-term cultivation of human tonsil follicular dendritic cells, Virch. Arch.B Cell Pathol 59: 95 (1990).
11. D.Vercelli, H.H.Jabara, B.W.Lee, N.Woodland, R.S.Geha and D.Y.Leung, Human recombinant interleukin-4 induces FcεR2/CD23 on normal human monocytes. *J.Exp.Med.* 167: 1406 (1988).
12. T.Bieber, A.Rieger, C.Neuchrist, J.C.Prinz, E.P.Rieber, G.Boltz-Nitulescu, O.Scheiner, D.Kraft, J.Ring and G.Stingl, Induction of Fc_{ϵ}R2 (CD23) on human epidermal Langerhans cells by human recombinant Interleukin-4 and τ-Interferon, *J.Exp.Med.* 170: 309 (1989).
13. J.C.Prinz, X.Baur, G.Mazur and E.P.Rieber, Allergen-directed expression of Fc-receptors for IgE (CD23) on human T lymphocytes is modulated by interleukin 4 and interferon, *Eur.J.Immunol.* 20: 1259 (1990).
14. T.Kawabe, M.Takami, M.Hosoda, Y.Maeda, S.Sato, M.Mayumi, H.Mikawa, K.Arai and J.Yodoi, Regulation of Fc_{ϵ}R2/CD23 gene expression by cytokines and specific ligands (IgE and anti-Fc_{ϵ}R2 monoclonal antibody, *J.Immunol.* 141: 1376 (1988).
15. A.Yokota, H.Kikutani, T.Tanaka, R.Sato, E.L.Barsumian, M.Suemura and T.Kishimoto, Two species of human Fc_{ϵ}Receptor II (Fc_{ϵ}RII/CD23): Tissue-specific and IL-4-specific regulation of gene expression, *Cell* 55: 611 (1988).
16. N.Endres, J.C.Prinz & E.P.Rieber, T cells express the B isoform of FcεRII after stimulation with allergen or PHA, *Immunobiol.* 181: 152 (1990).
17. U.Pirron, T.Schlunk, J.C.Prinz and E.P.Rieber, IgE-dependent antigen focusing by human B lymphocytes is mediated by the low affinity receptor for IgE, *Eur.J.Immunol.* 20: 1547 (1990).

FOLLICULAR DENDRITIC CELLS IN MALIGNANT LYMPHOMAS

- DISTRIBUTION, PHENOTYPES & ULTRASTRUCTURES -

Kunihiko Maeda, Mikio Matsuda, Masaru Narabayashi,
Ryuichi Nagashia, Noriyuki Degawa and Yutaka Imai

Department of Pathology, Yamagata University School of Medicine
2-2-2 Iida-Nishi, Yamagata, 990-23, Japan

INTRODUCTION

Follicular dendritic cells (FDC) are morphologically and functionally specialized cells which located within the reactive follicles of secondary lymphoid tissues. Recent years, several reports [1-9] described that FDCs also distributed in the lymphoid tissues involved with malignant lymphoma such as follicular lymphoma, diffuse lymphoma, and Hodgkin's disease. We expected that the comparison of the distribution pattern, morphology, immunological phenotypes and functional properties of FDC in malignant lymphoma with those of reactive tissues provide us important instances to understand the relationships between the functional properties or differentiation of FDC and the microenvironment organized by surrounding lymphoid cells. In the present study we had examined the morphology, surface phenotypes and distribution patterns of FDC in lymphoma including follicular lymphoma diffuse B-cell lymphoma and T-cell lymphoma and compared with those of reactive follicles.

MATERIALS AND METHODS

Table 1 indicated the monoclonal and polyclonal antibodies employed in this study. For the present, the specific markers for human FDC have not been established yet, but DRC-1 (R4/23) and DF-DRC1 are used as relatively reliable makers. In routinely processed paraffin sections, anti-CD35 or anti-CD21, anti-CD23 also used as markers for FDC.

Seven cases of follicular lymphoma, 20 cases of diffuse B-cell lymphoma and 5 cases of T-cell lymphoma were examined. Follicular lymphomas and diffuse B-cell lymphomas were subclassified based on NCI working formulation of malignant lymphoma subclassification [10] and T-cell lymphomas were Up-to-date Kiel classification [11].

RESULTS AND DISCUSSIONS

FDC in follicular lymphomas

Seven cases of follicular lymphoma were examined. They included a case of predominantly small cleaved cell type, 5 cases of mixed small cleaved and large cell type

Dendritic Cells in Fundamental and Clinical Immunology,
Edited by Kamperdijk *et al.*, Plenum Press, New York, 1993

Table1, Antibodies employed in the present study

Antibody (clone)	Isotype	Source & Dilution
R4/23 (DRC-1)	mouse IgM	DAKO (M709) 1:50
DF-DRC-1	mouse IgG1	Sera-Lab (MAS450) 1:50
anti-CD21 (BU32)* anti-CR-2 or C3bR	mouse IgG1	Binding Site (ME110) 1:100
anti-CD23 (BU38)* anti-FcεR-II	mouse IgG1	Binding Site (ME112) 1:100
anti-CD35 (Ber-MAC-DRC) * anti-CR-1 or C3dR	mouse IgG1	DAKO (M846) 1:10
anti-human IgG	rabbit polyclonal antisera	DAKO (A424) 1:800
anti-human IgA	rabbit polyclonal antisera	DAKO (A262) 1:400
anti-human IgM	rabbit polyclonal antisera	DAKO (A426) 1:800
anti-human IgD	rabbit polyclonal antisera	DAKO (A093) 1:100
anti-human IgE	rabbit polyclonal antisera	DAKO (A094) 1:100
anti-human C1q	rabbit polyclonal antisera	DAKO (A136) 1:50
anti-human C3d	rabbit polyclonal antisera	DAKO (A063) 1:50
anti-human C5	rabbit polyclonal antisera	DAKO (A055) 1:50

* available on the routinely processed paraffin sections.

and a case of large cell type. In all cases, distinct reticular positivity of DRC-1, DF-DRC-1, CD21 (CR2 or C3b-receptor), CD35 (CR1 or C3d receptor) were seen coincidentally with neoplastic follicles (Fig.1a, 1b). At ultrastructural level, the cells which had euchromatic slight indented nucleus with prominent one or a few nucleoli, entangled very intricate cytoplasmic processes and sometimes desmosome structures (Fig.2). These features were very similar to those of FDC in reactive follicles.

On immunoglobulin and complement components, the reticular depositions of IgM, IgE, C3,or C5 were detected partly in the neoplastic follicles of some cases (Fig.3a and 3b). Our recent result also indicated that the decay accelerating factor (DAF), which is one of complement regulatory factors and detected in the reactive follicles in association with FDC, were also recognized in the neoplastic follicles but the intensity of the reactivity was much less than reactive follicles (unpublished data). These results suggested that FDC in the neoplastic follicles still had the function to trap and retain immune complexes on their surface as well as in the reactive follicles, but the functional capability or regulatory mechanisms might be impaired. These functional difference between the reactive lymphoid follicles and the neoplastic follicles of follicular lymphoma was quite interesting for understanding on the functional diversity of FDC.

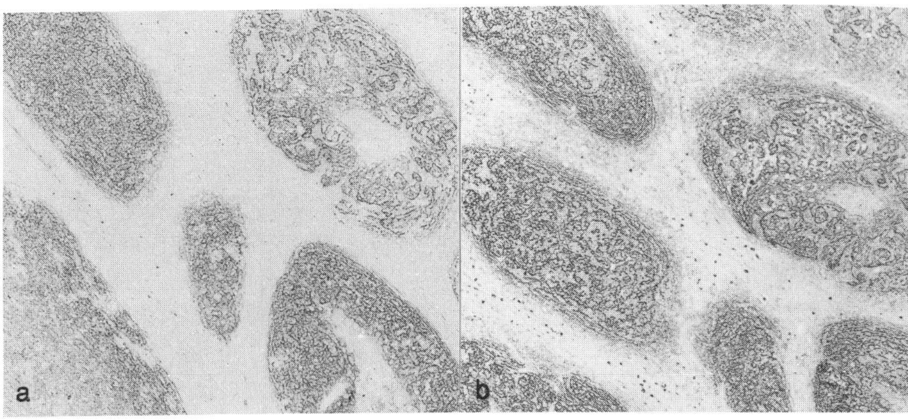

Fig.1 Light micrographs showing the reactivity of DRC-1(R4/23, 1a) and CD35 (CR-1 or C3dR, 1b) in the lymph node with small cleaved cell type of follicular lymphoma. (both X40)

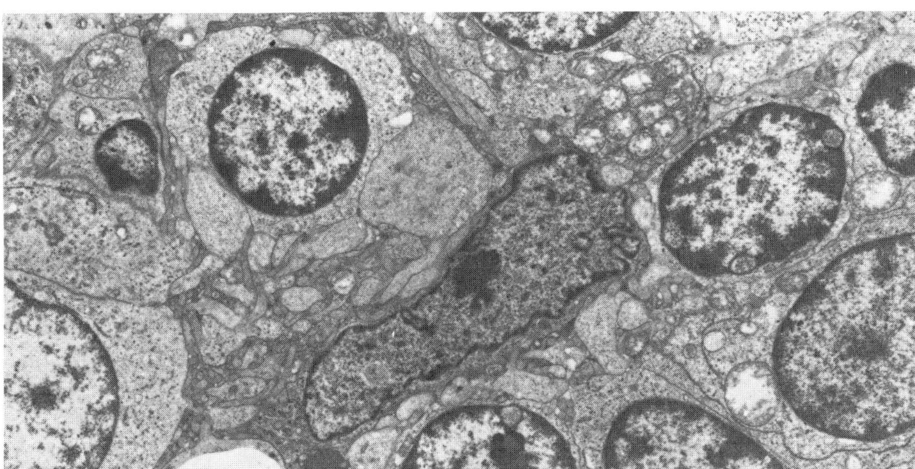

Fig.2 Electron micrograph showing FDC in the neoplastic follicle of follicular lymphoma. Note entangled, very intricate cytoplasmic processes of this cell. (X5,500)

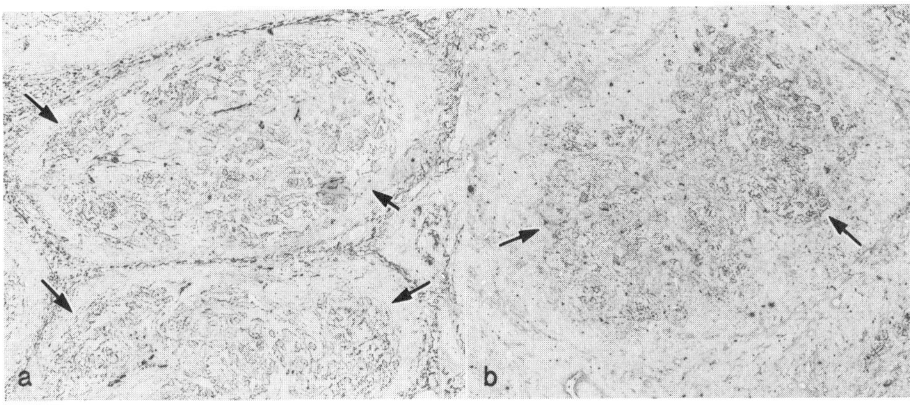

Fig.3 Light micrographs demonstrating C3d staining (3a) and IgM (3b) staining of lymph node involved follicular lymphoma. Arrows indicate reticular deposition of C3d or IgM. (both X40)

FDC in diffuse B-cell lymphomas

20 cases of diffuse B-cell lymphoma were investigated. These were including 13 cases of small cleaved cell-type, a case of mixed small and large cell type, and 6 cases of large cell type.. In diffuse lymphomas, two distinct distribution patterns of FDC were noticed. One of them was tentatively designated "vestigial pattern", which mean the vestigial foci of the pre-existing follicles. These pattern consisted of a few cells which had many microvilli-like cytoplasmic projections and intense surface labeling of DRC-1 and DF-DRC1 (Fig.4a, 4b). Another pattern was designated "spreading pattern", which mean the widely distribution of the cells. The cells constituted this arrangement revealed spindle or polygonal shape, well developed rER and euchromatic nuclei similar (so-called "fibroblastic appearance") and partial reactivity of DRC-1 and DF-DRC1 (Fig.5a, 5b). Table 2 indicated the relationships between these categories on the distribution of FDC and subclassification in the well examined 12 cases. The vestigial pattern were seen in various type of diffuse B-cell lymphoma but the spreading pattern were not seen in large cell type but prominent in the small cleaved cell type. Mori et al[3] and Scoazec et al[4] also described almost the same results. These observations suggested that the cells constituting "spreading pattern" might represent the developing or alterative process rather than the vestiges of pre-existing follicles. In other words, growth and proliferation of lymphoma cells, especially small cleaved type of cells, might induce the phenotypic differentiation of stromal cells toward FDC de novo. Cerny et al[12] and Imai et al[13] also suggested experimentally that populations of B-cells or microenvironment organized B-cells may contribute the development or differentiation of FDC. Small cleaved cells are thought to a neoplastic counterpart of centrocytes (germinal center cells exclusively located in light zone of germinal center, in which well developed FDC also localize). Thus, the microenvironment organized by small cleaved type of lymphoma cells or centrocytes might be necessary to the development of FDC.

Table 2 Distribution pattern of FDC in diffuse lymphomas

CASE	Subclassification	Vestigial Pattern	Spreading Pattern
1	small cleaved cell + vague nodularity	–	+++
2	small cleaved cell	++	+++
3	small cleaved cell	+	++
4	small cleaved cell + vague nodularity	+++	++
5	small cleaved cell	–	–
6	small cleaved cell	++	–
7	small cleaved cell	++	+++
8	small cleaved cell	++	+++
9	mixed	++	–
10	large cell (cleaved)	–	–
11	large cell (non-cleaved)	+	–
12	large cell (non-cleaved)	–	–

FDC in T-cell lymphomas

5 cases of T-cell lymphomas including 4 cases of AILD type and one case of lymphoblastic lymphoma with pseudonodular pattern were examined. Interestingly the wide-spreading FDC were observed in all cases of AILD type (Fig.6). This type of lym-

phoma also associated polyclonal plasma cells proliferation, arborizing proliferation of high endothelial venules and hypergammaglobulinemia or dysproteinemia. These unique features may suggest the functional peculiarity of the lymphoma cells. In contrast, lymphoblastic lymphoma did not have distribution of FDC, even if it revealed follicular appearance (pseudonodular pattern).

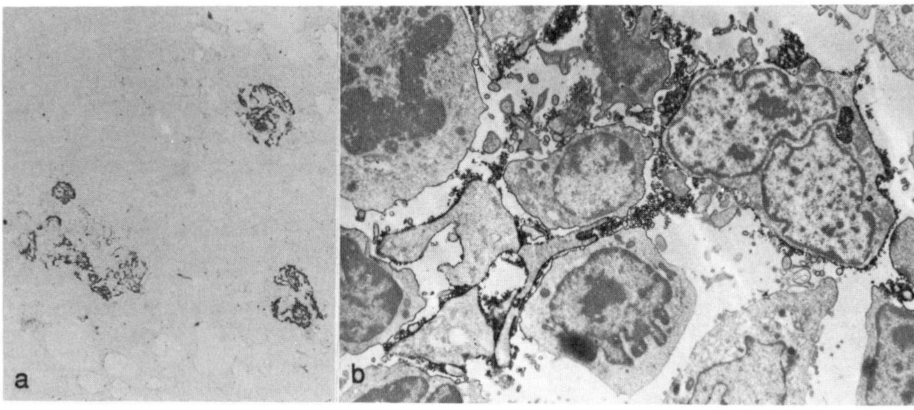

Fig.4, Vestigial pattern of distribution of DF-DRC1$^+$ cells in diffuse B-cell lymphoma. Fig.4a shows LM view and 4b shows EM figures of the cell constituting this pattern. (4a X30, 4b X3,650)

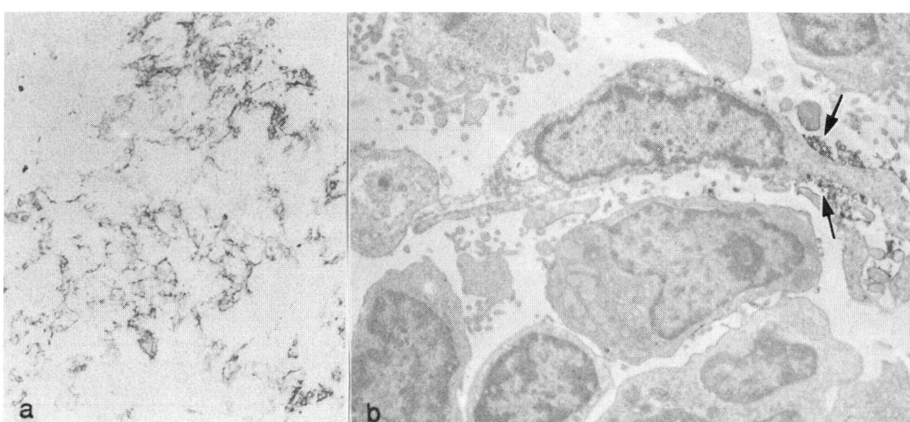

Fig.5, Spreading pattern of distribution of DF-DRC1$^+$ cells in diffuse B-cell lymphoma. Fig.5a shows LM view and 5b shows EM figures of the cell constituting this pattern. (5a X30, 5b X3,300)

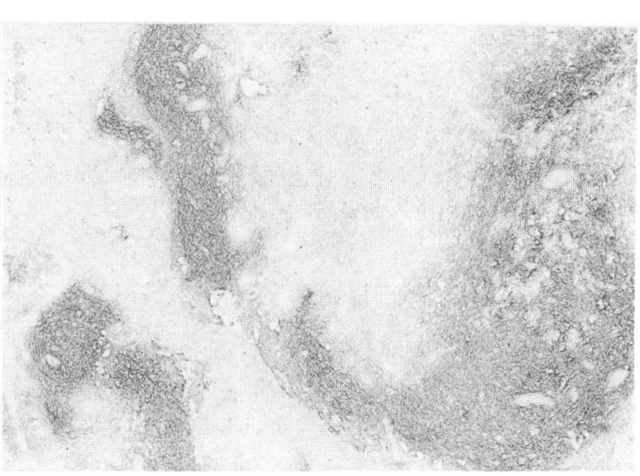

Fig.6 Light micrograph showing the distribution of DRC-1$^+$ cells in AILD type of T-cell lymphoma. (X30)

REFERENCES

1, J. Gerdes and H. Stein, Complement (C3) receptors on dendritic reticulum cells of normal and malignant lymphoid tissue. *Clin. Exp. Immunol.* 48: 348 (1982).

2, R. Manconi, A. Poletti, R. Volpe, S. Sulfaro and A. Carbone, Dendritic reticulum cell pattern as a microenvironmental indicator for a distinct origin of lymphoma of follicular mantle cells. *British J. Haematol.* 68:213 (1988).

3, N. Mori, K. Oka and M. Kojima, DRC antigen expression in B-cell lymphomas. *Am. J. Clin. Pathol.* 89:488 (1988).

4, J.-Y. Scoazec, F. Berger, J.-P. Magaud, J. Brochier, B. Coiffier and P.-A. Bryon, The dendritic reticulum cell pattern in B cell lymphomas of the small cleaved, mixed, and large cell types: An immunohistochemical study of 48 cases. *Human Pathol* 20: 124 (1989).

5, Y. Imai, M. Matsuda, K. Maeda, M. Narabayashi and A. Masunaga, Follicular dendritic cells in lymphoid malignancies –morphology, distribution and function–. *in*: "Lymphoid malignancy -Immunocytology and Cytogenetics-." M. Hanaoka, M.E. Kadin, A. Mikata and S. Watanabe, eds., Field & Wood, New York, 229 (1990).

6, S. Petrasch, C. Perez-Alvarez, J. Schmitz, M.H. Kosco and G. Brittinger, Antigenic phenotyping of human follicular dendritic cells isolated from nonmalignant and malignant lymphatic tissue. *Eur. J. Immunol.* 20:1013 (1990).

7, H. Tabrizchi, M.-L. Hansmann, M.-R. Parwaresch and K. Lennert, Distribution pattern of follicular dendritic cells on low grade B-cell lymphomas of the gastro intestinal tract immunostained by Ki-FDC1p: A new paraffin-resistant monoclonal antibody. *Modern Pathology*,3:470 (1990).

8, M. Narabayashi, K. Maeda, M. Matsuda and Y. Imai, Immunohistological study in the distribution and function of follicular dendritic cell (FDC) in follicular lymphoma. (in Japanese) *J. Jpn. RES Soc.* 30:333 (1990).

9, M.J. Alavaikko, M.-L. Hansmann, C. Nebendahl, M.R. Parwaresch and K. Lennert, Follicular dendritic cells in Hodgkin's disease. *Am. J. Clin. Pathol.* 95:194 (1991).

10, The Non-Hodgkin's Lymphoma Pathologic Classification Project, National cancer institute sponsored study of classifications of non-Hodgkin's lymphomas. Summary and description of a working formulation for clinical usage. *Cancer* 49:2112 (1982).

11, T. Suchi, K. Lennert, L.-Y. Tu, M. Kikuchi, E. Sato, A.G. Stansfeld and A.C. Feller, Histopathology and immunohistochemistry of peripheral T cell lymphomas: a proposal for their classification. *J. Clin. Pathol.* 40:995 (1987).

12, A. Cerny, R.M. Zinkernagel and P. Groscurth, Development of follicular dendritic cells in lymph nodes of B-cell-depleted mice. *Cell Tissue Res.* 254:449 (1988)

13, Y. Imai, K. Maeda, M. Yamakawa, Y. Karube, M. Matsuda, M. Dobashi, H. Sato and K. Terashima, Heterogeneity and cellular origin of follicular dendritic cells. *in this volume*, (1992).

LYMPHOID FOLLICLES IN CYNOMOLGUS MONKEYS AFTER INFECTION WITH SIMIAN IMMUNODEFICIENCY VIRUS

Piet Joling,[1] Peter Biberfeld,[2] Henk K. Parmentier,[1] Dick F. van Wichen,[1] Timo Meerloo,[1] Louk H.P.M. Rademakers,[2] Jürg Tschopp,[3] Jaap Goudsmit,[4] Henk-Jan Schuurman[1]

[1] Departments of Pathology and Internal Medicine, University Hospital, P.O. Box 85.500, 3508 GA Utrecht, The Netherlands
[2] Immunopathology Laboratory, Karolinska Institute, Stockholm, Sweden
[3] Institute of Biochemistry, University of Lausanne, Switzerland
[4] Human Retroviral Laboratory, Academic Medical Center, Amsterdam, The Netherlands

INTRODUCTION

In the first phase of infection by the Human Immunodeficiency Virus type 1 (HIV-1), persistent generalised lymphadenopathy can be the only feature of the disease, with or without associated clinical symptoms. The main histologic abnormality observed in the swollen lymph node is hyperplasia of follicles. The network comprising of follicular dendritic cells (FDC) appears to be damaged. Immunohistochemistry shows signs of fragmentations and indentations. HIV-1 components have been suggested to play a role in this process, as proteins,[1,2] virions[3] and mRNA[2,4] of the virus are concentrated in the germinal centers. We[5] and others[6] have provided evidence that FDC can be infected by HIV-1. Some authors have proposed a role for the immune response towards the virus. This was supported by the demonstration of cytotoxic cells, documented by the presence of mRNA encoding serine esterase Granzyme B, which occurs in granules of cytotoxic cells.[7]

For *in vivo* studies on HIV-1 infection, a number of animal models are available. Out of these, infection of Cynomolgus monkeys (*Macaca fascicularis*) by Simian Immunodeficiency Virus (SIV) closely resembles HIV-1 infection in man.[8] This prompted us to evaluate lymphoid tissue of monkeys after SIV infection. We focussed on the following aspects:

- histology of lymphoid tissue follicles;
- immunohistochemistry on tissue sections and FDC purified from tissue for the presence of FDC-markers and SIV-protein;
- immunohistochemistry on tissue sections for the presence of lymphocyte subpopulations with the phenotype of cytotoxic cells;
- ultrastructure of FDC.

MATERIAL AND METHODS

This study included five monkeys. Some clinical and laboratory data is presented in Table 1. The animals were infected with 10-10^5 animal doses of SIV_{sm}, isolate SMM3, that was provided by Dr. P. Fultz, Yerks Primate Research Center, Atlanta, USA. Four animals received experimental chemotherapy in the first period after infection, that included azidothymidine, '3-fluorothymidine, deoxy-4-thymidine, or dideoxyinosine. Animal Nr.5 was was sacrificed because of severe clinical disease; the other four animals were sacrificed according to the experimental protocol.

Spleen and lymph nodes were processed in four different ways. One part was fixed in formalin and processed for conventional histopathology. A second small part was processed for transmission electron microscopy, that was performed as described elsewhere.[3] A third part was snap-frozen for immunohistochemistry. The fourth part was subjected to an isolation procedure to obtain FDC in suspension. This procedure included: mincing in small fragments; collagenase enzyme digestion; sedimentation on a discontinuous gradient of bovine serum albumin; and density gradient centrifugation on a percoll gradient.[9] In cytological preparations stained by May Grünwald-Giemsa, mainly cells with the morphology of FDC, either solitary or in clusters with lymphocytes, were observed. There were almost no macrophage-like cells in the purified cell population. The cells were analysed by immunocytochemistry for FDC markers and SIV protein, and subjected to electron microscopy.

In immunohistochemistry, the following mouse monoclonal antibodies were applied; anti-DRC-1 (Dakopatts, Glostrup, Denmark) or anti-Ki-M4 (Behringwerke, Marburg-Lahn, Germany) directed to FDC; anti-Leu-2a (CD8) (Becton Dickinson, Mountain View, CA, USA) directed to T cells of cytotoxic phenotype; SF4 towards SIV *gag* p28 protein (Biotech Research Laboratories, Rockville, MD, USA). A rabbit antiserum was used to detect serine esterase Granzyme-B present in granules of T-cytotoxic cells and natural killer cells.[10] To demonstrate mesenchymal markers, a rabbit anti-desmin and a mouse monoclonal anti-vimentin antibody were applied (Eurodiagnostics, Apeldoorn, The Netherlands). Immunohisto- and cytochemistry was performed by a three-step immunoperoxidase method. After the first antibody, a rabbit anti-mouse immunoglobulin (only in case of mouse monoclonal antibody) was applied, and then a swine anti-rabbit immunoglobulin (both secondary reagents were used conjugated to horseradish peroxidase, Dakopatts). Enzyme visualization was done using 3-amino-9-ethylcarbazole and H_2O_2 as substrate, followed by slight counterstaining with hematoxylin. Two-color immunofluorescence was done using rabbit anti-desmin antibody on one hand, and monoclonal reagents anti-DRC-1 or anti-vimentin on the other. After first incubation with monoclonal antibody, fluorescein isothiocyanate-conjugated goat anti-mouse antibody (Nordic Immunological Laboratories, Tilburg, The Netherlands) was applied. This was followed by incubations with anti-desmin and tetramethyl rhodamine isothiocyanate-conjugated goat anti-rabbit antibody (Nordic).

RESULTS AND DISCUSSION

Histopathology and Immunohistochemistry of FDC

In histologic sections of spleen and lymph nodes, changes observed ranged from follicular hyperplasia to follicular fragmentation and involution, and lymphoid atrophy (Table 1). These changes are similar to those in lymph nodes of patients after HIV-1 infection.[1,2,4,5] The labeling patterns observed in frozen section immunohistochemistry by anti-DRC-1 and Ki-M4 were in accord with the histologic observations. In hyperplastic

Table 1. Clinical and Laboratory Data of the Cynomolgus Monkeys Investigated.

Code	Sex	Infection period (days)	Chemotherapy	Clinical symptoms	Blood parameters	Histology	SIV Ag in follicles isolated FDC	SIV Ag in follicles tissue section
1	f	366	None	LAS, splenomegaly, weight loss	CD4 0.65, CD4/8 1.35	Hyperplasia	+	+
2	m	239	d-4T 0-10 days	none	CD4 0.3 CD4/8 0.2	Hyperplasia, fragmentation	+	+
3	m	239	DDI 0-10 days AZT 9 weeks	LAS, splenomegaly	CD4 1.0 CD4/8 0.3	Hyperplasia, fragmentation	-	±
4	m	239	DDI 0-10 days FLT 9 weeks	none	CD4 1.3 CD4/8 1.5	Hyperplasia, fragmentation	-	±
5	f	82	Linomide -4 to +4 days	Weight loss, wasting syndrome		Hyperplasia, atrophy	+	+

Abbreviations: f, female; m, male; DDI, dideoxyinosine; AZT, azidothymidine; FLT, '3-fluorothymidine; d-4T, deoxy-4-thymidine; LAS, lymphadenopathy syndrome. CD4 counts are presented in 10^9/L.

follicles a fine reticular pattern with a varying extent of indendations and fragmentations was observed. In follicle atrophy (animal Nr.5), only remnants of the original FDC network were seen. The preparations of cells purified from tissue showed immunoreactivity of FDC-like cells for DRC-1 and Ki-M4 antigen. In two-color tissue section immunofluorescence for mesenchymal markers and DRC-1/Ki-M4 antigen, anti-desmin manifested a reticular immunolabeling pattern of follicular and interfollicular areas. Follicles showed an identical immunolabeling by anti-DRC-1/anti-Ki-M4, and by anti-vimentin. This data indicates that FDC in germinal centers express vimentin and desmin, and thus are mesenchymal cells. This tissue origin of FDC confirms similar data for FDC in man.[11]

Immunohistochemistry for SIV p28 *Gag* Protein

On tissue sections, SIV p28 immunoreactivity co-localized with that by anti-DRC-1 or anti-Ki-M4. Outside follicles, only solitary cells at scattered location manifested immunoreactivity for p28. Also on cytocentrifuge preparations of purified FDC, p28 immunoreactivity was observed for these cells, either solitary or in clusters with lymphocytes. This data indicates that SIV p28 concentrates in follicles, presumably in the form of immune complex trapped in the labyrinth of cytoplasmic extensions of FDC.

This immunolabeling of FDC in tissue sections and in cytocentrifuge preparations was observed for animals 1, 2, and 5. Purified FDC from animals 3 and 4 did not show any labeling by anti-SIV p28, and on tissue sections of these animals there was only a faint immunolabeling product observed in follicles. FDC in these preparations, both isolated and in tissue section, showed immunoreactivity for anti-DRC-1 and anti-Ki-M4. Animals 3 and 4 differed from the others, as these were treated during the first 9 weeks after infection with azidothymidine and '3-fluorothymidine (Table 1). This data suggests that there is in these animals a reduced synthesis of viral protein and subsequent concentration in follicles. This suggestion fits with the hypothesis that SIV protein contributes to follicle damage after SIV infection. Following this hypothesis, the removal of SIV protein from the germinal center may have a beneficial effect on the host immune system.

Immunohistochemistry for Lymphocyte Subpopulations

Serial sections were immunolabeled for anti-DRC-1, CD8, and anti-Granzyme B antibody. The density of CD8 antibody was high in interfollicular areas. Also in follicles CD8-positive cells were observed, especially in indented and fragmented areas where there was no immunolabeling for DRC-1. Areas with DRC-1 immunoreactive stromal cells showed only scattered some CD8-positive cells. Granzyme-B-positive cells were observed solitary in interfollicular areas. Fragmentations and indentations of germinal centers showed scattered a few positive cells, and there were no Granzyme-B-immunoreactive cells in DRC-1-positive areas. This data indicates that there are high proportions of CD8-positive cells in spleen and lymph node. Part of these cells show serine esterase Granzyme-B, and are thus considered as cytotoxic cells. But, such cells with the phenotype of cytotoxic cells do not occur in the network of FDC in germinal centers identified by DRC-1 immunolabeling. This data does not support the hypothesis that cytotoxic cells are involved in the destruction of the germinal center stroma after SIV infection. This conclusion confirms data in the human situation after HIV-1 infection.[12]

Electron Microscopy

The subcellular analysis focussed on FDC in germinal centers of spleen and lymph node. In tissue sections, these cells showed well-differentiated characteristics, with villous extensions with electron-dense deposits, presumably immune complexes. Within these complexes, virions resembling SIV particles were observed, in particular in animals 1, 2 and 5 (Table 1). The presence of SIV virions was confirmed by immunogold electron microscopy, demonstrating *gag* p28 positivity of particles between extensions of isolated FDC. The animals 1, 2, and 5 also showed p28 immunoreactivity of FDC in suspension and in tissue sections (see above).

A number of cell types in the germinal center manifested paracrystalline arrays in the rough endoplasmic reticulum, that resembled so-called tubulo-reticular structures.[13] These were seen in some atypical cells with a lymphoblastoid morphology, some small lymphocytes, and in centrocytes. Tubulo-reticular structures were not observed in macrophages. They were found in FDC, either with the morphology of differentiated cells and with the morphology of undifferentiated cells present in the mantle zone of follicles. Tubulo-reticular structures have been associated with the synthesis of α-interferon, either by the cell itself or by cells in the local microenvironment, as part of immunologic reactions to virus infection.[14] The structures have been documented in lymph nodes of patients after HIV-1 infection, but thus far not in FDC. Their presence in undifferentiated FDC in follicles of monkeys after SIV infection suggests that FDC already early during their differentiation from mesenchymal precursor cells contribute to the reaction towards virus infection.

CONCLUSIONS

- The histopathology of lymphoid organs of cynomolgus monkeys after SIV_{sm} infection closely resembles that of patients after HIV-1 infection. Similarities are documented for histology of paraffin sections, SIV *gag* p28 expression on FDC, and the presence of SIV virions observed by electron microscopy.
- The presence of SIV components supports the hypothesis that these components have a role in the destruction of germinal centers, in particular of the FDC meshwork. Apparently, cells with a cytotoxic phenotype are not involved in this destruction. FDC may contribute to the anti-viral response in the local environment, that is associated with α-interferon synthesis.

- The reduced presence of SIV components in lymphoid follicles in case of some types of chemotherapy during the first phase after SIV infection indicates the potential value of the experimental animal model to evaluate new modalities of treatment and prevention of viral infection.

ACKNOWLEDGEMENT

This work was supported by the Dutch Ministry of Health, as part of the National Program on AIDS Research, Grant Nr. RGO/WVC 88-79/89005.

REFERENCES

1. P. Biberfeld, A. Porwit-Ksiazek, B. Böttiger, L. Morfeldt-Månson, and G. Biberfeld, Immunohistopathology of lymph nodes in HTLV-III infected homosexuals with persistent adenopathy or AIDS, *Cancer Res.* 45 Suppl:4665s (1985).
2. H.-J. Schuurman, W.J.A. Krone, R. Broekhuizen, and J. Goudsmit, Expression of RNA and antigens of human immunodeficiency virus type-1 (HIV-1) in lymph nodes from HIV-1 infected individuals, *Am. J. Pathol.* 133:516 (1988).
3. L.H.P.M. Rademakers, H.-J. Schuurman, J.F. de Frankrijker, and A. van Ooyen, Cellular composition of germinal centers in lymph nodes after HIV-1 infection: evidence for an inadequate support of germinal center B lymphocytes by follicular dendritic cells, *Clin. Immunol. Immunopathol.* 62:148 (1992).
4. P. Biberfeld, K.J. Chayt, L.M. Marselle, G. Biberfeld, R.C. Gallo, and M.E. Harper, HTLV-III expression in infected lymph nodes and relevance to pathogenesis of lymphadenopathy, *Am. J. Pathol.* 125:436 (1986).
5. H.K. Parmentier, D.F. van Wichen, D.M.D.S. Sie-Go, J. Goudsmit, J.J. Borleffs, and H.-J. Schuurman, HIV-1 infection and virus production in follicular dendritic cells in lymph nodes. A case report, with analysis of isolated follicular dendritic cells, *Am. J. Pathol.* 137:247 (1990).
6. I. Stahmer, J.P. Zimmer, M. Ernst, T. Fenner, R. Finnern, H. Schmitz, H.-D. Flad, and J. Gerdes, Isolation of normal human follicular dendritic cells and CD4-independent *in vitro* infection by human immunodeficiency virus (HIV-1), *Eur. J. Immunol.* 21:1873 (1991).
7. O. Devergne, M. Peuchmaur, M.-C. Crevon, J.A. Trapani, M.-C. Maillot, P. Galanaud, and D. Emilie, Activation of cytotoxic cells in hyperplastic lymph nodes from HIV-infected patients, *AIDS* 5:1071 (1991).
8. P. Putkonen, K. Warstedt, R. Thorstensson, R. Benthin, J. Albert, B. Lundgren, B. Öberg, E. Norrby, and G. Biberfeld, Experimental infection of cynomolgus monkeys (*Macaca fascicularis*) with simian immunodeficiency virus (SIV_{sm}), *AIDS* 2:359 (1989).
9. H.K. Parmentier, J.A. van der Linden, J. Krijnen, D.F. van Wichen, L.H.P.M. Rademakers, A.C. Bloem, and H.-J. Schuurman, Human follicular dendritic cells: isolation and characteristics in situ and in suspension, *Scand. J. Immunol.* 33:441 (1991).
10. O. Krähenbühl, C. Rey, D. Jenne, A. Lanzavecchia, P. Groscurth, S. Carrel, and J. Tschopp, Characterization of granzymes A and B isolated from granules of cloned human cytotoxic T lymphocytes, *J. Immunol.* 141:3471 (1988).
11. L.H.P.M. Rademakers, D.F. van Wichen, and R.A. de Weger, Immunohistochemical localization of intermediate filament and contractile proteins in the tonsil with reference to follicular dendritic cells, *in* "Lymphatic Tissues and In Vivo Immune Responses," S. Ezine, S. Berrih-Aknin, and B. Imhof, eds., Marcel Dekker, New York 721 (1991).
12. H.K. Parmentier, D.F. van Wichen, P.J. Peters, J. Tschopp, R.A. de Weger, and H.-J. Schuurman, No histological evidence for cytotoxic T cells in destruction of lymph-node follicle centers after HIV infection, *AIDS* 5:778 (1991).
13. J. Luu, D. Bockus, F. Remington, M.A. Bean, and S.P. Kammar, Tubuloreticular structures and cylindrical confronting cisternae: a review, *Hum. Pathol.* 20:617 (1989).
14. P.M. Grimley, G.L. Davis, Y.-H. Kang, J.S. Dooley, J. Strohmaier, and J.H. Hoofnagle, Tubuloreticular inclusions in peripheral blood mononuclear cells related to systemic therapy with α-interferon, *Lab. Invest.* 52:638 (1985).

DESTRUCTION OF FOLLICULAR DENDRITIC CELLS IN MURINE ACQUIRED IMMUNODEFICIENCY SYNDROME (MAIDS)

Akihiro Masuda[1], Gregory F. Burton[1], Bruse A. Fuchs[2], Andras K. Szakal[3], and John G. Tew[1]

[1]Department of Microbiology and Immunology
[2]Department of Pharmacology and Toxicology
[3]Department of Anatomy, Division of Immunobiology
Medical College of Virginia/Virginia Commonwealth University
P.O. Box 678 Richmond, Virginia 23298 U.S.A.

INTRODUCTION

Follicular dendritic cells (FDCs) degenerate in individuals manifesting persistent generalized lymphadenopathy after infection with HIV. At the light microscopic level, this degeneration of FDC is noted by the destruction of the FDC network in lymph node follicles[1-3], and it is called "follicle lysis"[1]. Ultimately, FDC are depleted as indicated by a lack of labeling with FDC specific monoclonal antibodies[3]. At the electronmicroscopic level, it has been reported that HIV, observed as viral particles, were attached to the processes of FDC[2,4]. These findings support the concept that HIV infection leads to destruction of FDC and loss of FDC functions.

The major functional role of FDC is in the stimulation of B cells[5-7]. FDC traps antigens in the form of immune complex and retains them on their surfaces for long periods as intact molecules[6]. The retained antigens on the FDC interact with and stimulate B cells and thereby function in the induction and maintenance of humoral immune responses—especially of secondary responses—[7]. Because FDC play an important role in the immune system, we hypothesized that part of the immunodeficiency associated with AIDS is attributable to the loss of FDC and FDC function.

C57BL/6 mice infected with LP-BM5 murine leukemia virus develop a disease that has many of the features of the human acquired immunodeficiency syndrome[8]. Hence the abbreviation MAIDS (murine acquired immunodeficiency syndrome) is generally accepted. In LP-BM5 infected mice, follicular hyperplasia and its destructive change are observed[9]. However, the relationship between MAIDS and changes in FDC structure and function has not been studied.

The objectives of this study were 1) to determine if LP-BM5 leads to destruction of FDC in mice as HIV does in man. To examine this, we examined what happened to FDC and follicular B cells immunohistochemically using monoclonal antibodies to FDC and lymphocytes. And 2) to determine if LP-BM5 infection affects the capacity of FDC to trap and retain antigen. To examine this, we quantified FDC function by examining their ability to trap and retain radioiodinated antigen.

MATERIALS AND METHODS

Eight to 12-wk-old C57BL/6J mice were injected i.p. with one infectious dose of LP-BM5 MuLV stock obtained from Dr. John Billelo, VA Med. Center, Baltimore, MD. To assess the effect of MAIDS infection on antigen trapped prior to infection, mice were immunized with human serum albumin (HSA). Actively immunized mice were inoculated with virus 1 week after the final booster injection of 2.5μg of ^{125}I-HSA in the left hind foot pad. To assess the loss of ability to trap new antigen after MAIDS infection, mice were passively immunized by an injection of 0.5 ml of rabbit anti-HSA serum i.p. Subsequently, the mice were challenged in the left hind foot pad with 2.5μg of ^{125}I-HSA 1 day after passive immunization. At specific times after viral inoculation, mice were killed and various lymph nodes and the spleen were carefully removed and radioactivity was counted in an LKB Gamma Counter. Then the tissues were fixed for 6 hours in periodate-lysin-paraformldehyde solution containing 2% paroformaldehyde. The tissues were frozen in Tissue-Tek OCT compound (Miles Inc., Elkhart, IN), and 6μm cryostat sections were prepared. Cryostat sections were used for immunohistochemistry as described previously[10]. The monoclonal antibodies used for immunohistochemistry were FDC-M1 (rat IgG anti-mouse FDC)[11], anti-B220, anti-Thy1.2, anti-immunoglobulins and anti-IgD. For the localization of germinal center cells, PNA was utilized in conjunction with peroxidase histochemistry. To detect anti-HSA on tissue sections, biotinylated HSA and avidin-peroxidase-complexes method were used. For localization of ^{125}I-HSA, autoradiography was done in cryostat sections as described previously[12]. Anti-HSA antibody was assayed with anti-mouse IgG using a solid phase RIA[13].

RESULTS

Splenomegaly and lymphadenopathy

To confirm that LP-BM5 causes splenomegaly and lymphadenopathy, spleen and lymph node weights were determined in actively immunized mice. On day 30 after infection (day 37 after final boost), spleen and left draining popliteal lymph node (PLN) weighed 259.6 ± 39.8 and 16.5 ± 1.17 respectively, while those of control mice weighed 89.6±3.73 and 5.38±0.62 respectively.

Quantitative study of FDC function

To determine if FDC can function after MAIDS infection we examined antigen trapping and retention in infected mice. Results are summarized in figure 1.

Loss of antigen trapped prior to MAIDS infection. A major function of FDC is to retain antigen for months after immunization. We reasoned that if FDC were being damaged by the MAIDS virus, antigen retention by FDC may fall below normal levels. To test this prediction, actively immunized mice were inoculated with the virus 1 week after the final booster injection of ^{125}I-HSA. The reason for the 1 week delay on inoculation was to allow clearance of immune complexes by macrophages and other phagocytic cells and to clear radiolabeled HSA from the circulatory system. Mice were sacrificed 10, 21, 30, 44,

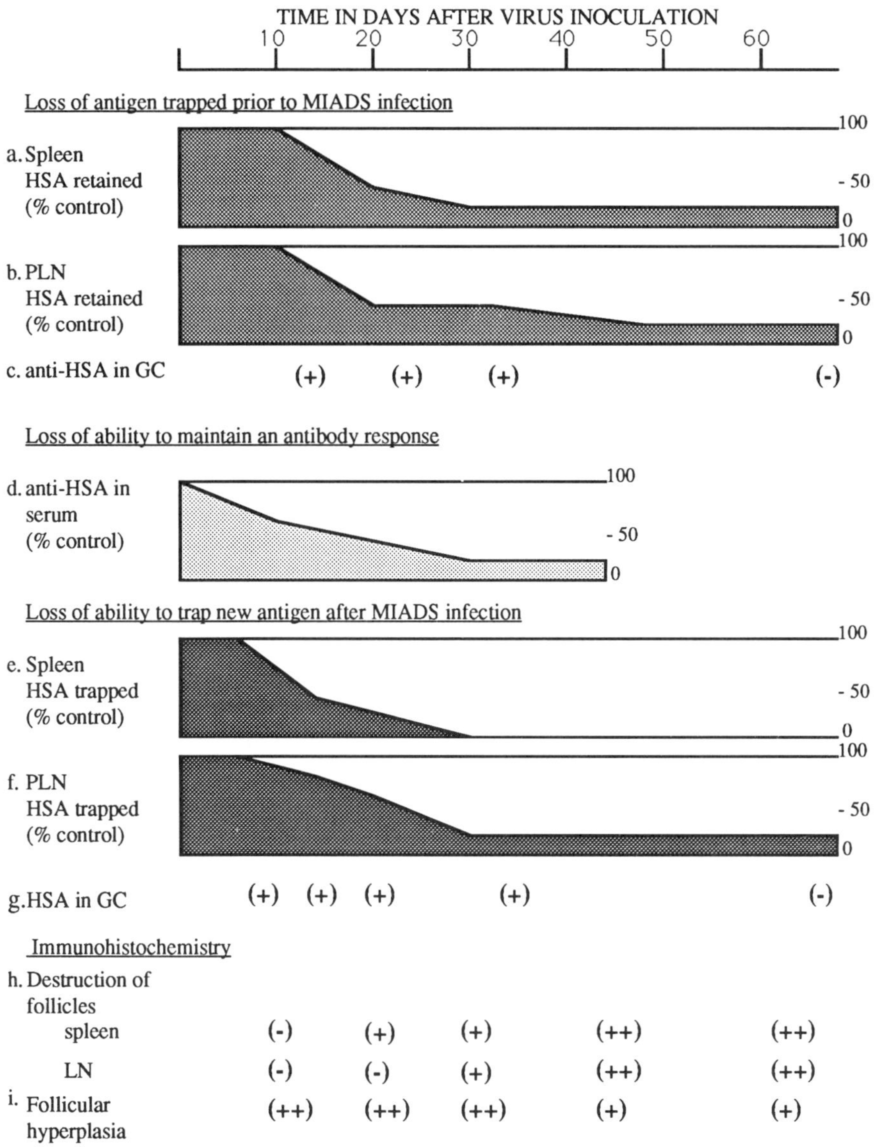

Figure 1. (a),(b),(c), & (d): Actively immunized mice were inoculated with virus 1 week after the final booster injection of ^{125}I-HSA in the left hind foot pad. Mice were sacrificed 10, 21, 30, 44, 64 days after inoculation and the amount of retained HSA was detemined in spleen(a) and left popliteal lymph node (PLN), (b) and amount of anti-HSA in serum was determined by solid phase RIA (d), and to determine if anti-HSA in immune complexes was trapped in germinal center (GC), immunohistochemistry with using biotinylated HSA was performed(c).

(e),(f), & (g): Infected group and non-infected group were given passive immunization. The day after passive immunization, ^{125}I-HSA was challenged. The mice were allowed to rest 7 days to allow the macrophage to degrade antigen and the amount of retained HSA in spleen(e) and PLN(f) was determined. And in order to determine if the HSA was trapped in GC, autoradiograohy was done(g). Each data point represents data from 4 to 10 mice.

(h)&(i): Degree of destruction and hyperplasia of follicle assesed by tissue sections stained with immunohistochemistry are shown by (-) to (++).

64 days after virus inoculation. A deterioration in ability to retain antigen trapped prior to MAIDS infection was observed in PLN at 3 weeks after virus inoculation. By day 44, the ability of FDC to retain immune complexes decreased to less than 20% of the controls. By 3 weeks in the spleen, retained antigen in LP-BM5 infected mice was less than 50% that of the controls. By day 60 after inoculation, autoradiography failed to detect FDC-trapped antigen. This finding was comfirmed by the use of biotinylated HSA.

Loss of specific serum antibody. FDCs are thought to be involved in maintenace and regulation of specific antibody responses[5]. To determine how LP-BM5 affects the specific antibody response, the amount of anti-HSA of IgG isotype was assessed by RIA. By day 21 after infection, decrease in anti-HSA was apparent.

Loss of ability to trap new antigen after MAIDS infection. To determine how well LP-BM5 infected mice can trap and retain new antigen, mice were infected and then immunized. We used passive immunization to assure that level of antibody in the LP-BM5 infected and normal control mice were identical. In this study we gave infected mice antigen to determine if they were capable of trapping newly injected antigen which contrasts with the previous experiment where mice were given antigen prior to inoculation. One day after passive immunization, both group of mice were challenged with ^{125}I-HSA in the left hind foot pads. The mice were allowed to rest for 7 days to allow the macrophages to degrade phagocytized immune complexes and then the level of retained antigen was quantified from the level of radioactivity persisting in the draining popliteal lymph node and spleen. ^{125}I-HSA was injected at day 7, 14, 22, 29, 40, 80 to the passively immunized mice after inoculation of LPBM-5. Three weeks after infection, PLN of LP-BM5 infected mice were able to trap and retain only about 60 % as much antigen as normal controls. This antigen trapping decreased to 20% by day 29. In spleen, almost no antigen was trapped at 4 weeks. At day 80, PLN still could trap about 25% of CPM of control (MAIDS: 2500 ± 460cpm, control: 8800 ± 1400cpm). However, no focal distribution of grains suggestive of antigen localization was observed on autoradiography sections of the infected mice.

Immunohistochemistry of FDC and lymph follicles in MAIDS infected mice

To determine whether MAIDS infection results in follicular destruction, we examined lymph nodes and spleens using immunohistochemistry. Follicular labeling of FDC in infected lymph nodes persisted for weeks and by 1 month the striking features was a follicular hyperplasia. Three weeks after inoculation at the time when the deterioration in FDC ability to retain antigen trapped prior to MAIDS infection was observed in both PLN and spleen, large extensive FDC networks were observed. At 4 weeks after inoculation, hyperplasia of follicles still observable. However, staining intensity of FDC-M1 was reduced and distortion of lymph follicle, including disruption and fragmentation was apparent and mantle zone of follicles showed reduction and destruction. In addition, decrease of T cell zone and immunoblastosis was observed. After that, the destruction of follicular organization of lymph nodes continued. By 60 days, follicles were appeared to be compressed by immunoblastosis occurinig extrafollicular area and the follicles showed severe damage. By 4 months the loss of FDC was apparent. However,in some smaller generally less affected nodes, FDC were still present.

DISCUSSION

The present study showed that FDC function was destroyed before the stage of morphological destruction of FDCs. Both antigen trapping and long term retention by FDC was affected by 3 weeks, indicating a loss of FDC function. In contrast a morphologically apparent destruction of FDC occurred only 10 days later. Because FDCs play important roles as accessory cells and have the ability to present antigen to B cells and provide an environment that promote B cell proliferation and maturation[5,7], part of the immunodeficiency associated with MAIDS may be attributable to the loss of FDC and FDC functions. In this model, as well as HIV, there is a possibility that loss of FDC function may be important in progression of the deseases. Similarly to MAIDS, the loss of FDC function in AIDS patient is thought to emerge at an early stage in which the morphological destruction of lymphoid follicles are not yet conspicuous. Ultrastructurally, however, in MAIDS infected mice FDC degenerates and dendritic fragmentation is in progress within 24 hour after inoculation and the trapping of injected horse radish peroxidase appears to be reduced[14]. However, there was no decrease of ability for trapping and retaining of antigen until about 10 days after inoculation in the experimental system we adopted in this report, nevertheless possibility exists that FDC function is already deteriorated within a week after inoculation.

We also observed a decrease of anti-HSA in MIADS mice challenged with antigen after infection. There is correlation between the proximity of plasma cells responsible for restoring serum antibody levels and antigen persisting on FDC after a year's time lag[15]. This supports a role for FDC-retaining antigen in the maintenance phase of the immune response. Thus, the loss of FDC-retained antigen in MAIDS may also contribute to the loss of the maintenance phase of the immune response.

The present study showed that the destruction of lymphoid follicles and FDC, which had also been reported to occur during lymphadenopathy in HIV infection, is also observable in MAIDS infection. Immunohistochemical studies show that infection by MAIDS virus does not initially eliminate FDC. By one month after infection, hyperplasia of FDC was apparent. However after 4 months of infection, the FDCs disappeared in most lymph nodes and the remaining FDCs were fragmented and faintly stained. Follicular B cells also disappeared and B cells were no longer localized in the follicles. Szakal et al[14] observed by electronmicroscopy that virus particles were trapped by FDCs, which became subsequently atrophic. In MAIDS, changes similar to that of HIV occurred at ultramicroscopic level.

Like HIV in man, LP-BM5 infection in mice leads to the destruction of FDC. In mice this process only requires a few months and represents an attractive model for the study of the immunodeficiency resulting from the functional disturbance and the destruction of FDC.

REFERENCES

1. G.S. Wood, C.F. Garcia, R.F. Dorfman and R.A. Warnke, The immunohistology of follicle lysis in lymph node biopsies from homosexual man, *Blood.* 66:1092 (1985).
2. M.A. Piris, C. Rivas, M. Morente, C. Rubio, C. Martin and H. Olivia, Persistent and generalized lymphadenopathy: a lesion of follicular dendritic cells? An immunohistologic and ultrastructural study, *Am J Clin Pathol.* 87:716 (1987).
3. P. Biberfeld, A. Porwit, G. Biberfeld, M. Harper, A. Bodner and R. Gallo, Lymphadenopathy in HIV (HTLV-III/LAV) infected subjects: the role of virus and follicular dendritic cells, *Cancer Detect Prev.* 12:217 (1988).
4. J.A. Armstrong and R. Horne, Follicular dendritic cells and viruslike particles in AIDS-related lymphadenopathy, *Lanset.* 2:370 (1984).
5. J.G. Tew, M.H. Kosco, G.F. Burton and A.K. Szakal, Follicular dendritic cells as accessory cells, *Immunol Rev.* 117:185 (1990).
6. T.E. Mandel, R.P. Phipps, A. Abbot and J.G. Tew, The follicular dendritic cell: long term antigen retention during immunity, *Immunol Rev.* 53:3 (1980).

7. J.G. Tew, M.H. Kosco and A.K. Szakal, The alternative antigen pathway, *Immunol Today*. 10:229 (1989).
8. D.E. Mosier, Animal model for retrovirus-induced immunodeficiency disease, *Immunol Invest*. 15:233 (1986).
9. P.K. Pattengale, C.R. Taylor, P. Twomey, S. Hill, J. Jonasson, T. Beardsley and M. Haas, Immunopathology of B-cell lymphomas induced in C57BL/6 mice by dualtropic murine leukemia virus (MuLV), *Am J Pathol*. 107:362 (1982).
10. K. Maeda, G.F. Burton, D.A. Padgett, D.H. Conrad, T.F. Huff, A. Masuda, A.K. Szakal and J.G. Tew, Murine follicular dendritic cells and low affinity Fc receptors for IgE (FcεRII), *J Immunol*. 148:2340 (1992).
11. M.H.Kosco, E.Pflugfelder, and D.Grey, Follicular dendritic cell-dependent adhesion and proliferation of B cells in vitro, *J Immunol*. 148:2331 (1992).
12. R.P. Phipps, T.E. Mandel, C.T. Schnizlein and J.G. Tew, Anamnestic responses induced by antigen persisting on follicular dendritic cells from cyclophosphamide-treated mice, *Immunology*. 51:387 (1984).
13. G.F.Burton, M.H.Kosco, A.K.Szakal, and J.G.Tew, Iccosome and the secondary antibody response, *Immunology* 73:271-276 (1991).
14. A.K. Szakal, B.S. Bhogal, B.S. Fuchs, A. Masuda, G.F. Burton and J.G. Tew, Antigen transport and follicular dendritic cells (FDC) in mouse AIDS. An ultrastructural study. *FASEB J* 5:A1369 (1990).
15. S.L. Donaldson, M.H. Kosco, A.K. Szakal and J.G. Tew, Localization of antibody-forming cells in draining lymphoid organs during long-term maintenance of the antibody response, *J Leukoc Biol*. 40:147 (1986).

CHANGES IN FOLLICULAR DENDRITIC CELL AND CD8+ CELL FUNCTION IN MACAQUE LYMPH NODES FOLLOWING INFECTION WITH SIV_{251}

Yvonne J. Rosenberg[1], Marie H. Kosco[2], Mark G. Lewis[1], Enrique C. Leon[1], Jack J. Greenhouse[1], Kyle E. Bieg[1], Gerald A. Eddy[1], and Philip M. Zack[3]

[1] Henry M. Jackson Foundation, Rockville, MD 20850, USA, [2] Basel Institute for Immunology, CH-4058 Basel, Switzerland, [3]Walter Reed Army Institute of Research, Washington, DC 20307, USA

INTRODUCTION

By virtue of their extensive meshlike network of dendritic processes expressing C3b, C3d and Fc receptors, follicular dendritic cells (FDC) found in lymphoid follicles are able to trap and retain unprocessed antigen in the form of Ag/Ab complexes (1-4). As such these cells provide the microenvironment for germinal centre reactions in which antigen specific B cells proliferate, differentiate into memory cells and antibody forming cell precursors and undergo class switching and affinity maturation. To date FDCs appear to serve two major functions. Firstly to present immunogen in the form of iccosomes to B cells and secondly as a source of of a costimulatory signal(s) required for B cell proliferation (Tew et. al. and Burton et.al., this volume). Although T cells are also required for germinal centre formation in vivo and for cluster formation and proliferation in vitro, little is known about FDC/T cells interactions. An understanding of these roles is achieved by dissecting the cellular and molecular interactions leading to the spontaneous cluster formation and proliferation observed in in vitro cultures of FDC-enriched lymphoid cells 5,6).

Many reports have indicated that during HIV and SIV infection, degeneration of the FDC occurs, ultimately resulting in follicular involution and lymphocyte depletion (7-9). The mechanisms underlying such destruction are not known but it is clear that subtle deleterious changes in lymph nodes (LN), and specifically within germinal centres (GC), occur prior to the dramatic loss of LN architecture. Studies with SIV-infected macaques have indicated that despite the well characterized decline in the CD4% and CD4/CD8 ratio in the blood during the chronic phase of infection, the cellular composition of the LN remains relatively constant while the blood CD4/CD8 ratio is >.5. Below this level, CD4% and CD4/CD8 ratios decreases, immunohistochemistry reveals SIV antigen expression by FDC in GC and CD8+ cells undergo a CD45RAhi->activated CD45RAlo transition and infiltrate the GC (10,11). Since early double labelling studies in HIV+ individuals have demonstrated groups of CD8+ (and CD4+) cells regularly present in close proximity to viral p24 antigen located on the surface of FDC (12,13), it is possible that either accumulation of viral antigens and virions and/or cytokines released from CD8+ cells may be partly responsible for loss of functional FDCs well as CD4+ cells in these nodes. For this reason, SIV-infected LN known to have high or low CD4% without or with activated $CD45RA^{lo}$ CD8+ cells , were used to prepare FDC-enriched populations in order to assess (i) the FDC/CD8+ cell association, (ii) the functional properties of these infiltrating CD8+ cells and (iii) the ability of such FDC to promote cluster formation and B cell proliferation.

MATERIALS AND METHODS

Rhesus macaques were infected with 1-10 ID_{50} SIV-251 and sacrificed at D112 (5568) or D205 (81C, 5606) at which time LN were removed for analysis. Control LN were obtained from pig-tailed macaques from the Regional Primate Centre, Seattle , WA.

Phenotypic Analysis

Cultured cells were double or triple stained with fluorescein isothiocyanate (FITC) coupled CD45RA (Gentrak, Wayne, PA), CD4 coupled to phycoerythrin and CD20 and CD8 labelled with either FITC or PerCP (Becton Dickenson, Mountain View, CA). Analysis was done on a FACSCAN (Becton Dickenson). Individual wells were assayed separately.

Isolatation of Cells and Culture Conditions

FDC-enriched cultures were prepared using mesenteric LN from uninfected or SIV-infected macaques according to the method of Kosco et. al. (5). Macaque LN were first chopped into small pieces before being subjected to three rounds of enzyme digestion. The low density , nonadherent fraction was cultured in 24 well plates at 1-4 x 10^6/ well/ml. At several time points cells were analysed by flow cytometry or incorporation of ^{3}(H)-TdR. In the latter case, 100ul aliquots from three wells were pulsed with ^{3}(H)-TdR for 6-8 hrs before harvesting.

Cytotoxic assay

K562, HUT 78, H9, AA2 and SIV_{E11S}-infected AA2 were labelled with ^{3}(H)-TdR for 6 hr and plated at 5 x 10^3 cells per 200ul well using the JAM assay of Matzinger (14). LN cells from FDC-enriched cultures or from normal PHA plus IL 2 stimulated cultures were added at E:T ratios of 1:1 to 50:1 for 6-14 hr. Contol labelled targets cultured in the absence of killers were harvested immediately and similarly at 6-14 hr using a Betaplate (Pharmacia LKB, Gaithersburg, MD).

Immunohistichemistry

LN cryosections were stained with anti-CD8 and DRC-1 anti-FDC MAb (Dako, Carpenteria, CA). The binding was visualized using avidin-biotin peroxidase according to the method described (15).

RESULTS AND DISCUSSION

As mentioned, during HIV and SIV infections major changes accompany the decline in CD4% and CD4/CD8 ratios in LN. Figure 1 compares the extent of migration of CD8+ cells into the germinal centres of SIV-251 infected LN that still contained normal levels of CD4% (30-55%) with a LN where CD4% had begun to decline. Very few CD8+ cells could be seen associated with the GC from nodes of monkeys where CD4% was unchanged (33-54%)(Fig. 1A). In contrast, in LN where CD4% had begun to decline (24-30%), marginal zones and GC were heavily infiltrated by CD8+ cells which also appeared to undergo division as evidenced by the presence of small foci (Fig.1B). In addition to a cellular infiltration, the reticular FDC staining pattern observed with anti-CD8 MAb also infers the FDC binding of free CD8 molecules, possibly released from the atypical activated CD45RAlo, CD8+ population which now predominates in these nodes. In order to ascertain whether these immigrant CD8+ cells were associated with damage to the FDC, we performed FDC staining on the GC sections adjacent to those stained with anti-CD8 MAb above. Figs. 1C and D demonstrate that advanced degeneration of the FDC as measured by their motheaten and fragmented appearance, occurs only in the GCs with marked CD8+ cell migration and suggest cause and effect relationships between decreased CD4%, CD8+ cell

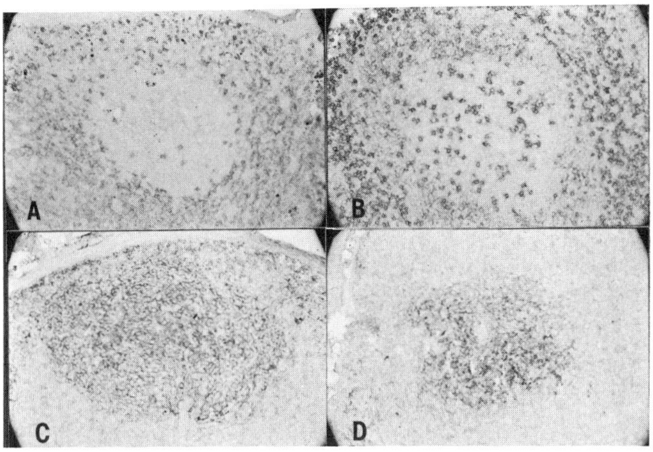

Fig. 1. Axillary LN cryosections showing GC from SIV_{251}-infected macaques immunostained with anti-CD8 MAb (Figs.1A and 1B) and the DRC-1 anti-FDC MAb (Figs. 1C and 1D). The LN in Figs. 1A and 1C contained normal levels of CD4%, showed no signs of infiltrating CD8+ cells and had a healthy reticular FDC network which occupied the entire GC. In comparison, the LN shown in Figs. 1B and 1D had reduced CD4% , had undergone a marked influx in CD8+ cells into GC and exhibited extensive disruption and fragmentation of the FDC network. (ABC, x 70)

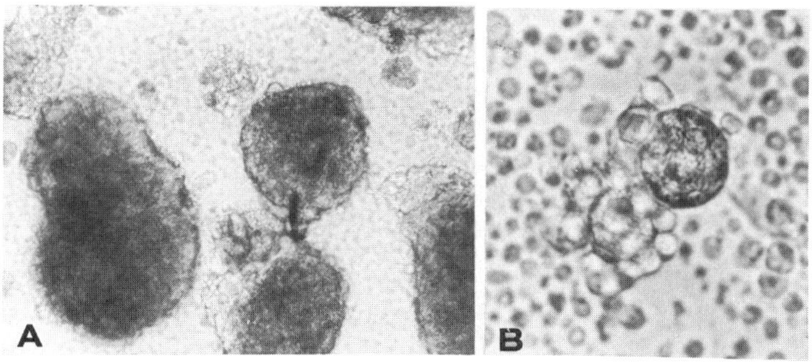

Fig. 2. Clusters of cells in D7 cultures of FDC-enriched cells from mesenteric LN of a rhesus macaque 112 days following infection with SIV_{251}. A. Huge clusters which are usually loosely bound to adherent cells on the plate and reform rapidly following dispersion. B Small aggregates (~5-10 cells) surrounding a larger cell also persist in these cultures. (ABC, x 150)

migration and destruction of FDC within lymphoid tissue.

To further analyse the possible association between FDC and CD8+ cells in late stage SIV infection, FDC-enriched populations prepared from LN of SIV- and SIV+ macaques were cultured and analysed both for their ability to form clusters and for the phenotype of the proliferating cells. As a reflection of ongoing enviromental stimulation, cultures containing isolated FDC from normal pig-tailed monkey donors contained both small aggregates of <10 cells as well large cellular clusters containing 100's of cells. The larger clusters usually appeared to be loosely bound to adherent cells on the bottom of the culture plate.The magnitude of the reactions varied with individual macaques. The FDC-enriched cultures derived from SIV-infected macaques LN clearly differed in several respects from those in which primed mice have served as the source of lymphoid cells and were much more dramatic than those derived from uninfected monkeys. For example, (i) while aggregates in cultures from primed mice contained 30-100 cells and some clusters derived from normal monkey LN are larger, those from SIV-infected nodes are numerous and consist of 1,000's of cells. (ii) Although CD8+ cells have rarely been found within clusters from immunized mice and were variable but low in LN culture from control monkeys, Fig.3 indicates that

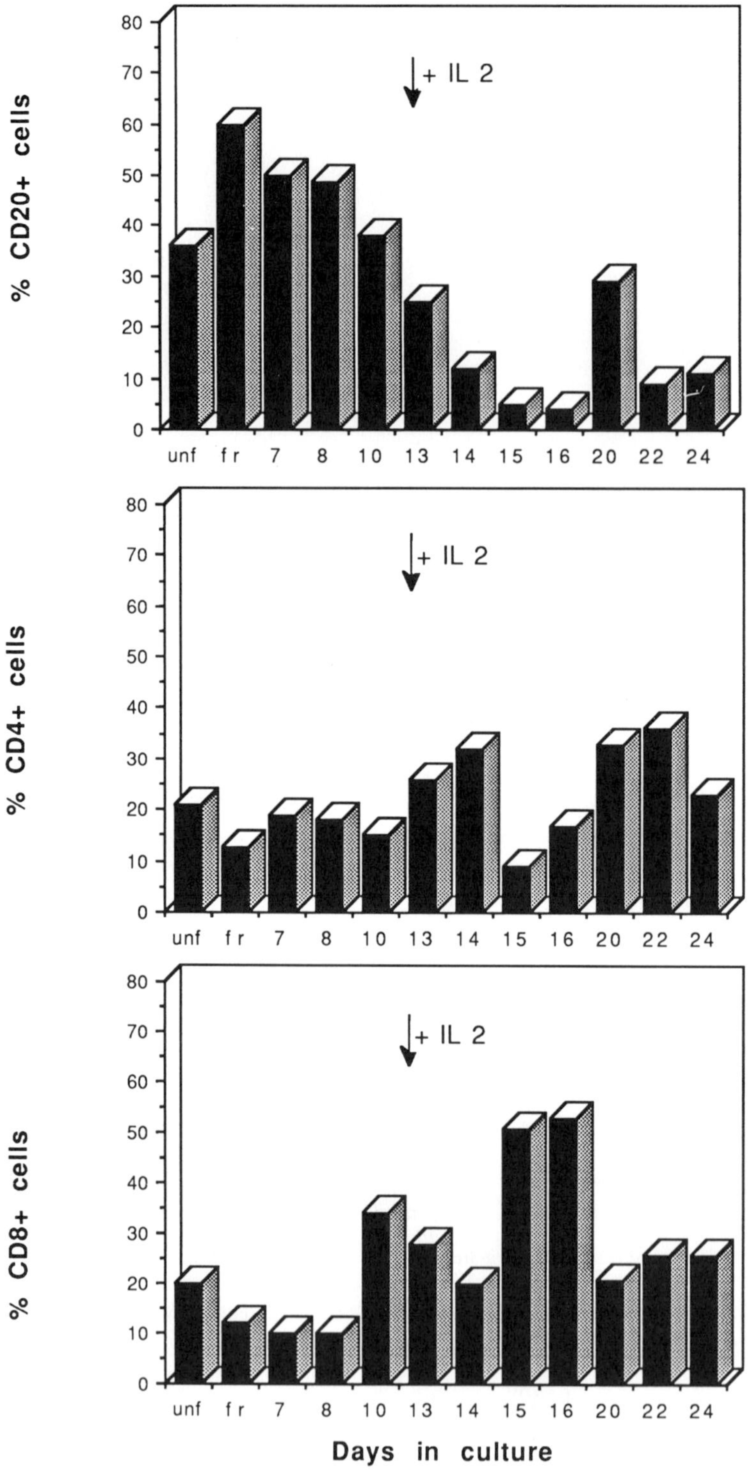

Fig. 3. Growth of CD20+, CD4+ and CD8+ lymphocyte populations during in vitro cultures of FDC-enriched cells derived from LN of SIV-251-infected macaques. IL 2 was added at D12 (arrow).unf = unfractionated, fr = fractionated.

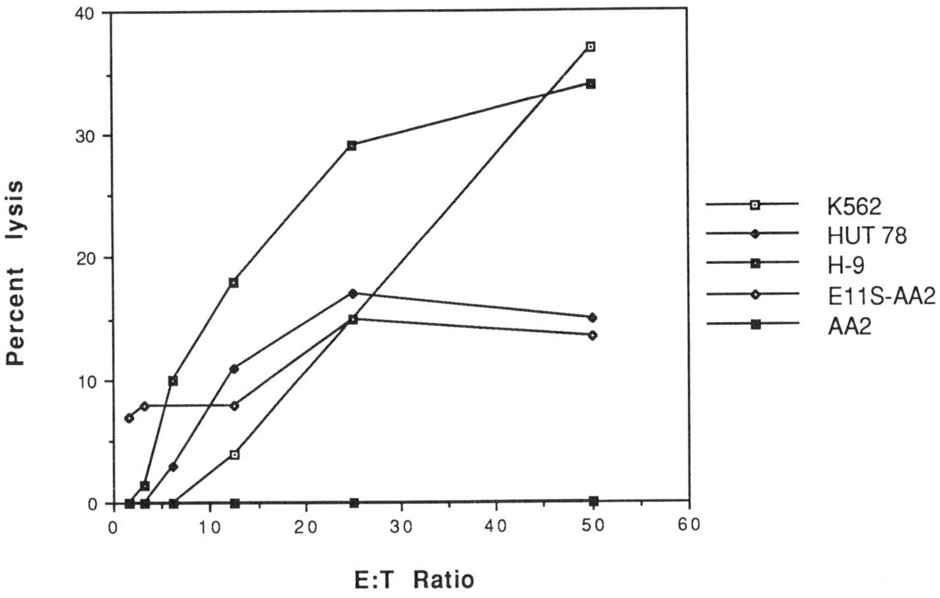

Fig. 4. Cytolytic activity of FDC-enriched LN cells from SIV_{251}-infected macaques following 14 days in culture. Cultured cells were tested against ^{3}(H)-TdR labelled K562, HUT 78, normal AA2 and SIV_{E11S}-infected AA2 at various E:T ratios. Percent lysis was calculated from CPM in parallel wells containing the same targets in the absence of killers.

CD8+ lymphocytes were maintained in cultures from infected donors in significant numbers (~10-20%). In addition, such cells were also activated measured by as CD45RAlo expression. Following addition of IL 2 at day 12, the presence of these CD8+ cells could be even further expanded in many wells with concomitant decrease in B and CD4+ cells. B cells were maintained at a level of >40% for at least 10 days in cultures from SIV+ monkeys. It should also be noted that low levels of CD56+ cells also persisted and could represent 20% of cells in wells receiving IL 2 (iii) Perhaps as a result of viral replication, the survival of clusters in cultures derived from infected macaques was much greater that controls. For example, the FDC-enriched cultures shown in Fig.2 incorporated >20,000 CPM ^{3}(H)-TdR daily for 9 days and although the numbers of clusters decreased in number with time, selected wells continued to proliferate (20,000-40,000CPM) until D28. Indeed growth was so vigorous that the original wells were split twice during the four week period. To date, the mean ^{3}(H)-TdR incorporation at D1 in three FDC cultures from SIV- versus SIV+ monkeys is 5,136 +/-1438 and 24,017 +/- 4109 respectively.

For the presence of activated CD8+ cells to be deleterious to the well being of FDC, such cells must either produce certain cytokines and/or possess cytotoxic potential.To date the presence of CD8+ cells containing cytotoxic granules in infected nodes is controversial (16,17). In preliminary experiments, cells from the D14 cultures described in Fig.3 were assayed for their ability to kill a variety of 3(H)-TdR labelled targets. Fig. 4 demonstrates a broad sprectrum of killing using such cells; being highest against K562 and H-9, moderate against HUT 78 and E11S-infected AA2 line and absent using uninfected AA2 targets. Such cytolytic activity is clearly not antigen-specific or MHC-restricted as reported for gag-specific CTL in SIV-infected macaques (18).Our results are similar however to the env-specific CD16+ cell mediated killing observed in blood in these monkeys (19) and indicate the potential of these FDC-associated cells to do harm. In control experiments, IL 2 expanded LN suspension cultures from a late stage SIV+ but not SIV- monkeys also exhibited killing of K562, HUT 78 and H-9 cells. Whether such killing was mediated by CD8+, CD8+CD56+ or CD8-CD56+ cells is currenly being investigated on different targets including autochthonous lymphocytes.

These results indicate that FDC-enriched cultures from monkey LN spontaneously undergo germinal centre reactions with the appearance of large in vitro clusters which probably reproduce the GCs originally present in situ and reflect the level of ongoing antigenic stimulation. In the present experiments, LN from monkeys in late stage SIV disease with decreased CD4%, reduced CD4/CD8 ratios, marked SIV antigen expression on FDC and increased levels of activated CD45RAlo CD8+ cells in their LN, were chosen for study in order to assess how the process of FDC degeneration, seen in Fig.1, affects their ability to support GC reactions in vitro. Surprisingly, LN FDC from late stage infected macaques appear to have an enhanced ability to support proliferation which suggests that FDC damage does not reduce its capacity to provide a costimulatory signal(s). It is not clear however what role precoating of the FDC with SIV viral proteins or virions or binding of new replicated virus during culture has in driving the system. In this context, Masuda et. al. (this volume) have demonstrated a markedly depressed ability of FDC from MAIDS infected mice to trap new antigen, which is central to the continued selection of high affinity B cells and consequently high affinity antibody profiles. This is particularly relevant to late stage HIV and SIV disease in which CD4+ cell decline, FDC degeneration and demonstrable budding of virions has been shown to correlate with low titres of anti-viral antibody in association with hypergammaglobulaemia (20,21). In situ, low numbers of specific anti-HIV antibody producing cells have been visualized in the interfollicular areas and medulla of LN from individuals with advanced disease but strong associations with the state of FDCs or levels of p24 or CD4+ cells were not apparent. This apparent paradox, which is known to occur in several infections and diseases might be predicted if B cells within the GC become activated as a result of HIV or SIV infection and receive a costimulatory signal delivered by the FDC in the absence of any selection mechanism required for maintaining the levels of protective high affinity. In this context, it is of interest that in late stage SIV-PBj infected macaques, a decline LN CD4% appears to correlate with an increase in the number of B cells in cell cycle as measured by 7AAD staining (not shown).

Finally, the present results raise another question regarding the association of FDCs and T cells. For example, the findings that CD8% (and CD4%) in these FDC-enriched cultures are maintained at significant levels indicate that, rather than simply being passenger lymphocytes, the activated CD8+ cells shown to infiltrate the GC in Fig. 1 may actually be bound to the FDC either via specific TcR recognition of antibody-complexed antigen or as a result of increased expression of adhesion molecules eg. ICAM-1/LFA-1 interactions. Thus by providing an environment within GC for CD8+ cells to differentiate into effectors, FDC may pay the price of their own destruction while the humoral immune response suffers due to an overall loss in the affinity of anti-HIV and -SIV antibody.

ACKNOWLEDGEMENTS

The authors wish to thank Dr. Suzanne Gartner for reading the paper and Jerry Coates for help with the histology.

REFERENCES

1. G.G.B. Klaus, J.H. Humphrey, A. Kunkl, and D.W. Dongworth, The follicular dendriric cell: its role in antigen presentation in the generation of immunological memory, *Immunol. Rev.* 53:3-28 (1980).
2. J.G. Tew, M.H. Kosca, G.F. Burton, and A.K. Szakal, Follicular dendritic cells as accessory cells. *Immunol. Rev.* 117:185-211.20 (1992).
3. J. Jacob, and G. Kelsoe, In situ studies of the primary immune response to (4hydroxy-3-nitrophenyl) actyl. II. A common clonal origin for periarteriolar lymphoid sheath-associated foci and germinal centres, *J. Exp. Med.* (in press).
4. Y.J. Liu, G.D. Johnson, J. Gordon, and I.C.M. MacLennan, Germinal centers in T-cell-dependent antibody responses. Immunol. Today 13: 17-21 (1992).
5. M.H. Kosco, E. Pflugfelder, and D. Grey, Follicular dendritic cell-dependent adhesion and proliferation of B cells in vitro, *J. Imm.* 148:2331-2339 (1992).
6. G. Koopman, H.K. Parmentor, H.J. Schuurman, W. Newman, C.J.L.M. Meijer, and S.T.Pals, Adhesion of human B cells to follicular dendritic cells involves both the lymphocyte function-associated antigen1/intercellular adhesion molecule1 and very late antigen 4/ vascular cell adhesion molecule 1 pathways, *J. Exp. Med.* 173:1297-1304 (1991).

7. P. Racz, K. Tenner-Racz, C. Kahl, A.C. Feller, P. Kern, and M. Dietrich, Spectrum of morphological changes of lymph nodes from patients with AIDS or AIDS-related complex. *Prog. Allergy* 37:81-181 (1986).
8. P. Biberfeld, K.J. Chayt, L.M. Marselle, G. Biberfeld, R.C. Gallo, and M.E. Harper, HTLV-III expression in infected lymph nodes and relevance to pathogenesis of lymphadenopathy. *Am. J. Pathol.* 125:436-442 (1986).
9. L.V. Chalifoux, N.W. King, and N.L. Letvin, Morphologic changes in lymph nodes of macaques with an immunodeficiency syndrome. *Lab. Invest.* 51:22-26 (1984).
10. Y.J. Rosenberg, A. Shafferman, B.D. White, S.F. Papermaster, E. Leon, G.A. Eddy, R. Benveniste, D.S. Burke, M.G. Lewis, Variation in the CD4+and CD8+populations in lymph nodes does not reflect that in the blood during SIVmne/e11s infection of macaques, Munksgaard (1992).
11. G. Janossy, A.J. Pinching, M. Bofill, J. Weber, J.E. McLaughlin, M. Ornstein, K. Ivory, J.R. Harris, M. Favrot, and D.C. Macdonald-Burns, An immunohistological approach to persistent lymphadenopathy and its relevance to AIDS. *Clin. Exp. Immunol.* 59:257-266 (1985).
12. K. Tenner-Racz, P. Racz, S. Gartner, J. Ramsauer, M. Dietrich, J-C. Gluckman, and M. Popovic, Ultrastructural analysis of germinal centers in lymph nodes of patients with HIV-1-induced persistent generalized lymphadenopathy: Evidence for persistence of infection. *Prog. AIDS Pathol.* 1:29-40 (1989).
13. C.D. Baroni,and U. Stefania, HIV and EBV expression in lymph node immunohistology and in situ hybridization, *in:* "Modern Pathology of AIDS and Other Retroviral Infections", S.A. Karger, Basel (1990).
14. P. Matzinger, The JAM Test : A simple assay for DNA fragmentation and cell death, *J. Immuno. Methods* 145:185-192 (1991).
15. V.M. Hirsch, P.M. Zack, A.P. Vogel, and P.R. Johnson, Simian immunodeficiency virus infection of macaques: End-stage disease is characterized by widespread distribution of proviral DNA in tissues, *J. Infect. Dis.* 163:976-988 (1991).
16. J.D. Laman, E. Claassen, N.V. Rooijen, and J.A. Wim, Immune complexes on follicular dendritic cells as a target for cytolytic cells in AIDS, *AIDS.* 3:543-548 (1989).
17. L. Vago, M.C. Antonacci, S. Cristina, C. Parravicini, A. Lazzarin, M. Moroni, C. Hegri, C. Uberti-Foppa, M. Musicco, G. Costanzi, Morphogenesis, evolution and prognostic significance of lymphatic tissue lesions in HIV infection. *Appl Pathol.* 7:298-309 (1989).
18. M.D. Miller, C.I. Lord, V.Stallard, G.P. Mazzara, N.L. Letvin, Gag-specific cytotoxic T lymphocytes in rhesus monkeys infected with the simian immunodeficiency virus ofmacaques, *J. Immunol.* 144:122-128 (1990).
19. H. Yanamoto, M.D. Miller, D.I. Watkins, G.B. Snyder, N.E. Chase, G.P. Mazzara, L. Gritz, D.L. Panicali, and N.L. Letvin, Two distinct lymphocyte populations mediate simian immunodeficiency virus envelope-specific target cell lysis, *J. of Immunol.* 145: 3740-3746 (1990).
20. P.U. Cameron, R.L. Dawkins, J.A. Armstrong, and E. Bonifacio, Western blot profiles, lymph node ultrastructure and viral expression in HIV-infectedpatients: a correlative study, *Clin. exp. Immunol.* 68:465-478 (1987).
21. A. Shirai, M. Cosentino, S.F. Leitman-Klinman, and D.M. Klinman, Human immunodeficiency virus infection induces both polyclonal and virus-specific B-cell activation, *J. of Clin. Invest.* 89:561-566 (1992).

RAPID AND SELECTIVE ISOLATION OF FOLLICULAR DENDRITIC CELLS BY LOW SPEED CENTRIFUGATIONS ON DISCONTINUOUS BSA GRADIENTS

Christian Marcoty, Ernst Heinen, Nadine Antoine, Rikiya Tsunoda* and L.J. Simar

Institute of Human Histology, University of Liège, Belgium,
Department of Anatomy, Fukushima Medical College, Japan*

INTRODUCTION

Germinal centers are the main sites for T-dependent humoral immune responses; these peculiar microenvironments are built up by the follicular dendritic cells (FDC; Papamichail et al., 1975; Klaus et al., 1980; Heinen et al., 1991). Some functions of the FDC have been elucidated: immune complex trapping via F_c or C_3b receptors (Herd and Ada, 1969; Radoux et al., 1984; Sellheyer et al., 1989), increase of the survival and the proliferation of lymphoid cells (Cormann et al., 1986). However, other roles remain unclear: relationship to the memory cell formation, to cell migration, etc. Isolation of pure FDC populations would allow the analysis of a lot of these functional aspects.

Attempts to purify FDC are hindered by the fact that they are located only in the germinal centers, ar scarce (2 % of the total cell population, Heinen et al. 1986), intermingled with other cells and connected by desmosomes. Lilet-Leclercq et al. (1984) prepared cell suspensions from human tonsils containing a few percent of FDC. These cells enveloped lymphoid cells in their cytoplasmic extensions and thus appeared as spherical clusters (FDC clusters). More efficient isolation procedures were developed by Ennas et al. (1989) and Sellheyer et al. (1989). Tsunoda et al. (1990) could obtain pure FDC clusters preparations. The procedure, involving sedimentation at unit gravity and exploiting the capacity of FDC to adhere to sustrates, is, however, prolonged and suffers from cell loss. Since FDC are rare, we intended further to improve the FDC clusters isolation procedure.

MATERIALS AND METHODS

Freshly removed tonsils were transported at 4°C in a physiological solution containing 0,4 % bovine serum albumin (BSA).

a) Enzymatic digestion of the tonsillar follicles (modified from Lilet-Leclercq et al., 1984)

The tonsils were cut in slices (1 mm thick) and the follicles were isolated under biomicroscopes. These follicles were then maintained under constant agitation for 30 min at 37°C in minimal essential medium (MEM) containing 0.05 % collagenase and

Dendritic Cells in Fundamental and Clinical Immunology,
Edited by Kamperdijk *et al.*, Plenum Press, New York, 1993

0.4 % BSA. The supernatant was collected and the follicles were then incubated 20 min at 37°C in MEM containing 0.05 % COLLAGENASE, 0.05 % dispase, 0.004 % DNase and 0.4 % BSA (all enzymes from Boehringer Mannheim). The supernatant was collected and this digestion step was repeated once.

b) Isolation of FDC clusters
See results

c) Observation of the cell populations
Each purification step was controled by phase contrast microcopy and by immunostaining with anti-DRC_1 antibody (DAKO) on cytospin preps.

RESULTS

a) Isolation of FDC clusters

Our major aim was to obtain, after the digestion steps, highly concentrated FDC-clusters by an efficient and rapid way without cell loss. Best results were given by centrifugations on discontinuous BSA gradients. Different parameters were checked: the diration and the speed of centrifugation, the number of phases and BSA concentration. We applied the following procedure (Fig.1). The cell suspension obtained from the follicles was adjusted to a maximum of 30.10^6 cells/ml (step A in Fig.1). 1.5 ml of this suspension was deposited in a 10 ml tube on the top of a gradient composed as follows:

- a lower phase containing 1.8 ml of 7.5 % BSA in PBS
- an upper phase containing 1.5 ml of 3 % BSA in PBS

The tubes were centrifuged at 10 g for 7 min in a Jouan CR 312 centrifuge (step B). The 3 % phase were discarded and the 7.5 % phases were recovered. The macrophages contained in this cell suspension were eliminated by adherence in culture dishes for 1 hour at 37°C in RPMI 1640 containing 10 % foetal calf serum (FCS) (step C). The non adherent cells were then recovered. Thereafter, the step B was repeated once or twice in smaller tubes (steps D and E). Afterwards, the recovered cells were allowed to settle on glass cover-slips (+/- 2.10^4 clusters per cover-slip) in culture wells for 3 to 6 hours at 37°C in RPMI 1640 + 5 % FCS (step F).

In these conditions, FDC-clusters adhered to the substrate; they preserved their spherical aspect and retained the lymphoid cells they enveloped. Free lymphocytes did not fix on the cover-slips. They were discarded by rinsing the culture wells with RPMI 1640 + 5 % FCS at 37°C (step G). Other structures like fragments of blood vessels, epithelial cells or collagen bundles were also washed off the preparation. It follows that only the FDC clusters remained attached.

b) Efficiency of the technique

The quality (size, number of follicles, abundance of collagen fibers) of the tonsils obtained from the clinics is extremely variable. This leads to great differences in the number and the purity of the FDC-clusters. Therefore, all the values and the symbols used in table 1 give only a general overview of our results.

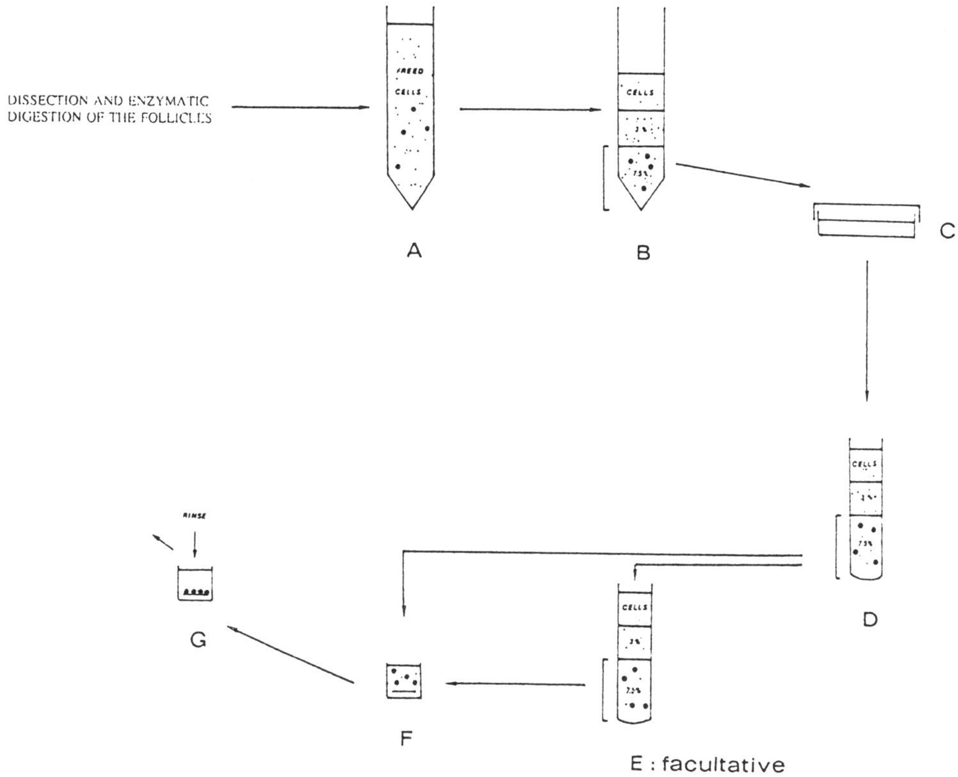

Fig. 1: Procedure for FDC isolation. See text for explanations.

Table 1. Percentages and absolute numbers of FDC-clusters and frequency of contaminating cells or structures after each step of the isolation procedure.

	% of FDC clusters	Absolute number of FDC clusters	Principal contaminating cells or structures			
			Macro phages	Epithelial cells	Free lymphocytes	Collagen bundles, fragments of blood vessels
Step A	0.03 -0.025	3.10^4 - 4.10^5	ND	ND	ND	ND
Step B	1 - 8	negligible loss	++	+	++++	++
Step C	20 - 10	negligible loss	-	+	++++	+
Step D	20 -50	slight loss	-	+	++	+
Step E	40 - 70	slight loss $2.7.10^4$ - $3.6.10^5$	-	+/-	-	+/-
Step G	almost 100	slight loss	-	-	+/-	-

ND: not determined

We calculated that the loss of FDC clusters ramining in the upper phase of the gradients was below 10 % of the total of the FDC clusters present before the application of the sedimentation procedure. Using both tonsils of a patient, we currently obtain 2.10^4 to 3.10^5 FDC clusters at the end of our isolation procedure. Trypan blue exclusion tests showed that almost 100 % of the recovered FDC-clusters were viable.

c) DRC$_1$ labelling

DRC$_1$ immunostaining was performed on isolated FDC clusters loyered on cytospin preps (after step D or E) or attached to cover-slips (after step G). Mostly all cell agregates were DRC$_1^+$. Few DRC$_1^-$ cells contamined them after step G. We notized differences in the intensity of the DRC$_1$ expression among the FDC-clusters (Fig.2).

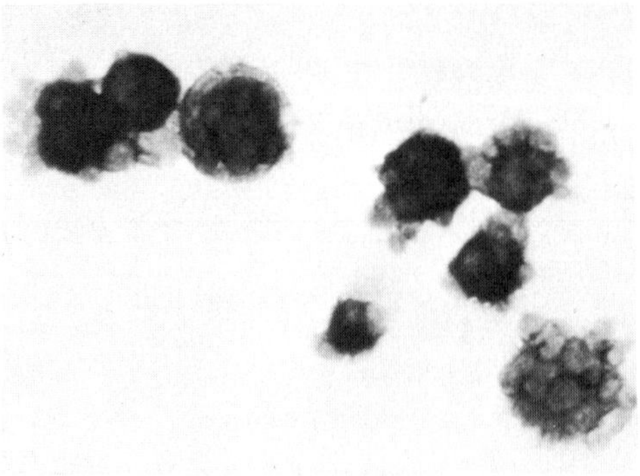

Fig.2. DRC$_1$ staining: all FDC clusters adhering on covel slips are positive. Notice differences in the staining intensity.

CONCLUSIONS

Our new technique for the isolation of the follicular dendritic cells is rapid, simple and efficient. We can in this way concentrate FDC clusters from 0.03 - 0.25 % up to 50 % (step D) or almost 100 % (step G).

Only 10 % of the FDC clusters are lost along the manipulations. Furthermore, the mildness of our procedure prevents cell alterations. This ensures the preservation of the physiological properties of the FDC, for example attraction and binding of lymphoid cells (data not shown).

The main point is now to develope techniques to prevent the rapid dedifferentiation of the FDC reported by Tsunoda et al. (1990). This problem could be solved by cultivating the FDC clusters inside extracellular matrix gels or in determining the factors produced by lymphoid and other cells acting on FDC.

REFERENCES

1. Cormann N., Lesage F., Heinen E., Schaaf-Lafontaine N., Kinet-Denoël C. and L. J. Simar. "Isolation of follicular dendrictic cells from human tonsils and adenoids . V. Effect on lymphocyte proliferation and differentiation.*Immunol. Lett.* 14:29-35 (1986).
2. Ennas M.G., Chilosi M., Scarpa A., Lantini M.S., Cadeddu G. and Fiore-Donati L. "Isolation of multi cellular complexes of follicular dendritic cells and lymphocytes: Immunophenotypical characterization electron microscopy and culture studies."*Cell. Tissue.Res.* 257:9 (1989).
3. Heinen E., Braun M., Coulie P.G., Van Snick J., Moeremans M., Cormann N., Kinet-Denoël C. and L.J. Simar. "Transfer of immune complexes from lymphocytes to follicular dendritic cells."*Eur. J. of Imm.* 16:167 (1986).
4. Heinen E., Cormann N., Tsunoda R., Kinet-Denoël C. and. L. J. Simar. "Ultrastructural and functional aspects of follicular dendritic cells in vitro." Racz P., Dijkstra C.D., Gluckman J.C. (eds):"Accessory cells in HIV and other retroviral infections." *Basel, Karger* 14:29-35 (1991).
5. Herd Z.L. and Ada G.L. "The retention of 125 I-immunoglobulins, IgG subunits and antigen - antibody complexes in rat footpads and draining lymph nodes."*Aust.J. Exp. Biol. Med. Sci.* 47:63 (1969).
6. Lilet-Leclercq C., Radoux D., Heinen E., Kinet-Denoël C., Defraigne J.O., Houben-Defresne M.P. and L.J. Simar. "Isolation of follicular dendritic cells from human tonsils and adenoids I. Procedure and morphological characterization." *J.Immunol. Meth.* 66:235-244 (1984).
7. Klaus G.G.B., Humphrey J.H., Kunkl A. and Dongworth D.W. "The follicular dendritic cell : its role in antigen presentation in the generation of immunological memory." *Immunol. Rev.* 53:3-28 (1980).
8. Papamichail M., Gutierrez C., Embling P., Johnson P., Holborow E. and Pepys M. "Complement dependence of localization of aggregated IgG in germinal centers."*Scand.J. Immunol.* 4:343-347 (1975).
9. Radoux D., Heinen E., Kinet-Denoël C., Tihange E. and Simar L.J. "Precise localization of antigens on follicular dendritic cells." *Cell Tiss. Res.* 235:267-274 (1984).
10. Sellheyer K., Schwarting R. and Stein H. "Isolation and antigenic profile of follicular dendritic cells." *Clin. Exp. Immnol.* 78:4331-436 (1990).
11. Tsunoda R., Nakayama M., Ongaki K., Heinen E., Cormann N., Kinet-Denoël C. and Kojima M. "Isolation and long-term cultivation of human tonsil follicular dendritic cells."*Virchows Arch. B cell Pathol.* 59:95-105 (1990).

FOLLICULAR DENDRITIC CELLS DO NOT PRODUCE TNF-α NOR ITS RECEPTOR

Isabelle Mancini, Alain Bosseloir, Elisabeth Hooghe-Peters (*), Ernst Heinen and Léon Simar

Institute of Human Histology, University of Liège, and Department of Pharmacology (*), Medical School, Free University of Brussels, Belgium

INTRODUCTION

The immunomodulatory role of TNF-α on the humoral response is complex: in vitro, TNF-α can enhance the proliferation of activated B lymphocytes and, in certain conditions, the Ig production. In vivo, it probably acts on the differenciation and growth of B lymphocytes (Kehrl et al., 1987).
Tsunoda et al. have detected the production of TNF-α in supernatants of mixed culture of lymphoid cells and FDCs; this TNF-α could be produced by FDCs. On the contrary, another studies indicated that FDCs do not express mRNA for TNF-α (Schriever et al., 1991).
In this work, by in situ hybridization, we checked the sites were TNF-α is produced in human tonsils, using ^{35}S labeled cDNA and RNA probes on cryosections. In order to localize the cells expressing the receptors for this cytokines, we developed a cytochemical procedure based on a biotinylated TNF-α probe. We also tested the effect induced by TNF-α on cultured tonsil cells. To do this, we separated germinal center cells from the other cells of the tonsils on the base of their capacity to fix peanut agglutinin (PNA).

MATERIALS AND METHODS

a) ISH using a cDNA probe

Tonsil cryosections were fixed in PBS containing 4% of paraformaldehyde and 20mM vanadyl ribonucleoside complex (VRC, Sigma) for 10 min at room temperature. The slides were prehybrized for 2 hr at 37°C in a solution containing 50% deionized formamide, 1xDenhardt's solution (Sigma), 2xSSC, yeast tRNA (200µg/ml, Boehringer Mannheim, FRG), poly(rA) (200µg/ml, Pharmacia, Uppsala, Sweden), heat-denatured salmon sperm DNA (150µg/ml, Sigma) and 200mM DTT. For hybridization, human TNF-α (hTNF-α cDNA, provided by Professor Fiers, University of Gand, Belgium) was ^{35}S labeled by random priming. The sections were then covered with 100µl of heat-denatured labeled DNA probe (10^5cpm/slide) in the prehybridization solution and kept for 16 hr at 37°C. After hybridization, the slides were washed for 1 hr in the same formamide buffer at 37°C

Dendritic Cells in Fundamental and Clinical Immunology,
Edited by Kamperdijk *et al.*, Plenum Press, New York, 1993

with two changes, then with SSC 1x, 0.5x for 1 hr each at room temperature. Afterwards, slides were dehydrated, air-dried and immersed in a Ilford K2 emulsion. After appropriate exposure times (10 to 20 days), the slides were developed in Kodak D19 developer and fixed in 30% sodium thiosulfate.

b) ISH using a riboprobe. (Deman et al., in press)

The slides were hybridized for 16 hr at 50°C in a RNase-free solution containing 50% deionised formamide, NaCl 0.6M, Tris-HCl 10mM pH 7.5, EDTA 1mM, 1% SDS, DTT 10mM, tRNA 0.25 mg/ml, 1xDenhardt's solution, 10% PEG and the ^{35}S-labeled RNA antisens probe (10^5cpm/slide).
Controls for the ISH have been realized by a pretreatment with RNase (10mg/ml, Boehringer Mannheim) or using the RNA sens probe.

c) Localization of TNF-α receptors

Cryosections of human tonsils were fixed in acetone for 10 min at room temperature. Slides were then covered with a solution of PBS-BSA (1%) containing human recombinant TNF-α conjugated with biotin (Boehringer Mannheim) at various dilutions (1 to 1000ng/ml) and incubated overnight at 4°C. The complexes "receptor-biotinylated TNF-α" were stained using the streptavidin-biotin-peroxidase reaction.

d) Cell culture in presence of TNF-α.

Fragments of the tonsil dissected under the bimicroscope were digested to obtain free cells, called "total population". The cells of the germinal centers were separated from the other cells of the tonsil by the "panning" technique, using the capacity of the germinal center cells to fix peanut agglutinin (Coico et al., 1983). Also, we obtain two populations: the "PNA^+" and the "PNA^-". Each cell population (total, PNA^+, PNA^-) were set up in 96-well flat-bottomed microtest plates (Nunc). The cells (10^5 cells/well) were kept in a final volume of 200μl Iscove medium (serum-free medium) supplemented with 1% Pokeweed mitogen (PWM, Gibco) and TNF-α at various dilutions: 12, 20, 25, 30ng/ml. Cultures were incubated at 37°C in a 5% CO_2 enriched atmosphere. The culture was continued for 3 days and DNA synthesis was mesured during the last 6 hr of the culture period by incorporation of ^{3}H thymidine using standard liquid scintillation counting techniques. Controls were cultivated in the same conditions, but in a TNF-α free medium.

RESULTS

a) Localization of the mRNA of TNF-α in human tonsil

After hybridization with ^{35}S labeled TNF-α cDNA and RNA, small clusters of positive cells were found scattered in three areas of the considered tonsils: the interfollicular zone (fig. 1), the subepithelial zone (fig. 2) and the conjunctive capsule. No staining was observed in the epithelium and the lymph follicles (table 1). In control sections hybridized with unrelated or RNA sens probe, no accumulation of silver grains could be detected. Furthermore, the treatment with ribonuclease before hybridization abolished the autoradiographic signal. Our results are based on experiments performed on cryosections from nine different tonsils.

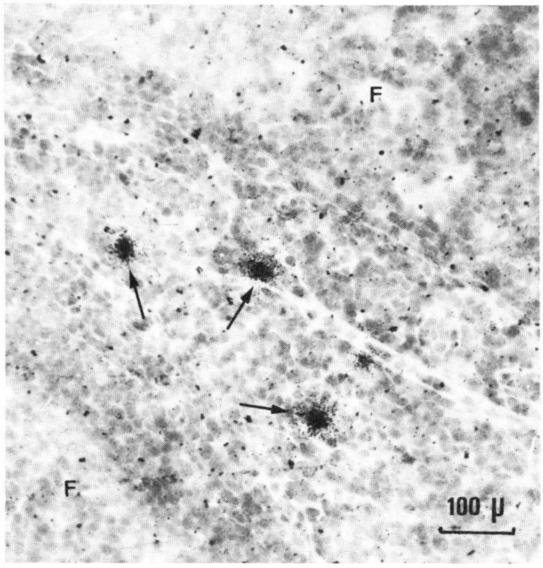

Figure 1. In situ hybridization of human tonsils. Exposure time 20 days. Probe specific activity $5x10^5$ cpm/μg. Section were counterstained with hematoxylin. The autoradiographic signal is located in the interfollicular zone. No labeling is found in the lymph follicles. F: lymph follicles.

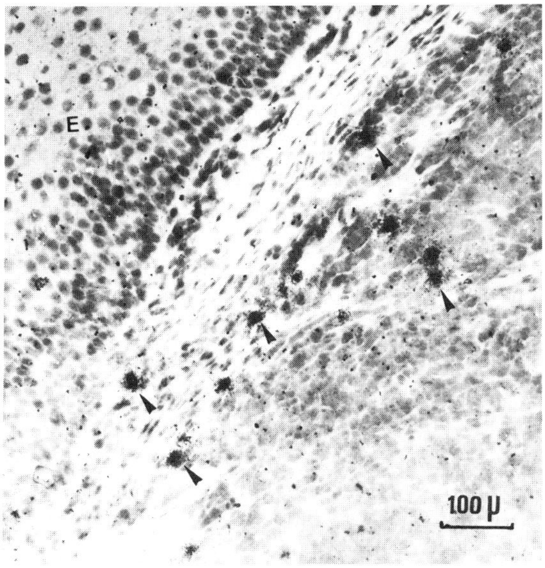

Figure 2. ISH of human tonsils. The epithelial zone is less of autoradiographic signal. Small clusters of strongly positive cells are found in the subepithelial zone. E: epithelium.

Table 1. Localization of mRNA in human tonsils using RNA and cDNA probes.
a) +++ to - represent estimation of the number of labeled cells in the considered zones.

Probe \ Zones (a)	Interfollicular	Subepithelial	Capsular	Epithelial	Follicular
RNA	++	+	+	-	-
cDNA	+++	++	+	-	-

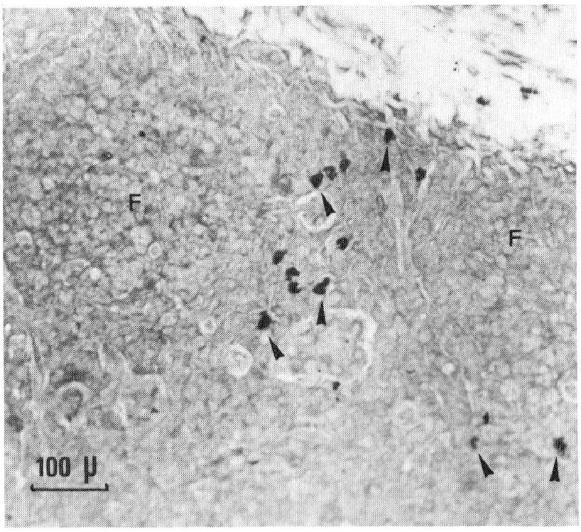

Figure 3. Localization of TNF-α receptors in cryosections of tonsils using a biotinylated TNF-α probe (100 ng/ml). The positive cells are located in the interfollicular zone. The follicles are negative. F: lymph follicles.

b) Localization of TNF-α receptors

Using biotinylated TNF-α, we have tried to localize the cells expressing TNF-α receptors in cryosections of five tonsils. We found an optimal labeling at the concentration of 200 ng/ml. TNF-α receptors were mainly detected in the interfollicular zones (fig. 3). The lymph follicles were not labeled. Epithelial and capsular areas were also marked but it could be an aspecific marking since it was also found among the controls. This aspecific labeling seems to be deposits on the borders of the sections.

c) Cell culture in presence of TNF-α

The DNA synthesis was measured for the three cell populations prepared from human tonsils (total, PNA^+, PNA^-) cultivated during three days in presence of TNF-α at various concentrations (fig. 4). The experiment had been repeated four times. The PNA^+ population, composed of germinal center cells, was not influenced by the presence of TNF-α in the culture medium, whichever its concentration was. The total population underwent a significative proliferation enhancement when cultivated in presence of TNF-α at the concentration of 20 ng/ml. At 30 ng/ml, the ^3H thymidine incorporation had nearly tripled. The PNA^- population was more sensitive to the presence of TNF-α in the medium than the others. A peak was observed at the concentration of 25ng/ml while the DNA synthesis is reduced at 30ng/ml.

DISCUSSION

In this work, we have analyzed the locations of TNF-α expression and of its receptors inside cryosections of human tonsils and tested the influence of TNF-α on the proliferative

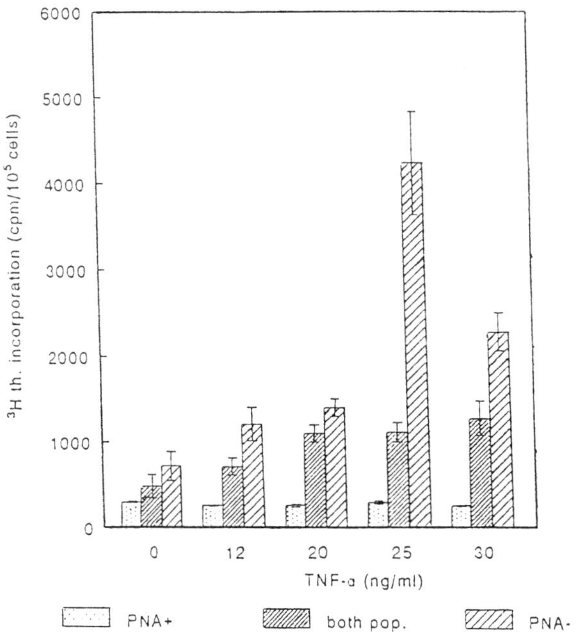

Figure 4. ^3H thymidine incorporation in the cells of different tonsilar populations after 3 days of culture in presence of 1% pokeweed mitogen (PMW) and of various concentrations of TNF-α.

activity of different tonsillar populations. It appears that TNF-α is synthetized in the interfollicular, subepithelial and capsular zones, but not in the lymph follicles. The TNF-α receptors are mainly located in the interfollicular zones of the tonsils but absent from the lymph follicles. We observed that TNF-α modulates the proliferation of the cells located
lymph follicles. We observed that TNF-α modulates the proliferation of the cells located
outside of the germinal centers, therefore we hypothetize that TNF-α probably acts on T cells because they are the most numerous in this area of the tonsil.

Our three different approaches to analyse the part played by TNF-α in human tonsils are concording: the TNF-α producing cells, the cells expressing its receptor and the cells responding to the action of TNF-α are located in the same areas of the tonsils: the interfollicular zones. In according with Schriever et al., 1991, we show that FDCs do not produce TNF-α.

At a first sight, it appears astonishing to note that a cytokine able to influence the B lymphocyte proliferation and the Ig secretion by these cells is not produced inside germinal centers and does not act on the proliferative activity of the germinal centers cells. This situation is less surprising when one compares the TNF-α localization with that of other cytokines inside the tonsils. IL-2, IL-4 and IL-6 have also been located in the interfollicular, subepithelial and capsular zones, but not in the lymphoid follicles (Bosseloir et al., 1989, Von Gaudecker et al., 1991). On the other hand, IL-1 and IL-5 appear to de produced in germinal centers (Lasky et al., 1989, Bosseloir et al., 1991).

On basis on our experiments, we can also conclude that germinal centers are areas where B lymphocytes are apparently isolated from given cytokines. Germinal center B cells can apparently only meet these cytokines during their migration in the perifollicular areas. We consider that during the development of the humoral immunity, B cells pass trough evolutive phases in which each one depends on the encounter with a given cytokines, but some of these phases must, on the contrary, be realized in the absence of certain cytokines.

REFERENCES

1. Kehrl, J., Miller, A. and Fauci, A. "Effect of tumor necrosis factor alpha on mitogen-activated human B cells." *J. Exp. Med.* 166:786 (1980).
2. Tsunoda, R. and Kojima, M. "immunocytological characterisation of the constituent cells of the secondary nodules in human tonsils." *Adv. Exp. Med. Biol.* 149:829 (1982).
3. Schriever, F., Freeman, G. and Nadler, M. "Follicular dendritic cells contain a unique gene repertoire demonstrated by single-cell polymerase chain reaction." *Blood* 77:787 (1991).
4. Bosseloir, A., Hooghe-Peters, E., Heinen, E., Cormann, N., Kinet-Denoël, C., Vanhaelst, L. and Simar, L.J. "Localisation of interleukin-6 mRNA in human tonsils by in situ hybridization." *Eur. J. Immunol.* 19:2379 (1989).
5. Von Gaudecker, B., Mielke, V., Ritter, M.A. and Sterry, W. "Subcellular localization of cytokines and their receptors in peripheral lymphoepithelial organs." *Lymphatic tissues and in vivo immune response. (Imhof, B.E., Berrih-Aknin, S., Ezine, S., Eds.)* 32 (1991).
6. Lasky, J.L. and Thorbecke, G.J. "Growth requirements of SJL lymphomas in vitro: effects of BCGF II" *Adv. Exp. Med. Biol.* 237:145 (1989).
7. Bosseloir, A., Hooghe-Peters, E., Heinen, E., Marcoty, C., Cormann, N., Kinet-Denoël, C. and Simar, L.J. "Localization of interleukin-4 and interleukin-6 messenger ribonucleic acid in human tonsils by ISH. *Lymphatic tissues and in vivo immune response. (Imhof, B.E., Berrih-Aknin, S., Ezine, S., Eds.)* 315 (1991).
8. Deman, J., Martin, M.T., Delvenne, P., Humblet, C., Boniver, J. and Defresne, M.P. "Analysis by ISH of cells expressing mRNA for TNF-α in developping thymus of mice. *Devel. Immunol.* in press.
9. Coico, R.F., Bhogal, B.S. and Thorbecke, G.J. *J. Immunol.* 131:2254 (1983).

SPLENIC LESIONS IN HYPOGAMMAGLOBULINAEMIA

J. Weston,[1] B.M. Balfour,[2] W. Tsohas,[1] N. English,[2] J. Farrant[2] and A.D.B. Webster[2]

[1]Department of Histopathology, Northwick Park Hospital, Harrow, Middlesex, HA1 3UJ, U.K.
[2]Division of Immunological Medicine, Clinical Research Centre, Harrow, Middlesex, HA1 3UJ, U.K.

INTRODUCTION

Common Variable Immunodeficiency (CVI) is a syndrome characterised by hypogammaglobulinaemia and defects in cellular immunity.[1] The differentiation of B-lymphocytes into plasma cells does not occur and the patients suffer from recurrent infections.[2] About 30% of CVI cases develop splenomegaly.[3] Some of these patients develop symptoms of hypersplenism which further increases their susceptibility to infection.[3] In most cases the mechanism is benign, but unknown. Splenectomy is often required to control hypersplenism and two such cases are presented here, together with a third patient whose spleen was removed for suspected lymphoma, which was not found. A spleen from a healthy donor was provided by a liver transplant team. Conventional histological and immunohistochemical techniques were used to study the distribution of macrophages, follicular dendritic cells (FDC) and B-lymphocytes. The principal finding was the presence of granulomas in all 3 CVI spleens.[4]

CLINICAL FINDINGS

The 3 patients were all young women. Their clinical details and the reasons for splenectomy are summarised in Table 1.

METHODS

Histology

The spleens were weighed and processed for routine histology. The area occupied by white pulp was estimated using an image analysis computer system and expressed as a percentage of the total area of the spleen (Table 2). The degree of germinal centre formation was expressed in a semiquantitative

Dendritic Cells in Fundamental and Clinical Immunology,
Edited by Kamperdijk *et al.*, Plenum Press, New York, 1993

manner as follows: - no germinal centres; + germinal centres present; ++ very large lymphoid follicles with prominent germinal centres. Granulomas were measured and their average diameter and anatomical location noted (Table 2).

Immunohistochemistry

Paraffin sections of formalin fixed tissue were labelled with a biotinylated polyclonal antibody against lysozyme, a macrophage marker, followed by extravidin:peroxidase. Frozen sections were labelled for nonspecific esterase (NSE), another marker of macrophages. FDC and B-lymphocytes were identified with mouse monoclonal antibodies and immunoperoxidase.

Table 1. Case Histories.

Case	Age at splenectomy	Duration of illness	Type of illness	duration of splenic enlargement	reason for splenectomy
1	34 years	4 years	sinusitis ear infections	6 months	? lymphoma
2	29 years	8 years	recurrent ear and chest infections Hepatitis C	4 years	hypersplenism
3	27 years	10 years	pneumonia diarrhoea bronchiect- -asis	4 years	hypersplenism

RESULTS

Histology

The results are summarised in Table 2. There was expansion of the white pulp and germinal centres in cases 1 and 2. Non-caseating sarcoid-like epithelioid granulomas were present in the spleens of all 3 cases of CVI. The number, size and site of the granulomas varied from case to case (Table 2 and Figure 1), but in general they were found in or near the white pulp. Spleen 3 contained many large granulomas which obliterated almost all the follicles. In spleen 2 the lesions were much smaller and spleen 1 contained only a few small granulomas.

Table 2. Summary of histological findings.

Case	Wt	Percentage occupied by white pulp	Germinal centres	Granulomas Average size	Granulomas Site
1	570g	38.5%	++	0.140mm	marginal zone
2	950g	24.5%	+	0.247mm	white pulp mantle zone
3	1100g	20.0%	not assessable	0.430mm	white pulp follicle centres
normal	130g	7.1%	-	-	-

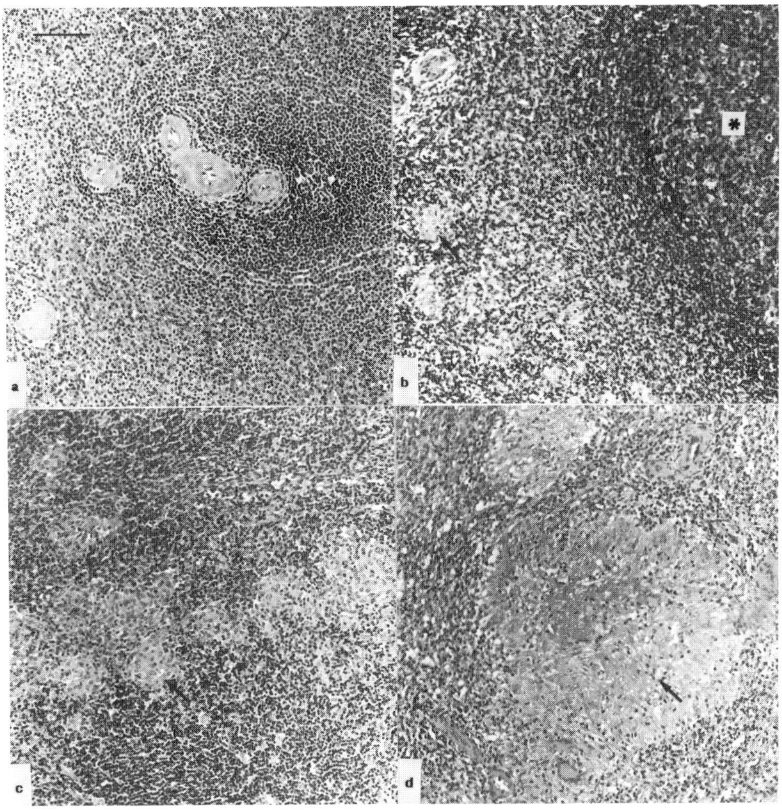

Figure 1. Normal spleen and 3 cases of CVI with granulomas (arrowed). **a** normal spleen. **b** case 1. **c** case 2. Note reactive follicle with germinal centre indicated by *. **d** case 3. Scale bar=0.1mm.

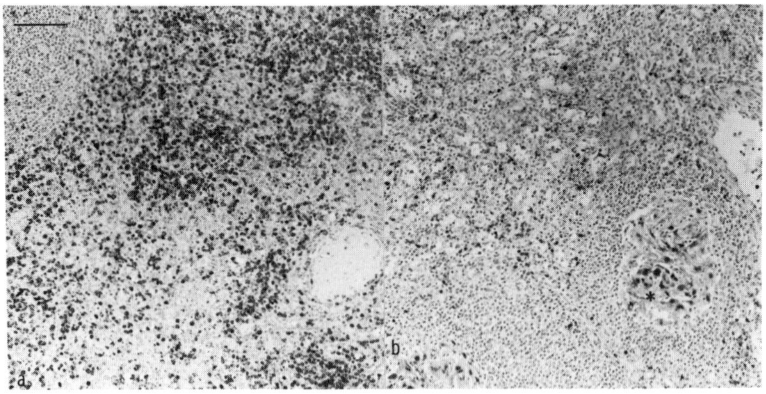

Figure 2. Normal spleen and spleen from a patient with CVI labelled for lysozyme. **a** normal. **b** case 3 with granuloma indicated by *. Positive cells in the granuloma and in the red pulp stain darkly. Note diminished reactivity of red pulp macrophages in this spleen compared to the normal spleen. Scale bar=0.1mm.

Immunohistochemistry

When sections of normal spleen were stained for lysozyme and NSE many positive cells were seen in the red pulp (Figure 2a). The reaction was very weak in spleen 3 (Figure 2b) and somewhat diminished in spleen 2, but only slightly less intense than normal in spleen 1. The epithelioid granuloma cells were positive for lysozyme and NSE confirming that they were macrophage-derived (Figure 2b). In spleen 2 however, some of the cells within the granulomas were not positive for lysozyme or NSE suggesting that the granulomas were composed of an admixture of epithelioid cells and fibroblasts.

In spleens 1 and 2 FDC were seen in lymphoid follicles and germinal centres. In spleen 3 islands of surviving cells remained in close association with small groups of CD 19 positive B-cells, but these islands were separated from each other by granulomas. The total number of B-cells in this spleen was much reduced (Figure 3).

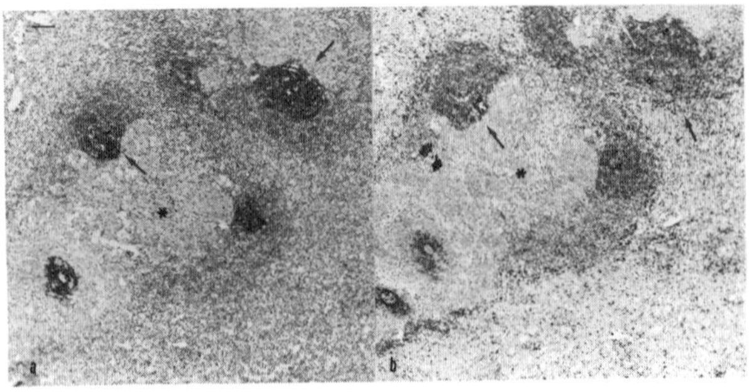

Figure 3. A follicle centre in the spleen of case 3 is largely obliterated by a large granuloma indicated by *. **a** FDC marker showing small groups of FDC (arrowed). **b** CD19 positive B-cells (arrowed) are only seen in association with FDCs. Scale bar=0.1mm.

DISCUSSION

To summarise, all 3 patients with CVI had splenic pathology with reactive changes and granulomas. The number and size of the lesions were proportional to the duration of the illness. The granulomas contained epithelioid cells derived from macrophages. In spleens 2 and 3 the red pulp macrophages had very weak lysozyme and NSE activity. These findings suggest an intrinsic abnormality in macrophage function leading to defective antigen handling and presentation.[5]

Neither splenomegaly nor granuloma formation occur in patients with X-linked agammaglobulinaemia (XLA), therefore, these changes cannot be attributed solely to chronic infection. Granuloma formation is enhanced by activated T-cells[6] and since these cells are found in large numbers in CVI but not in XLA patients,[7] they may be one of the prime causes of the lesion. Constant stimulation of the immune system by infections in the presence of activated T-cells could be one of the factors responsible for granuloma formation in a proportion of patients.

In case 3 FDC and B-lymphocytes were only found in small islands and this disruption of the normal architecture by granulomas must have interfered with the generation of memory cells and the selection of B-cells secreting high affinity antibody.

ACKNOWLEDGMENTS

We would like to thank J. MacKenzie for taking the photomicrographs and J. Bradbeer for his help with the image analysis.

REFERENCES

1. G.P. Spickett, A.D.B. Webster and J. Farrant, Cellular abnormalities in common variable immunodeficiency, *Immunodefic. Rev.* 2:199 (1990).
2. A. Saxon, J.V. Giorgi, E.H. Sherr and J.M. Kagan, Failure of B-cells in common variable immunodeficiency to transit from proliferation to differentiation is associated with altered B-cell surface molecule display, *J. Allergy Clin. Immunol.* 84:44 (1989).
3. A.D.B. Webster, J. Farrant, M. Hany, M. North, E. Toubi and R. Beatie, Clinical and cellular features of 'common variable hypogammaglobulinaemia' *Estratto dalla rivista EOS.* 11:32 (1991).
4. M.J. Crofts, M.V. Joyner, J.C. Sharp, J. Costello and D. Vergani, Sarcoidosis associated with combined immunodeficiency, *Postgrad. Med. J.* 56:258 (1980).
5. J.W. Mannhalter, G.J. Zlabinger and M.M. Eibl, Defective macrophage-T-cell interaction in common varied immunodeficiency, *in:* 'Progress in immunodeficiency research and therapy,' C. Griscelli and J. Vossen, ed., Excerpta medica, Amsterdam, 1:147 (1984).
6. A. McInnes and D.M. Rennick, Interleukin 4 induces cultured monocytes/macrophages to form multinucleated giant cells, *P.N.A.S.* 167:598 (1988).
7. J. Farrant, Personal communication.

ONTOGENIC STUDY ON THE BRONCHUS-ASSOCIATED LYMPHOID TISSUE (BALT) IN THE RAT, WITH SPECIAL REFERENCE TO DENDRITIC CELLS

Yoshifusa Matsuura, Nobuo Imazeki, Akira Senoo, and Yusuke Fuse

Department of Pathology II
National Defense Medical College
Tokorozawa, Saitama 359, Japan

INTRODUCTION

In several mammalian species, the bronchus–associated lymphoid tissue (BALT) appears as condensed lymphoid aggregates to the bronchial wall. Recently, a population of antigen–bearing cells with dendritic morphology has been identified in rat respiratory tract tissue (Sertl et al., 1986). We studied on the ontogenic development of BALT with special reference to dendritic cells (DC) of rat lung ultrastructurally and immunohistochemically.

MATERIALS AND METHODS

A total of 55 Sprague–Dawley (SD) rats were kept in an ordinary environment with food and water provided ad libitum. Furthermore, a total of 12 Specific Pathogen Free (SPF) SD rats were obtained. Animals were sacrificed by cervical dislocation or by the intraperitoneal injection of a lethal dose of barbital.

The lungs were quickly removed, dissected in a longitudinal plane and fixed in 8% paraformaldehyde for preparation of wax sections. Specimens were also processed for electron microscopy in a routine way. The first layer antibodies used in immunohistochemistry were as follows: MAS 026 (Sera–lab, Sussex, UK) to anti–rat leukocyte common antigen; MAS 027 (Sera–lab) to anti–rat Thy–1; MAS 043 (Sera–lab) to anti–rat Ia antigen; MCA 340 (Serotec, Oxford, UK) ; and anti–bovine S–100 protein (Dacopatts, Glostrup, Denmark), a marker of rat follicular dendritic cells (FDC). S–100–positive cells were also observed with a confocal scanning laser microscope (CSLM).

RESULTS

Late foetal period

Rats in 21 days gestation showed few leukocyte common antigen–positive cells and no BALT–like structure was observed.

One week of age

At 6 days of age, there was an aggregation of the fibroblastic reticulum cells (FRC)

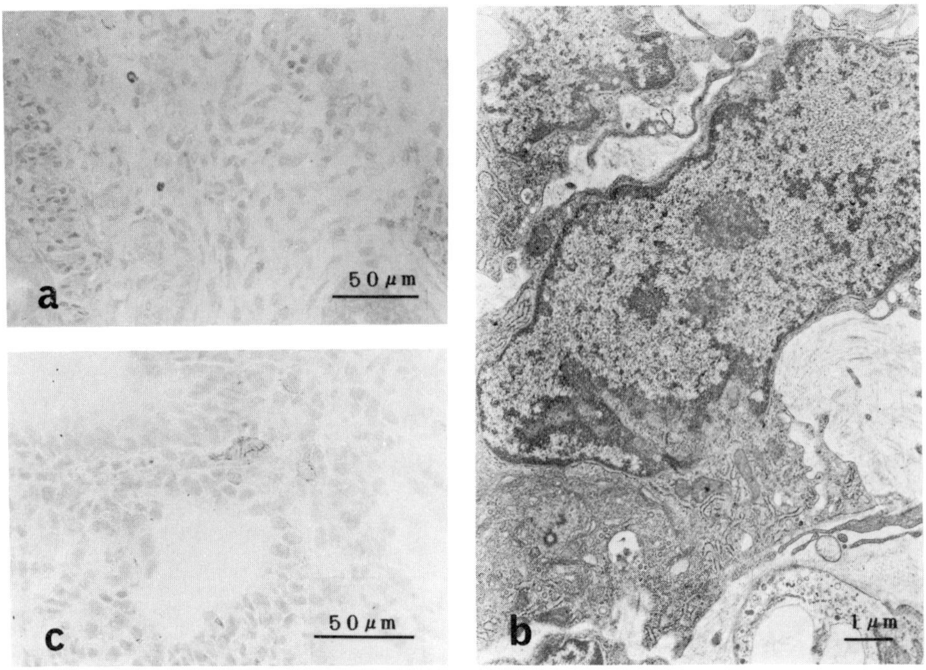

Figure 1.(a) Semi–thin section of the lung of a 6 day old rat stained with toluidine blue. Note the aggregation of reticular cells. A mast cell showing metachromasia of the granules is also observed. (b) Ultrastructure of the lung of a 6 day old rat. Subepithelial fibroblastic reticulum cells with extracellular bundles of collagen fibers are seen. (c) Immunoperoxidase staining for Ia antigen in the lung of a 6 day old rat. Ia–positive large spindle cells are observed in the interstitial tissue.

in the peribronchial interstitial tissue, which was supposed to be an initial BALT (Figure 1a,b). Ia–positive cells were also observed (Figure 1c).

Two weeks of age

By 10 days of age, the aggregation of lymphoid cells together with S–100–positive reticular cells were observed under the muscularis mucosae of the bronchus (Figure 2).

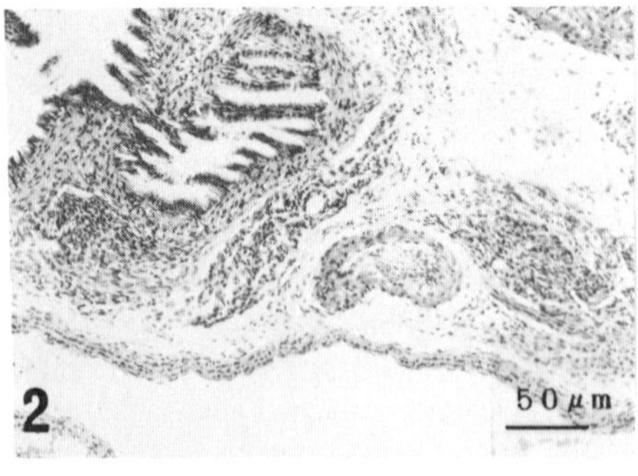

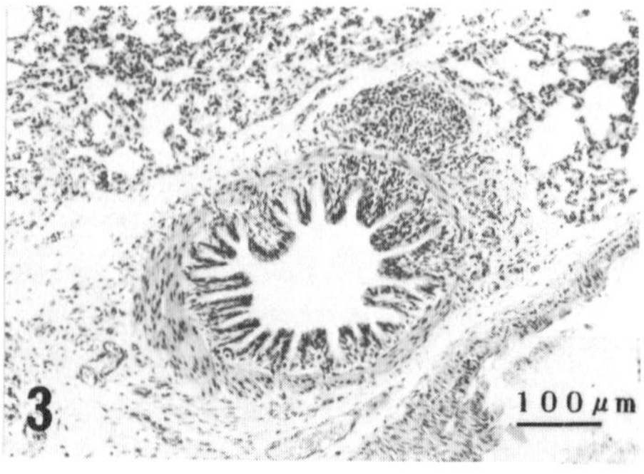

Figure 2. Haematoxylin and eosin (HE) staining of the lung of a 9 day old rat. Lymphoid aggregation together with reticulum cells is observed between the main bronchus and the accompanying vessels.
Figure 3. HE staining of the lung of a 20 day old rat. Many vessels and lymphocytes breaking through the muscle layer are observed in BALT.

Three to four weeks of age

By 20 days of age, many vessels, lymphocytes which broke through the muscle layer, and markedly S–100–positive cells were observed in BALT (Figure 3).

SPF rats

At 21–42 days of age, SPF rats developed less and smaller BALT and fewer S–100–positive cells compared with animals kept conventionally.

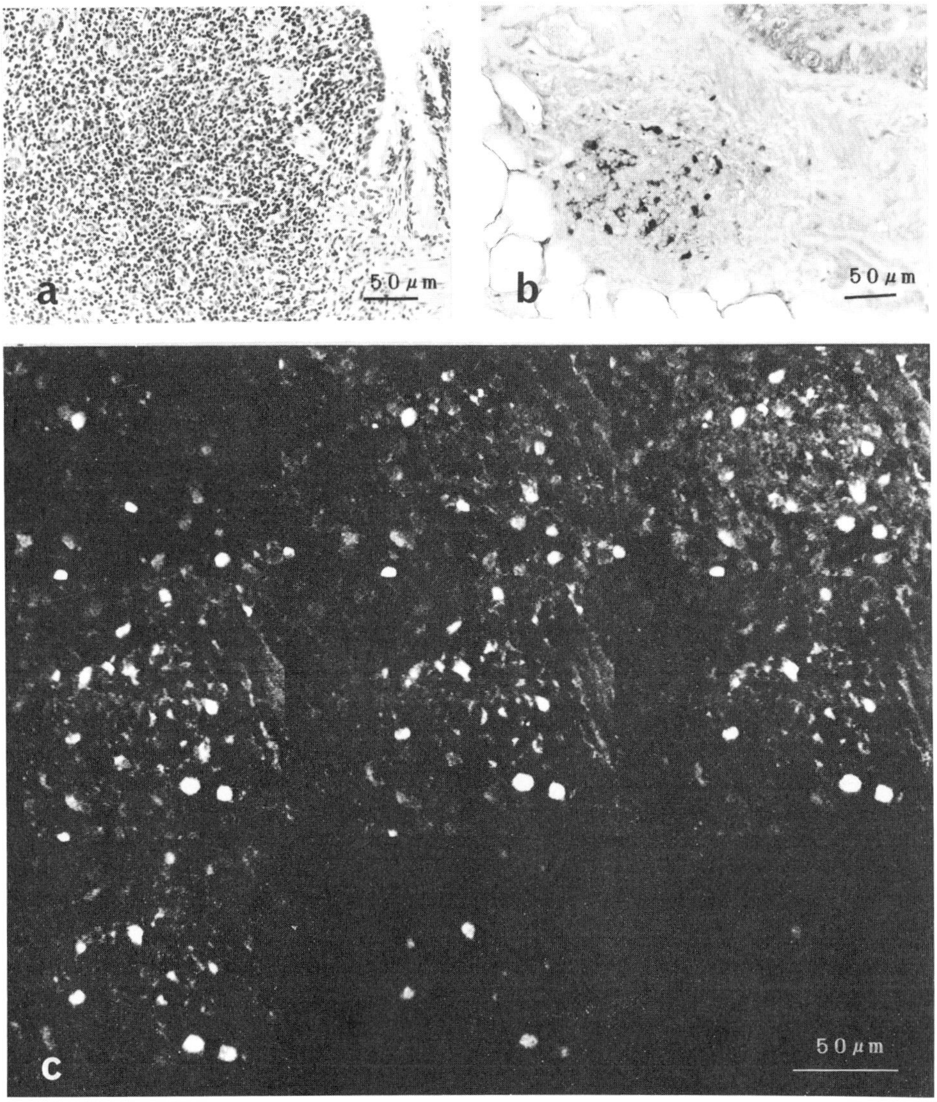

Figure 4.(a) HE staining of BALT in the lung of a 7 week old rat. (b) Immunoalkaline–phosphatase staining of a lymphoid follicle and S–100–positive dendritic cells in BALT of a 15 week old rat. (c) Tomographical view of S–100–positive cells in BALT of a rat observed in the same specimen as Figure 4a by immunofluorescent microscopy with CSLM. The distance between the shown sections is 1.2 μm.

Adult rats

In the adult animals, a B–cell area was present in central part and subepithelium of BALT, where S–100–positive cells with a dendritic form were observed (Figure 4a,b). Three–dimensional visualization of S–100 positive cells was performed (Figure 4c,d). Ultrastructure of central part of BALT showed reticular cells with dendritic cytoplasmic processes, which contact the lymphocytes (Figure 4e,f).

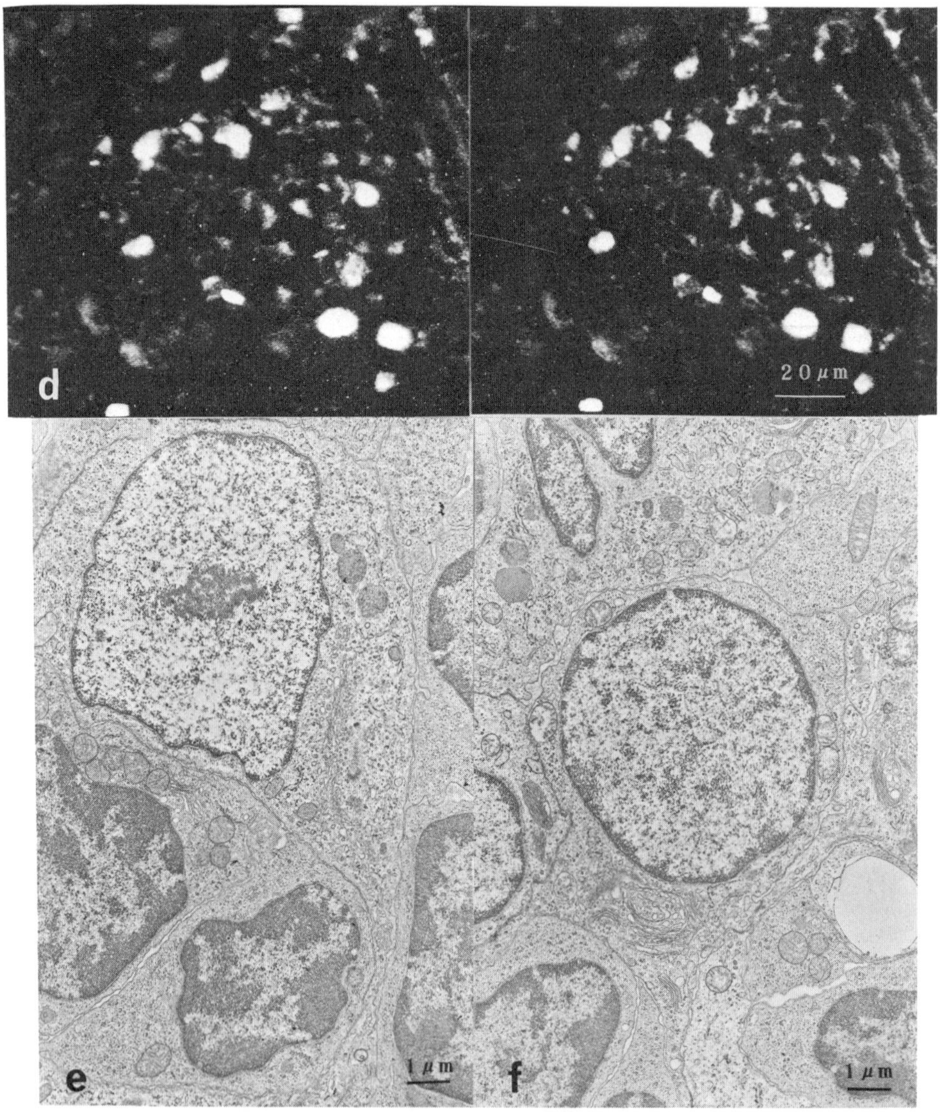

Figure 4. (d) Stereopair of S–100–positive cells in BALT of the rat. (e) Ultrastructure of the central part of rat BALT. A reticular cell shows the euchromatic nucleus with well–defined nucleolus and prominent fibrous lamina and the cytoplasm which has rough–surfaced endoplasmic reticulum, mitochondria, Golgi apparatus, and polysomes. Dendritic cytoplasmic process which contacts the lymphocytes is observed. (f) Ultrastructure of the central part of rat BALT that shows a reticular cell with prominent Golgi apparatus and dendritic cytoplasmic processes.

DISCUSSION

Recent years have seen several reports concerning with DC in the lung, but further studies on their morphology and functional state are needed. Sertl et al. (1986) reported that Langerhans' cell (LC)-like DC with antigen presenting capability reside in human lung. Subsequently, distribution of Ia-bearing cells in rat lung has been demonstrated (Holt et al., 1988; Schon-Hegrad et al., 1991).

S-100 protein is positive in human LC (Shamoto and Shinzato, 1991) and negative in human FDC (Imai et al.,1991), whilst is positive in rat FDC (Imai et al.,1991), and negative in T-cell area of rat spleen or outside of BALT (data not shown). DC observed in BALT seem to be different cells from LC because of S-100-positivity and lack for LC granules.

BALT develops postnatally in response to antigen stimulus and may play an important role in local defenses of respiratory tract. S-100-positive cells in BALT appeared to be identical FDC seen in the secondary follicles of lymphoid organs from their ultrastructural features, although they lacked coiled processes or extracellular dense material which are shown in active FDC (Fuse, et al., 1991). Present ontogenic study of DC in BALT also suggests that they might be derived from FRC, and may function to present antigen to lymphocytes.

ACKNOWLEDGMENT

The authors wish to thank Mr. Toshio Haneda and Mr. Makoto Kato, Olympus Optical Co., LTD. for their excellent technical assistance with CSLM.

REFERENCES

Fuse, Y., Senoo, A., and Imazeki, N., 1991, Ultrastructure of follicular dendritic cells, in: "Dendritic Cells in Lymphoid Tissues," Y. Imai, J.G. Tew, and E.C.M. Hoefsmit, ed., Elsevier, Amsterdam.

Holt, P.G., Schon-Hegrad, M.A., and Oliver, J., 1988, MHC class II antigen-bearing dendritic cells in pulmonary tissues of the rat, J. Exp. Med. 167:262.

Imai, Y., Matsuda, M., Maeda, K., Yamakawa, M., Dobashi, M., Satoh, H., and Terashima, K., 1991, Dendritic cells in lymphoid tissues. Their morphology, antigen profiles and functions, in: "Dendritic Cells in Lymphoid Tissues," Y. Imai, J.G. Tew, and E.C.M. Hoefsmit, ed., Elsevier, Amsterdam.

Schon-Hegrad, M.A., Oliver, J., McMenamin, P.G., and Holt, P.G., 1991, Studies on the density, distribution, and surface phenotype of intraepithelial class II major histocompatibility complex antigen (Ia)-bearing dendritic cells (DC) in the conducting airways, J. Exp. Med. 173:1345.

Sertl, K., Takemura, T., Tschachler, E., Ferrans, V.J., Kaliner, M.A., and Schevach, E.M., 1986, Dendritic cells with antigen-presenting capability reside in airway epithelium, lung parenchyma, and visceral pleura, J. Exp. Med. 163:436.

Shamoto, M., and Shinzato, M., The characterization of interdigitating reticulum cells: a comparison between Langerhans' cells, in: "Dendritic Cells in Lymphoid Tissues," Y. Imai, J.G. Tew, and E.C.M. Hoefsmit, ed., Elsevier, Amsterdam.

DRC_1 EXPRESSION ON NORMAL AND PATHOLOGICAL LYMPHOID CELLS

Nadine Antoine, Ernst Heinen, Christian Marcoty, Alain Bosseloir and Léon Simar

Laboratory of Human Histology
University of Liège, Belgium

INTRODUCTION

In 1983, Naiem et al., produced a murine monoclonal antibody recognizing tonsillar follicular dendritic cells (FDC). This anti-DRC_1 appeared highly selective for human FDC. It has proved of great value in the analysis of the distribution of FDC in reactive lymphoid tissues, in follicular lymphomas and in other diseases, namely in AIDS. Anti-DRC_1 (M 706, Dakopatts) strongly reacts with the FDC extensions in mantle zones (Naiem et al. 1983). However it also reacts, though much lesser with splenic marginal zone B cells and with some mantle B lymphocytes.

DRC_1 expression was also found in pathological conditions: EBV transformed B cells and blood B cells from patients with rhumatoid arthritis (Terashima et al. 1988).

In keeping with these results, the aim of the present work was to precise the exact distribution of DRC_1 antigen on various normal and pathological lymphoid cells but also to study the role played by this antigen, namely to determine if it connects lymphoid cells to follicular dendritic cells as it had been suggested by Louis et al. (1989).

DRC_1 expression was analysed by FACS after indirect immunolabelling on normal blood and on lymphoid cells prepared from leukemic patients.

MATERIAL AND METHODS

a) Cell separation: Lymphoid cells were separated on density gradients (Ficoll-Hypaque) from heparinized peripheral blood of healthy persons and from patients with chronic lymphocytic (CLL), acute lymphblastic (ALL) or prolymphocytic leukemias (PLL). They were then washed twice with PBS medium; T cells were prepared by "rosetting" with sheep red blood cells.

Dendritic Cells in Fundamental and Clinical Immunology,
Edited by Kamperdijk *et al.*, Plenum Press, New York, 1993

b) Immunolabeling: Cells were stained with individual monoclonal antibodies (Table 1) for 30 minutes at 4°C. Each antibody was diluted in PBS to its optimal working dilution. After this time, lymphocytes were washed three times in PBS, and labeled 30 minutes at 4°C with fluorescein-labeled rabit anti-mouse immunoglobulin Ig (1/200). Cells were then washed and stored at 4°C before flow cytometric analysis.

Table 1. Antibodies used in the present study and their dilutions.

TESTED ANTIBODIES	DILUTIONS
α - CD 19 (DAKO)	1/20
α - CD 3 (DAKO)	1/20
α - DRC_1 (DAKO)	1/20
α - CD 10 (DAKO)	1/10

c) Flow cytometric analysis: After indirect immunofluorescence labeling, the lymphocytes were analysed in a fluorescent activated cell sorter (FACS-SCAN Becton Dickinson) equipped with a 488 nm argon laser. Fluorescent signals above 520 nm were considered for at least 10.000 cells per analysis.

d) Adhesion test: 1. *Isolation and purification of FDC:* surgical removed palatine children tonsils were dissected and digested. FDC clusters were then purified by centrifugations on discontinuous BSA gradients as described by Marcoty et al. (see this volume).

2. *Adhesion*: the adhesion test was adapted from that of Louis et al. (1989). After isolation, FITC-labeled lymphocytes (30 minutes at 37°C with FITC, 50 μg/ml at pH 6.8 were incubated with FDC. After 45 minutes of incubation, the fluorescent cells adhering to FDC were counted and their mean number was calculated (n=30 FDC clusters).

RESULTS

a) Cytofluorimetric analysis

Phenotypic aspects of normal and pathological lymphoid cells are summarized in table 2.

Table 2. Immunophenotyping of normal or pathological lymphoid cells with anti-B, anti-T monoclonal antibodies and anti-DRC_1.

* number of T cells persisting after rosetting
** n = number of analyzed cases
B-CLL = B - chronic lymphocytic leukemia
B-ALL = B - acute lymphoblastic leukemia
B-PLL = B - prolymphocytic leukemia

B cell antigens / Cases analyzed	CD 19	CD 3	CD 10	DRC_1
Tonsillar B cells (n=10)	62 +/- 3 %	38 +/- 1,2 %	/	60 +/- 1,8 %
Normal blood B lymphocytes after "rosetting" (n=5)**	41 +/- 4,2 %	*14 +/- 2,8 %	/	45 +/- 2 %
B-CLL (n=7)	97,2 +/- 28 %	2 +/- 0,3 %	0 %	84 +/- 5,1 %
B-ALL (n=3)	64,5 +/- 5 %	21 +/- 2,4 %	63 +/- 4,8 %	0 %
B-PLL (n=1)	98 +/- 1,2 %	0,5 +/- 0,2 %	0 %	0 %

We observed DRC_1 expression (45 +/- 2 %) in the preparation of normal blood cells. This percentage was approximately that obtained with the CD 19 antigen (41 +/- 4,2 %). A double label flow cytometry allowed us to attest that only B lymphocytes express DRC_1. All T cells appeared DRC_1 negative. B cells from chronic lymphocytic leukemia also strongly expressed DRC_1 contrary to the blood B lymphocytes from acute lymphoblastic leukemias and a prolymphocytic leukemia which were DRC_1 negative.

b) Contacts between follicular dendritic cells (FDC) and B lymphocytes

When the adherence capacity of B leukemia cells with FDC was checked, only B lymphocytes from ALL were able to establish contacts with FDC isolated from human tonsils (Table 3.). Lymphocytes from B-CLL were unable to attach on FDC clusters.

Table 3. Adherence test of normal and pathological B lymphocytes on FDC

	Mean number of fluorescent lymphocytes on FDC clusters after 45 minutes of incubation
Tonsillar B cells (n=10)	3,7 +/- 1,6 %
Normal blood B lymphocytes (n=5)	0
B-CLL (n=7)	0
B-ALL (n=3)	8 +/- 2,5 %

CONCLUSIONS

The advent of flow cytometry and monoclonal antibodies has offered important supplemental informations that give better reproductibility and objectivity of leukemic diagnosis and characterization than current immunohistologic methods.

Our results support the hypothesis that the antibody α-DRC_1 is not specific for follicular dendritic cells. Indeed, the DRC_1 antigen seems to be a B cell marker. When double label flow cytometry was performed on normal peripheral blood and in some lymphoproliferative disorders (CLL), all CD 19 positive cells were also DRC_1 positive. All T cells analysed appeared DRC_1 negative.

DRC_1 labeling appears to be valuable to discriminate between CLL (DRC_1 positive) and PLL (DRC_1 negative) inasmusch as the distinction of the two entities generally is difficult (Degan, 1989).

In conclusion, the DRC_1 appears thus to be rather an intrinsic antigen expressed on various B cells populations: B lymphocytes from human tonsils follicles (Antoine et al. 1992), normal blood and malignant B lymphocytes from patients with CLL disease.

We also show here that the DRC_1 antigen is not a determinant factor for the adhesion between lymphoid cells and FDC: when the adherence of isolated FDC to B leukemia cells was checked, no clear relationship was found between the adherence capacity and DRC_1 expression; only B lymphocytes from B-ALL which were DRC_1 negative were able to react with FDC.

Further analyses are programmed to verify the diagnostic value of DRC_1 antigen and to define its exact function on intercellular cooperations.

REFERENCES

1. Antoine N., Beckers C., Marcoty C., Bosseloir A., Beguin Y., Schaaf-Lafontaine N. and Simar L.J. "Expression de l'antigène DRC_1 par les cellules leucémiques" *Revue Médicale de Liège XLVII* 65-69 (1992).
2. Degan M.G., "Membrane antigen analysis in the diagnosis of lymphoid leukemias and lymphomas"*Arch. Pathol. Lab. Med.*, 113: 606-618 (1989).
3. Louis E., Philippet B., Cardos B., Heinen E., Cormann N., Kinet-Denoël C., Braun M. and Simar L.J. "Intercellular contacts between germinal center cells. Mechanisms of adhesion between lymphoid cells an follicular dendritic cells" *Acta Oto-Rhino-Laryngologica Belgica* 43: 4, 297-320 (1989).
4. Marcoty C., Heinen E., Antoine N., Tsunoda R. and Simar L.J. (see this volume) " Rapid and selective isolation of follicular dendritic cells by low speed centrifugations on discontinuous BSA gradients.
5. Naiem M., Gerdes J., and Adulaziz Z., Stein H., Mason D.Y. "Production of a monoclonal antibody reactive with human dendritic reticulum cells and its use in immunohistological analysis of lymphoid tissues." *J. Clin. Path.* 36: 167-175 (1983).
6. Terashima K., Ukai K., Tajima K., Yuda F., and Imai Y. "Morphological diversity of DRC_1 positive cells: human follicular dendritic cells and their relatives" *Adv. Exp. Med. Biol.* 237: 157-163 (1988).

BINDING OF HIV-1 TO HUMAN FOLLICULAR DENDRITIC CELLS

Piet Joling,[1] Leendert J. Bakker,[2] Dick F. van Wichen,[1] Loek de Graaf,[2] Timo Meerloo,[1] Maarten R. Visser,[2] Jos A.G. van Strijp,[2] Jaap Goudsmit,[3] Jan Verhoef,[2] and Henk-Jan Schuurman[1]

[1] Departments of Pathology and Internal Medicine, University Hospital P.O. Box 85.500, 3508 GA Utrecht, The Netherlands
[2] Department of Clinical Microbiology, University Hospital, Utrecht
[3] Human Retroviral Laboratory, Academic Medical Center, Amsterdam The Netherlands

INTRODUCTION

Follicular dendritic cells (FDC) are among the first cells involved in tissue damage after infection with Human Immunodeficiency Virus type-1 (HIV-1). These cells, being of mesenchymal origin, contribute to the stroma of germinal centers in lymphoid organs. In the first phase after HIV-1 infection, these germinal centers are hyperplastic, with distinct fragmentations and indentations. At these locations, FDC give the impression of being damaged by the infection. An active role of HIV-1 components has been proposed in this process. HIV-1 proteins, complete virions and HIV-1 RNA have been observed in the germinal center, in higher density than at other locations of the lymphoid tissue.[1,2] Apparently, protein and virus particles are concentrated, presumably in the form of immune complex, in the labyrinth of cytoplasmic protusions of FDC in the germinal center. Studies on FDC purified from a lymph node of an HIV-1-infected patient,[3] and *in vitro* infection of FDC from normal tissue,[4] have provided support for the suggestion that the cells can be infected and actively produce virus.
The characteristics of the binding of HIV-1 to FDC are unresolved thus far. FDC show in part expression of the CD4 antigen, the "classical" receptor for HIV-1.[5] Alternatively, a role for receptors for complement components and for Fc fragment of immunoglobulin can be proposed, as these receptors are involved in immune complex binding. We therefore studied the binding of HIV-1 to FDC in suspension. The data shows a pivotal role of complement component C3 in this process.

MATERIAL AND METHODS

Isolation of Cells. FDC were purified in suspension from tonsil obtained at tonsillectomy of children (age range 3-10 years).[5] The tissue was minced in small fragments (about 1 mm^3); after removal of most leukocytes, a collagenase digestion (15 U/ml) was performed for 30 min at room temperature, followed by 1x*g* sedimentation on

Dendritic Cells in Fundamental and Clinical Immunology,
Edited by Kamperdijk *et al.*, Plenum Press, New York, 1993

a discontinuous gradient of 1.5, 2.5, 5.0, and 7.5% (w/v) bovine serum albumin (BSA), respectively. The interphase between 2.5 and 5% BSA was harvested. Then, cells were separated over a discontinuous Percoll (Pharmacia, Uppsala, Sweden) gradient of 1.070, 1.060 and 1.030 g/l, at 1200x*g* for 20 min. The FDC-enriched population was harvested from the 1.030-1.060 interphase. In cytocentrifuge preparations stained by May Grünwald-Giemsa, most cells showed the morphology of FDC, and were located either solitary or clustered with lymphoid cells. Cells with a macrophage-like cytology were almost absent, and a few solitary lymphocytes were observed. Human blood mononuclear cells (MNC) were isolated from blood of HIV-1-negative donors by conventional Ficoll-Isopaque density centrifugation.

HIV-1 and Binding Studies: HIV-1 virus was purified from the supernatant of SUP-T1 cells after infection by the HIV-1 IIIB strain.[6] The virus was conjugated to fluorescein isothiocyanate (FITC).[7] Culture supernatant was filtered (0.45 μm) and centriguged at 70,000x*g* for 1 hour. The pellet was incubated with FITC (0.1 mg/l, Sigma Chemical Co., St. Louis, MI, USA) in Tris buffer, pH 7.3, for 1 hour. Centrifugation was done on a sucrose gradient (layers of 22% and 50%, in 0.05 M Tris buffer) at 40,000x*g* for 17 hours at 4°C. The layer containing virus was dialysed against phosphate-buffered saline.

Fifty μl FITC-conjugated HIV-1 solution was incubated with $5x10^5$ cells. The incubation mixture was supplemented with 20% serum. The sources of serum are presented below. After incubation at 37°C for 30 minutes, the cells were washed and fixed in 1% paraformaldehyde.

Binding was analyzed using the following methods. Conventional electron microscopy and immunogold electron microscopy using ultracryotomy with mouse monoclonal anti-HIV-1 *gag* p24 antibody (Abbot Laboratories, North Chicago, IL, USA) or with rabbit

Table 1. Fluorescence signal (flow cytometry) of FDC or blood monocytes after incubation with FITC-labeled HIV-1 in the presence of various sera, and the effect of quenching of extracellular fluorescence using trypan blue.

Serum	Mean fluorescence	Quenching -	Quenching +
Follicular dendritic cells			
Background, no HIV-1/FITC	5.3	4.1	3.9
Culture medium	11.5	8.3	5.7
Control heated serum	8.2	5.9	6.4
Control fresh serum	66.7	20.3	8.2
Heated HIV-1 serum	n.d	5.9	5.0
Fresh HIV-1 serum	n.d	34.9	9.7
C3-deficient serum	8.3	n.d	n.d
C5-deficient serum	46.3	n.d	n.d
Blood monocytes			
Control heated serum	n.d	6.4	7.4
Control fresh serum	n.d	57.5	29.0

Quenching: - denotes no trypan blue, + incubation with trypan blue. n.d, not done anti-FITC antibody (Molecular Probes Inc, Eugene, OR, USA) was done as described elsewhere.[8] Fluorescence microscopy was performed after processing to cytocentrifuge preparations. Flow cytometry was performed on a FACStar flow cytometer (Becton Dickinson, Mountain View, CA, USA), equipped with an Argon 2W laser (excitation 488 nm), and Consort C30 software. Gatings were set on measurement of lymphocytes and larger cells. To discriminate between intracellular and extracellular (cell membrane) fluorescence, quenching was performed by incubation with 0.2 g/l trypan blue for 20 minutes in sodium acetate buffer, pH5.8. Before and after trypan blue incubation, the cells were washed in the same buffer. This was followed by fixation in 1% paraformaldehyde. The control experiment included the omission of trypan blue.

Sera. Control serum in binding tests was a pooled serum of healthy donors, that was heated for 30 minutes at 56°C to remove complement activity. This pooled serum was negative in anti-HIV-1 antibodies, documented by western blotting. The same pooled serum was used without heating ("fresh serum"). The same was done for serum pooled from HIV-1 positive individuals ("HIV-1 serum"). Serum from a patient with deficiency of complement component C3 was provided by the Department of Immunology, University Hospital for Children and Youth, Utrecht ("C3-deficient serum"). In control experiments, the addition of C3 restored complement activity of this serum. Serum deficient in complement component C5 ("C5-deficient serum") was obtained from Sigma Chemical Co. All sera were stored at -70°C until use.

RESULTS AND DISCUSSION

Flow Cytometry

Fluorescence profiles were prepared for FDC and blood MNC after incubation with HIV-1, with various serum supplements in the incubation mixture (Fig. 1, Table 1). Using heated pooled control serum in incubation with FDC, only a marginal increase in fluorescence signal was observed with respect to the control (FDC without HIV-1). The application of heated HIV-1 serum similarly showed only a marginal increase in fluorescence. FDC incubated with HIV-1/FITC in the presence of fresh pooled control serum manifested a substantial increase, indicative of HIV-1 binding to FDC. Using combinations of fresh pooled control serum and anti-HIV-1 serum resulted in an even higher signal in flow cytometry.

This data indicates that a factor in fresh serum, removed by heating, mediates the binding of HIV-1 to FDC. Anti-HIV-1 antibodies in HIV-1 serum show only a marginal effect on binding, but enhance this fresh serum-associated effect.

It is beforehand to suggest that complement components present in fresh serum mediate binding of HIV-1 to FDC. This was investigated using fresh serum lacking C3 or C5 (Table 1). In the presence of C3-deficient serum, there was no binding above the background values. In the presence of C5-deficient serum, the intensity of the fluorescence signal was similar to that in the presence of pooled fresh control serum. This data indicates that complement C3 mediates binding of HIV-1 to FDC.

Trypan blue fluorescence quenching was applied to evaluate whether the fluorescence is present intracellularly or on the cell surface. In the presence of the dye, extracellular fluorescence is quenched and no longer detectable. The mean fluorescence of FDC after binding HIV-1 in the presence of complement was reduced to about 30% of the maximum value (Table 1). As control, blood MNC were subjected to FDC binding.[7] A less reduction of fluorescence, to about 50% of the maximum signal, was observed after quenching. This data indicates that blood monocytes show a higher internalization of HIV-1 after

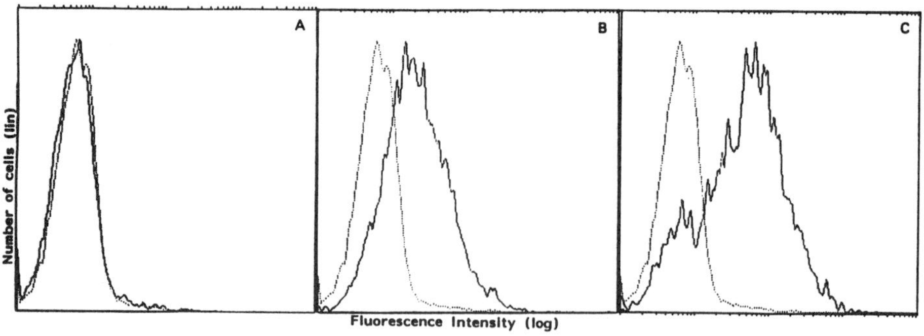

Figure 1. Fluorescence profiles of FDC after binding FITC-labeled HIV-1. In each of the figures shown, the background fluorescence of FDC not subjected to HIV-1 binding is indicated by a dotted line. **A**: In the presence of **heated serum** (either from controls or from HIV-1 positive individuals) there is no increase in fluorescence signal. **B**: In the presence of **fresh control serum** a higher fluorescence signal is seen indicative of HIV-1/FITC binding. **C**: In the presence of **fresh serum from HIV-1 positive individuals** an even higher signal is observed.

complement-mediated binding to the cell surface than FDC. This conclusion is made with the restriction of the experimental approach, that included incubation at 37°C for 30 minutes. When extrapolated to *in situ* data, this phenomenon is in accord with the extracellular localization of HIV-1 virions, that has been claimed for the labyrinth of cytoplasmic protrusions of FDC in histologic sections of the germinal center. The internalization of part of HIV-1 virions supports observations mentioned above, that FDC can be infected and then actively produce virus.[3,4]

Fluorescence Microscopy and Immunogold Electron Microscopy

Reading cytocentrifuge preparations of FDC after incubation with HIV-1 with various serum supplements confirmed the flow cytometric observations (Fig. 2). Only a faint fluorescence was observed for FDC-like cells, either solitary or in clusters with lymphocytes, when using heated serum either from controls or from HIV-1 positive patients. Preparations after incubation in the presence of fresh serum showed a high fluorescence of FDC-like cells. The fluorescence localized on the surface of the centrally situated FDC in clusters, extending between adherent lymphocytes. Lymphocytes

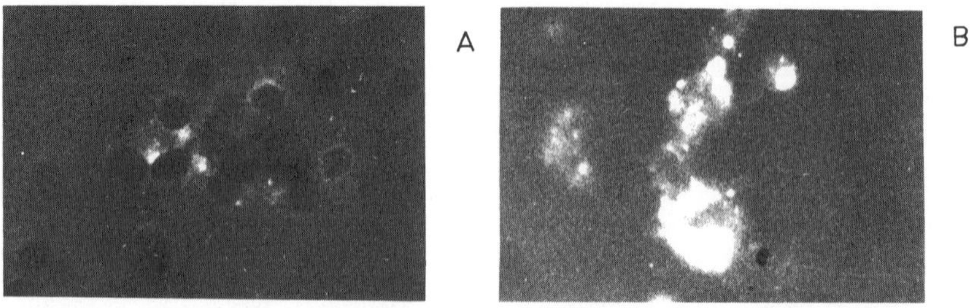

Figure 2. Immunofluorescence of FDC after binding FITC-labeled HIV-1, processed to cytocentrifuge preparations. **A** In the presence of heated serum, there is only a faint fluorescence observed of FDC-like cells. **B** In the presence of fresh serum from an HIV-1 positive individual, a strong fluorescence of FDC-like cells is observed. Magnification x264

themselves, either in clusters or present as solitary small-sized cells, showed only a faint fluorescence.

Electron microscopy of FDC after HIV-1 binding showed intact virions in the villous extensions of FDC-like cells, either solitary cells or cells located centrally in clusters with adherent lymphocytes. Accidentally virions were observed around free lymphocytes. In immunogold electron microscopy, labeling product of HIV-1 p24 or FITC localized on the cell surface of FDC-like cells, and on virions attached to the cell surface. Also intracellularly immunogold label was seen. It was not possible to distinguish which cytoplasmic component was labeled due to the poor preservation of the cell. This was not only related to a loss of architecture in ultracryotomy, but also to loss of architecture due to the isolation from tissue and subsequent incubation procedure with HIV-1.

CONCLUSIONS

- FDC *in vitro* bind HIV-1 in the presence of complement component C3. Anti-HIV-1 antibodies reinforce this binding, but do not mediate binding without complement. There is no apparent role for the "classical" virus-target cell interaction molecules, i.e. determinants on HIV-1 *env* gp 120 and the CD4 molecule, in this process. This may be related to the abundance of receptors for complement C3 on FDC, whereas there is only marginal expression of receptors for the Fc fragment of immunoglobulin (Tuijnman and Van Wichen, unpublished observations). The presence of CD4 molecules on FDC is subject of continuous discussion. In our hands, only a proportion (about 20%) of FDC in suspension are CD4-immunoreactive.[5]
- The majority of HIV-1 after incubation with FDC at 37°C for 30 minutes remains on the cell surface. This contrasts with blood monocytes after complement-mediated binding of FDC. In accord with their phagocytic characteristics, the major part of HIV-1 is internalized.
- The germinal center of lymphoid organs is a site, where *in vivo* HIV-1 concentrates in the first period after infection. From other studies is has been concluded that HIV-1 itself at this location contributes to the process of destruction of FDC. Approaches to eliminate HIV-1 from the germinal center, to treat infection and prevent viral spread, should take into account that HIV-1 binding to germinal center cells is mainly mediated by complement, and not by the interaction between HIV-1 *env* gp120 and the CD4 receptor.

ACKNOWLEDGEMENT

This work was supported by the Dutch Ministry of Health, as part of the National Program on AIDS Research, Grant Nr. RGO/WVC 88-79/89005.

REFERENCES

1. H.-J. Schuurman, W.J.A. Krone, R. Broekhuizen, and J. Goudsmit, Expression of RNA and antigens of human immunodeficiency virus type-1 (HIV-1) in lymph nodes from HIV-1 infected individuals, *Am. J. Pathol.* 133:516 (1988).
2. L.H.P.M. Rademakers, H.-J. Schuurman, J.F. de Frankrijker, and A. van Ooyen, Cellular composition of germinal centers in lymph nodes after HIV-1 infection: evidence for an inadequate support of germinal center B lymphocytes by follicular dendritic cells, *Clin. Immunol. Immunopathol.* 62:148 (1992).
3. H.K. Parmentier, D.F. van Wichen, D.M.D.S. Sie-Go, J. Goudsmit, J.J. Borleffs, and H.-J. Schuurman,

HIV-1 infection and virus production in follicular dendritic cells in lymph nodes. A case report, with analysis of isolated follicular dendritic cells, *Am. J. Pathol.* 137:247 (1990).
4. I. Stahmer, J.P. Zimmer, M. Ernst, T. Fenner, R. Finnern, H. Schmitz, H.-D. Flad, and J. Gerdes, Isolation of normal human follicular dendritic cells and CD4-independent *in vitro* infection by human immunodeficiency virus (HIV-1), *Eur. J. Immunol.* 21:1873 (1991).
5. H.K. Parmentier, J.A. van der Linden, J. Krijnen, D.F. van Wichen, L.H.P.M. Rademakers, A.C. Bloem, and H.-J. Schuurman, Human follicular dendritic cells: isolation and characteristics in situ and in suspension, *Scand. J. Immunol.* 33:441 (1991).
6. M. Popovic, M.G. Sargadharan, E. Read, and R.C. Gallo, Detection, isolation, and continuous production of cytopathic retroviruses (HTLV-III) from patients with AIDS and pre-AIDS, *Science* 224:497 (1984).
7. L.J. Bakker, H.S.L.M. Nottet, N.M. de Vos, L. de Graaf, J.A.G. van Strijp, M.R. Visser, and J. Verhoef, Antibodies and complement enhance binding and uptake of human immunodeficiency virus type 1 by human monocytes. *AIDS* 6:35 (1992).
8. T. Meerloo, H.K. Parmentier, A.D.M.E. Osterhaus, J. Goudsmit, and H-J Schuurman, Modulation of cell surface nolecules during HIV-1 infection of H9 cells. An immunoelectron microscopic study, *AIDS* 1992; in press.

FOLLICULAR DENDRITIC CELLS IN GERMINAL CENTER REACTIONS

John G. Tew[1], Gregory F. Burton[1], Leo I. Kupp[1] and Andras Szakal[2]

[1] Department of Microbiology and Immunology
[2] Department of Anatomy, Division of Immunobiology
Medical College of Virginia/Virginia Commonwealth University
P.O. Box 678 Richmond, Virginia 23298 U.S.A.

INTRODUCTION

Data implicating antigen persisting in lymphoid follicles in the production of germinal centers and B memory cells has been available for many years[1,2,3]. Follicular dendritic cells (FDC) represent the specialized cells which bind and retain the immunogen in the follicles for many months. The basic concept is that FDC with their associated antigens represent the accessory cell population responsible for the stimulation of antigen-specific B cells and subsequent germinal center formation. In recent years this concept has been further supported by computer assisted morphometric analysis and *in vitro* studies of germinal center formation using isolated FDC. A summary of recent results are listed in Table 1 (top of the next page). The items enumerated in Table 1 will be explained in subsequent sections of this paper in the sequence listed.

THE 1:1 RELATIONSHIP BETWEEN LOCALIZED ANTIGEN BEARING FDC AND THE DEVELOPMENT OF GERMINAL CENTERS

Within a few minutes after booster immunization it is possible to detect antigen moving from the subcapsular sinus of the draining lymph nodes into the outer cortex[4]. Within 24 hours the vast majority of antigen persisting in the lymph node has moved through these transport sites into a number of lymphoid follicles. When horseradish peroxidase (HRP) is used as the immunogen is possible to visualize these FDC bearing HRP-anti-HRP immunecomplexes in many follicles. The intertwining process of the FDC form a localized network or reticulum and multiple antigen retaining reticula (FDC-reticula) can be observed in sections of the draining lymph nodes of young adult mice[4]. By serially sectioning through lymph nodes and using morphometric analysis it was determined that the average young adult mouse develops 8 FDC-reticula per popliteal and 12 FDC-reticula per axillary lymph node

Dendritic Cells in Fundamental and Clinical Immunology,
Edited by Kamperdijk *et al.*, Plenum Press, New York, 1993

Table 1. Recent evidence implicating FDC in B cell stimulation and germinal center formation.

1. A 1:1 relationship exists between localized areas with antigen bearing FDC (FDC-reticula) and the development of germinal centers.
2. FDC deficient aged mice have reduced numbers of FDC-reticula but the FDC-reticula to germinal center ratio remains 1:1.
3. Germinal center (GC) formation in vitro is FDC dependent. FDC are necessary both for B cells to form large clusters and proliferate.
4. FDC not only provide GC B cells with antigen but also provide important costimulatory signals. For example, B cells stimulated through sIg with anti-μ do not form large clusters and or proliferate optimally in the absence of FDC.
5. FDC apparently provide a signal(s) which are able to render resting B cells capable of chemotaxis. Early in the GC reaction the GC B cells are capable of chemotaxis and this may help explain germinal center dissociation and homing of GC B cells to the bone marrow.

upon challenge. Germinal center development in these mice was also monitored morphometrically by utilizing the fact that germinal center B cells selectively bind peanut agglutinin[5]. The 8 and 12 FDC-reticula correlated nicely with a mean of 8 and 14 germinal centers respectively[6]. These correlations support the concept that one FDC-reticulum induces the development of one germinal center as suggested by previous reports[1,2,3,7,8].

FDC DEFICIENT AGED MICE HAVE REDUCED NUMBERS OF FDC-RETICULA BUT THE FDC-RETICULA TO GERMINAL CENTER RATIO REMAINS 1:1

As explained above, antigen transport leads to formation of FDC-reticula composed of FDC with their antigen bearing processes intertwining[4]. In recent studies with aged mice[6,9], we observed that the vast majority of antigen transport sites did not develop and that the antigen transport sites that did develop appeared atrophic[6]. This deficit corresponded with a deficit in the number of FDC-reticula by day 1-3 after antigenic challenge[6]. Follicular dendritic cells in the antigen retaining reticula that did develop were ultrastructurally atrophic, retained little antigen and produced no iccosomes.

These results prompted us to examine the capacity of these old mice to develop germinal centers. By using methods to distinguish between preexisting and newly developed (de novo) germinal centers, the previous impression of a paucity of germinal centers in old mice[10] was quantitatively confirmed. In old mice that developed germinal centers (60% did not develop any), the mean number of FDC-reticula was only 2 in popliteal and 4 in axillary lymph nodes. These FDC-reticula induced a mean of 2 and 4 germinal centers in the respective lymph nodes. Although there is a marked age-related depression in germinal center development, the ratio of FDC-antigen retaining reticula to germinal center numbers remained 1:1. As in young lymph nodes, the significance of this ratio is that it supports the concept that germinal center development requires functional antigen retaining FDC.

This concept that FDC are required for germinal center formation and normal antibody production is further supported by cell transfer studies[11]. Old mice with defective FDC have markedly depressed secondary antibody responses. The anamnestic response in these old mice can not be restored by reconstituting the animals with young memory T and B cells. In contrast, young control mice with normal FDC give normal anamnestic responses when given memory T and B cells. The old FDC are apparently unable provide immunogen in the form of immune complexes in a construct (presumably iccosomes) that is stimulatory for young lymphocytes[11].

GERMINAL CENTER FORMATION IN VITRO IS FDC DEPENDENT

Further evidence indicating a requirement for FDC in germinal center development comes from studies of "germinal center reactions *in vitro*"[12]. Using the technique developed by our laboratory to isolate FDC[13] a mixed population containing 85-90% B cells, 3-10% FDC and 1-5% T cells is obtained. This combination of cells form clusters of 20 to 50 lymphocytes per FDC within hours of culturing at 37°C. Adding 3H-thymidine at various intervals demonstrated that cells continue to proliferate for at least 5 days. Immunocytochemistry in conjunction with autoradiography revealed that the majority of cells undergoing proliferation were of the germinal center B cell phenotype. The necessity for FDC and T cells was assessed by depleting these populations from the isolated cells using magnetic beads coated with FDC specific or T cell specific monoclonal antibodies (FDC-M1, a rat anti-mouse FDC antibody; T 24 a rat anti-mouse thy 1 antibody). When FDC were depleted, no cellular clusters formed and incorporation of 3H-thymidine was minimal. Removal of T cells resulted in the formation of smaller clusters and dramatically decreased proliferation. Adding recombinant IL1, IL2, IL3 IL4, IL5, IL6 either alone or in combinations, or a source of T cell conditioned supernatant, did not replace the removal of FDC or T cells[12]. These data indicate FDC are required for cluster formation and that FDC-B cell interactions are necessary for B cell proliferation. T cells are also necessary but the nature of the T cell help is not yet clear.

FDC PROVIDE B CELLS WITH IMPORTANT COSTIMULATORY SIGNALS

We reasoned that FDC might provide co-stimulatory signals which would enhance the ability of antigen to stimulate B cell proliferation in the germinal centers. To test this, FDC were cultured with B cells activated by a sIg dependent (anti-μ) or independent (LPS) pathway and their proliferation measured using ^{3}H-thymidine incorporation. The addition of FDC markedly augmented B cell proliferation in a dose dependent fashion. Depletion of FDC from cultures abrogated the increased proliferation. Addition of highly purified FDC obtained from cell sorting resulted in B cell co-stimulation whereas addition of other sorted cells was without effect. The FDC accessory activity was apparent over the entire culture period and over a wide range of either polyclonal B cell activator. When B cells and activators were cultured in the absence of FDC, only about one-fourth of the cells remained viable after 3 days. In contrast, virtually all cells in cultures containing FDC, B cells and activator were viable supporting a role for FDC in maintaining B cell viability. Cultures containing FDC and B cells from nude mice proliferated normally in the presence of anti-μ+rIL-4 indicating that functional T cells were not important in this system. The costimulatory activity of the FDC could not replace either the anti-μ or IL-4 in this system and was not MHC restricted (Burton et.al unpublished). These data support the concept that FDC provide not only antigen but also facilitate B cell proliferation by means of other costimulatory interactions which contribute to make the microenvironment in the germinal center favorable for B cells to proliferate.

FDC APPARENTLY PROVIDE PRIMARY SIGNALS TO B CELLS WHICH CAN RENDER B CELLS CAPABLE OF CHEMOTAXIS

We have recently reported that germinal center B cells obtained 2 to 4 days after antigenic challenge are capable of chemotaxis[14]. This is at the early stage when germinal centers are dissociating and germinal center B cells may be found in the draining lymph and blood. Many of these cells are antibody forming cells and are homing to the bone marrow[14]. We reasoned that FDC might play a major role in providing signals necessary for these B cells to acquire the motile phenotype capable of chemotaxis. When resting B cells from nude mice were mixed with isolated FDC, the B cells acquired the mobile phenotype and were capable of chemotaxis. In this case, the signal could be transmitted by FDC alone and the addition of anti-μ and IL-4 was not necessary (Kupp et. al, Unpublished). The ability of B cells to migrate and respond to chemotactic signals may be important in germinal center dissociation which takes place early after antigen challenge. Furthermore, this chemotactic ability may allow the germinal center B cell to move after binding cells in the sinuses of the bone marrow into the bone marrow proper. Cells in the sinuses of the bone marrow are capable of binding galactose which is in rich abundance on the surface of the germinal center B cell[14]. We reason that agents in the bone marrow attract the germinal center B cell into the bone marrow proper. This concept was supported by the observation that germinal center B cells are capable of chemotaxis toward supernatant fluids from bone marrow cultures[15].

CONCLUDING COMMENTS

The basic results discussed here are summarized in Table 1. In short, the accumulated evidence supporting the concept that FDC are required for germinal center formation is compelling. In addition to providing specific immunogen to stimulate B cells bearing appropriate sIg, FDC provide costimulatory signals which facilitate B cell proliferation and other functions like B cell chemotaxis. This relationship between the FDC and B cell is only partially understood but it is becomming clear that FDC-B cell interactions are vital for normal humoral immune responses.

REFERENCES

1. R.G. White, V.I. French, and J.M. Stark, A study of the localization of a protein antigen in the chicken spleen and its relation to the formation of germinal centers, *J. Med. Microbiol.* 3:65 (1970).
2. G.L. Thorbecke, T.J. Romano, and S.P. Lerman, Regulatory mechanisms in proliferation and differentiation of lymphoid tissue, with particular reference to germinal center development, *in*: "Progress in Immunology," L. Brent and J. Holbrow, eds., North-Holland Publishing Co., Amsterdam (1974).
3. J.G. Thorbecke and S.P. Lerman, Germinal centers and their role in immune responses, *in*: "The Reticuloendothelial System in Health and Disease: Functions and Characteristics," S.M. Reichard, M.R. Escobar, and H. Friedman, eds., Plenum Press, N.Y. (1976).
4. A.K. Szakal, K.L. Holmes, and J.G. Tew, Transport of immune complexes from the subcapsular sinus to lymph node follicles on the surface of non-phagocytic cells, including cells with dendritic morphology, *J. Immunol.* 131:1714 (1983).
5. A.K. Szakal, J.K. Taylor, J.P. Smith, M.H. Kosco, G.F. Burton, and J.G. Tew, Kinetics of germinal center development in lymph nodes of young and aging immune mice, *Anat. Rec.* In press (1990c).
6. A.K. Szakal, J.K. Taylor, J.P. Smith, M.H. Kosco, G.F. Burton, and J.G. Tew, Morphometry and kinetics of antigen transport and developing antigen retaining reticulum of follicular dendritic cells in lymph nodes of aging immune mice, *Aging: Immunol. and Inf. Dis.* 1:7 (1988b).
7. A.K. Szakal, M.H. Kosco, and J.G. Tew, A novel *in vivo* follicular dendritic cell-dependent iccosome-mediated mechanism for delivery of antigen to antigen-processing cells, *J. Immunol.* 140:341 (1988a).
8. G.G.B. Klaus and A. Kunkl, The role of germinal centers in the generation of immunological memory, *in*: "Microenvironments in Haemmopoietic and Lymphoid Differentiation, CIBA Foundation Symposium 84," Pitman Medical, London (1981).
9. A.K. Szakal, J.K. Taylor, J.P. Smith, M.H. Kosco, G.F. Burton, and J.G. Tew, Kinetics of germinal center development in lymph nodes of young and aging immune mice, *Anat. Rec.* In press (1990c).
10. M.G. Hanna Jr, C.C. Congdon, and C.J. Wust, Effect of antigen dose on lymphatic tissue germinal centers, *Proc. Soc. Exp. Biol. Med.* 121:286 (1966).
11. G.F. Burton, M.H. Kosco, A.K. Szakal, and J.G. Tew, Iccosomes and the secondary antibody response, *Immunol.* 73:271-276 (1991).
12. M.H. Kosco, E. Pflugfelder, and D. Gray, Follicular dendritic cell-dependent adhesion and proliferation of B cells in vitro, *J.Immunol.* 148:2331 (1992).
13. C.T. Schnizlein, M.H. Kosco, A.K. Szakal, and J.G. Tew, Follicular dendritic cells in suspension: identification, enrichment, and initial characterization indicating immune complex trapping and lack of adherence and phagocytic activity, *J.Immunol.* 134:1360 (1985).
14. Leo I. Kupp, M.H. Kosco, H.A. Schenkein, and J.G. Tew, Chemotaxis of germinal center B cells in response to C5a, *Eur. J. Immunol.* 21:2697-2701 (1991)
15. J.G. Tew, R.M. DiLosa, G.F. Burton, M.H. Kosco, L.I. Kupp, A. Masuda, and A.K. Szakal, Germinal centers in the immune response, *Immunol. Rev.* 126:99-112 (1992).

PANEL DISCUSSION ON DENDRITIC CELL NOMENCLATURE

FOLLICULAR DENDRITIC CELLS AND DENDRITIC CELL NOMENCLATURE

John G. Tew

Department of Microbiology and Immunology
Medical College of Virginia/Virginia Commonwealth University
P.O. Box 678 Richmond, Virginia 23298 U.S.A.

In the 10 years since the last nomenclature committee on dendritic cells met (Tew, Thorbecke and Steinman, J. Reticuloendothelial Society 31:371, 1982), considerable evidence has accumulated supporting the concept that Langerhans cells, interdigitating cells, and veiled cells are the same cell in different stages of maturation. It also appears that the follicular dendritic cell (FDC) is quite different from these dendritic cells. In short, it seems to me that we are dealing with two distinct dendritic cell types and that this concept should be clearly stated. I have a suggestion on how this might be done in Table 1. The designation "T cell associated dendritic cell" or "B cell associated dendritic cell" in Table 1 does not mean to imply that these interactions are absolute and exclusive. Rather, it is clear that one type of dendritic cells typically interacts with T cells and the other typically interacts with B cells.

It is desirable to set up simple definitions for cells that are clear and unambiguous. Such definitions can eliminate a host of semantic problems and serve as a "gold standard" for cell identification. T cells and B cells are excellent examples. If a cell has a T cell receptor and rearranged genes for the T cell receptor it is a T cell. Some T cells may lack cell surface determinants like CD4 and CD8, some may be large and have a dendritic morphology, some T cells even express class II and are capable of presenting antigen to other T cells. Nevertheless, these cells are still T cells if they have the T cell receptor and the variable features simply represent heterogeneity within the T cell compartment. Similarly, if a cell has surface immunoglobulin and rearranged immunoglobulin genes it is a B cell. Variations in morphology and surface markers represent heterogeneity.

Unfortunately, it is not as easy to set up such simple definitions for the dendritic cells. However, the nomenclature committee that reviewed dendritic cells 10 years ago tried to define the essential or cardinal features of the cells in a few phrases (Tew, Thorbecke and Steinman, J. Reticuloendothelial Society, 31:371,

1982). I think this is still the most desirable approach. We need to define the cardinal features and acknowledge that, like other cells, considerable heterogeneity exists among dendritic cells. For example, the committee felt that FDC may be distinguished from other cells by three features which are considered diagnostic or cardinal: 1- they are dendritic, 2- they are restricted to lymphoid follicles and 3- they trap and retain immune complexes on their surfaces for long periods of time. These are the features that caught the attention of Szakal and Hanna as well as Nossal and colleagues in 1968 when both groups identified FDC as the cells responsible for retaining antigen. If this does not capture the essential or cardinal features of FDC, other well accepted and non-controversial requirements may be added (Table 2). However, there is heterogeneity among FDC and addition of other features could cause problems rather than solve problems.

Table 1. TWO DISTINCTLY DIFFERENT DENDRITIC CELL TYPES (Distinguishable by the lymphocytes they typically interact with and location in the body)

I. T cell associated dendritic cells (Widely dispersed)
   1. Thymus - Interdigitating cells or IDC
   2. Secondary lymphoid tissue with T cells -- IDC
   3. Blood-- Dendritic cells or veiled cells DC/VC
   4. Afferent lymph -- Veiled cells or VC
   5. Skin -- Langerhans cells or LC.
   6. Connective tissues of organs (e.g. heart and kidney) -- DC

II. B cell associated dendritic cells -- follicular dendritic cells or FDC (Highly localized)
   1. Follicles of all normal secondary lymphoid tissue
   2. Follicles in neoplastic tissues - e.g. follicular lymphoma
   3. Germinal centers in certain sites of chronic inflammation

Table 2. DISTINCTIVE FEATURES OF FDC THAT ARE GENERALLY ACCEPTED

A. Cardinal or Diagnostic Features
   1. Dendritic morphology
   2. Localized to lymphoid follicles (Germinal centers)
   3. Binds and retains immune complexes

B. Non-Controversial Features Accepted by a Number of Laboratories
   1. Lack of phagocytic activity
   2. Euchromatic nucleus (or nuclei) with variable shapes and lobes
   3. Presence of complement receptors
   4. Bears ICAM-1
   5. Clusters with B cells in vitro
   6. Monoclonal antibodies (It is believed that these react with all FDC but most also react with certain other non-FDC cell types)
      a. Human FDC -- DRC-1, Ki-M4, HJ2
      b. Mouse FDC -- FDC-M1
      c. Rat FDC -- MRC OX-2, KiM4R

LANGERHANS CELLS AS OUTPOSTS OF THE DENDRITIC CELL SYSTEM

A.H. Warfel, G.J. Thorbecke and D.V. Belsito

Departments of Dermatology and Pathology
and the Kaplan Cancer Center
New York University School of Medicine
New York, NY 10016, USA

Introduction

The T cell function associated dendritic cell (DC) system consists of various members located in different organs and differing in phenotypic expression, possibly on the basis of different states of "maturation" or as a direct result of microenvironmental influences. These DC, whether located in lymphoid or non-lymphoid tissues, are present only as a very minor cell population. In human blood, less than 0.1% of the white blood cells are DC.[1] In the epidermis, Langerhans cells (LC) represent 3 to 5% of the total population.[2] In spleen cell suspensions, the DC are about 1-1.6% of the cells.[3]

DC turnover and migration

DC arise in the bone marrow (BM) and travel via the blood to lymphoid and nonlymphoid tissues. Splenic DC in young mice have at least a 4 fold higher turnover rate than do macrophages.[4] It has been estimated, on the basis of continuous ^{3}H-thymidine labeling _in vivo_, that the turnover time for the total murine pool of splenic DC is 8-11 days, and that 1.5 - 2.5% are labeled after a single pulse of ^{3}H-thymidine.[4] Scant information is available for murine epidermal LC in this respect. It has been reported that $<0.01\%$ of these cells are labeled after a pulse of ^{3}H-thymidine.[5] However, this information is in direct contradiction to the observations on human FACS-sorted epidermal LC that have been found to be cycling at the rate of 1.3 - 3.3% in S and 1.0 - 2.5% in G2/M phase.[6] Continuous labeling with BuDR of LC in human skin grafts on nu/nu mice results in labeling of all the LC in 16 days.[7]

Another aspect of DC turnover that has been studied is the rate at which donor bone marrow precursor cells enter the recipient DC compartment. Repopulation of the depleted splenic DC in γ-irradiated inbred mice by F_1 hybrid donor cells is approximately 80-90% complete within 15 days.[4] Replacement of the epidermal LC compartment by donor cells appears to be much slower: approximately one third of the LC are of donor origin 1 month after bone marrow transplantation in both humans[8] and mice.[9,10] The combined data on bone marrow repopulation of irradiated recipients suggest that the precursor cells which repopulate the splenic DC and the epidermal LC compartments do so relatively independently. However, it should be realized that recipient

Dendritic Cells in Fundamental and Clinical Immunology,
Edited by Kamperdijk *et al.*, Plenum Press, New York, 1993

epidermal LC are not depleted by irradiation, whereas the DC isolated from spleen in cell suspensions, studied in all these experiments as "splenic DC", are γ-irradiation sensitive.

There may be some specificity in the migration patterns of DC, as observed during the first 24 h after injection of labeled DC, but the available information is still insufficient to allow firm conclusions. Intravenous infusion of epidermal LC in syngeneic mice leads to the preferential localization of these cells to the skin within 16 h after transfer. As much as 20% of injected radioactivity is located in skin, while approximately 7% is present in the spleen.[11] About 8% of splenic DC, injected iv into syngeneic or allogeneic mice, localize in spleen, where they are found mainly in the T-dependent areas of the white pulp. Splenic DC do not appear to home to lymph nodes (LN), although approximately 4% of them are found in skin within 24 h of iv transfer.[12,13] Unfortunately, the methods of skin sample preparation in these studies were apparently different, and the total weight of skin per mouse used in the calculations was not clear. Because of these differences, one cannot without further data assume that DC preferentially localize to the organ of origin. In this regard, it is of interest that when splenic DC are injected subcutaneously, they promptly appear in the draining LN, suggesting that the method of administration of the cells and the time of harvesting the tissues influence the observed fate of DC.

The emigration of DC from their organ location has primarily been studied for LC. On percutaneous sensitization with antigen, LC are known to migrate to other sites. Results from early work showed that following antigenic challenge, antigen (ferritin) containing, Birbeck granule positive LC migrate from the epidermis into the dermis, into lymphatic vessels, and finally into draining LN.[14-16] Moreover, normal animals passively or actively sensitized to 2,4-dinitro-1-chlorobenzene (DNCB), show an increase in the numbers of LC in the dermis and in dermal vessels within 3-6 hr after skin challenge with DNCB.[14,16]

Conclusions from more recent studies employing several different experimental approaches have confirmed and extended these earlier observations.[17-19] DC in the draining lymph nodes of skin painted with FITC increase in number; some are labeled with FITC and possess the characteristic Birbeck granules of LC.[17,20] A positive correlation exists between the numbers of DC present in a particular LN and the proliferative capacity of the cells within that node, thus reflecting the degree of contact sensitivity.[20] Draining LN from mice painted with picryl chloride or DNCB derivatives also show increased numbers of DC with time, and they are capable of initiating primary stimulation of lymphocytes _in vitro_ and of specifically sensitizing recipient mice.[21-23] The potent antigen presenting quality of the cells in the draining LN after skin painting with contact sensitizers[24] was demonstrated to be due to the migratory cells from the skin rather than to resident LN cells. In these experiments employing C3H skin grafted onto BALB/c mice, the H-2 restricted ability of the DC in the draining LN to present antigen to C3H and not to BALB/c T cells clearly showed their origin from the skin graft.[19]

In a recent study, there was a significant increase in the number of DC in the draining LN following antigenic challenge and up to 50% of these DC were found to contain high levels of antigen. Of interest, there also was an increase in total numbers of low density DC in contralateral and distant LN, but less than 3% of those cells carried detectable antigen.[25] Furthermore, some of these DC seen in the contralateral LN contain Birbeck granules.[26] This suggests that "systemic signals" may attract LC from non-sensitized skin areas into the contralateral and distant LN. TNF-α and IL-1 may be among the factors that cause this nonspecific migration of LC during inflammatory responses.[27,28] This effect of cytokines appears invariably associated with an induction of enhanced surface Ia expression on the LC.[27,29,30]

Under normal physiological conditions, this migration of LC from skin to draining LN also occurs in man. This was illustrated by the demonstration of cells expressing both S-100 and CD1 (i.e. epidermal LC-like) in the paracortex of peripheral LN, but not in mesenteric LN or spleen, where interdigitating DC stain for S100 but not for CD1.[31] The reason for this "normal" migration is unclear. The observation that LC continuously process environmental allergens regardless of whether an immune response will subsequently occur[32] suggests that factors induced during antigen processing may stimulate migration. Given the likelihood of frequent cutaneous exposure to potential allergens, it is therefore not unexpected that, even in the absence of overt immunologic responses, trafficking of LC from epidermis to LN is seen.

It is not known whether T cell function associated DC at each distinct anatomical site arise from the same or separate BM precursor cells. Multiple precursor cells could account for the apparent specificity of cell homing. However, a simpler explanation would be that there is a common precursor cell and that the blood DC migrates in a somewhat directed manner into several specific sites, where, under specific microenvironmental conditions, the DC undergo phenotypic changes. Little is known about the phenotypic properties that induce the precursors of LC to migrate in and out of the epidermis and reside there for undefined periods of time. It has been proposed that members of the β1 integrin family expressed on the LC surface, particularly VLA-5 and VLA-6, allow for migration across the basement membrane of the epidermis.[33] These receptors on LC were shown to be important in the adhesion of LC to fibronectin and laminin in vitro.[34]

DC or "veiled" cells have been found in the afferent, but not in the efferent lymphatics of LN.[35] Thus, once LC migrate to a draining LN they appear to remain there, although again little is known about the duration of their stay in LN. It has been shown that severing the afferent lymphatics to LN depletes the LN of macrophages and interdigitating DC, suggesting that the major route of entry of DC into the LN is indeed via the afferent lymphatics.[36,37] These findings do not, however, illuminate the percentage of afferent lymphatic DC that have first resided in the epidermis versus that derived from the blood without an intervening stage as LC.

There is increasing evidence showing that DC can undergo phenotypic changes in tissue culture. When resident LC and cultured LC (cLC) are compared, several distinct features are found to be changed. Some of these changes are apparent within 24-48 hr after LC isolation and culture. There is a striking resemblance between both the functional[39,40] and the phenotypic (Table 1) properties of such cLC and those of DC isolated from lymphoid organs or blood (LDC). Specifically, FcγR, ATPase, CD1a and F4/80, as well as Birbeck granules, disappear from cLC and are absent in LDC. In view of the trypsinization step required in the preparation of LC suspensions, the disappearance of markers from their surface can only be interpreted after a culture period sufficient to allow for re-expression of these markers; such enzymes are generally not used in the preparation of LDC. Expression of class I and class II MHC antigens, ICAM-1, LFA-3, IL-2R, B7 and CD40 are either enhanced or first appear on LC after culture, while all of these are present on LDC (Table 1).

The in vivo changes that LC appear to undergo upon migration to draining LN are similar to those exhibited in vitro. It is not clear whether this is primarily the result of a change in the microenvironment or more related to the difference between the sessile and the mobile state of the LC. This process of change is frequently referred to as "maturation", but there is no real evidence that the mobile state is

Table 1

COMPARISON OF PROPERTIES OF EPIDERMAL LC AND CULTURED LC WITH THOSE OF LYMPHOID DC

Parameter studied	Epidermal LC	Lymphoid DC	Cultured LC
Class I MHC	+ (43)	+ (44)	++ (43)
Class II MHC	+ (45,46)	++ (44)	++ (39)
FcγRII (CDW32)	+ (47,48)	- (44)	- (39)
ATPase	+ (49)	- (50)	- (39)
Birbeck granules	+ (51)	- (50)	- (39)
CDla (T6, human)	+ (52,53)	- (44)	± (54)
F4/80 (mouse)	+ (55)	- (55)	- (39)
CD25 (IL-2R)	- (54,56)	+ (44)	+ (54,56)
CD69 (AIM, human)	+ (57)	?	± (57)
CD54 (ICAM-1)	± (48,54)	++ (44)	++ (54)
CD58 (LFA-3)	+ (48,58)	++ (59,60)	++ (48)
B7 (BB-1)	± (61)	++ (61,62,63)	+ (59)
CD40	± (54)	+ (44)	+ (54)

more mature than the sessile one. The LDC obtained in suspensions of lymphoid tissues probably represent primarily the cells that are mobile rather than the more sessile interdigitating DC of the paracortex, which, particularly with respect to their endosomal compartment, more closely resemble sessile LC.[41,42]

It is of interest to speculate that sessile DC, such as LC in the epidermis, process antigens and thus require different functional properties, particularly with respect to their class II MHC molecule turnover[40], than do DC (LC) that are enroute to present antigen processed elsewhere to T cells. During migration, they must hold onto the processed peptide attached to their Ia, and thus stop renewing their class II MHC. In contrast, expression of adhesion molecules involved in the interaction with T cells, such as ICAM-1, B7 and LFA-3, must be upregulated.

The available evidence indicates that, both following <u>in vitro</u> culture and after <u>in vivo</u> migration out of the epidermal compartment, there is a profound decrease in the expression of FcγRII on LC. This suggests that the epidermal milieu supports the expression of FcγRII on resident LC or that the higher Ig content of plasma in dermis and LN down-regulates the FcγRII. The precise factor(s) responsible for this modulation of FcγRII on LC are unknown. Acute triggering of the FcγR inhibits all cytokine-induced regulation of Class II expression on LC.[64] Therefore, it may be of importance that LC lose FcγR expression as they migrate towards LN compartments. As shown in Fig. 1, upregulation of Ia antigen expression on murine LC in epidermal sheets by either IL-2 or IL-6 is completely abrogated by rabbit or human IgG, but not by human IgA. In additional studies it was shown that Fc but

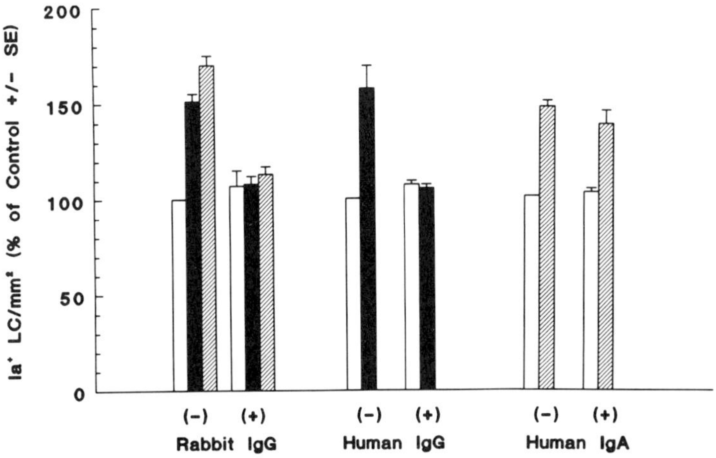

Figure 1. Effect of rabbit or human IgG on the IL-2- and IL-6-induced increases in the number of Ia^+ epidermal LC (from[64]). Skin biopsies were obtained from BALB/c mice. Epidermal sheets were incubated in medium containing IL-2 (250 U/ml, ▬), IL-6 (1000 U/ml, ▨) or medium alone (▭) in the presence (+) or absence (-) of 5 µg/ml of rabbit or human IgG or 5/µg ml of human IgA (control). Sheets were then stained for Ia antigen. Results are expressed as percent of the control specimen incubated in medium alone. Control values (100%) = 695 ± 18 Ia^+ LC/mm^2 of epidermis for rabbit IgG, 626 ± 50 for human IgG, and 686 ± 25 for human IgA (n = 3 to 7 per group).

not Fab fragments of IgG were inhibitory. Thus, loss of FcγR by LC as they migrate towards LN allows for modulation of Class II expression on LC by cytokine production locally within LN, irrespective of IgG or immune complexes present in that environment.

Effect of the epidermal microenvironment on LC

Whether because they attract LC or because they influence LC phenotypically through the microenvironment, keratinocytes may be crucial regulators of typical LC characteristics. LC are seen in all keratinizing epithelia, including keratinizing metaplastic epithelia[65] and thymus.[66-68]. To our knowledge, no reports have appeared in which DC <u>in vitro</u> were exposed to keratinocytes in attempts to influence their phenotypic properties. Once in the epidermis, phenotypic changes occurring locally may reprogram the properties of the cells.

In the epidermis, LC are surrounded by keratinocytes that are capable of producing many cytokines.[69] Most of these cytokines, including IL-1, IL-3, IL-4, IL-6, TNF-α, or GM-CSF, as well as the lymphokines IL-2 or IFN-γ, cause a significant enhancement of Ia expression on LC after <u>in vivo</u> administration[70] or <u>in vitro</u> when epidermal or whole skin sheets are incubated in their presence for 16 h at 37°C.[30] Not all these cytokines are thought to exert a direct effect on the LC, since both <u>in vivo</u> and <u>in vitro</u>, cyclosporin A[30] as well as transforming growth factor-β_1 (TGF-β_1)[70], counteract the effect of some cytokines, such as TNF-α, IL-1 and IFN-γ, but not of others, such as IL-2 and IL-6. This inhibitory effect of cyclosporin A

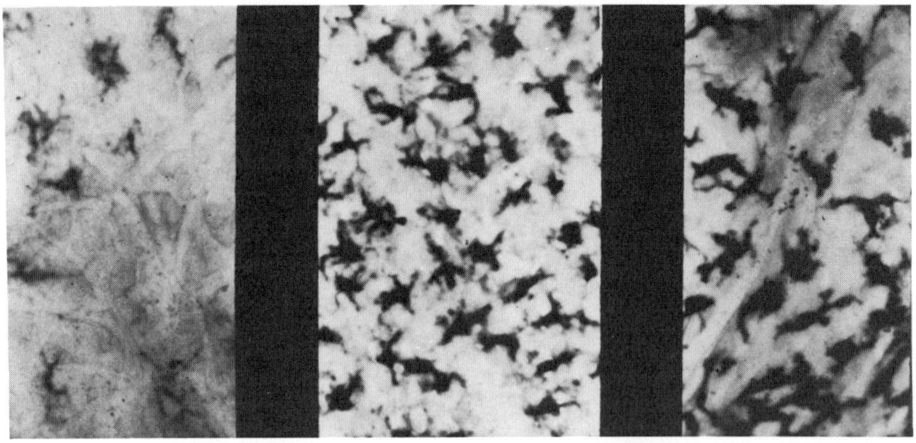

Figure 2. Effect of cytokines on Ia^+ epidermal LC. Epidermal sheets were prepared from skin of 20 mos. old BALB/c mice and incubated (16 h, 37°C) in RPMI-1640 containing 2% fetal calf serum and antibiotics, which was not further supplemented (left panel) or supplemented with 1 U/ml of ultrapure human IL-1 (Genzyme) (center panel) or 1000 U/ml of human rIL-6 (donated by Dr. H. Karasuyama, Tokyo U.) (right panel). Epidermal sheets were then washed and stained for Ia antigen using M5/114 (immunoperoxidase as in[30]). Note the cytokine-induced enhanced expression of Ia antigen on LC as compared to the medium control (X 248).

and/or TGF-β_1 could be due to interference with the synthesis of an intermediary cytokine by either keratinocytes or LC themselves. Thus, although expression of Ia on epidermal LC has been thought to be "constitutive", the possibility that it results from local epidermal cytokine production should also be considered.

In this regard, the effects of ultraviolet B light (UVB) on Ia expression are of interest.[29] Following low dose UVB (60 mJ/cm^2/day x 2 days), there is a profound decrease in the number of Ia^+ LC within the epidermis (720 $\pm$ 19 Ia^+ LC/mm^2 pre-UVB; 225 $\pm$ 59 Ia^+ LC/mm^2 post-UVB). Recovery of Ia^+ LC following exposure to this dose of UVB *in vivo* takes 7 days. However, the number of Ia^+ LC in previously irradiated epidermis can be normalized within 16 h by incubation in medium containing 0.5 U/ml of IL-1 (816 $\pm$ 49 Ia^+ LC/mm^2 in irradiated skin; 910 $\pm$ 64 Ia^+ LC/mm^2 in unirradiated skin). This suggests that, regardless of the mechanism by which UVB down-regulates Ia expression on LC, re-expression of Ia is greatly dependent on cytokines, especially IL-1, in the microenvironment. Thus, the prolonged recovery of Ia following UVB in the absence of exogenous cytokines could be due to the well-described induction of IL-1Ra production in keratinocytes by UVB.[71] Furthermore, it raises the possibility that the normal "constitutive" expression of Ia by epidermal LC relates to the known constitutive production of IL-1 and/or other cytokines by keratinocytes which surround LC in the epidermis.

With aging, there is a decrease in production of IL-1[72] and possibly of other cytokines by keratinocytes. Thus, if expression of Ia is dependent on constitutive cytokine production, an age-related decline in the density of Ia^+ LC would be expected. In fact, there is a profound decrease with age in the density of Ia+ LC in mice, such that by 18-20 mos. of age, the density of Ia+ LC in epidermis of BALB/c mice is 40 to 60% below that for normal young (2-5 mos. old) mice. When epidermis from aged mice is incubated in IL-1 or in several other cytokines, there is a significant upregulation of Ia antigen expression on LC as well as in the number of LC expressing Ia (Fig. 2). However, when compared to similar cytokine treatment of normal young epidermis,

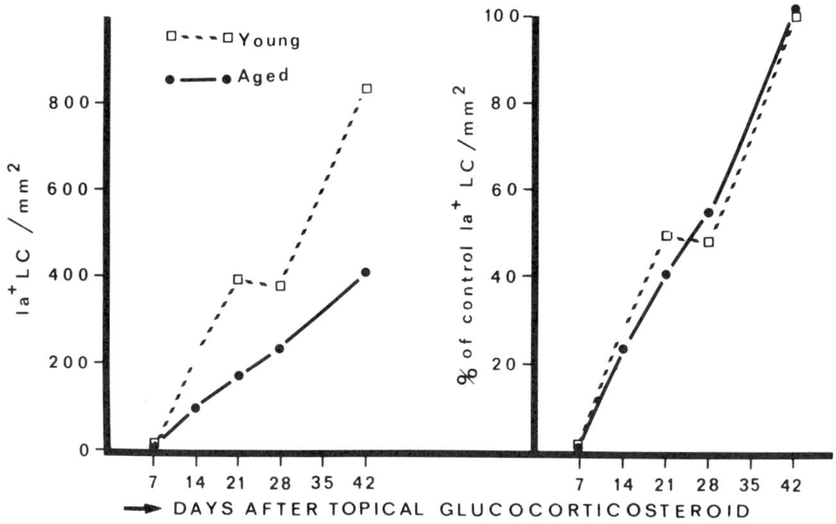

Figure 3. Recovery of Ia+ epidermal LC following glucocorticoid treatment of young (3-5 mos) and aged (18-20 mos) BALB/c mice. Animals were treated daily with 100 µl of 0.001% mometasone furoate in ethanol for 5 days on the dorsal skin. Control mice were similarly treated with ethanol alone. Biopsies of skin were taken prior to treatment and at various times following cessation of topical applications. Epidermal sheets were prepared from biopsy specimens, stained for Ia, and the number of Ia+ LC/mm^2 of epidermis quantitated. Data were expressed as the absolute numbers of repopulating Ia+ LC/mm^2 (left panel) or as the percentages of the pretreatment (baseline) density of Ia+ LC (right panel). Note that the rate of recovery of Ia+ LC to baseline (right panel) is identical in young and old, although like the baseline value, the absolute density obtained is lower in the aged than in the young mice. Treatment with ethanol alone has no effect (n = 3 mice per age group followed longitudinally).

most of these cytokines fail to normalize the density of Ia+ LC in epidermis from aged mice. Indeed, in skin from both young and aged mice, these cytokines as well as 2-mercaptoethanol induce a population of approximately 200-300 Ia− or Ia^{lo} LC/mm^2 of epidermis to express Ia.[30] This suggests that factors other than deficient local epidermal cytokine production contribute to the reduced expression of Ia antigen on LC from aged mice.

Among the possible explanations for a decline in LC density with age could be precursor cell defects. However, studies using topical glucocorticoids, which are cytolytic for LC[73], fail to demonstrate any significant differences in the rate of re-population of Ia^+ LC in the skin of aged and young mice. In these studies, treatment of skin with mometasone furoate 0.001% once daily for 5 days results in near ablation of Ia^+ LC locally without any evidence for a systemic effect.[74] Following cessation of treatment, the rate of recovery of Ia^+ LC in the epidermis of aged and young mice, expressed as percentages of the original numbers present, is identical with return to baseline in approximately 6 weeks (Fig. 3). These data suggest that innate deficiencies in precursor cell numbers are unlikely to be responsible for the decline in Ia^+ LC numbers seen with age.

Another possibility, the validity of which has not yet been investigated, is that the rate of mitotic activity in intra-epidermal LC is lower in aged than in young mice, possibly as a result of a local cytokine defect, such as GM-CSF, a cytokine known to promote DC generation from BM precursors.[75] Such a defect could result in a lower plateau of Ia expressing LC after recovery from glucocorticoid treatment in aged mice (see Fig. 3) in spite of a normal influx of precursor cells.

In conclusion, there are many aspects of the interrelationship between epidermal LC and LDC that require further investigation. If we assume the existence of a common precursor in the blood, then it needs to be established what percentage of these precursors reaches the peripheral LN with an intermediate stage as LC in the skin and what the relationship is to precursors that enter the spleen. It is not known whether there is an exchange of DC between these tissues back and forth: the DC origin and trafficking needs further study. Moreover, the regulatory influences of the microenvironments need to be identified in more detail, and it needs to be determined whether any of these influences should be regarded as causing terminal differentiation with respect to phenotypic and/or functional characteristics of the T cell function associated DC. Additional information is also required concerning the stimuli causing (directed ?) migration of LC and the surface properties involved in facilitating such migration into and out of the epidermis.

Acknowledgements: These studies were supported by USPHS Grant AG-04860 and AR-39749. AHW is the recipient of an NIAID AIDS Fellowship, Training Grant 89-AI-20.

REFERENCES

1. W.C. Van Voorhis, L.S. Hair, R.M. Steinman, and G. Kaplan, Human dendritic cells. Enrichment and characterization from peripheral blood, J. Exp. Med. 155:1172 (1982).
2. G. Stingl, S.I. Katz, E.M. Shevach, E. Wolff-Schreiner, and I. Green, Detection of Ia antigens on Langerhans cells in guinea pig skin, J. Immunol. 120:570 (1978).
3. R.M. Steinman and Z.A. Cohn, Identification of a novel cell type in peripheral lymphoid organs of mice I. Morphology, quantitation, tissue distribution, J. Exp. Med. 137:1142 (1973).
4. R.M. Steinman, D.S. Lustig, and Z.A. Cohn, Identification of a novel cell type in peripheral lymphoid organs of mice. III. Functional properties in vivo, J. Exp. Med. 139:1431 (1974).
5. I.C. Mackenzie, Labelling of murine epidermal Langerhans cells with H^3-Thymidine, Am. J. Anat. 144:127 (1975).
6. J.M. Czernielewski, P. Vaigot, and M. Prunieras, Epidermal Langerhans cells - a cycling cell population, J. Invest. Dermatol. 84:424 (1985).
7. J.M. Czernielewski and M. Dermarchez, Further evidence for the

self-reproducing capacity of Langerhans cells in human skin, J. Invest. Dermatol. 88:17 (1987).
8. M. Pelletier, C. Perreault, D. Landry, M. David, and S. Montplaisir, Ontogeny of human epidermal Langerhans cells, Transplantation 38:544 (1984).
9. S.I. Katz, K. Tamaki, and D.H. Sachs, Epidermal Langerhans cells are derived from cells originating in bone marrow, Nature 282:324 (1979).
10. K. Tamaki, and S.I. Katz, Ontogeny of Langerhans cells, J. Invest. Dermatol. 75:12 (1980).
11. P.D. Cruz, Jr., R.E. Tigelaar, and P.R. Bergstresser, Langerhans cells that migrate to skin after intravenous infusion regulate the induction of contact hypersensitivity, J. Immunol. 144:2486 (1990).
12. J.W. Kupiec-Weglinski, J.M. Austyn, and P.J. Morris, Migration patterns of dendritic cells in the mouse. Traffic from the blood, and T cell-dependent and -independent entry to lymphoid tissues, J. Exp. Med. 167:632 (1988).
13. J.M. Austyn, J.W. Kupiec-Weglinski, D.F. Hankins, and P.J. Morris, Migration patterns of dendritic cells in the mouse. Homing to T Cell-dependent areas of spleen, and binding within marginal zone, J. Exp. Med. 167:646 (1988).
14. I. Silberberg, R.L. Baer, S.A. Rosenthal, G.J. Thorbecke, and V. Berezowsky, Dermal and intravascular Langerhans cells at sites of passively induced allergic contact sensitivity, Cellular Immunol. 18:435 (1975).
15. I. Silberberg-Sinakin, G.J. Thorbecke, R.L. Baer, S.A. Rosenthal, and V. Berezowsky, Antigen-bearing Langerhans cells in skin, dermal lymphatics and in lymph nodes, Cellular Immunol. 25:137. (1976).
16. I. Silberberg-Sinakin, I. Gigli, R.L. Baer, and G.J. Thorbecke, Langerhans cells: role in contact hypersensitivity and relationship to lymphoid dendritic cells and to macrophages, Immunological Rev. 53:203 (1980).
17. S.E. Macatonia, S.C. Knight, A.J. Edwards, S. Griffiths, and P. Fryer, Localization of antigen on lymph node dendritic cells after exposure to the contact sensitizer fluorescein isothiocyanate, J. Exp. Med. 166:1654 (1987).
18. C.P. Larsen, R.M. Steinman, M. Witmer-Pack, D.F. Hankins, P.J. Morris, and J.M. Austyn, Migration and maturation of Langerhans cells in skin transplants and explants, J. Exp. Med. 172:1483 (1990).
19. M.L. Kripke, C.G. Munn, A. Jeevan, J-M. Tang, and C. Bucana, Evidence that cutaneous antigen-presenting cells migrate to regional lymph nodes during contact sensitization, J. Immunol. 145:2833 (1990).
20. I. Kimber, A. Kinnaird, S.W. Peters, and J.A. Mitchell, Correlation between lymphocyte proliferative responses and dendritic cell migration in regional lymph nodes following skin painting with contact-sensitizing agents, Int. Arch. Allergy. Appl. Immunol. 93:47 (1990).
21. S.C. Knight, J. Krejci, M. Malkovsky, V. Colizzi, A. Gautam, and G.L. Asherson, The role of dendritic cells in the initiation of immune responses to contact sensitizers. I. In Vivo exposure to antigen, Cellular Immunol. 94:427 (1985).
22. S.C. Knight, P. Bedford, and R. Hunt, The role of dendritic cells in the initiation of immune responses to contact sensitizers. II. Studies in nude mice, Cellular Immunol. 94:435 (1985).
23. A. Kinnaird, S.W. Peters, J.R. Foster, and I. Kimber, Dendritic cell accumulation in draining lymph nodes during the induction phase of contact allergy in mice, Int. Arch. Allergy Appl. Immunol. 89:202 (1989).

24. I. Silberberg, R.L. Baer, and S.A. Rosenthal, The role of Langerhans cells in allergic contact hypersensitivity. A review of findings in man and guinea pigs, J. Invest. Dermatol. 66:210 (1976).
25. S. Hill, A.J. Edwards, I. Kimber, and S.C. Knight, Systemic migration of dendritic cells during contact sensitization, Immunology 71:277 (1990).
26. S. Hill, S. Griffiths, I. Kimber, and S.C. Knight, Migration of dendritic cells during contact sensitization, Abst. 2nd Intl. Symp. on Dendritic cells in fundamental and clinical Immunology, Amsterdam, pg. 74 (1992).
27. E.N. Lundqvist and O. Back, Interleukin-1 decreases the number of Ia^+ epidermal dendritic cells but increases their expression of Ia antigen, Act. Derm. Venerol. (Stockh) 70:391 (1990).
28. M. Cumberbatch and I. Kimber, Dermal tumour necrosis factor-α induces dendritic cell migration to draining lymph nodes, and possibly provides one stimulus for Langerhans' cell migration, Immunology 75:257 (1992).
29. D.V.Belsito, J.M. Schultz, R.L. Baer, and G.J. Thorbecke, Effects of various cytokines on Ia antigen expression by Langerhans cells from normal, ultraviolet B-irradiated and glucocorticoid- treated young mice and normal aged mice, in: "The Langerhans Cell", J. Thivolet & D. Schmitt, eds., John Libbey Eurotext Ltd., London, 172:241 (1988).
30. D.V. Belsito, S.P. Epstein, J.M. Schultz, R.L. Baer, and G.J. Thorbecke, G.J., Enhancement by various cytokines or 2-ß-mercaptoethanol of Ia antigen expression on Langerhans cells in skin from normal aged and young mice. Effect of Cyclosporine A., J. Immunol. 143:1530 (1989).
31. M. Shamoto, S. Hosokawa, M. Shinzato, C. Kaneko, Comparison of Langerhans cells and interdigitating reticulum cells, Abst. 2nd Intl. Symp. on Dendritic cells in fundamental and clinical Immunology, Amsterdam, pg. 68 (1992).
32. W. Sterry, N. Kuhne, K. Weber-Matthiesen, J. Brasch, and V. Mielke, Cell trafficking in positive and negative patch-test reactions: demonstration of a stereotypic migration pathway, J. Invest. Dermatol. 96:459 (1991).
33. B. Le Varlet, C. Dezutter-Dambuyant, M.J. Staquet, P. Delorme, and D. Schmitt, Human epidermal Langerhans cells express integrins of the ß1 subfamily, J. Invest. Dermatol. 96:518.(1991).
34. B. Le Varlet, M.J. Staquet, C. Dezutter-Dambuyant, P. Delorme, and D. Schmitt, In vitro adhesion of human epidermal Langerhans cells to laminin and fibronectin occurs through ß1 integrin receptors, J. Leuk. Biol. 51:415 (1992).
35. R.H. Kelly, B.M. Balfour, J.A. Armstrong, and S. Griffiths, Functional anatomy of lymph nodes. II. Peripheral lymph-borne mononuclear cells. Anat. Rec. 190:5 (1978).
36. E.W.A. Kamperdijk, and E.C.M. Hoefsmit, Birbeck granules in lymph node macrophages, Ultramicroscopy 3:137 (1978).
37. E.C.M. Hoefsmit, B.M. Balfour, E.W.A. Kamperdijk, and J. Cvetanov, Cells containing Birbeck granules in the lymph and the lymph node. Z. Immunol. Forsch. 154:321 (1978).
39. G. Schuler, and R.M. Steinman, Murine epidermal Langerhans cells mature into potent immunostimulatory dendritic cells in vitro, J. Exp. Med. 161:526 (1985).
40. E. Pure, K. Inaba, M.T. Crowley, L. Tardelli, M.D. Witmer-Pack, G. Ruberti, G. Fathman, and R.M. Steinman, Antigen processing by epidermal Langerhans cells correlates with the level of biosynthesis of major histocompatibility complex Class II molecules and expression of invariant chain, J. Exp. Med. 172:1459. (1990).
41. J.M.S. Arkema, I.L. Schadee-Eestermans, D.M. Broekhuis-Fluitsma, and

E.C.M. Hoefsmit, Immunocytochemical chacterization of dendritic cells, Abst. 2nd Intl. Symp. on Dendritic cells in fundamental and clinical Immunology, Amsterdam, pg. 4 (1992).
42. T. Maruyama, E.C.M. Hoefsmit, and G. Kraal, A monoclonal antibody (MIDC-8) recognizes an epitope of a dendritic cell specfic endosomal determinant involved in presentation of protein antigens, Abst. 2nd Intl. Symp. on Dendritic cells in fundamental and clinical Immunology, Amsterdam, pg. 7 (1992).
43. M.D. Witmer-Pack, J. Valinsky, W. Olivier, and R.M. Steinman, Quantitation of surface antigens on cultured murine epidermal Langerhans cells: Rapid and selective increase in the level of surface MHC products, J. Invest. Dermatol. 90:387 (1988).
44. R.M. Steinman, The dendritic cell system and its role in immunogenicity, Annu. Rev. Immunol. 9:271 (1991).
45. G. Rowden, M.G. Lewis, and A.K. Sullivan, Ia antigen expression on human epidermal Langerhans cells, Nature 268:247 (1977).
46. L. Klareskog, U.M. Tjernlund, U. Forsum, and P.A. Peterson, Epidermal Langerhans cells express Ia antigens, Nature 268:248 (1977).
47. G. Stingl, E.C. Wolff-Schreiner, W.J. Pichler, F.G. Gschnait, W. Knapp, and K. Wolfe, Epidermal Langerhans cells bear Fc and C3 receptors, Nature 268:245 (1977).
48. G. Stingl, Dendritic cells of the skin, Dermatologic Clinics 8:673 (1990).
49. R.H. Cormane and G.L. Kalsbeek, ATP-Hydrolyzing enzymes in normal human skin, Dermatologica 127:381 (1963).
50. R.M. Steinman, and M.C. Nussenzweig, Dendritic cells: Features and functions, Immunological Rev. 53:127 (1980).
51. M.S. Birbeck, A.S. Breathnach, and J.D. Everall, An electron microscope study of basal melanocytes and high-level clear cells (Langerhans cells) in vitiligo, J. Invest. Dermatol. 37:51 (1961).
52. E. Fithian, P. Kung, G. Goldstein, M. Rubenfeld, C. Fenoglio, and R. Edelson, Reactivity of Langerhans cells with hybridoma antibody, Proc. Natl. Acad. Sci. USA 78:2541 (1981).
53. G.F. Murphy, A.K. Bhan, S. Sato, T.J. Harrist, and M.C. Mihm, Jr. Characterization of Langerhans cells by the use of monoclonal antibodies, Laboratory Invest. 45:465 (1981).
54. N. Romani, A. Lenz, H. Glassel, H. Stossel, U. Stanzl, O. Majdic, P. Fritsch, and G. Schuler, Cultured human Langerhans cells resemble lymphoid dendritic cells in phenotype and function, J. Invest. Dermatol. 93:600. (1989).
55. D.A. Hume, A.P. Robinson, G.G. Macpherson, and S. Gordon, The mononuclear phagocyte system of the mouse defined by immunohistochemical localization of antigen F4/80. Relationship between macrophages, Langerhans cells, reticular cells, and dendritic cells in lymphoid and hematopoietic organs, J. Exp. Med. 158:1522 (1983).
56. G. Steiner, E. Tschachler, M. Tani, T.R. Malek, E.M. Shevach, W. Holter, W. Knapp, K. Wolff, and G. Stingl, Interleukin 2 receptors on cultured murine epidermal Langerhans cells, J. Immunol. 137:155. (1986).
57. T. Bieber, A. Rieger, G. Stingl, E. Sander, P. Wanek, and I. Strobel, CD69, An early activation antigen on lymphocytes is constitutively expressed by human epidermal Langerhans cells, J. Invest. Dermatol. 98:771 (1992).
58. G. De Panfilis, G.C. Manara, C. Ferrari, and C. Torresani, Adhesion molecules on the plasma membrane of epidermal cells. III. Keratinocytes and Langerhans cells constitutively express the lymphocytes function - associated antigen 3, J. Invest. Dermatol. 96:512 (1991).

59. W. Egner, T.C.R. Prickett, and D.N.J. Hart, Dendritic cells, monocytes and macrophages: A comparison of adhesion molecule repertoire, Abst. 2nd Intl. Symp. on Dendritic cells in fundamental and clinical Immunology, Amsterdam, pg. 15 (1992).
60. U. O'Doherty, W.J. Swiggard, I. Kopeloff, Y. Yamaguchi, N. Bhardwaj, R.M. Steinman, Anti-human dendritic cell mAB production: Tolerizing mice to human leukocytes neonatally and with cyclophosphamide, Abst. 2nd Intl. Symp. on Dendritic cells in fundamental and clinical Immunology, Amsterdam, pg. 40. (1992)
61. A.A. Gaspari, personal communication (1992).
62. S.D. Norton, K.B. Urdahl, L. Zuckerman, R. Shefner, J. Miller, and Jenkins, M.K., Costimulating factors and signals relevant for antigen presenting cell function, Abst. 2nd Intl. Symp. on Dendritic cells in fundamental and clinical Immunology, Amsterdam, pg. 18 (1992).
63. M.K. Jenkins, personal communication (1992).
64. S.P. Epstein, R.L. Baer, and D.V. Belsito, Effect of triggering epidermal Fcγ receptors on the Interleukin-2- and Interleukin-6-induced upregulation of Ia antigen expression by murine epidermal Langerhans cells: The role of prostaglandins and cAMP, J. Invest. Dermatol. 97:461 (1991).
65. S. Hosokawa, M. Shinzato, C. Kaneko, and M. Shamoto, Studies on Langerhans cells in the tracheal squamous metaplasia of vitamin A deficient rats, Abst. 2nd Intl. Symp. on Dendritic cells in fundamental and clinical Immunology, Amsterdam, pg. 133. (1992).
66. E.C.M. Hoefsmit, and J.A.M. Gerver, Epithelial cells and macrophages in the normal thymus, in: "Biological Activity of Thymic Hormones", D.W. van Bekkum and A.M. Kruisbeck, eds., Kooyer, Rotterdam, pp. 63-68 (1975).
67. U.J.G. Van Haelst, Light and electron microscopic study of the normal and pathological thymus of the rat. III. A mesenchymal histiocytic type of cell, Z. Zellforsch. Mikrosk. Anat. 99:198 (1969).
68. I. Olah, C. Dunay, P. Rohlich, and I. Toro, A special type of cell, in the medulla of the rat thymus, Acta Biol. Hung 19:97 (1968).
69. T.A. Luger and T. Schwarz, Evidence for an epidermal cytokine network, J. Invest. Dermatol. 95:100S (1990).
70. S.P. Epstein, R.L. Baer, G.J. Thorbecke, and D.V. Belsito, Immunosuppressive effects of transforming growth factor ß: Inhibition of the induction of Ia antigen on Langerhans cells by cytokines and of the contact hypersensitivity response, J. Invest. Dermatol. 96:832 (1991).
71. T. Scwarz, A. Urbanska, F. Gschnait, and T.A. Luger, UV-irradiated epidermal cells produce a specific inhibitor of interleukin 1 activity, J. Immunol. 138:1457 (1987).
72. D.N. Sauder, Effect of age on epidermal immune function, Dermatologic Clinics 4:447. (1986).
73. W. Aberer, N. Romani, A. Elbe, and G. Stingl, Effects of physico-chemical agents on murine epidermal Langerhans cells and Thy-1 positive dendritic epidermal cells, J. Immunol. 136:1210 (1986).
74. D.V. Belsito, R.L. Baer, J.M. Schultz, and G.J. Thorbecke, Relative lack of systemic effects of mometasone furoate on Langerhans cells of mice after topical administration as compared with other glucocorticosteroids, J. Invest. Dermatol. 91:219 (1988).
75. W.E. Bowers, Dendritic cell differentiation in the bone marrow, Abst. 2nd Intl. Symp., Amsterdam-the Netherlands, pg. 115 (1992).

HETEROGENEITY OF DENDRITIC CELLS AND NOMENCLATURE

Elisabeth C.M. Hoefsmit, Joanne M.S. Arkema, Michiel G.H. Betjes, Carin E.G. Havenith, Ellen van Vugt, Robert H.J. Beelen, and Eduard W.A. Kamperdijk

Department of Cell Biology, Medical Faculty
Vrije Universiteit
NL-1081 BT Amsterdam

INTRODUCTION

The Dendritic Cell System (DCS) comprises Dendritic Cells (DC) isolated from lymphoid organs and related in vivo equivalents such as interdigitating cells (IDC), which are present in thymus dependent areas of lymphoid tissues, and in the medulla of the thymus. However relatives of DC have also been isolated from other tissues and compartments such as epidermal Langerhans cells (LC), and all these cells share the functional capacity to present antigen to T cells.

DENDRITIC CELLS

Dendritic cells (DC) have originally been isolated form mouse lymphoid organs, such as spleen, lymph nodes, and thymus[1]. DC, a minor population in crude cell suspensions, are separated from lymphocytes by transient glass adherence and density gradient centrifugation, and from macrophages by Fc-rozetting[2]. The cells are non-phagocytic, express MHC class II molecules[3], and actively stimulate the MLR[4]. Rat DC do not adhere to glass or plastics[5] and thus the purification of rat DC was modified by irradiation and a culture step to remove most of the lymphocytes[6]. In that study isolated DC were characterized, by dendritic morphology, expression of MHC class II molecules and acid phosphatase activity in a central spot in the nuclear concavity. The superior capacity of these cells to present soluble antigen to sensitized T lymphocytes compared to macrophages was confirmed[7].

Thymic DC

In crude cell suspensions of rat thymus, prepared by incubation of small tissue fragments to allow mobile cells to leave the epithelial reticulum, and by pressing the remaining fragments through a nylon gauze, different types of non-lymphoid, mononuclear cells could be identified by LM and EM (immuno-) cytochemistry[8]: Type I cells were monocyte-like. Many of them expressed MHC class II molecules, but only a few had myeloperoxidase granules. Acid phosphatase activity was present in a few scattered granules. Many of type II cells had Birbeck granules but all of them had acid phosphate activity in a central spot and expressed MHC class II molecules or Ia antigen on the plasma membrane. They frequently had phagolysosomes, which contained basophilic, presumably nuclear debris. Type III cells had acid phosphatase activity throughout the cytoplasm in cytocentrifuge preparations. They did not express Ia antigen and 50 % of them had endogenous peroxidatic activity in the endoplasmic reticulum (ER) like resident macrophages. Birbeck granules, characterizing epidermal LC were never seen in type III cells. These results suggested that type II cells represent the members of the DCS from the medulla, and type III cells the cortical macrophages, whereas type I cells were considered as precursors of type II and/or type III cells. Comparison of nucleo/cytoplasmic ratio and buoyant density on continuous Percoll gradients confirmed that the monocyte like type I cells were the relatively dense cells, whereas both type II and III cells were present in all the fractions of lower density. The volume of the cells increased with decreasing buoyant density, but no transitional forms between type II and III cells were present (E.H. unpublished results). As it is generally accepted that both macrophages and DC are bone marrow derived, the results suggest that the relatively dense type I cells are the precursors of both type II and type III cells and that the decreasing nucleo/cytoplasmic ratio and decreasing buoyant density reflect the lifetime or maturation of the cells.

In contrast to type II cells, DC enriched from crude rat thymic celsuspensions by density centrifugation, irradiation, and depletion of Fc-rozetting cells, compose relatively homogenous populations of cells. They were all characterized by expression of MHC class II molecules on the plasmamembrane and acid phosphatase activity in a spot near the nucleus. However the phagocytic activity of these cells was lower in comparison to the Type II cell in the crude suspensions. In addition these cells have the capacity to present antigen to T cells of the same magnitude as DC isolated from spleen[9]. We suppose that isolated thymic DC represent the low-density and FcR negative subpopulation of the type II cells from the crude thymic cell suspensions.

Lymph node DC

DC isolated from lymph nodes constitute homogenous populations too, irrespective the nature and the stage of the immune response in the node[10]. Only the numbers of DC that could be obtained were greater during the first 24 hours after antigen stimulation, than in the memory phase i.e. 2 weeks after stimulation. However during the induction phase of the immune response some DC contained Birbeck granules (BG), the characteristic cell organelles of the epidermal Langerhans cells[11]. All cells expressed MHC class II molecules and demonstrated acid phosphatase activity in a central spot near the nucleus. At the ultrastructural level the DC had an irregular outline, resembling veiled cells from skin draining lymph. However they never had phagocytosed recognizable extracellular material[10].

Peritoneal DC

DC appeared also to be present in the peritoneal cavity of man[12] and rat[13]. In male rats in a steady state about 1 % of the peritoneal cells are DC and have the same morphological and immunocytochemical characteristics as DC isolated from spleen and lymph nodes. These percentages of DC have also been isolated from other non-lymphoid tissues and seem to represent a steady state population of surveyors of the immune system.

DC in exudates and infiltrates

Remarkably increased numbers of DC could be detected in an acute[14] and a chronical[15] inflammatory peritoneal exudate, if estimated on the same criteria of morphology, acid phosphatase cytochemistry, expression of MHC class II molecules and function. In the acute inflammatory exudate the percentages of DC had increased from 1.0 to 3.3 and the absolute numbers with a factor 10, whereas the percentages of macrophages decreased[14,15]. In the chronical exudate the absolute numbers of DC increased to the same extent, but the percentages increased with a factor 2, whereas the percentages of macrophages remained the same[15]. The DC had the same phenotype and function as DC isolated from lymphoid organs. However, if the macrophages were added to the purified DC in the antigen presentation assay, they suppressed the T cell proliferation[14]. In the human system, the resident peritoneal macrophages harvested during laparoscopy have a low stimulatory capacity whereas the exudate macrophages, which form the main population of cells in continuous ambulatory peritoneal dialysis effluents[16], stimulate the allogenic MLR better than the monocytes of the same patients, after overnight culture[12].

DC could also be detected in normal bronchoalveolar lavages in the rat[17]. Two days after intratracheal instillation of BCG, DC increased about a factor 25 in the bronchoalveolar lavages. The alveolar macrophages had increased only less than three times. It was shown in *in vitro* antigen presentation studies, that DC represent the antigen presentation capacity, whereas alveolar macrophages suppressed the response in a concentration dependent manner [17].

These results suggest that DC not only function as surveyors of the immune system in a steady state but that they increase both in percentages and in absolute numbers during inflammatory reactions. Qualitatively they represent the most potent antigen presenting cells, whereas macrophages vary in their antigen presenting capacity, and may even suppress the response.

DC from human blood

Finally DC were also isolated from human blood, by an overnight adherence stap, density gradient centrifugation and elimination of Fc-receptor bearing cells by IgG panning. The antigen presenting capacity was tested in a MLR[18]. In immuno-EM studies it appeared that DC had not only MHC class II molecules on the plasma membrane, but also a prominent cluster of MHC class II positive vesicles near the nucleus[18]. To evaluate the intracellular location of lysosomal activity and MHC class II molecules immuno-EM was combined with ultrastructural cytochemistry. DC had as well vesicles with membrane bound MHC class II molecules, as vesicles demonstrating acid phosphatase activity. Both types of vesicles had a location near the nucleus. Only a few vesicles had both MHC class II molecules and acid phosphatase activity[19] or both MHC class II molecules and lysosome associated membrane proteins (Lamp-1)[20]. However the majority of the MHC class II labelled vesicles are probably storage vesicles[19]. In these studies it was also shown that DC differ from monocytes: Monocytes, which still contaminate the DC enriched cell suspensions, are identified by the presence of myeloperoxidase containing vesicles. They did not contain MHC class II labelled storage granules, and they did not concentrate their lysosomal vesicles near the centrioles[21]. Moreover, several monoclonal antibodies that are known to be relatively monospecific for monocytes and/or macrophages, probably because they recognize epitopes on macrophage lysosomal constituents, label DC too, but in a juxtanuclear position, such as EBM11 (CD68) in the human [22], ED1 in the rat, and MOMA-2 in the mouse [23]. These observations together suggest that the cocompartmentalization of the MHC class II storage vesicles and lysosomes in the central area of the cell reflects the specialized function of DC in antigen processing, in comparison to monocytes and macrophages.

INTERDIGITATING CELLS (IDC)

IDC originally described by Jan Veldman in the inner cortex of the lymph node, were classified as a specialized type of macrophage. They were identified on pure morphological grounds[24] in the medulla of the thymus and the in T cell dependent areas of peripheral lymphoid organs, . Most surprising was the observation, that the population of IDC changed during the different stages of the immune response[10], from monocyte like cells, small veiled cells, small IDC containing only a few small lysosomes near the nucleus, up to huge IDC, with many big phagolysosomes. These phagolysosomes contained the same particulate antigen that induced the immune response, in addition to necrotic lymphocytes. It was clear that *in vivo* IDC have the capacity to phagocytose to the same extend as macrophages. However IDC concentrate the lysosomes around the centrioles and the Golgi stacks; the smaller ones near the central area and the bigger ones more in a peripheral location. On the contrary, macrophages, like tingible body macrophages have a nucleus in the centrum of the cell and phagolysosomes containing necrotic lymphocytes throughout the cytoplasm[25]. The sequence of different IDC from small and monocyte-like cells, up to huge phagocytic IDC of decreasing nucleo/cytoplasmic index, appeared to reflect the sequence of non-lymphoid mononuclear cells that can be obtained from the afferent lymphatics, which drain the infiltrate from the injection site of the antigen at subsequent time points after injection[24]. This sequence of different IDC represents the different developmental stages in the life history of the IDC. Besides the obvious phagocytic capacity of the more developed IDC *in vivo*, nothing is known about their function, simply because they have not yet been isolated.

EPIDERMAL LANGERHANS CELLS (LC)

The evidence that LC belong to the DCS is not merely the morphological similarity of LC and IDC, or the presence of Birbeck granules in IDC of skin draining lymph nodes during the first day after antigen stimulation[26]. Studies of the kinetics of DC and LC (see review[27]), and the results of comparative studies of the *in vitro* function of isolated LC and DC[28] demonstrated that *in vitro* LC may loose the capacity to process protein antigens, but become potent stimulators of resting T cells[28].

We have recently found that in contrast to blood DC LC *in situ* contain prelysosomes, which could intensely be labeled for MHC class II molecules[20]. These prelysosomes demonstrate endocytotic activity of LC *in situ*. The presence of membrane bound MHC class II molecules in this compartment might indicate that these prelysosomes play a role in the processing of protein antigens to immunogenic complexes of peptides and MHC class II molecules.

To study the relation between LC and IDC, the non-lymphoid mononuclear cells in the afferent lymph from a subcutaneous depot of diphtheria toxoid were studied, at different time points after injection. A complete sequence of developmental stages of the cellular infiltrate passed. Monocyte like cells, veiled cells demonstrating macropinocytosis, and veiled cells demonstrating phagocytosis of cellular debris with and without Birbeck granules, including macrophages[24] were observed. Following these cells through the lymph node, it was observed that the small and motile veiled cells penetrated into the paracortex or T cell dependent area, and transform in situ into fully developed IDC, whereas hugh macrophages form a thick layer under the subcapsular sinus and populate the medullary sinus[24]. Assuming that both LC and DC are derived from bone marrow precursors, both LC and IDC seem relatively mature: The nucleo/cytoplasmic ratio is small, the cells extend long processus between the surrounding cells, and the cytoplasm comprises many cell organelles, including endosomes and lysosomes.

It was also shown, that the successive distribution of veiled cells and macrophages in the parathymic lymph node is the same as in skin draining nodes, although Birbeck granule containing cells were neither observed in the peritoneal cavity nor in the parathymic node[29]. These results suggested, that during an immune response in skin draining nodes, in addition to relative mature, epidermal LC, the cells from the inflammatory infiltrate are the IDC precursors[30], including those, which may have got Birbeck granules in the epidermis.

LC are also present in the respiratory epithelium of the human upper airways[31,32]. These cells were identified in situ as LC, by the expression of CD1a, and the possession of Birbeck granules. Some of these LC were only very small, had a relatively large nucleus, lacked the long cytoplasmic extensions and had only a very few cell organelles. We suggest that these small cells have newly arrived in the epithelium, and may represent intermediates between precursors and fully developed LC. The function of these small cells is not yet estimated. The increased numbers of LC in the mucosa of the upper airways may play a role in allergic reactions[33]

DISCUSSION AND CONCLUSIONS

DC are primarily identified in vitro. They form a relatively homogeneous population of cells on the base of morphology, cytochemistry, expression of MHC class II molecules and function. They are closely related to the macrophages. They are negatively selected from macrophages by adherence and Fc-receptor binding. They are present in increased numbers in inflammatory exudates and infiltrates, and play the most important role in T cell proliferation and differentiation.

In situ many different phenotypes of DC relatives are present. During the immune response different developmental stages of IDC are observed in the lymph node. They ranged from small monocyte like cells from the early inflammatory infiltrate to huge, phagocytic IDC in the memory phase of the response. Thymus IDC *in situ* usually demonstrate phagocytic activity. Isolated DC represent the non-phagocytic subpopulation of IDC. The function of phagocytic IDC is not known.

Epidermal LC are identified *in situ* and *in vitro*. The function of freshly isolated cells is not stable. The cells loose their Birbeck granules and acid organelles[34] demonstrating loss of endocytotic activity. The presence of MHC class II molecules in the late endosomes in LC *in situ*[20] demonstrate the endocytotic activity and suggest an antigen processing capacity. The LC precursor is not yet identified. As LC are presumably bone marrow derived, blood DC may be a candidate precursor. However blood derived DC differ from LC by the presence of MHC class II containing storage granules and the absence of a well developed endosomal system. It may well be that DC can modulate their function of antigen processing and presentation by regulation of both endocytotic activity and intracellular location of MHC class II molecules.

The small cell with a very few cell organelles and a single Birbeck granule that have been described in human respiratory epithelium, might be another candidate LC precursor.

It is clear that the DCS in vivo is a complex system, comprising many different phenotypes. As the fenotype, the kinetics, and functional capacity of the precursor cell and subsequent developmental stages in relation to the microenviromental conditions are not yet known, we should first collect these relevant data, before we change the nomenclature.

REFERENCES

1. R.M. Steinman and Z.A. Cohn, Identification of a novel cell type in peripheral lymphoid organs of mice. I. Morphology, quantisation, and tissue distribution, *J. Exp. Med.* 137:1142 (1973).
2. M. Crowley, K. Inaba, M. Witmer-Pack, and R.M. Steinman, The cell surface of mouse dendritic cells: FACS analyses of dendritic cells from different tissues including thymus, *Cell Immunol.* 118:108 (1989).
3. R.M. Steinman, and M.C. Nussenzweig, Dendritic cells: Features and Functions, *Immunol. Rev.* 53:127 (1980).
4. R.M. Steinman, and K. Inaba, Stimulation of the primary mixed lymphocyte reaction, *CRC Crit. Rev. Immunol.* 5:331 (1985).
5. W.E.F. Klinkert, J.H. LaBadie, J.P. O'Brien, C.F. Beyer and W.E. Bowers, Rat dendritic cells function as accessory cells and control the production of a soluble factor required for mitogenic responses of T lymphocytes, *Proc. Natl. Acad. Sci. USA,* 77:5414 (1980).
6. E.W.A. Kamperdijk, M.L. Kapsenberg, M. van den Berg, and E.C.M. Hoefsmit, Characterization of dendritic cells, isolated from normal and stimulated lymph nodes of the rat, *Cell Tissue Res.*,242:469 (1985).
7. E.W.A.Kamperdijk, M.A.M. Verdaasdonk, R.H.J. Beelen, Visual and functional expression of Ia antigen on macrophages and dendritic cells in ACI/Mal rats, *Transplant. Proc.* 19:3024 (1987).
8. A.M. Duijvestijn, R. Schutte, Y.G. Köhler, C. Korn, and E.C.M. Hoefsmit, Characterization of the population of phagocytic cells in thymic cell suspensions, *Cell Tissue Res.*, 231:313 (1983).
9. E.W.A. Kamperdijk, Joanne M.S. Arkema, Marina A.M. Verdaasdonk, Robert H.J. Beelen, and Ellen van Vugt, Rat thymic dendritic cells, *This Symposium.*
10. E.W.A. Kamperdijk, E.B.J. van Nieuwkerk, M.A.M. Verdaasdonk, and E.C.M. Hoefsmit, Macrophages in different compartments of the non-neoplastic lymph node, *in:* "Current Topics in Pathology, Vol. 84/1 Reaction Patterns of the Lymph Node Part 1 Cell Types and Functions, E. Grundmann and E. Vollmer, eds, Springer-Verlag, Berlin Heidelberg (1990) p 219.
11. E.C.M. Hoefsmit, B.M. Balfour, E.W.A. Kamperdijk and J Cvetanov, Cells containing Birbeck granules in the lymph and the lymph node, *Adv. Exp. Med. Biol.* 114:389 (1979).
12. M.G.H. Betjes, C.W. Tuk, and Robert H.J. Beelen, Phenotypical and functional characterization of dendritic cells in the human peritoneal cavity, *This Symposium.*
13. E. van Vugt, J.M.S. Arkema, M.A.M. Verdaasdonk, R.H.J. Beelen and E.W.A Kamperdijk, Morphological and functional characteristics of rat steady state peritoneal dendritic cells, *Immunobiol.* , 184:14 (1991).
14. E. van Vugt, M.A.M. Verdaasdonk, R.H.J. Beelen and E.W.A. Kamperdijk, Induction of an increased number of dendritic cells in the peritoneal cavity of rats by intraperitoneal administration of Bacillus Calmette-Guérin, *Immunobiol* in press.
15. E. van Vugt, M.A.M. Verdaasdonk, E.W.A. Kamperdijk and R.H.J. Beelen, Antigen presenting capacity of peritoneal macrophages and dendritic cells, *This Symposium.*
16. H.J. Bos, D.G. Struijk, C.W. Tuk, J.C. de Veld, T.J.M. Helmerhorst, E.C.M. Hoefsmit, L. Arisz, and R.H.J. Beelen. Peritoneal dialysis induces a local sterile inflammatory state and the mesothelial cells in effluent are related to the bacterial peritonitis incidence,*Nephron*, 59:508 (1991).
17. C.E.G. Havenith, A.J. Breedijk, and E.C.M. Hoefsmit, Effect of Bacillus Calmette-Guérin inoculation on numbers of Dendritic Cells in Brochoalveolar lavages of rats,*Immunobiol.* 184:336 (1992).
18. J.M.S. Arkema, D.M. Broekhuis-Fluitsma, P.A.J.M. de Laat and E.C.M. Hoefsmit, Human peripheral blood dendritic cells concentrate in contrast to monocytes intracellular HLA class II molecules in a juxtanuclear position, *Immunobiol.*, 181:335 (1990).

19. J.M.S. Arkema, I.L. Schadee-Eestermans, D.M. Broekhuis-Fluitsma and E.C.M. Hoefsmit, Localization of Class II Molecules in storage vesicles, endosomes and lysosomes in human dendritic cells, *Immunobiol.*, 183:396 (1991).
20. J.M.S. Arkema, I.L. Schadee-Eestermans, D.M. Broekhuis-Fluitsma and E.C.M. Hoefsmit, Immunocytochemical characterization of dendritic cells, *This Symposium*.
21. J.M.S. Arkema, I.L. Schadee-Eestermans, R.H.J. Beelen and E.C.M. Hoefsmit, A Combined method for both endogenous myeloperoxidase and acid phosphatase cytochemistry as well as immunoperoxidase surface labelling discriminating human peripheral blood derived dendritic cells and monocytes, *Histochemistry, 95:537 (1991).*
22. M.G.H. Betjes, M.C.Haks, C.W. Tuk, and R.H.J. Beelen, Monoclonal antibody EBM11 (anti-CD68) discriminates between dendritic cells and macrophages after short-term culture, *Immunobiol*, 183:79, (1991).
23. R.H.J. Beelen, E. van Vugt, J.E. Steenbergen, M.G.H. Betjes, Carin E.G. Havenith, and E.W.A. Kamperdijk, Dendritic cells isolated from rat and human non-lymphoid tissue are very potent accessory cells, *This Symposium.*
24. E.C.M. Hoefsmit, E.W.A. Kamperdijk, H.R. Hendriks, Robert H.J. Beelen and B.M. Balfour, Lymph node macrophages, *in*: "The Reticuloendothelial System", Vol.I, I. Carr and W.T. Daems eds., Plenum Publishing Corporation, New York, 1980.
25. E.W.A. Kamperdijk, E.B.J. van Nieuwkerk, J.M. Arkema, A.M. Duijvestijn, and E.C.M. Hoefsmit, *in*: "Accessory Cells In HIV and Other Retroviral Infections", P. Racz, C.D. Dijkstra, J.C. Gluckman, eds., Karger, Basel, (1991).
26. E.W.A. Kamperdijk, E.M. Raaymakers, J.H.S. de Leeuw, and E.C.M. Hoefsmit, Lymph node macrophages and reticulum cells in the immune response. I. The primary response to paratyphoid vaccine, *Cell Tiss. Res.* 192:1, (1978).
27. A.H. Warfel, G.J. Thorbecke and D.V. Belsito, Langerhans cells as outposts of the Dendritic cell system, *This Symposium.*
28. G.Schuler, F. Koch, C. Heufler, E. Kämpgen, G. Topar, and N. Romani, Murine epidermal Langerhans cells as a model to study tissue dendritic cells, *This Symposium.*
29. E.W.A Kamperdijk, M. van den Berg, and E.C.M. Hoefsmit, Lymph node accessory cells in the immune response.The primary response to paratyphoid vaccine in rat parathymic lymph nodes, *Cell Tissue Res.*237:39 (1984).
30. E.C.M. Hoefsmit, A.M. Duijvestijn, and E.W.A. Kamperdijk, Relation between Langerhans cells, veiled cells, and interdigitating cells, *Immunobiol.*161:255 (1982).
31. E.B.J. van Nieuwkerk, E.W.A. Kamperdijk, M.A.M. Verdaasdonk S. van der Baan, and E.C.M. Hoefsmit, Langerhans cells in the respiratory epithelium of the Human adenoid, *Eur. J. Cell Biol.*, 54:182 (1991).
32. W.J. Fokkens, D.M. Broekhuis-Fluitsma, E. Rijntjes, Th.M. Vroom, and E.C.M. Hoefsmit, Langerhans cells in nasal mucosa of patients with grass-pollen allergy, *Immunobiology* 1991.
33. W.J. Fokkens, C.A.F.M. Bruijnzeel-Koomen, Th.M. Vroom, E. Rijntjes, E.C.M. Hoefsmit G.C. Mudde, and Bruynzeel PLB. The Langerhans cell: an underestimated cell in atopic disease, *Clin. Exp. Allergy* 20:627 (1990).
34. H. Stössel, F. Koch, Eckhart Kämpgen, P. Stöger, A. Lenz, C. Heufler, N. Romani and G. Schuler. Disappaerance of certain acidic organelles (endosomes and Langerhans Cell granules) accompanies loss of antigen processing capacity upon culture of epidermal Langerhans cells, *J.Exp.Med.* 172:1471 (1990).

REPORT OF THE PANEL DISCUSSION

Elisabeth C.M. Hoefsmit

There was unanimous agreement of two different types of Dendritic Cells: the Follicular Dendritic Cells, and the Dendritic Cells, which comprise different phenotypes such as Langerhans Cells and Interdigitating Cells. Although this nomenclature suggest a different location of the same cells, there was no consensus of new names to discriminate both cell types. The confusion could best be minimized by excluding the Follicular Dendritic Cells from the Dendritic Cell System. The classical names should therefore be defined, and the cardinal features updated.

Main issues of the Dendritic Cell System, that have been discussed are the phenotype of the precursors of Dendritic Cells in bone marrow and in blood, the influence of growth factors on the phenotype, the sequence of maturation steps and the influence of cytokines, and specialized properties of antigen processing and presentation. It appeared that the nomenclature of the different phenotypes depends on the methods of investigation: whether the function of the cells can be studied *in vitro* such as in experimental immunology, or the only possible method is determination of the phenotype *in situ* as is usually the case in pathology.

Recent developments in immunoelectron microscopy of cryosubstituted and cryoembedded tissue fragments, enables us to define and compare intracellular organelles en phenotype of different types of DC as well in situ as in functional in vitro assay's.
It may be expected that these methods will contribute to bridge the gap between in vitro and in vivo studies in animal experiments and in human pathology. For that reason cardinal features can probably best be attributed to cells indicated by their classical names.

Under the auspices of the International Union of Immunological Societies and on initiation of the WHO/IUIS Nomenclature Committee the panel has formed a Nomenclature Subcommittee on Nomenclature of Dendritic Cells on the occasion of the 2nd International symposium on Dendritic Cells in fundamental and clinical Immunology, June 21-26 in Amsterdam.

The panel discussion aimed to reconsider the nomenclature on Dendritic cells and prepare a consensus report on the names and cardinal features of Dendritic Cells.

The subcommittee consists of:

G.J. Thorbecke
R.M. Steinman
G.J. Tew
P. Nieuwenhuis (liaison with IUIS)
E.C.M. Hoefsmit, chairperson

DENDRITIC CELLS IN TRANSPLANTATION

Jonathan M. Austyn

Nuffield Department of Surgery, University of Oxford
John Radcliffe Hospital, Headington, Oxford OX3 9DU

THE PASSENGER LEUKOCYTE CONCEPT

According to the passenger leukocyte concept, a subset of Ia-rich (MHC class II-positive) leukocytes, carried over to the recipient within transplanted organs, provides the primary stimulus for allograft rejection [1]. This concept has important implications for transplantation, since it implies that the bulk of a transplanted tissue is weakly or non-immunogenic. Thus, for some years, there has been growing interest in the idea that removal of passenger leukocytes from organs before transplantation may permit successful allografting with less recipient immunosuppression. If this could be achieved, the recipient should be less susceptible to the sometimes severe and even fatal side effects that are associated with current non-specific immunosuppressive therapies in the clinical setting.

Many of the early studies of experimental transplantation that lead to the passenger leukocyte concept focused on the rejection of allografts, or the induction of graft-versus-host reactions, after transplantation of tissues from chimeric animals. The key observation, made by several investigators, was that the replacement of bone marrow-derived cells within a tissue with those from another strain could elicit transplantation reactions, even though other graft elements were syngeneic to the recipient. In addition, it was noted that various Ia-negative cells were sometimes unable to induce immune responses after administration to allogeneic recipients, and it now seems that some of these can induce a state of specific immunological unresponsiveness, or tolerance, in some situations.

Subsequently, various attempts were made to reduce the immunogenicity of tissue allografts by manipulation of passenger leukocytes. Many of these centred on the observation that small tissues in a relatively dispersed form, such as parathyroid, thyroid and pancreatic islets of Langerhans, were rendered less immunogenic by prior organ culture, particularly in high O_2 concentrations or at low (24°C) temperatures. It was postulated that successful allografting in these situations was associated with selective destruction of a small subset of cells, the presumptive passenger leukocytes.

Dendritic Cells in Fundamental and Clinical Immunology,
Edited by Kamperdijk *et al.*, Plenum Press, New York, 1993

The passenger leukocyte concept was put on a more rational basis with the discovery of Ia+ dendritic cells as a trace cell type in lymphoid organs and the finding that these cells played a central and perhaps unique role in the initiation of T-dependent responses, of which allograft rejection is a classical example, and the growing awareness that related cells are present in most non-lymphoid organs that clinicians wish to transplant.

DENDRITIC LEUKOCYTES

Members of the family of dendritic cells, sometimes referred to as dendritic leukocytes (DL), are distributed throughout lymphoid and most non-lymphoid tissues (except for the bulk of the brain) and migratory forms also exist in blood and afferent lymph [2]. These cells are phenotypically and functionally diverse apparently in relation to their precise anatomical localization.

Function of lymphoid DC

Many studies have demonstrated that DL isolated from lymphoid tissues (lymphoid DC) have the capacity to present foreign antigens to resting T cells and deliver signals for T cell activation (this process being termed immunostimulation) thereby initiating immune responses in vitro [2,3]. Of central importance to the initiation of allograft rejection, one example is the capacity of DC to stimulate the allogeneic mixed leukocyte reaction (MLR) when they are cultured with allogeneic T cells; in this case alloreactive T cells recognize alloantigens (in many cases, foreign peptide - allogeneic MHC complexes) presented by the DC, and receive the necessary signals for activation and proliferation. In addition, alloreactive CTL are generated in the MLR, even in the absence of CD4+ T cells which suggests that DC can activate both CD4+ and at least some CD8+ T cells directly. In contrast, other Ia+ leukocytes from lymphoid tissues, such as resting B cells and macrophages, lack the capacity to initiate these and other primary T and T-dependent responses. These observations support the idea that DL may have the capacity to function as passenger leukocytes, and suggest quite strongly that the mere induction of Ia on non-DL graft elements may not be sufficient to trigger the rejection response.

Maturation of non-lymphoid DL

Further important insights into the family of DL came from work showing that Langerhans cells (LC) from skin epidermis could develop in culture into cells closely resembling lymphoid DC both in phenotype and function [2,3]. However, freshly-isolated LC (and presumably epidermal LC in situ) were unable to stimulate the MLR, despite their expression of the relevant alloantigens. It was then found that fresh LC could pinocytose and process native protein antigens, and that immunogenic peptide-MHC complexes were subsequently expressed at the cell surface for several days as the cells lost the capacity to process native antigens but acquired immunostimulatory activity [2-4]. In addition, fresh LC can phagocytose a variety of particulates, including certain yeasts and bacteria, but this capacity is also lost during culture (see C. Reis e Sousa et al.).

It is possible, therefore, to distinguish between fresh LC which can be thought of as immature DL, and cultured LC which, like lymphoid DC, can be considered as mature DL. The transition between these two stages of DL can thus be termed maturation, and for LC this process is controlled and influenced by different cytokines, particularly GM-CSF, TNFα and IL1 (see J. Roake et al.). This maturation process is accompanied by phenotypic remodelling of the cell surface, including loss of Fc receptors (CDw32) and mannose receptors, a drastic reduction in the rate of biosynthesis of MHC class II heterodimers and the invariant chain, and loss of intracellular acidic organelles, particularly early endosomes, any or all of which may be implicated in loss of antigen uptake and processing capabilities of cultured LC [4]. In addition, it is intriguing that cultured but not fresh LC express the B7/BB1 molecule (the ligand for CD28 and CTLA-4 on T cells) concomitant with their acquisition of immunostimulatory activity (C. Larsen, personal communication). Ia+ leukocytes with dendritic profiles are distributed throughout most other non-lymphoid tissues, and it has been suspected that at least some of these cells belong to the DL family [2]. In topologically external tissues, such as the lung and airways, the urogenital tract, and gut, their distribution somewhat resembles that of LC in skin, whereas in solid organs such as heart and kidney they are present in the interstitial spaces leading to the description of these cells as interstitial DC. DL have been isolated from several non-lymphoid tissues and shown to express immunostimulatory activity in culture. However, an important issue concerns the precise maturation stages of these non-lymphoid DL in situ, because this activity may be acquired during culture (analogous to maturation of LC).

It has been shown that freshly-isolated DL from mouse hearts and kidneys are unable to initiate primary responses in vitro but they acquire this capacity after overnight culture concomitant with changes in phenotype and other properties, suggesting that the cells in situ are functionally immature (see A. Rao et al.). This may well apply to other non-lymphoid DL, and it has been shown that DL in other sites can acquire and present antigens that were, for example, administered in aerosolized form into the lung or injected into the gut [4]. This activity is of course more characteristic of immature DL than of mature lymphoid DC. Therefore, immature DL may be situated primarily in non-lymphoid tissues, where they can acquire and process any foreign antigens gaining access to these sites, whereas mature DC may be localized in lymphoid tissues.

Migration of DL

There is a considerable flux of DL from non-lymphoid tissues into afferent lymph and thence to draining lymph nodes, particularly from transplants and sites of inflammation or infection [5]. It seems likely that these cells, which are termed veiled cells, are transporting foreign antigens in immunogenic form (peptide-MHC complexes) into the lymphoid tissue in order to interact with antigen-specific T cells and initiate immune responses. Indeed, DC from lymph nodes draining areas of skin to which contact sensitizing agents are applied can sensitize naive recipients after adoptive transfer. The phenotype of veiled cells, as for example isolated from pseudoafferent lymph draining the intestine in rats, is heterogeneous and it appears that the cells have already commenced their maturation process. The maturation and perhaps migration of DL out of non-lymphoid tissues is presumably stimulated by locally-produced cytokines (see J. Roake et al.).

Migratory DL can also be isolated from peripheral blood, but these appear to consist of two distinct forms. First, there are bone marrow-derived blood precursors which apparently seed the non-lymphoid tissues. Second, there are DL that may have already resided in non-lymphoid organs and acquired antigens, and which are now en route to the spleen. The latter migratory route was first shown in studies with purified splenic DC that were radio- or fluorochrome-labelled and transfused into mice; these cells migrated from the blood primarily into the spleen and homed to central white pulp (T areas) [5]. These observations were later complemented by studies of migration of DL after experimental transplantation (see below) which further suggested that this form of blood DC may originate from solid organs such as hearts. Therefore, DL may migrate from these sites via blood and lymph, and from body surfaces such as the skin perhaps only via lymph.

DENDRITIC CELLS IN TRANSPLANTATION

The properties of DL outlined above indicate they could function as passenger leukocytes in transplantation, and direct evidence to support this has now been obtained. Moreover, in some cases, studies on the basic immunobiology of DL has shed important insights into the initiation of allograft responses.

DL as passenger leukocytes

Lechler and Bachelor [6] obtained evidence that DL might be important passenger leukocytes for renal allograft rejection. They found that replacement of leukocytes in rat kidneys, by temporary residence in immunosuppressed secondary allogeneic recipients, resulted in long-term survival when the kidneys were retransplanted to untreated recipients syngeneic with the secondary hosts. However, rejection occurred when donor DC, but not other leukocytes tested, were administered at the time of retransplantation. Subsequently, more direct evidence was obtained by other investigators [7,8] who showed that pretreatment of mouse islets of Langerhans or thyroid tissue with a monoclonal antibody specific for DC and complement resulted in long-term survival after transplantation to allogeneic recipients; again administration of donor DC lead to rejection. It is important to note, however, that these findings only applied to particular strain combinations and they could not be reproduced in others. Part of the reason for this may be that the host's own DL may be able to acquire graft antigens and present them to recipient T cells, a pathway termed indirect sensitization, in contrast to direct sensitization by donor DL.

Maturation and migration of donor DL after transplantation

Early studies of experimental transplantation lead to the concept that sensitization to skin grafts occurs in recipient lymph nodes (central sensitization) whereas sensitization against fully-vascularized organ allografts such as heart transplants occurs within the organ itself (peripheral sensitization) [1,5]. However, studies on the behaviour of LC after skin grafting [9] and of DL after heart transplantation [10] have raised the possibility that sensitization can occur centrally in both cases.

In the case of skin grafts, it has been shown that after transplantation the epidermal LC rapidly increase in size and upregulate MHC class II expression, before they migrate into the dermis and enter dermal lymphatics [9]. Similar changes are observed in organ cultured skin where, in addition, the LC leave the tissue and accumulate in the tissue culture vessel; this process is accompanied by phenotypic and functional maturation of the LC. Presumably, in the case of a skin allograft, these cells would enter the bed of the graft and migrate to the draining lymph nodes to initiate the response. A rapid loss of DL from hearts has also been observed after allotransplantation, but these donor-derived cells were subsequently detected within recipient spleens, associated with CD4+ T cells in peripheral white pulp (B areas) [10]. Thus, while it is possible that DC can also migrate in lymph draining from the transplant, they can certainly enter the blood since the spleen lacks afferent lymphatics, a conclusion supported by other control studies (J. Roake et al., unpublished observations). This would appear to provide a route for central sensitization to fully-vascularized allografts, and similar observations have been made in a rat limb transplantation model [11].

Role of host DL in allograft rejection

While direct sensitization of a transplant recipient against donor DL is thought likely to provide the major stimulus for rejection, because of the high frequency of alloreactive T cells in the repertoire, it has been shown that host cells can present graft antigens through the indirect pathway and trigger skin graft rejection, although the cell type involved has not been defined precisely [12]. It seems likely that if host DL are involved in this process, then they gain access to graft antigens at a relatively immature stage, perhaps by migrating as blood precursors into the transplant. Here they may develop into immature DL with the capacity to internalize and process graft-derived soluble antigens and perhaps even cells and other debris from the graft, before migrating to lymphoid tissues to present foreign peptide - self MHC complexes to host T cells. This idea is also supported by the observation that mature host-strain DC are unable to gain access to skin and heart allografts in the mouse [13].

Potential approaches to manipulate DL during transplantation

Insights into the maturation, migration and function of DL, and increasing evidence for their role as passenger leukocytes in transplantation, suggest a number of approaches to manipulate DL in order to reduce or prevent the initiation of rejection. For example, it would clearly be advantageous to deplete these cells from non-lymphoid organs before transplantation. While it may be difficult to use specific monoclonal anti-DL antibodies for solid organs, one study has shown that perfusion of kidneys with anti-CD45 antibodies before clinical transplantation can result in fewer rejection episodes [14], conceivably by depleting DL or altering their function.

An alternative approach may be to use cytokines to promote migration of DL out of the organ, provided it could be maintained ex vivo for a suitable period of time. For smaller, dispersed tissues such as islets of Langerhans, it has been noted that culture alone can promote the egress of DL into the culture vessels, providing an alternative explanation for the reduced immunogenicity of some cultured tissues noted at the outset. In addition, anti-cytokine antibodies could conceivably be used to arrest the maturation of resident DL and/or their migration to host lymphoid tissues, and might even promote cell death for it is known, for example, that the viability of cultured LC is sustained by TNFα. It seems likely that other strategies will be suggested by further studies on the immunobiology DL.

REFERENCES

1. J.M. Austyn and R.M. Steinman, The passenger leukocyte - a fresh look, Transplant. Revs. 2:139-176 (1988)
2. R.M. Steinman, The dendritic cell system and its role in immunogenicity, Annu. Rev. Immunol. 9:271-296 (1991)
3. J.P. Metlay, E. Pure and R.M. Steinman, Control of the immune response at the level of antigen-presenting cells: a comparison of the function of dendritic cells and B lymphocytes, Adv. Immunol. 47:45-116 (1989)
4. J.M. Austyn, Antigen uptake and presentation by dendritic leukocytes, Seminars Immunol. 4: In Press (1992)
5. J. M. Austyn and C. P. Larsen, Migration patterns of dendritic leukocytes: implications for transplantation, Transplantation 49:1-7 (1990)
6. R.I. Lechler and J.R. Batchelor, Restoration of immunogenicity to passenger cell-depleted kidney allografts by the addition of donor strain dendritic cells, J. Exp. Med. 155:31 (1982)
7. D. Faustman et al., Prevention of rejection of murine islet allografts by pretreatment with anti-dendritic cell antibody, Proc. Natl. Acad. Sci. USA. 81:3864 (1984)
8. H. Iwai et al., Acceptance of murine thyroid allografts by pretreatment of anti-Ia or anti-dendritic cell antibody, Transplantation 47:45 (1989)
9. C.P. Larsen et al., Maturation and migration of Langerhans cells in skin transplants and explants, J. Exp. Med. 172:1483-1493 (1990)
10. C.P. Larsen, P.J. Morris and J.M. Austyn, Migration of dendritic leukocytes from cardiac allografts into host spleens: a novel pathway for initiation of rejection, J. Exp. Med. 171:370-314 (1990)
11. M.A. Codner et al., Migration of donor leukocytes from limb allografts into host lymphoid tissues. Ann. Plastic Surg. 25:353-359 (1990)
12. R.A. Sherwood et al, Presentation of alloantigens by host cells. Eur. J. Immunol. 16:569 (1986)
13. C.P. Larsen et al., Failure of mature dendritic cells of the host to migrate from the blood into cardiac or skin allografts. Transplantation 50:294-301 (1990)
14. Y. Brewer et al., Effect of graft perfusion with two CD45 monoclonal antibodies on the incidence of kidney allograft rejection. Lancet ii; 935-937 (1989)

DOWN-REGULATION OF MHC-EXPRESSION ON DENDRITIC CELLS IN RAT KIDNEY GRAFTS BY PUVA PRETREATMENT

B. v. Gaudecker, R. Petersen, M. Epstein, J. Kaden, and H. Oesterwitz

Anatomisches Institut der Universität Kiel
Olshausenstr. 40, D-2300 Kiel, Germany

INTRODUCTION

It is the aim of this investigation to elucidate the cellular mechanisms responsible for graft acceptance in rat transplantation models after PUVA pretreatment of the graft. PUVA-therapy is used successfully in the treatment of psoriasis and other dermatoses with local accumulations of dendritic cells, as well as skin-lymphomas. Following these experiences Oesterwitz et al. (1989) developed a PUVA-treatment of the graft before transplantation. In this respect PUVA means: i. v. injection of the photoactive substance 8-Methoxypsoralen into the donor and in vitro exposure of the kidney graft to long wave UVA-irradiation. This PUVA therapy significantly prolonged graft survival times in rats even across a strong MHC-barrier. It has been shown, that 40% of PUVA treated rat kidneys are accepted by the recipient for more than 100 days. MHC-class-II-positive dendritic cells present donor specific alloantigen to the recipient and induce rejection. PUVA treatment reduces the expression of MHC-class-II-molecules and may lead to tolerance of the recipient towards the graft.

MATERIAL AND METHODS

8-Methoxypsoralen (Gerot Pharmazeutica, Vienna, Austria) was given intravenously at a dosage of 0,1 mg/kg body weight via the vena penis dorsalis of the organ donor 10 min. before removal of the graft. UVA-irradiation was applied in a dosage of 11.32 J cm^{-2} to the graft ex vivo during hypothermic preservation in Euro-Collins solution. The time of irradiation was two hours. Thereafter the kidneys were transplanted orthotopically into allogenic recipients, which did not receive any immunosuppressive drugs. The rat strain combinations LEW($RT1^{l}$) as donor and DA($RT1^{av1}$) as recipient (LEW→DA) and the converse (DA→LEW) mainly were used in this investigation. MHC-class-II-positive cells were identified by light microscopical immunohistochemistry and counted. Using the monoclonal antibody (mab) OX3, which reacts with MHC-class-II of the $RT1^{l}$ locus in LEW-rats and the mab. OX6, which reacts with MHC-class-II in both LEW- and DA-rats, we were able to distinguish between MHC-class-II-positive cells of graft origin from those of the recipient. After immunohistochemical staining untreated control grafts were compared with kidneys after PUVA treatment during a period of one week and after 100 days. Using immuno-electron microscopy (Perry et al. 1992) the positive cell types were characterized according to ultrastructural criteria.

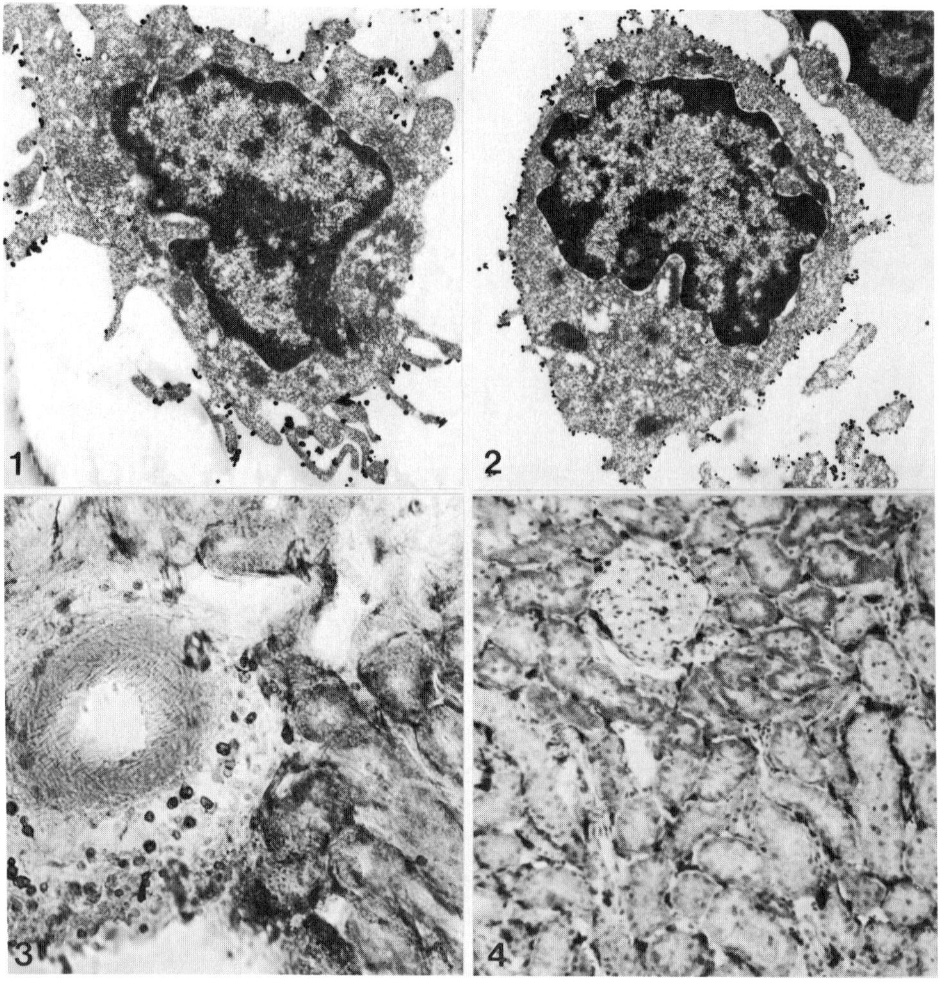

Fig. 1 and **Fig. 2:** Immuno-electron micrographs. Label is indicated by silver enhanced gold particles.
Fig. 1) LEW-rat incubated with OX6: Dendritic cell between kidney tubules. 20000x.
Fig. 2) DA→LEW, PUVA, one day after transplantation, incubated with OX3: Positive monocyte from the LEW recipient has invaded the graft. 14000x.
Fig. 3) LEW→DA, PUVA, one day after transplantation. Light micrograph from kryostate section incubated with OX6: MHC-class-II-positive cells from the recipient accumulate in the perivascular space around an artery. Positive dendritic cells (elongated) from the donor are located between the tubuli. 300x.
Fig. 4) DA→LEW, PUVA. Kryostate section incubated with OX3: Transplanted kidney grafts show no signs of rejection after five days. Dendritic cells from the LEW-recipient ($OX3^+$) are located between the tubuli of the DA-kidney-graft. 170x.

RESULTS

Untreated kidneys which were investigated after 2 hours of preservation in Euro-Collins solution without transplantation showed numerous MHC-class-II-positive cells in the interstitium between the glomeruli and tubuli. Most of these were dendritic cells with numerous cytoplasmic projections. They could be clearly identified in immuno-electron micrographs (Fig. 1). Additionally positive monocytes (Fig. 2), and macrophages were recognized. Considerable up-regulation of donor specific MHC-class-II-molecules was observed in untreated control grafts one day after the operation (possibly caused by interferon γ from

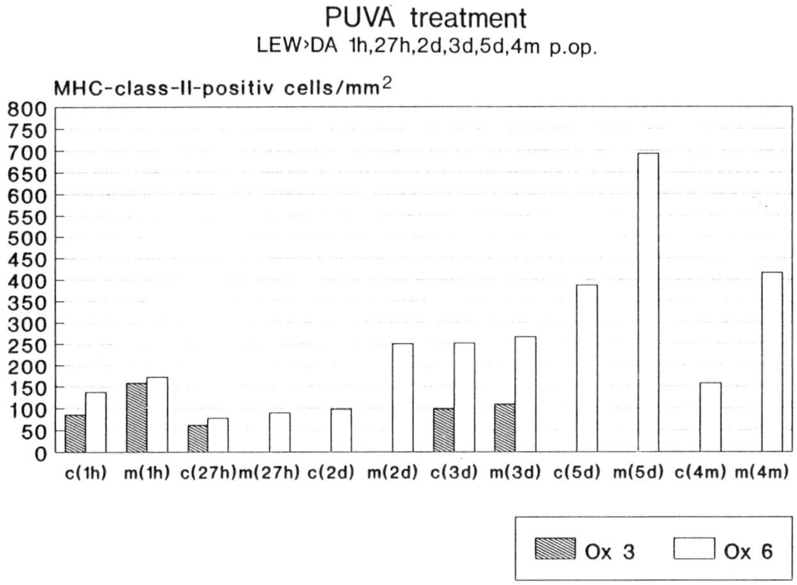

Fig. 5: Transplantation series LEW→DA with PUVA treatment. Quantitative evaluation of MHC-class-II-positive cells from the donor ($OX3^+$), and from the host ($OX6^+$-$OX3^+$) in the kidney grafts. During the first five days after transplantation MHC-class-II-positive cells of donor origin disappear from the graft. Many of them migrate into lymphatic organs of the recipient. On the other hand, MHC-class-II-positive cells from the host accumulate in the graft. The highest numbers of these cells are counted five days after transplantation. Many of them differentiate into dendritic cells, which home between the tubuli of the kidney graft.

the recipient). Control kidney grafts one day after transplantation investigated with OX6 revealed not only positive dendritic cells but also numerous MHC-class-II-positive round cells, which were predominantly located within the perivascular spaces around arteries and veins. If a DA-kidney was transplanted into a LEW-recipient, only these round cells were stained after incubation with the LEW specific mab. OX3. This proves that the round cells originate from the LEW-recipient and have invaded the graft. Invasion of the leukocytes proceeds during the next days after transplantation and causes rejection of the graft. Transplanted control kidneys do not survive more than five to seven days.

A significant down-regulation of donor specific MHC-class-II-molecules can be observed in animals that received a PUVA treated kidney. Positively reacting donor specific dendritic cells are considerably reduced in grafts investigated one day after transplantation. In LEW-grafts investigated 2 days after transplantation no MHC-class-II-positive donor cells could be observed. The specimens taken after three days gave variable results. Some of them revealed low numbers of donor specific MHC-class-II-positive cells, but in all PUVA-treated kidney grafts the amount of positive cells was reduced compared to untreated controls. After five days, and later, no donor specific antigen presenting cells (APC) were seen in the grafts of LEW→DA combinations (Fig. 5).

In all grafts, in which PUVA-treatment had been successful, the parenchyma of the kidney was unaltered (Fig. 3, 4). No signs of rejection were noticed even in grafts that had been transplanted for longer than 100 days. Reaction of the recipient however can be noticed: MHC-class-II-positive cells from the recipient initially invade a PUVA treated kidney graft in the same way as an untreated control graft. One and two days after transplantation one finds numerous positive round cells in the perivascular spaces (Fig 3). After five days and three months round cells have disappeared again from the graft, but increased numbers of recipient specific cells (OX6-OX3) were counted from the first to the fifth day after transplantation. These leukocytes however do not disturbe the parenchyma of the graft. By morphological criteria they were identified as dendritic cells and they were located around the glomeruli and in the narrow interstitium between the kidney tubuli (Fig 4). This means, that now donor specific dendritic cells have been replaced by APCs from the recipient.

In order to answer the question, of wether MHC-class-II-positive cells can home to the lymphatic organs of the host, we investigated spleens of the recipients in the combinations LEW→DA. OX3-positive donor cells were found in the spleens of the untreated controls and of the DA-recipients, which got a PUVA-treated LEW-kidney graft. These cells were predominantly located in increased numbers in the marginal zone of the white pulp. In spleens of untreated controls five days after transplantation a considerable number of these cells lined the periphery of the PALS. At the same time the kidney graft was widely disturbed by rejection. After PUVA-treatment, on the other hand, the number of positive cells was lower compared to untreated controls. After five days only very few OX3-positive cells were noticed and after three months they were absent from spleens of the recipient.

In the reciprocal transplantation series DA→LEW successful PUVA-treatment of kidney grafts was less frequent than in the combination LEW→DA. It is noteworthy that initially in DA-kidneys before transplantation the number of MHC-class-II-positive cells was considerably higher than in LEW-kidneys. The third day after transplantation seems to be critical, for several of the DA-grafts showed signs of rejection in spite of PUVA-treatment. Two LEW-, three SD-(SD→BDIX) and two DA-kidneys stayed in the host for longer than 100 days. The LEW- and SD-kidney grafts were accepted and fully functional when taken out and investigated. Both PUVA-treated DA-kidney grafts however were fully rejected.

Conclusion

PUVA-treatment of the donor leads to alteration of the alloantigen presenting cells of the graft but seems not to be harmful for the kidney parenchyma. The dendritic cells seem not to be eliminated, but their antigen presentation is obviously blocked, and the expression of their MHC-class-II-molecules is downregulated. It has been shown in vitro that additionally other adhesion molecules such as ICAM-1 can be inactivated on the surface of dendritic cells by UV-irradiation (Tang et al. 1991). APCs from untreated and PUVA-treated grafts are able to invade and to home to the lymphatic organs of the host. In this investigation they were observed in the spleen. There sensitization and clonal proliferation of host lymphocytes is induced and finally leads to rejection of the graft (Larsen et al. 1990). Dendritic cells from PUVA-treated grafts obviously lose their ability to present alloantigen to the lymphocytes of the recipient. The firm binding of the T-cell-receptor to the MHC-molecule seems to be inhibited (possibly due to a missing second signal). We conclude that APCs from PUVA-treated grafts become tolerogenic, since the sensitization of the immune system in the host towards the donor is widely reduced. So during the first critical days the graft is protected from rejection.

Leukocytes from the recipient initially invade the PUVA-treated graft in the same way as untreated control grafts, but they are anergic. The parenchyma of transplanted kidneys stays undisturbed during the first days. Later, between the fourth and fifth day after transplantation mononuclear cells from the recipient home in the interstitium of the graft. There they develop into APCs. So donor specific dendritic cells have been replaced by those of the host. This leads to permanent acceptance of the graft in the rat model (Hart et al. 1980).

If our observations in these transplantation experiments reflect a common principal, one might conclude, that also in clinical transplantation the acceptance of a graft after the first critical weeks is maintained by invasion of dendritic cells originating from the bone marrow of the patient . These cells home to the graft and protect it from being recognized as foreign by the immune system of the recipient. So PUVA treatment seems to result in specific tolerance to the graft. If this is true, PUVA-pretreatment of the graft in future might be a therapy which can be used routinely. In successful cases it might help to reduce or in some cases completely avoid the use of harmful immunosuppressive drugs. However, not all of the PUVA treated kidney grafts are accepted by the recipient, probably due to genetic differences leading to a higher or lower tolerance induction. It would be of great interest, to work out a test system which allows to predict the success of PUVA therapy in clinical transplantation

Acknowledgements

This investigation was supported by grants from the Deutsche Forschungsgemeinschaft. The authors appreciate the excellent technical assistance of Mrs. K. Lüdtke, Mrs. K. Schlender, Mrs. H. Siebke, and Mrs. H. Waluk.

References

Hart, D.N.J., Winearls, C.G., and Fabre J.W., 1980, Graft adaptation: Studies on possible mechanisms in long-term surviving rat renal allografts, *Transplantation* 30: 73.

Larsen, C.P., Morris P.J., and Austyn J.M., 1990, Migration of dendritic leukocytes from cardiac allografts into host spleens, *J. Exp. Med.* 171: 307.

Oesterwitz H., Kaden J., Gruner S., Schneider W., Eichler C., Schirrow R., May G., Scholz D., and Mebel M., 1989, PUVA-Behandlung von Organtransplantaten – experimentelle und klinische Ergebnisse, *Z. Urol. Nephrol.* 82: 1.

Perry M.E., Brown K.A., and Gaudecker von B., 1992, Ultrastructural identification and distribution of the adhesion molecules ICAM-1 and LFA-1 in the vascular and extravascular compartments of the human palatine tonsil, , *Cell Tissue Res* 268: : 317.

Tang A., and Udey M.C., 1991, Inhibition of epidermal Langerhans cell function by low dose ultraviolet B radiation. – Ultraviolet B radiation selectively modulates ICAM-1 (CD54) expression by murine Langerhans cells, *J. Immunology* 146, 3347.

CYTOKINE MEDIATORS OF NON-LYMPHOID DENDRITIC CELL MIGRATION

Justin A. Roake, Abdul S. Rao, Christian P. Larsen, Deborah F. Hankins, Peter J. Morris, and Jonathan M. Austyn

Nuffield Department of Surgery
University of Oxford
John Radcliffe Hospital
Headington
Oxford, OX3 9DU
United Kingdom

INTRODUCTION

The system of dendritic cells (DC) comprises a diverse lineage, distributed throughout the tissues of the body, in different stages of development and maturation. DC develop from precursors in the bone marrow but it has proved difficult to isolate these cells in marrow culture. Recently, however, proliferating MHC class II-negative precursors, presumably in transit from the bone marrow to the tissues, have been isolated from mouse blood[1]. The dendritic leukocytes (DL) resident in the tissues are, by definition, MHC class II-positive, but it is uncertain whether in most tissues there is also a significant population of resident MHC class II-negative precursors. In the thymus[2] and neonatal epidermis[3], at least, MHC class II-negative precursors have been demonstrated. Langerhans cells (LC) which have the capacity to endocytose and process antigens, and probably other non-lymphoid DL, appear to be immature lymphoid DC that migrate to the lymphoid tissues, via the lymph (or blood), and there as mature DC initiate primary T-dependent responses.

The factors that mediate DL migration are largely unknown but experiments in which LC were found to migrate from the epidermis into the dermis in organ culture[4], or within 24h of skin transplantation (that is, before the development of an adaptive immune response), suggest that locally produced inflammatory mediators may be important in LC migration. Similar mediators may also be important in promoting the migration of DL from other organs such as cardiac allografts[5].

There is good evidence that the DL that are normally resident in most transplanted tissues, and are carried over from donor to recipient during transplantation, provide the most potent route of sensitisation to an allograft and initiate rejection. It is likely that in order to function, these DL must migrate via the lymph, or blood, from the transplanted tissue into the recipient lymphoid tissues (See Austyn J.M.). We have started to define the mediators that may be responsible for DL migration, since greater understanding this process may lead to new methods by which organs may be depleted of DL before transplantation, or suggest new ways in which DL function may be inhibited, and thus contribute to the prevention of acute rejection in clinical transplantation.

Here we briefly review what is known about the effects of cytokines on various DL populations (summarised in the figure), and present evidence that lipopolysaccharide (LPS) and TNFα deplete heart, kidney and to a lesser extent epidermis, of MHC class II-positive DL *in vivo* through the promotion of cell migration, and further that in the heart the response to LPS is accompanied by recruitment of MHC class II-negative DL precursors.

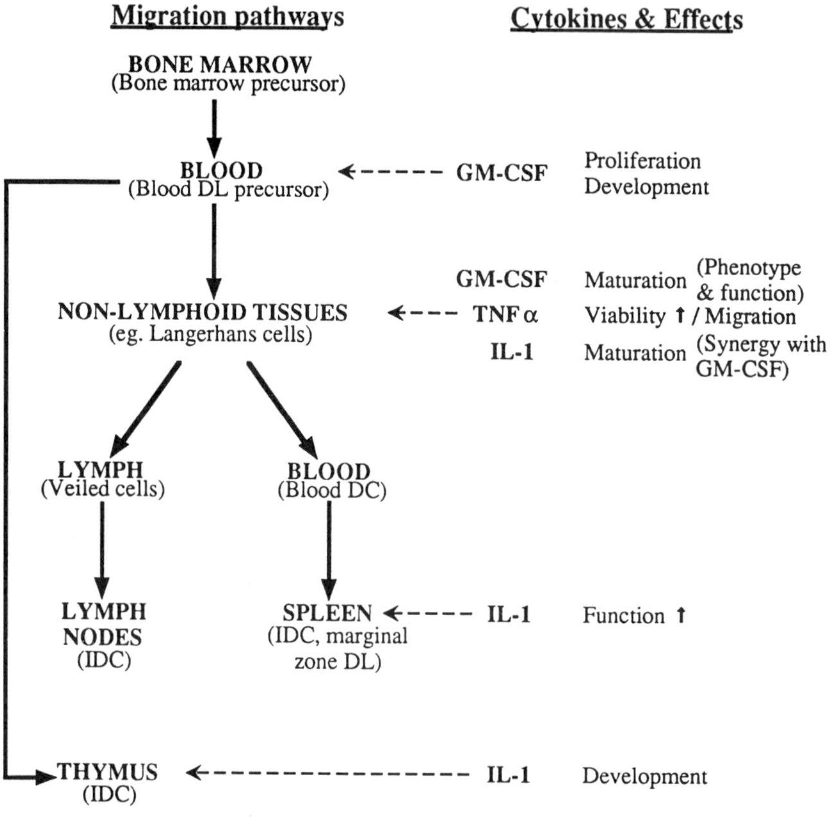

Legend: The influence of cytokines on dendritic leukocytes in the mouse. Development, as used here, refers to the generation of MHC class II-positive DL from MHC class II-negative precursors. The process of maturation refers to MHC class II-positive DL, and includes changes in phenotype and immunostimulatory function.

CYTOKINES IN THE DEVELOPMENT AND MATURATION OF DL

It is known that certain cytokines have important effects upon isolated DL populations *in vitro*. IL-1, which is known to enhance splenic[6] and thymic[2] DC immunostimulatory function, also supports the development of thymic DC from MHC class II-negative precursors. When suspensions of low density CD4-, CD8-, Ig- thymic cells, which had been depleted of MHC class II-positive DC by treatment with anti-MHC class II antibody and complement, were cultured in the presence of IL-1, MHC class II-positive immunostimulatory DC developed from MHC class II-negative precursors. Other cytokines (IL-2, IL-3 and GM-CSF) induced the development of few, or no, DC and it was not possible to detect MHC class II-negative DC precursors in spleen and lymph node, or the development of DC from bone marrow suspensions in the presence of IL-1[2]. Recently GM-CSF but not other CSFs was found to support the development of proliferating DC from MHC class II-negative precursors in mouse blood[1].

LC, in bulk epidermal cell cultures, undergo phenotypic and functional changes that seem to represent a maturation process. However LC enriched by panning (separated from keratinocytes in these cultures) required added GM-CSF in the form of purified natural or recombinant cytokine, keratinocyte conditioned medium, or conditioned medium from LPS stimulated macrophages, to promote their viability and functional maturation[7]; these events could be blocked by anti-GM-CSF antibody but not by anti-TNF of anti-IFN-γ. Other cytokines, including IL-1α, IL-2, IL-3, IL-4, G-CSF, M-CSF, rHuTNF and IFN-α/β/γ, were ineffective when tested individually. In similar, independent, studies GM-CSF was again shown to promote LC viability as well as function, and although IL-1α alone was not effective in promoting LC viability, it (but not IL-3) further enhanced LC immunostimulatory function twofold when used in combination with GM-CSF[8]. The ability of GM-CSF to maintain LC viability has been confirmed in studies by Koch et al., in which a panel of additional cytokines was tested for their ability to maintain the viability of LC in culture after enrichment by panning or sorting[9]. It was found that only murine TNFα also supported LC viability, and this did not appear to result from the induction of GM-CSF production because even in the presence anti-GM-CSF antibody TNFα supported LC viability. Interestingly recombinant human TNFα did not support LC viability at comparable doses and was only partially supportive at higher doses, so in this respect TNF exhibited a species preference.

The survival of lymph-borne veiled cells (VC) from pseudoafferent lymph in rats has also been shown to be dependent upon GM-CSF since supplementation of the culture medium with Con A-stimulated spleen cell supernatant (rich in GM-CSF) or recombinant GM-CSF but not other cytokines (rIL-1, rIL-2, rIFN-γ and an IL-3 rich supernatant) markedly increased their viability and this could be blocked by a GM-CSF-specific antiserum[10]. After 16h in culture VC developed increased immunostimulatory capacity, but this process did not appear to be dependent upon GM-CSF.

Whether GM-CSF, IL-1 and TNFα are similarly important *in vivo* is uncertain but when epidermal cells (EC) were examined for the expression of mRNA specific for GM-CSF, IL-1 and TNFα it was found that freshly-prepared EC expressed TNFα mRNA but little or no GM-CSF mRNA and no IL-1 mRNA[9]. However, after 3h and 24h of culture under conditions in which LC maturation occurs, the EC expressed significant levels of GM-CSF and IL-1 mRNA respectively. This suggested that the potential for the local production of these cytokines exists in the epidermal microenvironment. Normal human keratinocytes synthesise TNFα at low levels but after stimulation with lipopolysaccharide (*in vitro*) or UV light (*in vitro* & *in vivo*) TNFα synthesis was dramatically increased[11]. Keratinocytes have also been demonstrated to capable of releasing several other cytokines including IL-1α & β, IL-6, IL-8 and GM-CSF.

The full range of cytokine receptors expressed by members of the DC lineage has yet to be defined. It is known that lymphoid DC and many DL isolated from murine lung[12] express the IL-2 receptor (p55 chain IL-2R, CD25, p75 chain not examined) whereas freshly isolated LC, or fresh VC, are almost completely negative for IL-2R. After several hours in culture, however, both LC[13] and VC[14] acquire IL-2R expression in parallel with their functional maturation, but the significance of this is obscure since IL-2 has not yet been demonstrated to have any effect on DC. In the case of VC the induction of IL-2R appears to be at least partially dependent upon GM-CSF since supplementation with Con A-stimulated spleen cell supernatant or rGM-CSF induced more rapid and intense expression of IL-2R and this could be blocked by a GM-CSF-specific antiserum[14]. Although expression of other cytokine receptors by DC has not been reported, mRNA specific for TNF(I & II), IL-1 and GM-CSF receptors has been detected by PCR amplification in purified preparations of DC from human tonsil (D.N.J. Hart personal communication), and in view of the known effects of cytokines on enriched LC in culture it seems reasonable to assume that TNFα, IL-1 and GM-CSF receptors are expressed, or can be induced, on LC.

CYTOKINES AS MEDIATORS OF THE MIGRATION OF DL

There is only limited direct evidence concerning which cytokines may mediate DL migration. Systemic or intracutaneous administration of rIL-1β decreased the density of MHC class II-positive LC in epidermal sheets by up to 50% between 2-7 days after injection[15]. Since the level of expression of MHC class II on the remaining LC was enhanced, it seems likely that IL-1 may have induced migration of LC from the epidermis rather than simply down-regulation of MHC class II expression. In another study, intracutaneous injection of murine, but not human, TNFα into the ears of mice resulted in increased numbers of DC in the draining lymph nodes which, it was suggested, may have originated in the epidermis[16].

Following intravenous injection of LPS into mesenteric lymphadenectomized rats the number of VC released into the pseudoafferent lymph was increased[14] possibly because the cytokines (including TNFα, IL-6, IL-8, and in some studies IL-1) which LPS is known to induce *in vivo* promoted intestinal DL migration.

We have begun to define the mediators of the migration of other non-lymphoid DL. In the heart and kidneys of mice injected with a single sublethal dose of LPS there was a profound dose and time dependent reduction in the number of leukocytes expressing MHC class II as determined by immunoperoxidase labelling. The maximum depletion was seen by 2 days with almost complete recovery of MHC class II-positive cell numbers by 8 days. With ≥50μg of LPS the reduction in the heart was about 95%, and in the kidney depletion of similar magnitude also occurred. In the renal medulla, but not in the renal cortex or heart, the DL depletion was preceded by an early, transient, increase in the size of, and level of MHC class II expressed by the DL. In the epidermis of ear skin remote from the site of LPS injection there was a modest (25%) reduction in the number of LC by 2 days after injection of LPS.

These observations are consistent with either migration of the DL out of the heart, kidney and epidermis or profound down-regulation of MHC class II expression. We do not favour the latter explanation firstly because in the spleen and lymph nodes MHC class II expression by the interdigitating cells was unchanged indicating that LPS had not simply induced generalised down-regulation of MHC class II. Secondly isolated cardiac leukocytes and LC did not down-regulate MHC class II expression when cultured in medium containing LPS, a finding that is consistent with the observations of Witmer-Pack et al.[7]. Thirdly, on double labelling the epidermal LC no NLDC-145-positive, MHC class II-negative LC were present in the epidermis indicating that the apparent depletion of epidermal LC induced by LPS was not simply the result of down-regulation of MHC class II expression by these cells.

Depletion of the MHC class II-positive cells was accompanied by recruitment of other leukocytes into the tissues, the majority of which expressed macrophage (mAb F4/80 strong) or neutrophil (mAb 7/4) markers. When MHC class II-negative hearts from LPS treated donors were transplanted into fully allogeneic recipients, cells expressing donor MHC class II were found within the white pulp of the recipient spleen after 1 - 3 days, in a similar location, and expressing a similar phenotype to the donor cells (putative DC) which have previously been reported to migrate into the spleen following transplantation of hearts from untreated donors[5]. This indicated that MHC class II-positive cells, possibly DC, had developed from MHC class II-negative precursors in the heart, a view further supported by the observations that the hearts were rejected at the same rate as controls from untreated donors and that leukocyte-rich isolates from hearts of animals treated with LPS and cultured overnight were more stimulatory in a primary allogeneic MLR than isolates from normal hearts.

Between 1 - 4 days after injection of LPS, MHC class II-positive cells with dendritic morphology were present in the dermis of ear skin remote from the site injection, where they formed cord-like structures, presumably within dermal lymphatics. These "dermal cords" resembled those previously found after migration of epidermal LC into the dermis during skin organ culture or after transplantation[4]. However, double labelling studies indicated that the majority of cells comprising the cords formed after LPS treatment were MHC class II-positive, CD45-positive but NLDC145-negative or -weak whereas after organ culture the cells were NLDC145-positive, which in the skin is specific for LC. This suggested that after LPS treatment many of the cord-forming cells may have been recruited from the dermis or the blood.

The response to injection of recombinant human (rHu) TNFα was similar to that induced by LPS, including leukocyte recruitment into the heart and kidney, dermal cord formation and modest depletion of epidermal LC, but only 50% depletion of DC in the heart and kidneys was observed at 48h after injection of the maximum dose of rHuTNFα tested. Neutralising anti-TNF antibody injected prior to LPS did not prevent DC depletion from, or recruitment of Mø into, the heart or kidneys despite abolition of detectable serum TNF. However, it did inhibit neutrophil recruitment, epidermal LC depletion and the formation of dermal-cords. TNFα has been shown to mediate many of the effects of LPS in other experimental models *in vitro* and *in vivo*, so it was perhaps not surprising that recombinant TNFα reproduced may of the effects on DL populations that had been observed following injection of LPS. However TNFα is known to exhibit species preference in some of its effects and this may explain why it was not possible to produce the same magnitude of effect with respect to LPS-induced DL depletion, despite administration of relatively large doses of cytokine. That anti-TNF antibody was effective at inhibiting the neutrophil recruitment, epidermal LC depletion and dermal cord formation induced by LPS, but not the depletion of heart or kidney DL suggests that either other mediators were also important in the latter process, or that neutralisation of serum TNF was not sufficient, perhaps because the affinity of the antibody-TNF interaction was too low, or TNF at the tissue level was not sufficiently neutralised.

Intraperitoneal injection of rHuIL-1α did not deplete epidermal LC or MHC class II-positive DL from the heart or kidney cortex but the number of MHC class II-positive DL in the kidney medulla was reduced by about 50%. IL-1 also induced the formation of MHC class II-positive, NLDC145-negative dermal cords, and recruitment of macrophages and neutrophils into the heart and kidney. In control studies no effects were observed after intraperitioneal injection of similar doses of rHuIL-2.

Clearly the administration of IL-1α or TNFα produced different effects on the DL populations examined. TNF caused both depletion of DL and recruitment of leukocytes into the tissues whereas IL-1α did not, in most cases, cause depletion but it did result in leukocyte recruitment. IL-1α also caused formation of dermal cords, composed mainly of cells that were MHC class II-positive but NLDC145-negative or -weak, without measurable depletion of epidermal LC. We therefore favour the view that the cords formed after administration of IL-1α were composed mainly of cells recruited from the dermis or blood, whereas following administration of LPS or TNFα it seems likely that a proportion of the cord-forming cells may have originated in the epidermis since under these circumstances we observed depletion of epidermal LC.

In summary, it is our working hypothesis that systemic administration of LPS stimulates migration of MHC class II-positive DL from heart, kidney and epidermis which is accompanied by recruitment of MHC class II-negative DC precursors and other leukocytes from the blood and that these processes are mediated by the cytokines induced by LPS.

REFERENCES

1. K. Inaba *et al*, Identification of proliferating dendritic cell precursors in mouse blood, *J. Exp. Med.* 175:1157 (1992).
2. K. Inaba *et al*, The function of Ia$^+$ dendritic cell and Ia$^-$ dendritic cell precursors in thymocyte mitogenesis to lectin and lectin plus interleukin 1, *J. Exp. Med.* 167:149 (1988).
3. N. Romani, G. Schuler and P. Fritsch, Ontogeny of Ia-positive and Thy-1 positive leukocytes of murine epidermis, *J. Invest. Dermatol.* 86:129 (1986).
4. C.P. Larsen *et al*, Migration and maturation of Langerhans cells in skin transplants and explants, *J. Exp. Med.* 172:1483 (1990).
5. C.P. Larsen, P.J. Morris and J.M. Austyn, Migration of dendritic leukocytes from cardiac allografts into host spleens: a novel pathway for initiation of rejection, *J. Exp. Med.* 171:307 (1990).
6. S.L. Koide, K. Inaba and R. Steinman, Interleukin-1 enhances T-dependent immune responses by amplifying the function of dendritic cells, *J. Exp. Med.* 165:515 (1987).
7. M. Witmer-Pack *et al*, Granulocyte/Macrophage colony-stimulating factor is essential for the viability and function of cultured murine epidermal Langerhans cells, *J. Exp. Med.* 166:1484 (1987).
8. C. Heufler, F. Koch and G. Schuler, Granulocyte-macrophage colony-stimulating factor and interleukin-1 mediate the maturation of murine epidermal Langerhans cells into potent immunostimulatory dendritic cells, *J. Exp. Med.* 167:700 (1988).
9. F. Koch *et al*, Tumor necrosis factor α maintains the viability of murine epidermal Langerhans cells in culture, but in contrast to granulocyte/macrophage colony-stimulating factor, without inducing their functional maturation, *J. Exp. Med.* 171:159 (1990).
10. G.G. MacPherson, Properties of lymph-bourne (veiled) dendritic cells in culture: I. Modulation of phenotype, survival and function: partial dependence on GM-CSF, *Immunol* 68:102 (1989).
11. A. Kock *et al*, Human keratinocytes are a source for tumor necrosis factor α: evidence for synthesis and release upon stimulation with endotoxin or ultraviolet light, *J.Exp.Med.* 172:1609 (1990).
12. A.M. Pollard and M.F. Lipscomb, Characterization of murine lung dendritic cells: Similarities to Langerhans cells and thymic dendritic cells, *J. Exp. Med.* 172:159 (1990).
13. N. Romani *et al*, Cultured human Langerhans cells resemble lymphoid dendritic cells in phenotype and function, *J. Invest. Dermatol.* 93:600 (1989).
14. G.G. MacPherson, S. Fossum and B. Harrison, Properties of lymph-bourne (veiled) dendritic cells in culture: II. Expression of the IL-2 receptor: Role of GM-CSF, *Immunol* 68:108 (1989).
15. E.N. Lundqvist and O. Back, Interleukin-1 decreases the number of Ia+ epidermal dendritic cells but increases their expression of Ia antigen, *Acta Derm Venereol (Stockh)* 70:391 (1990).
16. M. Cumberbatch and I. Kimber, Dermal tumour necrosis factor-alpha induces dendritic cell migration to draining lymph nodes, and possibly provides one stimulus for Langerhans cell migration, *Immunol.* 75:257 (1992).

ISOLATION OF DENDRITIC LEUKOCYTES FROM NON-LYMPHOID ORGANS

Abdul S. Rao, Justin A. Roake, Christian P. Larsen, Deborah F. Hankins, Peter J. Morris and Jonathan M. Austyn

Nuffield Department of Surgery
University of Oxford
John Radcliffe Hospital
Headington
Oxford OX3 9DU.
United Kingdom

Members of the dendritic cell (DC) lineage are present in both lymphoid and non-lymphoid organs (Steinman 1991). DC resident in the interstitial connective tissue of non-lymphoid organs are thought to be important passenger cells which sensitise the host T-cells against graft antigens and initiate allograft rejection (Lechler and Batchelor 1982; Austyn and Steinman 1988; see Austyn J.M.). While freshly isolated, mature lymphoid DC are potent inducers of T-dependent immune responses (immunostimulation) (Austyn 1987), Langerhans cells (LC), the immature form of DC in skin, are unable to initiate this primary response, but they develop in culture in the presence of GM-CSF into cells resembling lymphoid DC with potent immunostimulatory capacity (Schuler et al. 1985; Schuler and Steinman 1985; Inaba et al. 1986; Heufler et al. 1987; Witmer-Pack et al. 1987).

Despite the important role of DC in direct alloantigen presentation, very little is known about the *in vivo* and *in vitro* functions of dendritic leukocytes (DL) from kidneys and hearts. We briefly review the isolation and *in vitro* functions of DC from a variety of different non-lymphoid organs, and present evidence that freshly isolated DL from mouse kidneys and hearts are poor stimulators of primary *in vitro* T-cell responses, but that after overnight culture they develop into potent immunostimulatory cells. In this respect, freshly isolated renal and cardiac DL, and presumably the cells *in situ*, more closely resemble fresh epidermal LC than lymphoid DC.

Dendritic Cells in Fundamental and Clinical Immunology,
Edited by Kamperdijk *et al.*, Plenum Press, New York, 1993

EPIDERMIS

Freshly isolated epidermal LC express MHC Class II molecules, which are upregulated during culture (Witmer-Pack et al. 1988). They also express the macrophage (Mø) marker F4/80 and FcγRII (CDw32), but both are lost during *in vitro* maturation along with Birbeck granules and other non-specific markers (Schuler and Steinman 1985; Heufler et al. 1987). *In vitro* proliferative assays have shown that fresh LC are weak stimulators of primary T-cell proliferative responses (oxidative mitogenesis and the allogeneic mixed leukocyte reaction), but that after culture these cells mature into potent immunostimulatory cells and come to resemble mature DC from lymphoid organs (Inaba et al. 1986).

LIVER

In liver, DC are localised primarily in the portal triads whereas Kupffer cells are situated in the sinusoids (Klinkert et al. 1982). Very low stimulatory activity was found in partially enriched population of nonparenchymal cells that was prepared by collagenase perfusion of rat liver followed by differential centrifugation of isolated cells, possibly due to the presence of inhibitory Mø. After fractionation on BSA, the low density fraction showed a five fold higher immunostimulatory activity than the high density fraction, which was mainly composed of parenchymal cells. If the low density fraction was γ-irradiated and cultured overnight in tissue culture dishes, most of the stimulatory activity was recovered in the non-adherent fraction, whereas the adherent fraction which was rich in Mø and fibroblast had no detectable activity. If the non-adherent fraction was further enriched for DC by EA-rosetting, most of the activity was then recovered in the EA^- population, which contained 85% of DC, whereas the immunostimulatory activity in the EA^+ population, which contained 15% DC, was much lower than expected possibly due to the presence of Fc receptor-positive and non-specific esterase-positive macrophages, which could inhibit T-cell proliferation (Klinkert et al. 1982). These and other observations suggest that in rat liver, most of the immunostimulatory activity resides in the cells that express MHC Class II molecules, are Fc receptor-negative, non-adherent, radio-resistant and negative for non-specific esterase (Klinkert et al. 1982).

LUNG

In human (Sertl et al. 1986; Nichod et al. 1989), rat (Klinkert et al. 1982; Holt et al. 1987; Rochester et al. 1988) and mouse (Pollard and Lipscomb 1990), MHC Class II positive parenchymal dendritic leukocytes are localised at the alveolar septal junctions, close to sub-epithelial alveolar macrophages, and this is also true for the airways where intraepithelial DC lie in close proximity to the macrophages. Analogous to epidermal LC, freshly isolated airway and lung parenchymal DC have relatively weak immunostimulatory capacity, but this activity along with increased expression of surface MHC Class II molecules is observed after overnight culture (Holt et al. 1990). This phenotypic and functional maturation can be inhibited by pulmonary alveolar macrophages (Holt et al. 1987).

GUT

Unfractionated cell suspensions from mouse intestinal lamina propria and Peyer's patches are able to stimulate T-cell proliferation both in the mixed leukocyte reaction and oxidative mitogenesis (Spalding et al. 1983; Pavli et al. 1990). Depletion of fibronectin-adherent Mø from unfractionated cells isolated from both the small and the large intestinal lamina propria leads to a loss of high dose inhibition, which is also reversed by indomethacin. Further enrichment of the fibronectin non-adherent fraction by density gradient centrifugation, resulted in a 20-50 fold enhancement of MLR stimulatory activity. Like lymphoid DC, these MHC Class II positive stimulatory cells are non-phagocytic, have low density with poor plastic adherence, and lack typical Mø, T-cell and B-cell markers. These observations suggest that freshly isolated DC from intestinal lamina propria and Peyer's patches, are similar to those from spleen. It has been shown that GM-CSF causes maturation of epidermal LC in culture (Witmer-Pack et al. 1987), and as human intestinal lamina propria cells constitutively produce GM-CSF, this could account for *in vivo* maturation of lamina propria DC, or alternatively functional maturation might be induced during isolation (Pavli et al. 1990).

KIDNEYS

Immunocytochemical analysis of cryosections from normal mouse kidney shows the presence of interstitial MHC Class II positive cells, both in the renal cortex and the medulla. Double-labelling of these sections shows that almost all of the MHC Class II positive cells are CD45-positive, and most of them are also F4/80-positive. We have purified dendritic leukocytes from mouse kidneys using enzyme digestion, followed by overnight culture in tissue culture dishes. The non-adherent fraction was then subjected to density gradient centrifugation, and the low density fraction was recovered. Further enrichment of the non-adherent fraction was achieved with immunomagnetic bead extraction or fluorescence-activated cell sorting (FACS). The freshly isolated bulk population (Day 0), was found unable to stimulate the allogeneic mixed leukocyte reaction or syngeneic oxidative mitogenesis, but, similar to epidermal LC, this activity was acquired after overnight culture (Day 1). Both CD45-positive and negative sorted fractions on Day 0 were unable to stimulate oxidative mitogenesis, whereas on Day 1 the immunostimulatory activity was found only in the CD45-positive fraction. When MHC Class II positive cells were isolated on Day 0 and Day 1 with B21-2 (anti-Ia^d) coated immunomagnetic beads, both the MHC Class II enriched and depleted fractions on Day 0 failed to stimulate T-cell responses, whereas after overnight culture the stimulatory activity was entirely restricted to the MHC Class II enriched fraction. Further, no inhibitory activity was observed if freshly isolated bulk populations or the CD45-positive and negative sorted populations from Day 0, were titrated into oxidative mitogenesis with spleen stimulators. This indicates that the lack of immunostimulation on Day 0 is not due to the presence of an inhibitory cell population masking the activity of accessory cells, unlike liver (Klinkert et al. 1982), lung (Holt et al. 1990; Pollard and Lipscomb 1990), and gut (Pavli et al. 1990), but is probably due to the lack of immunostimulatory capacity of freshly isolated renal dendritic leukocytes. Analogous to LC, functional maturation of kidney DL, may be accompanied by up-regulation of MHC Class II surface antigens, but unlike epidermal LC, Fc receptor II (2.4G2) and antigen F4/80 may be unchanged, although this awaits confirmation. Freshly isolated MHC Class II positive cells were also able to phagocytose FITC conjugated zymosan, as has been shown for freshly isolated epidermal LC (see C. Reis e Sousa et al.), but this activity was somewhat reduced after overnight culture as the cells apparently undergo maturation (Rao, Roake et al. unpublished observations).

HEARTS

Immunofluorescence and immunoperoxidase labelling of cryosections from normal rat hearts with an anti-MHC Class II monoclonal antibody demonstrated positively stained structures, with an orientation parallel to the cardiac myocytes and a dendritic morphology (Hart and Fabre 1981). Double-staining of these interstitial MHC Class II positive cells with a panel of monoclonal antibodies showed that the cells were also positive for the leukocyte common antigen and for the MHC Class I molecules, but were not stained for W3/13 (which labels T lymphocytes and granulocytes). Further, histochemical characterization showed that the majority of the MHC Class II positive cells were negative for non-specific esterase, acid phosphatase, β-glucuronidase and ATP-ase activity. A few MHC Class II positive cells were positive for non-specific esterase, and these were later shown to be cardiac Mø (Hart and Fabre 1981). The presence of DC and Mø in normal rat hearts has also been demonstrated by using specific anti-Mø monoclonal antibody (Spencer and Fabre 1990).

Immunocytochemical analysis of cryosections from normal mouse hearts also shows the presence of MHC Class II positive cells, which on double-labelling are also positive for the leukocyte common antigen (CD45). Most of the functional data we have obtained with mouse heart DL are similar to those of mouse renal DL namely that the freshly isolated cells are poor stimulators of primary T-cell responses but that this activity develops after overnight culture. However, like LC, functional maturation may be accompanied by down-regulation of Fc receptor and F4/80 expression whereas unlike LC MHC Class II expression may remain unchanged, although this needs to be confirmed (Rao, Roake et al. unpublished observations).

SUMMARY

These observation suggest that dendritic leukocytes from several different non-lymphoid organs *in situ* are functionally immature and that in this respect they more closely resemble epidermal LC than mature lymphoid DC. The exception appears to be the interstitial dendritic leukocytes from small and large intestinal lamina propria and Peyer's patches, where functional maturation could be attributed to constitutively secreted GM-CSF by lamina propria cell *in situ*, or alternatively to the isolation procedure which might lead to functional maturation of gut DC. After overnight culture, and possibly following organ transplantation, interstitial dendritic leukocytes may mature into potent activators of antigen-specific T-cell proliferation (immunostimulation). Further studies are needed to characterize dendritic leukocytes in solid non-lymphoid organs, and these may lead to new strategies for overcoming graft rejection by inhibiting the maturation of dendritic leukocytes after transplantation

REFERENCES

Austyn, J. M. (1987). "Lymphoid dendritic cells."*Immunology* . **62**: 161.

Austyn, J. M. and R. M. Steinman (1988). "The passenger leukocyte - a fresh look."*Transplantation.Rev* .**2**: 139.

Hart, D. N. J. and J. W. Fabre (1981). "Demonstration and characterization of Ia-positive dendritic cells in the interstitial connective tissues of rat heart and other tissues, but not brain."*J.Exp.Med.* **153**: 347.

Heufler, C., F. Koch and G. Schuler (1987). "Granulocyte-macrophage colony-stimulating factor and interleukin-1 mediate the maturation of murine epidermal Langerhans cells into potent immunostimulatory dendritic cells."*J.Exp.Med* .**167**: 700.

Holt, P. G., M. A. Schon-Hegard and P. G. McMenamin (1990). "Dendritic cells in respiratory tract."*Int.Rev.Immunol.* **6**: 139.

Holt, P. G., M. A. Schon-Hegard and J. Olivier (1987). "MHC class II antigen-bearing dendritic cell in pulmonary tissue of the rat. Regulation of antigen presentation activity by endogenous macrophage population."*J.Exp.Med* **167**: 262.

Inaba, K., G. Schuler, M. D. Witmer, J. Valinsky, B. Atassi and R. M. Steinman (1986). "The immunologic properties of purified Langerhans cells: distinct requirements for the stimulation of unprimed and sensitized T lymphocytes."*J .Exp .Med* **169**: 1169.

Klinkert, W. E. F., J. H. Labadie and W. E. Bowers (1982). "Accessory and stimulating properties of dendritic cells and macrophages from various rat tissues."*J .Exp .Med.***156**: 1.

Lechler, R. I. and J. R. Batchelor (1982). "Restoration of immunogenicity to passenger cell-depleted kidney allografts by the addition of donor strain dendritic cells."*J.Exp.Med.***155**: 31.

Nichod, L. P., M. F. Lipscomb, J. C. Weissler and G. B. Toews (1989). "Mononuclear cells from human lung parenchyma support antigen induced T-cell proliferation."*J .Leuk .Biol* . **45**: 336.

Pavli, P., C. E. Woodhams, W. F. Doe and D. A. Hume (1990). "Isolation and characterization of antigen-presenting dendritic cell from mouse intestinal lamina propria."*Immunology* . **70**: 40.

Pollard, A. M. and M. F. Lipscomb (1990). "Characterization of murine lung dendritic cells: Similarities to Langerhans cells and thymic dendritic cells."*J.Exp.Med.***172**: 159.

Rochester, C. L., E. M. Goodell, J. K. Stoltenborg and W. E. Bowers (1988). "Dendritic cells from rat lung are potent accessory cells."*Am.Rev.Respir.Dis.* **138**: 121.

Schuler, G., N. Romani and R. M. Steinman (1985). "A comparison of murine epidermal Langerhans cells with spleen dendritic cells."*J.Invest.Dermatol.* **85**: 99.

Schuler, G. and R. M. Steinman (1985). "Murine epidermal Langerhans cells mature into potent immunostimulatory dendritic cells in vitro."*J.Exp.Med.***161**: 526.

Sertl, K., T. Takemura, E. Tschachler, V. J. Ferrans, M. A. Kaliner and E. M. Shevach (1986). "Dendritic cells with antigen-presenting capability reside in airway epithelium, lung parenchyma, and visceral pleura."*J.Exp.Med.* **163**: 436.

Spalding, D. M., W. J. Koopman, E. J.H., J. R. McGhee and R. M. Steinman (1983). "Accessory cells in murine Peyer's patch. 1. Identification and encrichment of a functional dendritic cell." *J.Exp.Med.* **157**: 1646.

Spencer, S. C. and J. W. Fabre (1990). "Characterization of tissue macrophages andinterstitial dendritic cell as distinct leukocytes normally resident in the connective tissue of rat heart."*J.Exp.Med.* **171**: 1841.

Steinman, R. M. (1991). "The dendritic cell system and its role in immunogenecity."*Annu.Rev.Immunol.* **9**: 271.

Witmer-Pack, M. D., W. Olivier, J. Valinsky, G. Schuler and R. M. Steinman (1987). "Granulocyte/macrophage colony-stimulating factor is essential for the viability and function of cultured murine epidermal Langerhans cells."*J.Exp.Med.* **166**: 1484.

Witmer-Pack, M. D., J. Valinsky, W. Olivier and R. M. Steinman (1988). "Quantitation of surface antigens on cultured murine epidermal Langerhans cells: rapid and selective increase in the level of surface MHC products."*J.Invest.Dermatol..* **90**: 387.

RT1B/D+ NON-LYMPHOID DC IN EARLY GVHD AND Hg-INDUCED AUTOIMMUNITY OF RAT SALIVARY AND LACRIMAL GLANDS

Åke Larsson[1], Kouji Fujiwara[2], and Michael Peszkowski[1]

[1]Department of Oral Pathology, School of Dentistry, Lund University, Malmö, Sweden
[2]Department of Pathology, Saga Medical School, Saga, Japan

INTRODUCTION

Low dose injections of $HgCl_2$ into Brown Norway (BN), but not other rat strains, are followed by T cell-expansion, resulting in immunopathologic events[1]. A pathogenetic role of MHC class II-autoreactive T cells was suggested, since T helper cells from Hg-injected BN rats BN(Hg) are able to transfer the immune disease into normal BN[2,3]. In BN(Hg), lymphocyte infiltrates spontaneously appeared in peripheral tissues, e.g. oral mucosa and salivary glands[4,5], which may be related to the (migratory?) properties of the autoreactive T cells. To test this hypothesis, an experimental model, in which activation and proliferation of autoreactive T cells occurs by ways other than chemical induction, would be of interest.

Experimental graft-versus-host disease (GVHD) may offer such a model[6,7]. In selected parent to offspring (P->F1) combinations of non-irradiated animals, the initiating mechanism for GVHD induction ("afferent" phase[8]) is recognition of MHC class-II alloantigens of the host lymphoid tissues by "naive" $CD4^+$ donor T cells[9]. This will give rise to a clonal expansion of donor T cells with properties of primed T cells, which may now enter peripheral circulation. Theoretically, these T cells may show peripheral tissue migratory properties similar to those of the autoreactive T cells thought to be induced in BN(Hg). Furthermore, potential targets of such cells in peripheral tissue may be constitutively class II-expressing dendritic cells (DC), which may thereby initiate the development of peripheral lesions.

The purpose of the present study was to immunohistochemically characterize the population of class II expressing DC-like, and some other related, cells in lacrimal and salivary glands of control rats and to perform a comparative study of the lesions appearing in these glands in BN(Hg) and (BN->(BNxLEW)F1) GVHD.

MATERIAL AND METHODS

Sexually mature, inbred Brown Norway (BN) rats ($RT1^n$) from The Wallenberg Laboratory, Lund, Sweden were used. (BNxLEW)F1 hybrids were obtained in our own

Animal Care Center, by mating female BN and male Lewis (LEW) rats ($RT1^l$) from Mϕllegaard Breeding Center, Copenhagen, Denmark. Animals three to five months old, were kept in separate cages and fed standard diet and tap water at libitum.

For neuroleptanalgesia, etorphine-acepromazine (Immobilon^R; 12.5μg/kg body weight) and its antidote diprenorphine (Revivon^R; 45μg/kg body weight) were used[10,11].

In BN(Hg), age matched male and female BN rats received $HgCl_2$ in H_2O (1mg/kg body weight) subcutaneously at days 0, 2, 5, 7 and 9 and were sacrificed at day 13[4,5,12]. Untreated animals were used as controls.

In GVHD, BN rat spleens, removed under deep neuroleptanalgesia (Immobilon, 3ml/kg bw), were minced in RPMI 1640 (KEBO AB, Sweden) and filtered through nylon sieve. Acquired cell suspension was centrifuged 3x10 min at 700 g, the cells counted by the dye exclusion test with trypan blue and resuspended in RPMI to a final concentration of 0.5-1x10^8/ml (modified from Fujiwara et al.[13]).

The GVH reaction was induced in (BNxLEW)F1 hybrids by one i.v.(tail vein) injection of 1 ml of BN spleen cell suspension. F1 rats injected with equal amount of untreated F1-rat spleen cells, and untreated (BNxLEW)F1 served as controls. The donor and recipient animals were sex-matched in all experiments.

Half of the (BNxLEW)F1 recipients of BN spleen cells were sacrificed 7 days after injection (early GVHD). Remaining animals were killed at first oral signs of GVHD (between day 11 and 14), noted as perioral and labial redness with some degree of labial ulceration. The control animals (F1->F1) were sacrificed at day 14.

In both experiments, submandibular, parotid and lacrimal glands were dissected, embedded in Tissue-Tek^R O.C.T compound (Miles Inc, Elkhart, IN, USA) and snapfrozen in precooled isopentane at -80°C. Frozen sections were formalin fixed and stained with hematoxylin-eosin for routine histology.

Table 1. Murine anti-rat monoclonal antibodies (Serotec, Oxon, England) and their specificity.

OX 34	CD2, pan T
W3/25	CD4, T helper, macrophages, dendritic cells
OX 8	CD8, T supressor and cytotoxic
R73	TCR αβ
OX 22	CD45RC (190-220 kD), "naive" T
OX 33	CD45 (240 kD), pan B
OX 6	RT.1B (homol to human HLA-DQ), monomorphic
OX 17	RT.1D (homol to human HLA-DR), monomorphic
OX 3	RT.1B^l, polymorphic
ED 1	Macrophages, monocytes, veiled cells

For immunohistochemistry, six μm consecutive frozen sections were air-dried, fixed for 10 min in acetone and washed in phosphate buffered saline (PBS) for 20 min. Endogenous peroxidase activity was blocked by treatment with 0.3% H_2O_2 in PBS for 30 min and followed by 1% normal rabbit serum (Dakopatts, Copenhagen, Denmark) in PBS for 30 min, to block nonspecific background staining. Primary monoclonal antibodies (mAbs), as specified in Table 1, in PBS were applied for 60 min at room temperature. Following a rinse in PBS, the sections were incubated with peroxidase-conjugated rabbit antimouse IgG (Dakopatts) in PBS containing 5% of normal rat serum (Dakopatts). As chromogen 0.05% 3,3'-diaminobenzidine tetrahydrochloride (DAB) with 0.01% H_2O_2 was used (30 min). The sections were lightly counterstained with Mayer's hematoxylin and mounted[5].

RESULTS

The different glands showed the same staining pattern, although changes were consistently most prominent in lacrimal glands. In the GVH animals, at 7 days post-induction, spleen was enlarged to almost its double size, as an indicator of immuno-stimulation rather than suppression[9].

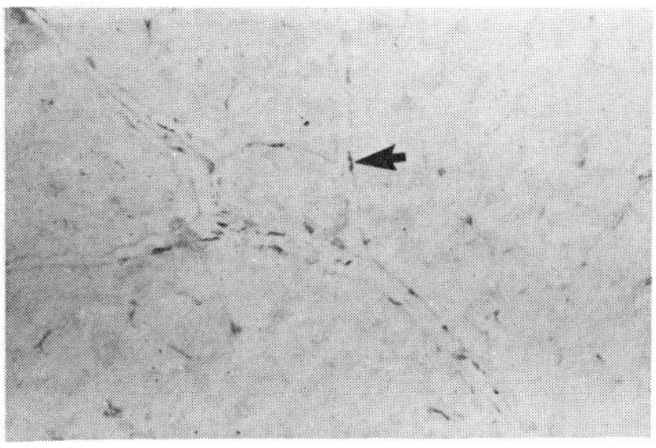

Fig. 1. (All sections show lacrimal gland). Control section stained with W3/25. A few stained cells are present in the interstitial c.t. septa (arrow) with only a few cells at surrounding acini. A similar stain pattern was seen with ED1. x 100

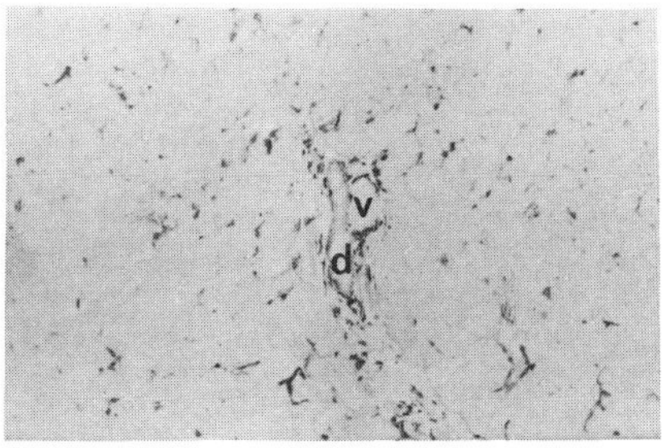

Fig. 2. Control section stained with OX6. Stained cells with a DC-like appearance are present in interstitial c.t., especially at the d/v plexus (d, v) but also at surrounding acinar structures. OX3 and OX17 showed a similar staining pattern. x 50

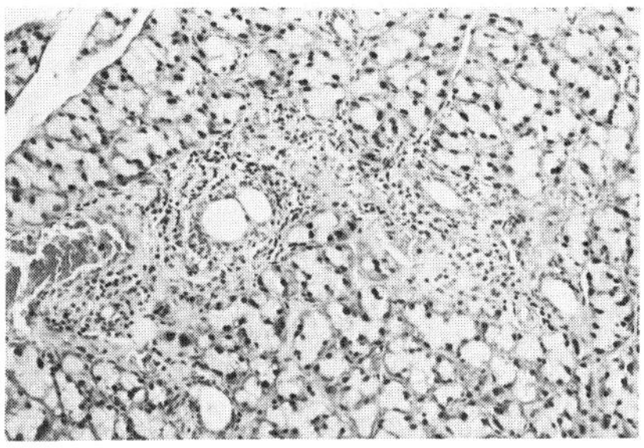

Fig. 3. Symptomatic GVH, section stained with Htx-eosin. A focal collection of lymphocytic cells occurs at the d/v complex. x 100

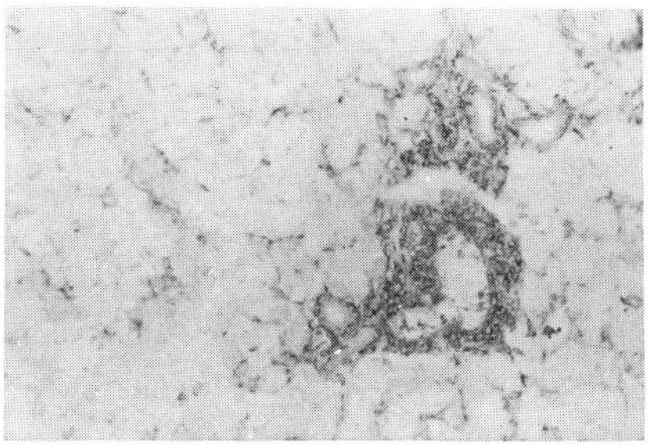

Fig. 4. BN(Hg), section stained with W3/25. A heavy staining is seen at the d/v complex (cf Fig. 1), with a few stained cells at surrounding acini. x 100

In control glands, isolated lymphocytic $CD2^+$, $TCR\alpha\beta^+$ and $CD45RC^+$ cells could only occasionally be detected. A small but regularly appearing number of $CD4^+$ (Fig. 1) and $ED1^+$ cells was generally present within the interlobular connective tissue (c.t.) septa, carrying the excretory ducts with accompanying blood vessels ("ductal/vascular" (d/v)-plexus). In contrast, only a few $CD4^+$ and $ED1^+$ cells could be seen in the sparse c.t. present around acini. Staining with OX6(RT1B) was similar to OX17(RT1D) (Fig. 2), with DC-like cells appearing within the c.t. septa and around acini. These cells were particularly prominent at the d/v plexus. In the F1 hybrids, similar staining patterns were obtained with OX3 (polymorphic RT1B, reactive with $RT1B^l$ but not $RT1B^n$).

In BN(Hg), an increased number of $CD4^+$ and $ED1^+$ cells was found in focal collections of lymphocyte cells at the d/v plexus (Fig. 3). These foci contained T($CD2^+$), but no B($CD45^+$) and only isolated "naive" ($CD45RC^+$) cells. Such foci were never observed outside the d/v plexus but in contrast to controls, a small number of $CD4^+$ (Fig.

4) and $ED1^+$ cells also appeared in the sparse c.t. around the acini. The lymphocyte foci at the d/v plexus stained heavily with $OX6(RT1B^+)$ (Fig. 5) and $OX17(RT1D^+)$. Some of the stained cells were small and lymphocyte-like. Also the ductal epithelium and the vascular endothelium often stained (Fig. 6). In addition, there was an increased number of $OX6^+(RT1B^+)$ and $OX17^+(RT1D^+)$, DC-like cells around the acini. No differencies were observed between RT1B and RT1D staining pattern.

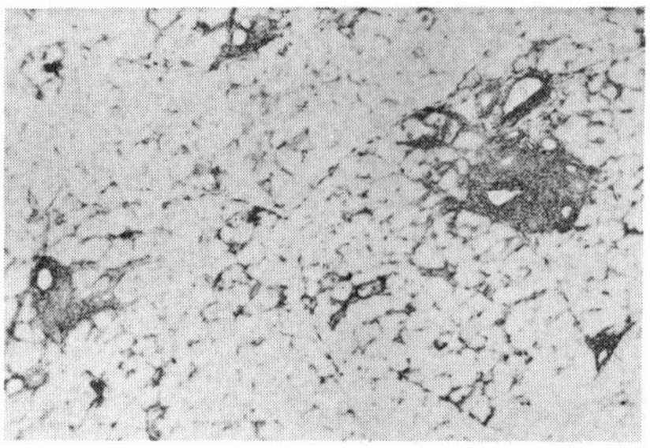

Fig. 5. BN(Hg), section stained with OX6, cf Fig. 2. A dense focus of stained cells is seen at the d/v complexes and also an increased number of DC-like cells at surrounding acini. x 50

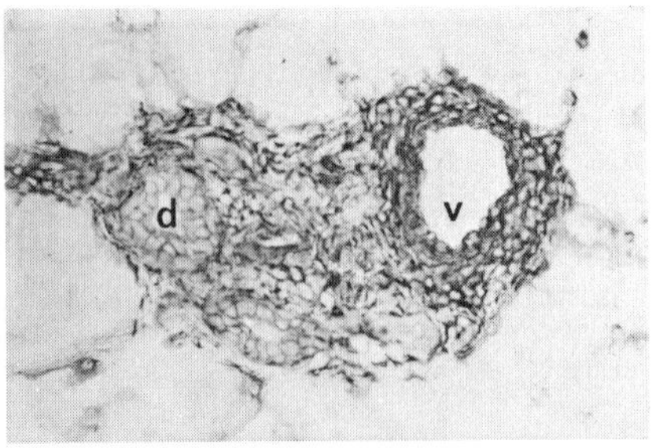

Fig. 6. BN(Hg), section stained with OX17. High magnification of d/v complex shows a dense infiltrate of cells and also a positive staining of ductal epithelium (d) and vascular lining (v). OX6 showed a similar staining pattern. x 200

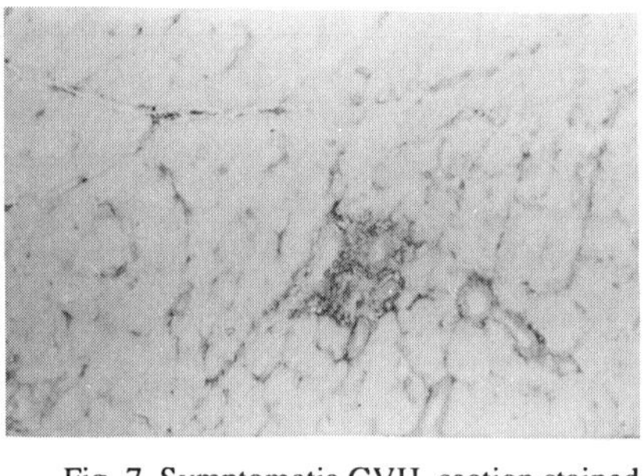

Fig. 7. Symptomatic GVH, section stained with W3/25. Staining pattern is similar to BN(Hg) (cf Fig. 4). x 50

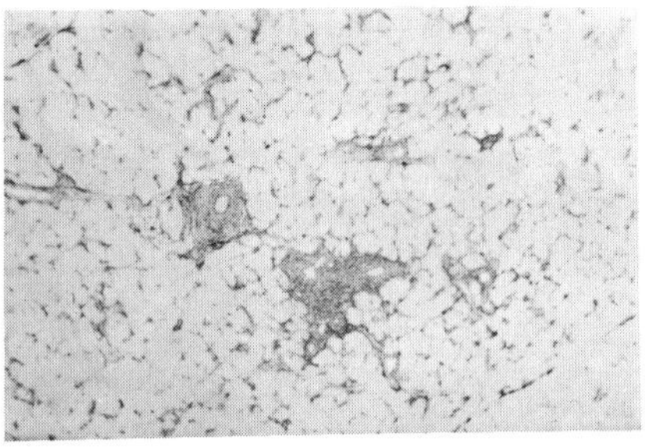

Fig. 8. Symptomatic GVH, section stained with OX6 staining pattern is similar to BN(Hg) (cf Fig. 5). x 50

In GVHD, stained sections showed a reaction remarkably similar to that seen in BN(Hg) (Fig. 7). Sections from early (7 days) GVH-animals showed fewer $RT1B^+$ and $RT1D^+$ DC-like cells than the BN(Hg)-sections. This difference was however no longer observed in sections from rats with oral signs (11-14 days) of GVHD (Fig. 8). With OX3 ($RT1B^l$/LEW class II specific) antibody, the staining pattern was similar to that of OX6/OX17, although the number of the $OX6/17^+$ cells seemed to slightly exceed that of $OX3^+$ cells in the foci.

DISCUSSION

Our findings indicate that a population of $RT1B/D^+$, $CD4^+$ and $ED1^+$ macrophages [14,15] may well be involved in the development of the early lymphocyte foci at the d/v complex[16], but that constitutively class II (RT1B/D^{+}) expressing DC may also contribute to (initiate?) this process. The similarities between BN(Hg) and (BN->(BNxLEW)F1) GVHD have been a striking finding in the present study. With support from the literature on BN(Hg)[2,3] and (P->F1) GVHD[9,17,18], we suggest that common pathogenic mechanisms are involved.

The initiating mechanisms for (P->F1) GVHD induction in the non-irradiated animal model is recognition of MHC class II antigens of the host lymphoid tissues, mainly spleen and lymph nodes, by donor T cells[9,18]. Injected donor spleen cells enriched for T cells by passing over nylon wool[19] or nylon sieve (preliminary data) would be expected to act in the F1 hybrid as "naive" cells, with corresponding migration pattern, i.e. preferential migration to spleen and lymph nodes[9]. Thus, no peripheral tissue pathosis would be expected to occur in the "afferent" phase of (P->F1) GVHD. In no or insignificant MHC class I differences, as in the present study, an MHC class II disparity will give rise to a clonal expansion of allostimulated CD4^{+} T-helper donor cells[18], as also indicated by spleen enlargement. Following this expansion, activated T cells with the properties of primed T cells may potentially contribute to peripheral pathosis in the "efferent" GVHD phase.

We have somewhat arbitrarily chosen 7th post-induction day of (P->F1) GVHD for analysis of the early "efferent" phase. We have found (preliminary observations) that the majority of the infiltrating T lymphocytes lack a "naive"-T phenotype (i.e. they are CD45RC^{-})[20], supporting our thesis that they may represent activated, primed T cells. Similar findings were made in BN(Hg). In GVH animals, we have not been able to detect any significant differences between OX6/17 (monomorphic RT1B/D) and OX3 (polymorphic RT1B^{l}) staining patterns, indirectly suggesting that class II expressing DC-like cells in the lesional tissue are mainly of host ((BNxLEW)F1) origin. Slightly more OX6/17^{+} cells, as compared with OX3^{+} may be due to class II expressing, activated T cells[21]. In rat salivary and lacrimal glands, there is no constitutive class II expression on ductal epithelium and no obvious lymphocyte infiltration in normal tissue. Furthermore, class II induction in this epithelium during the "afferent" phase of GVH, preceding the stage of lymphocyte infiltration, seems unlikely, althought this may require further study. At present, we interpret the class II expression of glandular epithelium both in BN(Hg) and GVHD as a result of induction, following the lymphocyte infiltration[22]. In this process, DC, normally present in lacrimal and salivary gland tissue and constitutively expressing class II[23-28], may become targets in an initial "recall" antigen recognition event. This will result in T cell reactivation and cytokine production (γ-IFN?, TNF?) followed by class II induction of adjacent ductal epithelium and endothelium, with recruitment of more lymphocytes into the area. Part of the process may involve a gradual population of the tissue by more DC-like class II-expressing cells of host origin in a way which may mimic the events in grafted tissue[29].

According to this concept, lacrimal and salivary gland lesions in BN(Hg) and GVHD are initiated by the local migration of primed, class II autoreactive T cells, and presence of constitutively class II expressing DC (as a specific antigenic determinant). In BN(Hg), it can not be excluded that Hg, retained in and expressed on DC (with class II), may contribute to reactivation of primed T lymphocytes. However, the resulting foci would not differ from those in GVHD. Further studies are in progress to confirm this concept.

ACKNOWLEDGEMENTS

This study was supported by grants from the Swedish Medical Research Council (24B-09024).

REFERENCES

1. E. Druet, C. Sapin, G. Fournie, C. Mandet, E. Gunther, and P. Druet, Genetic control of susceptibility to mercury-induced immunonephritis in various strains of rat, *Clin Immunol Immunopathol.* 25:203 (1982).
2. L. Pelletier, R. Pasquier, F. Hirsch, C. Sapin, and P. Druet, Autoreactive T cells in mercury-induced autoimmune disease: in vitro demonstration, *J Immunol.* 137:2548 (1986).
3. L. Pelletier, R. Pasquier, J. Rossert, M-C. Vial, C. Mandet, and P. Druet, Autoreactive T cells in mercury-induced autoimmunity. Ability to induce the autoimmune disease, *J Immunol.* 140:750 (1988).
4. J. Aten, C.B. Bosman, J. Rozing, T. Stijnen, P.J. Hoedemaeker, and J.J. Weening, Mercuric chloride-induced autoimmunity in the Brown Norway rat. Cellular kinetics and major histocompatibility complex antigen expression, *Am J Pathol.* 133:127 (1988).
5. G. Warfvinge. "Experimental Oral Contact Hypersensitivity", (Dissertation), Malmö, Sweden: University of Lund (1990).
6. E. Gleichmann, H-W. Vohr, C. Stringer, J. Nuyens, and H. Gleichmann, Testing the sensitization of T cells to chemicals. From murine graft-versus-host (GVH) reactions to chemical-induced GVH-like immunological diseases, *in*: "Autoimmunity and Toxicology. Immune Disregulation Induced by Drugs and Chemicals", M.E. Kammuller, N. Bloksma, and W. Seinen, eds., Elsevier Science Publishers B.V. (Biomedical Division), Amsterdam (1989).
7. M. Goldman, P. Druet, and E. Gleichmann, T_H2 cells in systemic autoimmunity: insights from allogeneic diseases and chemically-induced autoimmunity, *Immunol Today.* 12:223 (1991).
8. J.L.M. Ferrara, and H.J. Deeg, Graft-versus-host disease, *N Engl J Med.* 324:667 (1991).
9. A.G. Rolink, A. Strasser, and F. Melchers, Autoimmune diseases induced by graft-vs.-host disease, *in*: "Graft-vs.-Host Disease. Immunology, Pathophysiology, and Treatment", S.J. Burakoff, H.J. Deeg, J. Ferrara, and K. Atkinson, eds., Marcel Dekker, Inc., New York and Basel (1990).
10. A.V. Fisker, I. Stage, and H.P. Philipsen, Use of etorphine-acepromazine and diprenorphine in reversible neuroleptanalgesia of rats, *Lab Anim.* 16:109 (1982).
11. E.E. Ahlfors, and Å. Larsson, Light microscopic study of experimental cell-mediated immunity in rat oral mucosa, *Scand J Dent Res.* 92:549 (1984).
12. L. Pelletier, F. Hirsch, J. Rossert, E. Druet, and F. Druet, Experimental mercury-induced glomerulonephritis, *Springer Semin Immunopathol.* 9:359 (1987).
13. K. Fujiwara, N. Sakaguchi, and T. Watanabe, Sialoadenitis in experimental graft-versus-host disease. An animal model of Sjögren's syndrome, *Lab Invest.* 65:710 (1991).
14. W.A. Jefferies, J.R. Green, and A.F. Williams, Authentic T helper CD4 (W3/25) antigen on rat peritoneal macrophages, *J Exp Med.* 162:117 (1985).
15. C.D. Dijkstra, B.A. Dopp, P. Joling, and G. Kraal, The heterogeneity of mononuclear phagocytes in lymphoid organs: distinct macrophage subpopulations in the rat recognized by monoclonal antibodies ED1, ED2 and ED3, *Immunology.* 54:589 (1985).
16. H.M. Shulman, Pathology of chronic graft-vs.-host disease, *in*: "Graft-vs.-Host Disease. Immunology, Pathophysiology, and Treatment", S.J. Burakoff, H.J. Deeg, J. Ferrara, and K. Atkinson, eds., Marcel Dekker, Inc., New York and Basel (1990).
17. T. Ghayur, T.A. Seemayer, and W.S. Lapp, Histologic correlates of immune functional deficits in graft-vs.-host disease, in: "Graft-vs.-Host Disease. *Immunology*, Pathophysiology, and Treatment", S.J. Burakoff, H.J. Deeg, J. Ferrara, and K. Atkinson, eds., Marcel Dekker, Inc., New York and Basel (1990).
18. C.S. Via, Kinetics of T cell activation in acute and chronic forms of murine graft-versus-host disease, *J Immunol.* 146:2603 (1991).
19. F.J. Guillen, J. Ferrara, W.W. Hancock, et al., Acute cutaneous graft-versus-host disease to minor histocompatibility antigens in a murine model, *Lab Invest.* 55:35 (1986).
20. F. Powrie, and D. Mason, Phenotypic and functional heterogeneity of $CD4^+$ T cells, *Immunol Today.* 9:274 (1988).
21. J.F. Kaufman, C. Auffray, A.J. Korman, D.A. Shackelford, and J. Strominger, The class II molecules of the human and murine major histocompatibility complex, *Cell.* 36:1 (1984).
22. M. Feldmann, Regulation of HLA class II expression and its role in autoimmune disease, *in*: "Autoimmunity and Autoimmune Disease", D. Evered, and J. Whelan, eds., John Wiley & Sons Ltd, Chichester (1987).

23. R.M. Steinman, The dendritic cell system and its role in immunogenicity, *Annu Rev Immunol.* 9:271 (1991).
24. W.L. Ford, and H.J. Deeg, Bone marrow transplantation, with emphasis on GVH reactions, *Transplant Proc.* XV:1517 (1983).
25. J.M. Austyn, Lymphoid dendritic cells, *Immunology.* 62:161 (1987).
26. P.D. King, and D.R. Katz, Mechanisms of dendritic cell function, *Immunol Today.* 11:206 (1990).
27. J. Sprent, and M. Schaefer, Antigen-presenting cells for unprimed T cells, *Immunol Today.* 10:17 (1989).
28. D.N.J. Hart, and J.W. Fabre, Demonstration and characterization of Ia-positive dendritic cells in the interstitial connective tissues of rat hart and other tissues, but not brain, *J Exp Med.* 153:347 (1981).
29. E.J. Weringer, and A.A. Like, Identification of T cell subsets and class I and class II antigen expression in islet grafts and pancreatic islets of diabetic BioBreeding/Worcester rats, *Am J Pathol.* 132:292 (1988).

IN-VITRO INFECTION OF PERIPHERAL BLOOD DENDRITIC CELLS WITH HUMAN IMMUNODEFICIENCY VIRUS-1 CAUSES IMPAIRMENT OF ACCESSORY FUNCTIONS

Jihed Chehimi, Kesh Prakash, Vedapuri Shanmugam, Stephanie J. Jackson, Santu Bandyopadhyay, and Stuart E. Starr

The Division of Infectious Diseases and Immunology, The Joseph Stokes, Jr. Research Institute of The Children's Hospital of Philadelphia, Pennsylvania, 19104

INTRODUCTION

Dendritic cells (DC) although a minor bone-marrow derived cell population, play an important role in initiating T-cell-mediated immune responses (Steiman and Inaba,1989). The involvement of DC in acquired immune deficiency syndrome (AIDS) has been suggested by several reports (Eales et al.,1988; Patterson and Knight,1987; Langhoff et al.,1991). Patterson and Knight (1987), using a preparation containing 2 to 48% DC, detected viral budding from the plasma membrane 5 days after HIV-IIIB infection. Surprisingly, Langhoff et al (1991) reported that DC produced ~ 10-fold more virus, than did T lymphocytes or fresh adherent monocyte populations when infected with several strains of HIV. Our studies were designed to test the relative ability of different strains of HIV-1 displaying different cellular tropism, to infect human peripheral blood DC and to determine the effect of such infection on DC accessory functions.

MATERIALS AND METHODS

Purification of DC from human peripheral blood. Peripheral blood mononuclear cells (PBMC), 5 x 10^6/ml in RPMI 1640 containing 10% heat-inactivated human AB serum were incubated in plastic tissue culture flasks for 90 min. Nonadherent cells were removed, and adherent cells were washed to remove floating nonadherent cells. Flasks were then incubated for 18 h at 37°C in 5% CO2. Released nonadherent cells were incubated for 30 min at 4°C with a mixture of monoclonal antibodies (MAbs): OKT3 (anti-CD3), B36.1 (anti-CD5); B52.1 (anti-CD14); B73.1, Leu11b and 3G8 (anti-CD16); THB5 (reacting with Epstein-Barr virus receptor) and BC1 (anti CD20). Cells were then washed 3 times with ice cold PBS and indirectly rosetted with CrCl3-treated goat anti-mouse Ig-coated sheep erythrocytes. The preparation was then centrifuged over a Ficoll gradient for 30 min. The interphase containing most of the $CD3^-$, $CD5^-$, $CD14^-$, $CD16^-$, $CD20^-$ cells was washed 3 times and incubated for 10 min at 37°C with low-toxicity baby rabbit complement diluted 1:3, or panned on goat anti-mouse immunoglobulin coated plastic dishes. Cells were then washed 3 times and residual monocytes were removed by repeated pannings (1 to 3) using human immunoglobulin-coated plastic dishes. Purified DC preparations were analyzed by flow cytofluorometry.

Detection of CD4 mRNA by Northern (RNA) blot analysis. Total RNA was extracted from purified DC as described by Chomczynski and Sacchi (1987), and hybridization was done as described previously (Chehimi et al.,1991). The probe used for the detection of the CD4 mRNA was the full-length 3.0-kb cDNA, pT4B.

Dendritic Cells in Fundamental and Clinical Immunology,
Edited by Kamperdijk *et al.*, Plenum Press, New York, 1993

Viruses. Culture supernatants of the following HIV-1 strains were used: III-B, WMJ1, SF2, ARV2; SF162, 89.6, and clone HXB2. Production and titration of viral stocks were done as described previously (Chehimi et al.,1991).

HIV-infection of DC. Purified DC (~ 5 x 10^5) in 15-ml conical plastic centrifuge tubes were inoculated with different cell-free HIV isolates or equivalent amounts of heat-inactivated virus (3h at 56°C), at multiplicities of infection of 0.1-0.5 based on 50% tissue culture infective doses. After 3 to 4 h of incubation at 37°C, DC were washed several times with serum-free RPMI, and twice rapidly with 0.05% trypsin to remove any surface bound virus. DC were then resuspended in RPMI-1640 containing culture media supplemented with 15% Lymphocult-T-LF and media was changed after 18h incubation. In selected experiments DC were incubated for 1 h at 37°C with 25 μg/ml of MAb 3G8 (anti-FcγRIII) or 25 μg/ml of MAb Leu-3a prior to inoculation with HIV. In these experiments, the MAbs were present throughout the culture period. Cultures were fed with fresh media containing the appropriate mAbs every 3-4 days. In selected experiments PHA-IL-2 stimulated PBMC and monocytes, were included in parallel with DC for the purpose of comparison. At different time intervals culture supernatants and cells were harvested for p24 core Ag determinations. Cells were washed three times in phosphate-buffered saline (PBS) and lysed.

Detection of HIV DNA and HIV RNA in infected DC. DNA from uninfected and 89.6-infected DC was extracted, digested with the restriction endonuclease SacI, electrophoresed in a 1% agarose gel, transferred to a nitrocellulose membrane and hybridized to a nick-translated ^{32}P-labelled full-length 9.1-kb HIV probe derived from the clone HXB2 as described previously (Chehimi et al., 1991). Total RNA was isolated from uninfected and 89.6-infected DC, 89.6-infected monocytes and HIV-IIIB-infected PBMC as described (Chomczynski and Sacchi,1987). RNA (10 μg) was separated on a 1% agarose-2.2 M formaldehyde gel, and transferred to nitrocellulose. Prehybridization and hybridization were carried out as described above using a Bgl II fragment (0.58-kb) consisting of a portion of the HIV env gene sequences.

Functional assays. Mock-infected and HIV-infected DC were cultured in RPMI-1640 containing 10% human AB serum and 15% Lymphocult-T-LF. Then the cells were washed two times and used as stimulator cells in all functional assays. For ConA stimulation, purified syngeneic $CD8^+$T cells were used as responding cells. ConA (5 μg/ml) stimulated cultures were pulsed after 72 hrs, with 2 μCi [^{3}H]-TdR for 4 h. Syngeneic and allogeneic MLR were done as described previously by Young and Steinman (1988). Plastic nonadherent PBL obtained from a normal HIV-seronegative donor, containing ≥ 70% T cells were used as responders. Responding cells were cultured for 6 days with irradiated syngeneic or allogeneic stimulator cells in 96-well plates in 200 μl medium containing 15 μg/ml recombinant soluble CD4 to prevent HIV transmission to responding T cells. For assays of HSV-1 specific lymphocyte proliferation, cultures consisting of responding PBL and syngeneic irradiated-DC were incubated with UV-irradiated-HSV-1 strain NS (15 μl/well, giving a final antigen dilution of 1:100). Recombinant soluble CD4 (15 μg/ml) was present throughout the assay. After 144 h cultures were pulsed with [^{3}H]-TdR for 16 h, and harvested. The incorporation of [^{3}H]-TdR was measured using a liquid scintillation counter and the results were expressed as mean cpm ± S.D of triplicate cultures.

RESULTS AND DISCUSSION

The purpose of the present study was to determine whether human peripheral blood DC could be infected in-vitro with different strains of HIV-1 and to study the effect of such infection on DC accessory functions. Purified DC preparations contained ≥ 80% HLA-DR positive cells as determined by flow cytometry and 1-2 % expressed markers associated with T or B lymphocytes, NK cells or monocytes . Since the flow cytofluorometry data suggested that DC preparations might contain small numbers of contaminating cells, Northern blot analyses were done to determine if CD4 mRNA could be detected. While DC did not express any detectable CD4 mRNA (Figure 1, lane A), $CD4^+$ T cells expressed a 3.0 kb transcript (Figure 1, lane B). Fresh monocytes expressed a similar 3.0-kb transcript (not shown).

Peripheral blood DC maintained in lymphocyte-conditioned media were susceptible to HIV-1 strains displaying different patterns of cellular tropism. The duotropic (monocyte-tropic replicating in T cells) strain 89.6 and SF162 yielded consistently higher extracellular p24 Ag levels, 8 and 16 days postinfection than did lymphocyte-tropic strains IIIB, SF2 and WMJ1. The clone HXB2 yielded higher p24 Ag levels than did its parent strain IIIB (Table 1).

Table 1. Replication of different strains of HIV-1 in DC

HIV strains	p24 Ag (pg/ml) in culture supernatants at various time post-infection Day 8	Day 16
III-B	52	18
SF2	45	21
WMJ1	71	28
HXB2	699	403
SF162	520	175
89.6	498	479

Purified DC were infected with different strains of HIV-1 and cultured in RPMI 1640 containing 10% FBS, supplemented with 15% Lymphocult-T-LF. Culture medium was changed after 24 hrs and levels of p24 Ag were determined by ELISA.

In contrast, identical concentrations of Leu3a had no effect on the amounts of cell-free or cell-associated p24 Ag detected after similar numbers of DC were inoculated with SF162, 89.6 or HXB2 (Table 2). MAb 3G8 (anti-FcγRIII) had no inhibitory effect on HIV replication in any of the cell populations tested. These results suggest that an alternate non-CD4, non-FcγRIII-receptor mechanism of viral entry exists for DC in the peripheral blood.

To determine the ability of anti-CD4 MAb (Leu 3a) or anti-FcγRIII (3G8) to inhibit HIV infection, we performed blocking experiments in which similar numbers of DC, monocytes and PBMC were incubated with Leu3a or 3G8 MAbs prior to and after infection with various strains of HIV. In all experiments the presence of Leu3a substantially reduced the ability of HIV to replicate in monocytes and PBMC (Not shown).

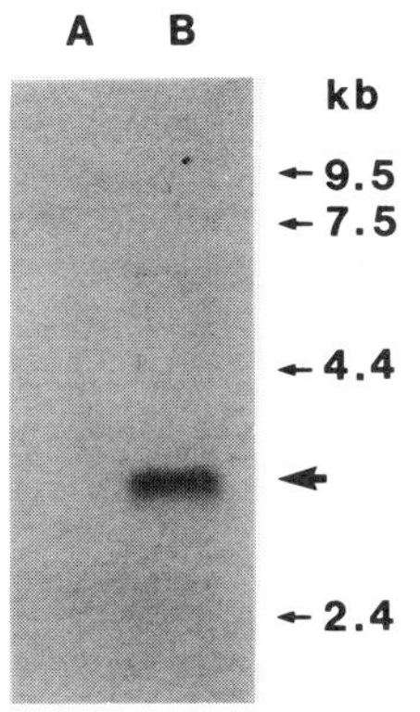

Figure 1. Northern blot analysis of expression of CD4-specific mRNA in DC. Lanes: A, DC; B, CD4+ cell line. Total RNA (10 μg) was electrophoresed through a 1% agarose-formaldehyde gel, transferred to nitrocellulose, and hybridized with a nick-translated full-length 3.0-kb CD4 cDNA.

Table 2. Blocking experiments with anti-CD4 and anti-CD16 MAbs

Virus strains	p24 Ag (pg/ml) in culture supernatant in the presence of:			Cell associated p24 Ag (pg/ml) in the presence of:		
	Media	CD4	CD16	Media	CD4	CD16
SF162	550	535	490	410	470	429
HXB2	595	565	619	489	ND	509
89.6	417	398	402	480	ND	505

DC were incubated with anti-CD4 MAb Leu3a (25 μg/ml) or anti-FcRγIII MAb 3G8 (25 μg/ml). After 1 h incubation cells were exposed to SF162, HXB2, or 89.6. Cells were continuously exposed to indicated MAbs throughout the culture period.

To determine whether HIV DNA could be detected in infected DC, high molecular-weight DNA was extracted at 7 days post-infection from HIV-89.6-infected DC, cultured in the continuous presence of Leu3a MAb. While no bands were detected with DNA obtained from uninfected DC (Fig 2, lane 1), a single band of approximately 9.3-kb was detected in DNA extracted from a T cell line infected with HIV-IIIB (lane 2) and from DC infected with 89.6 (lane 3)

In order to examine the synthesis of viral RNA in infected DC, total RNA was extracted at 7 d post-infection, transferred to nitrocellulose after electrophoresis and hybridized to a 0.58-kb Bgl II fragment derived from the env gene of HXB2. As shown in Figure 3, no bands were detected in uninfected DC (lane 1), whereas the three HIV-specific RNA species, 9.3-kb (genomic RNA as well as mRNA), 4.3-kb and 2.0-kb were present in DC infected with HIV-89.6 (lane 2) and PBMC-infected with 89.6 used as a control (lane 3). Similar results were obtained when a 9.1-kb probe was used (not shown). These results show that (a) less viral RNA is detected in DC as compared to PBMC and (b) the 4.3-kb mRNA is synthesized in greater amounts than the 2.0-kb and 9.3-kb HIV-specific RNA species.

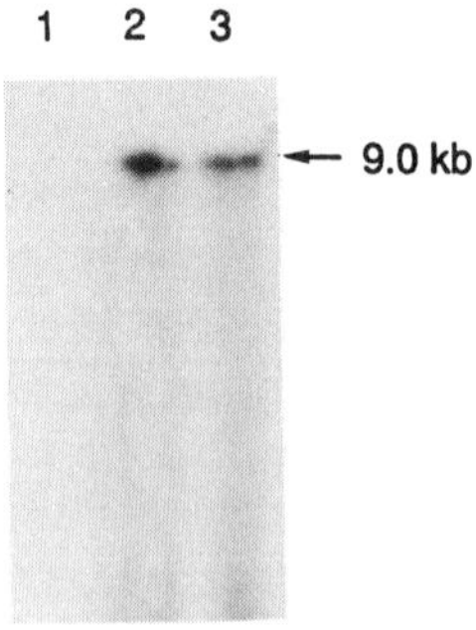

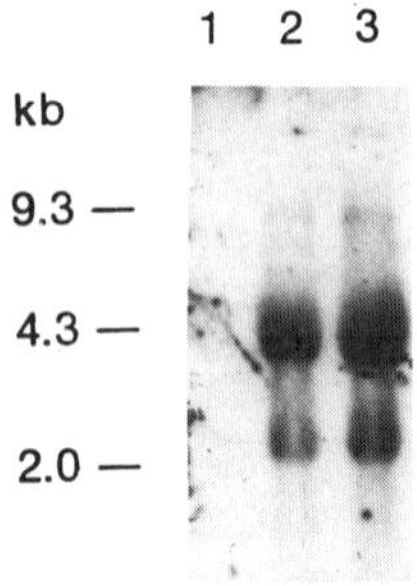

Figure 2 and 3. Southern blot analysis of HIV DNA (left panel) and Northern blot analysis of HIV RNA (right panel) in DC infected with HIV-89.6.

The ability of DC to stimulate proliferative responses before and after exposure to HIV in-vitro was examined using both autologous and heterologous MLR, as well as antigen and mitogen-stimulated lymphocyte proliferation. When SF162 or IIIB-infected DC were used as accessory cells, $CD8^+$ T cell proliferation in response to Con A was reduced compared to that observed with mock-infected DC. In the presence of DC exposed to heat-inactivated HIV, $CD8^+$ T cell proliferation was not reduced suggesting that reduction of DC function required live virus (Table 3).

Table 3. Effect of HIV-1 infection of DC on ConA presentation to syngeneic $CD8^+$ T cells.

Assay Mixture	^{3}[H] thymidine uptake, cpm
Mock infected DC + CD8 cells	135 ± 39
Mock infected DC + Con A	272 ± 43
Mock infected DC + CD8 cells + Con A	1520 ± 185
HI-IIIB-infected DC + CD8 cells + Con A	1510 + 119
IIIB-infected DC + CD8 cells + ConA	1052 + 110 (30)
HI-SF162-infected DC + CD8 cells + Con A	1480 ± 165
SF162-infected DC + CD8 cells + Con A	629 ± 204 (59)

$CD8^+$ cells were cultured for 72 hr with or without ConA in the presence of mock-infected, heat-inactivated (HI)-exposed, and HIV-infected DC. Number in parenthesis indicate the percent reduction determined as: [(cpm in cultures containing mock-infected DC, CD8 cells and ConA minus cpm in culture containing infected DC, CD8 cells and ConA)/cpm in cultures containing mock-infected DC, CD8 cells and ConA] x 100.

Similarly, autologous and heterologous MLR responses elicited when 89.6 or HXB2-infected DC were used as stimulator cells. The cells were 40 to 80% lower than responses in the presence of noninfected DC (Table 4). DC infected with 89.6 and pulsed with HSV-1 were 32 to 62% less effective than HSV-1-pulsed non-infected DC in stimulating T cell proliferation (Table 4).

This study clearly indicates that peripheral blood DC can be productively infected in-vitro by various HIV-1 isolates. Viral replication was demonstrated by p24 Ag production, through the presence of HIV DNA and HIV RNA within infected DC. Infection of DC was non-cytopathic and reduction in DC number starting at day 3 in both infected and noninfected preparations (not shown) do not support the suggested idea that DC are killed in-vitro by HIV. Infectious virus was recovered from filtered cell-free supernatants of 89.6-infected DC. Filtered cell-free supernatant containing approximately 500 pg/ml of p24 Ag contained infectious virus which could infect CEM-174 cells (not shown). The reduction in DC number starting day 4 in uninfected as well as in infected preparations, contradicts the suggested idea that DC are killed in vitro by HIV. Such infection was not blocked by anti-CD4 or anti-FcγRIII MAbs and may involve an alternate mechanism of viral entry.

Two independent studies recently described the susceptibility of peripheral blood DC to infection with HIV (Patterson and Knight,1987; Langhoff et al.,1991). Patterson and Knight described budding of HIV-IIIB from partially enriched DC preparations five days after infection. Langhoff and coworkers reported a higher level of HIV production in DC than in activated T cells or fresh monocytes. In our hands, not all HIV strains replicated in DC, which were more susceptible to macrophage-tropic than to lymphocyte-tropic strains. Such differences might be explained by the different methods used to purify DC.

Table 4. Stimulation of the MLR and HSV-1 specific lymphocyte proliferation by uninfected and HIV-infected dendritic cells.

Stimulator cells	^{3}[H]-thymidine uptake (cpm) in the presence of responder cells during : MLR		HSV-1 specificlymphocyte proliferation
	Responder cells from: donor 1	donor 2	Autologous responder cells
donor 1			
Cultured DC	3960 ± 419	38940 ± 1001	10303 ± 445
89.6-infected DC	2150 ± 355 (46)	21400 ± 635 (45)	6990 ± 497 (32)
HXB2-infected DC	680 ± 102 (83)	15407 ± 890 (60)	ND
donor 2			
Cultured DC	22183 ± 845	2408 ± 250	7840 ± 510
89.6-infected DC	8970 ± 498 (60)	949 ± 106 (61)	2975 ± 305 (62)
HXB2-infected DC	6490 ± 397 (71)	699 ± 93 (71)	ND

DC were inoculated with strains 89.6 and HXB2 for 4 days, and used as accessory cells in six days MLR or HSV-1 specific lymphocyte proliferation.

An important issue raised by our study and others, is wether DC are infected in-vivo. In earlier studies, autologous and heterologous MLR, were found to be depressed in patients with AIDS (Eales et al,1988). Recently, Macatonia et al (1991) reported depletion and dysfunction of DC in HIV-infected persons, as well as the detection of HIV in DC using in-situ hybridization. Since DC specific MAbs were not available to identify the cells that hybridized with HIV-specific probe, these results dealing with a population comprising ≤1% of peripheral blood, must be interpreted with caution.

While HIV infection impairs the ability of DC to stimulate T lymphocyte proliferation in response to mitogenic and antigenic stimulation, further studies will be needed to define the precise role of DC infection in the pathogenesis of AIDS.

Acknowledgements: This work was supported by Public Health Service grant AI-31368 from the National Institutes of Health. We thank Giorgio Trinchieri for providing many of the monoclonal antibodies used

REFERENCES

Chehimi, J., Bandyopadhay, S., Prakash, K., Perussia, B., Hassan, N.F., Campbell. D., Kornbluth, J., and Starr, S.E., 1991, In-vitro infection of natural killer cells with different human immunodeficiency virus type 1 isolates, J Virol. 65:1812.

Chomzynski, P., and Sacchi, N., 1987, Single-step method of RNA isolation by acid guanidium thiocyanate-phenol-chloroform extraction, Anal Biochem. 162:156.

Eales, L.J., Farrant, J., Helbert, M., and Pinching, A.J., 1988, Peripheral blood dendritic cells in person with AIDS and AIDS-related complex:loss of high intensity class II antigen expression and function, Clin Exp Immunol. 71:425.

Langhoff, E., Terwilliger, E.F., Bos, H.J., Kalland, K.H., Poznansky, M.C., Bacon, O.M.L., and Haseltine, W.A., 1991, Replication of human immunodeficiency virus type 1 in primary dendritic cell cultures, Proc Natl Acad Sci. 88:7998.

Macatonia, S.E., Lau, R., Patterson, S., Pinching, A.J., and Knight, S.C., 1990, Dendritic cell infection, depletion and dysfunction in HIV-infected individuals, Immunology. 71:38.

Patterson, S., and Knight, S. C., 1987, Susceptibility of human peripheral blood dendritic cells to infection by human immunodeficiency virus, J Gen Virol. 68:1177.

Young, J.W., and Steinman, R.M, 1988., Accessory cell requirement for the mixed leukocyte reaction and polyclonal mitogenesis, as studied with a new technique for enriching blood dendritic cells, Cell Immunol. 111:167.

SIMIAN IMMUNODEFICIENCY VIRUS (SIV) INDUCED ALTERATIONS OF THYMUS IDCs*

V. Krenn[2], J. Müller[1], W. Mosgöller[3], S. Czub[1], C. Schindler[1], C. Stahl-Hennig[4], C. Coulibaly[4], G. Hunsmann[4], HK. Müller-Hermelink[1]

[1]Inst. of Pathology, University of Würzburg(FRG)
[2]Fellow of the FWF(Erwin Schrödinger scholarship),Histol-Embryol Inst.,University of Vienna(A)
[3]Histol-Embryol Inst.,University of Vienna(A)
[4]Inst. of Virology, German Primate Center,Göttingen(FRG)
*supported by the Bundesministerium für Forschung und Technologie(FRG)

INTRODUCTION

Although large amounts of information have been accumulated regarding the role of dendritic cells during HIV infection (Armstrong and Horne 1984, Tenner-Racz et al. 1985, Tschachler et al. 1987) no attention has been payed so far to the role of thymic interdigitating dendritic cells (IDCs). The goal of this study was to investigate whether thymic IDCs are altered morphologically at defined timepoints in the early and late phase of SIVmac (simian immunodeficiency virus) infected rhesus monkies (Macaca mulatta).

The macaque modell was used because it largely represents all aspects of HIV induced disease of humans (King et al., 1983, Letvin and King, 1990). The morphology of the rhesus monkey thymus was analysed by means of light and transmission electron microscopy (TEM). Additionally, a morphometric analysis was carried out to detect possible differences in ultrastructural morphology of IDCs from infected and non-infected control animals.

MATERIAL AND METHODS

Animals and virus

A total of 11 juvenile rhesus monkeys (Macaca mulatta) was investigated. All animals were infected and housed individually as previously described (Stahl-Hennig et al. 1990) To analyse the role of IDCs during SIVmac infection 7 healthy juvenile monkeys were sacrified in the early course of infection and the findings were compared with 2 juvenile non-infected animals and 2 juvenile animals suffering from AIDS. 8 juvenile chinese monkeys were infected with 100 MID50 (monkey infectious doses infecting 50% of recipients) of the SIVmac 251-32H virus stock solution (Cranage et al. 1990) and one indian animal was infected intrathecally with 1ml SIVmac. 239 containing 10^3-10^4 TCID50(tissue culture infective dose infecting 50% of tested cultures). All SIV-infected animals showed seroconversion (ELISA) and positive in vitro virus reisolation from the peripheral blood (Kneitz et al, in preparation).

Electron microscopy

In order to prevent artificial alterations of the thymus the organ was explanted immediately after exsanguination. From at least two different regions of the thymus one cross section was cut into small cubes of about 1mm side length, which were fixed in 2,5% glutaraldehyde buffered with 0,05% phosphat buffer. Subsequently the tissue cubes were postfixed by immersion in 2% OsO4(Sigma) buffered with 0,1 %phospate at pH 7,2. After dehydration in graded ethanol they were embedded in epon 812(Roth) via propylenoxide. Semithin sections were stained with Giemsa and were examined by light microscopy. Thin sections were cut with an OMU2 (Reichert Austria) and contrasted with lead citrate in PBS. The ultrathin sections were examined using an EM 10 Zeiss transmission electron microscope.

Evaluation of ultrathin sections

Representative areas of the cortex and the medulla were selected for ultrathin sections. Ultrastructural analysis of the thymus was done by large scale examination of each two ultrathin sections of the cortex and of the medulla. Each ultrathin section was completely photographed at a magnification of 3140x, and the prints were assembled to a large photocollage representing the semithin section and measuring up to 2m and more. In these conditions IDCs could unequivocally be identified and localized within the different thymic compartments. IDCs laying in the perivascular space were excluded from further analysis.

Morphometric analysis of the ultrastructure of IDC. A total of 36 randomly chosen cells from two infected (31 and 66wpi; weeks post infectionem) and two non infected animals displaying the cytocentre were photographed at a magnification of 4500x. Based on these EM photo micrographes, the average area of the cell section and the perimeter were measured using a personal computer and technics described by Young et al. (1987). These morphological data concerning thymic IDCs of two infected and two non-infected animals were compared by Student's t-test (Sokal and Rohlf, 1981) in order to investigate possible differences in the two populations.

RESULTS:

Ultrastructure of IDC in uninfected control animals

IDCs were diffusely distributed in the cortico-medullary boundary and the medulla. The shape of the nucleus was elongated with indentations on the side facing the cytocentre. The nucleus contained large amounts of euchromatin and in some cases a prominent nucleolus. The abundant cytoplasm was highly branched, electron lucent and contained numerous organelles: mitochondria, smooth and rough endoplasmatic reticulum in form of irregulary distributed narrow cisternae (tubulovesicular system), a well developed Golgi field, and some small electrone dense granules. The cell membrane exhibited the characteristic cell membrane invaginations which contain electrone dense material. Birbeck granules were not detected in the cytoplasm of IDCs.

Ultrastructure of IDC in infected animals

Rhesus monkies at 1,3 and 6 wpi showed no alteration of the thymus whereas the animals 12 and 24wpi showed a considerable narrowing of the thymus cortex without alterations of the medulla. IDCs in these animals exhibited the same morphology as seen in uninfected control animals (Fig. 1a) Two animals (31 and 66wpi) showed weight loss but no opportunistic infections. These animals showed a shrinkage of the thymic cortex, thymic medulla and severe alterations of IDCs. These alterations of IDCs consisted of an almost complete loss of the cytoplasmatic processes and a reduction in the amount of the cytoplasmaic organelles (Fig 1b). Compared to IDCs of uninfected animals the smooth as well as the rough endoplasmatic reticulum , the Golgi complexes and the electron dense granules were reduced in number. Animals with AIDS (122 and 124 wpi) showed a marked involution with a severe depletion of lymphocytes and epithelial cells. In these animals thymic IDCs were completely lost.

Morphometric analysis of IDCs

18 randomly selected EM crossections of altered IDCs in 2 infected animals (31wpi and 66wpi) showed a significant smaller cytoplasmic area P:< 0,01 when compared to the same number of sections from 18 IDCs of two control animals. The average cytoplasmic area of the IDCs from infected animals (31wpi and 66wpi) was $\bar{x}=229\pm96\mu m^2$, in the uninfected animals it was $\bar{x}=346\pm106\ \mu m^2$. The cytoplasmic protrusions were clearly reduced in number and of a blunt shape as observed at the ultrastructural level. In order to compare the surface of cell membranes of IDCs we measured the average length of the perimeter of the cell membrane as displayed on randomly selected sections through the cytocentre. There was a significant decrease (P:<0,01) of the average cell perimeter of IDCs of infected animals. The average membrane perimeter of the two infected animals (31wpi and 66wpi) was $\bar{x}=270\pm80\mu m$ and of the two uninfected animals was $\bar{x}=431\pm164\ \mu m$.

DISCUSSION

In this paper we present evidence for an involvement of thymic IDCs in SIVmac infection. First ultrastructural alterations of the IDCs ocurred in two animals 31 and 66wpi. In these animals there was a loss of the cytoplasmic processes, and the morphometric analysis showed a significant reduction of the cell surface and a significant reduction of the cell membrane perimeter.

As these animals showed only a weight reduction but no opportunistic infections, we consider these alterations to be SIV-induced. Alterations of IDCs in SIV infected monkies differed clearly from alteration of IDCs in acute accidental thymic involution. This was shown by Duijvestijn et al.(1982)

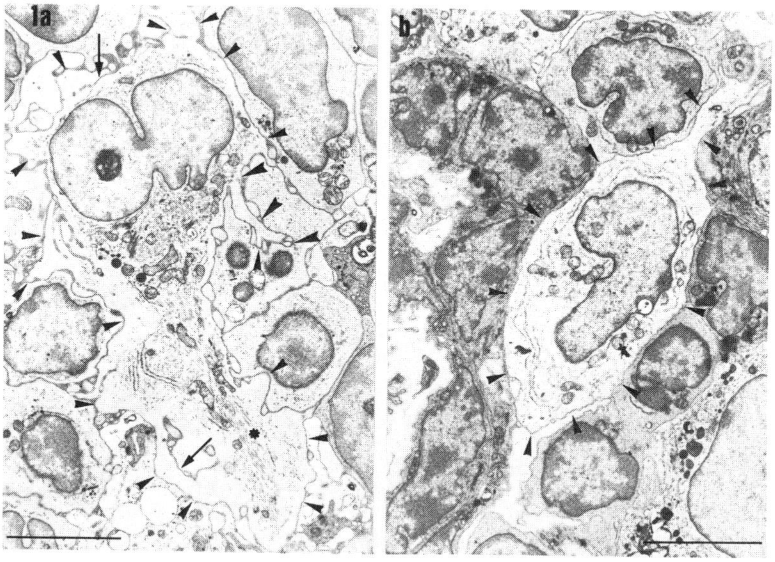

Fig.1a: Electron micrograph of an IDC from an uninfected control animal exhibiting the typical features : elongated nucleus with indentations, abundant electron lucent and higly branched cytoplasm (arrowheads). The cytoplasm contains several organelles: Mitochondria, Golgi complexes, smooth and rough endoplasmatic reticulum(tubulovesicular-complex, asterix) some electron dense vesicles. Arrows point at the characteristic membrane invaginations which contain electron dense material. 1b: Electron micrograph of a "degenerated" IDC from an SIV-infected animal (66wpi). In comparison with the IDC of the control animal there is a reduction of the cytoplasmic processes and of the organelles. (arrow heads point the periphery of the shrunken cytoplasm). Inserted bars: 5 μm.

who demonstrated an increase in the density and strong phagocytic activity of IDCs during the acute cortison induced thymic involution of rats. But in SIVmac infected animals, IDCs appeared to be "degenerating" and finally this resulted in a dramatic numberical reduction in animals with AIDS. Consequently, it appears that the alterations and the numeric reduction of the thymic IDCs represent a specific event occuring during SIVmac induced thymic atrophy. Ultrastructurally no virus could be detected close or within the IDCs. This is well in line with the findings of Spiegel et al. (1992) who could find no evidence of HIV infection in human IDCs of the lymph node by means of in situ hydridisation. We think that indirect mechanisms may play a role in SIV-induced alteration of IDCs: This may consist of a humoral or cellular autoimmune attach (Amadori et al., 1989) or an altered maturation of the IDC precursors in the damaged thymic microenvironment.

Our findings showed that the ultrastructure and the distribution of the IDCs of rhesus monkies and of humans (Kaiserling et al. 1974, v Gaudecker 1991) is nearly identical. The macacque modell is therefore well suited to study the involvement of IDCs in human AIDS. IDCs are considered to be involved in the negative selection of the T-lymphocytes (Kyewski et al. 1986, Inaba et al. 1991). One might speculate therefore that the loss of cytoplasmic processess and the reduction of organelles of IDCs could disturb the process of negative selection and this might give rise to autoimmune phenomena as observed in SIV and HIV infections (Savino et al. 1986, Kumar et al. 1989,).

REFERENCES

Amadori,A. Zamarchi,R. Ciminale,V. del Mistro,A. Siervo, S. Alberti,A. Colombatti,M. Chieco-Bianchi,L. Hiv-1 specific B-cell activation. A major constituent of spontaneous B-cell activation during HIV-1 infection. J.Immunology.143:2146-2152(1989)

Armstrong,A. Horne,R . Follicular dendritic cells and virus like particles in AIDS-related lymphadenopathy. Lancet ii:370-372(1984).

Duijvestijn,A. Köhler,Y. Hoefsmit, E. Interdigitating cells and macrophages in the acute involuting rat thymus. Cell Tiss. Res.219:291-301(1982)

Cranage,M. Cook,N. Johnstone,P. Greenaway,P. Kitchin,P. Stott,E. Almond,N. Baskerville,A. SIV infection of rhesus macaques: in vitro titration of infectivity and development of an experimental vaccine, in Schellekens H and Horzinek M (Eds): Animal models in AIDS. Amsterdam;Elsevier,103-114(1990)

Inaba,M. Inaba,K . Hosono,M. Kumamoto,T. Ishida,T. Muramatsu,S. Masuda,T. Susumu,I. Distinct mechanism of neonatal tolerance induced by dendritic cells and thymic B cells. J. Exp.Med. 173:549-559,(1991)

Kaiserling,E. Stein,H. Müller-Hermelink,HK. Interdigitating reticulum cells in the human thymus. Cell. Tissue Res.:155:47-55(1974)

King,N. Hunt,R. Letvin,N. Histopathologic changes in macaques with an acquired immunodeficiency syndrom (AIDS). Am. J. Pathol. 113:382-388(1983)

Kumar,K. Resnick,L. Loewenstein,D. Berger,J. Eisdorfer,C. Brain-reactive antibodies and the AIDS dementia complex. J.Acquired Immune.Defic. Syndroms 2:469-471(1989)

Kyewski, B. Fathman, C. Rouse, R. Intrathymic presentation of circulating non MHC antigens by medullary dendritic cells. J.Exp. Med.163:231-246(1986)

Letvin,N. King,N. Immunologic and pathologic manifestations of the infection of rhesus monkeys with simian immunodeficiency virus of macaques. J.Acquired Immune. Defic. Syndroms 3:1023-1040(1990)

Savino,W.Dardenne,M.Marche,C.Trophilme,D.Dupuy,Pekovic,D.Lapointe, N.Bach,J. Thymic epithelium in AIDS. An immunohistologic study. Am.J.Pathol.122:302-307(1986)

Sokal,R. Rohlf,J. Biometry, 2nd edition,W.H.Freeman and Company,New York. (1981)

Spiegel, H. Herbst, H. Niedobitek, G. Foss, H-D. Stein, H. Follicular dendritic cells are a mayor reservoir for human immunodeficiency virus type 1 in lymphoid tissues facilitating infection of $CD4^+$ T-helper cells.Am. J. Pathol.140:15-21(1992)

Stahl-Hennig, C. Herchenröder, O. Nick, S. Evers, M. Stille-Siegener, M. Jentsch, K-D. Kirchhoff, F. Tolle, D. Gatesman, T. Lüke, W. Hunsmann, G.Experimental infection of macaques with HIV2-ben, a novel HIV-2 isolate. AIDS 4:611-617(1990)

Tenner-Racz, K. Racz, P. Dietrich, M. Kern,P. Altered follicular dendritical cells and virus like particles in AIDS and AIDS-related lymphadenopathy. Lancet i:105-106(1985)

Tschachler, E. Groh, V. Popovic, M. Mann, D. Konrad, K. Safai, B. Eron, L. DiMarzo-Veronese, F. Wolff, K. Stingl, G. Epidermal Langerhans cells as primary targets for HIV infection. Invest. Dermatol.88:233-237(1987)

v.Gaudecker, B. Functional histology of the human thymus. Anat. Embryol:183:1-15(1991)

Young, S Royer, S. Groves, P. Kinnamon, J. Three-dimensional reconstructions from serial micrographs using the IBM PC. J.Electr.Microsc.Tech. 6:207-217(1987)

MURINE LEUKAEMIA VIRUS INFECTIONS AS MODELS FOR RETROVIRAL DISEASE IN HUMANS

Mary S. Roberts, Jennifer J. Harvey, Steven E. Macatonia and Stella C. Knight

Antigen Presentation Research Group
Clinical Research Centre
Watford Road
Harrow
Middlesex HA1 3UJ
U.K.

INTRODUCTION

The characterisation of animal retroviral infections permits the study of early immunological events after infection and may lead to a better understanding of human retroviral diseases such as AIDS. In retrovirus infections, either stimulation or suppression of T cells can be induced by DC. Early in HIV-infection, antigen-presenting cells may stimulate protective immune responses to the virus. However, presentation by DC of antigens other than HIV is impaired and precedes any direct effect on T cell function[1]. In contrast, another human retrovirus, HTLV-1, is not immunosuppressive. In HTLV-1 infected patients with tropical spastic paraparesis, the lymphoproliferation characteristic of the disease is due to stimulation by DC, some of which are infected by the virus[2]. Here we tested the function of DC and T cells from healthy and retrovirally infected mice in a primary mixed leukocyte reaction (MLR). Comparison was made between the strongly immunosuppressive Rauscher leukaemia virus (RLV) and the weakly immunosuppressive Moloney leukaemia virus (MLV).

MATERIALS AND METHODS

Mice and Inoculation

Young, adult, female Balb/c mice and CBA mice were obtained from the specific pathogen free unit at the Clinical Research Centre. Balb/c mice were inoculated intravenously (tail vein) with a 0.1ml volume of saline (controls) or a suspension containing RLV[3] or MLV[4]. The virus suspensions contained $1x10^5$ focus forming units (ffu)/ml.

Dendritic Cells in Fundamental and Clinical Immunology,
Edited by Kamperdijk *et al.*, Plenum Press, New York, 1993

Cell Preparations

Spleens and inguinal, axillary and salivary lymph nodes were taken from CBA mice and from Balb/c mice 2 weeks post-inoculation. Single cell suspensions, prepared by pressing the tissues through wire mesh, were washed and cultured at 5-$10x10^6$ cells/ml of medium (RPMI 1640, Dutch Modification, Flow Laboratories, with 100IU/ml penicillin, 100 μg/ml streptomycin, $1x10^{-5}$M 2-mercaptoethanol and 10% FCS) in tissue culture flasks overnight at 37°C. Non-adherent cells were then layered onto 2ml metrizamide (Nygaard, Oslo, 14.5g plus 100ml medium) and centrifuged for 10min at 600xg. Cells at the interface were washed once and resuspended in medium to yield 60-70% enriched DC as identified morphologically under the light microscope. Pellet cells from lymph node suspensions were passed through nylon wool columns to obtain >90% pure T cells[5].

Mixed Leucocyte Reaction (MLR)

Cultures, 20μl hanging drops in inverted Terasaki plates[6], contained $25\text{-}100x10^3$ T cells plus 1000 syngenic or allogeneic DC. After 3 days, the cultures were pulsed with ^{3}H thymidine (Amersham International, Amersham, Bucks) 2Ci/mM at 1 μg/ml for 2hr and harvested by blotting onto filter discs. Uptake of ^{3}H thymidine was measured by liquid scintillation counting of the acid-precipitable fraction of the cells from triplicate cultures.

RESULTS

T Cell Function

T cells from RLV and MLV infected mice responded well to

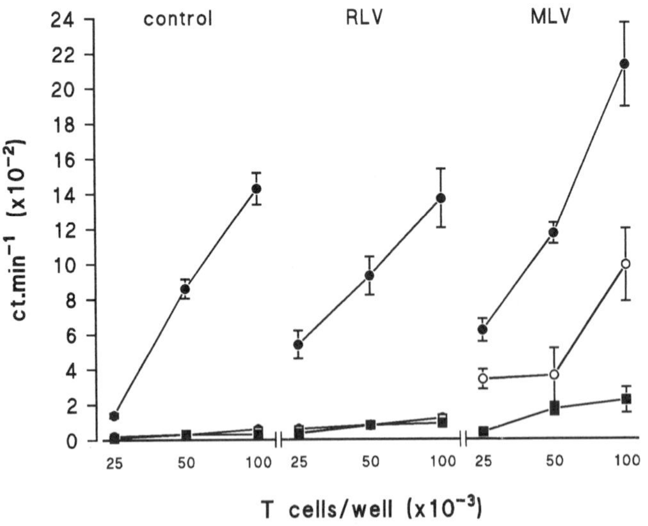

Figure 1. Function of T cells from virally-infected Balb/c mice in an MLR. Cultures contained Balb/c T cells alone (■—■), or with 1000 syngeneic DC(○—○) or 1000 allogeneic (CBA) DC (●—●).

stimulation by allogeneic DC from uninfected mice (figure 1). T cells from RLV-infected mice had a low background turnover, and this was not increased by addition of syngeneic DC from infected mice. In contrast, syngeneic stimulation by DC of T cells from MLV-infected mice was very high. Three experiments gave similar results.

DC Function

DC from RLV-infected mice stimulated healthy allogeneic T cells weakly in comparison to control DC. DC from MLV-infected mice stimulated as strongly as those from control animals. Figure 2 shows one of three experiments giving similar results.

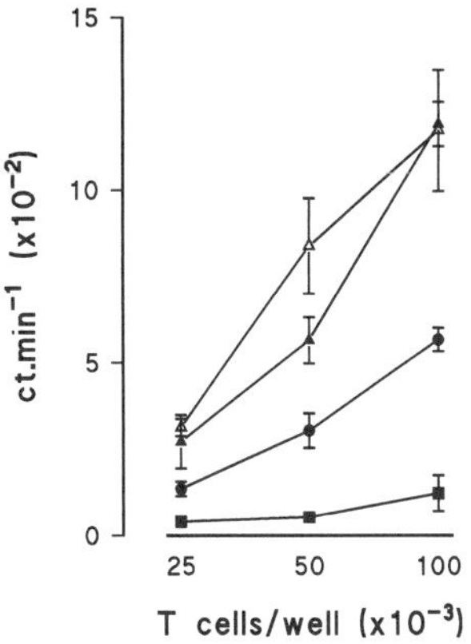

Figure 2. Function of lymph node DC from virally-infected mice in an MLR. Cultures contained CBA T cells alone (■—■), or with 1000 allogeneic (Balb/c) DC from saline treated controls (▲—▲), MLV-infected mice (△—△), or RLV-infected mice (●—●).

DISCUSSION

In early infection of mice with RLV, the ability of DC to stimulate a primary MLR is severely compromised with little effect on T cell function. This parallels the function of cells in asymptomatic HIV seropositive individuals[1]. Since RLV is strongly immunosuppressive, it may provide a useful small animal model of HIV-infection. In contrast to the impaired function of DC in RLV infection, MLV did not affect the stimulatory capacity of DC in allogeneic responses. DC from infected animals did, however, stimulate higher than normal proliferation in syngeneic T cells. High levels of autologous stimulation are also seen in cells of HTLV-1 infected individuals[2]. MLV is comparatively weakly immunosuppresive and may thus resemble HTLV-1 in its pathology. It was therefore of interest to compare and contrast the effects of these two viruses on immune function in infected mice.

T cells from RLV and MLV infected mice responded as well as those from controls to stimulation by allogeneic DC. In contrast, DC function in RLV infection was impaired. The immunosuppression caused by some retroviral infections may thus occur through a defect in DC-mediated antigen presentation. Whether a T cell defect develops at a later stage in RLV infection (as in HIV infection) is unclear.

Other groups have observed aberrations in antigen presentation in murine retroviral infection. For example, Mizuochi et al,[7] describe a decreased ability of splenic antigen-presenting cells to stimulate antigen-specific proliferation of Th clones during infection of C57/BL mice with the LP-BM5 virus (the virus causing murine AIDS 'MAIDS'). The defect was detected just one week post-infection. Using a higher dose of virus, LP-BM5 infected C57/BL mice show a defect in secondary proliferative responses of CD4+ T cells to soluble antigen and mitogens, but not until greater than 5 weeks after infection[8]. These observations again suggest that a defect in antigen presentation may occur prior to a defect in T cell responsiveness.

The high syngeneic stimulation of T cells by DC from MLV-infected mice may parallel the marked T cell proliferation stimulated by DC in HTLV-1 infected individuals with tropical spastic paraparesis[2]. HTLV-1 is also associated with adult T cell leukaemia. Persistent stimulation by DC may predispose cells for malignant transformation in the leukaemic form of HTLV-1 associated disease. Such a connection could be investigated in murine retrovirally induced T cell leukaemias such as that induced by MLV.

The advantages of small animal models for human retroviral infections are obvious. Retroviral disease develops quickly and therapeutic interventions can be tested against the same genetic background. Such models may help to identify methods of circumventing the defects in immune function and, in particular, antigen presentation caused by retroviruses.

REFERENCES

1. S.E. Macatonia, R. Lau, S. Patterson, A.J. Pinching, and S.C. Knight, Dendritic cell infection, depletion and dysfunction in HIV-infected individuals, Immunology. 71:38 (1990).
2. S.E. Macatonia, K. Cruickshank, P. Rudge, and S.C. Knight, Dendritic cells from patients with tropical spastic paraparesis are infected with HTLV-1 and stimulate autologous lymphocyte proliferation, AIDS Hum Retrov. (in press).
3. F.J. Rauscher, A virus-induced disease of mice characterised by by erythrocytopoiesis and lymphoid leukaemia, J Natl Cancer Inst. 29:515 (1962).
4. J.B. Moloney, Biological studies on a lymphoid-leukaemia virus extracted from sarcoma 37. 1. Origin and introductory investigations, J Natl cancer Inst 24:933 (1960).
5. S.C. Knight, Lymphocyte proliferation assays, *in*: "Lymphocytes: a Pratical Approach," G.G.B Klaus, ed., IRL Press, Oxford (1987).
6. M.H. Julius, E. Simpson, and C.A. Herzenberg, A rapid method for the isolation of functional thymus-derived murine lymphocytes, Eur J Immunol. 3:645 (1973).

7. T. Mizuochi, J. Mizuguchi, T. Uchida, K. Ohnishi, M. Nakanishi, Y. Asano, T. Kakiuchi, Y. Fukushima, K. Okuyama, H.C. Morse III, K. Komuro, A selective signalling defect in helper T cells induced by antigen-presenting cells from mice with murine acquired immunodeficiency syndrome, J Immunol. 144:313 (1990).
8. A. Cerny, A.W. Hugin, K.L. Holmes, and H.C. Morse III, CD4+ T cells in murine acquired immunodeficiency syndrome: evidence for an intrinsic defect in the proliferative response to soluble antigen, Eur J Immunol. 20:1577 (1990).

LANGERHANS CELLS AND INTERDIGITATING CELLS IN HIV-INFECTION

Andrea M.R. von Stemm[1], Julia Ramsauer[3], Klara Tenner-Racz[4], Heidemarie F. Schmidt[1], Irma Gigli[2], Paul Racz[1]

[1]Department of Pathology
[2]Department of Immunology
Institute for Tropical Diseases
[3]Department of Dermatology
[4]Department of Haematology
A.K. St. Georg
2000 Hamburg, Germany

INTRODUCTION

The role of Langerhans cells (LC) and dendritic cells (DC) in HIV infection has been a subject of interest during the last several years. It has been reported by several authors that LC are important targets for HIV-1, thus, playing a major role in the dissemination of the infection[1-3]. Other researchers, however, have failed to demonstrate the virus in LC[4-7]. Similarly, infection of DC from peripheral blood of HIV-1 infected individuals has been stated[8,9], but this could not be confirmed by others[10].

Because of these conflicting reports, we studied skin and lymph node (LN) specimens of HIV-1-infected patients in an effort to clarify whether or not LC and interdigitating cells (IDC) of the LNs are productively infected by the virus.

MATERIAL AND METHODS

This study involved samples from 55 HIV-1 infected patients from Hamburg (Germany), San Diego (CA,USA) and Zaire. From the 57 skin samples 49 were biopsies removed for diagnostic purposes, and 8 were obtained from autopsies. From two patients we recieved 2 biopsies each. The postmortem specimen were obtained within 6 hours after death. The clinical stages of the patients are listed on Table 1. Four biopsies from HIV seronegative persons served as controls. In 5 cases (all CDC stage III) LN biopsies from the same patients were also obtained.

All samples were fixed in 4% formaldehyde and embedded in paraffin. In situ hybridization was done on 5µm sections according to a published protocol[11]. A very sensitive ^{35}S-labeled antisense RNA probe synthesized from 5 DNA templates was used. This probe is able of detecting 30-300 copies of target RNA[12].

Combined immunohistochemistry and in situ hybridization was performed on 10 skin biopsies and 5 LNs. Immunostaining of paraffin sections was performed using the monoclonal antibodies (MAB) OPD4 and KiM1P applying the alkaline phosphatase anti-alkaline phosphatase technique. Langerhans cells and IDC were stained with a polyclonal antibody to the S-100 protein and the reaction was visualized with peroxidase anti-peroxidase. All antibodies were obtained from Dakopatts, Copenhagen, Denmark. To inhibit the ribonuclease activity in the antibodies, RNasin (Boehringer,Mannheim, Germany) was added to each antiserum (0.3 units/µl). After immunstaining, in situ hybridization was performed omiting treatment with proteinase K.

Table 1

Classification stage	No. of cases with positive signal for HIV-RNA
CDCD III	1 / 12
CDC IV KS	3 / 14
CDC IV OI	1 / 3
CDC IV KS and OI	0 / 6
Unclassified African Cases	1 / 20
HIV-negative controls	0 / 4
Total	6 / 59

RESULTS

The method of in situ hybridization applied in this study provides excellent morphology and yields high hybridization signals. Multiple skin sections from each of the 59 specimen studied (12 sections from HIV seronegative and 253 sections from HIV seropositive individuals) were probed for viral RNA. Under bright field and incident polarized mercury arc illumination we found no positive cells in the epidermis. In 6 cases one to three cells positive for HIV-1 transcripts were detected in the dermis (Fig.1). Each contained at least 50 silver grains over background. No preferential localization of these cells was noted. Positive cells were found in 3/22 KS biopsies, but not within the KS lesion. Positive cells were also found in the skin of 3/35 patients without KS (Table 1). The number of positive cells was too small to characterize them by performing colocalization studies in serial sections using immunostaining for the S-100 protein, OPD4 and KiM1P.

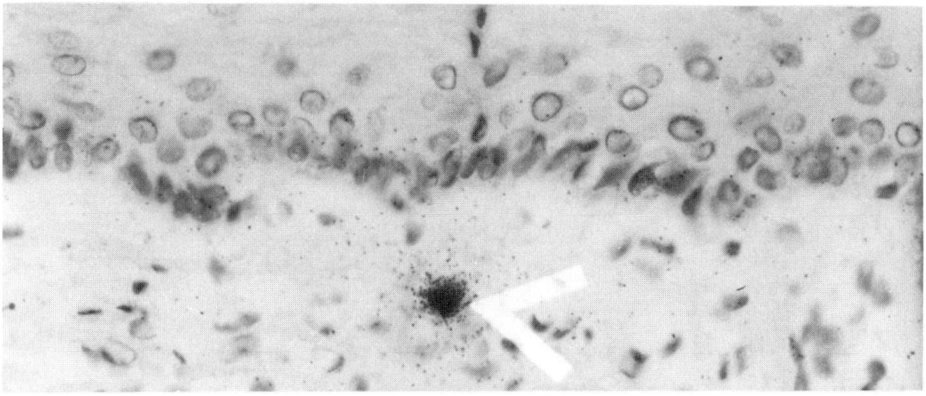

Fig. 1. HIV-positive cell in the upper dermis. (x 250)

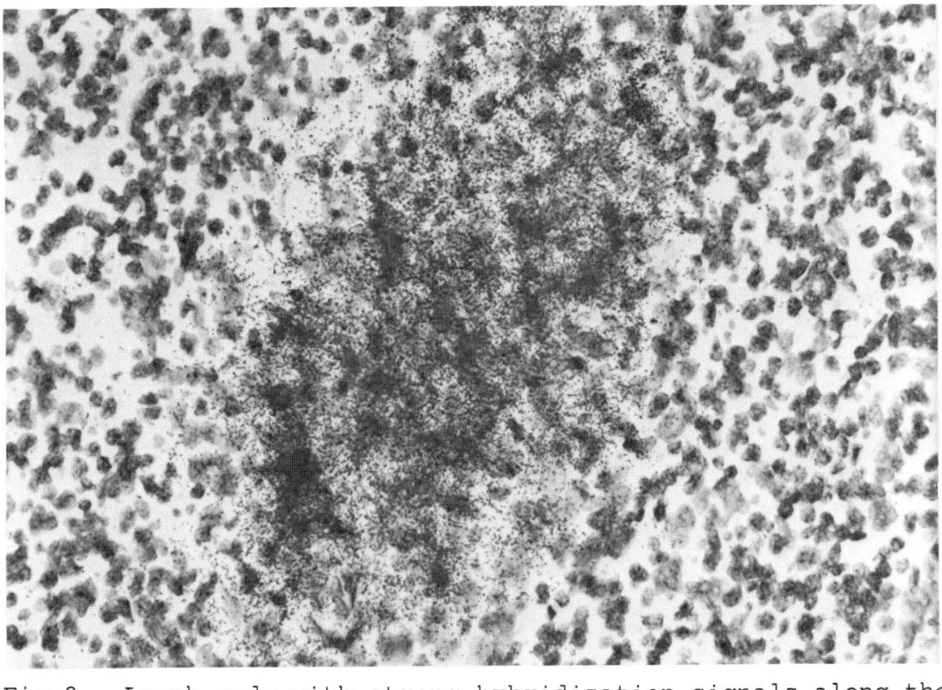

Fig.2. Lymph node with strong hybridization signals along the FDC network (x 370).

Examination of lymph nodes showed explosive follicular hyperplasia. The germinal centers exhibited strong hybridization signals demonstrating the high sensitivity of the probe. The reactivity was localized mainly along the network of follicular dendritic cells (Fig.2) with few RNA positive cells in the extrafollicular parenchyma. Combined immunohistochemistry and in situ hybridization demonstrated that none of the S-100 positive cells, characteristic for IDC, produced positive signals for HIV-1 transcripts (Fig.3). Moreover IDC were not in close contact to infected cells.

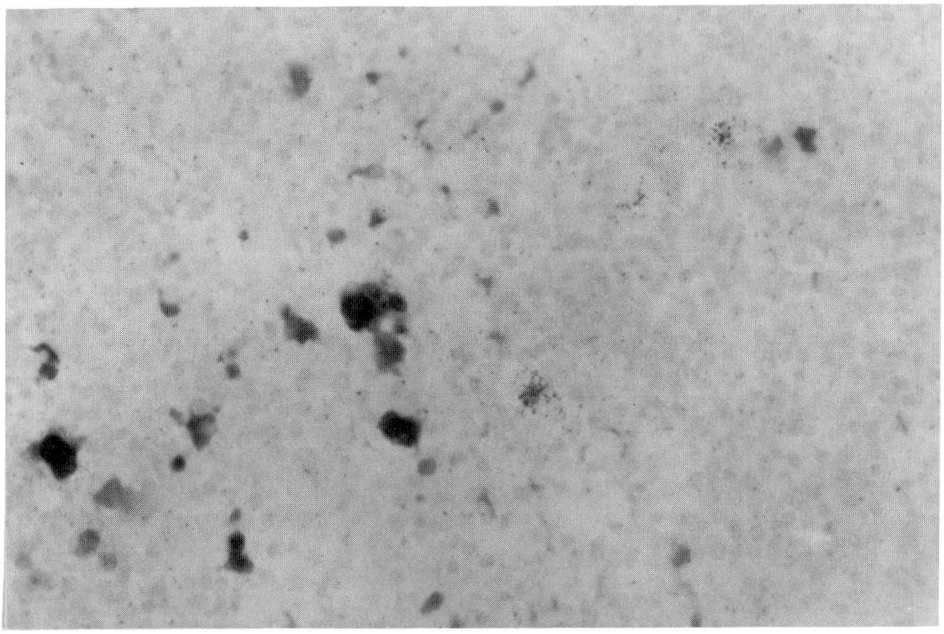

Fig. 3. Two HIV-RNA positive cells in the paracortex of lymph nodes. S-100 positive cells do not coexpress HIV.(x370)

DISCUSSION

The results presented here are in agreement with recent data in the literature showing that the number of infected LC and IDC is very low. Using in situ hybridization in combination with confocal laser microscopy Mahoney and coworkers found no convincing evidence for infection of LC by HIV-1[7]. Similarly, Kalter et al[6] detected a solitary cell positive for HIV-RNA in only one skin section from one patient out of 22 HIV seropositive subjects. Based on this finding they suggested that a role for LC as a principal viral reservoir or vector for transmission is highly unlikely. In contrast, using immunohistochemical techniques Tschachler and collegues demonstrated in 7/40 skin biopsies positive LC using monoclonal antibodies to the HIV-1 protein p17. However, using anti-p24 antibodies only 1/40 biopsies showed positive LC. The authors concluded that many LC were infected[1].

The above observations are in disagreement with results from other laboratories. In an earlier study, we noted that only very few LC were positive for the p24 antigen[13]. Becker and coworkers could not detect any HIV antigen in the epithelium from oral mucosa of 26 HIV positive patients[4]. Kanitakis et al[5] investigated 44 skin biopsies from HIV-1 infected patients and found no reactivity with antibodies to the p24 and the gp120 viral proteins. They showed, however, that a monoclonal antibody to the p17 protein cross-reacts with basal epidermal cells. Parravicini et al[14] and we[13] made similar observations. This cross-reactivity could account for the high number of p17 positive cells observed in the epidermis in the work cited above.

Our finding that IDC appear to be negative for HIV-1 transcripts is consistent with those of Spiegel et al.[15] who performed colocalization studies using antibodies to CD1a on frozen sections and did not find double labeling with HIV-1 RNA. Based on this data it may be concluded that the hypothesis that infected LC migrate to the LNs where they transform into IDC and infect surrounding lymphocytes is rather unlikely. We believe that there is only little evidence that LC and IDC are productively infected by HIV-1. To elucidate the extent of latent infection of LC and IDC further investigations using polymerase chain reaction in combination with in situ hybridization on sections are in progress.

ACKNOWLEDGEMENT

The skilful technical assistance by Birgit Gühlk and Corinna Thomé are gratefully acknowledged.
This is a partial presentation of an ongoing work with Dr. A.S. Fauci and C.H. Fox, National Institute of Health, Bethesda, Maryland and Dr. P. Gigase and Dr. R. Colebunders, Institute for Tropical Medicine, Antwerp.
This study was supported by the Bundesministerium für Forschung und Technologie by grant FKZ BGA III - 006-89/FVP 6, and the Körber Foundation, Hamburg.

REFERENCES

1. E. Tschachler, V. Groh, M. Popovic, D.L. Mann, K. Konrad, B. Safai, L. Eron., F. diMarzo Veronese, K. Wolff and G. Stingl, Epidermal Langerhans cells-a target for HTLV-III/LAV infection, **J. Invest. Dermatol**. 88:233 (1987).
2. L.R. Braathen, G. Ramirez, R.O. Kunze and H. Gelderblom, Langerhans' cells as primary target cells for HIV infection, **Lancet** ii:1094 (1987).
3. M. Landor, Z. Harish and R. Rubinstein, Human immunodeficiency virus transmission: is the integument a barrier?, **Am J Med**. 87:489 (1989).
4. J. Becker, P. Ulrich, R. Kunze, H. Gelderblom, A. Langford and P. Raichert, Immunohistochemical detection of HIV structural proteins and distribution of T-lymphocytes and Langerhans cells in the oral mucosa of HIV infected patients, **Virchows Archiv A Pathol Anat Histopathol**. 412:413 (1988).

5. J. Kanitakis, C. Marchand, H. Su, J. Tivolet, G. Zambruno, D. Schmitt and L. Gazzolo, Immunohistochemical study of normal skin of HIV-1-infected patients shows no evidence of infection of epidermal Langerhans cells by HIV, **AIDS Res Human Retrovir**. 5:293 (1989).
6. D.C. Kalter, J.L. Greenhouse, J.M. Orenstein, S.M. Schnittman, H.E. Gendelman and M.S. Meltzer, Epidermal Langerhans cells are not principal reservoirs of virus in HIV disease, **J Immunol**. 146:3396 (1991).
7. S.E. Mahoney, M. Duvic, B.J. Nickoloff, M. Minshall, L.C. Smith, C.E.M. Griffiths, S.W. Paddock and D.E. Lewis, **J Clin Invest**. 88:174 (1991).
8. S. Pattersen, S.E. Macatonia, J. Gross and S.C. Knight: Infection of bone marrow-derived dendritic cells in HIV, in:" Modern Pathology of AIDS and other Retroviral Infections", P. Racz, A.T. Haase, J.C. Gluckman, ed., Karger, Basel (1990).
9. E. Langhoff, E.F. Terwilliger, H.J. Bos, K.H.Kalland, M.C. Poznansky, O.M. Bacon and W.A. Haseltine, Replication of human immunodeficiency virus type 1 in primary dendritic cell cultures, **Proc Natl Acad Sci USA**. 88:7988 (1991).
10. P.U. Cameron, U. Forsum, H. Teppler, A. Granelli-Piperno and R. Steinmann, During HIV-1 infection most blood dendritic cells are not productively infected and can induce allogeneic CD4 + T cells clonal expansion, **Clin Exp Immunol**. 88:226 (1992).
11. C.H. Fox, S. Koenig and A.S. Fauci, In situ hybridization for molecular diagnosis, in: Protocols for In Situ Hybridization. Rockville, MD: Lofstrand Labs. (1989).
12. C.H. Fox, K. Tenner-Racz, P. Racz, A. Firpo, P.A. Pizzo and A.S. Fauci, Lymphoid germinal centers are reservoirs of human immunodeficiency virus type 1 RNA, **J Inf Dis**. 164:1051 (1991).
13. J. Ramsauer, K. Tenner-Racz, W. Meigel and P. Racz, Immunohistochemical study of Langerhans' cells in the skin of HIV-1-infected patients with and without Kaposi's sarcoma, in:" Accessory Cells in HIV and other Retroviral Infections", P. Racz, C.D. Dijkstra, J.C. Gluckman, ed., Karger, Basel (1991).
14. C. Parravicini, D. Klatzmann, P.Jaffray, G. Constanzi and J.C. Gluckman, Monoclonal antibodies to the human immunodeficiency virus p18 protein cross-react with normal human tissues, **AIDS**. 2:171 (1988).
15. H. Spiegel, H. Herbst, G. Niedobitek, H.-D. Foss and H. Stein, Follicular dendritic cells are a major reservoir for human immunodeficiency virus type 1 in lymphoid tissues facilitating infection of CD4 + T-helper cells, **Am J Pathol**. 140:15 (1992).

DENDRITIC CELLS IN HIV-1 AND HTLV-1 INFECTION

Stella C. Knight, Steven E. Macatonia,
Kennedy Cruickshank, Peter Rudge and
Steven Patterson

Northwick Park Hospital &
Clinical Research Centre
Watford Road
Harrow Middlesex HA1 3UJ

SUMMARY

Individuals with HIV-1 infection show two major immunological effects - the early persistent stimulation of immune responses (e.g. of CD8 cells and antibody production) and later catastrophic immunodeficiency. Peripheral blood dendritic cells (DC) become infected with HIV-1[1-6] and infected cells can stimulate responses to virus[7]. By contrast infected DC show a reduced capacity to stimulate either primary or secondary responses to other antigens[2,3,8]. We have proposed that the block, particularly in primary responses, may be instrumental in the development of immunodeficiency[8,9].

In HTLV-1 infected patients with tropical spastic paraparesis (TSP) a major feature of disease is 'spontaneous' T cell proliferation thought to underlie development of inflammatory neurological disease[10]. We have now shown that some DC in addition to T cells are infected in TSP and that DC stimulate the persistent T cell activity. Here we demonstrate this using cells from an informative family where the daughter was normal, the father an HTLV-1 seronegative carrier and the mother had TSP. DC from all individuals stimulated normal allogeneic T cells in a mixed leukocyte reaction (MLR) and T cells responded well to normal allogeneic DC. T cells from the daughter showed little stimulation with autologous DC, those from the father showed significant but low stimulation, and T cells from the TSP patient gave a response to autologous DC which exceeded that to allogeneic DC. Taken together, the studies of DC and both HIV-1 and HTLV-1 indicate that infection of DC may play a central role in development of T cell abnormalities in human retrovirus-induced diseases.

INTRODUCTION

Following early observations that DC from peripheral blood were susceptible to infection with HIV-1 *in vitro*[1] and that some

Dendritic Cells in Fundamental and Clinical Immunology,
Edited by Kamperdijk *et al.*, Plenum Press, New York, 1993

Langerhans' cells were infected *in vivo*[12] studies from our laboratory have sought to answer three questions concerning peripheral blood and HIV-1 infection. Firstly, we have asked whether peripheral blood from seropositive patients are infected with HIV-1 *in vivo* and compared the level of infection with that in other cell types. Secondly, we have tested whether infected DC are able to stimulate primary responses to the HIV-1 in autologous T cells and finally we have looked at the capacity of DC exposed to HIV-1 to stimulate responses to antigens other than HIV-1. Results on these different aspects are already published[1-5,7-9] and in this article the salient features of these studies are given. The work has now asked the same three questions concerning infection of DC with HTLV-1[11]. Here we also review this work and illustrate the major finding with data on T cell and DC function using cells from a family in which 2 members were positive for HTLV-1 and one of these had the disease TSP.

INFECTION OF PERIPHERAL BLOOD DC WITH HIV-1 AND HTLV-1

Early studies of exposure of DC to HIV-1 *in vitro* showed by electron microscopy within 3-5 days, productive infection of some DC indicating a greater susceptibility to infection than in other freshly isolated leukocytes[1]. These cells are non-dividing indicating that productive infection can occur in the absence of cell turnover. The sub-population of DC infected was not the 'majority' cell type isolated from peripheral blood (Type 1 DC with irregular shape and many projections and some cytoplasmic vacuoles). Instead, productive infection was only seen in a proportion of DC having a morphology somewhat similar to that of Langerhans' cells (type II DC) and in occasional fully developed veiled cells (type III). In order to see infection, therefore, the technique used to obtain DC should permit the separation or development of the type II and III DC which may be equivalent to more mature stages of DC. Infection of DC was also shown in *in vitro* infected DC using a double labelling histological technique where cells contaminating enriched DC populations were labelled with antibodies and *in situ* hybridisation for HIV-1 performed on the same preparation[2]. *In vivo* infection has also been demonstrated in cells identified by the histological double labelling technique[3,4] and this shows a much higher proportion of DC infected than of either lymphocytes or macrophages. The polymerase chain reaction using nested gag primers indicates that the DC enriched populations from infected patients often have 1-2 orders of magnitude greater levels of infection than lymphocyte populations (9, Patterson *et al*, in preparation). The contrast between these observations and those which fail to show infection in DC[13] seems likely to be related to the lower proportion of DC isolated by other authors.

DC infected by HTLV-1 are seen in cells isolated from the peripheral blood of patients with asymptomatic infection with HTLV-1 or those with the disease TSP. The proportion of infected DC (0.5 -5%) is similar to that of T cells infected[11]. The cells were again identified by purification of DC in addition to immunohistology and *in situ* hybridisation. DC from non-infected individuals are also susceptible to infection with HTLV-1 *in vitro* (Ali, Patterson, Rudge, Cruickshank, Dalgleish and Knight, in preparation).

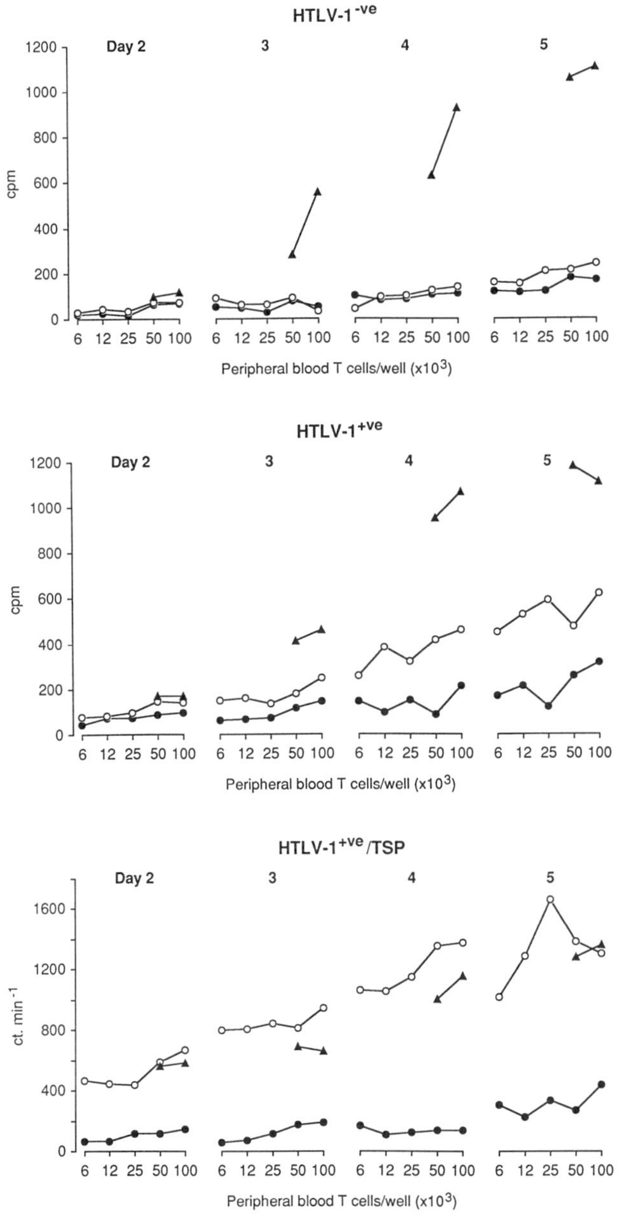

Figure 1a, 1b, 1c. Blood samples were taken from a woman with HTLV-1 positive TSP, her 'carrier' HTLV-1 positive husband and HTLV-1 negative daughter and one unrelated control. Enriched dendritic cells were isolated from 24hr non-adherent mononuclear cells using 13.7% w/v metrizamide gradients. T cells were purified from pellet cells by E-rosetting. Different numbers of T cells from family members were cultured in 20µl hanging drops (●) and autologous (○) or normal allogeneic (▲) dendritic cells added. Stimulation was measured from uptake of ^{3}H-thymidine[2,3,11]. If T cells were purified after removal of dendritic cells all T cell samples showed low turnover, even those from the TSP patient. Cells from all individuals gave good responses to normal allogeneic DC but only those from infected individuals showed significant autologous stimulation and this was very high in the TSP patient.

STIMULATION OF RESPONSES TO HIV-1 AND HTLV-1 BY DC

Using cells from non-immune donors, DC have been exposed *in vitro* to live HIV-1, to gp120 or to a small envelope peptide of HIV-1. These antigen pulsed cells then stimulated primary proliferative responses and the development of cytotoxic T cells (CTL) that kill virus infected target cells. Thus DC can stimulate primary responses to the virus and the technique can be used to identify primary T cell epitopes in humans which may have protective potential for vaccination purposes[7]. With high levels of virus infection the DC do not have this stimulatory capacity[15]. DC taken from infected individuals do not cause marked proliferative responses in autologous T cells. However, ongoing CTL responses are reported in infected individuals but the nature of the antigen presenting cells fuelling these responses is not known. DC and M0 from HIV-1 infected individuals without disease stimulated B and T cell cultures to produce antibodies to gp120 and p24 antigen (Roberts *et al*, in preparation), indicating an ongoing role for antigen presenting cells fuelling protective immune responses in these individuals.

In HTLV-1 infected patients with TSP a major immunological effect is the ongoing T cell activity[10]. When DC are removed from the samples and the T cell purified the high background turnover of T cells in TSP is lost and cells show low background turnover. Figure 1 demonstrates how this stimulation is initiated by the presence of DC causing an autologous MLR. Adding back DC from HTLV-1 positives who are asymptomatic caused low but significant levels of autologous stimulation whereas those with TSP showed autologous stimulation often greater than that seen in allogeneic MLR (Fig. 1).

The DC thus appear to be fuelling the T cell activity which in turn may be involved in the development of disease. The autologous stimulation is blocked by antibodies both to class II molecules and to HTLV-1 itself suggesting that persistently infected DC may be presenting antigen to T cells.

STIMULATION OF RESPONSES TO ANTIGENS BY INFECTED DC

Defective stimulation by DC of both primary and secondary responses in T cells has been identified when they have been exposed to HIV-1 *in vitro* or *in vivo*[2,3,8]. A defective recruitment of T cells into the memory pool has been proposed as a cause of the slow decline of immune responses in HIV-1 infection as memory T cells are lost through various mechanisms (9, for references). By contrast, there is no evidence that the capacity of DC to stimulate mixed leukocyte reactions or Concanavalin A responses is lost following infection of DC with HTLV-1.

In conclusion, the infection of DC with HIV-1 or with HTLV-1 and the effects produced as a consequence on T cell function may be an important route of immunomodulation in the production of disease.

REFERENCES

1. S. Patterson and S.C. Knight. Susceptibility of human

peripheral blood dendritic cells to infection by human immunodeficiency virus. *J. Gen. Virol.* 68:1177(1987).
2. S.E. Macatonia, S. Patterson and S.C. Knight. Suppression of immune responses by dendritic cells infected with HIV. *Immunology* 67:285(1989).
3. S.E. Macatonia, R. Lau, S. Patterson, A.J. Pinching and S.C. Knight. Dendritic cell infection, depletion and dysfunction in HIV-infected individuals. *Immunology* 71:38(1990).
4. R.A. Hughes, S.E. Macatonia, I.F. Row, A.C.S. Keat and S.C. Knight. The detection of human immunodeficiency virus DNA in dendritic cells from the joints of patients with aseptic arthritis. *Br. J. Rheumatol.* 29:166(1990).
5. S. Patterson, J. Gross, P.A. Bedford and S.C. Knight. Morphology and phenotype of dendritic cells from peripheral blood and productive and non-productive infection with human immunodeficiency virus type 1. *Immunology* 72:361(1991).
6. E. Langhoff, E.F. Terwilliger, H.J. Bos, K.H. Kalland, M.C. Poznansky, O.M.L. Bacon and W.A. Haseltine. Replication of human immunodeficiency virus type 1 in primary dendritic cells cultures. *Proc. Nat. Acad. Sci. USA* 88:7998(1991).
7. S.E. Macatonia, S. Patterson and S.C. Knight. Primary proliferative and cytotoxic T cell responses to HIV induced *in vitro* by human dendritic cells. *Immunology* 74:399(1991).
8. S.E. Macatonia, M. Gompels, A.J. Pinching, S. Patterson and S.C. Knight. Antigen-presentation by macrophages but not by dendritic cells in HIV infection. *Immunology*, in press.
9. S.C. Knight, S. Macatonia, S. Patterson. Loss of dendritic cell function, in: "Immunology of HIV Infection", G. Janossy, B. Autran, F. Miedema, ed., Karger, Basel, (1992).
10. Y. Itoyama, S. Minato, J.L. Kiral, I. Goto, H. Sato, K. Okochi and N. Yamamoto. Altered subsets of peripheral blood lymphcoytes in patients with HTLV-1 associated myelopathy (HAM). *Neurology* 38:816(1988).
11. S.E. Macatonia, J.K. Cruickshank, P. Rudge and S.C. Knight. Dendritic cells from patients with tropical spastic paraparesis are infected with HTLV-1 and stimulate autologous lymphocyte proliferation. *AIDS Hum. Retrov.* in press.
12. E. Tsachler, V. Groh, M. Popovic, D.L. Mann, K. Konrad, B. Sofai, L. Eron, F. Veronese, K. Wolff and G. Stingl. Epidermal Langerhans' cells. A target for HTLV-111/LAV infection. *J. Invest. Dermatol.* 88:233(1987).
13. P.U. Cameron, U. Forsum, H. Teppler, A. Granelli-Piperno and R.M. Steinman. During HIV-1 infection most blood dendritic cells are not productively infected and can induce allogeneic $CD4^+$ T cells clonal expansion. *Clin. exp. Immunol.* 88(2):226(1992).
14. S.C. Knight, S. Patterson and S.E. Macatonia. Stimulatory and suppressive effects of infection of dendritic cells with HIV-1. *Immunol. Lett.* 30:213(1991).

DENDRITIC CELLS IN ALLERGIC AND CHRONIC INFLAMMATORY RESPONSES

Leonard W. Poulter and George Janossy

Academic Department of Clinical Immunology
Royal Free Hospital School of Medicine, London NW3 2QG

INTRODUCTION

Dendritic cells represent a distinct population of non-lymphoid immunocompetent cells that exhibit characteristic morphologic, physiologic, and phenotypic features, united by their characteristic of highly efficient antigen presentation to T cells (1). The poorly adherent dendritic splenic macrophages described some twenty years ago (2) are now recognised as just one of a family of dendritic cells including veiled cells, interdigitating cells, Langerhans cells and thymic dendritic cells. Even considering this limited selection, it is clear that although exhibiting unifying features each of these dendritic cell types has unique features of its own and many of these features locally develop in their microenvironment.

Initial functional investigations of such cells relied on *in vitro* studies where the characteristics of poor adherence to plastic or glass could be used to separate these cells from tissue macrophages (3). We were fortunate in developing a monoclonal antibody RFD1 following mouse immunisation with a human thymic cell suspension that appeared to see an epitope on interdigitating cells that was not present on monocyte or tissue macrophages (4). This RFD1 antibody reacts with a similar but not fully overlapping population which expresses HLA-DQ. Subsequent studies have revealed that all the different cell categories listed among the dendritic family express RFD1 positivity. The antibody reacts with an unstable epitope that appears to be associated with the MHC Class II. Nevertheless it has a totally different and more restricted expression, as not all MHC Class II$^+$ cells are positive for RFD1. In terms of function, however, it has been shown that RFD1 blocks T cell transformation as efficiently as anti HLA-DR (5). These were the primary observations that initiated the widespread use of this particular antibody in immunohistological investigations in both normal and diseased tissues. The fetal development of the RFD1$^+$ dendritic cells has also been described (6). Our particular interest has been the distribution of dendritic cells and their role in allergic and chronic inflammatory disease.

DISTRIBUTION OF DENDRITIC CELLS

With immunohistological techniques RFD1$^+$ dendritic cells have been shown to be present in the cellular infiltrates of <u>all</u> chronic inflammatory diseases investigated, although their relative numbers vary from being a major non-lymphoid cell present, as in the inflamed synovium of patients with rheumatoid arthritis (7), to a relatively minor population seen in Crohns Colitis (8). It has been possible to relate numbers of dendritic cells to disease activity in some diseases such as sarcoidosis, where cell numbers can be quantified in bronchoalveolar lavage and increasing disease severity is seen associated with

Dendritic Cells in Fundamental and Clinical Immunology,
Edited by Kamperdijk *et al.*, Plenum Press, New York, 1993

increasing numbers of such cells (9). In chronic inflammation any local increase in numbers of dendritic cells is parallelled by the accumulation of T lymphocytes with which they always appear in intimate association. Such an association mirrors that seen in the T cell areas of lymphoid tissue. This finding is considered strong circumstantial evidence that antigen presentation occurs locally in these inflamed tissues. Indeed, when such cells have been isolated it has been possible to demonstrate their highly efficient antigen presenting capacity *in vitro* (10). Such interaction thus represents a local cell mediated immune response. This is most clearly seen in cases of leprosy. In the lesions of tuberculoid leprosy, characterised by cell mediated hypersensitivity reactions with few or no mycobacteria, large numbers of $RFD1^+$ dendritic cells are associated with extensive T cell infiltrates (11). At the other end of the leprosy spectrum the lesions of lepromatous leprosy are hyperbacillary with sparce $RFD1^+$ and few T cells. Such patients appear anergic.

Tuberculoid leprosy and sarcoidosis mentioned above are granulomatous diseases. Although accumulations of dendritic cells may be seen early in the inflamed areas, the granulomata once formed appear to contain few dendritic cells and predominant epithelioid cells (12). Within the lymphocyte mantle of the mature granuloma $RFD1^+$ cells may however be found, although they do not appear in as large a number as in the non-granulomatous lymphocyte dominated chronic inflammation seen in arthritic synovium (7) or in psoriatic skin (13).

In the inflammation associated with allergic disease $RFD1^+$ cells again occur. In the bronchial wall of asthmatic patients 3-5 fold increases in $RFD1^+$ cells may be recorded (14). In lesions of atopic dermatitis (AD) the dermis contains raised numbers of $RFD1^+$ cells including considerable numbers of dendritic $CD1^+$ Langerhans cells. The increases in these phenotypically identified cell subsets however appear to occur without any significant rise in the overall number of mononuclear non-lymphoid cells identified with the CD68 antibody (15). As seen in chronic inflammatory disease, the dendritic cells present in allergy related lesions are apposed to large numbers of T lymphocytes (16); by these criteria the immunopathology of the bronchi in asthma and the skin in AD reflects the situation seen in those conditions classically described as chronic inflammatory disease. Such similarities should not be surprising as conditions such as asthma and AD may have acute exacerbations promoted by allergens but nevertheless clinically represent chronic conditions with underlying immune disorders that may persist for years. In conclusion, all inflammatory reactions involving lymphocytic infiltration are characterised by the presence of dendritic cells. The immunopathologic picture is one of a tissue reaction promoting development of local cellular interaction reflecting that normally present in the paracortical areas of the lymphoid tissue.

THE FUNCTION OF DENDRITIC CELLS IN INFLAMMATORY REACTIONS

Dendritic cells are antigen presenters designed to induce specific T cell activity. Allergic and chronic inflammatory responses are a reflection of immune dysfunction. The question thus arises as to whether the dendritic cells are functioning 'normally' within an aberrant immune system or whether the dysfunction is caused by aberrantly functioning dendritic cells. Both situations may occur and the dissection of the causes and effects is difficult without experiments with isolated calls *in vitro*.

There are, however, some functional and immunophenotypic clues which help in approaching these issues. From our studies of dendritic cells in leprosy it became clear that these cells, despite their recognised poor phagocytic capacity, may be parasitized by the mycobacteria (17). One consequence of infection was to down- regulate the expression of HLA-DR on these cells (17). Clearly, this effect may have dramatic consequences on the antigen presenting capacity of these cells and therefore contribute directly to local anergy. Although not fully understood, the recruitment and/or differentiation of dendritic cells within peripheral tissues is likely to be mediated by cytokines. In this regard it has been demonstrated that IFN-gamma may be influential in promoting dendritic cell activity (18); thus regulation of dendritic cell/T cell interaction may be viewed as a bidirectional phenomenon.

Chronicity of inflammation is by and large regarded as a problem of immune regulation. There is no doubt that dendritic cells involved in such reactions could be functioning "normally". In the case of allergic diseases interest has recently focused on the expression by dendritic cells of CD23 antigen (low affinity receptor for IgE). This molecule is normally expressed by granulocytes, B cells and some macrophages. In lesions of AD, however, expression of CD23 antigen occurs on dermal dendritic cells including $CD1^+$ Langerhans cells that migrate into the dermis (15). Further studies of patch test reactions in AD patients has also revealed that this increased CD23 expression on dendritic cells can occur within 48 hours of allergen contact (19). It has been postulated that the emergence of this receptor on dendritic cells could influence their antigen presenting function; particularly in allergic disease where IgE/allergen complexes are present (20). In asthma, raised expression of HLA-DR is evident on dendritic cells. This increase together with the increased numbers of DC within the bronchial wall result in quantitative increases in HLA-DR. This phenomenon has been documented by optical density measurements on tissue sections (21). Of some significance is the fact that these increases in asthmatics show a strong correlation with the degree of bronchial hyperactivity exhibited by these patients (21). Whether this increased HLA-DR expression reflects increased antigen presenting capacity remains to be determined.

In summary it can be said that in certain examples of chronic inflammation the dendritic cells involved may be caught up as "innocent bystanders" and be functioning in a normal fashion within a dysregulated system. Examples are to be found, however, where there is convincing evidence that the function of the DC may itself be altered and thus be directly contributing to the aberrant immunologic response that forms the basis for the inflammatory disease.

THE BALANCE OF DENDRITIC CELLS/MACROPHAGES

The division of non-lymphoid accessory cells into dendritic antigen- presenting cells and phagocytic macrophages that function as inducers and effector cells of T cell immune responses, respectively, does not reflect the full complex picture of cellular diversity. Several workers have reported the presence of another subset of macrophages which is capable of a suppressor function. These cells are abundantly found within the lung, where they down-regulate the induction of cell mediated immunity (22). Indeed, it has been proposed that control of T cell responses by these suppressor cells may promote tolerance induction in the developing fetus, particulary at the mucosal surfaces and inside the gut wall where such cells are identifiable early during development (23). These cells appear to act by down-regulating the action of dendritic antigen presenting cells which are also seen at mucosal surfaces (24,25). Although most evidence of the action of these cells has come from studies in rodents, we have reported the presence of similar cells in humans (26). Interestingly, these suppressor cells are also MoAb $RFD1^+$ but also express the epitope seen by another MoAb, RFD7 (26). This RFD7 is absent on dendritic cells. The suppressor $RFD1^+,RFD7^+$ macrophages exhibit characteristics of phagocytic cells rather than antigen presenting cells (27). Using both RFD1 and RFD7 MoAb it has been possible to investigate the balance of dendritic cells ($RFD1^+,RFD7^-$), suppressor macrophages ($RFD1^+,RFD7^+$), and effector macrophages ($RFD1^-,RFD7^+$) within normal and inflamed tissues. It has been found that these three subsets are consistently balanced at specific locations, and that this balance differs at different sites (28). Mucosal tissue is characterised by the presence of a major population of suppressor cells with dendritic cells only forming a small proportion of total macrophage like cells. In the skin, on the other hand, all three subsets are evenly balanced (28).

Of considerable interest are the observations that when the number of dendritic cells increases, as seen in the BAL of patients with sarcoidosis, the suppressor macrophages concomittantly increase their numbers. In other inflammatory conditions, such as colitis and asthma, however, a rise in dendritic cells is associated with a significant <u>decline</u> in suppressor macrophages (28,29). Such observations indicate that changes in dendritic cell numbers in inflammation should be viewed in the context of the overall change in balance with other functionally distinct macrophage subsets.

CONCLUSIONS

The use of MoAbs to identify dendritic cells *in situ* and to help isolate them *in vitro* by their phenotype has proved of real value in revealing the distribution of the cells in allergic and chronic inflammatory responses. Some caution must be exercised, however, because a given phenotype may not be unique to one cell type (see above), nor may the phenotype be stable on any given cell lineage. *In vitro* studies in our laboratory have revealed that IFN-gamma (18), corticosteroids (30) and TGF-beta (unpublished observation) can change the phenotype of macrophage subsets in culture. Others have described changes in dendritic cell phenotype *in vivo* associated with the promotion of peritoneal inflammation (31). It is fascinating in the light of these observations to raise the possibility that cell functions may also change along with cell phenotype. Could the balance of dendritic cells, suppressor cells and phagocytes seen within an area of inflammation be a product of the local environment? Despite these remaining questions it is clear that dendritic cells not only play an integral part in promoting inflammatory reactions by their function of antigen-presentation but may also themselves be responsible for the specific derangement of the immune system in these clinical conditions.

ACKNOWLEDGEMENT

Dr L. W. Poulter and Prof. G. Janossy acknowledge the collaboration of their clinical colleagues Dr C. Burke and Dr M. Rustin. The experiments described were performed by post graduate students, Dr C. Munro, Dr M. Spiteri, Dr C. Power, Dr C. Buckley and Dr M. Allison.

REFERENCES

1. M. C. Nussenzweig, R. M. Steinman. Contribution of dendritic cells to stimulation of the murine syngeneic mixed leucocyte reaction. *J Exp Med.* 151:1196 (1980).
2. R. M. Steinman, Z. A. Cohn. Identification of a novel cell type in peripheral lymphoid organs of mice. I.Morphology, quantitation, tissue distribution. *J Exp Med.* 137:1142 (1973).
3. S. C. Knight, J. Farrant, A. Bryant, A. J. Edwards, S. Burman, A. Lever, J. Clarke, A. D. B. Webster. Non-adherent, low-density cells from human peripheral blood contain dendritic cells and monocytes, both with veiled morphology. *Immunology* 57:595 (1986).
4. L. W. Poulter, D. A. Campbell, C. Munro, G. Janossy. Discrimination of human macrophages and dendritic cells using monoclonal antibodies. *Scand J Immunol.* 24:351 (1986).
5. L. W. Poulter, O. Duke. Dendritic cells of the rheumatoid synovial membrane and fluid, *in* "Quatrieme Cours d'Immuno-rhumatologie et Seminaire Internationale d'Immunopathologie Articulare". J Clot, J Sony, ed., (1983).
6. G. Janossy, M. Bofill, L. W. Poulter, G. Burford, C. Navarette, A. Zeigler, E. Kelemen. Separate ontogeny of two macrophage-like accessory cell populations in the human foetus. *J Immunol.* 136:4354 (1986).
7. L. W. Poulter, G. Janossy. The involvement of dendritic cells in chronic inflammation. *Scand J Immunol.* 21:401 (1985).
8. M. C. Allison, S. Cornwall, L. W. Poulter, A. P. Dhillon, R. E. Pounder. Macrophage heterogeneity in normal colonic mucosa and in IBD. *Gut* 29:1531 (1988).
9. G. Ainslie, L. W. Poulter, R. M. Dubois. Relation between immunocytological features of bronchoalveolar lavage fluid and clinical indices in sarcoidosis. *Thorax* 44:501 (1989).
10. J. P. Metlay, E. Pare, R. M. Steinman. Distinct features of dendritic cells and anti Ig activated B cells as stimulators of the primary mixed leucocyte reaction. *J Exp Med.* 169:239 (1989).

11.L. A. Collings, M.F.R. Waters, L. W. Poulter. The involvement of dendritic cells in the cutaneous lesions associated with tuberculoid and lepromatous leprosy. *Clin exp Immunol.* 62:458 (1985).
12.C. S. Munro, D. A. Campbell, L. A. Collings, L. W. Poulter. Monoclonal antibodies distinguish macrophages and epithelioid cells in lesions of sarcoidosis and leprosy. *Clin exp Immunol.* 68:282 (1987).
13.L. W. Poulter, R. Russell-Jones, S. Hobbs. The significance of antigen presenting cells in psoriasis, *in:* "Immunodermatology," D.MacDonald, ed., Butterworth, London (1985).
14.L. W. Poulter, A. Norris, C. Power, A. Condez, H. Barnes, B. Schmekel, C. Burke. T cell dominated inflammatory reactions in the bronchioles of asymptomatic asthmatics are also present in the nasal mucosa. *Postgrad Med J.* 67:747 (1991).
15.C. Buckley, C. Ivison, M. Rustin, L. W. Poulter. Altered cellular expression of the FcIgEII receptors in atopic dermatitis. *J Invest Derm.* in press (1992).
16.C. B. Zachary, L. W. Poulter, D. Macdonald. Cell mediated immune responses in atopic dermatitis. The relevance of antigen presenting cells. *Brit J Derm.* 113:10 (1985).
17.L. W. Poulter, A. Condez. The effect of intracellular parasitism on cell phenotype. *Clin exp Immunol.* 74:344 (1988).
18.L. W. Poulter, G.A.W. Rook, J. Steel, A. Condez. Influence of I-25-(OH)2 vitamin D3 and gamma interferon on the phenotype of human peripheral blood monocyte-derived macrophages. *Infect Immun.* 55:2017 (1987).
19.C. Buckley, C. Ivison, L. W. Poulter, M.H.A. Rustin. CD23 expression in contact sensitivity reactions: a comparison between aero-allergen patch test reactions in atopic dermatitis & the nickel reaction in non-atopic individuals. Submitted (1992).
20.J. Brostoff, C. Carini. Production of IgE complexes by allergen challenge in atopic patients and the effect of sodium chromoglycate. *Lancet* ii:1268 (1979).
21.L. W. Poulter, C. Power, C. Burke. The relationship between bronchial immuno-pathology and hyperresponsiveness in asthma. *Eur Resp J.* 3:792 (1990).
22.P. G. Holt. Down-regulation of immune responses in the lower respiratory tract: the role of alveolar macrophages. *Clin exp Immunol.* 63:261 (1986).
23.P. G. Holt, C. McMenamin. Review: Defence against allergic sensitization in the healthy lung: the role of inhalation tolerance. *Clin Exp Allergy.* 19:155 (1989).
24.A. M. Pollard and M. F. Lipscomb. Characterization of murine lung dendritic cells: similarities to langerhans cells and thymic dendritic cells. *J Exp Med.* 172:159 (1990).
25.P. G. Holt, M. A. Schon-Hegrad, M. J. Phillips, P. G. McMenamin. Ia-positive dendritic cells form a tightly meshed network within the human airway epithelium. *Clin Exp Allergy* 19:597 (1989).
26.M. A. Spiteri and L. W. Poulter. Characterisation of immune inducer and suppressor macrophages from the normal human lung. *Clin exp Immunol.* 83:157 (1991).
27.M. A. Spiteri, S. W. Clarke, L. W. Poulter. The isolation of phenotypically and functionally distinct alveolar macrophages from human broncho-alveolar lavage. *Eur Resp Jour.* in press (1992).
28.C. Hutter and L. W. Poulter. The balance of macrophage subsets may be customised at mucosal surfaces. *FEMS Immunol Microbiol.* In press (1992).
29.M. C. Allison and L. W. Poulter. Changes in phenotypically distinct mucosa macrophage populations may be a prerequisite for the development of inflammatory bowel disease. *Clin exp Immunol.* 85:504 (1991).
30.L. Marianayagam and L. W. Poulter. Corticosteroids can alter antigen expression on alveolar macrophages. *Clin exp Immunol.* 85:531 (1991).
31.F.A.M. Josien, Wijffels, Zegerine de Rover, Nico van Rooijen, G. Kraal, R.H.J. Beelen. Chronic inflammation induces the expression of dendritic cell markers not related to functional antigen presentation on peritoneal exudate macrophages. *Immunobiol.* 184:83 (1991).

PULMONARY DENDRITIC CELL POPULATIONS

Patrick G. Holt

The Western Australian Research Institute for Child Health
Princess Margaret Hospital
Perth, Western Australia 6001

INTRODUCTION

Epithelial surfaces in the respiratory tract represent a uniquely fragile interface between the immune system and the outside environment. The lumenal surface of the respiratory tract is exposed virtually continuously to low levels of potentially pathogenic microbial antigens, as well as to a myriad of inert antigens in the form of organic dusts and aerosols of both plant and animal origin. These antigens include a wide range of low molecular weight highly soluble proteins, a proportion of which are highly likely to penetrate into the epithelium.

On purely teleological grounds, it can be argued that higher animals require a finely tuned and efficient antigen surveillance system, functional at the level of the epithelial surfaces in both the lower and upper respiratory tract. Ideally, such a system should have in-built controls to limit local "presentation" of antigen to T-cells within the epithelium itself - which carries substantial attendant risk of T-cell mediated damage to delicate surrounding tissues - to situations which pose a genuine threat to the host. Data emerging from a number of laboratories point to local populations of Dendritic Cells (DC) present within the airway epithelium and alveolar septal walls, as prime candidates for the role of principal local antigen presenting cells (APC), and further suggest that tight control of the APC functions of these DCs by adjacent cell populations is of major importance in regulation of local immunological homeostasis.

CHARACTERISATION OF PULMONARY DC POPULATIONS IN TISSUE SECTIONS

Dendritic and/or Langerhans Cells (LC) were initially described in lung biopsies from patients with chronic inflammatory and fibrotic diseases[1,2]. However, they are now recognised as normal residents of epithelial microenvironments throughout the respiratory tract in disease-free individuals[3-5], and can also be found as a discrete (albeit minor) subset amongst the cell population accessible by bronchoalveolar lavage (BAL)[4,6].

Interest in the potential significance of DC in immune regulation in the respiratory tract has grown steadily in the last few years, particularly following demonstration of the previously unrecognised high population density of these cells within the airway epithelium. In particular, it was shown that the use of conventional (viz. vertical) planes of section to prepare airway tissues for immunohistochemical analysis led to a completely false classification of intraepithelial cell populations expressing class II MHC (Ia) antigen[7]. Due to the large size and highly pleiomorphic shape of mature DCs (e.g. Fig. 1b) vertical sections through the airway epithelium often reveal only a portion of the body or parts of processes of individual cells (Fig. 1a), which has led many authors to conclude erroneously that Ia expression is heterogeneous at these sites, and is a function of many cell types within the normal epithelium. However, by use of a plane of section through the epithelium and parallel to the underlying basal lamina, it has been possible to demonstrate that in the normal (inflammation-free) airway epithelium virtually 100% of Ia staining can be accounted for by cells with the characteristic morphology of DCs (Fig. 1e,f) both in experimental animals[7] and in humans[8]. These DC display a distribution pattern and density (in the order of 600 per mm^2 epithelial surface in the large airways)[9] which is comparable to the epidermal LC network (Fig. 1c,d).

Ultrastructural analysis of pulmonary DC populations indicates marked heterogeneity with respect to Birbeck granule expression[3,4,10], which may be explicable on the basis of variations in degree of activation[11]. At the light microscopic level, heterogeneity in expression of surface markers has also been reported in a number of studies comparing DC-like cells from different pulmonary tissue microenvironments[3,4,9,12,13] and within individual pulmonary DC populations[9,12,15], and these differences are also likely to reflect local activation (see below).

EFFECTS OF EXOGENOUS STIMULI ON PULMONARY DCs

Recent evidence from several laboratories demonstrate the potential responsiveness of pulmonary DC populations to inflammatory stimulation. In the human, inhalation of chemical irritants in the form of tobacco smoke increases the density of DCs in the bronchial mucosa and alveolar parenchyma, and effects changes in CD1 and Birbeck granule expression[4]. Moreover, human smokers display increased numbers of DCs in BAL fluids[6], suggesting inflammation-induced recruitment. Inflammation associated with allergic rhinitis is also accompanied by increased influx of $CD1^+$ DCs into the nasal mucosa[10].

In the rat model, intratracheal challenge with killed BCG induces DC influx into the alveoli[14] - immunocytochemical staining of BAL-derived cells from these animals revealed a characteristic DC morphology (small size, highly ruffled Ia^+ cell membrane, single supra-nuclear spot of Acid Phosphatase activity) consistent with the influx of young DC from the bone-marrow[14].

The DC population in the rat airway epithelium is also capable of responding dynamically to local challenge with microbial stimuli, and expands rapidly within 24 hrs of exposure to aerosolised LPS[9]. DC density returns to baseline levels within 24 hrs, suggesting that net traffic of these cells from the inflamed epithelium to draining lymph nodes (and hence net flux of associated antigens) increases transiently during the acute phase of the inflammatory response[9]. Chronic airways inflammation in rats induced by housing on dusty bedding (pine shavings) also increases DC density, intensity of Ia-immunostaining on individual cells and integrin expression[9]. It is also noteworthy that the distribution of DC within the respiratory tree of normal SPF rats (notably the inverse relationship between airway diameter and intraepithelial DC density)[9,13] broadly reflect

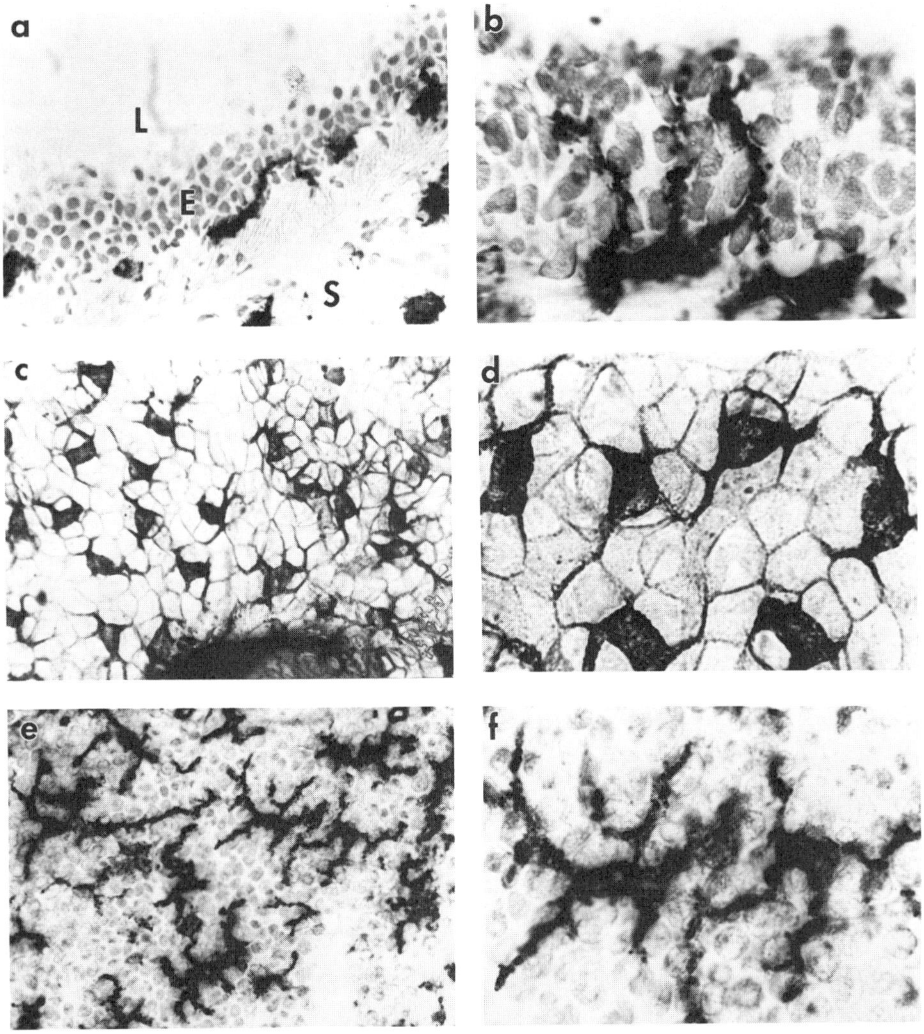

Figure 1. Immunoperoxidase staining for Ia in rat tracheal epithelium and epidermis. Ethanol-fixed sections were immunoperoxidase stained with the monoclonal antibody Ox6 as described previously[19]. Epidermal sheets from rat ears were fixed and stained in parallel. **a:** Transverse section of tracheal epithelium. L = lumen; E = epithelium;S = submucosa. x 40 obj. **b:** Transverse section of tracheal epithelium. x 100 obj. **c:** Epidermal sheet. x 40 obj. **d:** Epidermal sheet. x 100 obj. **e:** Tangential section of tracheal epithelium. x 40 obj. **f:** Tangential section of tracheal epithelium. x 100 obj.
Reproduced with permission from Int. Arch. Allergy Appl. Immunol. 91:155 (1990).

patterns of deposition of inhaled particulate, suggesting that the overall "tonus" of the DC network in the respiratory tract in the steady-state may be regulated via the degree of stimulation provided by irritants in inspired air.

Parenteral administration of the pro-inflammatory cytokine IFNg has also been demonstrated to increase the density of Ia^+ DC within the lung parenchyma[14] and airway epithelium[13], and the presence of significant numbers of DC within the vascular bed of IFNg-treated animals suggests that the underlying mechanism involves recruitment from peripheral blood, as opposed to stimulation of Ia expression on local Ia^- DC precursors[14].

FUNCTIONAL CAPACITY OF PULMONARY DCs

The capacity of lung-derived DC to present antigen-specific activation signals to immune T-cells was first demonstrated in studies on cell populations derived from collagenase digestion of parenchymal tissue from rat lung[17], and similar findings have since been reported for other species and also for airway-derived DC (reviewed in Holt[18]). The antigen presenting cell (APC) activity of the lung DCs was also demonstrated to be inhibited by endogenous tissue macrophages in the rat[17,19] and also in humans[20,21].

Both airway[13] and lung parenchymal DC[22] have also been shown to bind inhaled antigen *in situ* and process it into immunogenic form for subsequent presentation to T-cells *in vitro*. However, recent studies indicate that, analogous to DC in the epidermis and lymph, resident lung and airway DC in the rat function poorly as APC when freshly isolated, and acquire this capacity during overnight culture[26] (Fig. 2). Moreover, this functional maturation process can be inhibited by co-culture with alveolar macrophages (AM) across a semi-permeable membrane (Fig. 2), and can be partially reproduced by TNFα[23].

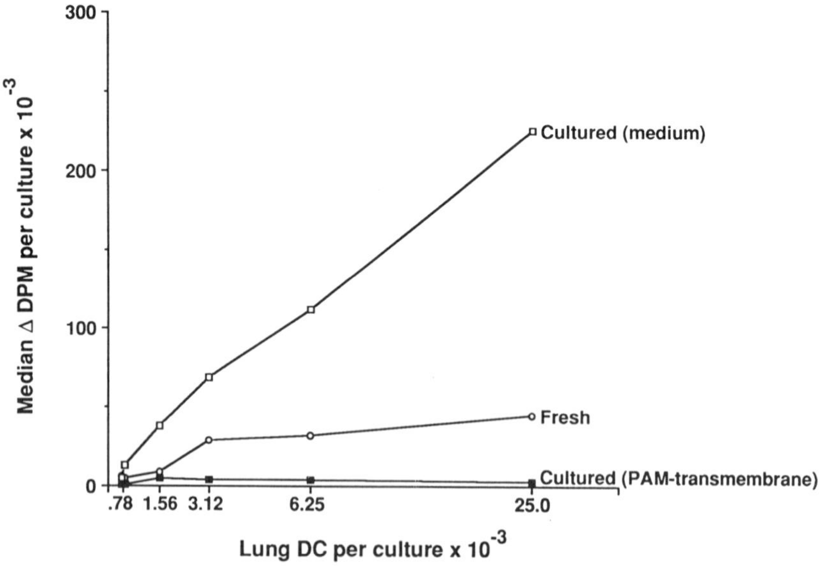

Figure 2. MLR stimulatory activity of rat lung parenchymal DC assayed immediately after preparation (fresh), or after overnight preculture in medium, or in a dual chamber culture vessel separated from pulmonary alveolar macrophages (PAM) by a semi-permeable membrane.

As part of the latter study, *in vivo* AM depletion of groups of rats was carried out employing the liposome "suicide" technique. As detailed in[24], this procedure involves intratracheal inoculation of liposomes containing the cytotoxic drug dichloromethylene diphosphonate, and results in depletion of up to 90% of resident AM within 48-72 hrs, without detectable inflammatory side-effects. Pulmonary DC harvested from AM-depleted animals demonstrate high levels of APC activity when freshly isolated, equivalent to that normally achieved only by overnight culture[23], suggesting that the APC function(s) of the DC may be downregulated *in situ* in the steady state by secreted products of AM.

In support of the feasibility of such a control mechanism operating *in vivo,* electron microscopic examination of perfused-fixed rat lungs demonstrates that the majority of AM are normally localised in clefts at alveolar septal junctions, separated from underlying interstitial tissue containing the parenchymal DC population by an attenuated covering of Type 1 epithelium as thin as 0.1 μm[23]. Similarly in the airways, the intraepithelial DCs are closely juxtaposed to underlying submucosal macrophages, separated by only the epithelial basal lamina[12,19,23].

CONCLUSIONS

On the basis of the studies cited above, it is reasonable to postulate an important "sentinel" role for pulmonary DC, analogous to that demonstrated previously for epidermal LC, viz. their prime function appears to be sampling inhaled antigens impinging on the airway surface, for subsequent presentation to the T-cell system after migration to regional lymph nodes. During their residence *in situ* in the respiratory tract, their capacity to present antigen to T-cells is actively downregulated by adjacent macrophages via secreted products (including TNFα), and this may be viewed as a "protective" mechanism to limit T-cell-mediated damage to host epithelial tissues in the lung and airways. The rapid responsiveness of the pulmonary DC network to inflammatory stimuli suggest inherent ability to upregulate the functional capacity of the network in the face of challenge by microbial pathogens, including (on the basis of data in Havenith[14]) antigen presentation activity.

It may thus be speculated that pulmonary DC play a major role in the aetiology and pathogenesis of a variety of respiratory tract diseases with an immunological component, in particular virus infections and allergic diseases triggered by airborne environmental allergens.

REFERENCES

1. O. Kawanami, F. Basset, V.J. Ferrans, P. Soler, and R.G. Crystal, Pulmonary Langerhans cells in patients with fibrotic lung disorders, *Lab .Invest.* 44:227 (1981).
2. S. Hamman, D. Brockus, S. Remmington, and M. Bartha, The widespread distribution of Langerhans cells in pathological tissues. An ultrastructural and immunohistochemical study, *Human Path.* 17:894 (1986).
3. K. Sertl, T. Takemura, E. Tschachler, V.J. Ferrans , M.A. Kaliner,and E.M. Shevach, Dendritic cells with antigen-presenting capability reside in airway epithelium, lung parenchyma, and visceral pleura, *J. Exp. Med.* 63:436 (1986).
4. P. Soler, A. Moreau, F. Basset, and A.J. Hance, Cigarette smoking-induced changes in the number and differentiated state of pulmonary dendritic cells/Langerhans cells, *Am. Rev. Respir. Dis.* 139:1112 (1989).
5. S. Richard, S. Barbey, A. Pfister, P. Scheinmann, F. Jaubert, and C.Nezelof, Mise en evidence de

cellules de Langerhans dans l'epithelium bronchique humain, *C. R. Acad. Sci. Paris Serie III* 305:35 (1987).

6. M.A. Casolaro, J-F. Bernaudin, C. Saltini, V.J. Ferrans, and R.G. Crystal, Accumulation of Langerhans' cells on the epithelial surface of the lower respiratory tract in normal subjects in association with cigarette smoking, *Am. Rev. Respir. Dis.* 137:406 (1988).
7. P.G. Holt, M. Schon-Hegrad, J. Oliver, B.J. Holt, and P. McMenamin, A contiguous network of dendritic antigen presenting cells within the respiratory epithelium. *Int. Arch. Allergy Appl. Immunol.* 91:155 (1990).
8. P.G. Holt, M.A. Schon-Hegrad, M.J. Phillips, and P.G. McMenamin, Ia-positive Dendritic Cells form a tightly meshed network within the human airway epithelium, *Clin. Exp. Allergy* 19:597 (1989).
9. M.A. Schon-Hegrad, J. Oliver, P.G. McMenamin, P.G. Holt, Studies on the density, distribution and surface phenotype of intraepithelial class II MHC antigen (Ia)-bearing dendritic cells (DC) in the conducting airways, *J. Exp. Med.* 173:1345 (1991).
10. W.J. Fokkens, T.M.A. Vroom, E. Rijnije, and P.G.H. Mulder, CD1$^+$ (T6) HLA-DR-expressing cells, presumably Langerhans cells in nasal mucosa, *Allergy* 44:167 (1989).
11. D. Hannau, M. Fabre, J-P. Lepoittevin, J-L. Stampf, E. Grosshaus, and C. Benezra, La formation des granules de Langerhans semble liee a lactivite AT Pasique membranaire des cellules de Langerhans epidermiques, *C. R. Acad. Sci. Paris Serie III* 301:167 (1985).
12. P.G. Holt, and M.A. Schon-Hegrad, Localization of T cells, macrophages and dendritic cells in respiratory tract tissue: Implications for immune function studies, *Immunol.* 62:349 (1987).
13. J.L. Gong, K.M. McCarthy, J. Telford, T. Tamatani, M. Miyasaka, and E.E. Schneeberger, Intraepithelial dendritic cells: a distinct subset of pulmonary dendritic cells obtained by microdissection. *J. Exp. Med.* 175: 797 (1992).
14. K.E.G. Havenith, A.J. Breedijk, and E.C.M. Hoefsmit, Effect of BCG inoculation on numbers of dendritic cells in bronchoalveolar lavages of rats, *Immunobiol. (*1992 in press).
15. W. Xia, E.E. Schneeberger, K. McCarthy, and R.L. Kradin, Accessory cells of the Lung. II. Ia$^+$ pulmonary dendritic cells display cell surface antigen heterogeneity, *Am. J. Respir. Cell Mol. Biol.* 5:276 (1991).
16. R.L. Kradin, K.M. McCarthy, W. Xia, D. Lazarus, and E.E. Schneeberger, Accessory cells of the lung. I. Interferon-gamma increases Ia$^+$ dendritic cells in the lung without augmenting their accessory activities, *Am. J. Respir. Cell Mol. Biol.* 4:210 (1991).
17. P.G. Holt, A. Degebrodt, C. O'Leary, K. Krska, T. Plozza, T-cell activation by antigen presenting cells from lung tissue digests: Suppression by endogenous macrophages, *Clin. Exp. Immunol.* 62:586 (1985).
18. P.G. Holt, Regulation of antigen presenting cell function(s) in lung and airway tissues, *Europ. Resp. J.* (1992 in press).
19. P.G. Holt, M.A. Schon-Hegrad, and J. Oliver, MHC class II antigen-bearing dendritic cells in pulmonary tissues of the rat: Regulation of antigen presentation activity by endogenous macrophage populations, *J. Exp. Med.* 167:262 (1988).
20. L.P. Nicod, M.F. Lipscomb, G.B. Toews, and J.C. Weissler, Separation of potent and poorly functional human lung accessory cells based on autofluorescence, *J. Leukocyte Biol.* 45:458 (1989).
21. M.A. Spiteri, and L.W. Poulter, Characterisation of immune inducer and suppressor macrophages from the normal human lung, *Clin. Exp. Immunol.* 83:157 (1991).
22. P.G. Holt, J. Oliver, C. McMenamin, and M.A. Schon-Hegrad, Studies on the surface phenotype and functions of Dendritic Cells in parenchymal lung tissue of the rat, *Immunol.* 75:582 (1992).
23. P.G. Holt, J. Oliver, C. McMenamin, N. Bilyk, G. Kraal, and T. Thepen, The antigen presentation functions of lung Dendritic Cells are downmodulated *in situ* by soluble mediators from pulmonary alveolar macrophages, (submitted).
24. T. Thepen, C. McMenamin, J. Oliver, G. Kraal, and P.G. Holt, Regulation of immune responses to inhaled antigen by alveolar macrophages (AM): differential effects of AM elimination *in vivo* on the induction of tolerance versus immunity, *Europ. J. Immunol.* 21:2845 (1991).

BLOOD DENDRITIC CELLS ARE HIGHLY ADHERENT TO UNTREATED AND CYTOKINE-TREATED CULTURED ENDOTHELIUM

K. Alun Brown, Frances LeRoy, Penelope A. Bedford,[1] Stella C. Knight[1] and Dudley C. Dumonde

Department of Immunology, UMDS, St Thomas' Hospital, London SE1 7EH and [1]Antigen Presentation Research Group, CRC, Middlesex, UK

INTRODUCTION

Blood dendritic cells are potent stimulators of T cell proliferation[1] and their entry into tissues may give rise to the many forms of interdigitating cells such as skin Langerhans cells[2]. The binding of lymphocytes, monocytes and polymorphonuclear leucocytes to vascular endothelium is dependent in part upon the surface expression of the CD11/CD18 β2 integrin family[3], adhesion molecules that are also present on the surface of dendritic cells[4]. Treatment of endothelium with inflammatory cytokines increases their adhesiveness for leucocytes by inducing or enhancing the expression of other surface adhesion determinants (e.g. ICAM-1) that recognise corresponding receptors (i.e. CD11a) on the leucocytes[5].

The aim of this study was to investigate the adhesive properties of blood dendritic cells for cultured endothelial cells. Experiments were designed to compare the binding properties of dendritic cells with those of lymphocytes and monocytes isolated from the same subject using a quantitative monolayer adhesion assay with human umbilical cord vein endothelial cells as the adherence substrate. Adhesion experiments were also performed with endothelial monolayers treated with the inflammatory cytokine, tumour necrosis factor (TNF).

MATERIALS AND METHODS

Isolation of Lymphocytes, Monocytes and Dendritic Cells

Heparinised blood (10 u/ml) from healthy subjects was diluted with an equal volume of RPMI, layered onto ficoll-paque (Pharmacia) and centrifuged at 600 *g* for 30 minutes.

Dendritic Cells in Fundamental and Clinical Immunology,
Edited by Kamperdijk *et al.*, Plenum Press, New York, 1993

The mononuclear cell layer was removed and washed twice. The cells were placed in tissue culture flasks (5-10 x 10^6 ml^{-1}) and incubated at 37°C overnight. The non adherent cells were removed and centrifuged over hypertonic metrizamide gradients (Nyegaard, 13.7% w/v). The adherent cells were vigorously removed with a cell scraper (Costar). More than 90% of these adherent cells were monocytes. The pellet cells from the metrizamide gradients were gently resuspended and used as a source of lymphocytes. The low density dendritic cell interface from these gradients were further enriched by incubation on human IgG-coated petri dishes to remove contaminating Fc receptor positive monocytes. Approximately 70% of cells in these preparations showed dendritic cell morphology when viewed under a light microscope.

Lymphocytes, monocytes and dendritic cells from each subject were suspended in 1 ml HBSS containing 100 µCi (3.7 MBq) of $Na2^{51}Cr04$ for 1 h at 37°C. The radiolabelled cells were resuspended in RPMI with 10% autologous serum.

Culture of Endothelial Cells from Umbilical Cord Veins

Each vein of a human umbilical cord was filled with RPMI medium and 1 mg/ml collagenase solution and incubated for 15 min at 37°C. Thereafter, the collagenase solution was aspirated and centrifuged at 600 X *g* for 10 min. The endothelial cell pellet was resuspended in medium (RPMI plus 20% pooled human AB serum, 2mM glutamine, 200 U/ml penicillin, 100 U/ml streptomycin), added to a culture flask the floor of which had previously been treated with 1% (w/v) gelatin, and incubated in a 10% CO_2-humidified atmosphere at 37°C. When the endothelial cells had formed a confluent monolayer, the medium was expelled and the cells treated with 1 ml 0.05% trypsin and 0.025% EDTA in phosphate-buffered saline (PBS) for 1-3 min. The endothelial cells were resuspended in medium to 1 X 10^5 cells/ml and 100 µl added to each well of a collagen-coated (0.15% Type IV) 96-well microtitre plate. Confluent monolayers of endothelial cells were obtained after 4 days of culture at 37°C in a humidified atmosphere of 10% CO_2 in air. Identification of the endothelial cells was confirmed by their reactivity with an anti-human factor VIII antiserum and by their characteristic morphology using light microscopy.

Adherence Assay To each well were added 200 µl of labelled lymphocytes, monocytes or dendritic cells (1 X 10^6 cells/ well). Each test was performed in quadruplicate in randomly allotted wells. After incubation at 37°C for 1 h the non-adherent leucocytes were aspirated and the endothelial monolayers washed on five occasions with RPMI, without serum, at 37°C to remove loosely adherent cells. The endothelium was osmotically disrupted and the lysate counted in an auto-gamma scintillation counter. The percentage of leucocytes adhering to endothelium was calculated as follows:

$$\% \text{ Adherence} = \frac{\text{cpm in 200 µl lysate - cpm background}}{\text{cpm in 200 µl original MNC suspension - cpm background}} \times 100$$

To investigate the binding of leucocytes to TNF-treated endothelium the monolayers were incubated with 1-100 U/ml of rTNFα for 4 h prior to the assay.

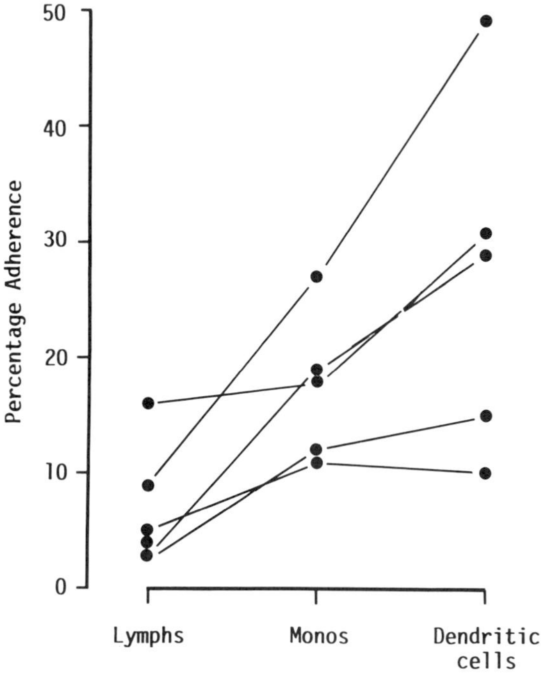

Summary of experiments in which blood lymphocytes, monocytes and dendritic cells from 5 normal subjects were added to endothelial monolayers.

Figure 1. Adherence of lymphocytes, monocytes and dendritic cells to untreated endothelium

RESULTS

Figure 1 shows the comparative adherence to untreated endothelium of blood lymphocytes, monocytes and dendritic cells from five healthy subjects. In terms of their ability to bind to endothelium the list in increasing order of adhesiveness is lymphocytes<monocytes<dendritic cells.

In the eighteen experiments summarised in Table 1, dendritic cells were significantly more adherent than lymphocytes (mean 118% increase; $p<0.001$) and monocytes (mean 50% increase; $p<0.05$) to untreated endothelium. There was considerable variation in the attachment to endothelium of dendritic cells (range 10-49%) and this was due to a difference in the adhesive properties of endothelial cells obtained from various umbilical cord veins rather than a difference between the adherence characteristics of dendritic cells from one subject to another.

Table 1 also shows that pretreatment of endothelial monolayers with TNF increased the number of adherent lymphocytes, monocytes and dendritic cells. There was a greater number of dendritic cells binding to TNF-treated endothelial cells in comparison with monocytes (mean 43% increase; $p<0.001$) and lymphocytes (mean 49% increase; $p<0.001$).

Table 1. Binding of blood dendritic cells to untreated and TNF-treated cultures of human endothelium

	%ADHERENCE	
	UNTREATED ENDOTHELIUM	TNF-TREATED ENDOTHELIUM
LYMPHOCYTES	11 ± 5	17 ± 7
MONOCYTES	16 ± 8	21 ± 9
DENDRITIC CELLS	24 ± 12	33 ± 13

Results are the mean (± standard deviation) of 18 experiments.
In each experiment lymphocytes, monocytes and dendritic cells from a normal subject were added to endothelial monolayers prepared from one umbilical cord vein.
Dendritic cells were significantly more adherent than lymphocytes ($p<0.001$) and monocytes ($p<0.01$) to untreated endothelium and more adherent than lymphocytes and monocytes ($p<0.001$) to TNF-treated endothelium.

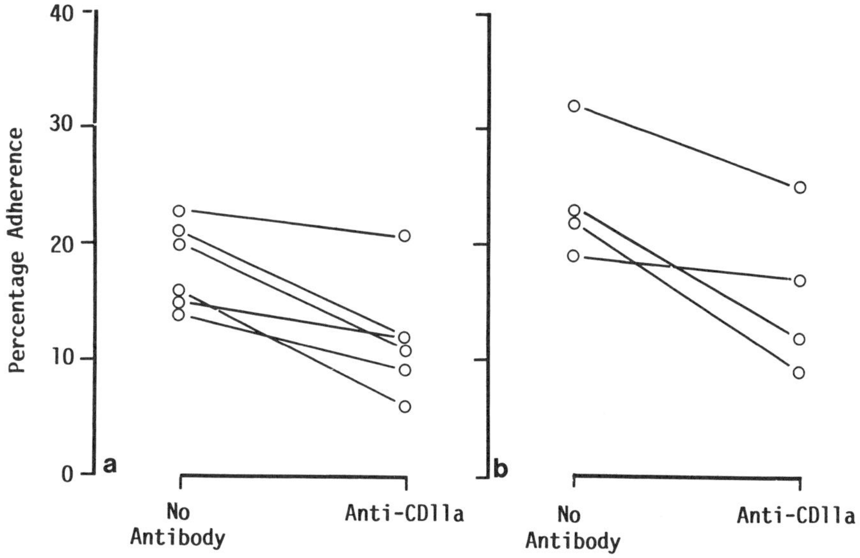

Blood dendritic cells from 6 normal subjects were pretreated with an anti-CD11a antibody prior to their addition to untreated and TNF-treated monolayers.

Figure 2. Anti-CD11a antibodies inhibit binding of dendritic cells to (a) untreated and (b) TNF-treated endothelium

To determine if the binding of dendritic cells was dependent upon the expression of CD11a, cells were pretreated with anti-CD11a antibodies prior to their addition to untreated and TNF-treated endothelial cells. From the experiments illustrated in Figure 2, the anti-CD11a antibody produced a mean 36% and 35% inhibition of dendritic cell adhesion ($p<0.01$) to untreated endothelium and TNF-treated endothelium respectively.

DISCUSSION

Small numbers of dendritic cells are present in normal blood and these bone-marrow derived cells are believed to be the precursors of tissue dendritic cells[2]. The blood dendritic cells are very potent antigen presenting cells *in vitro*[1] and lack typical markers of monocytes and lymphocytes[4]. Our investigation shows that preparations of low density cells enriched in blood dendritic cells, as determined by phenotypic and functional studies[7], are more adherent than lymphocytes and monocytes to untreated and TNF-treated endothelial monolayers. Binding to blood vessel walls is a prerequisite for leucocyte extravasation and the abnormal adhesiveness of dendritic cells for vascular endothelium is likely to facilitate their entry into tissue and maturation into peripheral dendritic cells. Blood dendritic cells express CD11a on their surface[4] and this integrin recognises both the constitutively expressed ICAM-2 and cytokine-upregulated ICAM-1 on the surface of endothelial cells. In the present study antibody blocking experiments demonstrated that the binding of dendritic cells to untreated and TNF-treated endothelium was dependent in part upon the expression of LFA-1. Studies are in progress to investigate the contribution of other adhesion molecules in dendritic cell-endothelial interaction. The increased propensity to adhere to cytokine-stimulated endothelial cells could result in a recruitment of blood dendritic cells into inflammatory lesions and the perpetuation of antigen presentation. Such antigens may be of local origin or transported by infiltrating dendritic cells since these cells are known to process, carry and present viral antigens to lymphocytes[8] and carry complement complexes on their surface[9]. Indirect support for this view is that there is currently no evidence of dendritic cell proliferation though we do not exclude the possibility that the number of dendritic cells at an inflammatory site may be maintained or increased by the differentiation of local cells.

We suggest that an augmented binding of blood dendritic cells to vascular endothelium may be a key event in the recruitment of dendritic cells to sites of T-cell reactivity.

ACKNOWLEDGEMENTS

This work was supported by the Arthritis and Rheumatism Council and The Psoriasis Association.

REFERENCES

1. J.M. Austyn, Lymphoid dendritic cells, *Immunology*. 62:16. (1987).

2. C.D.L. Reid, P.R. Fryer, C. Clifford, A. Kirk, J. Tikerpoe, and S.C. Knight, Identification of hematopoietic progenitors of macrophages and dendritic Langerhans cells (DL-CFU) in human bone marrow and peripheral blood, *Blood*. 76:1139 (1990).
3. T.A. Springer, Adhesion receptors of the immune system, *Nature*. 846:425 (1990).
4. R.M. Steinman, The dendritic cell system and its role in immunogenicity, *Ann. Rev. Immunol*. 9:271 (1991).
5. J. Pober, Cytokine-mediated activation of vascular endothelium, *Am. J. Pathol*. 133:426 (1988).
6. A.J. Stagg, B. Harding, R.A. Hughes, A. Keat, and S.C. Knight, The distribution and 'functional' properties of dendritic cells in patients with seronegative arthritis, *Clin. Exp. Immunol*. 84:66 (1991).
7. S.C. Knight, J. Farrant, A. Bryan, et al, Non-adherent low density cells from human peripheral blood contain dendritic cells and monocytes, both with veiled morphology, *Immunology*. 57:595 (1986).
8. S.C. Knight, S.E. Macatonia, M.S. Roberts, J.J. Harvey, and Patterson, S., *in*: "Viruses in the Cellular Immune Response", B. Thomas, ed., Marcel Dekker Inco., New York (1992).
9. R., Würzner, H. Xu, A. Franzke, M. Schulze, J.N. Peters, and O. Götze, Blood dendritic cells carry terminal complement complexes on their cell surface as detected by newly developed neoepitope-specific monoclonal antibodies, *Immunology*. 74:132 (1991).

ANTIGEN SPECIFIC T CELL PRIMING *IN VIVO* BY INTRATRACHEAL INJECTION OF ANTIGEN PRESENTING CELLS

Carin E.G. Havenith, Annette J. Breedijk, Wim Calame, Robert H.J. Beelen and Elisabeth C.M. Hoefsmit

Department of Cell Biology
Division of Electron Microscopy
Medical Faculty
Vrije Universiteit
van der Boechorststraat 7
1081 BT Amsterdam, The Netherlands

INTRODUCTION

In the lung alveolar macrophages (AM) are the first cells to encounter pathogens and airborne antigens. In general, AM function poorly as accessory cells and even inhibit T cell proliferation *in vitro*[1,2]. In previous experiments it was found that the increased numbers of dendritic cells (DC) in the bronchoalveolar lavages after intratracheal instillation of Bacillus Calmette-Guérin (BCG) were responsible for the antigen presentation *in vitro*[2]. The AM suppressed the T cell proliferation in a concentration dependent manner[2]. Many cell types are capable of presenting antigen to T cells that already have undergone priming *in situ*, whereas the DC has been shown capable of delivering exogenous antigens directly to naive T cells *in vitro*[3].

In the present study splenic DC and AM were pulsed with antigen *in vitro*, and subsequently intratracheally instilled to test whether these two cell populations have the capacity to sensitize T cells in the draining lymph nodes of the lung.

MATERIALS AND METHODS

Animals

Inbred male ACI/Ma1 rats were obtained from Harlan CPB (Zeist, The Netherlands). All animals were kept under routine laboratory conditions. ACI/Ma1 rats are high responders for the copolymer glutamine-tyrosine (GT, ICN ImmunoBiologicals, Lisle, Israel), in an Ia-dependent manner[4].

Preparation of (antigen-pulsed) DC and AM

Bronchoalveolar lavage cells were collected by repeated bronchoalveolar lavages as described in Havenith et al.[2]. The cells were washed with phosphate-buffered saline (PBS) and resuspended in culture medium (complete RPMI, 10% heat-inactivated FCS, 20 mM L-glutamine, penicillin 50 U/ml, streptomycin 50 μg/ml, 50 μM ß-mercaptoethenol) at a concentration of 1 to 2 x 10^6 cells/ml and maintained overnight in Teflon Beakers. To some of these cultures 100 μg GT/ml was added to pulse the cells overnight with this antigen. After overnight culture the AM were collected by gently scraping with a rubber policeman.

Dendritic cells were enriched from a cell suspension of the spleen (incubation with collagenase and DNAse), using a overnight adhesion culture step followed by a Nycodenz gradient as described in Havenith et al.[5]. During overnight culture, to some of these cultures 100 μg GT/ml was added.

Cytocentrifuge preparations were made and the cellular compositions of the populations were evaluated by immuno-and enzymecytochemistry as described elsewhere[2]. The (un)pulsed DC populations consisted of 60 to 75% of DC (Ia-positive cells, with an acid phosphatase spot in the vicinity of the nucleus), whereas the AM populations consisted of more than 97% of Ia-negative cells. The contaminating cells of the AM populations were mainly T cells, whereas DC never exceeded 0.5% of the total cells.

GT-primed T Cell Proliferation Assay

Two weeks after immunization in the footpads with GT emulsified in 100 μl Freund's complete adjuvant, GT-primed T cells were isolated from the draining lymph nodes, using a Nylon wool column and an incubation step with the anti-Ia monoclonal antibody Ox6 and Dynabeads[5]. In the antigen presentation assay (un)pulsed DC or AM were added to 1 x 10^5 GT-primed T cells with or without GT (100 μg/ml) in a total volume of 200 μl culture medium in flat-bottomed 96-wells microtiter plates. T cell proliferation was determined by measuring ^{3}H-thymidine incorporation[5]. Data, mean values of triplicate wells, are expressed in counts per minute (cpm).

In Situ Priming with Antigen-pulsed DC or AM

The DC and AM populations were intratracheally instilled in recipient rats, after they were lightly anaesthetized with 0.2 ml Hypnorm (Janssen, Goirle, The Netherlands), intramuscularly injected. The cells, doses 0.5 to 1 x 10^6 cells, were injected in a volume of 0.1 ml PBS. As controls a GT-solution (10 μg/0.1 ml) and killed cells (three times freezing in liquid nitrogen followed by thawing) were used.
Four days later the draining paratracheal lymph nodes were removed and teased into a cell suspension. Antigen-specific T cell proliferation was determined by culturing 3 x 10^5 lymph node cells in a total volume of 200 μl (culture medium: 5% heat-inactivated normal rat serum instead of FCS) with or without antigen (100 μg/ml) and by measuring ^{3}H-thymidine incorporation. Data, mean values of triplicate wells, are expressed in counts per minute (cpm).

RESULTS

Accessory Cell Function of (un)pulsed DC or AM *In Vitro*

The accessory cell function of the DC and AM populations, pulsed or unpulsed with the antigen GT, were tested in a GT-primed T cell proliferation assay. Splenic DC, pulsed with the antigen GT during overnight culture, stimulated the primed T cells to higher levels in comparison to unpulsed DC (Table 1). AM, pulsed or unpulsed, did not induce antigen-specific T cell proliferation. Co-culture of unpulsed DC-T cell cultures with GT-pulsed AM showed that the AM suppressed the T cell proliferation (Table 1). After killing of the GT-pulsed AM, the suppressive activity disappeared, and even higher responses were found than the response of DC-T cell cultures alone (Table 1).

Table 1. Testing of GT-pulsed dendritic cells or alveolar macrophages in a GT-primed T cell proliferation assay.

Antigen presenting cell population(s)[1]	Incorporation of ^{3}H-TdR by GT-primed T cells (x 10^3 cpm)[2] no GT added	GT added
DC	45.9	211.0
GT-pulsed DC	200.4	244.1
AM	<0.1	<0.1
GT-pulsed AM	<0.1	<0.1
DC + GT-pulsed AM	0.8	6.7
DC + GT-pulsed AM***	105.6	285.3

[1]DC, splenic dendritic cells; GT-pulsed DC, splenic DC cultured overnight with 100 μg GT/ml; AM, bronchoalveolar lavage cells cultured overnight in Teflon beakers; GT-pulsed AM, AM cultured overnight with 100 μg GT/ml; ***, three times frozen in liquid nitrogen followed by thawing.

[2]1 x 10^5 GT-primed T cells were cocultured with 2 x 10^4 antigen presenting cells; T cell proliferation was determined by measuring ^{3}H-thymidine (^{3}H-TdR) incorporation. Data, mean values of triplicate wells, are expressed in counts per minute (cpm).

Accessory Cell Function of (un)pulsed DC or AM In Vivo

The accessory cell function *in vivo* of the DC and AM populations was tested by instillation of the cells in the bronchoalveolar lumen. The draining paratracheal lymph nodes were taken 4 days later, and tested for *in vivo* priming, by boosting *in vitro* with the antigen GT. Intratracheal instillation of unpulsed DC or AM and also of free antigen (10 μg GT/0.1 ml) gave no significant antigen-specific responsiveness in the draining lymph nodes (Table 2). Whereas instillation of GT-pulsed DC or AM resulted in the development of an antigen-specific T cell proliferation in the paratracheal lymph nodes (Table 2). However after killing the GT-pulsed DC or AM, different results between the two cell populations were found. Killed GT-pulsed DC gave no antigen-specific T cell priming, whereas killed GT-pulsed AM, still were able to induce T cell priming (Table 2).

Table 2. T cell priming *in vivo* in paratracheal lymph nodes by GT-pulsed dendritic cells or alveolar macrophages instilled intratracheally.

LNC isolated 4 days after injection of[2]	Incorporation of ^{3}H-Thymidine by 3x10^5 LNC (x 10^3 cpm)[1]	
	no antigen	100 ug/ml GT
DC	1.2	3.4
GT-pulsed DC	1.5	281.4
GT-pulsed DC ***	0.5	4.5
AM	0.4	2.8
GT-pulsed AM	0.3	185.0
GT-pulsed AM ***	0.5	112.0
free antigen (100 μg/ml)	1.0	9.3

[1]Proliferation of lymph node cells (LNC), mean of triplicate wells, was expressed in counts per minute (cpm).

[2]0.5-1.0 x 10^6 cells intratracheally instilled. Abbreviations see Table 1.

DISCUSSION

This study demonstrates that viable DC, pulsed with an antigen during overnight culture, instilled in the bronchoalveolar lumen, are able to induce an antigen-specific T cell priming in the draining lymph nodes of the lung. So, increased numbers of bronchoalveolar DC *in vivo*, which appear after induction of an inflammatory reaction[2], may be involved in antigen-specific T cell priming for airborne antigens. Also instillation of GT-pulsed AM leads to the induction of a immune response. However the finding that, in contrast to killed antigen-pulsed DC, killed antigen-pulsed AM still leads to antigen-specific T cell priming suggests that antigen-specific priming *in vivo* by AM involve different mechanisms than the presently supposed mechanisms for DC.

DC are potent accessory cells in the primary syngeneic and allogeneic MLR[6] and are known to home to the lymphoid organs[7,8]. Also Inaba et al.[3] demonstrated that DC are capable of delivering the antigen directly to naive T cells *in situ*. On the contrary, the AM, suppress antigen-or mitogen stimulated lymphocyteproliferative responses *in vitro*[1,2]. This is again demonstrated in the *in vitro* experiments in this study. The fact that Ia-negative AM, alive or dead, leads to the initiation of a response suggests that it is not the AM itself that presents the antigen to resting T cells *in vivo*, but that antigen becomes accessible to other APC. The DC, which are present in different compartments of the lung[9,10], are the most likely cells to pick up the antigen and migrate to the draining lymph nodes. The here presented *in vitro* antigen presenting experiments implicate that the antigen released from killed GT-pulsed AM can be presented by DC.

REFERENCES

1. P.G. Holt, Downregulation of immune responses in the lower respiratory tract: the role of alveolar macrophages, *Clin. Exp. Immunol.* 63:261 (1986).

2. C.E.G. Havenith, A.J. Breedijk, and E.C.M. Hoefsmit, Effect of Bacillus Calmette-Guérin inoculation on numbers of dendritic cells in bronchoalveolar lavages of rats, *Immunobiol.* 184:336 (1992).
3. K. Inaba, J.P. Metlay, M.T. Crowley, and R.M. Steinman, Dendritic cells pulsed with protein antigens in vitro van prime antigen-specific, MHC-restricted T cells in situ, *J. Exp. Med.* 172:631 (1990).
4. D.Armerding, D.H. Katz, and B. Benacerraf, Immune response genes in inbred rats. I. Analysis of responder status to synthetic polypeptides and low doses of bovine serum albumin, *Immunogenetics* 1:79 (1979).
5. C.E.G. Havenith, A.J. Breedijk, M.A.M. Verdaasdonk, E.W.A. Kamperdijk, and R.H.J. Beelen, An improved and rapid method for the isolation of rat lymph nodes or spleen T lymphocytes for T cell proliferation assays, *J. Immunol. Methods* (in press).
6. R.M. Steinman and M.D. Witmer, Lymphoid dendritic cells are potent stimulators of the primary mixed leukocyte reaction in mice, *Proc. Natl. Acad. Sci.* 75:5132 (1978).
7. J.M. Austyn, J.W. Kupiec-Weglinski, D.F. Hankins, and P.J. Morris, Migration patterns of dendritic cells in the mouse. Homing to T cell-dependent areas of spleen, and binding within marginal zone, *J. Exp. Med.* 167:646 (1988).
8. S. Fossum, Lymph-borne dendritic leukocytes do not recirculate, but enter the lymph node paracortex to become interdigitating cells, *Scan. J. Immunol.* 27:97 (1988).
9. M. Schon-Hegrad, J. Oliver, P.G. McMenamin, and P.G. Holt, Studies on the density, distribution, and surface phenotype of intraepithelial class II major histocompatibility complex antigen (Ia)-bearing dendritic cells (DC) in the conducting airways, *J. Exp. Med.* 173:1345 (1991).
10. K. Sertl, T. Takemura, E. Tschachler, V.J. Ferrans, M.A. Kaliner, and E.M. Shevach, Dendritic cells with antigen-presenting capability reside in airway epithelium, lung parenchyma, and viceral pleura, *J. Exp. Med.* 163:436 (1986).

HISTOLOGY AND IMMUNOPHENOTYPE OF DENDRITIC CELLS IN THE HUMAN LUNG

Jan Maarten W. van Haarst[1], Harm J. de Wit[1], Hemmo A.Drexhage[1] and Henk C.Hoogsteden[2]

Dept. of Immunology, Erasmus University (1) and Dept. of Pulmonary Medicine, University Hospital Dijkzigt (2)

INTRODUCTION

Monocytes, dendritic cells (DC) and macrophages (MΦ) play an important role in immune responses. Immunohistologically dendritic cells are characterized by their morphology (dendritic shape), enzyme histochemistry (spot or absence of acid phosphatase) and strong MHC class II expression. Other markers have been described on dendritic cells as well, for instance CD1a on Langerhans cells. The major function of dendritic cells is antigen presentation and the initiation of immune responses. Dendritic cells have been described in both the human lung[1,2,4] as well as in the rat[5] and in the mouse[1] lung. We performed experiments to confirm the data on the presence and distribution of dendritic cells in the human lung using the earlier mentioned immunohistological criteria. In addition an array of other markers was used to further characterize the dendritic cells of the human lung. Moreover we compared the marker pattern of lung dendritic cells with those of blood monocytes, of dendritic-like cells maturated from blood monocytes and of alveolar macrophages.

MATERIALS AND METHODS

Cells and tissues. Normal lung tissue was obtained from patients undergoing (partial) lung resection for bronchial carcinoma. Lung tissue was gently inflated with optimal cutting temperature cryoembedding material (O.C.T.tissue-Tek) 1:1 diluted in Phosphate Buffered Saline (PBS) and embedded in O.C.T.tissue-Tek. Specimens were frozen at -80°C. Cryostat sections (6 μm) were brought on poly-L-lysine coated slides and air dried. Broncho-alveolar lavage (BAL) cells were obtained from volunteering individuals undergoing routine gynecological surgery. Blood monocytes of healthy laboratory staff were isolated from heparin blood. The isolation procedure involved ficoll and percoll density gradient separation steps which yielded a purity of 70-95% monocytes as determined by non specific esterase staining. Blood derived dendritic-like cells (15-35% purity) were obtained by culturing these monocytes (16 hr, 37°C, 5% CO_2) under non-adherent conditions (polypropylene tubes) after stimulation with metrizamide (described by Kabel[6]). Cytocentrifuge preparations were prepared from the various cell suspensions.

Immunostaining procedures. The cytocentrifuge preparations and the histological sections were fixed in acetone for 10 min, and thereafter incubated with normal rabbit serum 10% in PBS for 10 min before incubating for 1 hr with the appropriate monoclonal antibody (MoAb). See table for MoAbs used. After washing three times with PBS the slides were incubated for 30 min. with rabbit-anti-mouse antiserum conjugated with horseradish peroxidase (RaMHRP) diluted in PBS with 10% normal human serum. The slides were rinsed three times in PBS and stained with di-amino-benzidine (DAB). Selected slides were stained for either acid phosphatase or non-specific esterase. Counterstaining was performed with haematoxilin.

For double immunostaining procedures the slides were incubated with a directly biotinylated second MoAb after the incubation with RaMHRP. Slides were rinsed three times in PBS and incubated with streptavidine-biotin complex conjugated with alkaline phosphatase. Slides were sequentially stained with 3-amino-9-ethylcarbazole (AEC)(peroxidase) and fast blue (alkaline phosphatase).

RESULTS

Bronchial and bronchiolar epithelium. Cells with a typical dendritic morphology were found in the epithelium of the bronchi and bronchioli in all patients. These cells stained for MHC class II, L25 and a part of these cells reacted strongly with CD1a (fig.1). Reactions with RFD1 varied considerably among the patients.

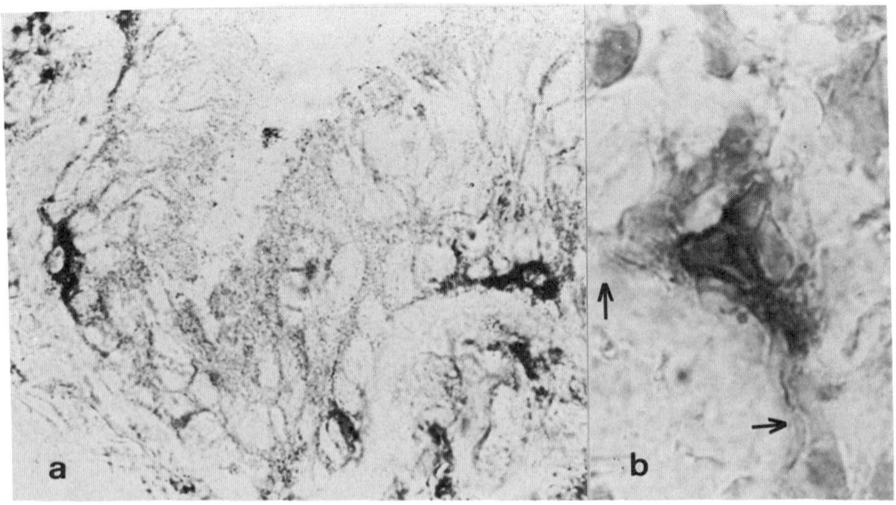

fig.1. Dendritic cells in the bronchial epithelium just above the basal membrane (arrows); 400x (a), and detail 1000x (b).

Subepithelial tissue. A heterogeneous infiltrate of mononuclear cells was found in the subepithelial tissue of the bronchus and bronchiolus of all patients, although the number of infiltrating cells varied considerably. Beside CD3-positive T-cells the infiltrating cells mainly consisted of irregular shaped cells positive for MHC class II and L25. A part of these cells reacted with CD14. Sporadically a CD1a-positive cell was observed. Only few macrophages were detected as determined by shape and positivity for KIM7(CD68), RFD7, RFD9 and strong acid phosphatase staining. B-cells (CD19) were only sporadically observed.

Bronchus associated lymphoid tissue (BALT). BALT was encountered in two of the twelve patients. Characteristic dendritic cells could be detected with OKIa, L25, RFD1 and CD1a in a small outer zone of the BALT. This outer zone could be identified as the T-cell zone on the basis of positivity for CD3.

Alveolar walls. In the alveolar walls recognition of dendritic cells was hampered by the fact that alveolar epithelial cells as well as endothelial cells showed class II-expression. Sporadically a cell with a dendritic morphology, negative for acid phosphatase and with a strong MHC class II-expression could be identified in the alveolar wall. Reactivity with OKT6 (CD1a) could not be observed in the alveolar septa.

Broncho-alveolar lavage. In the BAL less than 0.5% of the cells showed prominent dendritic protrusions, were strongly positive for MHC class II and were negative for or had a spot of acid phosphatase. These cells were negative for CD1a. There was however a population of large rounded cells, negative for MHC class II and acid phosphatase, but positive for CD1a (0-4% of the population).

In table 1 the marker pattern of the epithelial bronchial dendritic cells is compared to that of blood monocytes, blood-derived dendritic-like cells and alveolar macrophages (BAL/alveolar spaces).

Table 1. Comparison of marker patterns (++: strongly positive, +: positive, ±: weak, -:negative, +/-: partly positive).

	bronchial/ BALT-DC	blood-der. DC	blood monocyte	alv. MΦ
acid phosphatase	-	-	-	+
n.spec.esterase	-	±	+	+
OKIa(HLA-DR)	+	+	+	+
L25	+	+	-	±/-
OKT6(CD1a)	+/-	-	-	-
RFD1	+/-	+	-	+/-
My4(CD14)	-	+/-	+	-
KIM7(CD68)	-	-	+	++
RFD7	-	-	-	+
RFD9	-	-	-	+
OKT3(CD3)	-	-	-	-
B4(CD19)	-	-	-	-

DISCUSSION

This report clearly shows the presence of dendritic cells in the epithelium of the bronchial/bronchiolar wall and in the BALT of the human lung with a marker pattern resembling Langerhans cells and blood-derived dendritic-like cells. In the subepithelial tissue of the bronchial/ bronchiolar wall similar cells were observed. Their morphology was however less characteristic, and there was more overlap with markers of the macrophage lineage.

Our data are somewhat contrasting to those of others. The presence of a large number of OKT6(CD1a)-positive cells in the alveolar wall has been described by Soler[2], while Sertl[1] reported a lack of positivity of pulmonary dendritic cells for OKT6(CD1a). The differences between our and these studies might be attributed to differences in immunohistological techniques, collection of tissue specimens and/or the selection of patients. Soler used for instance fixed materials in contrast to us.

Unlike the description of a considerable number of dendritic cells in the alveolar septa of the human lung by Sertl[1] we could only sporadically recognize such cells in the alveolar wall. It must be noted that due to the strong MHC class II-expression in the alveolar wall (endothelium, pneumocytes) an enumeration of these cells in our specimens was impossible.

The virtual absence of characteristic dendritic cells in the BAL and in the alveolar tissue makes the alveolar compartment unsuitable for the initiation of immune responses. The abundant presence of macrophages in this compartment, which have been described to inhibit the stimulation of proliferative responses by antigen presenting cells[3], strengthens this view. It must be noted however that the population of rounded CD1a-positive cells in the BAL might be precursors for dendritic cells. It is also possible that these cells derive from the Langerhans-like cells in the bronchial epithelium. The bronchus and bronchiolus showing the clear presence of characteristic dendritic cells are better equipped for mounting an immune response than the alveolar compartment. The development of lymphoid tissue (BALT) in the bronchus/bronchiolus in a few patients supports this notion.

REFERENCES

1. K. Sertl et al. Dendritic cells with antigen-presenting capability reside in airway epithelium, lung parenchyma and visceral pleura, *J Exp Med.* 163:436-451 (1986).

2. P.Soler et al. Cigarette smoking-induced changes in the number and differentiated state of pulmonary dendritic cells/Langerhans cells, *Am Rev Resp Dis.* 139:1112-1117 (1989).

3. P.G.Holt et al. T-cell activation by antigen-presenting cells from lung tissue digests:suppression by endogenous macrophages, *Clin Exp Immunol.* 62:586-593 (1985).

4. P.G.Holt et al. Ia-positive dendritic cells form a tightly meshed network within the human airway epithelium, *Clin Exp Immunol.* 19:597-601 (1989).

5. P.G.Holt et al. A contiguous network of dendritic antigen-presenting cells within the respiratory epithelium, *Int Arch Allergy appl Immunol.* 91:155-159 (1990).

6. P.J.Kabel et al. Accessory cells with a morphology and marker pattern of dendritic cells can be obtained from elutriator purified blood monocyte fractions. An enhancing effect of metrizamide in this differentiation. *Immunobiol.* 179:395-411 (1989).

ACQUISITION OF CHLAMYDIAL ANTIGEN BY DENDRITIC CELLS AND MONOCYTES

Andrew J. Stagg,[1] Arthur Stackpoole,[2] William J. Elsley,[1] and Stella C. Knight[1]

[1]Antigen Presentation Research Group
[2]Section of Translantation Biology
Clinical Research Centre
Watford Road
Harrow Middlesex HA1 3UJ

INTRODUCTION

In some genetically susceptible individuals infection of the genitourinary tract with *Chlamydia trachomatis* appears to trigger an acute seronegative arthritis. Synovial fluids from such patients contain very large numbers of dendritic cells (DC)[1] and there is evidence that these cells may be presenting chlamydial antigen acquired *in vivo* (submitted for publication). These observations have led us to investigate the role of DC from healthy individuals in processing chlamydial antigen and stimulating T cell responses *in vitro*.

Low density cells isolated from human peripheral blood comprise approximately 60% monocytes (MO), 30% dendritic cells and a small number of lymphocytes[1]. They are potent antigen presenting cells (APC) for both primary and secondary T cell responses to complex chlamydial antigens *in vitro*. For primary T cell responses, assayed using a 20μl hanging drop culture technique[2], there is an absolute requirement for DC in the APC population although a role for contaminating MO cannot be excluded. In particular it remains controversial whether or not DC, which are poorly phagocytic in standard *in vitro* tests, are able to acquire and process an antigen as complex as a whole bacterium.

Using three colour fluorescence activated cell sorter (FACS) analysis we have looked directly at the ability of DC, MO and lymphocytes to bind whole biotinylated chlamydial elementary bodies (EB). Cell populations were identified with monoclonal antibodies and using permeabilised and non-permeabilised cells, surface and internalised antigen were distinguished.

MATERIALS AND METHODS

Preparation of Biotinylated Chlamydia

C. trachomatis (serovar L1) EB were harvested from BGMK cells after 48hr infection and partially purified by centrifugation through 25% v/v Isopaque. After washing they were biotinylated using a commercial kit (Amersham, U.K.) following the manufacturers' protocols. They were stored at 4°C until use.

Isolation of Cells

Mononuclear cells were isolated by centrifugation on Ficoll-Paque (Pharmacia Fine Chemicals) from defibrinated venous blood obtained by venepuncture of healthy laboratory personnel. The cells were washed twice in HEPES-buffered RPMI 1640 containing 2% fetal calf serum (FCS) and incubated overnight at 5×10^6/ml in $25cm^2$ tissue culture flasks (Falcon Co., Cockeysville, M.D.) in complete medium (Dutch modification of RPMI 1640 containing 10% FCS, 2mM L-glutamine 100 U/ml penicillin and 100μg/ml streptomycin). Non-adherent cells were centrifuged over hypertonic metrizamide (made as 14.5g plus 100ml complete medium) at 600g for 10 min. LDC were collected from the interface and the pellet cells used as a source of lymphocytes.

Detection of Bound Chlamydial Antigen

Biotinylated EB were incubated with LDC or lymphocytes at 37°C for various lengths of time and washed in the cold with medium containing sodium azide. In some experiments the cells were reincubated for a further 30 min and then washed again. For cell identification, the cells were labelled with fluorochrome conjugated monoclonal antibodies to B cells (CD19) to monocytes (CD14) and to HLA-DR. Biotinylated antigen on the surface of cells was detected by incubation with fluorochrome-labelled avidin followed by FACS analysis. For detection of total antigen cells were permeabilised before the addition of the avidin reagent. This was achieved by incubating in 70% ethanol for 20 min at 4°C.

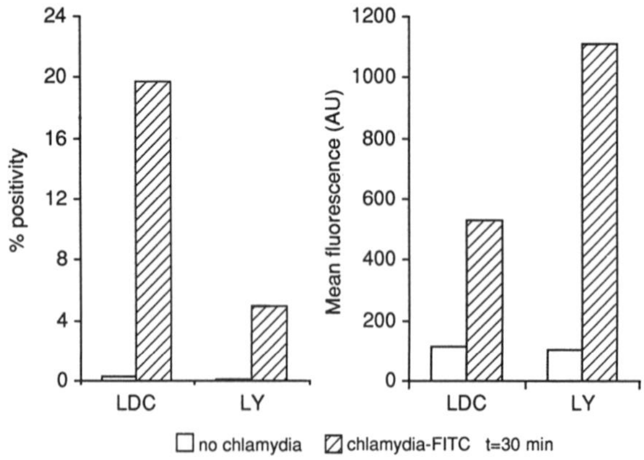

Figure 1. Binding of biotinylated EB by LDC and lymphocytes

RESULTS

Biotinylation of EB

Up to 90% of an EB preparation could be strongly labelled with avidin FITC following biotinylation (data not shown).

Detection of Surface Chlamydial Antigen

Initially, the ability of LDC and lymphocytes to bind EB was compared. A smaller percentage of lymphocytes bound EB but those that did bound larger amounts (Fig. 1).
Double labelling indicated that it was a small proportion of $CD19^+$ B cells within the lymphocyte preparation that bound EB (data not shown).

Further studies concentrated on the acquisition of EB by LDC. Antigen was detectable on the surface of LDC within 5 mins incubation and maximal levels were reached after 30-60 min (Fig. 2).

Following ethanol permeabilisation many additional LDC were found to be positive for chlamydial antigen and mean levels of fluorescence were also increased indicating the internalisation of large amounts of antigen (Fig. 3).

Evidence for Antigen Internalisation by DC

LDC are a mixture of cell types and in particular contain large numbers of monocytes. To look specifically at antigen uptake by DC three colour FACS analysis was used. DC were identified as $HLA\text{-}DR^+$

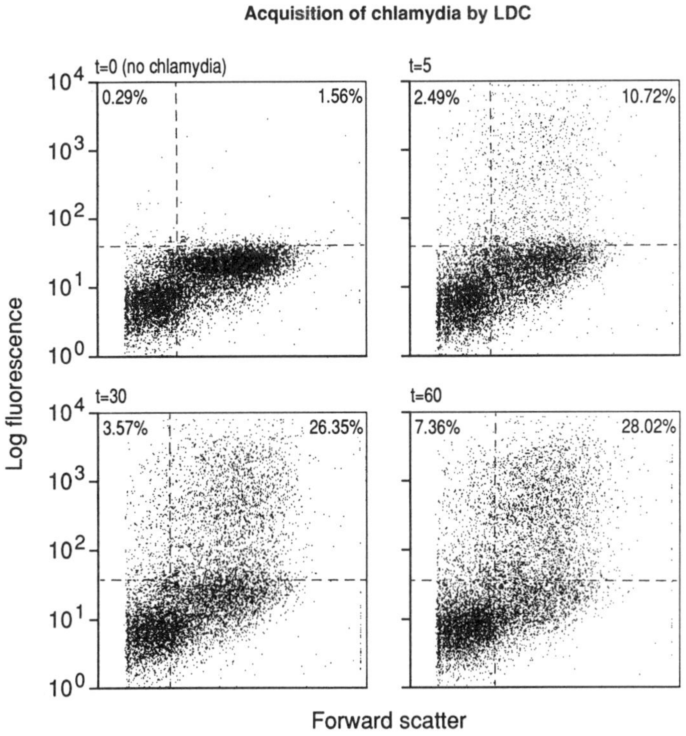

Figure 2. Kinetics of EB acquisition by LDC

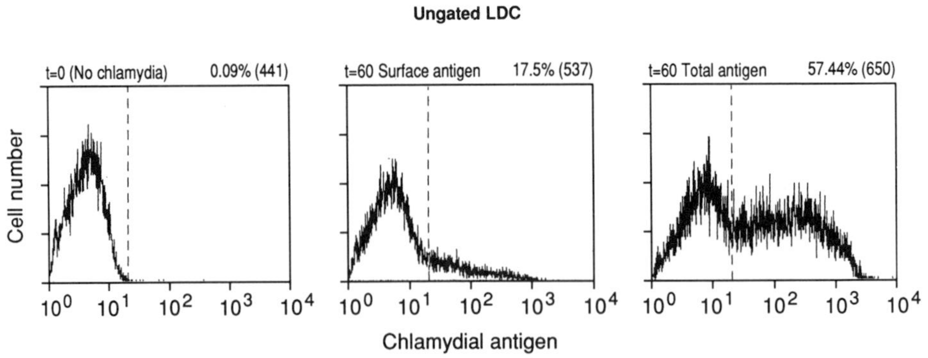

Figure 3. Permeabilisation revealed many LDC with internalised chlamydial antigen.

$CD14^-$ $CD19^-$ cells and acquisition of EB by this population was compared to that of $CD14^+$ HLA-DR^+ monocytes. As before, permeabilisation of the DC population before adding the avidin reagent resulted in additional antigen positive cells being detected (32% vs 11%) and an increase in mean fluorescence (540 vs 454) (Fig. 4). The percentage of DC bearing chlamydial antigen was markedly less than for monocytes (surface antigen positive = 30%, total antigen = 79%) but there was clear evidence that some DC are able to bind and internalise the complex bacterial antigen.

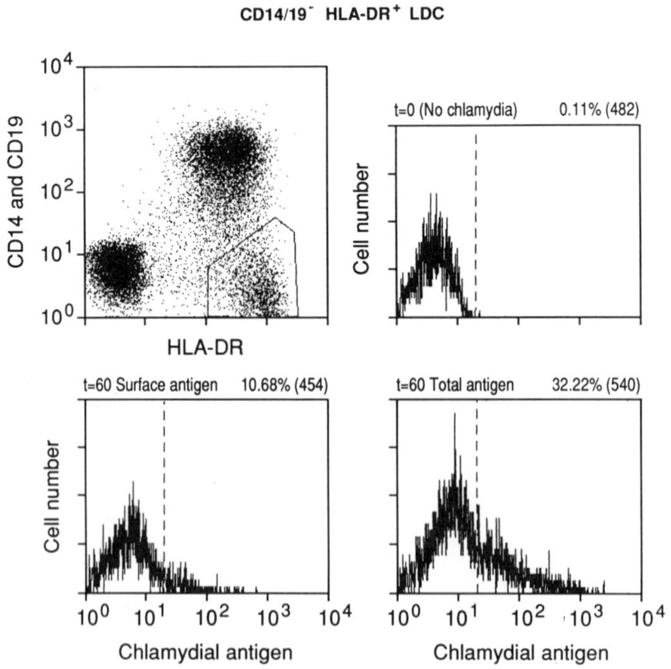

Figure 4. Three colour FACS analysis reveals internalisation of EB by DC

DISCUSSION

In this study we have demonstrated binding and internalisation of a particulate bacterial antigen by a proportion of human peripheral blood DC. In the mouse, studies on the ability of DC to process and present complex antigens have yielded conflicting results. Kapsenberg *et al*[3] and Guidos *et al*[4] reported that splenic DC were poor in comparison to phagocytic macrophages at presenting particulate antigen. In contrast, Kaye and colleagues[5] demonstrated that highly purified DC were able to present whole mycobacteria to paired T cells and they speculated that protein degradation was occurring at an extra-lysosomal site, possibly at the cell surface. In terms of sensitivity to glutaraldehyde fixation or inhibition by chloroquine presentation by DC and macrophages was similar or identical[6]. It is conceivable that conflicting data from different groups results from the use of different isolation protocols resulting in selective isolation of subtypes of DC. It is now clear that the functional properties of cells of the DC lineage alter dramatically during maturation[7].

Using multicolour FACS analysis we have studied directly the ability of human APC subpopulations to bind and internalise chlamydial EB. DC were identified as HLA-DR^+ LDC lacking monocytes and B cell markers. Following permeabilisation mean fluorescence levels increased and additional antigen positive DC were detected indicating that internalisation had occurred. Nonetheless, antigen was only detected in a minority of DC compared with over 80% of the monocytes. Human peripheral blood DC are heterogeneous by electron microscopy[8] perhaps reflecting the presence of cells of varying maturity. It is conceivable therefore that only a subpopulation of DC were able to internalise antigen or that they are only able to do it at certain stages in their development. It is notable that in the rat, internalisation of allogeneic lymphocytes by lymphoid interdigitating cells has been described *in vitro*[9] but cells with a similar phenotype from peripheral lymph were not active when tested *in vitro*.

We are currently extending these studies to compare the fate of antigen in DC and monocytes and look at the functional activity of antigen bearing cells.

Acknowledgement

This work was supported by the Arthritis and Rheumatism Council, U.K.

REFERENCES

1. A.J. Stagg, B. Harding, R.A. Hughes, A. Keat and S.C. Knight. The distribution and functional properties of dendritic cells and patients with seronegative arthritis. *Clin. exp. Immunol.* 84:66(1991).
2. S.C. Knight. Lymphocyte proliferation assays, *in*: "Lymphocytes: A Practical Approach", G.G.B. Klaus, ed., IRL Press, Oxford (1987).
3. M.L. Kapsenberg, M.B.M. Teunissen, F.E.M. Stiekema and H.G.Keizer. Antigen-presenting cell function of

dendritic cells and macrophages in proliferative T cell responses to soluble and particulate antigens. *Eur. J. Immunol.* 16:345(1986).

4. C. Guidos, M. Wong and K.-C. Lee. A comparison on the stimulatory activities of lymphoid dendritic cells and macrophages in T proliferative responses to various antigens. *J. Immunol.* 133:1179(1984).
5. P.M. Kaye, B.M. Chain and M. Feldmann. Non-phagocytic dendritic cells are effective accessory cells for anti-mycobacterial responses *in vitro*. *J. Immunol.* 134:1930(1985).
6. B.M. Chain, P.M. Kay and M. Feldmann. The cellular pathway of antigen presentation in biochemical and functional analysis of antigen processing in dendritic cells and macrophages. *Immunology*, 58:271(1986).
7. G. Schuler and R.M. Steinman. Murine epidermal Langerhans' cells mature into potent immunostimulating dendritic cells *in vitro*. *J. exp. Med.* 161:526(1985).
8. S. Patterson, J. Gross, P. Bedford and S.C. Knight. Morphology and phenotype of dendritic cells from peripheral blood and their productive and non-productive infection with human immunodeficiency virus type 1. *Immunology*, 72:361(1991).
9. S. Fossum and B. Rolstad. The roles of interdigitating cells and natural killer cells in the rapid rejection of allogeneic lymphocytes. *Eur. J. Immunol.* 16:440(1986).

EXPERIMENTAL CUTANEOUS LEISHMANIASIS: LANGERHANS CELLS INTERNALIZE *LEISHMANIA MAJOR* AND INDUCE AN ANTIGEN-SPECIFIC T-CELL RESPONSE

Heidrun Moll

Institute of Clinical Microbiology
University of Erlangen-Nürnberg
Wasserturmstrasse 3
W-8520 Erlangen, Germany

INTRODUCTION

Human leishmaniasis constitutes a diverse group of diseases caused by protozoa of the genus *Leishmania*. The parasites are transmitted by sandflies and the infection is initiated by introduction of promastigotes into the skin at the site of the vector's bite. In the mammalian host, *Leishmania* parasites exist as obligatory intracellular amastigotes residing in mononuclear phagocytes. The diseases vary in severity from the self-healing cutaneous leishmaniasis, characterized by a localized cutaneous lesion at the site of primary infection, to the progressive visceral leishmaniasis, where the parasites disseminate from the original skin lesion to local lymph nodes and then to spleen, liver and bone marrow. The clinical manifestation depends primarily on the species of parasite, but also on the genetic make-up of the host. The T cell-mediated immunity is crucial for the outcome of infection because activated T cells release various lymphokines that promote either the spreading or the elimination of parasites[1,2].

The entry of *Leishmania* into host cells occurs via attachment to cell surface receptors and subsequent internalization by phagocytosis. Receptor-ligand binding involves the receptor for complement component C3bi, CR3, and the mannose-fucose receptor on macrophages[3]. Infected macrophages express parasite antigen on their surface and thus serve as antigen-presenting cells for induction of a specific T-cell response.

In the mammalian skin, there are two appropriate cell types that would be in contact with inoculated *Leishmania* parasites - dermal macrophages and epidermal Langerhans cells (LC). LC express CR3 (Refs. 4, 5) and they are potent antigen-presenting cells bearing high levels of major histocompatibility complex (MHC) class II antigens. In contrast to macrophages, however, which are known to play a central role in the course of infection[6], the function of LC in leishmaniasis has so far not obtained adequate attention.

We used *Leishmania major*, the cause of human cutaneous leishmaniasis in the Old World, for analysis of the following issues in the mouse model: (1) the interaction of LC with *L. major*, (2) the efficacy of LC to present *L. major* antigen to T cells, and (3) the ability of LC to transport *L. major* antigen from the skin into lymphoid tissues for stimulation of T cells.

LANGERHANS CELLS CAN BE INFECTED BY *L. MAJOR*

In order to investigate the interaction of LC with *L. major* in vitro, single-cell suspensions of epidermal cells were prepared from ear skin of BALB/c mice by trypsinization procedures as previously described[7]. Freshly isolated epidermal cells were incubated with *L. major* amastigotes for 24 h. Subsequently, the cultures were harvested for removal of extracellular parasites and identification of infected epidermal cells by cytochemistry.

Double labeling combining immunoenzymatic labeling (ABC method[8]) to detect *L. major* antigens and immunofluorescence staining for expression of MHC class II (Ia) molecules revealed that only Ia^+ epidermal cells but not Ia^- cells contained *L. major* antigen. Because LC are the only Ia^+ cells in the normal epidermis, this finding demonstrated that epidermal LC can be parasitized by *L. major*. Furthermore, ultrastructural studies showed that LC accomodated intact amastigotes which were secluded within phagosomes (Figure 1). The ingestion of *L. major* by LC was blocked by anti-CR3 antibodies but could not be reduced by a soluble inhibitor of the mannose-fucose receptor. This indicated that in contrast to the uptake of *Leishmania* by macrophages, which requires the combined action of CR3 and the mannose-fucose receptor[3], the internalization by LC is mediated primarily by CR3.

Only freshly isolated LC were able to ingest *L. major*. When parasites were added to cultured epidermal cells, the rate of infection was inversely correlated with the time of pre-incubation. No uptake could be detected after overnight culture of epidermal cells prior to addition of parasites (Table 1).

L. major-containing LC could not only be identified after culture in vitro but were also detectable in situ in the lesions of mice that had been intradermally infected with *L. major* promastigotes. This was demonstrated in skin sections by mixed labeling[8] of LC with an antibody to nonlymphoid dendritic cells (NLDC-145, Ref. 10) and an antiserum to *L. major*. $NLDC\text{-}145^+$ LC expressing *L. major* antigen could only be traced in the dermal infiltrate, but not in the epidermis, and constituted less than 1% of nonlymphoid mononuclear cells in this area.

Table 1. The ability of LC to ingest *L. major* decreases with the time of culture.[1]

Time of culture before addition of *L. major* (hours)	infected LC (% of control ; $\bar{x} \pm$ SD)
0 (control)	100
2	59 ± 12
5	32 ± 11
9	8 ± 5
16	0 ± 0
24	3 ± 1

[1] *L.major* amastigotes were added to epidermal cell cultures at various time points. After additional 24 h, the cultures were harvested, extracellular parasites were removed, and the rate of infection of LC was determined by fluorescence microscopic analysis after staining with acridine orange and ethidium bromide[9].

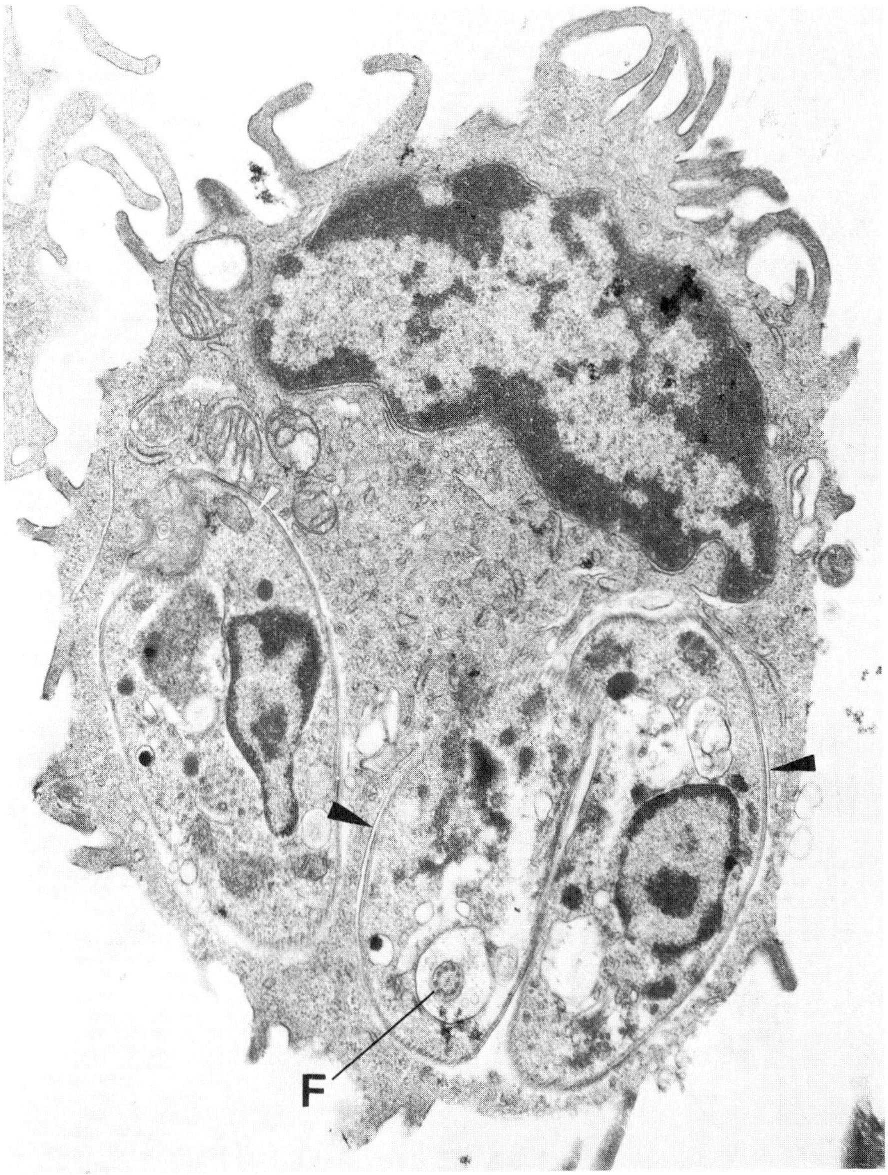

Figure 1. Electron microscopic documentation of a *L. major*-infected LC containing two amastigotes within one phagosome and a single organism within another phagosome. Note the unit mebrane of the phagosome (arrowheads) and the flagellum (F). Magnification = x15,000. (Courtesy of Klemens Rappersberger, M. D.)

The maximum level of infection in culture was 25% of total LC and the majority of infected LC had only ingested one parasite (mean: 1.4 amastigotes). Most interestingly, time course analysis revealed that the average number of organisms per infected cell did not increase after 24 h of culture, and by one week, *L. major*-containing LC were no longer detectable. Thus, LC are able to restrain parasite replication. This is in contrast to macrophages which are permissive to parasite infection unless they are activated by cytokines such as as interferon-γ[11].

The above findings showed that the parasite load of LC in vitro and in vivo is much lower than that of macrophages. This suggests that endocytosis of *L. major* by LC is not primarily aimed at antigen scavenging, the characteristic of macrophages. In addition, LC did not support intracellular parasite multiplication and are therefore unlikely to account for the massive expansion of amastigotes usually observed in the infected skin. It is conceivable, however, that the major function of LC, which are known to be amongst the most active antigen-presenting cells[12], is the presentation of *L. major* antigen to T cells. To test this possibility, we investigated the ability of LC to stimulate *L. major*-primed T cells in vitro.

LANGERHANS CELLS ARE POTENT STIMULATORS OF AN ANTIGEN-SPECIFIC T-CELL RESPONSE TO *L. MAJOR*

To assess the efficiency of epidermal antigen-presenting cells, we used unselected epidermal cell populations which contain 2% to 4% LC. The results demonstrated that epidermal cells are able to present *L. major* antigen in vitro to T cells from primed mice[13]. The heterogeneous epidermal cells were remarkably efficient because the doses required for optimal responses were equivalent to merely 250 to 2000 LC per culture. The T cell-stimulatory effect of epidermal cells could indeed be attributed to LC because removal of those cells completely abrogated the response. LC stimulated both T-cell proliferation and lymphokine production; the reaction was antigen-specific and could be induced by antigen in *L. major* lysate as well as by antigen derived from viable parasites[13]. The finding that only freshly isolated LC, but not cultured LC, were able to present *L. major* antigen was in line with our observation that only fresh LC could internalize parasites. This is reminiscent of earlier reports showing that freshly explanted LC can process native antigen for presentation to primed T cells, whereas cultured LC have lost this ability but are highly active mediators of stimuli that do not require antigen processing[14-16]. Freshly isolated LC would reflect the properties of intracutaneous LC, and LC cultured for several days would be the in vitro equivalents of lymphoid dendritic cells derived from LC that have migrated to lymph nodes.

L. MAJOR-BEARING LANGERHANS CELLS MIGRATE FROM THE INFECTED SKIN TO THE DRAINING LYMPH NODE

Evidence is accumulating that LC are capable of transporting antigens from peripheral non-lymphoid tissues via the lymphatics into lymphoid tissues for presentation to T cells[17,18]. Therefore, it was of interest to investigate whether dendritic cells carrying parasite antigen can be localized in lymph nodes draining the site of cutaneous infection with *L. major*. For this purpose, frozen sections of skin and lymph nodes from infected mice were analyzed by immunocytochemistry. In infected skin, the number of LC residing in the epidermis decreased rapidly within a few days after infection, whereas a gradually increasing number of LC appeared in the dermal compartment. Using mixed-labeling techniques[8], dendritic cells displaying *L. major* antigen were found not only in the dermis but also in the regional lymph node. In the early phase of infection, expression of parasite antigen in the lymph nodes was confined to dendritic cells and could not be detected on macrophages. Those lymph node cells had the potency of presenting parasite antigen to T cells because they stimulated antigen-specific T-cell proliferation in vitro and, after transfer to naive recipients, they induced DTH reactivity to *L. major* antigen. These findings indicate that LC capture *L. major* antigen in the skin and transport it to the regional lymph node for initiation of the T-cell immune response.

CONCLUDING REMARKS

In summary, our results suggest an important role of LC during cutaneous leishmaniasis. They support the concept that, upon inoculation of *L. major* into the skin, viable parasites are ingested not only by macrophages but also by LC that have migrated into the dermis. After appropriate processing of the organisms, LC would interact with antigen-specific T cells in two different ways. On the one hand, they are capable of carrying parasite antigen from the skin to the draining lymph node and presenting it to quiescent T cells of the circulating pool. On the other hand, LC may present parasite antigen to effector T cells infiltrating the cutaneous lesion and regulate the local immune response.

Several implications arise from these findings which are important for our understanding of the pathogenesis of cutaneous leishmaniasis. The first aspect relates to the well-documented phenomenon that resistance and susceptibility to infection can be attributed to $CD4^+$ T-cell populations with different patterns of lymphokine activity[1,2]. Release of interferon-γ appears to promote protection, whereas interleukin 4 has been suggested to facilitate survival of parasites. The possibility exists that different types of antigen-presenting cells (i.e. LC, lymphoid dendritic cells, macrophages and B cells) favor the expansion of disparate T-cell subsets on the basis of providing different costimulatory signals. The second point relates to the role of different parasite antigens. The processing machinery of LC may differ from that of other antigen-presenting cells and favor the presentation of distinct antigens or particular epitopes of a given antigen. A more detailed knowledge of this issue will be important for choosing the route of administration of potentially protective antigens. Finally, it will be of interest to analyze the mechanisms underlying the disparate permissiveness of resident macrophages and LC to infection with *L. major* or other *Leishmania* species. This may help to understand the varying disease patterns caused by different species of parasites. Current work in our laboratory addresses these issues. The above considerations emphasize that experimental cutaneous leishmaniasis provides an excellent model to study the role of LC in vivo during infectious diseases.

ACKNOWLEDGMENTS

It has been my privilege to study the role of LC in leishmaniasis in collaboration with the gifted students Christine Blank, Harald Fuchs and Antje Will. I am also very grateful to Dr. Klemens Rappersberger, University of Vienna Medical School, who performed the electron microscopical analyses. This work was supported by the Deutsche Forschungsgemeinschaft (SFB 263/A2).

REFERENCES

1. H. Moll and M. Röllinghoff, Resistance to murine cutaneous leishmaniasis is mediated by T_H1 cells, but disease-promoting $CD4^+$ cells are different from T_H2 cells, *Eur. J. Immunol.* 20:2067 (1990).
2. R. M. Locksley and P. Scott, Helper T-cell subsets in mouse leishmaniasis: induction, expansion and effector function, *in*: "Immunoparasitology Today", C. Ash and R. B. Gallagher, eds., Elsevier Science Publishers Ltd., Cambridge, UK (1991).
3. J. M. Blackwell, A. B. Ezekowitz, M. B. Roberts, J. Y. Channon, R. B. Sim, and S. Gordon, Macrophage complement and lectin-like receptors bind *Leishmania* in the absence of serum, *J. Exp. Med.* 162: 324 (1985).
4. G. Stingl, E. C. Wolff-Schreiner, W. J. Pichler, F. Gschnait, W. Knapp, and K. Wolff, Epidermal Langerhans cells bear Fc and C3 receptors, *Nature* (London) 268:245 (1977).
5. N. Romani, G. Stingl, E. Tschachler, M. D. Witmer, R. M. Steinman, E. M. Shevach, and G. Schuler,

The Thy-1 bearing cell of murine epidermis: a leukocyte distinct from Langerhans cell and perhaps related to NK cells, *J. Exp. Med.* 161:1368 (1985).

6. W. Solbach, H. Moll, and M. Röllinghoff, Lymphocytes play the music, but the macrophage calls the tune, *Immunol. Today* 12:4 (1991).
7. G. Schuler and F. Koch, Enrichment of epidermal Langerhans cells, *in*: "Epidermal Langerhans Cells", G. Schuler, ed., CRC press, Boca Raton (1991).
8. R. Gillitzer, R. Berger, and H. Moll, A reliable method for simultaneous demonstration of two antigens using a novel combination of immunogold-silver staining and immunoenzymatic labeling, *J. Histochem. Cytochem.* 38:307 (1990).
9. J. Y. Channon, M. B. Roberts, and J. M. Blackwell, A study of the differential respiratory burst activity elicited by promastigotes and amastigotes of *Leishmania donovani* in murine resident peritoneal macrophages, *Immunol.* 53:345 (1984).
10. G. Kraal, M. Breel, M. Janse, and G. Bruin, Langerhans cells, veiled cells and interdigitating cells in the mouse recognized by a monoclonal antibody, *J. Exp. Med.* 163:981 (1986).
11. C. F. Nathan, H. W. Murray, M. E. Wiebe, and B. Y. Rubin, Identification of interferon-γ as the lymphokine that activates human macrophage oxidative metabolism and antimicrobial activity, *J. Exp. Med.* 158:670 (1983).
12. G. Stingl and E. M. Shevach, Langerhans cells as antigen-presenting cells, *in*: "Epidermal Langerhans Cells", G. Schuler, ed., CRC press, Boca Raton (1991).
13. A. Will, C. Blank, M. Röllinghoff, and H. Moll, Murine epidermal Langerhans cells are potent stimulators of an antigen-specific T cell response to *Leishmania major*, the cause of cutaneous leishmaniasis, *Eur. J. Immunol.* 22:1341 (1992).
14. G. Schuler and R. M. Steinman, Murine epidermal Langerhans cells mature into potent immunostimulatory dendritic cells in vitro, *J. Exp. Med.* 161:526 (1985).
15. N. Romani, S. Koide, M. Crowley, M. Witmer-Pack, A. M. Livingstone, G. C. Fathman, K. Inaba, and R. M. Steinman, Presentation of exogenous antigens by dendritic cells to T cell clones. Intact protein is presented best by immature, epidermal Langerhans cells, *J. Exp. Med.* 169:1169 (1989).
16. J. W. Streilein and S. Grammer, In vitro evidence that Langerhans cells can adopt two functionally distinct forms capable of antigen presentation to T lymphocytes, *J. Immunol.* 143:3925 (1989).
17. S. E. Macatonia, S. C. Knight, A. J. Edwards, S. Griffiths, and P. Fryer, Localization of antigen on lymph node dendritic cells after exposure to the contact sensitizer fluorescein isothiocyanate. *J. Exp. Med.* 166:1654 (1987).
18. C. P. Larsen, R. M. Steinman, M. Witmer-Pack, D. F. Hankins, P. J. Morris, and J. M. Austyn, Migration and maturation of Langerhans cells in skin transplants and explants, *J. Exp. Med.* 172:1483 (1990).

ENDOCYTOSIS OF POTENTIAL CONTACT SENSITIZERS BY HUMAN DENDRITIC CELLS

Marie Larsson and Urban Forsum

Department of Clinical Microbiology
Faculty of Health Sciences
University of Linköping
S-581 85 Linköping
Sweden

INTRODUCTION

Several lines of evidence indicates that a population of lymphoid cells occurring in most organs of the body with a dendritic morphology is of prime importance in the initial stages of immune responses (1). Immunohistochemical distribution studies using various cell surfaces phenotypic markers has often been used to argue the probable importance of dendritic cells in tissues for the initial immune response (2,3). It has however proven difficult to compare these cells from all tissues functionally in vitro due to low yields and alteration of the phenotype when the cells are enzymatically dislodge from the tissues. Most studies have thus focused on the Langerhans cell of the skin, spleen dendritic cells and negatively selected dendritic cells from human peripheral blood.

The probable involvement of Langerhans cells in delayed hypersensitivity reactions and contact sensitization have led to a substantial literature on numbers of Langerhans cells in the epidermis in these conditions as well as the phenotypic changes that can be observed(4,5). In man significant changes in the morphology and density of Langerhans cells occur in irritant dermatitis as measured by immunohistochemistry on cryostat sections (6,7). Studies of enzymatically dislodged Langerhans cells in the mouse have clearly demonstrated their capacity to upregulate cytokine mRNA selectively and act as MHC-restricted antigen-presenting cells for contact sensitizers(8,9). Langerhans cells can endocytosis rhodamine-ovealbumin into intracellular granules (10), details on the process is however lacking.

A direct cell differentiation relationship between peripheral blood dendritic cells and Langerhans cells have not so far been proven but is however very suggestive based on phenotypic properties(11).With the view of establishing a reliable source of Langerhans cells equivalents for in vitro studies of contact and irritant dermatitis in man we are studing negatively selected dendritic cells from human buffy coats for inducible traits that are similar to those of Langerhans cells. In this paper we show that dendritic cells exposed to irritants shows an enhanced capacity for uptake of potential contact sensitizing substances indicating that the endocytotic process is facilitated by the irritant exposure.

MATERIALS AND METHODS

Isolation of mononuclear cell populations. Buffy coats from human blood were used to isolate mononuclear cells on Ficoll-Hypaque discontinuous gradients. Mononuclear cells were separated into E-rosette positive and negative fractions by rosetting with

Dendritic Cells in Fundamental and Clinical Immunology,
Edited by Kamperdijk *et al.*, Plenum Press, New York, 1993

neuraminidase treated sheep erythrocytes. The E-rosette negative fractions were incubated in RPMI medium including 5% fetal calf serum (FCS), glutamine, gentamicine and 2-mercaptoethanol in standard concentrations. Dendritic cells were negatively selected according to previously published methods (12) by incubating the E-rosette negative fractions for 40 h before removal of monocytes/macrophages by plastic adherence followed by panning on human IgG coated plastic dishes (Gammaglobulin-Kabi, Kabi pharmaceuticals, Stockholm). Low density cells were selected from the non-adherent Fc-receptor negative cells by metrizamide gradient (14.5% w/v) centrifugation at 600 g for 10 min. These populations contained 50-85 % large irregular-shaped cells with a phenotype of dendritic cells when analysed using a flow cytometer (EPICS Profile 4235637) with CYTO-STAT^R /COULTER CLONE^R single colour or two colour anti-CD2/CD19 (T11-RD1(IgG1)/B4-FITC(IgG1)), CD14 (Mo2(IgM)), CD56 (NKH-

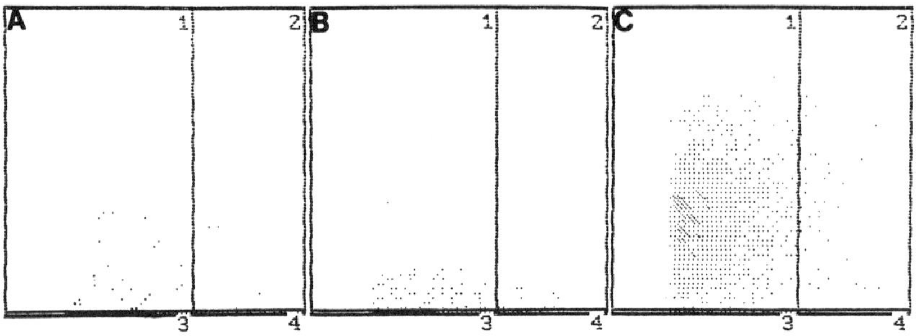

Figure 1. EPICS analysis of a dendritic cell enriched population. X-axis forward light scatter and Y-axis fluorescence intensity. The population was stained with (A) anti-CD 14 (2% of the cells CD 14 positive). (B) anti-CD 45R (5% of the cells CD 45 R positive) and (C) anti-HLA-DR (93% of the cells HLA-DR positive).

1(IgG1)) and CD3/34,29 Kd (T3-RD1(IgG1)/I3-FITC(IgG2a)) monoclonal antibodies (Coulter). In addition the cells were monitored at 37°C on an incubated stage inverted microscope (Nikon) equipped with a colour video camera, a time lapse video cassette recorder and monitor attached. DNP-albumin (DNP/albumin molecular ratio 35) and TRITC-albumin (adsorbance ratio 575/280 nm = 0.6)were produced according to published methods(13,14).
Confocal microscopy was done using a PHOIBOS confocal scanning laser microscope built around a Nikon Labphot microscope using software running on a Silicone graphics computer. Stacks of images were scanned and stored until futher processing in the computer and representative images recorded by color photography of the videomonitor picture.

RESULTS

The dendritic cell enriched populations were characterized by flow cytometry analysis and light microscopy at 37°C (figure 1). The populations typically contained 50-85% dendritic cells as defined by the criteria of typical veiled appearance at 37°C and a characteristic flow cytometry profile (large, macrophage, B-,T- and NK-cell marker negative and HLA-DR marker positive cells).

Viability of dendritic cells after exposure to 0.01-0.001% SLS remained high (>85%) for several hours, while other cells (mainly B- and NK-cells) within minutes became trypane blue positive.

Endocytosis of DNP-albumin (0.13 mg/ml) and TRITC-albumin (0.13 mg/ml) into dendritic cells with and without exposure to 0.001% SLS was then studied in confocal microscopy. Scaning stacks of images a clear intracellular location of DNP-albumin was noted before and after (more intense) SLS treatment (figure 2). TRITC-albumin was only seen intracellulary after SLS-treatment.

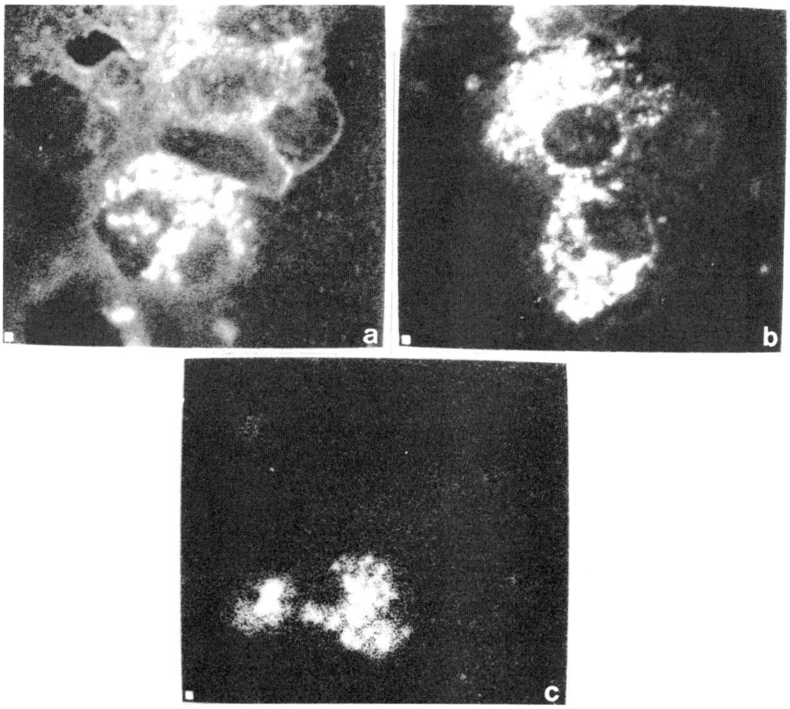

Figure 2. Confocal microscopy images of SLS (0.001%) exposed (A) and unexposed (B) dendritic cells after endocytosis of DNP-albumin (0.13 mg/ml) and SLS (0.001%) exposed (C) dendritic cells after endocytosis of TRITC-albumin (0.13 mg/ml). The images were obtained at the same level in the cells.

DISCUSSION

A key question for the concept of a network of dendritic cells in various tissues is to identify the dendritic cell population of various tissues as regards phenotypic similarities and differences. It may be assumed that the cell surface phenotype of dendritic cells from various tissues will vary according to the state of differentiation and activation of the cell in the tissue in question, partly as a response to physiological stimuli. Langerhans cells of the skin for example are Fc-receptor positive, F4/80 positive, membrane ATP-ase positive, non-specific esterase positive and have Birbeck granules, while dendritic cells from the peripheral blood are Fc-receptor negative, F4/80 negative, membrane ATP-ase negative, non-specific esterase negative and have no Birbeck granules (15). The precise equivalent of tissue dendritic cells in the circulation can not be determined, therefore, until traits have been identified that can induce peripherial blood dendritic cells to be phenotypically similar to tissue dendritic cells. Studies by video monitoring the cell suspension have revealed dendritic cells to be very motile cells with dendrites and veils (12). This behaviour could explain the extensive cytomembrane folding capacity found in skin by Langerhans cells. Falck et al suggest that Birbeck granules of the Langerhans cells are due to cytomembrane-sandwiching phenomenon which depends on the cell membranes coming in contact with each other (16). They also suggest that the dendrites of Langerhans cells catch antigen and thereby initiate membrane endocytosis (15). When human skin is exposed to irritants, cytomembrane-sandwiching is induced in Langerhans cells (17). Based on these and our findings it may be postulated that irritants enhance the endocytotic capacity of the Langerhans cells and that the same is true for dendritic cells. It may thus be possible to use negatively selected dendritic cells from peripherial blood in in vitro models for studing irritants and their importance in DTH reactions.

ACKNOWLEDGMENT

This study was supported by the Swedish Medical Research Council (09483).The authors wish to thank Dr Chris Andersson for valuable discussions on the biology of irritants, mrs Florence Sjögren for performing flow cytometry analysis and mrs K Holmgren-Peterson for help with the Phoibos instrument.

REFERENCES

1. R.M. Steinman, The dendritic cell system and its role in immunogenicity, *Ann Rev Immunol* 9:271-296 (1991).
2. U. Forsum, K. Claesson, E. Hjelm, A. Karlsson-Parra, L. Klareskog, A. Scheynius and U. Tjernlund, Class II transplantation antigens: Distribution in tissues and involvement in disease, *Scand J Immunol* 21:389-396 (1985).
3. P.D. King, and D.R. Katz, Mechanisms of dendritic cell function, *Immunol Today* 11(6):206-211 (1990).
4. A. Scheynius, T. Fischer, U. Forsum, and L. Klareskog, Phenotypic characterization in situ of inflammatory cells in allergic and irritant contact dermatitis in man, *Clin Exp Immunol* 55:81-90 (1984).
5. C. Avnstorp, E. Ralfkiaer, J. Jorgensen, and G. Lange-Wantzin, Sequential immunophenotypic study of lymphoid infiltrate in allergic and irritant reactions, *Contact Dermatitis* 16:239-245 (1987).
6. C.M. Willis, C.J.M. Stephens, and J.D. Wilkinson, Selective expression of immune-associated surface antigens by keratinocytes in irritant contact dermatitis, *J Invest Derm* 96:505-511 (1991).
7. C.M. Willis, C.J.M. Stephens, and J.D. Wilkinson, Differential effects of structurally unrelated chemical irritants on the density and morphology of epidermal CD1+ cells, *J Invest Derm* 95:711-716 (1990).
8. S. Aiba, and S.I. Katz, The ability of cultured Langerhans cells to process and present protein antigens is MHC-dependent, *J Immunol* 146:2479-2487 (1991).
9. H. A. Enk, and S.I. Katz, Early molecular events in the induction phase of contact sensitivity, *Proc Natl Acad Sci USA* 89:1398-1402 (1992).
10. E. Puré, K. Inaba, M.T. Crowley, L. Tardelli, M.D. Witmer-Pack, G. Ruberti, G. Fathman, and R.M. Steinman, Antigen processing by epidermal Langerhans cells correlates with the level of biosynthesis of major histocompatibility complex class II molecules and expression of invariant chain, *J Exp Med* 172:1459-1469 (1990).
11. N. Romani, M. Witmer-Pack, M. Crowley, S. Koide, G. Schuler, K. Inaba, and R.M. Steinman, Langerhans cells as immature dendritic cells, Chapter 7, *Epidermal Langerhans Cells* CRC Press (1991).
12. P.S. Freudenthal, and R.M. Steinman, The distinct surface of human blood dendritic cells, as observed after an improved isolation method, *Proc Natl Acad Sci USA* 87: 8698-7702 (1990).

13. T. Skogh, O. Stendahl, T. Sundqvist, and L. Edebo, Physicochemical properties and blood clearance of human serum albumin conjugated to different extents with dinitrophenyl groups, *Int Archs Allergy appl Immun* 70:238-244 (1983).
14. Antibodies, Cold Spring Harbor Laboratory press, Labeling Antibodies, Chapter 9, E. Harlow and D. Lane (ed), p 357 (1988).
15. H. Stössel, F. Koch, E. Kämpgen, P. Stöger, A. Lenz, C. Heufler, N. Romani, and G. Schuler, Disappearance of certain acidic organells (endosomes and Langerhans cell granules) accompanies loss of epidermal Langerhans cells, *J Exp Med* 172:1471-1482 (1990).
16. J. Bartosik, A. Andersson, S. Axelsson, B. Falck, and A. Ringberg, Direct evidence for the cytomembrane derivation of Birbeck granules: The membrane-sandwich effect, *Acta Derm Venereol* 65:157-160 (1985).
17. A. Mikulowska, A. Andersson, J. Bartosik, and B. Falck, Cytomembrane-sandwiching in human epidermal Langerhans cells: A novel reaction to an irritant, *Acta Derm Venereol* 68: 254-256 (1988).

DENDRITIC CELLS AND "DENDRITIC" MACROPHAGES IN THE UVEAL TRACT

John V. Forrester[1], Paul G. McMenamin[2] Janet Liversidge[1]and Lynne Lumsden[1]

[1]Dept. of Ophthalmology
University of Aberdeen
[2]Department of Anatomy and Human Biology
University of Western Australia

INTRODUCTION

The eye and the brain are considered sites of immune privilege(1) and it has been suggested that the lack of MHC Class II antigen presenting cells in neural tissue is responsible for this. However, neural tissue participates in many immune and autoimmune responses both clinically and experimentally(2) and clearly this must involve presentation of autoantigen to antigen specific T cells. In the CNS there is evidence that microglial cells may act as APC(3) during immune responses but, although similar cells occur in the retina, there is as yet no evidence that they may initiate immune responses to autoantigens. Aberrant expression of MHC Class II antigen by ocular cells such as retinal Muller cells, retinal pigment epithelial cells and ciliary body cells has been shown and under certain conditions, each of these cell types can be induced to present antigen albeit weakly(4-9). Conversely in other circumstances such cells may downregulate immune responses.

Despite initial negative reports, recent studies have shown that the normal iris and ciliary body contains cells staining positively for macrophages and the phenomenon of anterior chamber immune deviation (ACAID) has been attributed to them(10). Suppressive macrophages, however, have been described in sites not usually associated with immune privilege such as the lung(11). More recent studies have shown that there is a network of dendritic cells in the iris but their relationship to iris macrophages is unclear(12). There has to date, been no systematic study of dendritic cells and macrophages in the posterior segment of the eye (ie the retina and choroid) despite the fact that the major ocular autoantigens (such as retinal S antigen and interphotoreceptor retinol binding protein) are known to be located at this site.

MATERIALS AND METHODS

Tissue Harvest and Preparation

Adult Lewis rats (200-300g) were perfused with cold phosphate buffered saline(PBS) to remove circulating cells from the tissues and then perfused- fixed with 100% cold ethanol. The eyes, lymph nodes and spleen were harvested and frozen-embedded in OCT compound in liquid N_2. 5-8mm sections were cut and stained using a streptavidin-biotin-peroxidase labelling system.

Flat mount preparations of choroid tissue were also prepared by dissecting fresh PBS-perfused eyes in tissue culture medium(see below) and mounting the tissue face down on filter paper or electron microscopy grids.

Tissue culture

Explants of ciliary body/choroid (2x4 mm) were dissected from enucleated eyes of PBS-perfused Lewis rats and cultured in 96 well flat-bottomed trays for 48-96 hours in Minimum essential medium containing 10% FCS. Non-adherent cells were then harvested and prepared as cytospins, which were air dried overnight and then fixed in 100% ethanol prior to immunostaining.

Immunocytochemistry

Monoclonal antibodies used in the primary incubations included the following: 0X6 (MHC Class II), ED1,2,3 (macrophage markers)(13), ED7,8 (CD 11a,b), W3/25 (T cells and dendritic cells weakly), 0X8 (CD8), and OX 19/52 (pan T cell marker). Secondary sheep antimouse antibody (Amersham, Carshalton) was used at a concentration of 1/50.

Microscopy

Tissue and cell preparations were examined by conventional light and fluorescence microscopy using a Leitz orthoplan microscope, and by confocal scanning laser microscopy using a Biorad dual laser system. Lastly, videomicroscopy of live cells harvested from explant cultures was performed in a warm chamber attachment to an Olympus inverted microscope.

Functional studies

Dentritic cells were prepared from rat spleens using a modification of the method for isolating murine dendritic cells described by Knight et al.(13) Splenocytes as a single cell suspension were prepared from the spleens of naive Lewis rats. After removal of the adherent cells, the non-adherent cells (10^7/ml) were centrifuged on a metrizamide cushion and the low density cells (dendritic cells) from the interface were harvested, washed, and placed in culture overnight in growth medium at 5×10^6 cells per ml. The yield of dendritic cells isolated by this method was $1\text{-}5 \times 10^6$ DC's per rat spleen, the majority of non-lymphocyte cells were 0x6+ve and had dendritic morphology (APAAP staining of cytospins).

T cells were prepared from the pellet of high density splenocytes obtained after metrizamide separation of DC's. T cells were purified by passage over nylon wool followed by incubation over night before inclusion in the proliferation assays with the dendritic cells. Flow cytometry of "T" cells prepared by this method were typically >85% 0X8/W3-25 positive T cells. Monocytes and macrophages were absent, but the presence of a very bright population (1-2%) of OX 6 positive cells (peak channel 235) were present in some preparations suggesting that these "T" cells also contained residual dendritic cells, with up to 15% OX33B cells.

For the proliferation assays S-antigen was prepared as previously described (14) and interphotoreceptor retinol binding protein (IRPB) was prepared by a modification of Fongs' method (15) with further purification on a Mannose-agarose (Sigma) column to remove any residual Con-A. Dendritic cells, 2×10^4 - 2×10^1 per well were set up in quadruplicate 200μl microtitre well cultures with 2×10^5 "T" cells per well and IRBP (5 or 10μg/ml) or S-antigen (10 or 20μg/ml) for 120 hours and proliferation measured by (^{3}H) thymidine incorporation using the Matrix 96 system (Packard).

RESULTS

Dendritic cells in the uveal tract

Conventional frozen sections from perfused-alcohol fixed Lewis rat eyes showed typical dendritic cells(DC) in iris, ciliary body and choroid but no MHC Class II positive cells were observed in the retina (Fig 1a,b). Dendritic cells were observed in close apposition to the retinal pigment epithelium and also in the suprachoroidal layer and around vessels. DC were particularly frequent at the pars plana/peripheral retinal region and were also occasionally seen in the posterior chamber of the eye close to the ciliary processes.

Flat mount preparations of the iris and choroid revealed a dense network of connecting DC with a density distribution of about 350 cells per mm^2. DC were distributed throughout the thickness of each tissue often reaching 30mm in length. Confocal scanning laser microscopy demonstrated the extent of their processes which often assumed great lengths.

Macrophages in the uveal tract

In serial sections of tissues prepared for study of DC, cells staining positively with the ED series of monoclonal antibodies were observed, including ED1, ED2, ED3 and ED7and 8(Fig 1c,d). ED1 positive cells were small and rounded in appearance whereas ED2 positive cells were more elongated and resembled DC although they were smaller. Double staining for ED2 and MHC Class II indicated that ED2 positive macrophages and DC were two distinct populations of cells. ED positive cells had a similar distribution to DC ie near the peripheral retina and around blood vessels. Flat mount and confocal microscopy showed that ED2 positive cells were smaller and had finer processes than DC but had a similar density and tended to occur in clumps. ED3 positive cells(Fig.2a) were much less frequent than other cell types.

Few other cell types were observed. Occasional OX62 positive cells were observed but no T or B cells were observed using conventional antisera.

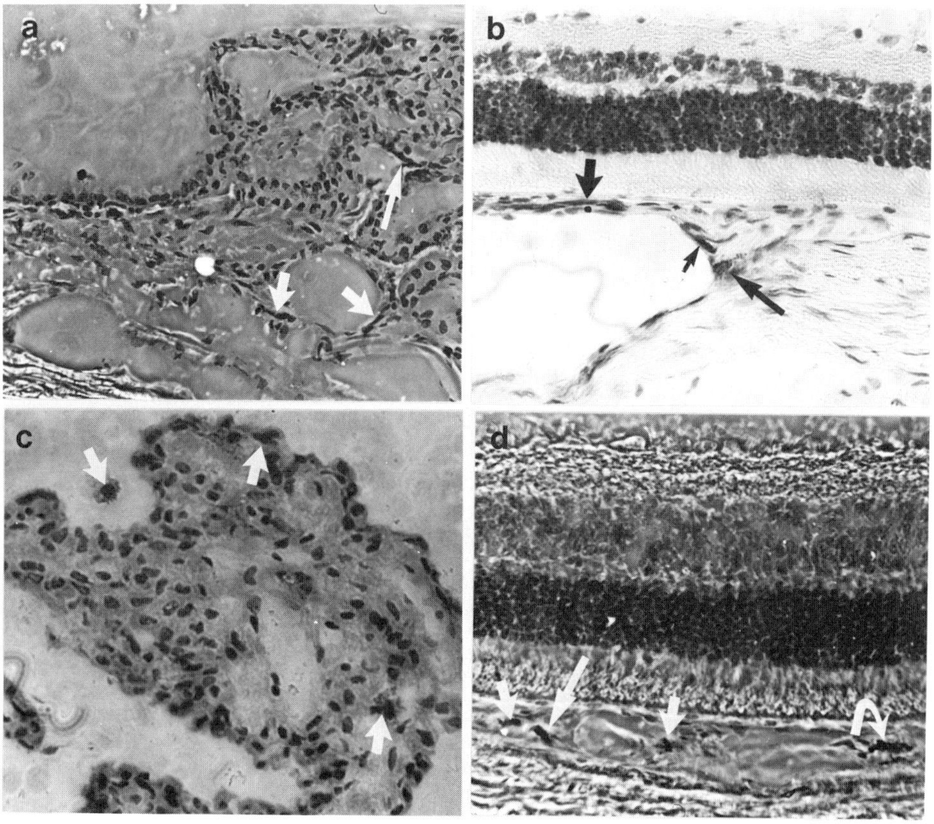

Figure 1. OX6 staining dentritic cells in (a)ciliary body, (b)choroid of rat eye. ED2 staining macrophages in (c)ciliary body (d)peripheral retina.

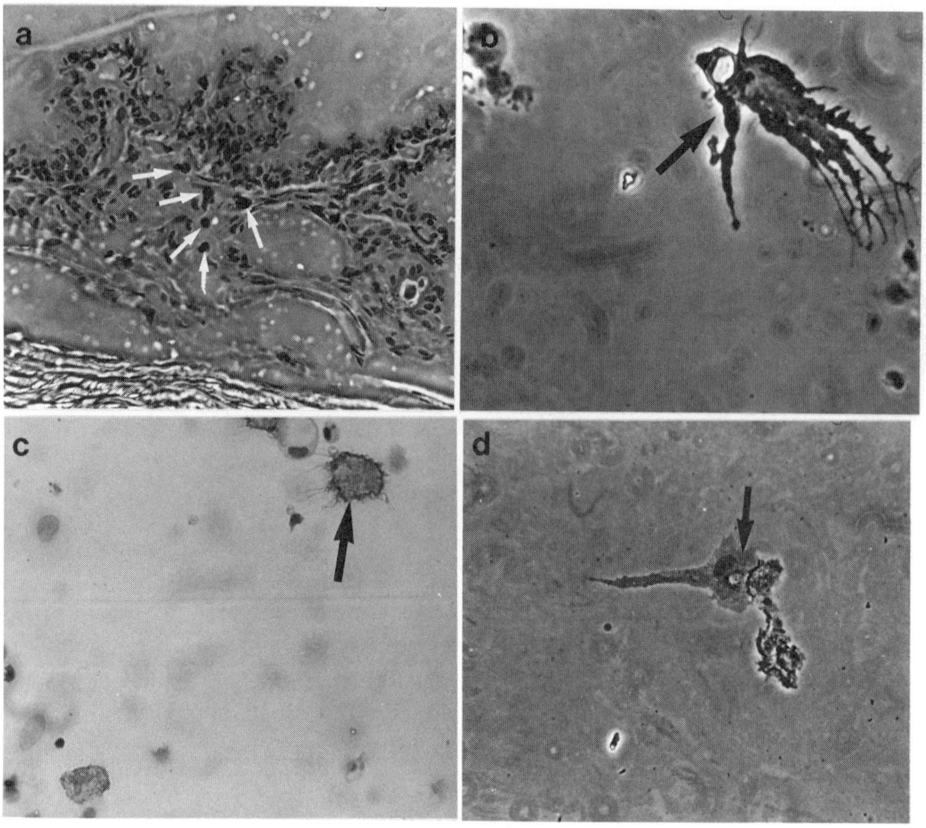

Figure 2. (a)E3 staing macrophage in choroid of rat-eyes. (b-d)Cytospin preparations of choroidexplant-cells showing OX6 staing dendritic dells (b,c) and an ED2 staining macrophage (arrow) in contact with an ED2 negative dendritic cells (d).

Choroid explant culture

Two to four days after explant culture of tissue blocks of choroid/ciliary body, cells that had migrated from the explants onto the culture dish were harvested and stained for MHC Class II and macrophage markers. Large MHC Class II positive cells were observed with very long cell processes, which were assumed to be DC(Fig 2b,c). In addition they appeared motile and formed clusters with smaller non-staining cells and with cells some of which appeared to be ciliary epithelium, while others were motile macrophage-like cells. Smaller ED2 positive staining cells were also observed, often in contact with ED negative DC, which were highly motile on videomicroscopy (Fig 2d). Round shaped ED1 positive cells were fewer in number and did not appear to be particularly motile. Few other cell types were observed.

Functional studies

Dentritic cells were found to be effective presenters of retinal antigens to naive T cells, inducing primary proliferative responses to both S-antigen and IRBP. The magnitude of the response was dependent upon the DC:T cell ratio and the amount of antigen present (Fig 3). A DC:T cell ratio of 1:20,000 was sufficient to elicit a significant, primary proliferative response to S-antigen ($p<0.005$ for 20µg/ml S- antigen) by naive T cells (Fig 3a). The ratio required for primary proliferation to 10µg/ml IRBP ($p<0.02$) was 1:2000 (Fig 3b).

Efficient presentation of both antigens was lost at APC responder ratios of 1:20. Control cultures did not proliferate in the absence of antigen and T cells or dendritic cells alone did not proliferate in the presence of retinal antigens.

DISCUSSION

This study has shown that there are at least three populations of "immune" cells in the normal uveal tract: dendritic cells, ED1 blood borne positive macrophages and ED2 positive resident macrophages. In addition, there is a small population of ED3 positive macrophages. No cells expressing these antigens were observed in the retina of the Lewis rat.

Functional studies indicate that dendritic cells have the potential to present ocular antigens to naive T cells (Fig 3) and it is suggested that uveal tract APC may be important in induction of autoimmune responses to retinal antigens because of the intimate association of these cells with the retinal pigment epithelium, a cell whose major function is to maintain the integrity of the photoreceptor cell by phagocytosing effete tips of the photoreceptor outer segments. The ability.of dendritic cells to induce primary immune responses is well documented (16-18), however their ability to process and present antigen diminishes with time in culture and after 24 hours processing function is reduced and eventually lost (19) In the functional studies of spleen-derived dendritic cells described here we show that even after overnight culture before inclusion in the assays splenic-derived dendritic cells can still effectively process and present retinal antigens, as few as 10 dendritic cells per microtitre well inducing proliferation of T cells to S-antigen.

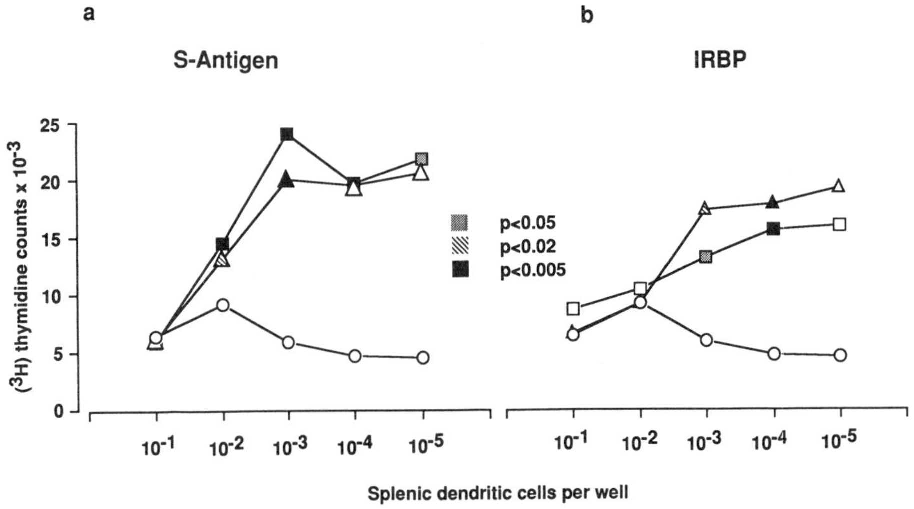

Figure 3. Dendritic cells, 10^{-1} - 10^{-4} per well were set up in quadruplicate 200ml cultures in round bottomed microtitre plates with 2 x 10^5 T cells and retinal S-antigen or IRBP. After 120 hours proliferation was measured by (^{3}H) thymidine incorporation, 0.5mci/well. Results are expressed as mean total counts x10^{-3} (6 min) $\pm$ 1SD. Significance values were calculated using the students T test.

The precise role of uveal tract DC in ocular immune responses remains to be elucidated. Previous studies have suggested that macrophages in the iris may be responsible for ACAID(see introduction), thereby explaining the immunologically privileged status of the anterior chamber. No similar studies have yet been performed in the posterior segment of the eye and there is no information available on the interactions between uveal macrophages and DC. It is likely, however that these cells have a significant role in maintaining immunological unresponsiveness in the eye.

Acknowledgments

This work was supported by the Leverhulme Trust, The Wellome Trust andthe Guide Dogs for the Blind

REFERENCES

1. J.W. Streilein, Immune regulation and the eye: a dangerous compromise. *FASEB Journal* . 75:199-208 (1987)
2. J.V. Forrester, J. Liversidge, H.S. Dua, H. Towler and P.G. McMenamin, Comparison of clinical and experimental uveitis. *Cur Eye Res.* 9(Suppl):75-84 (1990)
3. J.D. Sedgwick and R. Dorries, The immune system response to viral infect of the CNS. *Seminars in the Neuroscients*, 3:93-100 (1991)
4. G. Botttazzo, I. Todd, R. Mirakian, A.Belfiore and R. Pujol-Borrell, Organ-specific autoimmunity: a 1986 overview. *Immunological Rev.* 94:137-169 (1986)
5. C-C Chan et al, Expression of Ia antigen on retinal pigment epithelium in experimental autoimmune uveoretinits. *Cur. Eye. Res.* 5:325 (1986)
6. J.M. Liversidge, H.F. Sewell and J.V. Forrester, Interactions beween lmphocytes and cells of the blood-retina barrier: mechanisms of T lumphocyte adhesion to human retinal capillary endothelial cells and retinal pigment epithelial cells in vitro. *Immunology* 71:390-396 (1990)
7. A A Gaspari, M.K. Jenkins and S. Katz, Class II MHC-bearing keratinocytes induce antigen-specific unresponsiveness in hapten-specific Th1 clongs. *J. Immunol.* 141:2216-2220 (1988)
8. H. Helbig, R.C. Gurley, A G Palestine, R B Nussenblatt and R R. Caspi, Dual effect of ciliary body cells on T lymphocyte proliferation. *Ero. J. Immunol.* 20:2457-2463 (1990)
9. R.R. Caspi, R.G. Roberge and R.B. Nussenblatt, Organ resident non-lymphoid cells suppress proleferation of autoimmune T-helper lymphocytes. *Science* 237:1029-1032 (1987)
10. J.S.P. Williamson, D. Bradley and J.W. Streilein, Immunoregulatory properties of bone marrow-derived cells in the ris and ciliary body. *Immunology.* 67:96-102 (1989).
11. P.G. Holt, M.A. Schon-Hegrad and J. Liversidge, MHC Class II antigen-bearing dendritic cells in the pulmonary tissues of the rat. Regulation of antigen presentation activity by endogenous macrophage populations. *J. Exp. Med.* 167:262-274 (1988)
12. T.L. Knisley, T.M. Anderson, M.E. Sherwood, T.J. Flotte, D.M. Albert and R.D. Granstein, Morphological and ultrastructural examination of $I\text{-}A^+$ cells in the murine iris. *Invest. Ophth. Vis.Sci.* 32:2423-2431 (1991)
13. S.C. Knight, J. Mertin, A. Stackpool and J. Clark. Induction of immune responses *in vivo* with small numbers of veiled (dendritic) cells. *Proc. Natl. Acad. Sci.* 80:6032-6039 (1983)
14. J. Liversidge, A.W. Thomson, H.F. Sewell and J.V. Forrester, EAU in the guinea pig: inhibition of cell-mediated immunity and Ia antigen expression by Cyclosporin A. *Clin. Exp. Immunol.* 69: 591-600 (1987)
15. S-L Fong, G.I. Liou, R.A. Landel, R.A. Alvarez and C.D. Bridges, Purification and characterisation of a retinol binding glycoprotein synthesised and secreted by bovine, neural retina. *J. Biol. Chem.* 259:6534-6541 (1984)
16. S.E. Macatonia, P.M. Taylor, S.C. Knigh and B.A. Askonas, Primary stimulation by dendritic cells induces antiviral proliferative responses and cytotoxic T cell responses in vitro. *J. Exp. Med.* 169:1255-1264 (1989)
17. T. Sornasse, V. Flamand, G. De Becker, H. Bazin, K. Tielemans, J. Urbain, O. Leo and M. Moser, Angtigen-pulsed Dendritic cells can effeciently induce an antibody response in vivo. *J. Exp. Med.* 175:15-21 (1992)
18. N. Bhardwaj, S.M. Freidman, B.C. Cole and A.J. Nisanian, Dendritic Cells are potent antigen-presenting cells for microbial superantigens. *J. Exp. Med.* 175:267-273 (1992)
19. R.M. Steinman, The dentritic cell system and its rôle in immunogenicity. *Ann. Rev. Immunol.* 9:271-296 (1991)

MACROPHAGES AND DENDRITIC CELLS IN RAT COLON

IN EXPERIMENTAL INFLAMMATORY BOWEL DISEASE

Emmelien P. van Rees, Marsetyawan Soesatyo, Marja van der Ende, and Taede Sminia

Dept. Cell Biology, Div. Histology, Faculty of Medicine, Vrije Universiteit, Van der Boechorststraat 7, 1081 BT Amsterdam, The Netherlands

INTRODUCTION

The etiology of chronic inflammatory bowel diseases (IBD), like Crohn's disease (CD) and ulcerative colitis (UC), remains unclear. A recurrent theme is a failure to down-regulate a local hyperreactive immune response. Autoimmunity plays a role as circulating antibodies reactive against a protein in colonic epithelial cells are found in patients with ulcerative colitis (Takahasi et al., 1990). Local non-lymphoid cells like macrophages, dendritic cells and epithelial cells could be important in IBD in the pathogenesis and in the uptake and processing of antigen in affected tissue, leading to secondary bacterial infection.

Several studies indicate the involvement of macrophages and dendritic cells in IBD. Differences are found in numbers and heterogeneity of macrophages and dendritic cells between tissue obtained from patients and controls (Wilders et al., 1984; Allison et al., 1988; Seldenrijk et al., 1989). Furthermore, lamina propria macrophages obtained from mice have a suppressive effect on immune responses, whereas dendritic cells isolated from Peyer's Patches are very stimulating (Pavli et al., 1990).

Morris et al. have developed an animal model for chronic IBD, induced by a single rectal administration of 2,4,6-trinitrobenzene sulfonic acid (TNBS) in ethanol (Morris et al., 1989). The rats develop both clinical and histopathological symtoms of IBD, lasting for at least 8 weeks. In this model, segmental ulceration and granulomatous inflammation are common.

In the present study intestinal macrophages and dendritic cells were being studied in this animal model for IBD, using moab's ED1, ED2 and ED3 (Dijkstra et al., 1985) and Ox6.

MATERIALS AND METHODS

Induction of IBD

Colonic inflammation was induced in Wistar rats according to Morris et al. (1989) with small modifications. A flexible catheter was inserted 6-8 cm in the colon of anaesthetized rats, the tip of the catheter approximately at the splenic flexure. Through the catheter rats received a single dose of 30 mg TNBS dissolved in 40% ethanol. The total administrated volume was 0,25 ml. Control rats were treated with ethanol alone.

Experimental Design

At various times after induction of IBD (day 1, 14, 21 and 28) rats were killed by CO_2 intoxication. The colon was removed, opened and examined. Visible damage was scored on a 0-5 scale, according to Morris et al. (1989). The colon was removed and frozen in liquid nitrogen. A two-step immunoperoxidase method was used on cryostat sections (Dijkstra et al, 1983), moab's ED1, ED2, and ED3 were being applied. Moab Ox6 (Serotec) was used for recognizing class II antigens. Quantification was carried out on at least 10 microscopic fields of three sections for each specimen.

RESULTS

General observations

Almost all animals that were treated with TNBS in ethanol developed areas of bowel wall thickening, inflammation and ulcers. The ulceration included the muscularis mucosae. The mucosa adjacent to the ulcers showed extensive crypt distortion and granulomas were seen in some animals.

The rats treated with ethanol alone also showed symptoms of colitis, but they were recovered from the acute inflammation induced by ethanol in 2 weeks; whereas the TNBS-treated animals still suffered from chronic colitis a month after induction of IBD (Table 1).

Macrophage subpopulations

The effect of TNBS-induced colitis on local subpopulations macrophages, as recognized by mab's ED1, ED2 and ED3 was evaluated on day 1, day 14 and day 28 after the administration of TNBS. In the normal colon, ED1 marked a considerable number of macrophages, spread over the lamina propria. The number of macrophages recognized by ED1 had increased after induction of IBD (see Table 2), but no differences were found between the acute and the chronic phase of the inflammation.

In untreated rats some ED2+ macrophages are present, in the basal part of the crypts. In diseased animals the number of cells recognized by ED2 had sharply increased. Furthermore, the distribution pattern had changed as in IBD rats most of the ED2+ cells were found in the upper parts of the crypts (Fig. 1).

In untreated rats, ED3 only recognizes macrophages in lymphoid organs. After induction of IBD however, many ED3+ cells were demonstrable in the lamina propria. After the initial acute inflammation this number decreased, but still a considerable number of ED3+ cells remained present.

Ia-positive non-lymphoid cells

After induction of IBD, the number of Ia-positive dendritic cells in the lamina propria increased (Table 2), evenly spread over the lamina propria. The intensity of class II expression on each cell did not change.

Surprisingly, the epithelium did not express class II antigens (Fig. 2). Neither acute nor chronic inflammation induced Ia molecules on rat colonic epithelial cells in vivo.

DISCUSSION

In colonic tissue obtained from untreated rats, lamina propria

Table 1. *Damage score of the colon on several points of time after induction of experimental IBD.*

Treatment	Days after	Scores
Ethanol	1	5, 5, 5
Ethanol	8	5, 1
Ethanol	14	5, 1
Ethanol	21	0, 0, 0
Ethanol	28	0, 0, 0
TNBS + Eth.	1	5, 5, 5
TNBS + Eth.	8	5, 5, 1
TNBS + Eth.	14	5, 5
TNBS + Eth.	21	5, 5
TNBS + Eth.	30	5, 5, 5

Criteria for scoring according to Morris et al (1989), with modifications. 0, No damage; 1, localized hyperemia ≤ 1 cm but no ulcers; 2, one ulcer; 3, one ulcer and area of inflammation; 4, two or more sites of ulceration/inflammation; 5, two or more sites of ulceration/inflammation or one site > 1 cm length.

Table 2. *Non-lymphoid cells on several points of time after induction of experimental IBD, as recognized by moab's ED1,2,3 and Ox6.*

	Control	Day 1	Day 14	Day 28
ED1	+	++	++	++
ED2				
basal part of crypts	±	+	+	+
upper part of crypts	-	++	++	++
ED3	-	++	+	+
Ox6				
dendritic cells	++	+++	+++	+++
epithelial cells	-	-	-	-

-, not seen; ±, 1-5 cells per microscopic field (x20); +, 5-10 cells per field; ++, 10-20 cells per field; +++, more than 20 cells per field.

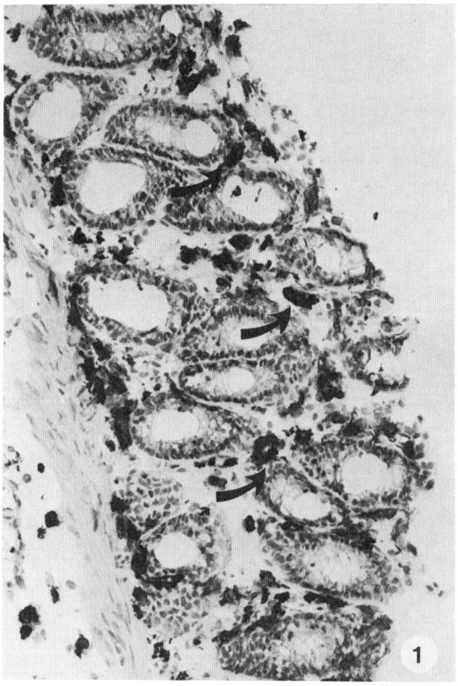

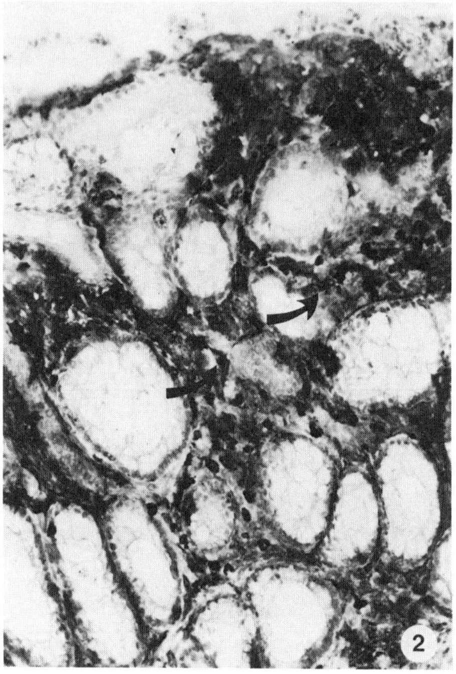

Fig. 1. *ED2+ macrophages (arrows) in the colon on day 1 after induction of experimental IBD.*

Fig. 2. *Ia+ dendritic cells (arrows) on day 1 after induction of experimental IBD. The epithelial cells are not being recognized by moab Ox6.*

macrophages are ED1+/ED2-/ED3-. Macrophages isolated from lamina propria have a suppressive function in vitro (Pavli et al., 1990). After induction of IBD, the number of macrophages stained by ED1 increased, but the number of cells marked by ED2 showed a larger increase. The net result is that there is a relative loss of macrophages that are only recognized by ED1. It is possible that the lamina propria macrophages have lost their suppressive function after expressing the antigen recognized by ED2, and have become stimulatory instead. This would implicate a contrasuppressive regulatory role for the ED2 antigen. ED2 is a marker for a membrane antigen present predominantly on fixed tissue macrophages (Dijkstra et al., 1985), consisting of three protein chains of 175, 160, and 95 kDA (Barbé et al., 1990). In ontogenetic development ED2 expression is always observed for the first time 1 to 2 weeks after the first macrophages (ED1+/ED2-) have populated the tissue (Van Rees et al., 1988ab; 1991). As the appearance of ED2 always precedes immunocompetence ED2 could have a contrasuppressive role in ontogeny as well.

ED3+ macrophages are usually only found in lymphoid organs. Outside the lymphoid organs, ED3 expression is only seen in several experimental autoimmune diseases, especially in the late and quiet effector stages (Verschure et al., 1989; Noble et al., 1990). This could indicate a suppressive role played by the ED3-determinant. In the present study however, many ED3+ cells are found in the initial, acute phase of the inflammation. Recently it has been shown that ED3 is a marker for sialoadhesin, which functions as a macrophage restricted lymphocyte adhesion molecule (Van den Berg et al., 1992). The signification of the appearance of the ED3 determinant in acute and chronic inflammation remains to be established.

Expression of HLA-DR by colonic epithelial cells in healthy humans has been described, and was enhanced in IBD-patients (Mayer et al., 1991, Malizia et al., 1991). In rats however, class II expression is only found in the small intestine and not in the colon. In the present study even under inflammatory conditions no Ia+ epithelial cells were found. However, there is no consensus on an antigen presenting function for enterocytes (Russell and Harmatz, 1991).

After induction of IBD, the number of Ia+ dendritic cells in the lamina propria had increased. This probably contributes, together with a relative decrease of suppressive lamina propria macrophages, to a state of local chronic inflammation. The balance between stimulatory and suppressive signals from dendritic cells and macrophages in IBD requires further attention. By manipulating subpopulations non-lymphoid cells in the animal model which is described in the present study, the role of non-lymphoid cells in IBD can be studied in vivo.

REFERENCES

Allison, M.C., Cornwall, S., Poulter, L.W., Dhillon, A.P., and Pounder, R.E., 1988, Macrophage heterogeneity in normal colonic mucosa and in inflammatory bowel disease. *Gut* 29:1531.

Barbé, E, Damoiseaux, J,G.M.C., Döpp, E.A., and Dijkstra, C.D., 1990, Characterization and Expression of the antigen present on resident rat macrophages recognized by monoclonal antibody ED2. *Immunobiol.* 182:88.

Dijkstra, C.D., and Döpp, E.A., 1983, Ontogenetic development of T and B lymphocytes and non-lymphoid cells in the white pulp of the rat spleen. *Cell Tissue Res.* 229:351.

Dijkstra, C.D., Döpp, E.A., Joling, P., and Kraal, G., 1985, The heterogeneity of mononuclear phagocytes in lymphoid organs: distinct macrophages ubpopulations in the rat recognized by monoclonal antibodies ED1, ED2 and ED3. *Immunology* 54:589.

Malizia, G., Calabrese, A., Cottone, M., Raimondo, M., Trejdosiewicz, L.K., Smart, C.J., Oliva, L., and Pagliaro, L., 1991, Expression of

leukocyte adhesion molecules by mucosal mononuclear phagocytes in inflammatory bowel disease. *Gastroenterology* 100:150.
Morris, G.P., Beck, P.L., Herridge, M.S., Depew, W.T., Szewczuk, M.R., and Wallage, J.L., 1989, Hapten-induced model of chronic inflammation and ulceration in the rat colon. *Gastroenterology* 96:795.
Noble, B., Ren, K., Taverne, J., Diporro, J., Van Liew, J., Dijkstra, C.D., Janossy, G., and Poulter, L.W., 1990, Mononuclear cells in glomeruli and cytokines in urine reflect the severity of experimental proliferative imune complex glomerulonephtitis. *Clin. exp. Immunol.* 80:281.
Pavli, P., Woodhams, C., Doe, W.D.F., and Hume, D.A., 1990, Isolation and characterization of antigen-presenting dendritic cells from the mouse intestinal lamina propria. *Immunology* 70:40.
Russell, G.J., and Harmatz, P.R., 1991, Major histocompatibility complex class II expression on enterocytes: to present or not to present. *Gastroenterology* 100:274.
Seldenrijk, C.A., Drexhage, H.A., Meuwissen, S.G.M., Pals, S.T., and Meijer, C.J.L.M., 1989, Dendritic cells and scavenger macrophages in chronic inflammatory bowel disease. *Gut* 30:484.
Takahasi, F., Shah, H.S., Wise, L.S., and Das, K.M., 1990, Circulating antibodies against human colonic extract enriched with a 40 kDa protein in patients with ulcerative colitis. *Gut* 31:1016.
Van den Berg, T.K., Breve, J.J.P., Damoiseaux, J.G.M.C., Döpp, E.A., Kelm, S., Crocker, P.R., Dijkstra, C.D., and Kraal, G., 1992, Sialoadhesin on macrophages: its identification as a lymphocyte adhesion molecule. *J. Exp. Med.* (in press)
Van Rees E.P., Dijkstra, C.D., Van der Ende, M.B., Janse, E.M., and Sminia, T., 1988a, The ontogenetic development of macrophage subpopulations and Ia-positive dendritic cells in gut-associated lymphoid tissue of the rat. *Immunology* 63:79.
Van Rees E.P., Voorbij, H.A.M., and Dijkstra, C.D., 1988b, Neonatal development of lymphoid organs and specific immune responses in situ in diabetes-prone BB rats. *Immunology* 65:465.
Van Rees, E.P., Van der Ende, M.B., and Sminia, T., 1991, Ontogeny of macrophage subpopulations and Ia-positive dendritic cells in pulmonary tissue of the rat. *Cell Tiss. Res.* 263:367.
Verschure, P.J., Van Noorden, C.J.F., and Dijkstra, C.D., 1989, Macrophages and dendritic cells during rthe early stages of antigen-induced arthritis in rats: Immunohistochemical analysis of cryostat sections of the whole knee joint. *Scand. J. Immunol.* 29:371.
Wilders, M.M., Drexhage, H.A., Kokje, M., Verspaget, H.W., and Meuwissen, S.G.M., 1984, Veiled cells in chronic idiopathic inflammatory bowel disease. *Clin. Exp. Immunol.* 55:461.

VACCINATION WITH TUMOR ANTIGEN-PULSED DENDRITIC CELLS INDUCES IN VIVO RESISTANCE TO A B CELL LYMPHOMA

V. Flamand, T. Sornasse, K. Thielemans*, C. Demanet*, O. Leo, J. Urbain and M. Moser

Laboratoire de Physiologie Animale, Université Libre de Bruxelles, Rue des Chevaux, 67; B-1640 Rhode-St-Genèse and *Division of Hematology-Immunology, Medical School, Vrije Universiteit Brussel, B-1090; Belgium.

INTRODUCTION

The role of the immune system in tumor surveillance is still a matter of speculation. Immune responses to murine experimental tumors have led to the unequivocal identification of tumor-associated antigens (TAA) which can become the target of rejection mechanisms (1). However, studies of 27 different tumors, all of strictly spontaneous origin, have revealed no evidence of tumor immunogenicity (2). It is therefore still unclear whether the lack of an efficient immune response able to eradicate the tumor is due to a lack of expression of TAA or to a suppressor mechanism that inhibits an ongoing immune response.

In the case of B cell tumors that express a membrane and secreted form of immunoglobulin (Ig), the idiotypic determinants associated with the surface Ig can serve as a target for host immunity and the secreted Ig can be isolated for use as the soluble form of TAA. The BCL1 lymphoma, which arose spontaneously in a female Balb/C mouse, has been extensively studied and shown to be analogous to human prolymphocytic leukemia (3). BCL1 is a very aggressive lymphoma which progresses too fast to allow generation of antiidiotype responses during tumor growth. Recent advances in immunology and, in particular, the discovery of a "costimulatory signal" required for optimal stimulation of T helper (Th) cells, suggests that tumor cells may not efficiently present TAA to Th cells. Therefore, the failure to eradicate the tumor could be due to inadequate presentation of tumor associated antigen.

Based on this assumption, we tried to induce an anti-tumor response by injecting BCL1 idiotype on specialized antigen-presenting-cells, i.e. dendritic cells (DC). It has indeed been shown that resting T lymphocytes require DC to initiate growth and differentiation. Previous data, obtained in our laboratory, demonstrated the potency of mouse DC pulsed with protein antigens as inducing cells in T-dependent antibody responses *in vivo* (4, 5). Inaba et al. (6) showed that dendritic cells can prime antigen-specific, MHC restricted T cells *in vivo.*

RESULTS AND DISCUSSION

Antigen-pulsed DC induce tumor resistance in vivo. Dendritic cells were enriched from the spleens of Balb/c mice by multi-step procedures over a period of 1 day in culture. Although purified DC have lost the capacity to process, they can be efficiently pulsed with protein antigen during the overnight culture (5). To assess the role of DC in inducing tumor resistance *in vivo*, Balb/c mice were injected with 3.6 10^5 idiotype-pulsed syngeneic DC (DC-BCL1) and inoculated with 1000 tumor cells either the same day or 5 days later. All untreated mice, that only received tumor cells, developped a tumor, as assessed by the level of idiotype in sera, whereas only 1 mouse treated with DC-BCL1 had high idiotype in serum (Figure 1a). The survival time of 4 groups correlated with the level of idiotype, i.e. all control mice and 1 DC-BCL1 injected mouse died by day 70 (Figure 1b).

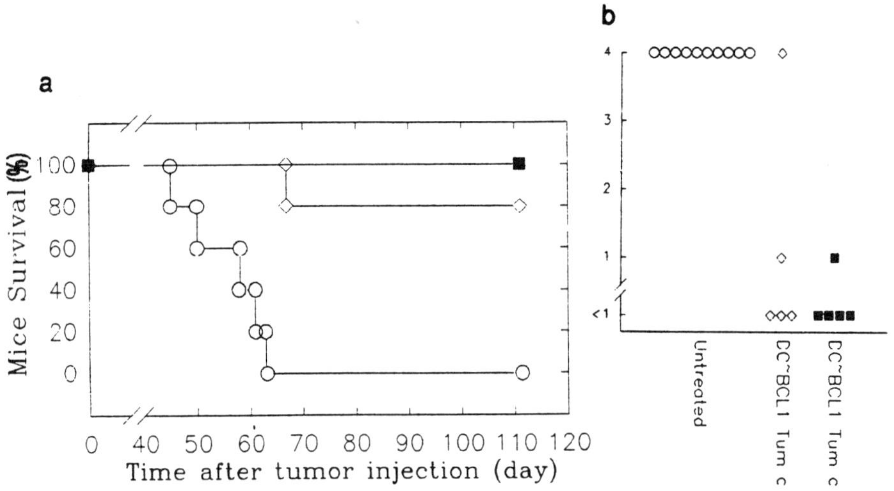

Figure 1: Effect of injection of BCL1-pulsed dendritic cells on survival of tumor-inoculated animals. Fig.1a: 2 groups of 5 mice were pretreated with 3.6 10^5 DC that were pulsed with 100μg of BCL1 during overnight culture. The animals were inoculated with tumor cells either the same day (diamonds) or 5 days later (squares). Control mice only received tumor cells (circles). All mice were checked daily for survival. Fig.1b: Mice were bled 28 days after tumor inoculation and the concentration of idiotype was measured in the sera by a standard ELISA assay using plates coated with monoclonal antiidiotype antibody. Data are expressed as serum dilution giving 50% of binding. Mice with undetectable levels of idiotype (< 100ng/ml) are scored as < 1.

Data collected from 5 experiments are summarized in Table 1 and confirm that a single injection of antigen-pulsed DC resulted in long-lasting tumor resistance in 83% of mice. By contrast, 2 injections of BCL1-pulsed B cells or 3 intraperitoneal immunizations of 50μg of idiotype emulsified in complete Freund adjuvant (CFA) had little effect on tumor incidence *in vivo.* Mice treated with BCL1 in saline, or with DC pulsed with irrelevant IgM, λ (MOPC 104) have roughly the same tumor incidence as untreated mice. Since only DC induce an immune response that prevents tumor cell growth, we reasoned that subcutaneous administration of soluble idiotype emulsified in CFA may be more efficient than intraperitoneal injection, since epidermal Langerhans cells of the skin have been shown to capture the antigen, migrate and mature into cells able to activate resting T cells (7). The results in Table 1 show indeed that 87% of mice that received 3 subcutaneous injections of BCL1 in CFA are protected against subsequent tumor inoculation. Experiments are in progress to determine whether soluble antigen administered subcutaneously in the absence of adjuvant results in the same protective effect. These data suggest that different modes of antigen administration may result in distinct responses, depending on the antigen-presenting cells that process and present the antigens.

Tab. 1. Effect of various modes of immunization on tumor incidence (day 160) in vivo.

Treatment group	Mice number with tumor/total (%)
untreated	18/20 (90%)
BCL1 idiotype in saline	14/17 (82%)
IgM, 1 (MOPC104)-pulsed DC	15/17 (88%)
BCL1-pulsed B cells	6/9 (66%)
BCL1-pulsed DC	5/28(17%)
BCL1 in CFA i.p. 3X	5/8 (62%)
BCL1 in CFA s.c. 3X	1/8 (13%)

The induced tumor resistance correlates with an antiidiotype response. Although the mechanism of resistance against BCL1 tumor cells is still poorly understood, most experiments suggest that an antiidiotype serological response is the main effective mechanism (8). Therefore, we analyzed the sera of mice injected with antigen-pulsed DC for the presence of idiotype-specific antibodies. The data of Figure 2 show the presence of a specific humoral response in mice pretreated with DC-BCL1, but not in control mice. The antibodies detected are idiotype specific, since they bind BCL1 but not an irrelevant IgM, 1 (MOPC104). Moreover, a significant amount of IgG2a antibodies was detected in the same mice. In all experiments performed, we observed a strict correlation between mice survival and secretion of antiidiotype antibodies of IgG2a isotype (data not shown).

Our data are in accordance with previous results emphasizing the role of antiidiotype antibodies in tumor protection. In particular, study of switch variants revealed that IgG2a antibodies were at least 10-fold more effective than other isotypes in tumor protection (9).

The presence of IgG2a antibodies in animals pretreated with antigen-pulsed DC most probably results from preferential activation of TH1 cells following DC injection. Although a direct role for T cells has not been ascribed in this system, splenic T cells that proliferated specifically in response to idiotypic IgM have been described in mice

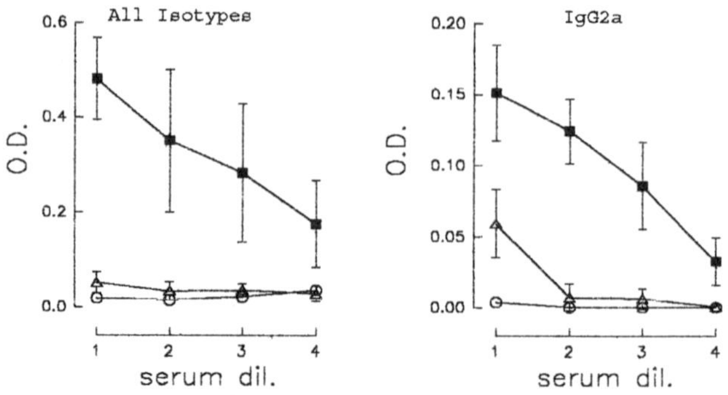

Figure 2. Injection of BCL1-pulsed DC induces an antiidiotype response in vivo: a group of 5 Balb/c mice were primed with BCL1-pulsed DC and boosted with 50μg of soluble idiotype 5 days later (squares). Control groups include untreated mice (circles) or mice injected with idiotype in saline (triangles). All mice were bled 14 days after antigen boost and the sera were tested individually. Antiidiotype antibodies were analyzed by a standard ELISA assay using plates coated with idiotype (BCL1, IgM, λ). Total specific response was revealed by rat anti-κ monoclonal antibodies and IgG2a antibodies were detected in the presence of rat antibodies specific for mouse IgG2a.

immunized with idiotype in CFA (10). Only low levels of cytotoxic T cells were found, suggesting that helper rather than cytotoxic T cells induce tumor resistance, either directly by activating a specific humoral response or directly by secretion of interferon-g. IFN-g has been indeed shown to cause suppression of IL-5-induced *in vitro* proliferation of the $CD5^+$ BCL1 lymphoma.

An hypothesis for tumor-induced suppression. One of the problem facing the active immunotherapy of cancer is the immunosuppressive nature of most malignancies. Recent experiments have shown that cultured BCL1 produces significant amounts of IL-10 and that expression of IL-10 is constitutive in splenic B cells from mice carrying the CD5 BCL1 lymphoma (11). Increased production of IL-10 probably occurs as a result of tumor growth and subsequently contributes to the survival of the tumor by interfering with immune regulation. Since IL-10 has been shown to inhibit TH1 functions (12), it is possible that the immune response becomes TH2-biased leading to secretion of TH2-type interleukins (IL-4 and IL-5) that are growth factors for BCL1 tumor cells *in vitro* (13), as represented in Figure 3.

Recent results have pointed out the importance of the balance IFN-γ/IL-10 in the regulation of CD5 B cells: administration of anti-IL-10 antibodies in mice results in interferon-γ elevation in serum that ultimately leads to selective depletion of CD5 B cells (14). Since DC seem to be potent activator of TH1 cells, they may prevent the TH1/TH2 imbalance, leading to an efficient immune response. The new finding that IL-10 does not inhibit antigen presentation by DC (15, 16) may be relevant in our system and may explain why antigen-pulsed DC induce long-lasting protection.

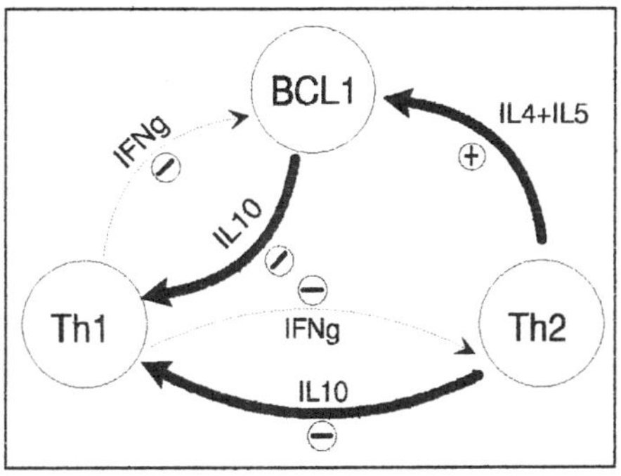

Figure 3. *hypothesis for tumor-induced suppression.*

Our results also imply that the effector cells are present in tumor bearing-animals and suggest that the major limitation in the induction of effective tumor responses is the activation of T helper cells.

Finally, since the ultimate goal in immunotherapy is to eradicate established tumors, experiments are in progress to use our immunization procedure for tumor-bearers.

REFERENCES

1. R.J. North, The murine antitumor immune response and its therapeutic manipulation, *Adv. Immunol.* 35:89 (1984).
2. H.B. Hewitt, E.R. Blake and A.S. Walder, A critique for the evidence for active host defense against cancer, based on personal studies of 27 murine tumours of spontaneous origin. *Brit.. J. Cancer*, 33:241 (1976).
3. K.A. Krolick, P.C. Isakson, J.W. Uhr and E.S. Vitetta, BCL1, a murine model for chronic lymphocytic leukemia: use of surface immunoglobulin idiotype for the detection and treatment of tumor, *Immunol. Rev.* 48:81 (1979).
4. M. Francotte and J. Urbain, Enhancement of antibody response by mouse dendritic cells pulsed with tobacco mosaic virus or with antiidiotypic antibodies raised against a private rabbit idiotype, *P.N.A.S. USA* 82:8149 (1985).
5. T. Sornasse, V. Flamand, G. De Becker, H. Bazin, F. Tielemans, K. Thielemans, J. Urbain, O. Leo and M. Moser, Antigen-pulsed dendritic cells can efficiently induce an antibody response in vivo, *J. Exp. Med.* 175:15 (1992).
6. K. Inaba, J.P. Metlay, M.T. Crowley and R.M. Steinman, Dendritic cells pulsed with antigens in vitro can prime antigen-specific, MHC-restricted T cells in situ, *J. Exp. Med.*, 172:631 (1990).
7. S.E. Macatonia, A.J. Edwards and S. C. Knight, Dendritic cells and the initiation of contact sensitivity to fluorescein isothiocyanate, *Immunology* 59:509 (1986).
8. A. J. T. George, S. G. Folkard, T. J. Hamblin, and F.K. Stevenson, Idiotypic vaccination as a treatment for a B cell lymphoma, *J. Immunol.*, 141:2168 (1988).
9. M.S. Kaminski, K. Kitamura, D.G. Maloney, M.J. Campbell and R. Levy, Importance of antibody isotype in monoclonal anti-idiotype therapy of a murine B cell lymphoma. A study of hybridoma switch variants, *J. Immunol.* 136:1123 (1986).
10. A.J.T. George, A.L.Tutt, and F. K. Stevenson, Anti-idiotypic mechanisms involved in suppression of a mouse B cell lymphoma, BCL1, *J.Immunol.*: 138:628 (1987).

11. A. O'Garra, R. Chang, N. Go, R. Hastings, G. Haughton and M. Howard. Ly-1 B (B-1) cells are the main source of B cell-derived interleukin 10, *Eur. J. Immunol.* 22:711 (1992).

12. D. F. Fiorentino, M.W. Bond, and T.R. Mosmann, Two types of mouse T helper cells. IV. Th2 clones secrete a factor that inhibits cytokine production by Th1 clones *J. Exp. Med.*, 170: 2081 (1989).

13. A. O'Garra, D. Barbis, J. Wu, P. D. Hodgkin, J. Abrams, and M. Howard, The BCL1 B lymphoma responds to IL-4, IL-5, and GM-CSF, *Cellular Immunology*: 123:189 (1989).

14. H. Ishida, R. Hastings, J. Kearney, and M. Howard, Continous anti-IL10 antibody administration depletes mice of Ly-1 B cells but not conventional B cells. *J. Exp. Med.*, 175:1213 (1992).

15. D. F. Fiorentino, A. Zlotnik, P. Vieira, T. R. Mosmann, M. Howard, K. W. Moore, and A. O'Garra, IL-10 acts on the antigen-presenting cell to inhibit cytokine production by Th1 cells, *J. Immunol.*, 146:3444 (1991).

16. L. Ding, and E. M. Shevach, IL-10 inhibits mitogen-induced T cell proliferation by selectively inhibiting macrophage costimulatory function, *J. Immunol.*, 148: 3133 (1992).

STUDIES ON LANGERHANS CELLS IN THE TRACHEAL SQUAMOUS METAPLASIA OF VITAMIN A DEFICIENT RATS

Satoru Hosokawa[1,2], Masanori Shinzato[1], Chiyuki Kaneko[1], and Mikihiro Shamoto[1]

[1]Division of Pathological Cytology, Fujita Health University School of Medicine, Toyoake, Aichi 470-11, Japan
[2]Eisai Co. Ltd., Kawashima-cho, Hashima, Gifu 501-61, Japan

INTRODUCTION

It has been generally accepted that Langerhans cells (LCs) are dendritic leukocytes derived from bone marrow (1). They migrate into epithelium which have the capacity to keratinize such as skin and other stratified squamous mucosal epithelium. They appear also in pathological conditions, such as epidermal cysts in dermis, ovarian dermoid cysts and squamous cell carcinomas. Therefore, some relationships have been postulated between LCs and keratinocytes (2). The present study was undertaken to reveal the developmental history of LCs and to reveal the relationships between LCs and keratinocytes in the tracheal squamous metaplasia of vitamin A deficient rats.

MATERIALS AND METHODS

Male Sprague-Dawley rats (Japan SLC, Shizuoka, Japan), initially 3 weeks old were fed a vitamin A-deficient diet (Oriental Yeast Co., Tokyo, Japan). The animals were sacrificed during the 12th to 18th week of the diet under pentobarbital anesthesia. The trachea were quickly removed and embedded according to Sato's AMeX(acetone, methylbenzoate and xylene) method (3). Anti-rat Ia monoclonal antibody(anti-Ia)(OX6; Sera-lab, U.K.) and anti-protein kinase C type II (PKCII) monoclonal antibody, kindly provided by Prof. H. Hidaka, Nagoya University, were used for the immunohistochemical demonstration of LCs (4,5). Some of the trachea were also fixed in a mixture of 2.5% gultaraldehyde and 2.0% paraformaldehyde for electron microscopy.

RESULTS

LCs and indeterminate cells were seen both immunohistochemically and electron microscopically in the metaplastic tracheal epithelium by vitamin A deficiency.

In the early stage of metaplasia, where basal cells begin to proliferate, small numbers of Ia and PKCII positive cells appeared among the hyperplastic basal cells(Fig.1). Electron microscopically they were immature cells with euchromatic oval or slightly intended nuclei and many free ribosomes. There were few cytoplasmic organella and no Birbeck granules.

Dendritic Cells in Fundamental and Clinical Immunology,
Edited by Kamperdijk *et al.*, Plenum Press, New York, 1993

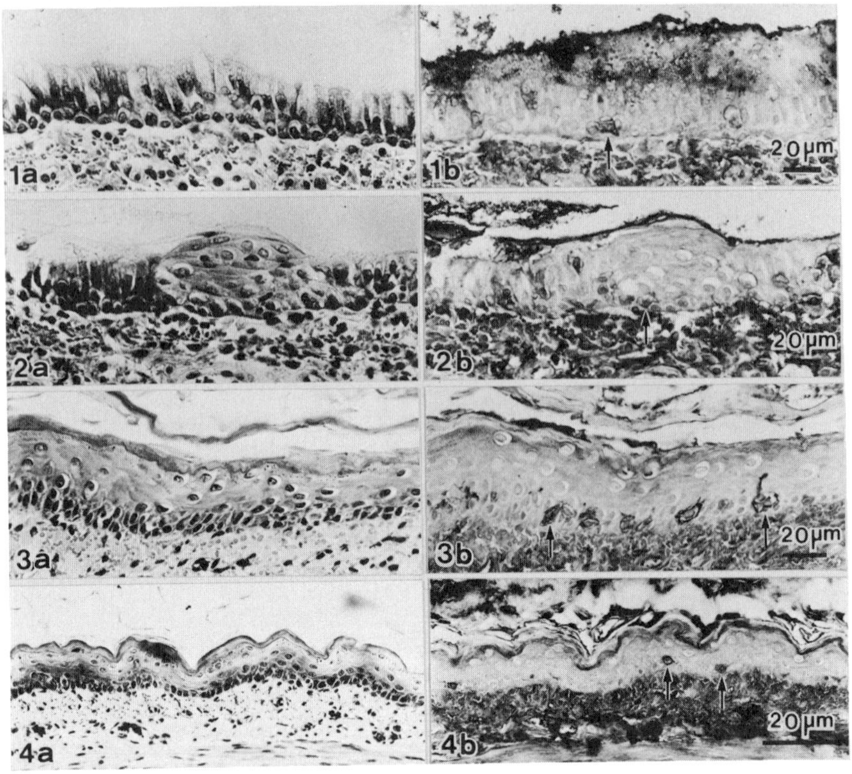

Fig.1-4. The PKC-II positive cells in the tracheal squamous metaplasia. Fig.1: Early stage of squamous metaplasia., Fig.2: Stage of stratification. Fig.3.: Stage of cornification. Fig.4: Final stage of metaplasia., **a**: H-E, **b**: anti-PKCII. Arrows indicate the PKC-II positive cells.

In the stage of stratification where proliferated basal cells differentiated into the flattered epithelium, small numbers of Ia and PKCII positive cells could be recognized (Fig.2). Ultrastructually these cells had ovoid or elongated nuclei, dark cytoplasm and few cytoplasmic processes. There were a few BGs and lysosome-like granules which have reported by Shamoto et al. (6) as atypical granules (AGs), mainly in the Golgi area. No BGs connected with the plasma membrane could be seen (Fig.6).

In the stage of cornification , a lot of Ia and PKCII positive dendritic cells could be recognized (Fig.3). Ultrastructually most of these were LCs with BGs. In the early stage of cornification, where the thin layer of desquamating cornified epithelium was observed, the LCs were found to have partially intended nuclei with prominent nucleoli and well-developed Golgi complexes. There were many BGs and AGs in the Golgi area, and some of theses granules were connected with Golgi apparatus (Fig.7). The AGs at this stage of LCs showed a varied morphology in size and shape. Some of the AGs were connected with BGs, and transitional forms between them were frequently observed. In these LCs, Birbeck granules which were connected with plasma membrane could be seen.

In the epithelium, where the squamous metaplasia had matured, most of the dendritic cells were the typical mature type of LCs with lobulated nuclei, clear cytoplasm and prominent dendritic processes (Fig.8). The numbers of BGs and AGs decreased when compared to the LCs mainly recognized in the early stage of cornification, and were distributed all over the cytoplasm. In the final stage of metaplasia where the basal cells also differentiate into the flatter epithelium, the numbers of LCs decreased (Fig.4). In the area where only basal cells differentiated into the flattened epithelium without a cornified layer, LCs with BGs were also seen (Fig.9)

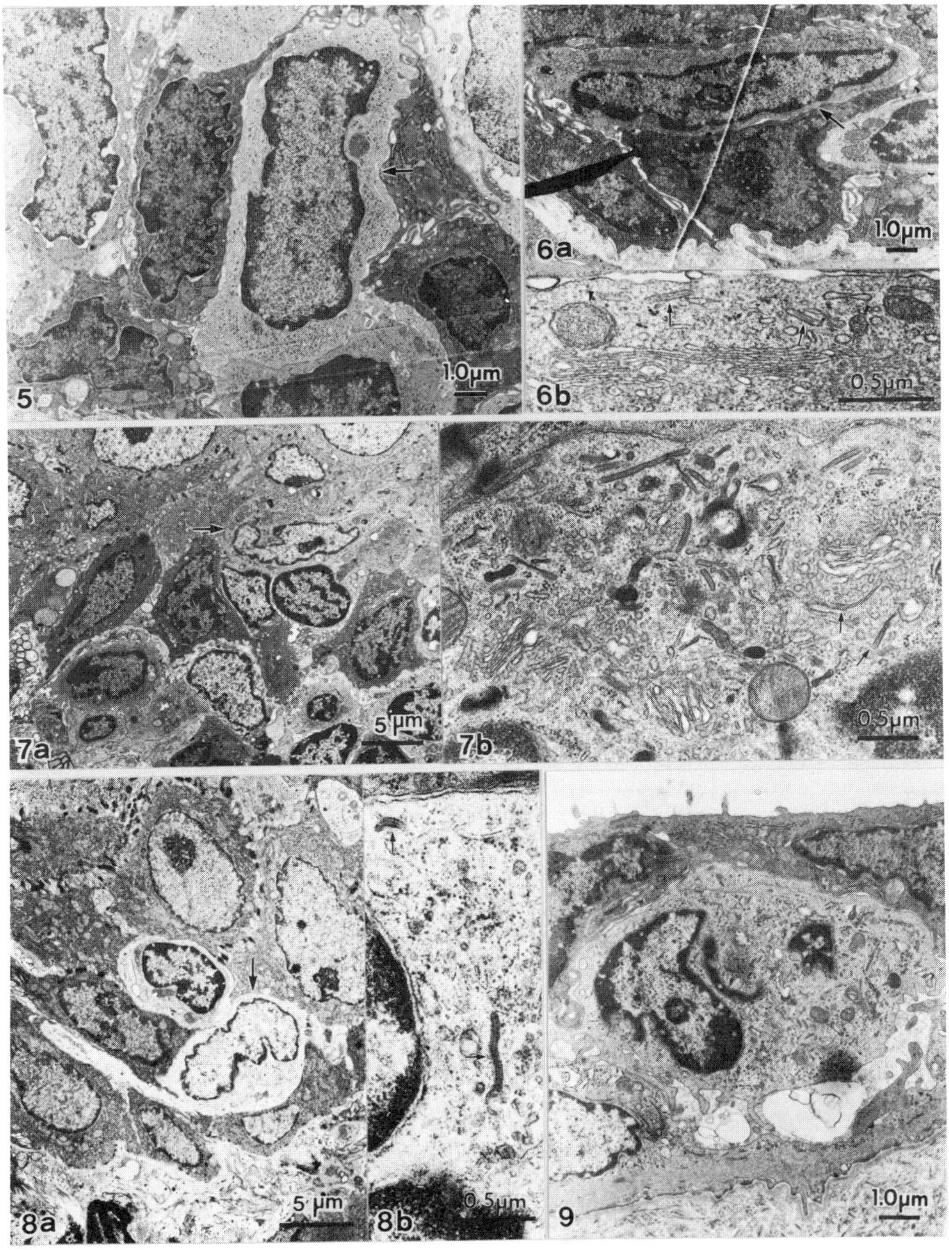

Fig.5. An immature cell (arrow) in the early stage of metaplasia. The cell possess slightly intended nucleus and many free ribosomes, but few cytoplasmic organella.
Fig.6. a: An immature LC (arrow) with elongated nucleus and dark cytoplasm in the stage of stratification. **b**: Higher magnification of **a**. BG (arrows) and AGs(arrowheads) can be seen in the Golgi area.
Fig.7. a: An LC (arrow) in the early stage of cornification. **b**: Higher magnification of **a**. Many BGs and AGs can be seen. Arrow indicate the BGs connected with Golgi apparatus.
Fig.8. a: A mature type of LC (arrow) with clear cytoplasm and lobulated nucleus in the epithelium where metaplasia was complete. **b**: Higher magnification of **a**. Arrows indicate the BGs.
Fig.9. An LC in the tracheal epithelium where only basal cells differentiate into the flatter epithelium.

DISCUSSION

The present study has showed that the tracheal squamous metaplasia lead to the migration of LCs possibly from bone marrow (7). The LCs with BGs first appeared in the stage of stratification. The Langerhans cells at this stage still possessed the characteristics of immature cells and was very few in number. The LCs increased in number along with the development of the metaplasia, and mostly could be seen during the stage of cornification. But they decreased in the final stage of metaplasia where basal cells also showed a transformation into flattened squamous epithelium. It is noteworthy that LCs with BGs could be seen even in the lesion where only the tracheal basal cells differentiated into flattened squamous cells. There have been reports suggesting some relationships between the distribution of LCs and the degree of keratinization (8-10). From the present study, it was considered that LCs which originated in bone marrow mature in squamous epithelium, and their distribution and morphology was considered to be dependent on the state of keratinocytes but not the presence of a cornified layer.

As for the formation of BGs, two main hypotheses have been postulated concerning the origin and function of BGs. One is that they derive from the Golgi apparatus with a secretory function (11-13). The other is that they develop from the plasma membrane as an endocytic organelle (14-17). Up to date, however, no clear conclusion has been made. In the present study, small numbers of BGs and AGs appeared in the Golgi area of the immature LCs, but BGs attached to the plasma membrane were observed only in the later developmental stage of LCs. BGs were closely connected with Golgi apparatus and many AGs resembling BGs have been observed. Transitional forms between BGs and AGs were also seen. These data strongly suggest that the formation of BGs occur in the Golgi area. Our present results support the hypothesis that BGs were derived from Golgi apparatus.

We recently reported that there are two types of different dendritic cells in the superficial and hilar lymph nodes, i.e. CD1a and S-100 positive LCs and S-100 positive but CD1a negative IDCs. No LCs could be seen in the mesenteric lymph nodes which were located in the deeper portion and spleens (18). Thus our previous paper indicated the close relationships with LCs and epidermal keratinocytes. Also in this study, the formation of BGs was closely related to the metaplastic keratinocytes. It was speculated that LCs need to make contact with keratinocytes in order to form BGs.

SUMMARY

LCs in the tracheal squamous metaplasia of vitamin A deficient rats were studied. The first appearance of LCs with Birbeck granules (BGs) was in the stage of stratification. The number of LCs increased along with the development of metaplasia, but decreased in number in the later stage, where basal cells also differentiate into flattened epithelium. In the area where only basal cells differentiated into flatter epithelium, without a cornified layer, LCs with BGs could also be found. These findings suggest that LCs which originate in bone marrow mature in squamous epithelium. The distribution and morphology of LCs are dependent on the state of kerachinocytes, but not the presence of a cornified layer.

REFERENCES

1. S.I. Kats, K. Tamaki, D.H. Sachs, Epidermal langerhans cells are derived from cells originating in bone marrow., *Nature* 282:324 (1979.)
2. J. Schweizer, Langerhans' cells, epidermal growth control mechanisms and type of keratinization., in: The Epidermis in Disease, R. Marks, E. Christophers, Eds., MTP press, England, pp.481 (1981).
3. Y. Sato, K. Mukai, S. Watanabe et al., The AMeX method. A simplified technique of tissue processing and paraffin embedding with improved preservation of antigens for immunostaining., *Am J Pathol*, 125:431 (1986).
4. A.N. Barkley, Different reticular elements in rat lymphoid tissue identified by localization of Ia, Thy-1 and MRCOX2 antigen., *Immunol* 44:727 (1981).
5. Y. Koyama, T. Hachiya, M. Hagiwara et al., Expression of protein kinase C isozyme in epidermal Langerhans cells of the mouse., *J Invest Dermatol*, 94:677 (1990).

6. M. Shamoto, Langerhans cell granule in Letterer-Siwe disease. An electron microscopic study., *Cancer*, 26:1102(1970).
7. Y.G. Wong, R.C. Buck, Langerhans cell in epidermoid metaplasia., *J Invest Dermatol*, 56:10 (1971).
8. P.A. Riley, Esterase in epidermal dendritic cells of the mouse. A study of the histochemical properties and distribution of activity of the enzyme in relation to patterning in the tail., *Br J Dermatol*, 78:388 (1966).
9. J. Schweizer, F. Marks, A developmental study of the distribution and frequency of Langerhans' cells in relation to formation of patterning in mouse tail epidermis., *J Invest Dermatol*, 69:198 (1977).
10. L.H. Hutchens, R.W. Sagebiel, M.A. Claeke, Oral epithelial dendritic cells of the rhusus monkey: histologic demonstration, fine structure and quantitative distribution., *J Invest Dermatol*, 56:325 (1971).
11. A.S. Zelickson, Granule formation in the Langerhans cell., *J Invest Dermatol*, 47:498 (1966).
12. G. Niebauer, W.S. Krawczyk, R.L. Kidd, Osmium zinc iodide reactive sites in the epidermal Langerhans cells., *J cell Biol*, 43:80 (1969).
13. M. Shamoto, M. Hoshino, T. Suchi, Cells containing Langerhans cell granules in human lymph nodes of "atypical hyperplasia" with fatal outcome and leukemic reticuloendotheliosis., *Acta Path Jap*, 26:311 (1976).
14. K. Hashimoto, Langerhans's granule. An endocytotic organelle., *Arch Dermatol*, 104:148 (1971).
15. M. Takigawa, K. Iwatsuki, M. Yamada et al., The Langerhans cell granule is an absorptive endocytic organelle., *J nvest Dermatol*, 85:12 (1985).
16. S. Takahashi, K. Hashimoto, Derivation of Langerhans cell granules from cytomembrane., *J Invest Dermatol*, 84:469 (1985).
17. A. Ray, D. Schmitt, C. Deztter-Dambuyant et al., Reappearance of CD1a antigenic sites after endocytosis on human Langerhans cells evidenced by immunogold relabeling., *J Invest Dermatol*, 92:217 (1989).
18. M. Shamoto, M. Shinzato, S. Hosokawa et al., Langerhans cells in the lymph nodes: mirror section and immunoelectron microscopic studies. *Virchows Arch[B]* 61:337 (1991).

DEPLETION OF LANGERHANS CELLS FOLLOWING CARCINOGEN TREATMENT IS PARTLY DUE TO ANTIGENICITY

Gregory M. Woods, Imogen H. Liew and H. Konrad Muller

Dept of Pathology
University of Tasmania
Box 252C, GPO, Hobart
Tasmania, Australia, 7001

INTRODUCTION

Langerhans cells (LC) are skin dendritic cells that trap foreign antigens[1] and migrate to the local lymph nodes where they present antigen to initiate an immune response. Maintenance of LC density is an essential component in the generation of an immune response as immunization through skin deficient in LC (either naturally depleted such as mouse tail skin or physically depleted by chemicals, UV light or "tape stripping") results in immunological unresponsiveness[2]. We have documented LC depletion following carcinogen/tumour promoter treatment and reported that sensitization with 2,4-dinitrofluorobenzene (DNFB) through such depleted sites induced immunological tolerance by activating antigen specific T suppressor cells[3]. The immunosuppressive ability of carcinogens/tumour promoters suggests that LC depletion is a phenomenon unique to these compounds but depletion of LC following antigen (2,4,6 trinitrochlorobenzene; TNCB) application has also been demonstrated[4]. Experiments by our group have clearly shown a rapid and substantial migration of LC from the epidermis following either antigen (TNCB) or carcinogen (7,12 dimethylbenz(a)anthracene; DMBA) treatment. These observations raise the possibility that depletion of LC following carcinogen/tumour promoter application is due to a normal immunological response. LC recognise the carcinogens as antigens and respond accordingly by migrating to the draining lymph node.

The aim of this work was to test the hypothesis that "Langerhans cells recognise carcinogens and tumour promoters as antigens which causes their depletion due to the triggering of migration".

MATERIALS AND METHODS

Carcinogen, tumour promoter, tumour initiator, carcinogen analogue & antigen sensitization

The dorsal skin of naive BALB/c mice (6-12 weeks of age) was shaved and treated with the tumour initiators (20 μl 10% urethane in acetone, 20 μl 0,25% benz(a)anthracene in acetone), tumour promoters (20 μl 0.005% TPA in acetone), antigens (20 μl of 1.0% TNCB in acetone/olive oil), complete carcinogens (1% DMBA in liquid paraffin/lanolin) and the non-carcinogenic analogue of DMBA (1% anthracene in liquid paraffin/lanolin). For carcinogens and carcinogen products 3 weekly applications were applied. For the antigen TNCB only one application was required.

Assessment of contact sensitization

Contact sensitivity responses were determined by reapplying the respective sensitizing agent to the dorsal surface of the right ear 5 days after the final sensitization. Ear thickness was measured 24 hours later with a spring loaded engineer's micrometer. Contact sensitivity was expressed as percent increase in ear thickness-

$$\frac{\text{thickness of challenged ear} - \text{thickness of unchallenged ear}}{\text{thickness of unchallenged ear}} \times 100\%$$

LC identification

LC were identified in epidermal sheets by staining for MHC class II antigen. The epidermal sheets were incubated overnight with mouse monoclonal anti-Ia^d (TIB-120) at 4%C. The sheets were washed in PBS and incubated with peroxidase conjugated rabbit antimouse immunoglobulins and the reaction product visualized with diaminobenzidine in PBS containing 0.02% H_2O_2. Specimens were mounted in 10% glycerol in PBS for microscopic examination. LC were examined by light microscopy and the number present in 8 random fields was counted and calculated as the mean number of LC per mm^2.

RESULTS

The ability of the carcinogen and carcinogen products to induce a contact sensitivity response is diagramatically represented in Figure 1. The carcinogen DMBA and the tumoür promoter TPA both induced a contact sensitivity response. In contrast, the tumour initiators BA and urethane as well as anthracene (the DMBA carcinogenic analogue) failed to induce contact sensitivity responses. The ability of these agents to deplete Langerhans cells from the dorsal trunk (Figure 2) skin reflected the ability to induce contact sensitivity. The chemicals causing contact sensitivity also caused LC depletion from the epidermis whereas no LC depletion was observed with the initiator and carcinogen analogue which failed to induce contact sensitivity (Table 1).

DISCUSSION

These experiments demonstrated that the complete carcinogen (DMBA) and the tumour promoter (TPA) were antigenic, compared to tumour initiators which did not induce an immune response. This antigenicity could explain their ability to induce Langerhans cell migration. Langerhans cells, by initiating an immune response to the carcinogen or tumour promoter, migrate to the draining lymph nodes and induce a contact sensitivity response to these agents. Following this 'antigen' induced migration the treated area of skin is transiently depleted of Langerhans cells which could lead to immune suppression should contact with a second antigen occur. It is also highly probable that the carcinogen damages (mutates) other cells within the epidermis which could eventually develop into tumour cells that escape immune detection due to the reduced density ofLangerhans cells.

The response to DMBA is consistent with the work by Klemme and colleagues[7] who also demonstrated a contact sensitivity response. Although the response to DMBA was weak (even after 3 weekly treatments) there is still strong evidence that antigen presentation occurred. The uptake of antigen is most likely to be the crucial signal that leads to the migration, irrespective of the ultimate contact sensitivity response. This weak response could be due to its antigenicity level or alternatiely, as DMBA requires metabolism into active carcinogenic product via cytochrome P-450-dependent monooxygenases and epoxide hydrolase[8], the immune response could be directed

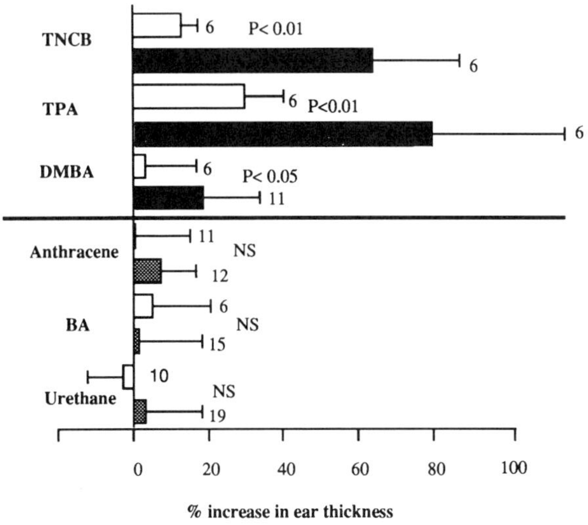

Contact sensitivity response following 3 weekly applications of carcinogen or carcinogen products.
Results represent mean ± standard deviation from the number of experiments indicated.
Unshaded boxes represent the vehicle controls.
Results analysed using Student's unpaired t-test.
TNCB received one application.
Results above solid line represent a significant contact sensitivity response.

Figure 1. Contact sensitivity following carcinogen sensitization

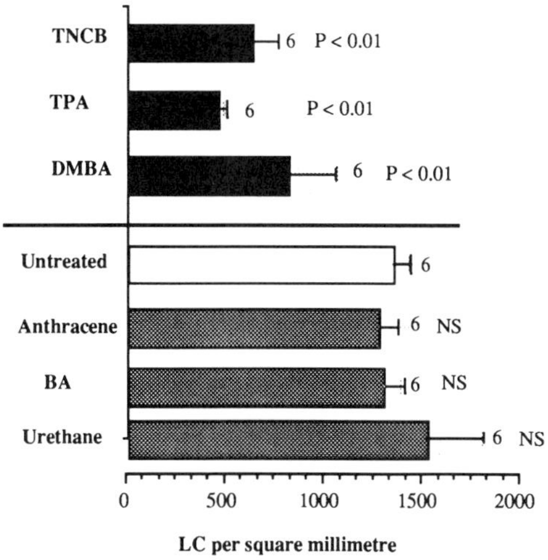

Langerhans cells were enumerated by class II staining following 3 weekly applications of carcinogen or carcinogen products.
Results represent mean ± standard deviation of the percent increase in ear thickness from the number of mice indicated.
Results analysed using Student's unpaired t-test compared to untreated control.
TNCB received one application.
Results above solid line represent a significant reduction in Langerhans cell density compared to untreated controls.

Figure 2. Langerhans cell density following carcinogen sensitization

Table 1. Summary of the effect of carcinogens on contact sensitivity and Langerhans cell depletion

Compound	Class of compound	Induction of contact sensitivity	Langerhans cell depletion
TNCB	antigen	strong	strong
DMBA	complete carcinogen	weak	moderate
TPA	tumour promoter	weak	moderate
BA	tumour initiator	none	none
Urethane	tumour initiator	none	none
Anthracene	carcinogen analogue	none	none

against these metabolites rather than the parent compound. This may also provide an explanation for the lack of immune reactivity to the DMBA structural analogue anthracene which is not metabolized to a carcinogenic product. To address this question the ability of metabolites to induce Langerhans cell depletion and contact sensitivity requires investigation.

CONCLUSIONS

These results indicate that depletion of LC from carcinogen treated skin is in part due to the recognition of these compounds as antigens. In response to antigens/carcinogens there was a significant migration with the treated area becoming deficient in LC as the normal rate of replenishment did not match the migration. The consequence of such depletion is that any *new* antigen appearing during this stage would not be presented correctly resulting in the generation of antigen specific T suppressor cells. This may explain how highly antigenic/carcinogenic induced tumours may avoid immune elimination.

We therefore propose that carcinogens or tumour promoters act at two levels:-

1: On initiated tumour cells promoting them into cancer cells.

2: On LC as they are initially treated as an antigen which initiates migration and ultimately to an impaired local immune response due to the reduction in LC density.

ACKNOWLEDGEMENTS

This work was supported by the Tasmanian Cancer Committee.

REFERENCES

1. W.B. Shelley and L. Juhlin. Selective uptake of contact allergens by the Langerhans cell, *Arch. Dermatol.* 113:187-192 (1977).
2. G.B. Toews, P.R. Bergstresser, and J.W. Streilein. Epidermal Langerhans cell density determines whether contact hypersensitivity or unresponsiveness follows skin painting with DNFB, *J. Immunol.* 124:445-453 (1980).
3. G.M. Halliday and H.K. Muller. Induction of tolerance via skin depleted of Langerhans cells by a chemical carcinogen, *Cell. Immunol.* 99:220-227 (1986).
4. P.R. Bergstresser, G.B. Toews, and J.W. Streilein, Natural and perturbed distributions of Langerhans cells: response to UV light, heterotopic skin grafting and dinitrofluorobenzene sensitisation, *J.Invest. Dermatol.* 75:73-77 (1980).
5. T.J. Slaga. Tumor promotion and skin carcinogenesis, *in:* "Mechanisms of Tumor Promotion. Volume II," T.J. Slaga, ed., CRC Press, Inc. (1984).
6. G.M. Halliday, G.M. MacCarrick, and H.K. Muller, Tumour promoters but not initiators deplete Langerhans cells from murine epidermis, *Br.J.Cancer* 56:328-330 (1987).
7. J.C. Klemme, H. Mukhtar, and G.A. Elmets, Induction of contact hypersensitivity to dimethylbenz(a)anthracene and benzo(a)pyrene in C3H/HeN mice, *Cancer Res.* 47:6074-6078 (1987).
8. P. Sims and P. L. Grover, Epoxides in polyaromatic hydrocarbon metabolism and carcinogenesis, *Cancer Res.* 20:165-274 (1974).

A PROPORTION OF PATIENTS WITH PREMATURE OVARIAN FAILURE SHOW LOWERED PERCENTAGES OF BLOOD MONOCYTE DERIVED DENDRITIC CELLS CAPABLE OF FORMING CLUSTERS WITH LYMPHOCYTES

Annemieke Hoek[1,2], Yvonne van Kasteren[1], Meeny de Haan-Meulman[2], Joop Schoemaker[1], and Hemmo A. Drexhage[2]

[1] Department of Gynaecology and Obstetrics, Free University, Amsterdam, The Netherlands
[2] Department of Immunology, Erasmus University Rotterdam, The Netherlands

INTRODUCTION

Premature ovarian failure (POF) is characterized by an amenorrhea before the age of 40 years, estrogen deficiency and elevated levels of gonadotropins[1]. An autoimmune mechanism in the etiology of a proportion of POF patients is indicated by: 1. the presence of autoantibodies against ovarian cells[2,3] and FSH receptors[4,5], 2. a higher association of POF with other autoimmune endocrinopathies[6,7,8], 3. lymphocytic and plasma cell infiltrates have been found in ovaries of POF patients[9,10]. Abnormalities in the cellular immune compartment have been described in these patients: increase in T-helper cells[11], elevated percentage of peripheral blood HLA-dr+ T-cells[12].

Besides T-lymphocytes and macrophages, the monocyte derived dendritic cells (DC) play a prominent role in endocrine autoimmune reactions[13,14]. DC are of crucial importance in the presentation of autoantigens during the elicitation of endocrine autoimmune reactions[15]. DC accumulation and clustering in the endocrine gland is seen in the initial stages of diseases such as thyroiditis and the insulitis[13,14]. Clustering with DC precedes and is essential for T-cell proliferation[16]. The clustering of DC with allogeneic lymphocytes reflects the initial steps of a non-antigen dependent homotypic clustering which is the initial step in antigen presentation. It can hence be easily assumed that the monocyte derived DC and macrophages are involved in the autoimmune proces of POF.

This report describes an investigation on the capability of DC isolated from peripheral blood monocytes of POF patients to form clusters with allogeneic lymphocytes.

PATIENTS AND CONTROLS

Heparinized blood was obtained from twenty eight POF patients, mean age

31.5 years (range: 25-38 years). All patients had a secondary amenorrhea and a normal karyotype, 46XX. Since effects of sexsteroids have been described on the function of the immune system[17], eight POF patients were tested under different endocrine conditions viz, treatment with estrogen substitution (to suppress the gonadotropin levels and elevate the estrogen levels).

As controls were studied: thirty one healthy women age between 22 and 43 years and forty nine healthy laboratory personnel (male and female). Eleven healthy postmenopausal women without estrogen substitution therapy, (mean age: 57.8 years; range: 47-75 years) and six postmenopausal women during estrogen substitution therapy (mean age: 53.6 years; range: 51-58) were studied.

ASSAYS AND METHODS

Hormone Assay. Plasma FSH and LH samples were analyzed in duplicate by a commercially available immunometric assay (Amerlite, Amersham, UK).

The Isolation of Peripheral Blood Monocytes. Peripheral blood mononuclear cells from patients, postmenopausal women and healthy controls were isolated by Ficoll-Isopaque density gradient centrifugation followed by Percoll gradient centrifugation[18]: Percoll 1.063 (Pharmacia, Diagnostics AC, Uppsala, Sweden). After centrifugation (40 min, 450 g) the cells were collected from the interface, washed twice in medium (10 min, 500 g) and counted: the suspension now contained 70-95% NSE-positive cells[19].

The maturation of DC from blood monocytes. Dendritic cells were prepared from peripheral blood monocytes according to the method described by Kabel[20]. Metrizamide (Serva, Heidelberg, FRG) was dissolved in RPMI supplemented with 10 % fetal calf serum. Cells from the isolated monocytic fractions were exposed to metrizamide in suspension culture for 30 minutes. Thereafter, the cells were washed (culture fluid was added slowly to prevent osmotic lysis of the cells), and further cultured under non-adhering conditions for 16 hours in polypropylene tubes (5% CO_2 and 37°C, 100% humidity). This procedure yields 40-80% cells with a dendritic morphology, showing class II MHC positivity, a decreased expression of the monocytic CD14 determinant, a decreased phagocytic capability, but an enhanced stimulator capability in the MLR.

Clustering of dendritic cells with allogeneic lymphocytes. The cluster assay as described by Kabel[20] was performed with modifications. $5x10^4$ dendritic cells prepared from peripheral blood monocytes exposed to metrizamide were allowed to cluster with $5x10^3$ allogeneic lymphocytes isolated from healthy controls (4 hours, 37°C 5% CO_2) in 250 μl flat-bottomed wells. The lymphocyte isolation was performed according to standard procedures with Ficoll-Isopaque and Nylon wool adherence (Leuko-Pak, Fenwall Laboratories, Ill, USA). Formed clusters were counted using an inverted microscope and values were expressed as the number of clusters per 6 microscopic fields (x250). A cluster was defined as an accumulation of 4-25 cells.

RESULTS

Endocrine Data. The patients with POF showed a hypergonadotropic hypoestrogenic hormone profile in untreated conditions. Estrogen substitution resulted in a statistic significant lowering of plasma gonadotropin levels in POF patients, it had relatively little effect on the gonadotropin levels in postmenopausal women.

Clustering of Dendritic Cells with Allogeneic Lymphocytes. The percentage of monocytes of the total of peripheral blood leucocytes in POF patients (mean 8%, s.d. 3%) and postmenopausal women (mean 7%, s.d. 1%) were within the normal range of the controls (mean 8%, s.d. 2%).

The percentage of DC isolated from these peripheral blood monocytes in patients with POF (mean 49%, s.d. 16%) did also not differ from postmenopausal women (mean 44%, s.d. 10%) or healthy laboratory personnel (mean 50%, s.d. 17%). Estrogen substitution did not influence these values.

Figure 1 shows that, the clustering capability of DC of postmenopausal women (108, s.d. 23) irrespective of estrogen substitution therapy (117, s.d. 22), was in the same range of that of the healthy controls (127, s.d. 25). The cluster capability of POF patients' DC was lower than that of healthy controls; namely a mean of 90 clusters (s.d. 46, $p<0.004$) for POF patients without estrogen substitution therapy and a mean of 99 clusters (s.d. 28, $p<0.009$) for POF patients with estrogen therapy. Estrogen substitution therapy had no effect on the number of clusters. The blood DC of 36% of the POF patients showed a defective capability to form clusters in vitro.

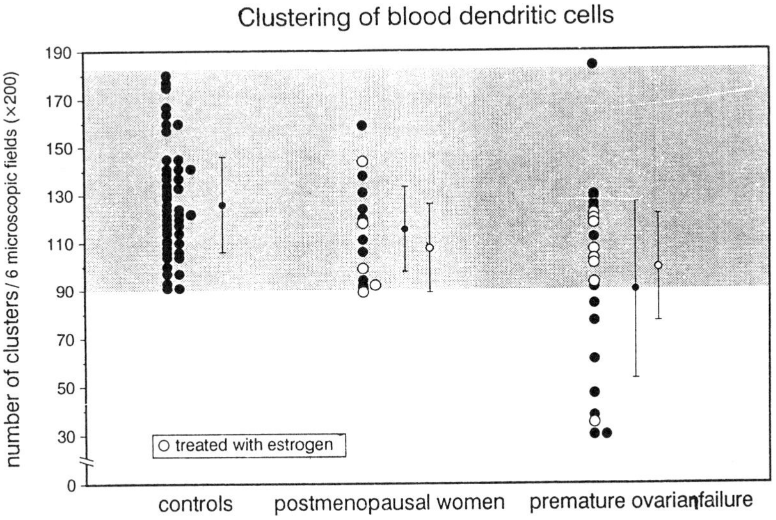

Figure 1. Number of clusters of DC with allogeneic lymphocytes (six microscopic fields, X200). The capability of blood DC from patients with POF to form clusters with allogeneic lymphocytes is decreased (wilcoxon $p<0.003$), compared to controls and postmenopausal women. 36% of patients with POF show a decreased cluster capability of peripheral blood DC. The hatched area represents normal values.

DISCUSSION

The results of this study demonstrate that peripheral blood monocyte derived DC show a decreased capability to form clusters with allogeneic lymphocytes. The lowered DC cluster capability of POF patients can not be ascribed to the hypergonadotropic hypoestrogenic hormone profile, since postmenopausal women with the same gonadotropin and estrogen profile did not show these defective DC cell functions; changes in the gonadotropin and estrogen profile (estrogen substitution therapy) had also no effect.

In Graves' disease a retroviral related low molecular weight factor (<25kd), the so-called transmembrane factor (TM-factor), with immunosuppressive properties[21,22,23] is present in the sera of these patients together with disturbances in dendritic cell clustering. However, correlation between the presence of this TM-factor and the disturbed DC functions could not be established[24].The decreased DC clustering capability can not be ascribed to the presence of the TM-factor in the sera of patients with POF, since the TM-factor is not detectable in the serum of patients (results not described). If neither the gonadotropin and estrogen hormone profile, nor the TM-factor is responsible for the abnormalities of peripheral blood derived DC function in patients with POF, what could be the mechanism?

In this respect it is of interest that similar findings have been reported, in another endocrine autoimmune endocrinopathy, namely type-1 diabetes. Martin et al[25] described that, a subpopulation of monocytes in peripheral blood of patients with diabetes mellitus, had a decreased expression of the adhesion molecule ICAM-1, whereas Ziegler et al[26] demonstrated in patients with recent onset type-1 diabetes, a decline of class-II positive Langerhans cells in the skin. Both authors explained their data by a assuming a redistribution of "active" monocytes and DC cells from the periphery (blood and skin) to the target gland (pancreas), where indeed an accumulation of active DC has been reported[14]. It is not known whether DC accumulate in the ovaries of POF patients, since POF ovarian tissue is not readily available and has, to our knowledge not been studied for the presence of such cells. If there is a redistribution of monocytes and DC from the periphery to the ovaries in POF patients, this could be caused by the following mechanisms for which there are indications in the literature: - overexpression of autoantigens in the ovary[27]. - venular defects in the target gland to be affected by the autoimmune disease[28]. - intrinsic defects in the function of monocytes and DC.

In conclusion: this study is the first report on a lowered percentage of monocyte derived dendritic cells capable of forming homotypic clusters in the peripheral blood of POF patients. The gonadotropin and estrogen hormonal profiles of these patients were not responsible for these decreases. Since similar defects have been found in well established autoimmune endocrinopathies, our observation strengthens the concept that autoimmune mechanisms play a role in a considerable proportion of POF patients.

REFERENCES

1. M. de Morales-Ruehsen and G.S. Jones. *Fert Steril.* 18:440-461 (1967).
2. F. Sotsiou, G.F. Botazzo, D. Doniach. *Clin Exp Immunol.* 39:97-111 (1980).
3. J.L. Luborsky, J. Visintin, S. Boyers, T. Asari, B. Caldwell, A. deCherney. *J Clin Endocrinol Metab.* 70:69-75 (1990).
4. V. Chiauzzi, S. Cigorraga, M.E. Escobar, M.A. Rivarola, E.H. Charreau. *J Clin Endocrinol Metab.* 54:1221-1228 (1982).
5. M.M. van Weissenbruch, A. Hoek, I. van Vliet-Bleeker, J. Schoemaker, H.A. Drexhage. *J Clin Endocrinol Metab.* 73:360-367 (1991).
6. M. Elder, N. Maclaren, W. Riley. *J Clin Endocrinol Metab.* 52:1137-1142 (1981).
7. A.M. Vasques, F.M. Kenney. *Obstet Gynecol.* 41:414-418 (1973).
8. M.E. Escobar, S.B. Cigorraga, V.A. Chiauzzi. *Acta Endocrinol.* 99:431-436 (1982).
9. D.D. Sedmak, W.R. Hart, R.R. Tubbs. *Int J Gynecol Pathol.* 6:73-81 (1987).
10. E. Gloor, J. Hurlimann. *Am J Clin Pathol.* 81:105-109 (1984).
11. M.H. Mignot, H.A. Drexhage, M. Kleingeld, E.M. van de Plassche-Boers, B.R. Rao, J. Schoemaker. *Eur J Obstet Gynecol Reprod Biol.* 30:67-72 (1989).
12. S.L. Rabinowe, K.L. George, V.A. Ravnikar, R.G. Dluhy, S.A. Dib. *Fert Steril.* 51:450-454 (1989).
13. P.J. Kabel, H.A.M. Voorbij, M. de Haan-Meulman, R. van der Gaag & H.A. Drexhage. *J Clin Endocrin Metab.* 66:199-207 (1988).
14. H.A.M. Voorbij, P.H.M. Jeucken, P.J. Kabel, M. de Haan-meulman, H.A. Drexhage. *Diabetes.* 38:1623-1629 (1989).
15. S.C. Knight, J. Farrant, J. Chan, A. Bryant, P.A. Bedford, C. Bateman. *Clin immunol Immunopathol.* 48:277-289 (1988).
16. J.M. Austyn, D.E. Weinstein & R.M. Steinman. *Immunology.* 63:691-696 (1988).
17. C.J. Grossman. *Endocrine Rev.* 5:435-455 (1984).
18. H. Pertoft, A. Johnsson, B. Wärmegård & R. Seljelid. *J Immunol Meth.* 29:133-137 (1979).
19. H. Mullink, M. von Blomberg, M.M. Wilders, H.A. Drexhage & C.L. Alons. *J Immunol Meth.* 29:133-137 (1979).
20. P.J. Kabel, M. de Haan-Meulman, H.A.M. Voorbij, M. Kleingeld, E.F. Knol & H.A. Drexhage. *Immunobiol.* 179:395-411 (1989).
21. E.M. v.d. Plassche-Boers, M. Tas, M. de Haan-Meulman, M. Kleingeld and H.A. Drexhage. *Clin Exp Immunol.* 73:348-354 (1988).
22. A.J.M. Balm, B.M.E. von Blomberg-vd Flier, H.A. Drexhage, M. de Haan-Meulman & G.B. Snow. *Laryngoscope.* 94:223-227 (1984).
23. I.B. Tan, H.A. Drexhage, R.J. Scheper, B.M.E. von Blomberg-van der Flier, M. de Haan-Meulman, G.B. Snow & A.J.M. Balm. *Arch Otolaryngol Head Neck Surg.* 112:942-945 (1986).
24. M. Tas, M. de Haan-Meulman, P.J. Kabel, H.A. Drexhage. *Clin Endocrinol.* 34:441-448 (1991).
25. S. Martin, H.D. Rothe, D. Tschope, B. Schwippert, H. Kolb. *Immunology.* 73:123-125 (1991).
26. A.G. Ziegler, E. Standl. *Diabetologica.* 31:632-635 (1988).
27. E.K. Muechler, K. Huang, E. Schenk. *Int J Fertil.* 36:99-103 (1991).
28. G. Manjo, I. Joris, E.S. Handler, J. Desemone, J.P. Mordes, A.A. Rossini. *Am J Pathol.* 128:210-215 (1987).

THYROID HORMONES AND THEIR IODINATED BREAKDOWN PRODUCTS ENHANCE THE CAPABILITY OF MONOCYTES TO MATURE INTO VEILED CELLS. BLOCKING EFFECTS OF α-GM-CSF

P. Mooij[1], M. de Haan-Meulman[1], H.J. de Wit[1] and H.A. Drexhage[1]

[1]Department of Immunology, Erasmus University Rotterdam
P.O. Box 1738, 3000 DR Rotterdam, The Netherlands

INTRODUCTION

Dendritic cells play a crucial role in the initiation of immune responses. The precursor of these cells present in the blood has not yet been characterised. However, we and others, Peters et al, have reported that 30-40% of cells with a dendritic/veiled morphology, MHC class-II positivity and weak to absent acid phosphatase positivity can be derived from almost pure blood monocyte fractions[1] after a pulse with 14.5% metrizamide and an overnight culture under nonadhering conditions[2]. These monocyte-derived dendritic/veiled cells are more capable than the original monocytes to act as stimulator cells in a MLR, are less phagocytic and express less CD14. Since metrizamide is an iodinated compound structurally related to the thyroid hormones T_3 and T_4 it was speculated that these thyroid hormones, their iodinated breakdown products and may be iodine itself might play a role in enhancing the monocyte-dendritic cell transition. In this study we investigated the effect of several iodinated compounds such as T_4, T_3, reverse T_3 (rT_3), their degradation products and MIT and DIT on the maturation of human peripheral blood monocytes into dendritic/veiled cells.

MATERIALS AND METHODS

The isolation of human peripheral blood monocytes. Peripheral blood monocytic cell fractions were obtained via counterflow elutriation as described previously in detail[3]. In brief, mononuclear cells were seperated from 450 ml whole blood via Percoll centrifugation (20 min., 1000g, room temperature). Thereafter, the mononuclear cells were injected into an elutriation system (Beckman J21 centrifuge with a JE-6 elutriaton rotor). PBS with 13 mM trisodium citrate and 5 mg human albumin per ml was used as elutriation medium. To seperate the different cell populations, a constant flow rate of 20 ml/min. was maintained, while the rotor speed decreased from 4000 to 0 rpm. The fraction collected at 2500 rpm contains 93-97% pure monocytes as judged by non-specific esterase reactivity. This fraction was used.

Dendritic Cells in Fundamental and Clinical Immunology,
Edited by Kamperdijk *et al.*, Plenum Press, New York, 1993

Incubation of human peripheral blood monocytes with hormones or iodinated compounds. Metrizamide (14.5% w/v, Serva, Heidelberg) and the other iodinated compounds T4, T3, rT3, 3,3'T2, 3',5'T2, 3,5T2, 3T1, 3'T1, T0, MIT and DIT (Sigma) were dissolved in either RPMI 1640 (Gibco) supplemented with 10% heat inactivated (30 mins., 56 °C) fetal calf serum (Integro bv, Zaandam, The Netherlands), 100 U/ml penicillin (Seromed, Biochrom, Berlin, Germany) and 0.1 mg/ml streptomycin (Seromed, Biochrom, Berlin, Germany) or in serum free medium (ISCOVE'S, Gibco) supplemented with SF-1 (Costar) and antibiotics. 1.10^6 monocytes/ml were incubated for 30 min. with the iodinated compounds or hormones at several concentrations (2.10^{-18}M - 2.10^{-6}M) under nonadhering conditions (polypropylene tubes, 37°C, 5% CO_2, 100% humidity). Thereafter, the cells were washed 3 times and further cultered for a period of 16 hours under the same conditions. To determine the influence of α-GM-CSF (Genzyme) on the dendritic cell maturation, 5 μg/ml of this monoclonal antibody was added after a 2 hour culture period. α-G-CSF was used as control.

Determination of the percentage of monocyte derived dendritic/veiled cells. After the 16 hour incubation period, cells were spun down and studied in light microscopy at a magnification of x400. Large cells, with veiled cytoplasmic processes were enumerated. The number of veiled cells counted in the wet preparations were verified using cytocentrifuge preparations.

50 μl of cellsuspensions at a concentration of $0.5.10^6$/ml were cytocentrifuged (3 minutes, 1100 rpm) using a cytospin (Nordic Immunological Laboratories, The Netherlands). Cytocentrifuge preparations were airdried for at least 1 hour and fixed in 100% acetone for 10 minutes. Thereafter, indirect immunoperoxidase staining was performed using either the monoclonal antibody OKI/a in a dilution of 1:100 (MHC-II, Orthodiagnostics) or RFD-1 in a dilution of 1:1000 (Class-II MHC associated antigen, Poulter[4]). After rinsing the slides an anti mouse IgG peroxidase labeled in 1% normal human pool serum (1:50) was applied as conjugate. The cells were then incubated for 30 minutes with acid phosphatase, dehydrated and embedded in DePeX. Cells with long cytoplasmic protrusions, a reniform nucleus and strong MHC class-II staining with absent or weak acid phosphatase reactivity were considered to be dendritic/veiled cells.

Cluster assay. To verify whether the monocyte derived cells with a veiled morphology were functionally active a cluster assay was performed. 5.10^4 monocyte derived veiled cells/100μl RPMI 1640 with 10% FCS and antibiotics were allowed to form clusters with 5.10^3/100μl allogeneic peripheral blood lymfocytes from a seperate donor for a period of 4 hours in flat bottemed wells at 37°C, 5% CO_2, 100% humidity. Cell clusters were counted per 6 microscopic fields (magn. 160x) using an inverted microscope. This type of cluster formation has been shown to be partly dependent on the adhesion molecules LFA-1/ICAM-1, on the temperature and on the presence of calcium[5]. The clustering of dendritic cells with T cells is essential for mitogenesis[6]. We defined a cluster as an accumulation of 4-25 cells, mostly involving various dendritic/veiled cells and allogeneic lymphocytes.

RESULTS

The enhancing effect of iodinated compounds and hormones on the monocyte-dendritic cell transition, and the blocking effects of α-Gm-CSF. Considering the maximum effect and at which dosage this effect was reached, T4, T3, rT3 and 3,3'T2 proved to be the most potent agents to enhance monocyte-veiled cell transition in comparison to metrizamide. However, 3T1, 3'T1, 3,5T2, 3',5'T2, MIT and DIT were less potent; T0 (the tyrosine backbone molecule) and KI were completely unable to enhance the maturation of dendritic cells from monocytes (see table 1 and fig 1). The veiled cells derived from the blood monocytes were functionally active in that they were capable to form clusters with allogeneic peripheral blood lymfocytes (see table 1). A good correlation was found (r=0.67) between the percentage of veiled cells counted directly in suspension (wet preparations) and the percentage of MHC class-II positive, acid phosphatase negative dendritic cells counted in the cytocentrifuge preparations . The maturation-enhancing effect of the iodinated compounds and hormones could be blocked completely by the monoclonal antibody α-GM-CSF (5 μg/ml being optimal). α-G-CSF did not block the enhanced monocyte-veiled cell transition (data not shown).

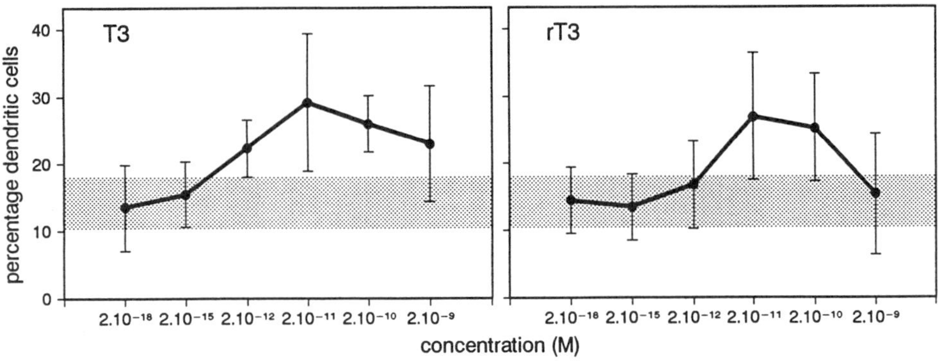

Figure 1. Dendritic cell maturation.
Human bloodmonocytes were incubated for 30 min. with T_3 and rT_3 at 37 °C. After washing and a further overnight culture under nonadhering conditions the percentage of veiled cells was determined. The hatched area represents the percentage veiled cells spontanuously derived from blood monocytes.

DISCUSSION

This study shows that a short pretreatment of blood monocytes (30 min.) with the thyroid hormones T_4 and T_3, with rT_3 and with the iodinated degradation product 3,3'T_2 enhance the capability of monocytes to mature into dendritic/veiled cells when cultured overnight under nonadhering conditions. This enhanced capability is mediated by GM-CSF, since α-GM-CSF blocked the effect.

Whether these monocyte-derived dendritic/veiled cells are fully active as splenic dendritic cells and capable of stimulating naive T cells[6] need to be elucidated. It is however clear that they do form clusters with allogeneic blood lymfocytes.

One can only speculate about the mechanisms which are involved in the enhancement of monocyte to dendritic/veiled cell transition. The active thyroid hormone T_3 may act via its nuclear receptor to stimulate cell metabolism in general. However, this does not explain why rT_3 (an inactive form of T_3 for which there is no nuclear receptor) and T_4 (a prohormone) have similar effects. The various degradation products, showing a gradual decrease in iodine content, also more or less show a decrease in their capability to stimulate the monocyte to dendritic/veiled cell transition. Tyrosine without iodinated groups lacks an effect, as do I^- ions. These observations point in the direction that iodonium groups (I^+) may be irritants for monocytes and it is tempting to speculate that these active redox groups are capable of interfering with the H_2O_2 generating system in the monocytes. Wathever the direct trigger mechanisms for the monocytes may be, our data show that GM-CSF is involved as a second factor. Whether this involves a higher production of GM-CSF by the cells or an upregulation of the GM-CSF receptor needs further clarification.

Table 1. Dendritic/veiled cell maturation.

	maximum effect	dosage	clusters/6 micr fields x160*
culture fluid	14%	-	239
metrizamide	26%	2.10^{-1}M	201
T0	13%	2.10^{-11}M	241
3T1	23%	2.10^{-10}M	250
3'T1	23%	2.10^{-11}M	270
3,5T2	21%	2.10^{-10}M	n.d.
3',5'T2	24%	2.10^{-10}M	230
3,3'T2	28%	2.10^{-11}M	247
T3	29%	2.10^{-11}M	245
rT3	27%	2.10^{-11}M	230
T4	25%	2.10^{-11}M	235
MIT	27%	2.10^{-6}M	n.d.
DIT	27%	2.10^{-9}M	n.d.

The maximum percentage of dendritic/veiled cells derived from monocytes after a 30 min. incubation with several iodinated compounds or hormones and an overnight culture period. The optimal dosages of the stimulus at which this maximal percentage is reached is also given.
* Functional assay for dendritic cells. The obtained dendritic cells were allowed to form clusters with allogeneic lymphocytes (normal values>125). n.d.: not done.

REFERENCES

1. J.H. Peters, S. Ruhl, and D. Friedrichs, Veiled accessory cells deduced from monocytes, *Immunobiol.* 176:154 (1987).
2. P.J. Kabel, M. de Haan-Meulman, H.A.M. Voorbij, M. Kleingeld, E.F. Knol and H.A. Drexhage, Accessory cells with a morphology and marker pattern of dendritic cells can be obtained from elutriator-purified blood monocyte fractions. An enhancing effect of metrizamide in this differentiation, *Immunobiol.* 179:395 (1989).
3. M. de Boer and D. Roos, Metabolic comparison between basophils and other leucocytes from human blood, *J. Immunol.* 136:3447 (1986).
4. L.W. Poulter, D.A. Campbell, C. Munro and G. Janossy, Discrimination of human macrophages and dendritic cells by means of monoclonal antibodies, *Scand. J. Immunol.* 24:351 (1986).
5. R.A. Scheeren, G. Koopman, S van der Baan, C.J.L.M. Meijer and S. Pals, Adhesion receptors involved in clustering of blood dendritic cells and T lymphocytes, *Eur. J. Immunol.* 21:1101 (1991)
6. J.M. Austyn, D.E. Weinstein and R.M. Steinman, Clustering with dendritic cells precedes and is essential for T-cell proliferation in a mitogenesis model, *Immunol.* 63:691 (1988).

RELATIONSHIP BETWEEN DENDRITIC CELLS AND FOLLICULO-STELLATE CELLS IN THE PITUITARY: IMMUNOHISTOCHEMICAL COMPARISON BETWEEN MOUSE, RAT AND HUMAN PITUITARIES

W. Allaerts[1], P.H.M. Jeucken[1], F.T. Bosman[2] and H.A. Drexhage[1]

Departments of [1]Immunology and [2]Pathological Anatomy, Erasmus University, Rotterdam, The Netherlands

INTRODUCTION

Interactions between the endocrine and immune systems have been described in terms of a variety of phenomena, occurring at different organismal levels, including the anterior pituitary (AP). Vankelecom *et al.* (1) previously ascribed a prominent role in immune-endocrine interaction to the folliculo-stellate (FS) cells, based on the production by FS cells of the cytokine interleukin-6 (IL-6). FS cells constitute a network of stellate-shaped non-hormone secreting cells within the AP (2). We previously demonstrated that FS cells, that are positive for the neuroectodermal marker S-100 (3,4), play a role in intercellular communication with hormone producing cells, generally resulting in the attenuation of secretory responses following stimulation of these cells (5,6). Otherwise, the S-100 protein has also been shown to be marker of dendritic Langerhans cells in the skin (7), and of the so-called interdigitating reticulum cell of the lymph node (8). We recently demonstrated in the normal pituitary of the mouse and the rat, a network of lymphoid dendritic cells (DC), positive for a lymphoid DC specific aminopeptidase in the mouse and expressing MHC class II determinants in both species (9).

In this report we describe the presence of lymphoid cell markers on these DC's in the normal mouse, rat and human pituitaries. Lymphoid DC markers and the S-100 protein are compared in three different rat strains, namely the Wistar, Lewis and autoimmune-prone BB/R rat. Human pituitaries were obtained by surgical obduction from patients that died from different non-endocrine causes.

The simultaneous occurrence during development in the AP of IL-6 producing (1,10) and interferon-γ responsive FS cells (11), and lymphoid DC expressing MHC class II determinants and various other lymphoid markers, elicits the question whether FS cells and pituitary DC belong to the same cell lineage, and/or whether both cell groups are functionally related.

Dendritic Cells in Fundamental and Clinical Immunology,
Edited by Kamperdijk *et al.*, Plenum Press, New York, 1993

MATERIALS AND METHODS

Human Material and Animals

Human pituitaries were obtained by surgical obduction from 4 male and 2 female subjects, that died from different non-endocrine causes. Pituitary fragments were embedded in Tissue-Tek (Miles Inc., Elkhart, Illinois) and stored frozen at -20°. Whole pituitaries from Balb/c mice and anterior pituitary lobes from BB/R, Lewis and Wistar rats were excised and prepared for immunohistochemistry or cell culture and separation techniques.

Immunohistochemistry

Frozen tissue was cut in 4-6 μm sections using a Leitz 1720 cryostat (Wild Leitz, Wetzlar, Germany). The sections were dried for 2-3 h at room temperature and then fixed with either acetone for 10 min (Merck, Darmstadt, Germany) or with Zamboni fluid for 2 h (12,13). Sections were stained with monoclonal antibodies (Moab) directed against murine, rat or human lymphoid markers (Table 1), as well as with a polyclonal rabbit antiserum to bovine S-100 (Dakopatt's, Glostrup, Denmark). Single immunostaining procedures consisted of an indirect peroxidase-labeled antibody method or peroxidase-anti-peroxidase method. Double immunostaining was performed using a combination of one of the former methods with an indirect alkaline phosphatase-labeled antibody method, using specific blocking steps with normal animal sera of the same species as used for raising the conjugated antibody.

Cell Preparation and Separation Techniques

Anterior pituitary lobes from BB/R and Wistar rats were cut into small tissue blocks with a razor blade, and further enzymatically dissociated into single cells according to the method of Denef *et al.* (13). The suspension of dissociated cells was separated into cell populations with differing proportions of FS cells, lactotrophs and gonadotrophs by velocity sedimentation at unit gravity in a linear gradient of bovine serum albumin (BSA; Sewa, Heidelberg, Germany) (13). The sedimentation pattern of FS

Table 1. Pattern of immunohistochemical staining with monocyte-dendritic cell-macrophage specific monoclonal antibodies.

Moab	Antigen	Spleen/ lymph node DC	Pituitary cells: stellate-shaped	Pituitary cells: non stellate-shaped
Mouse				
BMDM 1	aminopeptidase	++	++	—
M5/114	class II, MHC	++	++	—
M1/42	class I, MHC	++	++	—
30G12	T200 antigen	+	—	+
Rat				
OX-6	class II, MHC	++	++	—
Human				
CD11a	LFA-1 antigen (leucocyte function antigen)	++	—	±
CD13	aminopeptidase N	+	+	+
CD14	pi-linked protein (monocyt antigen)	++	++	—
CD45	T200 antigen (leucocyte common antigen)	++	—	+
OKIa	class II, MHC	++	++	—
L25	unknown	++	++	—
ZP214	PDGF-α chain	+	+	++

CD: cluster of differentiation, nomenclature as defined during Leukocyte Typing Conference.
++: numerous cells with distinct positive staining; +: few cells with distinct positive staining; ±: weak reactivity; —: no reactivity.

cells, lactotrophs and gonadotrophs (14) was similar to previous gradient preparations (5). Cells were suspended in serum-free defined culture medium (15) and allowed to reassociate into aggregates on a Bellco orbital Shaker (Vineland, New Jersey) in a humidified CO_2(1%)-air incubator at 37°C (6,16). Cytospin preparations using the Cytofuge centrifuge (Nordic Immunological Laboratories, Tilburg, The Netherlands) from freshly prepared gradient fractions or from re-aggregated cell populations, were immunostained with a Moab against rat MHC class II determinants (OX-6 antiserum; Sera-lab, Sussex, U.K.) to determine the sedimentation pattern of class II expressing cells. A magnetic cell separation system (MACS; Miltenyi Biotec GmbH, Bergisch Gladbach, Germany) was used to prepare MHC class II-enriched and MHC class II-free cell populations.

RESULTS

a. Phenotypical Analysis of Lymphoid DC and FS Cells in the Anterior Pituitary of Mouse, Rat and Human

Stellate-shaped cells positive for MHC class II determinants were found throughout the mouse, rat (Figure 1A) and human (Figure 2A) anterior pituitaries (Table 1). In the rat, MHC class II expressing DC were found to be more numerous in the BB/R strain pituitaries, than in Wistar or Lewis rat pituitaries. However, the number of MHC class II expressing cells was only a fraction of the number of S-100

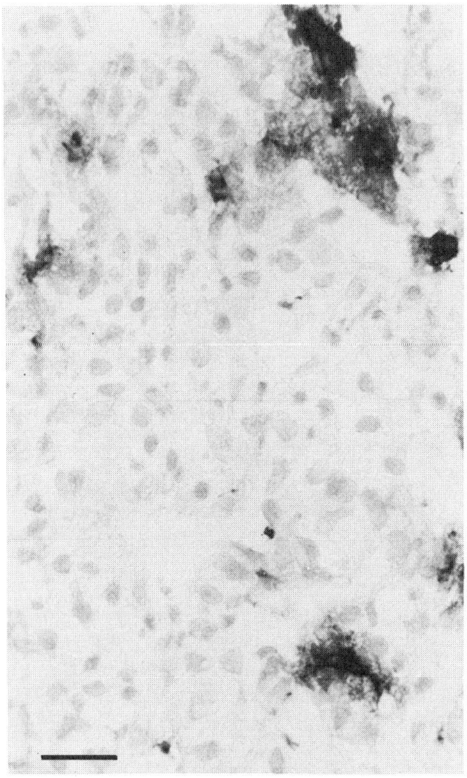

Figure 1A. Cryo-section of an adult female Wistar rat pituitary immunostained with the OX-6 Moab (MHC class II determinant). Stellate-shaped cells expressing MHC class II determinants bearing cytoplasmic extensions are found throughout the anterior pituitary.

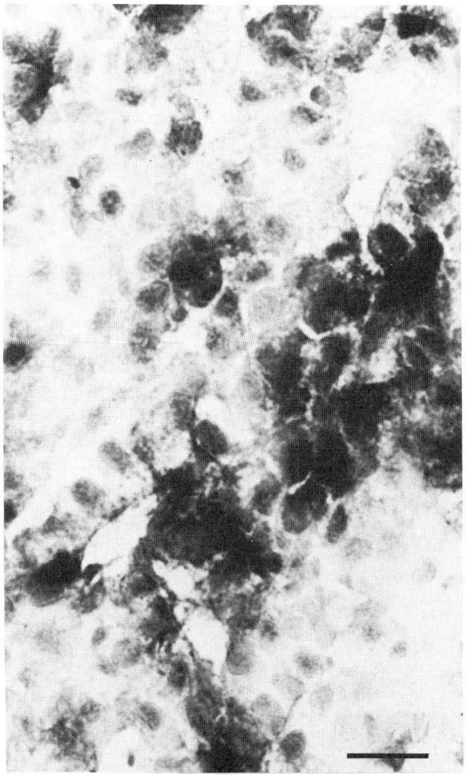

Figure 1B. Cryo-section of an adult female Wistar rat pituitary immunostained with the polyclonal anti-S-100 protein-serum, the common marker of pituitary folliculo-stellate cells. Sections are post-fixed with Zamboni fixative for 4 h (space bar = 20 μm).

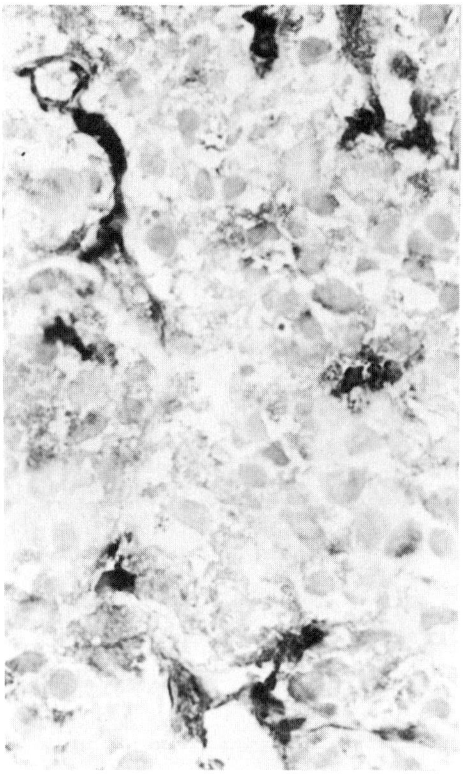

Figure 2A. MHC class II positive stellate cells in a human anterior pituitary obtained by surgical obduction.

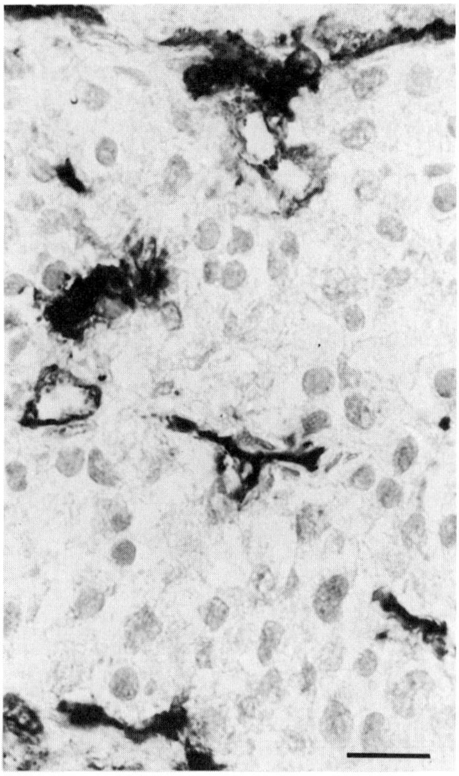

Figure 2B. CD14 antigen in a human pituitary obtained by surgical obduction. Typically a meshwork pattern surrounding cell chords is found (space bar = 20 μm).

positive FS cells (Figure 1B). The immunoreactivity to the S-100 protein in FS cells, and the number of MHC class II expressing DC showed a markedly synchronized increase in the ageing pituitary of female rats (17). Whereas in the Wistar and Lewis strains, a plateau level of S-100 expressing FS cells was reached in about 6 weeks, in the BB/R rats both S-100 and MHC class II expressing cell numbers increased still thereafter (17).

Double immunolabeling of rat pituitary cells with an antiserum to the S-100 protein and the OX-6 Moab (MHC class II) was unsuccessful because of the incompatibility of either antisera or fixation procedures. Frozen sections fixed with acetone alone showed poor S-100 immunoreactivity, whereas strong fixatives containing picric acid (Zamboni fluid) were detrimental to MHC class II immunoreactivity.

DC markers found in the pituitary of Balb/c mice were a DC aminopeptidase BMDM-1 (9) and MHC class I and class II antigens (Table 1). In the human pituitaries obtained from surgical obduction, stellate shaped cells positive for CD14 (Figure 2B), CD13 and L25 antigens were found that are reminiscent of lymphoid DC. Scattered stellate cells positive for the platelet-derived-growth factor α-chain (PDGF-α) were found within the parenchymal cords, separated by PDGF-α positive fibrous septal and endothelium-lined capillaries. Due to the lag time before obduction was performed (between 6 and 24 h), a possible effect of autolytic processes on the latter immunoreactivity can't be excluded on behalf of the present data.

b. Functional similarities between DC and FS cells in the rat

Using a sedimentation gradient at 1 *g* of Wistar rat pituitary cells, we found that MHC class II expressing cells were exclusively located in the upper two gradient fractions, corresponding to the fractions most enriched in S-100 positive FS cells (14). In the other gradient fractions, also in the bottom fractions that contain some 5-15% FS cells in clusters, MHC class II expressing cells were absent. MHC class II immunoreactivity remained when cells were cultured in 3D-aggregates. Using the MACS-separation technique, cell populations containing up to 61% MHC class II positive cells were prepared from adult female Wistars. In the latter cell populations, only 11% were found to be immunoreactive to S-100 protein. In the fraction devoid of MHC class II expressing cells, approximately 8% of the cell population was S-100 positive.

DISCUSSION

Although direct immunocytochemical evidence is lacking, based on double labeling of one cell type for both the S-100 protein, the common immunocytochemical marker for pituitary FS cells, and various markers for lymphoid DC, especially the markers for MHC class II determinants, there is increasing evidence in favor of a close relationship between the two differently defined cell groups. The similarities with respect to morphology, topographical distribution and sedimentation characteristics between FS cells and pituitary DC (present data) most conspicuously point towards a functional relationship between the two cell groups. Our data in the rat demonstrate a lower number of MHC class II expressing cells than the number of S-100 positive FS cells. The sedimentation pattern of MHC class II expressing cells was similar to the FS cell sedimentation pattern in a velocity gradient at 1 *g*, although S-100 positive cell clusters in the bottom fraction did not stain for MHC class II. This finding corroborates the findings of Vankelecom *et al.* (1,11), who found that S-100 immunoreactivity in different gradient preparations is highly correlated with IL-6 release, but not in as much with the magnitude of the inhibition of CRF-induced ACTH release by IFN-γ, which inhibition the authors believe is mediated by FS cells. The latter arguments strongly support the notion of functional heterogeneity within the group of FS cells, that previously has been shown to be heterogeneous with respect to immunocytochemical (18,19) and ultrastructural characteristics (20).

Since S-100 immunoreactivity was also demonstrated in lymphoid DC such as the Langerhans cells in the skin (7), and in view of the numerous stellate-shaped cells bearing DC markers in the pituitary of mouse, rat and man (present data), a subpopulation of the S-100 positive FS cells may belong to the lymphoid DC group. This would mean that these FS cells derive from three distinct Anlage, to with the neuroectodermal and (oral) ectodermal Anlage (19) and the (lymphoid) mesenchymal Anlage.

The multiple ontogenetic origin of FS cells, and their intermingling with lymphoid DC, makes it preferable to speak about a dendritic cell folliculo-stellate cell (DC-FSC) 'family' at the anterior pituitary level. Since these cells form a morphological network and since a functional interrelationship between the cell groups is suggested by the synchronized increase of S-100 protein and MHC class II determinants expressing cells during development, it is indeed worthwhile to speak about a cell 'family'. The increased MHC class II expression and S-100 immunoreactivity in the pituitaries of autoimmune-prone BB/R rats (17) raises further questions concerning a possible role of DC-FSC in the development of autoimmune endocrine disorders.

ACKNOWLEDGMENTS

This work is supported by NWO-MEDIGON (grant number 900-543-089). We thank Prof. Dr. W. van Ewijk and Dr. P.J.M. Leenen (Erasmus University Rotterdam) for their generous gift of Moab against mouse lymphoid markers. Mr. T.M. van Os and Mrs. H.J. Elsenbroek-de Jager are acknowledged for the photographic and typographic work, respectively.

REFERENCES

1. H. Vankelecom, P. Carmeliet, J. van Damme, A. Billiau, and C. Denef, Production of interleukin-6 by folliculo-stellate cells of the anterior pituitary gland in a histiotypic cell aggregate culture system, *Neuroendocrinology* 49:102 (1989).
2. E. Vila-Porcile, Le réseau des cellules folliculo-stellaires et les follicules de l'adénohypophyse du rat (pars distalis), *Z. Zellforsch.* 129:328 (1972).
3. D. Cocchia and N. Miani, Immunocytochemical localization of the brain-specific S-100 protein in the pituitary gland of adult rat, *J. Neurocytol.* 9:771 (1980).
4. T. Nakajima, H. Yamaguchi, and K. Takahashi, S100 protein in folliculo stellate cells of the rat of the pituitary anterior lobe, *Brain Res.* 191:523 (1980).
5. M. Baes, W. Allaerts, and C. Denef, Evidence for functional communication between folliculo-stellate cells and hormone-secreting cells in perifused anterior pituitary cell aggregates, *Endocrinology* 120:685 (1987).
6. W. Allaerts and C. Denef, Regulatory activity and topological distribution of folliculo-stellate cells in rat anterior pituitary cell aggregates, *Neuroendocrinology* 49:409 (1989).
7. D. Cocchia, F. Michetti, and R. Donato, Immunochemical and immunocytochemical localization of S-100 antigen in normal human skin, *Nature* 294:85 (1981).
8. K. Takahashi, H. Yamaguchi, J. Ishizeki, T. Nakajima, and Y. Nakazato, Immunohistochemical and immunoelectron microscopic localization of S-100 protein in the interdigitating reticulum cells of the human lymph node, *Virchows Arch. [Cell Pathol.]* 37:125 (1981).
9. W. Allaerts, P.H.M. Jeucken, L.J. Hofland, and H.A. Drexhage, Morphological, immunohistochemical and functional homologies between pituitary folliculo-stellate cells and lymphoid dendritic cells, *Acta Endocrinol. (Copenh.)* 125:92 (1991).
10. P. Carmeliet, H. Vankelecom, J. van Damme, A. Billiau, and C. Denef, Release of interleukin-6 from anterior pituitary cell aggregates: development pattern and modulation by glucocorticoids and forskolin. *Neuroendocrinology* 53:29 (1991).
11. H. Vankelecom, M. Andries, A. Billiau, and C. Denef, Evidence that folliculo-stellate cells mediate the inhibitory effect of interferon-γ on hormone secretion in rat anterior pituitary cell cultures, *Endocrinology* (in press).
12. L. Zamboni and C. de Martino, Buffered picric acid-formaldehyde: a new, rapid fixative for electron microscopy. *J. Cell Biol.* 35:148A (1967).
13. C. Denef, E. Hautekeete, A. de Wolf, and B. van der Schueren, Pituitary basophils from immature male and female rats: distribution of gonadotrophs and thyrotrophs as studied by unit gravity sedimentation, *Endocrinology* 103:724 (1978).
14. W. Allaerts, P.H.M. Jeucken, A.M.I. Tijssen, D.W. Koppenaal, and J. de Koning, Tissue-like integrity and primability to LHRH of rat anterior pituitaries *in vitro*, *submitted to J. Endocrinol* (1992).
15. M. Baes and C. Denef, Evidence that stimulation of growth hormone release by epinephrine and vasoactive intestinal peptide is based on cell-to-cell communication in the pituitary, *Endocrinology* 120:280 (1987).
16. B. van der Schueren, C. Denef, and J.-J. Cassiman, Ultrastructural and functional characteristics of rat pituitary cell aggregates, *Endocrinology* 110:513 (1982).
17. W. Allaerts, P. Jeucken, and H.A. Drexhage, The dendritic cell-folliculo stellate cell-family in the anterior pituitary in normal Wistar and autoimmune-prone BB/R rats, *in prep.*
18. F. Nakagawa, B.A. Schulte, and S.A. Spicer, Glycoconjugate localization with lectin and PA-TCH-SP cytochemistry in rat hypophysis, *Am. J. Anat.* 174:61 (1985).
19. O. Tachibana and T. Yamashima, Immunohistochemical study of folliculo-stellate cells in human pituitary adenomas. *Acta Neuropathol.* 76:458 (1988).
20. A. Scarbati, C. Zancanaro, S. Cinti, and F. Osculati, Marginal and folliculo-stellate cells of the pituitary gland of the rat. A comparative morphometric study in lactating animals, *Acta Anat.* 131:47 (1988).

DENDRITIC CELLS IN TUMOR GROWTH AND ENDOCRINE DISEASES

H.A. Drexhage, P. Mooy, A. Jansen, J. Kerrebijn, W. Allaerts and M.P.R. Tas

Dept. of Immunology, Erasmus University, Rotterdam, the Netherlands

The role of dendritic cells and macrophages in the contol of tumour growth

It is well established that dendritic cells and macrophages play an important role in the defence against tumour cells. Using monoclonal antibodies (moabs) specific for monocytes, macrophages, and dendritic cells many authors have observed these cells infiltrating neoplastically transformed tissues.

Over the past few years we have carried out investigations on the presence and histological localization of dendritic cells and macrophages in various neoplastic conditions, most notably in head and neck squamous cell carcinoma. In these studies the scavenger macrophages were identified on the basis of their morphology (round, ruffled, large cells), enzyme histochemistry (strong acid-phosphatase positivity) and marker pattern (diffuse CD68 positivity). Cells were regarded as dendritic cells on the basis of a clear dendritic morphology, negativity (or weak positivity) for acid phosphatase, strong positivity for MHC-class II molecules, and positivity for the moab markers RFD1 and L25. In the head and neck cancers studied cells meeting our criteria for dendritic cells were found inside nests of malignant epithelial cells (a few single cells), but most predominantly in the stromal compartment of the tumours, sometimes in contact with T lymphocytes, suggesting their role in a local antigen presentation and stimulation of tumour immunity. These dendritic cell-lymphocyte clusters sometimes formed extensive areas of lymphoid cell infiltration showing a certain degree of architecture (CD4:CD8=4:1).

Dendritic cells are thought to play a role in tumour defence by presenting tumour-associated antigens to virgin T cells to enable the cells to respond and build up a specific T-cell mediated immune response to tumour cells. This putative function of the dendritic cell was actually proven by Guyere et al. (1), who showed that immunity to a syngeneic sarcoma could be induced in rats by the transfer of dendritic lymph cells exposed to tumour antigens. Also in this proceedings there is prove of the tumour antigen presenting function of dendritic cells (Flamand et al). With regard to the infiltration of dendritic cells in tumours, Nomori et al. (2) showed that patients with a dense infiltration of dendritic cells in primary sites of

Dendritic Cells in Fundamental and Clinical Immunology,
Edited by Kamperdijk *et al.*, Plenum Press, New York, 1993

head and neck cancers survived longer than those without such infiltration. In the same study it was shown that there was no such relationship between prognosis and density of lysosome-positive macrophages at the site of the tumour. In papillary thyroid carcinoma the presence of S-100 positive dendritic cells has also been associated with a favourable prognosis (3) : Irrespective of other morphological and clinical features, no single instance of death resulting from cancer occurred among 23 papillary carcinoma patients with dense dendritic cell infiltrates, while 9 of 53 (17%) of the remaining patients ultimately died from thyroid cancer (3).

Scavenger macrophages were also numerous in our head and neck carcinomatous lesions studied, particularly inside the parenchyma of the tumours, suggesting a role in tumour cell destruction.

With regard to the effects of scavenger macrophages on cancer cells the data from the literature are confusing. It is clear that macrophages can exert cytotoxic/cytolytic effects on cancer cells both in vitro and in vivo; but significant tumoricidal effects have only been found when the macrophages had been activated by lymphokines or by bacterial products, like MDP incorporated into lysosomes (4). Mediators operative in such enhanced tumoricidal action are cytokines such as Tumour Necrosis Factor (TNF, 5). It must be noted, however, that some tumour cells may become insensitive to cytolysis by the lymphokine-activated macrophages because they are capable of building up resistance to TNF by exposing less TNF receptors on their surface (6).

Despite this cytotoxic potential of macrophages, Allen and Hogg (7) found no histological evidence for an in vivo cytotoxic role for macrophages in specimens of colorectal tumours. They did find a significant alteration of the mononuclear phagocyte population within the tumour, particularly in metastasizing tumours, i.e. more cells and more activated cells expressing the C3b receptor. In murine sarcomas and carcinomas the content of tumour infiltrated macrophages was even negatively correlated to tumour take, and the cells were not found to be cytotoxic, but even to stimulate the growth of the sarcoma cells in vitro (8). A growth promoting role of tumour-infiltrated macrophages has also been suggested by Alexander et al. (9), who found a preferential growth of blood-borne cancer cells at sites of trauma. The authors liked to believe that this was due to the effect of polypeptide growth factors (EGF, PDGF, etc.) released by the macrophages accumulated at the traumatic site to promote tissue repair.

Since macrophages and dendritic cells are capable of anti-tumor activity (be it depending on the circumstances) it is of importance that defects in the function of these cells have been found in patients with squamous cell carcinomas. A defect in tumoricidal capability of macrophages was reported by Cameron et al. (10). A clear impairment in the chemotactic capability of blood monocytes has been reported by us on several occasions (fig. 1), and this defect in monocyte chemotaxis in patients with squamous cell carcinomas of the head and neck has been confirmed by others (12). The latter authors additionally found that the number of chemoattractant (formylpeptide) binding sites on patient monocytes was increased despite the defective function (10). The defects in monocyte chemotaxis could be ascribed to the presence in serum of a tumor-derived low molecular weight factor (LMWF; <25 kD) capable of suppressing monocyte polarization (11, an early event prior to chemotaxis). This immunosuppressive factor produced by the tumor shares structural homology with p15E, the capsular protein of murine and feline leukemogenic retroviruses.

Bugelski et al. (13) have reported that in several animal models the recruitment of monocytes/macrophages into primary sites or metastases is decreased as tumor mass or the metastases enlarge. This resulted in a decrease in the density of tumor infiltrated macrophages. Whether our findings of a disturbed chemotactic capability of blood monocytes in carcinoma patients also results in a decrease in the density of human infiltrated macrophages and dendritic cells into the tumor site needs further investigation.

Defects in the clustering of peripheral blood dendritic cells of carcinoma patients has also been detected (fig. 1), and it is likely that these defects are also caused by the tumour-derived p15E like factors since isolated p15E like factors are capable to interfere with the cluster capability of healthy donor blood dendritic cells (14). The dendritic cell defects in the cancer patients are in line with earlier findings of disturbed cell-mediated immune functions in carcinoma patients, in which dendritic cells play a role. There are numerous reports on poor

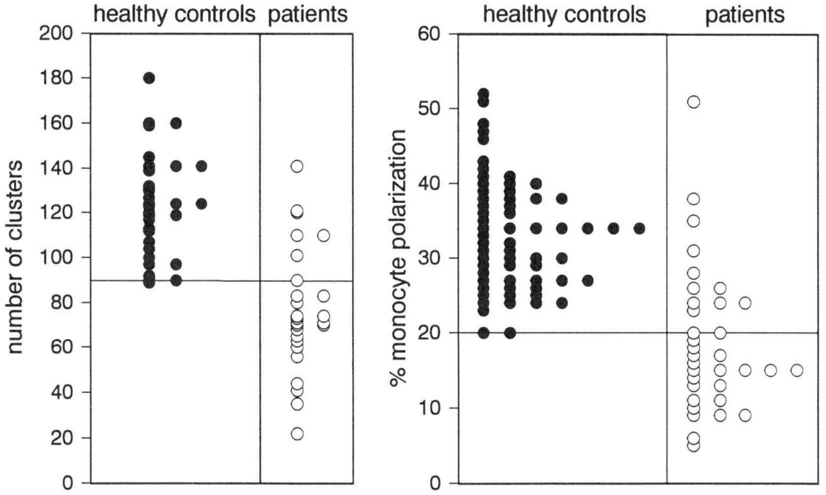

Fig. 1 The number of clusters of dendritic cells with allogeneic lymphocytes/six microscopic fields (x 200), and the fMLP induced polarization of blood monocytes in healthy donors and head and neck carcinoma patients. Note the lower number of dendritic cells capable of clusterformation and the lower number of chemoatractable monocytes in the carcinoma patients. For techniques see ref. 14.

responsiveness to DNCB sensitization (15), and diminished *in vitro* lymphocyte mitogenic responses (16). Whether the defective cluster capability of peripheral blood dendritic cells in carcinoma patients correlates with a defective infiltration and/or defective function of such cells in the tumor itself needs further investigation.

The p15E-like tumor factor that influences the monocyte chemotaxis and to a lesser extent the dendritic cell clustering, has been known since 1981. In that year it was demonstrated by Cianciolo et al. (17) that the inhibitory effect on monocyte chemotaxis was due to a LMWF produced by both animal and human malignant cells and that the effect of the LMWF could be adsorbed by any of three different moabs to the retroviral capsular protein, p15E. Later it became known that the suppressive effects of the p15E-like factors - and of a 17-amino-acid peptide synthesized from the highly conserved region of p15E (CKS-17) - are not restricted to monocyte chemotaxis, but that these factors have also suppressive effects on to monocyte chemotaxis, but that these factors have also suppressive effects on various other immune functions, including monocyte mediated killing (18), the respiratory burst of human monocytes (19), and the natural killer cell activity of NK cells (20).

The presence and the role of dendritic cells in endocrine tissues, most notably the anterior pituitary, and in endocrine autoimmune diseases

Recently we made some observations in studies on the presence of dendritic cells in endocrine tissues, which make an "endocrine function" of dendritic cells likely. In the normal anterior pituitary (human, rat, mouse) we surprisingly found networks of MHC class II positive dendritic cells with a marker pattern indentical to that of Langerhans cells and of lymphoid dendritic cells (spleen, thymus, lymphnodes). The pituitary dendritic cells lacked, however, markers specific for "immunologically active" dendritic cells (human: RFD-1, mouse: MP23). Characteristic for the pituitary dendritic cells was their positivity for markers such as BMDM-1 (mouse), CD14, and CD1 (human). A function in antigen-presentation of the pituitary dendritic cells can, however, not easily be assumed (there is not even a lymphatic drainage from the pituitary).

The localization and distribution pattern of the pituitary dendritic cells appeared similar to that of the earlier described S-100 positive pituitary folliculo-stellate cells (FS cells) (21). From subsequent studies isolating the cells from the pituitary it was further confirmed that the pituitary dendritic cells were indeed closely related, if not identical to the FS cells: they were found in the same isolate, and showed overlap in double staining techniques. Moreover,the cells had the same distribution pattern and quantity in aging Wistar and BB rats, and FS cells have been described to be positive for cytokines and growth factors which we and others also identify in monocytes and dendritic cells (IL-6, FGF, PDGF).

The function of the FS cells in the pituitary has been well documented, namely regulation of endocrine responses. We reestablished that fractionated pituitary dendritic/stellate cells buffer excessive hormonal responses of pituitary endocrine cells (22-23) in culture systems of reaggregated pituitary cells. Growth responses can also be measured by BrdU incorporation, and dendritic/stellate cells have been described to regulate the growth responses of pituitary endocrine cells by virtue of their growth factor/cytokine production and their physiological aggregation with the surrounding endocrine cells, while regulating the ionic composition (Ca^{2+}) of the local milieu of the aggregates (21).

Our study on stellate/dendritic cells in the pituitary examplifies a possible growth and function regulatory role of dendritic cells on neighbouring endocrine cells, and it can be envisaged that dendritic cells accumulated in endocrine organs other than the pituitary have such regulating functions. This is further suggested by the following observations:

(a) Dendritic cells can be found in the normal thyroid of man and animals in relatively low numbers. However, during iodine-deficiency-induced-metabolic-changes there is a clear accumulation of dendritic cells in the thyroid in parallel with the severity of the induced iodine deficiency and the enhanced metabolic activity of the thyrocytes. The accumulated dendritic cells form clusters with themselves and neighbouring thyrocytes. In iodine-deficient Wistar rats the thyroidal dendritic cell reaction is followed by the appearance of thyroid autoantibodies in the circulation as well as some minor T-lymphocytic infiltration in the thyroid. Similar observations - namely dendritic cell accumulation and thyroid autoantibody production have been found in humans during iodine-deficiency (24, 25).

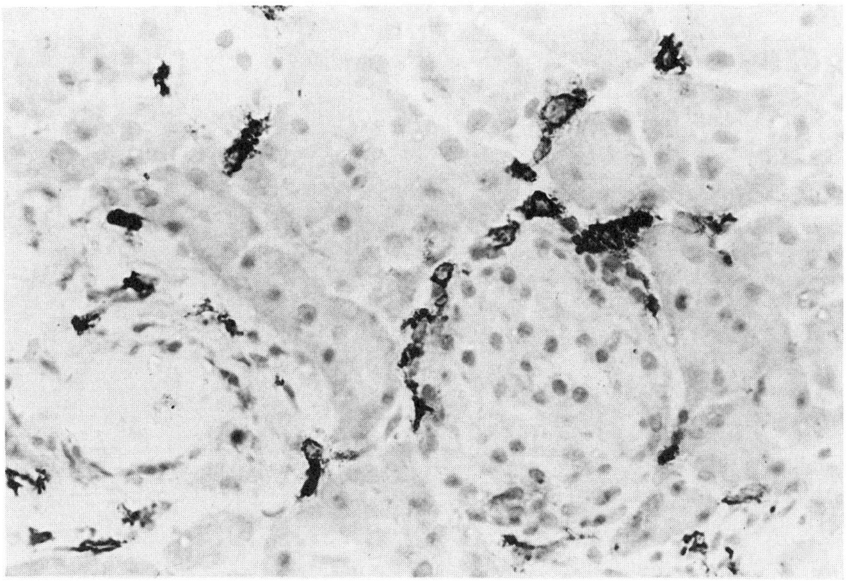

Fig. 2 MP 23 positive dendritic cells accumulating around an islet of Langerhans in a female NOD mouse. This accumulation of MP 23^+ dendritic cells is the earliest sign of a developping islet specific autoimmune reaction.

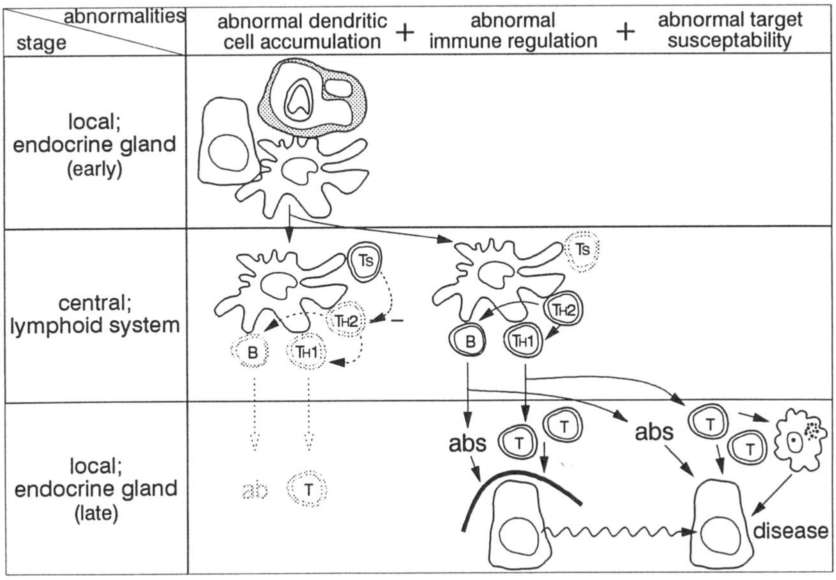

Fig. 3 A scheme showing that endocrine autoimmune diseases are multistage processes involving early and late events, both local and central. See text for explanation. The early local accumulation of dendritic cells is hypothesized to be meant for endocrine regulation. If there is also an abnormallity in immune regulation an excessive production of autoantibodies and autoreactive T cells will be possible. The role of an abnormal target susceptibility is particularly nicely demonstrated in the OS-chicken (34).

(b) Dendritic cells can be found in the normal ovary. The cells are situated particularly in the vicinity of steroid producing cells, namely in the theca layer, forming a little network and disappearing after luteinization. Thereafter their number increases in the corpus luteum (own observations, 26).
(c) It is known that cytokines produced by monocytes/dendritic cells (f.i. IL-6, TNFα, TGFβ) have damping effects on the function and growth of thyroid cells and steroid-producing cells (27).

In autoimmune diseases of endocrine organs such as Graves' disease, Hashimoto goitre and type 1 diabetes increased numbers of dendritic cells can be found in the affected organs (28-30). In fact an increase in the number of dendritic cells and a clustering of such cells is the *very first sign* of a developing endocrine autoimmune reaction (31-31). This local accumulation of dendritic cells precedes the reaction of the draining lymph nodes, the production of endocrine autoantibodies by these lymph nodes and further signs and symptoms of the endocrine disease (31). It is therefore likely that the early and locally accumulated dendritic cells take up autoantigens, travel with these to the draining lymph nodes and stimulate the immune system precipitating the endocrine autoimmune disease. In support of this view insulin-granulae containing dendritic cells have been observed in pancreatic lymph nodes (In 't Veld, personal communication). Knight et al. (33) indeed proved that dendritic cells are strong stimulators of endocrine autoimmune reactions by showing that very small numbers of dendritic cells pulsed with thyroglobulin were capable in inducing thyroiditis after transfer to donor Wistar-rats.

Recently we found that the NOD-mouse is a particular good model for the study of the early enhanced accumulation of dendritic cells in endocrine tissues to be affected by autoimmune disease, since there are monoclonal antibodies available with relatively good specificity for mouse lymphoid dendritic cells (f.i. MP23) and mouse classical macrophages (f.i. BM8). The early accumulation of MP23$^+$ dendritic cells in the NOD-insulitis is striking (fig. 2).

In conclusion: Dendritic cells may be considered as normal regulator cells in endocrine responses and they are the initiator cells of endocrine autoimmune reactions. It can be hypothesized that dendritic cells accumulate in endocrine glands primarely to regulate an endocrine abnormality. When this glandular accumulation of regulator cells - that are also capable of antigen presentation - takes place in the context of existing defects in immune tolerance one can easily envisage that due to the higher and professional autoantigen presentation dysbalances between effector and suppressor autoreactive systems are created and excessive autosensitization takes place resulting in an endocrine autoimmune disease (fig.3)

Acknowledgements: The work of the department is supported by various NWO-grants.

REFERENCES

1. L.A. Guyere, R. Barfoot, S. Denham, J.G. Hall, Immunity to a syngeneic sarcoma induced in rats by dendritic lymph cells exposed to the tumour either in vivo or in vitro, *Br J Cancer* 55:17-20 (1987).
2. H. Nomori, S. Watanebe, T. Nakajima, Y. Shimosato, T. Kameya, Histiocytes in nasopharyngeal carcinoma in relation to prognosis, *Cancer* 57:100-105 (1986).
3. S. Schröder, W. Schwarz, W. Rehpenning, T. Löning, W. Böcker, Dendritic/Langerhans cells and prognosis in patients with papillary thyroid carcinomas, *Am J Clin Path* 89:295-300 (1988).
4. I.J. Fidler, 1988 Metastasis, Ciba Foundation Symposium 141: 211. Wiley, Chichester.

5. M. Yamazaki, T. Okutomi, Augmentation of release of cytotoxin from murine bone marrow and peritoneal macrophages by tumor transplantation, *Cancer Res* 49:352-356 (1989).
6. L. Remels, L. Fransen, K. Huygen, P. de Baetselier, Immunological aspects of tumor-macrophage interactions, *Adv Exp Med Biol* 233:49-59 (1988).
7. C. Allen, N. Hogg, Elevation of infiltrating mononuclear phagocytes in human colorectal tumors, *J Natl Cancer Inst* 78:465-470 (1987).
8. L. Milas, J. Wike, N. Hunter, J. Volpe, I. Basic, Macrophage content of murine sarcomas and carcinomas: associations with tumor growth parameters and tumor radiocurability, *Cancer Res* 47:1069-1075 (1987).
9. P. Alexander, P. Murphy, D. Skipper, Preferential growth of blood-borne cancer cells at sites of trauma - a growth promoting role of macrophages, *Adv Exp Med Biol* 233:245-251 (1988).
10. D.J. Cameron, B. Stromberg, The ability of macrophages from head and neck cancer patients to kill tumour cells, *Cancer* 54:2403-2408 (1984).
11. G.J. Cianciolo & R. Snyderman, Monocyte responsiveness to chemotactic stimuli is a property of a subpopulation of cells that can respond to multiple chemoattractants, *J Clin Invest* 67:60-68 (1981).
12. R.J. Walter & J.R. Danielson, Characterization of formylpeptide chemoattractant binding on neutrophils and monocytes from patients with Head and Neck cancer, *JNCI* 78:61-67 (1987).
13. P.J. Bugelski, R. Kirsch, C. Buscarino, S.P. Corwin & G. Poste, Recruitment of exogenous macrophages into metastases at different stages of tumor growth. *Cancer Immunol Immunother* 24:93-98 (1987).
14. M. Tas, M. de Haan-Meulman, P.J. Kabel & H.A. Drexhage, Defects in monocyte polarization and dendritic cell clustering in patients with Graves' disease. A putative role for a non-specific immunoregulatory factor related to p15E. *Clin Endocrinol* 34:441-448 (1991).
15. J. Bier & U. Nicklish, Investigations of unspecific immunoreactivity in patients with head and neck carcinoma, *Arch Otorhinolaryngol* 232:145-163 (1981).
16. W.J. Catalona, W.F. Sample & P.B. Chretien, Lymphocyte reactivity in cancer patients: correlation with tumor histology and clinical stage, *Cancer* 31:65-71 (1973).
17. G.J. Cianciolo, J. Hunter, J. Silva, J.S. Haskill & R. Snyderman, Inhibitors of monocyte responses to chemotaxins are present in human cancerous effusions and react with monoclonal antibodies to the P15(E) structural protein of retroviruses. *J Clin Invest* 68:831-844 (1981).
18. E.S. Kleinerman, L.B. Lachman, R.D. Knowles, R. Snyderman & G.J. Cianciolo, A synthetic peptide homologous to the envelope proteins of retroviruses inhibits monocyte-mediated killing by inactivating interleukin 1. *J Immunol* 139:2329-2337 (1987).
19. R.A. Harrell, G.J. Cianciolo, T. D. Copeland, S. Oroszlan & R. Snyderman, Suppression of the respiratory burst of human monocytes by a synthetic peptide homologous to envelope proteins of human and animal retroviruses, *J Immunol* 136:3517-3520 (1986).
20. D.T. Harris, G.J. Cianciolo, R. Snyderman, S. Argov & H.S. Koren, Inhibition of human natural killer cell activity by a synthetic peptide homologous to a conserved region in the retroviral protein, p15E. *J Immunol* 138:889-894 (1987).
21. W. Allaerts, P. Carmeliet, C. Denef, New perspectives in the function of pituitary folliculo stellate cells, *Molec Cell Endocrinol* 71:73-81 (1990).
22. H. Vankelecom, P. Carmeliet, J. van Damme, A. Billiau, C. Denef, Production of interleukin-6 by folliculo stellate cells of the anterior pituitary gland in a histiotypic cell aggregate culture system, *Neuroendocrinol* 49:102-106 (1989).
23. W. Allaerts, C. Denef, Regulatory activity and topical distribution of folliculo stellate cells in rat anterior pituitary cell aggregates, *Neuroendocrinol* 49:409-418 (1989).
24. M.M. Wilders-Truschnig, P.J. Kabel, H.A. Drexhage, A. Beham, G. Leb, O. Eber, J. Heberstreit, D. Loidolt, G. Dohr, G. Lanzer, G.J. Kreys, Intrathyroidal dendritic cells, epitheloid cells and giant cells in iodine deficient goiter, *Am. J. Pathol.* 135:219-225 (1989).
25. M.M. Wilders-Truschnig, H.A. Drexhage, G. Leb, O. Eber, H.P. Brezinschek, G. Dohr, G. Lanzer, G.J. Kreys, Chromatographically purified IgG of endemic and sporadic goiter patients stimulates FRTL-5 cell growth in a mitotic arrest assay, *J Clin Endocrinol Metab* 70:444-452 (1990).
26. J.A. Loukides, R.A. Loy, R. Edwards, J. Honig, I. Visintin, M.L. Polan, Human follicular fluids contain tissue macrophages, *J. Clin. Endocrinol. Metab.* 71:1363-1367 (1990).
27. K. Bendtzen, Immune hormones (cytokines); pathogenic role in autoimmune rheumatic and endocrine diseases. *Autoimmunity* 2:177-189 (1989).
28. P.J. Kabel, H.A.M. Voorbij, M. de Haan, R.D. van der Gaag, H.A. Drexhage, Intrathyroidal dendritic cells in Graves' disease and simple goitre, *J Clin Endocrinol Metab* 65:199-207 (1988).

29. P.J. Kabel, H.A.M. Voorbij, M. de Haan-Meulman, S.T. Pals, H.A. Drexhage, High endothelial venules present in lymphoid cell accumulations in thyroid affected by autoimmune disease. A study in men and BB rat of functional activity and development. J. Clin. Endocrinol, *Metab* 68:744-751 (1989).
30. H.A.M. Voorbij, P.H.M. Jeucken, P.J. Kabel, M. de Haan, H.A. Drexhage, Dendritic cells and scavenger macrophages in the pancreatic islets of prediabetic BB rats, *Diabetes* 38:1623-1629 (1989).
31. H.A.M. Voorbij, P.J. Kabel, M. Haan de, P. Jeucken, R.D. Gaag van der, M.H. de Baets, H.A. Drexhage, Dendritic cells and class II MHC expression on thyrocytes during the autoimmune thyroid disease of the BB rat, *Clin Immunol Immunopathol* 55:9-22 (1990).
32. L.A. O'Reilly, P.R. Hutchings, P.R. Crocker, E. Simpson, T. Lund, D. Kioussis, F. Takei, J. Baird, A. Cooke, Characterization of pancreatic islet cell infiltrates in NOD-mice: effect of cell transfer and transgene expression, *Eur J Immunol* 21:1171-1180 (1991).
33. S.C. Knight, J. Farrant, J. Chan, A. Bryant, P.A. Bedford, C. Batman, Induction of autoimmunity with dendritic cells: studies on thyroiditis in mice, *Clin Immunol Immunopathol* 48:277 (1988).
34. G. Wick, H.P. Brezinschek, K. Hala, H. Dietrich, H. Wolf, G. Kroemer, The obese strain of chickens: an animal model with spontaneous autoimmune thyroiditis, *Adv Immunol* 47:433-500 (1989).

INDEX

Acid phosphatase activity, 17, 141, 571, 577
Adherence assay, 563
Adhesion molecules, 495
CD43, 71
ICAM-1, 65, 87, 93, 333, 377, 387
ICAM-2, 65
ICAM-3, 65
LFA-1, 65, 87, 93, 333, 387, 637
LFA-2, 65
LFA-3, 65, 87, 93
VCAM-1, 387
VLA-4, 387
Allergy, 551
Antibody, idiotype specific, 611
Antigen presentation, 35, 41, 117, 123, 129, 141, 179, 571, 599
Antigen processing, 11, 23
Antigen trapping, 411
Antitumor activity, 643
Apoptosis, 387
Atopic dermatitis, 551
Autoimmunity, 643
Autoradiography, 411
Azidothymidine (AZT), 533

Bare lymphocyte syndrome, 135
Birbeck granule, 293, 617
in epidermal LC, 219
in lymph node IDC, 219, 315
in thymic IDC, 141, 153
Bone marrow, 251
human, 263
mouse, 269
Broncho alveolar lavage (BAL), 571

Carcinogens, 623
Carcinoma, 643
Cell migration, 305, 315, 321, 469, 481, 489, 507, 571
Cell surface markers, 147
Cell transfer, 383
Cellular clustering, 53, 65, 93, 179, 225, 365, 449, 461, 629, 633, 643
Centrifugal elutriation, 185
Chemotaxis, 461
monocytes, 643
Chlamydia trachomatis, 581
Chloroquine, 35
Chronic inflammatory disease, 551
Complement receptor
C3b (CR1), 377
C3d (CR1/2), 377
Contact sensitivity, 623
Contact sensitization, 315
Costimulating signals
B7, 87, 159
CD28, 87, 159
Cryosections, 425
Cyclic AMP, 275, 287
Cyclophosphamide, 165
Cyclosporine A, 59, 153, 225
Cytokines, 243
IFN-τ, 47, 99, 111, 205, 333, 371
IL-1, 251, 275, 501
IL-1
receptors
mRNA, 81
IL-2, 47, 75, 87, 99, 371
IL-3, 99, 251
IL-4, 59, 75, 99, 231, 251, 275, 281, 333, 371, 393
IL-5, 111
IL-6, 47, 59, 99, 191, 275, 333
IL-7, 251
IL-10, 611
TNF-α, 251, 425, 563
Cytotoxicity assay, 173, 417

Dendritic cell nomenclature, 467, 469, 481, 487
Dendritic cell precursor, 263
Dendritic cells
antigen pulsing, 299
cultured, 173
derived from monocytes, 629
from blood, 17, 71, 75, 81, 93, 521, 563, 577, 581
from gut, 321, 507
from heart, 489, 501, 507
from intestines, 605
from kidney, 489, 501, 507

from liver, 507
from lung, 507, 557, 577
from monocytes, 287
from oral mucosa draining lymph node, 111
from peritoneal cavity, 117, 129, 141
from peyer's patches, 47
from respiratory tract, 557
from spleen, 11, 47, 99, 141, 159, 571
from thymus, 47, 141, 147, 159
from tonsil, 65, 81, 105
from uveal tract, 599
in osteopetrotic mice, 293
monocyte derived, 275
Desmosomes, 353, 359

Emperipolesis, 365
Endocytosis, 11, 333, 593
Endosomes, 35
proteases, 23
Endothelial cells, 563
Epithelial cells, 159
Epithelium
squamous, 617

Fibroblastic reticulum cell, 339, 443

Follicular dendritic cells (FDC)
freshly isolated, 333
from human blood, 425
from human spleen, 437
from human tonsil, 353, 359, 365, 393, 449, 455
from mouse lymph nodes, 371
heterogeneity, 339
in culture, 333, 417
isolation, 425
origin, 339
subtypes, 353, 359
Folliculo-stelliate (FS) cells, 637

Glands
mucosal, 513
oral, 513
salivary, 513
Granulocyte colony stimulating factor (G-CSF), 251
Granulocyte macrophage colony stimulating factor (GM-CSF), 81, 93, 191, 243, 251, 257, 269, 275, 281, 293, 501, 633
Grave's disease, 643

Hashimoto goitre, 643
HIV-1 virus, 359
Hybridization, 425

Iccosomes, 333, 387
IgE
low affinity receptor (FcER11/CD23), 393
receptor
Fc2ER/CD23, 231
FcER, 237
IL-6 receptor
high affinity, 191

Immune complex trapping, 377, 383
Immunoglobulins
IgA, 59
Immunogold labelling, 17, 35, 205, 237, 455
Immunophenotype, 117
Immunosuppression, 495, 551
Immunotargetting, 299
In situ hybridization, 431
Interdigitating cells (IDC), 35, 153, 311
from lymph node, 501, 539
from spleen, 501
from thymus, 141, 501

Keratinocytes, 469, 501, 617
Kidney
from rat, 495

Langerhans cells (LC), 617
epidermal, 17, 29, 205, 209, 219, 469, 489, 501, 507, 539, 557, 587, 623
epidermal
cultured, 587
freshly isolated, 587
hapten modified
cultured, 209
freshly isolated, 209
monocyte derived, 213
Liposomes, 345
Lipopolysaccharide (LPS), 501
Lung, 443
Lymphadenopathy, 359, 411
Lymphocytes, 581
B-cells, 11, 81, 87, 299, 387, 449, 461
CD23+, 339
T-cells, 135, 417, 489, 521, 545, 611
CD4+, 75, 87, 135, 417, 489, 521, 545
mRNA, 521
CD8+, 135, 417, 489, 521
T-helper cells
T_{H1}, 47, 99, 611
T_{H2}, 47, 59, 99, 611
Lymphoma
B-cell, 399
T-cell, 399
Lysosomes, 11, 17
proteases, 23

Macrophage colony stimulating factor (M-CSF), 243, 251, 293
Macrophages, 17, 71, 551
alveolar, 305
elimination, 345
from intestines, 605
from lung, 571
from peritoneal cavity, 129
from spleen, 99
peritoneal, 41, 305, 321
Mannose receptor, 191
Mannose-6-phosphate receptor, 17
MHC class II, 23, 251, 495, 507
expression, 29, 93, 129, 135, 141, 599
invariant chain, 29

labelling, 17
molecules, 87
MHC expression, 185
Mice
nude, 371, 383
SCID, 339, 383
Mitogenic potention, 53, 521
Mixed leucocyte reaction (MLR), 65, 205, 263, 269, 489, 533, 545, 557
Monkey, 527
Monoclonal antibodies, 117, 165, 293, 371, 387, 399, 405, 411, 513, 637
Monoclonal antibody
CD14, 213
CD1a, 213, 577
DRC1, 449
EBM11 (CD68), 123
ED1, 123, 599, 605
ED2, 599, 605
ED3, 599, 605
ED7, 179, 599
ED8, 179, 599
ED9, 179
F4/80, 293, 507
MIDC8, 35, 293
MOMA2, 123
NLDC145, 159, 293
OX6, 599, 605
OX7, 251
OX12, 251
RFD1, 551, 577
RFD7, 551, 577
Monocytes, 71
cultured, 173
freshly isolated, 173
from blood, 191, 281, 563, 577, 581, 633, 643

Non specific esterase, 577
Northern blot analysis, 521
Nuclear lamins, 287

Phagocytosis, 141
antigens, 199
candida albicans, 213
Polyclonal antibodies, 399
Premature ovarian failure, 629
Prostaglandins, 333
PUVA irradiation, 495

Rheumatoid arthritis, 551

S-100 protein, 311, 443, 637
Sarcoidosis, 551
Signal transduction, 87
Simian immunodeficiency virus, 405
Splenectomy, 437
Squamous cell carcinoma, 643
Superantigen, 41
Suppression, 123

T-cell
deletion, 159
differentiation, 111
proliferation, 209
regulation
V region gen expression, 135
repertoire selection, 159
response, 587
stimulation, 287, 533, 545
Thyroid
carcinoma, 643
hormone, 633
TNF, 643
TNFα receptor, 431
TNP-Ficoll, 345
Tolerance induction, 105, 159, 495
Tuberculoid
dermatitis, 551
leprosy, 551
Tumor surveillance, 611
Type I diabetis, 643

Ultrastructure, 129, 141, 153, 219, 293, 315, 333, 339, 353, 359, 399, 443, 617
follicular dendritic cells, 405
interdigitating cells, 527
thymus, 527

Veiled cells, 327, 489, 501, 633
from dogs, 225
Vesicular complex, 219
Vitamin A, 617